Editor in Chief: W. Plessas, Graz

Supplement 10

Few-Body Problems in Physics '98

*Proceedings of the 16th European Conference
on Few-Body Problems in Physics,
Autrans, France, June 1–6, 1998*

Edited by
B. Desplanques, K. Protasov, B. Silvestre-Brac, J. Carbonell

SpringerWienNewYork

Dr. Bertrand Desplanques
Dr. Konstantin Protasov
Dr. Bernard Silvestre-Brac
Dr. Jaume Carbonell
Institut des Sciences Nucléaires
Grenoble, France

Printing was supported by Bundesministerium für Wissenschaft und Verkehr, Wien

ISBN-13:978-3-7091-7409-8   e-ISBN-13: 978-3-7091-6798-4
DOI: 10.1007/978-3-7091-6798-4

© 1999 Springer-Verlag/Wien
Softcover reprint of the hardcover 1st edition 1999

Typesetting: Camera-ready by authors

Printed on acid-free and chlorine-free bleached paper
SPIN: 10690263

With 217 Figures and 1 Frontispiece

ISSN 0177-8811
ISBN-13:978-3-7091-7409-8   Springer-Verlag Wien New York

Organizing Committee

J. Carbonell (scientific secretary)

B. Desplanques (chairman)

C. Gignoux

A.M. Guglielmini

S. Kox

L. Lévy

K. Protasov (scientific secretary)

D. Rebreyend

J. Riffault

J.M. Richard

B. Silvestre-Brac (treasurer)

P. Valiron

L. Wiesenfeld

International Advisory Committee

J. Adamowski (Krakow)

J. Arvieux (LNS, CEA Saclay)

V.B. Belyaev (Dubna)

A. Boudard (DAPNIA, CEA Saclay)

M. Charlton (University College)

R.E. Chrien (BNL)

C. Cioffi Degli Atti (Perugia)

F. Close (Rutherford Appleton
Laboratory)

T.E.O. Ericson (CERN and Uppsala)

M. Fabre de la Ripelle (IPN Orsay)

A. Fonseca (Lisboa)

J. Friar (Los Alamos)

B. Frois (DAPNIA, CEA Saclay)

W. Glöckle (Bochum)

H. Grosse (Wien)

R. Guardiola (Valencia)

J.W. Humberston (University College)

K. Killian (Jülich)

C. Leluc-Lechanoine
(University of Geneva)

J. Macek (University of Knoxville
& Oak Ridge)

W. Plessas (Graz)

E. Pollak (Weizmann Institute)

D.O. Riska (Helsinki)

M. Rosina (Lubljana)

W. Sandhas (Bonn)

C. Schaerf (Roma)

E.A. Strokovski (Dubna)

T. Tel (Budapest)

J. Tjon (Utrecht)

W. Van Oers (TRIUMF)

S. Wiggins (CALTECH)

T. Yamazaki (Tokyo)

E. Zavattini (Trieste)

Sponsors

Institut National de la Physique Nucléaire et de la Physique des Particules (IN2P3 – CNRS)

Institut des Sciences Nucléaires (ISN), Grenoble

Département d'Astrophysique, de Physique des Particules, de Physique Nucléaire et d'Instrumentation Associée, Service de Physique Nucléaire (SPhN/DAPNIA – CEA)

Université Joseph Fourier (UJF), Grenoble

Institut National des Sciences de l'Univers (INSU – CNRS)

Ministère de l'Education Nationale, de la Recherche et de la Technologie (MENRT)

Ministère des Affaires Etrangères (MAE)

Conseil Général du département de l'Isère

Conseil Régional de la région Rhône-Alpes

Xerox

Foreword

The sixteenth *European Conference on Few Body Problems in Physics* has taken place from June 1 to June 6, 1998, in Autrans, a little village in the mountains, close to Grenoble. The Conference follows those organized in Peniscola (1995), Amsterdam (1993), Elba (1991), Uzhgorod (1990)... The present one has been organized by a group of physicists working in different fields at the University Joseph Fourier of Grenoble who find in this occasion a good opportunity to join their efforts. The core of the organizing committee was nevertheless located at the Institut des Sciences Nucléaires, whose physicists, especially in the group of theoretical physics, have a long tradition in the domain.

The Few Body Conference has a natural tendency to be a theoretical one – the exchange about the methods used in different fields is the common point to most participants. It also has a tendency to be a hadronic physics one – the corresponding physics community, perhaps due to the existence of experimental facilities devoted to the study of few body systems, is better organized. In preparing the scientific program, we largely relied on the advices of the International Advisory Committee, while avoiding to follow these trends too closely. Thus, the contact with physics problems has been mainly ensured by experimental talks. A large place has been devoted in plenary sessions to non-hadronic physics talks and we tried to incorporate new topics at the borderline of solid state physics (quantum dots) or in relation with the many body problem (few body systems in some medium).

The incursion of the many body problem into the few body one may look surprising. One should not forget that a few body system represents an idealized concept. From the field theory point of view, the description of such a system requires an infinite number of degrees of freedom. This led N. Isgur to postulate at the International Few Body Conference held in Williamsburg in 1995 that the Few Body conferences were going to end. Hopefully, when facing a new complicated problem, the first reaction of a physicist is to reduce it to a finite number of degrees of freedom, effective however. The end is not for tomorrow.

We are not sure that the scientific program has achieved the goals we assigned to it. Some of the physicists we contacted to give a talk decline our invitation. Quite politely, they presented reasonable excuses, but we sometimes got the feeling that they were hesitating to join a community of physicists largely unknown to them. Also, some of the contacted speakers had to cancel their participation in the Conference. These difficulties have especially affected the program of the sessions on the mathematical aspects of the few body problem and on few body systems near threshold. We strongly believe that these topics are full part of Few Body Conferences. The difficulty in making a coherent program in these domains is perhaps the indication that work is necessary there. In any case, we would like to express here our gratitude to those members of the International Advisory Committee who help us in preparing the scientific program, to the chairs of the plenary sessions who have the difficult task to give an introduction and an overview of their session and finally to the speakers who did a nice job.

The written versions of the talks given in plenary or parallel sessions are gathered in the present volume. We put together the contributions relative to each of the six major topics discussed at the conference. As far as possible, we tried to follow some logics in the order of presentation. Plenary session talks come first while parallel session talks appear close to each other when dealing with the same problem. Apart from that, one should not give too much importance to the fine structure. In most cases, the name of the author who presented the contribution appears first after the title.

The physicists of the former Soviet Union have always been very active in the field of the few body problem. Making possible their partipation in the Conference as well as that of physicists of other countries in a difficult economical situation has been a major concern for us. Managing this participation has been quite uneasy. While we largely anticipated some support, the uncertainty as to its amount obliged us to discard many applications of physicists whose participation in the Conference would have been quite useful. We apppreciated that the requests for financial support from the Organizing Committee were quite moderate, allowing one to help a larger number of physicists.

Many physicists expressed their opinion that the Conference went quite smoothly. The Organizing Committee is too aware of the imperfections to claim full responsability for that. We believe that the participants largely contributed to make the Conference interesting and successful. Our warmest thanks are due to all of them.

A Conference like the present one requires a sizeable support. This was provided for some part by scientific institutions: Institut National de la Physique Nucléaire et de la Physique des Particules(IN2P3-CNRS), Département d'Astrophysique, de Physique des Particules, de Physique du Noyau et de l'Instrumentation Associée (DAPNIA-CEA), Institut des Sciences Nucléaires (ISN-Grenoble), Université Joseph Fourier (UJF-Grenoble), and Institut National des Sciences de l'Univers (INSU-CNRS). Another part was provided by national or local government agencies as well as private companies: Ministère des Affaires Etrangères (MAE), Ministère de l'Education Nationale, de la Recherche et de la Technologie (MENRT), Conseil Général de l'Isère, Région Rhone-Alpes and Rank-Xerox.

We are also indebted to Anne-Marie Guglielmini and Jocelyne Riffault, who were in charge of the administrative aspects of the secretariat, to Alexander Samarine who took care of the Web page, to Christian Favro whose expertise was quite useful for a correct insertion of some figures in the Proceedings and to many other persons whose help will remain anonymous. Special thanks are due to Joël Chauvin, the Director of the ISN, for his continuous encouragements.

Grenoble, October 25, 1998   For the Organizing Committee

Bertrand Desplanques

Contents

Resolution methods for few-body problems

Mathematical aspects

Relativity

Few-body dynamics: Atomic and mesoscopic systems

Threshold effects and stability limit

Few-body dynamics: Nuclear and particle systems

Few-Body Systems Suppl. 10, 1–10 (1999)

Few-
Body
Systems
© by Springer-Verlag 1999

Quantum Monte Carlo Methods in Few-Body Physics

J. Carlson *

Theoretical Division, Los Alamos National Laboratory, Los Alamos, NM
87545, USA

Abstract. Monte Carlo methods are being used to bridge the gap between
few- and many-body quantum systems. In electronic systems, Quantum Monte
Carlo methods can be used to study the stability and orientation of various arrangements of atoms and molecules. In nuclear physics, Monte Carlo methods
provide a unique opportunity to study light nuclei with realistic nuclear interactions. In this article, we discuss why and how Monte Carlo methods have
been adapted to few-body physics, and a variety of the recent applications of
such methods.

We highlight applications in nuclear physics, where studies of the nuclear
interaction are extremely important. Three-nucleon interactions and relativistic effects are being studied in different nuclei with a variety of techniques.
Nuclei beyond A=4 offer unique opportunities to study the isospin dependence
of the three-nucleon interaction, as well as studies of bound states with important contributions from odd-parity partial waves. We have also begun to study
the parity-violating NN interaction, where some interesting new few-nucleon
experiments are planned and underway.

1 Introduction

Thanks to recent progress in computational methods and facilities, it is now
possible to examine many problems in few-nucleon and few-electron quantum
systems by direct solution of the Schrödinger equation. For very few degrees of
freedom, this can be done directly in either momentum- or coordinate space. In
nuclear physics, examples are recent applications of the Faddeev-Yakubovsky
and Correlated Hyperspherical Harmonics (CHH) methods to solve the four-
nucleon problem. As the number of particles grows, though, it is no longer
possible to directly enumerate the important quantum states of the system.
At this point Monte Carlo methods become valuable tools for solving for the
ground-state and low-lying excitations of few-body systems.

---

* *E-mail address:* carlson@lanl.gov

In this article, I begin by illustrating the scaling of quantum problems with the number of degrees of freedom. I then briefly describe available quantum Monte Carlo (QMC) methods, and their advantages and limitations. Finally I give a few short examples of recent studies in electronic and nuclear systems. In electronic systems, QMC methods have been used to study the binding energies of fairly complex molecules, and to provide serious tests for commonly applied mean-field theories.

In nuclear physics a primary motivation is to study the nuclear interaction. Here we are using a simplified picture of the nucleus as a system of interacting nucleons. In the last several years, such a picture of nuclei has gained credence through Chiral Perturbation theory [1], as well as a long history of detailed and precise comparisons with a rich body of experimental data [2].

The few-body community has worked diligently to understand three-nucleon bound states, scattering below- and above-breakup, and capture reactions. To a large extent a reliable picture and understanding of nuclear properties has been achieved, though with some notable exceptions such as the A_y problem [3]. In the last several years significant progress has also been made in the four-nucleon sector; several contributions to this volume present important progress in below-breakup scattering in A=4. Beyond A=4, though, one must currently resort to Monte Carlo techniques.

2 Why Monte Carlo?

A natural question, particularly in a few-body context, is why one should resort to Monte Carlo methods at all. Clearly explicit methods, those that employ a finite basis, have been very successful in describing few-body quantum problems. In electronic systems, full configuration-interaction (CI) calculations are possible in few-body problems. This method corresponds to directly diagonalizing the full Hamiltonian with Lanczos techniques; and has been highly successful in describing these systems. In nuclear physics, Faddeev-Yakubovsky and Correlated Hyperspherical Harmonics methods have been successfully applied to the three- and now four-nucleon systems, both in the bound-state and scattering regimes.

However, the general quantum problem scales exponentially with the number of particles. In electronic systems, the dominant Coulomb interaction is independent of the electrons spins, and hence the wave function can be written as a single function of the coordinates of all the particles. Denoting the number of important single-particle states as M, and the number of electrons by N, the number of wave function components (ignoring symmetry considerations) grows as N^M. There has been a tremendous progress in recent years, particularly in atomic systems where special symmetries can be effectively employed [4]. Despite the progress in theory and computational facilities, diagonalizing a general problem becomes rapidly prohibitive as the number of particles or basis states increases.

The scaling of the nuclear problem is even more daunting, both because of the strong short-range repulsion and the spin- and isospin-dependence of the interaction. The nuclear wave function can be decomposed into a sum over spatial functions ψ times products of individual spin- and isospin-states:

$$\Psi = \sum_{i=1}^{2^N} \sum_{j=1}^{N_t} \chi_\sigma(i) \, \chi_\tau(j) \, \psi_{ij}(\mathbf{R}). \tag{1}$$

The number of spin states grows as 2^A, while the number of isospin states grows somewhat more slowly due to charge conservation and (approximate) isospin conservation. To determine the overall scaling with system size, one must enumerate the important spatial states, often by putting the wave function on a grid of combined radial and angular variables. To obtain a crude estimate, we assume the nuclear wave function extends for 10 fermi in each direction with a required grid spacing of 0.2 fm, giving a spatial dimensionality on the order of 50^{3A}. Thus in an explicit spatial basis the wave function ^8Be would have approximately 10^{22} times the number of components of the alpha particle. The spin-isospin dependence renders the problem even more difficult.

Monte Carlo methods are used to reduce the effective dimensionality. Two methods are commonly employed in quantum calculations, Variational (VMC) and Green's function (or Diffusion) Monte Carlo (GFMC). VMC simply employs the Raleigh-Ritz variational principle to minimize the expectation value of the Hamiltonian

$$E_\alpha = \frac{\langle \Psi_T\{\alpha\} | H | \Psi_T\{\alpha\} \rangle}{\langle \Psi_T\{\alpha\} | \Psi_T\{\alpha\} \rangle} \tag{2}$$

as a function of a set of variational parameters $\{\alpha\}$. In electronic systems, it is popular to minimize the variance of the energy rather than the energy itself. Such methods can and have been used to study bound- and low-energy scattering states. VMC is rather similar to the stochastic variational method described by Varga in this volume. However, in the stochastic Variational Method the wave function is expanded in a complete, or nearly complete, basis, and the wave function is written as a linear combination of the basis states. This should work well for systems of moderate dimensionality.

For larger systems one must take a more restricted basis. One possibility is to use a small basis of correlated wave functions, each of which has the form:

$$|\Psi_{T,i}\rangle = \left[\mathcal{S} \prod_{i<j<k} F_{ijk} \right] \left[\mathcal{S} \prod F_{ij} \right] |\Phi_i\rangle. \tag{3}$$

Atomic physics applications are discussed in [5], while the nuclear physics wave functions are described in some detail in ref. [6]. In such a wave function, an anti-symmetrized product of single-particle states is incorporated into the state $|\Phi\rangle$. This product controls the quantum numbers of the state and much of the long-distance physics. The pair (F_{ij}) and three-body (F_{ijk}) correlation operators involve short-range correlations between the particles.

This approach has been quite successful in nuclear physics. The F_{ij} describe the correlations induced between the nucleons by the strong spin-dependent nuclear interaction. The two-nucleon correlations are obtained by solving parametrized two-nucleon Schrödinger equations, the three-nucleon correlations are primarily induced by the three-nucleon interactions.

The Monte Carlo method is only weakly dependent on system size; the computational cost scales basically as the time to calculate a wave function at a single set of spatial coordinates. In an electronic system this grows roughly proportional to the cube of the number of particles, where in nuclear physics it is proportional to the number of spin-isospin states times the number of pair and three-body operators. Hence, to date Monte Carlo methods in nuclear physics are still limited to few-nucleon ($A \leq 8$) systems. In principle Monte Carlo methods could be used to sample the spin-isospin degrees of freedom as well, though an effective scheme to do this has not yet been found.

While the Variational Monte Carlo wave functions we employ are accurate in many respects, the relative binding of different orientations of atoms or the relative binding of different nuclei often involves extremely sensitive cancellations. The p-shell nucleons are often rather loosely bound compared to the available breakup thresholds, and hence is placing extraordinary demands on the accuracy of the wave function. Similarly, the binding energies of molecules can be quite small on the atomic scale. Different spatial symmetries of atoms can yield almost the same total energy. It is important to enumerate these low-lying states and describe their relative binding, and hence one must obtain binding energies within a few per cent, which requires even much greater accuracy on the scale of the whole system. Monte Carlo methods have been able to successfully achieve this accuracy in a number of cases.

Green's function Monte Carlo methods are used to improve the estimates of the energies and other expectation values. The exact ground state of a quantum system can be obtained by projecting out the ground state via

$$|\Psi_0\rangle = \exp[-H\tau]|\Psi_T\rangle = \prod \exp[-H\Delta\tau]|\Psi_T\rangle, \tag{4}$$

which can be obtained by writing down the imaginary-time equivalent of the time-dependent Schrödinger equation. On the right-hand side we have written the propagator as a product over short-time steps $\Delta\tau$.

Over the past several years the efficiency of the numerical algorithms have improved dramatically. In electronic systems pseudo-potentials have been used with great efficiency to solve for a variety of quantum problems [5]. By using a pseudo-potential, one eliminates the energy scale associated with the inner electronic degrees of freedom, hence rendering calculations much more feasible. In nuclear systems, the dominant time scale is governed by the very repulsive short-range NN interaction. One recent source of improvement is using exact two-nucleon propagators in the expression used for $\exp[-H\tau]$. Previously we had used a simple (Feynman) product form: $\exp[-H\tau] = \exp[-V\tau/2]\exp[-T\tau]\exp[-V\tau/2]$. We now use

$$\exp[-H\Delta\tau] = \prod_i \exp[-H_i^0 \Delta\tau] \frac{\mathcal{S}\left[\prod_{i<j} \exp[-H_{ij}\Delta\tau]\right]}{\prod_{i<j} \exp[-H_{ij}^0 \Delta\tau]}, \tag{5}$$

where H_i^0 is the free-particle imaginary time propagator $H_i^0 = \exp[-T_i\tau]$, a simple Gaussian. The pair propagators must be solved for numerically [6]; including them explicitly in the algorithm allows us to make much larger time steps ($\Delta\tau$). Essentially this corresponds to exactly including multiple scattering terms between isolated pairs, and hence integrating out some of the highest-energy degrees of freedom in the NN interaction. Additional efficiencies gained in the last few years, including much more efficient treatment of three-nucleon interactions, have been crucial in allowing us to study 7- and 8-nucleon systems.

Of course, the reduced dimensionality of Monte Carlo problems comes at a price. Monte Carlo methods are valuable in calculating low-energy or thermal quantities like the properties of the low-lying states in the system or low-energy scattering properties. Even here, though, there are some difficulties. An exact treatment of fermionic systems with power-law scaling in computer time has yet to be achieved. In electronic calculations one typically employs a fixed-node scheme, where the nodes of the trial function are assumed to be exact. This yields an upper bound to the true ground-state energy which can be quite accurate. However, the fixed-node errors appear to be larger than the errors resulting from the pseudo-potential in modern calculations.

Nuclear physics calculations employ transient estimations, which are in principle exact but require exponential increase in computer time with increasing τ. Present-day calculations are limited to a total imaginary time of approximately 0.1 MeV^{-1}. However, we have employed variational wave functions in which the long-range degrees of freedom can be varied considerably; this is important to reduce the errors to acceptable levels. We are presently working on generalizations of the fixed-node method for nuclear problems.

In spite of these difficulties QMC methods are the most accurate available for ground states and low-lying states of such systems. A more serious difficulty is in applying QMC methods to the general quantum scattering problem. While in special cases (low-energy or semi-classical regime where Glauber theory applies) progress has been excellent, there is little understanding of how to attack the general problem. Quantum scattering above breakup threshold typically involves delicate interferences that are difficult to treat with Monte Carlo methods.

3 Electronic Systems

It is often important to know the lowest-lying states of various arrangements of atoms. In the simplest cases these estimates can be made in mean-field theory by Hartree-Fock methods. However, if the energy differences are very small one must resort to more accurate methods. One popular method is density functional theory, where one writes the total energy as a functional of the energy density. Another choice is to try to solve the quantum few- to many-body problem exactly with Monte Carlo techniques.

Density functional theory, in principle, is exact. However, in practice one typically resorts to local or gradient-corrected energy functionals, where the energy is written as a function of the local electronic density (LDA) and possibly its gradient (GCLDA). While this approximate method is quite accurate in most cases, it is useful to compare results to Monte Carlo methods in a few challenging cases.

One interesting problem that has been addressed recently is the electronic and structural properties of silicon clusters [7]. QMC calculations were performed for various arrangements of from two to twenty silicon atoms in various arrangements. Mean-field theory was used to select possible low-energy configurations of the atoms, and QMC used to calculate the actual binding energies for these systems. Experimental energies are available for clusters of up to 7 atoms; QMC results agree with these results within approximately ten per cent. Hartree Fock, on the other hand, can be off by a factor of two, and LDA results can be wrong by up to twenty per cent.

Extending the calculations up to larger systems, the QMC calculations determine unambiguously the role of correlation energy in the ordering for different structures, including a new lowest energy structure of Si_{20}. They also predict a different ground state for Si_{13}^- than determined by LDA. The LDA calculations overbind these larger systems by approximately 25 per cent. It is possible that the QMC algorithms will, by providing a number of useful such benchmarks, allow one to pin down the parameters in higher-order corrections to LDA much as the original electron-gas calculations have been used to determine the parameters in LDA itself. As computational facilities improve and computer time becomes less of an issue, it may be possible to compute electronic structures very accurately on a case-by-case basis.

4   Light Nuclei and the Nuclear Interaction

In nuclear physics the situation is rather different than in electronic problems. The nuclear interaction is, of course, neither so well-known nor so simple as the Coulomb force. Fortunately, it has been extremely well-studied experimentally [2]. In recent years several high-quality analyses of the experimental have been undertaken, these analyses determine the nucleon-nucleon phase shifts quite accurately in the regime of interest in the nuclear few- and many-body problem. Several models of NN interaction are available which provide high-quality fits to the experimental database. In our studies we employ the Argonne V18 interaction.

Beyond the two-nucleon interaction, two important effects can be studied in the Hamiltonian: relativity and three-nucleon interactions. Each of these has also been investigated with QMC algorithms. Preliminary studies of kinematic relativistic effects, those required by Poincare invariance, indicate a small repulsive effect in light nuclei [8]. Additional non-localities due to the relativistic corrections to one-pion-exchange are also being investigated. Of course, some of the relativistic effects can be transformed into three-nucleon interactions.

Explicit three-nucleon interactions can also be important because of the relatively low-lying excitations of the nucleon, in particular the Delta resonance. The longest range of the three-nucleon interaction is due to two pion exchanges with the excitation of an intermediate Delta resonance. Additional repulsive terms are also required in order to reproduce the saturation properties of nuclear matter. For simplicity, we use the simple Urbana model IX TNI, which consists of the Fujita-Miyazawa two-pion-exchange three-nucleon interaction plus a repulsive term with the range of two pion exchanges on each leg. The strength of this repulsive term is adjusted to reproduce the saturation density of nuclear matter, while the two-pion-exchange term is adjusted to reproduce the three- and four-nucleon bindings.

The resulting spectra is displayed in Fig. 1. The binding of light p-shell nuclei is very difficult to calculate due to the small energy scale of the binding as compared to the interaction itself. Expectation values of some of the terms in the Hamiltonian are displayed in Table 1 for a variety of nuclei. In ^{7}Li, the kinetic and potential energies are on the order of two-hundred MeV, while the binding compared to breakup into the three plus four-body channel is only a few MeV. Hence extremely accurate calculations are required. Also, we note that the one-pion-exchange contribution to the total two-body potential energy V_{ij} is quite large.

Overall the bindings and spectra are well-reproduced in the calculations. The remaining discrepancies are of the order of one to two MeV in nuclei near N=Z, while they are somewhat larger for neutron-rich nuclei like ^{8}He. In all cases, though, the discrepancies are smaller than the expectation value of the three-nucleon interaction.

Table 1. Total energy and contributions in A=2, 4, 6 and 7 (MeV)

Nucleus	T	V_{ij}	V_{ijk}	v_{ij}^{π}	$V_{ijk}^{2\pi}$	E(GFMC)	E(expt)
^{2}H	20	-22	0	-21	0	-2.2	-2.2
^{4}He	112	-136	-6	-99	-12	-28.3	-28.3
^{6}Li	150	-180	-7	-129	-14	-31.3	- 32.0
^{7}Li	186	-223	-9	-152	-17	-37.4	- 39.2

Without the three-nucleon interaction, the situation is considerably worse. Recent studies indicate that for the AV18 interaction alone, the He isotopes beyond A=4 are not bound, while the Li isotopes are near the limits of stability. The ^{8}Be is particularly ill-described, lying significantly above the two-alpha threshold.

Upon including the three-nucleon interaction, ^{8}Be is well described. On the other hand, ^{8}He moves approximately half way to the experimental value. We

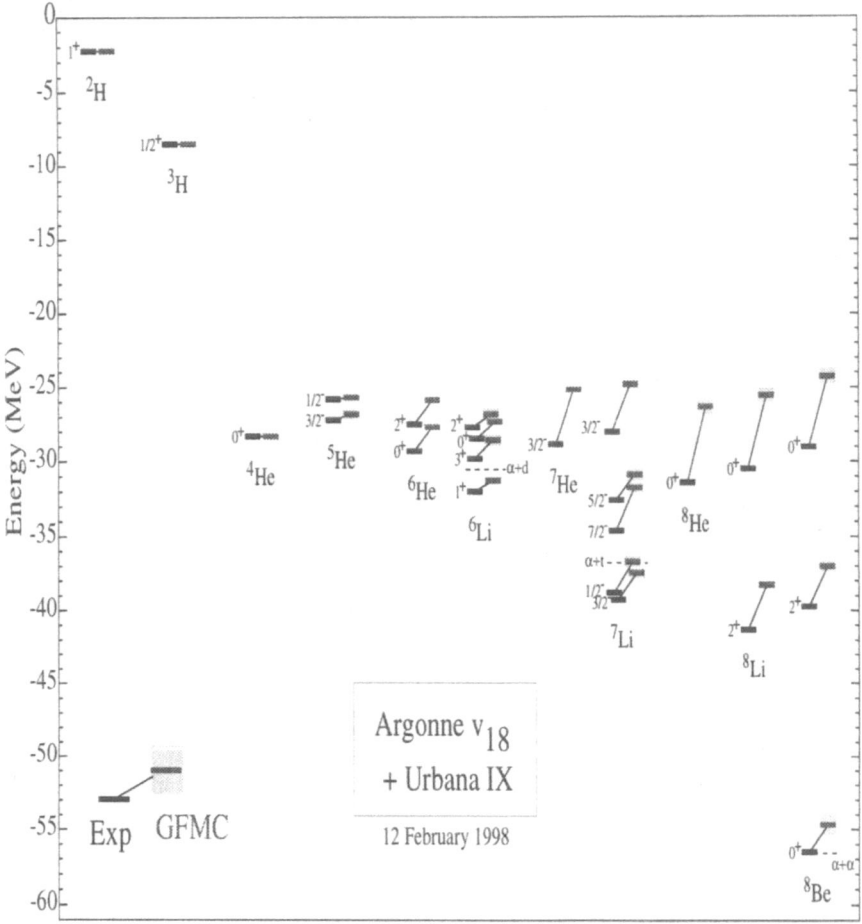

Figure 1. Energy spectra for light nuclei, calculated vs. experimental

have fit the three-nucleon interaction to isospin symmetric or nearly isospin-symmetric systems (^{3}H, ^{4}He, nuclear matter), and hence perhaps it is not surprising that these nuclei are better described than the isospin-rich systems.

The ordering of the excitations in light nuclei are correctly reproduced. However, it does appear that the L·S splittings are too small compared with experiment. In ^{5}He, only about 2/3 of the experimental splitting is recovered in the calculations. Without the TNI, even less of this splitting is obtained, approximately 1/2 of the total.

Each of these problems were also observed in previous calculations of neutron drops [9]. As compared with Skyrme-model calculations empirically fit to nuclei, the QMC results showed too little binding and too small of $L \cdot S$ splittings. Given the simple nature of the three-nucleon force used to date and the fact that it had only been fit near N=Z, the remaining discrepancies are not terribly surprising.

A host of other properties of these nuclei have also been calculated. These include electromagnetic elastic and transition form factors [10], one- and two-nucleon distributions, spin-polarization and momentum distributions.

5 The Future

Quantum Monte Carlo methods are the most accurate available for treating systems with more than a very small number of degrees of freedom. Important progress has been made in studies of both electronic and nuclear systems. In electronic systems the primary algorithmic need is for an efficient way to optimize the nodal positions in fixed-node calculations. While the present methods are very accurate, they do require significant optimizations. With future algorithmic improvements, it appears likely that a wide variety of molecular systems will be amenable to accurate calculations.

In nuclear systems there is a significant experimental and theoretical effort being made to understand the three-nucleon interactions and relativistic effects in light nuclei. If this is accomplished, few-nucleon systems can be successfully used to probe a variety of interesting and important physics. One example is the weak capture in A=4, the source of so-called *hep* neutrinos from the sun. Although this cross-section is extremely small, its precise value is important because it can influence analysis of possible signals of neutrino oscillations at Super-Kamiokande [11]. Another example is the analysis of recent and present-day parity-violating experiments in A=5. Combined, these experiments could shed considerable light on the weak parity-violating πNN coupling constant. However, a precise treatment of the strong-interaction physics is required for such an analysis. Quantum Monte Carlo methods will continue to play an important role in studies of the both the strong and weak interaction in nuclei.

Acknowledgement. The nuclear physics calculations have been supported by the National Energy Research Supercomputer Center, the IHPCI program at Los Alamos National Laboratory, and the Mathematics and Computer Science Division of Argonne National Laboratory. This work is supported by the U. S. Department of Energy.

References

1. S. Weinberg: Phys. Lett. **B251**, 288 (1990); Nucl. Phys. **B251**, 288 (1991)

2. J. Carlson and R. Schiavilla: Rev. Mod. Phys. **70**, 743 (1998)

3. H.D. Witala, W. Glöckle and H. Kamada: Phys. Rev. **C49**, R14 (1994)

4. J. Sapirstein: Rev. Mod. Phys. **70**, 55 (1998)

5. D.M. Ceperley and L. Mitas: Adv. in Chem. Phys. Vol **XCIII**, eds. I. Prigogine and S. A. Rice, Wiley, N.Y., 1 (1996)

6. B.S. Pudliner, V.R. Pandharipande, J. Carlson, S.C. Pieper, and R.B. Wiringa: Phys. Rev. **C56**, 1720 (1997)

7. J.C. Grossman and L. Mitas: Phys. Rev. Lett. **74**, 1323 (1995)

8. J.L. Forest, V.R. Pandharipande and J.L. Friar: Phys. Rev. **C52**, 568 (1995); J.L. Forest, et al., Phys. Rev. **C54**, 646 (1996)

9. B.S. Pudliner, A. Smerzi, J. Carlson, V.R. Pandharipande, and D.G. Ravenhall: Phys. Rev. Lett. **76**, 2416 (1996)

10. R.B. Wiringa and R. Schiavilla: Phys. Rev. Lett., to appear, (1998)

11. J. Bahcall and P. Krastev: Phys. Lett **B**, to appear (1998)

Few-Body Systems Suppl. 10, 11–18 (1999)

Few-
Body
Systems
© by Springer-Verlag 1999

Recent applications of the stochastic variational method

K. Varga[1] *, J. Usukura[2], Y. Suzuki[3]

[1] Physics Division, Argonne National Laboratory, 9700 South Cass Avenue, Argonne, Illinois 60439, USA

[2] Graduate School of Science and Technology, Niigata University, Niigata 950-2181, Japan

[3] Department of Physics, Niigata University, Niigata 950-2181, Japan

Abstract. This paper overviews the most recent developments and applications of the stochastic variational method for different physical systems.

1 Introduction

The variational method is a powerful and conceptually simple tool to solve the Schrödinger-equation. The wave function of the physical system is approximated by a linear combination of some appropriate "basis/trial" functions. The trial function may depend on some (continuous) parameters as well as on some quantum numbers (spins, isospins, orbital momentum, etc.) As the number of particles increases the selection of the most adequate trial function becomes more and more complicated. In principle one should optimize the wave function, but this way is not tractable (partly because the large number of parameters involved, partly because of the discrete variables (quantum numbers)).

The stochastic variational method (SVM) [1-13] attempts to find the most suitable basis functions by a simple trial and error gambling strategy. Several possible parameter and quantum number sets are probed and those which considerably improve the variational energy are selected to be basis states. The main idea is best explained by a simple example. We consider a simple yet nontrivial case: the Coulombic three-body system of two electrons and a positron.

---

* *On leave from:* Institute for Nuclear Research of the Hungarian Academy of Sciences (ATOMKI), 4001 Debrecen, PO Box 51, Hungary

Table 1. The energy of Ps$^-$ (in a.u.) for K basis states that are selected randomly. "Exact" energy is -0.262005. The energies Ei are obtained by starting from different random points. The energy of "Best 100" is the one calculated by sorting the best 100 basis states from 400 basis states, where 50 random trials are probed at each step of the basis selection. N_{eval} is the number of matrix elements evaluated during the optimization.

	$K = 100$	$K = 200$	$K = 400$	Best 100
$E1$	-0.2617619	-0.2619798	-0.2620032	-0.2619995
$E2$	-0.2617281	-0.2619793	-0.2620026	-0.2619978
$E3$	-0.2617918	-0.2619669	-0.2620016	-0.2619956
$E4$	-0.2618826	-0.2619675	-0.2620008	-0.2619953
$E5$	-0.2616285	-0.2619824	-0.2620024	-0.2619987
N_{eval}	5050	20100	80200	80200

The basis function, a correlated Gaussian [14, 15], is assumed in the form:

$$f(A, \mathbf{x}) = \exp\{-\frac{1}{2}\mathbf{x}A\mathbf{x}\} = \exp\{-\frac{1}{2}\sum_{i=1}^{2}\sum_{j=1}^{2}A_{ij}\mathbf{x}_i \cdot \mathbf{x}_j\} \tag{1}$$

where $\mathbf{x}_1$ and $\mathbf{x}_2$ are the relative coordinates between the two electrons, and between the center-of-mass of the two electrons and the positron, respectively. The matrix elements A_{ij} are the parameters of the trial function. This basis function belongs to $L = 0$ orbital motion, and electrons have antiparallel spins. As only Coulomb interaction is considered between the particles there is no additional quantum number needed to describe the systems and the trial function takes the above written simple form. The wave function is approximated as linear combination of the basis functions:

$$\Phi(\mathbf{x}) = \sum_{k=1}^{K} c_k f(A_k, \mathbf{x}). \tag{2}$$

The ground state energy of this system is fairly accurately known by different calculations: $E = -0.262005$ a.u.. In Table 1 we illustrate what happens if completely random parameters are generated for various basis dimension $K = 100, 200$ and 400. Five different random sets are generated (energies $E1, E2, E3, E4$ and $E5$ to show how much the results scatter depending on the random parameters. The last line shows the number of function evaluations needed to reach these results. This measures the necessary computer time. Table 1 shows that even a completely random basis can lead to accurate results provided that the dimension is big enough. It also shows that the energy difference between different random sets is not too big and it actually decreases by increasing the basis size. Different random parameters give very similar results and not all of the random states are equally important. One may get almost the same result by omitting some of the states. One can, for example, sort out

Table 2. Energy of Ps⁻ (in a.u.) by different optimization strategies. The basis size is set to $K = 100$. In the case of the full optimization by the Powell algorithm 7200 diagonalizations were required. The refining cycle is repeated only once.

Method	Powell	SVM($n = 1$)	SVM($n = 10$)	Refining ($n = 10$)
$E1$	-0.26200016	-0.26176191	-0.26199427	-0.26200231
$E2$	-0.26199947	-0.26172812	-0.26199382	-0.26200351
$E3$	-0.26200164	-0.26179182	-0.26199733	-0.26200312
$E4$	-0.26200135	-0.26188261	-0.26199778	-0.26200301
$E5$	-0.26200193	-0.26162811	-0.26199836	-0.26200271
N_{eval}	35466150	11615	86150	805050
Time	7200	27	43	195

a smaller basis from the $K = 400$ random basis. In the last column, under the title "best 100" we have selected the 100 basis states which gives the lowest energy amongst the 400 randomly chosen one.

In the stochastic variational method, random parameter sets are generated and we select the most appropriate set by comparing the energy gain. The basis is always built up step by step, increasing the dimension one-by-one, by adding a new set to the the previously fixed basis. The new set is chosen from several random candidates. Table 2 compares the energy obtained by the stochastic selection to that of the Powell optimization. The Powell optimization is a direct deterministic optimization of the parameters [16]. The fact that the Powell optimization gives slightly different results by starting from different random points shows that the omnipresence of the local minima makes such an optimization difficult. Table 2 also shows the result of SVM when we select amongst n random candidate. The $n = 1$ case means no selection but inclusion of random basis states, which is the same as in the case of Table 1 and it is included only for comparison. In the case of $n = 10$ random candidate, the results are very close to that of the Powell optimization, but the number of function evaluation, that is the computational load, is more than two order of magnitudes smaller. After we reached a given dimension, we may stop increasing the basis size and we may try to replace the already selected basis states with a better choice. This is a refining of the basis and it actually beats the results given by the Powell optimization, while the computational burden is still much smaller. The final result of the SVM is reached by repeating the refining steps and increasing the basis size until necessary/possible. Our most accurate calculation for this system is compared to the results of other calculations in Table 3.

2 Excited states

The Mini-Max theorem tells us that by diagonalizing the Hamiltonian in a K-dimensional basis one gets an upper bound not only for the ground state but also for the excited states as well. It may happen that the basis set found for the ground state is fairly good to predict the energies of the excited states with

Table 3. Energy and different separation distances for the $(e^+e^-e^-)$ Coulombic three-body system as a function of the basis dimension K. The virial ratio η is defined by $\eta = |1 + \langle V \rangle/(2\langle T \rangle)|$. Atomic units are used.

	SVM $(K{=}100)$	SVM $(K{=}200)$	SVM $(K{=}600)$	Hylleraas [17]
$-E$	0.26200465	0.2620050648	0.262005070226	0.2620050702328
$\langle r_{+-} \rangle$	5.489	5.48962	5.489633252	5.489633252
$\langle r_{--} \rangle$	8.548	8.54856	8.548580655	8.548580655
$\langle r_{+-}^2 \rangle^{\frac{1}{2}}$	6.958	6.95832	6.95837	6.95837
$\langle r_{--}^2 \rangle^{\frac{1}{2}}$	9.652	9.65284	9.65291	9.65291
η	0.46×10^{-4}	0.34×10^{-6}	0.54×10^{-10}	0.23×10^{-10}

the same conserved quantum numbers (angular momentum, parity etc.) as the ground state. By optimizing only the ground state, however, the energies of the excited states will not necessarily converge.

As the ith eigenvalue E_i obtained by diagonalization is the upper bound of the energy of the ith excited state, one may try to optimize this upper bound to get an accurate estimate for the energy of the excited state. In practice we repeat the same procedure as before, but now the basis selection is governed by the requirement that the ith eigenvalue, not the ground-state energy E_1, should be improved. To get the ith eigenvalue we need at least an i-dimensional basis to start with, but practical numerical considerations suggest that it is better to start with a basis in which all the lower eigenvalues ($k = 1, ..., i-1$) are already "stable". This means that we need a first guess for the lower eigenvalues, otherwise it may happen that when improving the first excited state, for example, we pick up such components that lower the ground state.

As the energies of the excited states of the Helium atom are known to a high accuracy, we will test our strategy in this case. First we optimize the ground state of the He atom on a $K = 100$-dimensional basis. The energy of the ground state is very accurate and even the energy of the first excited state is acceptable. Starting from this $K = 100$-dimensional basis, we increase the basis size one by one, picking up basis states which improve the energy of the first excited state. This procedure quickly improves the energy of the first excited state, while the energy of the ground state also improves a little bit (the important thing to note is that it does not get worse) because the basis size is increased. After reaching the basis size of $K = 200$ we switch to the second excited state and so on.

The above procedure gives us a basis where all the excited states are accurate up to a certain digit. We can of course create bases which give an accurate energy for an individual state, while the energy of the other states might be poor. In that case we simply carry out the stochastic search for a given state starting from a first guess basis (like the $K = 100$-dimensional basis in the previous case). This optimization may include refinement cycles as well, if ne-

Table 4. Energies in atomic units of the ground state and the first four ^1S excited states of the Helium atom. In column A the basis is optimized successively for all the states as described in the text ($K = 500$). In column B the basis is optimized separately for each state, leading to five different bases ($K = 600$) tailored for the respective states. The "exact" values are taken from [18].

State	A	B	"Exact"
E_1	-2.9037243758	-2.9037243769	-2.90372437698
E_2	-2.1459737740	-2.1459740452	-2.14597404605
E_3	-2.0612718887	-2.0612719880	-2.06127198974
E_4	-2.0335865085	-2.0335866779	-2.03358671702
E_5	-2.0211312479	-2.0211768312	-2.02117685157

cessary. The results we obtained in this way are compared to the "exact" (i.e., the best calculation in the literature) values [18] in Table 4. One can thus get as accurate energies for the excited states as for the ground state.

3 Coulombic few-body systems

The SVM has been widely applied for a number of Coulombic few-body problems. The real advantage of the method is that one can cope not only with 3 but 4-5-6 particles as well. It has been applied to few-electron atoms and for small molecules [6, 7]. It made it possible to prove the existence of several positronic atoms, for the first time [19, 20]. An extensive review can be found in ref [2]. Some of our results are shown in Table 5 for illustration.

Table 5. Energies of different Coulombic systems in atomic units. K is the basis dimension.

System	State	K	SVM	Other method	K	Ref.
Ps_2	$^1S^e$	800	-0.516003778	-0.516002	400	[21]
Ps_2	$^1P^o$	800	-0.334408112			
Li	$^1S^e$	600	-7.478058	-7.47806032	1589	[23]
Li	$^1P^o$	1000	-7.410151	-7.410156521	1715	[23]
Li	$^1D^e$	1000	-7.335520	-7.335523540	1673	[23]
$^\infty$HPs	$^1S^e$	1200	-0.7891964	-0.7891794		[22]
Be	$^1S^e$	500	-14.6673	-14.667355	1200	[20]
Li^-	$^1S^e$	600	-7.50012	-7.50076	∞	[24]
$Li + e^+$	$^1S^e$	1000	-7.53218			
$Li + Ps$	$^1S^e$	600	-7.73855			
$Be + e^+$	$^1S^e$	1000	-14.692			

The H_3^+ is a bound system but the system of three positrons and two electrons is found to be unbound [3]. It is intriguing to check at which mass ratio m/M the $(M^+, M^+, M^+, m^-, m^-)$ system looses the stability. The binding energy as a function of the mass ratio is shown in Fig. 1. The calculation shows that the stability is lost at around $m/M = 0.22$. Around that point the five-body system dissociates into a (M^+, M^+, m^-, m^-) four-body system (which is bound for any mass ratio m/M) and a M^+ particle. It seems that up to that point the heavy particles can form a slowly moving stable frame which will break beyond that mass ratio. This can be very nicely seen in Fig. 2, where the average distances between the particles are shown as a function of the mass ratio.

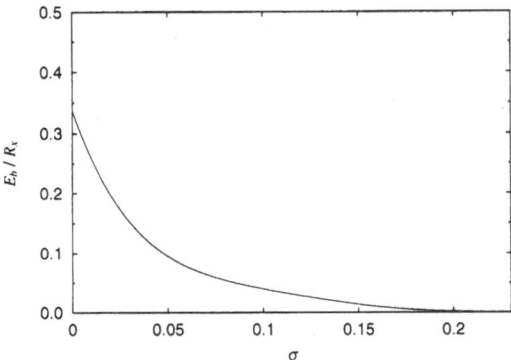

Figure 1. The binding energy of $(M^+, M^+, M^+, m^-, m^-)$ as a function of mass ratio. The energy is measured in Rydberg.

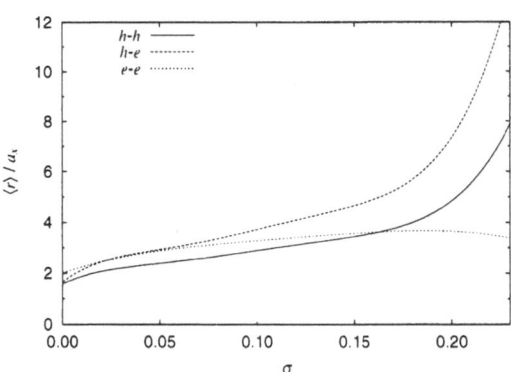

Figure 2. The average radius of $(M^+, M^+, M^+, m^-, m^-)$ as a function of mass ratio the distance is measured in Bohr.

Table 6. Energies of few electron quantum dots calculated by SVM and by a large scale shell model (SM) basis. Atomic units are used. The harmonic oscillator frequency, $\omega = 0.5$. The last column characterizes the shell model space used in the calculation.

$N\ (L, S, \pi)$	E (SVM)	K	E (SM)	$\hbar\omega$
3 $(1, 1/2, -)$	4.01324	100	4.01324	39
4 $(0, 0), +)$	6.35025	300	6.3506	14
5 $(1, 1/2, -)$	9.00331	500	9.0032	6

4 Confined few-electron systems

In this section we consider a quadratically confined three-dimensional few-electron system. This system can practically be viewed as an artificial atom, where the Coulombic attraction to the nucleus is replaced by a harmonic oscillator potential. Needless to say, the harmonic oscillator shell model type solutions are very succesful here because one only have to diagonalize the Coulombic repulsion between the electrons as the "residual interaction". We compare our results to a large scale shell model calculation [25] in Table 6 The energies are in very good agreement.

5 Summary

The stochastic variational method has proved to be useful in various fields of physics, including atomic, molecular, solid state, nuclear and subnuclear physics. This paper only reviewed a small part of the applications. Other contributions to this volume will show its usefulness in studies related to the structure of the baryons [26, 27]. Its main application is in nuclear physics, which has not been covered here but the interested reader can find examples in refs. [10-13]. We would like to extend its applicability to larger systems and for more complicated interactions. Such developments are under way.

Acknowledgement. This work has been supported by OTKA grant No. T 17298 (Hungary) and Grants-in-Aid for Scientific Research (Nos. 08044065 and 10640255) of the Ministry of Education, Science and Culture (Japan). The work of KV is also supported by the U. S. Department of Energy, Nuclear Physics Division, under contract No. W-31-109-ENG-39.

References

1. V.I. Kukulin and V.M. Krasnopol'sky: J. Phys. **G3**, 795 (1977)

2. Y. Suzuki and K. Varga: In: *Stochastic variational approach to quantum mechanical few-body problems*, Springer series on Lectures and Notes in Physics, in press.

18

3. K. Varga and Y. Suzuki: Phys. Rev. **C52**, 2885 (1995); Phys. Rev. **A53**, 1907 (1996)

4. Y. Suzuki, J. Usukura and K. Varga: J. Phys. **B31**, 31 (1998)

5. K. Varga and Y. Suzuki: Comp. Phys. Commun. **106**, 157 (1997)

6. K. Varga, J. Usukura and Y. Suzuki: Phys. Rev. Lett. **80**, 1876 (1998)

7. J. Usukura, K. Varga and Y. Suzuki: Phys. Rev. **A**, in press (1998)

8. Y. Suzuki, K. Varga and J. Usukura: Nucl. Phys. **A631**, 91c (1998)

9. K. Varga, Y. Suzuki and J. Usukura: Few-Body Systems **24**, 81 (1998)

10. K. Varga, Y. Ohbayasi and Y. Suzuki: Phys. Lett. **B396**, 1 (1997)

11. K. Varga, Y. Suzuki and R. G. Lovas: Nucl. Phys. **A371**, 447 (1994)

12. K. Arai, Y. Suzuki and K. Varga: Phys. Rev. **C51**, 2488 (1995)

13. K. Varga, Y. Suzuki and I. Tanihata: Phys. Rev. **C52**, 3013 (1995)

14. S.F. Boys: Proc. R. Soc. London. **A258**, 402 (1960)

15. K. Singer: Proc. R. Soc. London. **AA258**, 412 (1960)

16. W.H. Press et al.: In *Numerical Recipes in FORTRAN*, 2nd edition. New York: Cambridge University Press 1992

17. Y.K. Ho: Phys. Rev. **A48**, 4780 (1993)

18. R. Krivec, M.I. Haftel and V.B. Mandelzweig: Phys. Rev. **A44**, 7158 (1991)

19. G.G. Ryzhikh and J. Mitroy: Phys. Rev. Lett. **79**, 4124 (1997)

20. G.G. Ryzhikh, J. Mitroy and K. Varga: J. Phys. **B31**, L265 (1998)

21. A.M. Frolov, S.I. Kryuchkov and V.H. Smith, Jr.: Phys. Rev. **A51**, 4514 (1995)

22. A.M. Frolov and V.H. Smith, Jr.: J. Phys. **B29**, L433 (1996); Phys. Rev. **A55**, 2662 (1997)

23. Zhong-Chao Yan and G.W.F. Drake: Phys. Rev. **A52**, 3711 (1995)

24. C. Froese Fisher, J. Phys. **B26**, 855 (1993)

25. P. Navratil: Private communication.

26. L.Ya. Glozman, Z. Papp, W. Plessas, K. Varga and R. F. Wagenbrunn, Nucl. Phys. **A623**, 90 (1997); Phys. Rev. **C57**, 3406 (1998)

27. L.Ya. Glozman, W. Plessas, K. Varga and R.F. Wagenbrunn: Phys. Rev. D, in press (1998)

Few-Body Systems Suppl. 10, 19–26 (1999)

Few-
Body
Systems
© by Springer-Verlag 1999

Adiabatic hyperspherical expansion and three-body halos

A.S. Jensen[1], A. Cobis[1], D.V. Fedorov[1], E. Garrido[2] and E. Nielsen[1]

[1] Inst. of Physics and Astronomy, Aarhus University, DK-8000 Aarhus C
[2] Inst. de Estructura de la Materia, CSIC, Serano 123, E-28006 Madrid

Abstract. The three-body problem is formulated by use of hyperspherical coordinates and an adiabatic expansion of the Faddeev equations. For large distances and short-range potentials the equations simplify substantially and allow high accuracy solutions. The method can be formulated in two dimensions where the three-body systems have different properties. Applications for halo nuclei, Borromean systems, Efimov states, the hypertriton and helium plasma in astrophysical surroundings are briefly discussed.

1 Introduction

The discovery of the large interaction cross section of ^{11}Li and the subsequent interpretation in terms of an extended spatial neutron distribution triggered an enormous activity in the field of the so called neutron halos [1]. Halos are systems where one or two-particle separation energies are very small and the density distributions stretch out substantially into classically forbidden regions. It soon became clear that such weakly bound neutron dripline nuclei gain energy by correlating the motion of the valence neutrons in orbits outside a central core which resembles an ordinary nucleus. A description in terms of few-body models seemed appropriate and turned out as a tremendous success [2]. In particular the dominating configuration of ^{11}Li is two neutrons outside the ^{9}Li-core, i.e. a three-body system interacting via short-range two-body potentials.

Investigations of these weakly bound dripline nuclei revealed general properties related to the large-distance asymptotic behavior of the corresponding wave function. Spatially extended systems, called halos, arise when the binding energy and the orbital angular momentum are small [3]. The details of the potentials are unimportant in this limit [2]. Universal relations between radial moments and binding energies are established. They are distinctly different for two-body halos, for three-body halos with bound subsystems and for Borromean (without bound subsystems) three-body halos [2]. A description of the various three-body halos is obtained by solving the Faddeev equations. However, the decisive large-distance behavior must be found rather accurately. The wave functions and energies of the Efimov states are the extreme examples of these requirements [4].

2 Method

We use Jacobi coordinates $\mathbf{x}_i, \mathbf{y}_i$ and hyperspherical coordinates ρ, $\Omega = \{\alpha_i, \Omega_{xi}, \Omega_{yi}\}$, $\rho^2 = x_i^2 + y_i^2$, $x_i = \rho \sin \alpha_i$, $y_i = \rho \cos \alpha_i$, $i = 1, 2, 3$ [2, 5]. The wave function Ψ is for each ρ expanded on the adiabatic basis set $\{\Phi_n\}$:

$$\Psi(\rho, \Omega) = \frac{1}{\rho^{5/2}} \sum_{n=1}^{\infty} f_n(\rho) \Phi_n(\rho, \Omega) = \frac{1}{\rho^{5/2}} \sum_{n=1}^{\infty} f_n(\rho) \sum_{i=1}^{3} \frac{\phi_n^{(i)}(\rho, \Omega_i)}{\sin(2\alpha_i)} , \qquad (1)$$

where the three angular Faddeev components $\phi_n^{(i)}$ are solutions of

$$\hat{\Lambda}^2 \frac{\phi_n^{(i)}}{\sin(2\alpha_i)} + \frac{2m}{\hbar^2} \rho^2 V_{jk}(r_i) \Phi_n = \lambda_n(\rho) \frac{\phi_n^{(i)}}{\sin(2\alpha_i)} \qquad (2)$$

for two-body interactions V_{jk} and the angular eigenvalues $\lambda_n(\rho)$ depending on ρ. The radial expansion coefficients are then the solutions of

$$\left(-\frac{\partial^2}{\partial \rho^2} + \frac{\lambda_n + \frac{15}{4}}{\rho^2} - Q_{nn} - \frac{2mE}{\hbar^2} \right) f_n(\rho) = \sum_{n' \neq n} \left(Q_{nn'} + 2P_{nn'} \frac{\partial}{\partial \rho} \right) f_{n'}(\rho) , \quad (3)$$

where the P and Q-terms couple the different adiabatic solutions. The boundary conditions for large ρ are for bound and continuum states respectively

$$f_n(\rho) \propto \exp(-\kappa \rho) \qquad f_n^{(n')}(\rho) \to \delta_{n,n'} F_n^{(-)}(\kappa \rho) - S_{n,n'} F_n^{(+)}(\kappa \rho) , \qquad (4)$$

where $\kappa^2 = 2m|E|/\hbar^2$ and $S_{nn'}$ is the S-matrix and F is expressed in terms of the Hankel functions H_{K_n+2} $(K_n(K_n + 4) = \lambda_n(\rho = \infty))$

$$F_n^{(\pm)}(\kappa \rho) = \sqrt{\frac{m\rho}{4\hbar^2}} H_{K_n+2}^{(\pm)}(\kappa \rho) \to \sqrt{\frac{m}{2\pi\kappa\hbar^2}} \exp\left[\pm i\kappa\rho \pm \frac{i\pi}{2}(K_n + \frac{3}{2}) \right] . \qquad (5)$$

The S-matrix poles are found and computed by the complex scaling method, i.e. by solving Eq. (3) with complex energies $E = E_r - i\Gamma/2$ and the boundary condition $f_n \propto \exp(+i\rho\sqrt{2mE}/\hbar)$. Ignoring intrinsic spins the calculations proceed by partial wave expansion of the angular wave functions, i.e.

$$\Phi_n^{(i)}(\rho, \Omega_i) = \sum_{\ell_x \ell_y L} \frac{\phi_{n\ell_x \ell_y L}^{(i)}(\rho, \alpha_i)}{\sin(2\alpha_i)} \left[Y_{\ell_x m_x}(\Omega_{x_i}) \otimes Y_{\ell_y m_y}(\Omega_{y_i}) \right]^{LM_L} . \qquad (6)$$

Expressing all Faddeev components in Eq. (2) in one set of Jacobi coordinates and projecting on one set of angular momenta defines the operator R by

$$R_{ij}^{\ell_x \ell_y \ell_x' \ell_y' L} \left[\frac{\phi_{n\ell_x' \ell_y' L}^{(j)}(\rho, \alpha_j)}{\sin(2\alpha_j)} \right] \equiv \int d\Omega_{x_i} d\Omega_{y_i} \left[Y_{\ell_x \ell_y}^{LM_L}(\Omega_{x_i}, \Omega_{y_i}) \right]^*$$

$$\times \ \frac{\phi^{(j)}_{n\ell'_x\ell'_y L}(\rho,\alpha_j)}{\sin(2\alpha_j)} Y^{LM_L}_{\ell'_x\ell'_y}(\Omega_{x_j},\Omega_{y_j}) \ ,$$

where $Y^{LM_L}_{\ell_x\ell_y}$ are the coupled angular wave functions.

3   Large-distance solutions

For large distances the short-range potentials are zero unless $\alpha_i < \alpha^{(i)}_0 \propto \rho^{-1}$. By expansion to leading order in ρ^{-1} we only obtain finite contributions of the transformed wave functions in Eq. (7) when $\ell_x = \ell'_x = 0$, i.e.

$$R^{0L0LL}_{ij}\left[\frac{\phi^{(j)}_{n0LL}(\rho,\alpha_j)}{\sin(2\alpha_j)}\right] = (-1)^L \frac{\phi^{(j)}_{n0LL}(\rho,\varphi_k)}{\sin(2\varphi_k)} \ , \tag{7}$$

where $\tan\varphi_k = \sqrt{m_k(m_i+m_j+m_k)/(m_im_j)}$. The equations of motion then simplify substantially and in particular the higher angular momenta do not contribute to leading order in the s-wave equations which therefore can be solved independently in this limit. We now assume zero total angular momentum and spinless particles although the following procedure is generally applicable. The Faddeev equations in Eq. (2) can, for s-waves, be written as [5]

$$\left(\frac{\partial^2}{\partial\alpha_i^2} - \rho^2 v_i(\rho\sin\alpha_i) + \nu^2(\rho)\right)\phi^{(i)}_0(\rho,\alpha_i) = \rho^2 v_i(\rho\sin\alpha_i)$$

$$\times \sum_{j\neq i}\frac{1}{\sin(2\varphi_k)}\int_{|\varphi_k-\alpha_i|}^{\pi/2-|\pi/2-\varphi_k-\alpha_i|}d\alpha_j \phi^{(j)}_0(\rho,\alpha_j) \ , \tag{8}$$

where $\nu^2(\rho) = \lambda(\rho)+4$, $v_i(x) = \frac{2m}{\hbar^2}V_{jk}(x/mu_{jk})$ and $m\mu^2_{jk} = \frac{m_jm_k}{m_j+m_k}$.

The solution outside the potential, i.e for $\alpha_i > \alpha^{(i)}_0$ is then

$$\phi^{(i)}_0(\rho,\alpha_i) = A^{(i)}_0\sin\left[\nu\left(\alpha_i - \frac{\pi}{2}\right)\right] \tag{9}$$

and for $\alpha_i < \alpha^{(i)}_0$, where the integrand must be of the form in Eq. (9), we find the equations and the corresponding solutions

$$\left(\frac{\partial^2}{\partial\alpha_i^2} - \rho^2 v_i(\rho\sin\alpha_i) + \nu^2(\rho)\right)\phi^{(i)}_0(\rho,\alpha_i) =$$

$$2\rho^2 v_i(\rho\sin\alpha_i)\frac{\sin(\nu\alpha_i)}{\nu}\sum_{j\neq i}A^{(j)}_0\frac{\sin\left(\nu(\varphi_k-\frac{\pi}{2})\right)}{\sin(2\varphi_k)} \ , \tag{10}$$

$$\phi^{(i)}_0(\rho,\alpha_i) = \phi^{(i)}_{0h}(\rho,\alpha_i) - \frac{2}{\nu}\sin(\nu\alpha_i)\sum_{j\neq i}\frac{A^{(j)}_0\sin\left(\nu(\varphi_k-\frac{\pi}{2})\right)}{\sin(2\varphi_k)} \ , \tag{11}$$

where the homogeneous solution $\phi^{(i)}_{0h}(\rho,\alpha_i)$ for finite $v_i(0)$ simply is

$$\phi^{(i)}_{0h}(\rho,\alpha_i) = B^{(i)}_0\sin(\kappa\alpha_i) \qquad \kappa_i = \sqrt{\nu^2(\rho)-\rho^2 v_i(0)} \ . \tag{12}$$

For infinite $v_i(0)$, $\phi_{0h}^{(i)}$ depends on the details of the divergence of the potential. Matching at $\alpha_i = \alpha_0^{(i)}$ provides the quantization condition. The eigenvalues ν_n^2 converge towards the hyperspherical spectrum as ρ increases. Due to couplings the asymptotic values are now approached over a distance defined by scattering lengths, which might be very much larger than the interaction ranges.

For finite $v_i(0)$ and three identical bosons we find the eigenvalue equation

$$\frac{\kappa}{\nu}\cos(\alpha_0\kappa)\left[\frac{8}{\sqrt{3}}\sin(\nu\,\pi/6)\sin(\nu\alpha_0) - \nu\sin\left((\alpha_0 - \pi/2)\nu\right)\right]$$
$$= \sin(\alpha_0\kappa)\left[\frac{8}{\sqrt{3}}\sin(\nu\,\pi/6)\cos(\alpha_0\nu) - \nu\cos\left((\alpha_0 - \pi/2)\nu\right)\right],\qquad(13)$$

where α_0 is the matching point. If the scattering length a_s is large compared to ρ we get the Efimov equation by expansion

$$\sin(\nu_E\pi/6) - \nu_E\sqrt{3}\cos(\nu_E\pi/2) = \frac{\rho}{a_s}\sin(\nu_E\pi/2)\qquad(14)$$

with the limiting Efimov solution $\nu_E^2 = \lambda_E + 4 = -1.0125$.

4   Applications

The general procedure is now defined. We first solve the angular eigenvalue problem for small distances where the details of the potential are important and for large distances where the asymptotic equations are used. We minimize the basis size needed for small distances such that the results of the asymptotic solutions are reproduced at distances as small as possible. Finally we solve the coupled set of radial equations.

The key quantities are clearly the angular eigenvalue spectrum. We show in Fig. 1 an example for an asymmetric atomic helium trimer. The eigenvalues diverge for small ρ due to the strongly repulsive core in the interaction between two He-atoms. The lowest of these eigenvalues has an attractive pocket at small distance and diverges parabolically towards $-\infty$ for large ρ. Due to the very small two-body binding energy the lowest level remains almost constant up to about 1000 a.u. where the divergence sets in. The overlap region above which the asymptotic solutions can be used is around 100 a.u. The higher lying eigenvalues asymptotically approach the hyperspherical spectrum defined by $K(K+4)$ where K must be an even number because the total orbital angular momentum is zero.

Halos are systems with sizes extending well beyond the range of the effective interaction. A precise definition could be that half of the probability distribution is in this classically forbidden region [3]. The large distances are then essential for halos and our method is well suited. Accurate three-body calculations require a large number of hyperspherical harmonics in the basis as illustrated in Fig. 2 for the Borromean nucleus ^{6}He (n+n+^{4}He). The asymptotics of our large-distance equations are not reached before $K_{max} = 100$. An

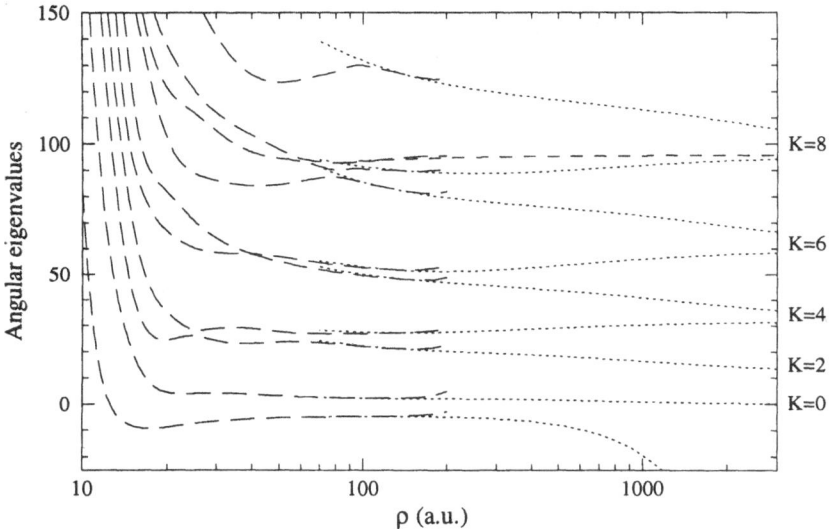

Figure 1. The lowest angular eigenvalues λ_n for the atomic $^3\text{He}^4\text{He}_2$-trimer as functions of ρ. The logarithmic scale on the ρ-axis uses the atomic length unit. The two-body interaction is the LM2M2 potential from [6]. The long-dashed curves are the angular eigenvalues obtained by diagonalization for the relatively small distances, the dotted curves are the large-distance s-wave solution and the short dashed curve is the extrapolation of the 9'th eigenvalue (non-zero ℓ) obtained by using the form $8(8+4) + c\rho^{-2}$. The hyperspherical quantum numbers K are shown to indicate the asymptotic spectrum $K(K+4)$ for $\rho \to \infty$. The lowest eigenvalue diverges as $-\rho^2$ corresponding to the bound dimer state.

even larger basis is needed if the asymptotic results are unavailable and larger distances still required.

The prototypes of three-body halo nuclei are ^{6}He and ^{11}Li (n+n+ ^{9}Li). Their *continuum structure* is so far not well established. We find a number of low-lying S-matrix poles in contrast to other investigations [7]. They are related to the large-distance structure treated accurately in our method. The wave functions for real energies are shown in Fig. 3 for ^{11}Li where the Efimov conditions nearly are fulfilled.

Breakup processes of halo nuclei reveal details of structure and reactions. *Neutron momentum distributions* are most sensitive but our three-body model and the sudden approximation reproduce the measurements rather accurately [8], see Fig. 4. The narrow distribution is a reflection of the spatial extension and the attractive final state interactions.

Weakly bound *two-dimensional structures* exhibit universal behavior for the qualitatively different potentials in Fig. 5 [9]. Two three-body bound states appear with energies proportional to the two-body energy. Borromean systems can only exist for two-body potentials with a short-range repulsive barrier. This weakly bound state is necessarily confined to distances within the barrier and

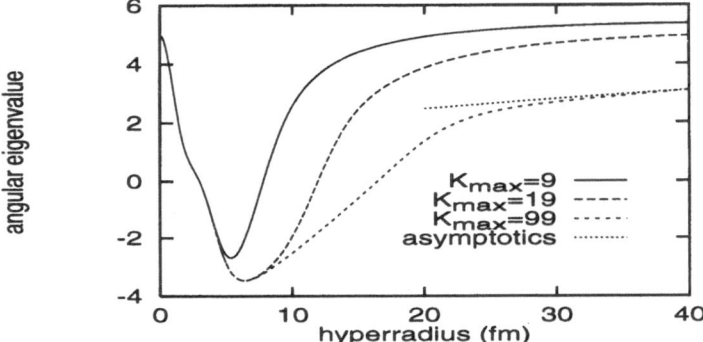

Figure 2. Lowest angular eigenvalue for ^{6}He ($J^\pi = 1^-$) calculated for different basis size and compared with results obtained from the asymptotic large-distance equations.

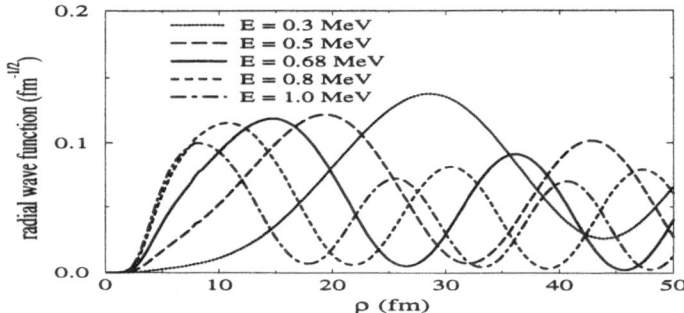

Figure 3. Absolute value of radial wave functions for ^{11}Li (n+n+^{9}Li) as functions of ρ for energies near the real part of the S-matrix pole $E_r = 0.68$ MeV for $J^\pi = \frac{3}{2}^+$.

therefore it depends on the details of the potential. The Efimov and Thomas effects are not present in two dimensions.

The *hypertriton* is the simplest strange halo system [10]. It is not Borromean since the neutron-proton system is bound. The Λ-particle is about 10 fm outside the deuteron. The large-distance properties of the Λ-nucleon potential are crucial and details correspondingly less important. To reproduce the binding energy we can constrain the singlet s-wave scattering length to be within 10% of 1.85 fm.

Triple α-correlations show up in the first excited 0^+-state of ^{12}C [11]. Two α-particles are only 0.092 MeV from being bound in ^{8}Be. The Efimov conditions are fulfilled, except for the Coulomb repulsion and the missing binding of 0.092 MeV. In a dense helium plasma the electrons screen the $\alpha - \alpha$ interaction at large distance. Then ^{8}Be becomes more bound and the tail of the long-range Coulomb interaction disappears. The Efimov conditions are approached. In Fig. 6 the energies are shown as function of the plasma density. At the two-body

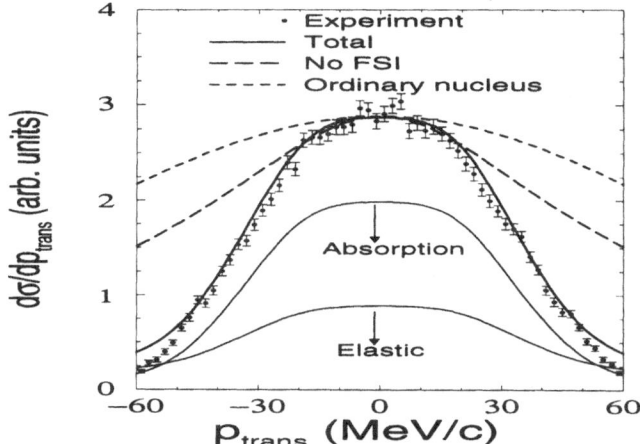

Figure 4. The transverse neutron momentum distribution after fragmentation of ^{6}He on ^{12}C at 240 MeV/u. The contributions are shown from both absorption and elastic scattering of one neutron on the target. The result without the final state interaction is also shown. The data are from Aleksandrov et al. Nucl. Phys. A633, 234 (1998).

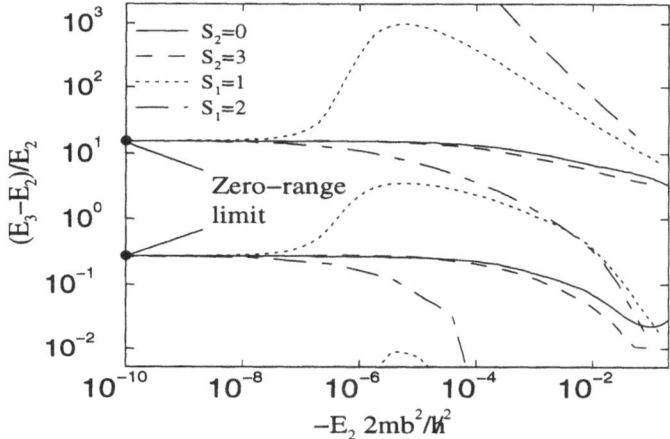

Figure 5. Ratio of three- to two-body energies as function of the two-body energy for different two-body potentials $V(r) = \frac{\hbar^2}{2mb^2}\left[S_1 \exp\left(-\frac{1}{2}r^2/b^2\right) + S_2 \exp\left(-2r^2/b^2\right)\right]$. The unspecified strength parameter S_i is used to vary the two-body binding.

threshold the Efimov states should appear, but the boundary conditions in the numerical calculations must then allow exponentially decreasing functions at extremely large distances beyond say 500 fm.

Atomic helium trimers exhibiting characteristics of both halo and Efimov states will be discussed in a separate contribution to this conference [12].

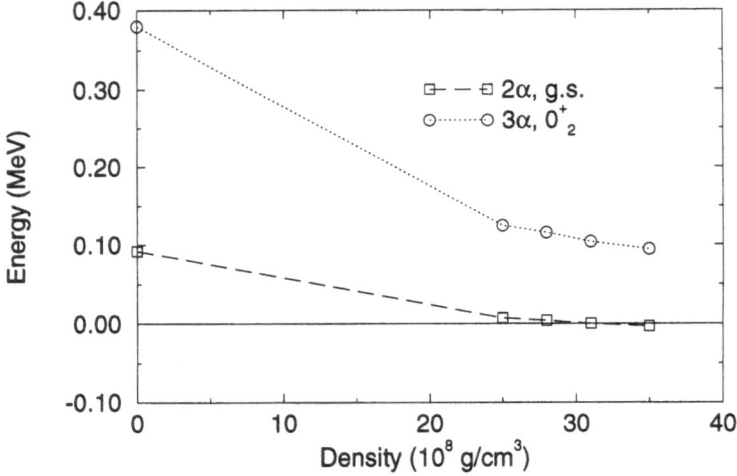

Figure 6. The energies of the ground state of the 2α system (long dashed) and the first excited 0^+ state in the 3α system (short dashed) as function of plasma density. The lines are drawn through the points to guide the eyes.

References

1. P.G. Hansen, A.S. Jensen, B. Jonson: Ann. Rev. Nucl. Part. Sci. **45**, 591 (1995)

2. D.V. Fedorov, A.S. Jensen, K. Riisager: Phys. Rev. **C49**, 201 (1994); **C50**, 2372 (1994)

3. D.V. Fedorov, A.S. Jensen, K. Riisager: Phys. Lett. **B312**, 1 (1993)

4. D.V. Fedorov, A.S. Jensen: Phys. Rev. Lett. **71**, 4103 (1993); D.V. Fedorov, A.S. Jensen, K. Riisager, ibid. **73**, 2817 (1994)

5. A.S. Jensen, E. Garrido, D.V. Fedorov: Few-Body Systems **22**, 193 (1997)

6. R.A. Aziz, M.J. Slaman: J. Chem. Phys. **94**, 8047 (1993).

7. A. Cobis, D.V. Fedorov, A.S. Jensen: Phys. Rev. Lett. **79**, 2411 (1997); Phys. Lett. **B424**, 1 (1998).

8. E. Garrido, D.V. Fedorov, A.S. Jensen: Europhysics Lett. in press.

9. E. Nielsen, D.V. Fedorov, A.S. Jensen: Phys. Rev. **A56**, 3287 (1997)

10. A. Cobis, D.V. Fedorov, A.S. Jensen: J. Phys. **G23**, 401 (1997)

11. D.V. Fedorov, A.S. Jensen: Phys. Lett. **B389**, 631 (1996)

12. E. Nielsen, D.V. Fedorov, A.S. Jensen: J. Phys. **B**, in press.

Few-Body Systems Suppl. 10, 27–36 (1999)

Few-
Body
Systems
© by Springer-Verlag 1999

Correlated Hyperspherical Harmonic Methods and Applications

A. Kievsky

Istituto Nazionale di Fisica Nucleare, Via Buonarroti 2, 56100 Pisa, Italy

Abstract. The Correlated Hyperspherical Harmonic Method is used to describe bound and scattering states in few-body systems. Special attention is given to the three-nucleon problem in which the presence of correlation factors substantially accelerate the convergence of the expansion. For scattering states the complex form of the Kohn variational principle is used to determine the S- or T-matrix. Using this formalism the method can be extended to energies above the three-body breakup threshold. Using realistic NN interactions comparisons to experimental data are given.

1 Introduction

In the theoretical description of bound and scattering states in few-body systems two important points have to be addressed. First of all, the method of solution. Among different possibilities, variational principles represent a powerful tool for the description of the dynamics of these systems. Here, particular interest deserves the Correlated Hyperspherical Harmonic (CHH) basis, due to the flexibility it provides in constructing the structure of the system of interest. The second point is related to the possibility of improving on our knowledge of the interaction through the study of observables. In this context nuclear systems are good examples because the interaction between nucleons is not yet known within the desired accuracy. In particular, three-body forces can be tested only by calculations of, at least, three-nucleon states and their comparison with the available experimental data.

In order to cover the richness of these systems the use of correlated bases, first introduced for bound states, has been extended to scattering states. In the present work the merit of both the CHH basis and the Kohn variational principle for the description of scattering states in three- and four-body systems will be shown. Applications to N-d scattering using realistic potentials will be presented. Moreover, the possibility of using the Kohn variational principle for energies above the deuteron breakup threshold and with the Coulomb potential taking into account will be discussed.

2 The CHH expansion

2.1 Bound State Calculations

The CHH expansion has been introduced in refs. [1, 2] for the description of the ground state of three- and four-nucleon systems. Recently, the Kohn variational has been used to extend the method to scattering states at energies below the three-body breakup [3, 4]. Following the aforementioned references and concentrating in the three–nucleon case for simplicity, the bound state wave function is written as a sum of three Faddeev–like amplitudes

$$\Psi = \psi(\mathbf{x}_i, \mathbf{y}_i) + \psi(\mathbf{x}_j, \mathbf{y}_j) + \psi(\mathbf{x}_k, \mathbf{y}_k) \ , \tag{1}$$

where $\mathbf{x}_i, \mathbf{y}_i$ are Jacobi coordinates. Each i–amplitude has total angular momentum JJ_z and total isospin TT_z. Using LS coupling it can be decomposed into channels

$$\psi(\mathbf{x}_i, \mathbf{y}_i) = \sum_\alpha^{N_c} F_{\alpha,i} \phi_\alpha(x_i, y_i) \mathcal{Y}_\alpha(jk, i) \tag{2}$$

$$\mathcal{Y}_\alpha(jk, i) = \left\{ \left[Y_{\ell_\alpha}(\hat{x}_i) Y_{L_\alpha}(\hat{y}_i) \right]_{\Lambda_\alpha} \left[s_\alpha^{jk} s_\alpha^i \right]_{S_\alpha} \right\}_{JJ_z} \left[t_\alpha^{jk} t_\alpha^i \right]_{TT_z}, \tag{3}$$

where x_i, y_i are the moduli of the Jacobi coordinates, $\mathcal{Y}_\alpha$ is the angular-spin-isospin function for each channel and $F_{\alpha,i}$ is a correlation factor.

The two-dimensional amplitude ϕ_α is expanded in terms of the Hyperspherical Harmonic basis:

$$\phi_\alpha(x_i, y_i) = \rho^{\ell_\alpha + L_\alpha} \left[\sum_K u_K^\alpha(\rho)^{(2)} P_K^{\ell_\alpha, L_\alpha}(\phi_i) \right] \ , \tag{4}$$

where the hyperspherical variables are defined by the equations $x_i = \rho \cos \phi_i$ and $y_i = \rho \sin \phi_i$, and $^{(2)} P_K^{\ell, L}(\phi)$ is an hyperspherical polynomial.

Different choices can be made for the correlation factor F. If it is taken as a pair correlation function, i.e. $F_{\alpha,i} = f_\alpha(x_i)$, then we have the Pair Correlated Hyperspherical Harmonic (PHH) basis [1]. More complex forms have been studied, for example, specific products of correlation functions [1, 2], i.e. for three particles we have $F_{\alpha,i} = f_\alpha(x_i) g_\alpha(x_j) g_\alpha(x_k)$. In all cases the intention is to take into account the correlations introduced by the strong repulsion of the potential at small distances. In practical applications the PHH basis has been used for $A = 3$ systems whereas the CHH basis has been employed for $A = 4$ systems. When the system interacts through a potential having a hard core, only the CHH can be applied.

The unknown quantities in the ground state wave function are the hyper-radial functions $u_K^\alpha(\rho)$. They can be obtained by applying the Rayleigh-Ritz variational principle. A possible choice is to expand the hyperradial functions in terms of Laguerre polynomials times an exponential tail:

$$u_K^\alpha(\rho) = \sum_m A_{K,m}^\alpha L_m^{(5)}(z) \exp(-z) \ , \tag{5}$$

where $z = \gamma\rho$ and γ is a nonlinear variational parameter. Let $|\alpha, K, m >$ be a totally antisymmetric element of the expansion basis, where α denotes the channel including the angular-spin-isospin dependence and the correlation function, and K, m are the indices of the hyperspherical and Laguerre polynomials, respectively. In terms of the basis elements the wave function (1) can be written as

$$\Psi = \sum_{\alpha, K, m} A_{K,m}^{\alpha} |\alpha, K, m > . \tag{6}$$

The problem is to determine the linear coefficients $A_{K,m}^{\alpha}$. They are obtained together with the energy of the system by solving the following generalized eigenvalue problem

$$\sum_{\alpha', K', m'} A_{K',m'}^{\alpha'} < \alpha, K, m|H - E|\alpha', K', m' >= 0 . \tag{7}$$

In ref. [5] the convergence properties of the expansion with respect to the indices K, m and the nonlinear parameter γ have been analyzed. Here some results obtained with the CHH or PHH basis are presented. For three nucleons the PHH basis has been used and different potential models have been taken into account: the Argonne NN potential (AV14), including the Coulomb potential (AV14+C) and the three-body force of Tucson-Melbourne (AV14+TM). The CHH basis has been used for describing the ground state of three helium atoms (trimer). To this aim the potential LM2M2 of Aziz and Slaman [6] and the new potential of Tang, Toennies and Yiu (TTY) [7] have been applied. For applications to four nucleons see ref. [8]. In Table 1 the corresponding binding energies are presented and compared to results obtained with other techniques.

Table 1. Binding energies for the three-nucleon system and for the helium trimer using different realistic potential models

method	potential	B(MeV)	potential	B(MeV)	potential	B(MeV)
PHH	AV14	7.684	AV14+C	7.033	AV14+TM	8.486
	ref. [9]	7.684	ref. [9]	7.033	ref. [10]	8.486

method	potential	B(mK)	potential	B(mK)
CHH	LM2M2	126.3	TTY	126.3
	ref.[11]	125.2		

An extremely good agreement is obtained in all cases showing the flexibility of the correlated basis to describe the structure of the system.

2.2 Scattering Calculations

The CHH or PHH basis can be used to describe scattering states. The complex Kohn variational principle is applied to directly obtain the S- or T-matrix. The description of the method for three- and four–nucleon systems is given in refs.

[3, 4, 5] for energies below the three-body breakup threshold, and in ref. [12] for energies above the three-body breakup. Here the method is outlined briefly.

The wave function corresponding to an N-d scattering state is written as a sum of two terms

$$\Psi = \Psi_C + \Psi_A \ . \tag{8}$$

The Ψ_C term describes the system when the three–nucleons are close to each other. For large interparticle separations and energies below the deuteron breakup it goes to zero, whereas for higher energies it must describe three outgoing particles [12]. The second term Ψ_A is the solution of the Schrödinger equation in the asymptotic nuclear region, where the two incident clusters are well separated. It can be written as a linear combination of functions of the form

$$\Omega_{LSJ}^{\lambda}(\mathbf{x}_i, \mathbf{y}_i) \qquad = \sum_{l_\alpha = 0,2} w_{l_\alpha}(x_i) \mathcal{R}_L^{\lambda}(y_i) \times$$
$$\left\{ \left[[Y_{l_\alpha}(\hat{x}_i) s_\alpha^{jk}]_1 \otimes s^i \right]_S \otimes Y_L(\hat{y}_i) \right\}_{J J_z} [t_\alpha^{jk} t^i]_{T T_z} \ , \tag{9}$$

where $w_{l_\alpha}(x_i)$ is the deuteron component in the state $l_\alpha = 0, 2$ and L is the relative angular momentum of the deuteron and the incident nucleon. The superscript λ indicates the regular ($\lambda = R$) or the irregular ($\lambda = I$) solution. They can be combined to form a general asymptotic state

$$\Omega_{LSJ}^{+}(\mathbf{x}_i, \mathbf{y}_i) = \Omega_{LSJ}^{0}(\mathbf{x}_i, \mathbf{y}_i) + \sum_{L'S'} {}^{J}\mathcal{L}_{LL'}^{SS'} \Omega_{L'S'J}^{1}(\mathbf{x}_i, \mathbf{y}_i) \ , \tag{10}$$

with the following asymptotic functions

$$\Omega_{LSJ}^{0}(\mathbf{x}_i, \mathbf{y}_i) = u_{00} \Omega_{LSJ}^{R}(\mathbf{x}_i, \mathbf{y}_i) + u_{01} \Omega_{LSJ}^{I}(\mathbf{x}_i, \mathbf{y}_i) \ , \tag{11}$$
$$\Omega_{LSJ}^{1}(\mathbf{x}_i, \mathbf{y}_i) = u_{10} \Omega_{LSJ}^{R}(\mathbf{x}_i, \mathbf{y}_i) + u_{11} \Omega_{LSJ}^{I}(\mathbf{x}_i, \mathbf{y}_i) \ . \tag{12}$$

The matrix elements u_{ij} are selected according to the four different choices of the matrix $\mathcal{L} = K, K^{-1}, S, -\pi T$ [13].

The three-nucleon scattering wave function for an incident state with relative angular momentum L, spin S and total angular momentum J is

$$\Psi_{LSJ}^{+} = \sum_{i=1,3} \left[\Psi_C(\mathbf{x}_i, \mathbf{y}_i) + \Omega_{LSJ}^{+}(\mathbf{x}_i, \mathbf{y}_i) \right] \ , \tag{13}$$

and its complex conjugate is Ψ_{LSJ}^{-}. A variational estimate for the matrix $\mathcal{L}$ can be obtained from the generalized Kohn variational principle

$$[{}^{J}\mathcal{L}_{LL'}^{SS'}] = {}^{J}\mathcal{L}_{LL'}^{SS'} - \frac{2}{\det(u)} \langle \Psi_{LSJ}^{+} | H - E | \Psi_{L'S'J}^{+} \rangle \ . \tag{14}$$

The trial parameters in the wave function Ψ_{LSJ}^{+} are varied in order to obtain a stationary value of the above functional.

For energies below the three-body breakup threshold the Ψ_C term goes to zero when the hyperradius $\rho \to \infty$, and it has been expanded in terms of the polynomial basis introduced before:

$$\Psi_C = \sum_{\alpha, K, m} A_{K,m}^{\alpha} |\alpha, K, m> . \tag{15}$$

With this choice of Ψ_C, the variation of the diagonal functionals with respect to the linear parameters leads to the following linear system

$$\sum_{\alpha', K', m'} A_{K', m'}^{\alpha'} < \alpha, K, m|H - E|\alpha', K', m' >= D_{LSJ}^{\lambda}(\alpha, K, m) , \tag{16}$$

with the two different inhomogeneous D terms corresponding to $\lambda \equiv 0, 1$

$$D_{LSJ}^{\lambda}(\alpha, K, m) = \sum_{j} < \alpha, K, m|H - E|\Omega_{LSJ}^{\lambda}(\mathbf{x}_j, \mathbf{y}_j) > . \tag{17}$$

The first order solution of the matrix $\mathcal{L}$ is obtained by solving the following algebraic equations

$$\sum_{L'', S''} {}^{J}\mathcal{L}_{LL''}^{SS''} X_{L'L''}^{S'S''} = Y_{LL'}^{SS'} , \tag{18}$$

with the coefficients X and Y defined as

$$X_{LL'}^{SS'} = \langle \Omega_{LSJ}^{1} + \Psi_{LSJ}^{1}|H - E|\Omega_{L'S'J}^{1} \rangle , \tag{19}$$

$$Y_{LL'}^{SS'} = \langle \Omega_{LSJ}^{0} + \Psi_{LSJ}^{0}|H - E|\Omega_{L'S'J}^{0} \rangle , \tag{20}$$

where Ψ_{LSJ}^{λ} is the solution of the set of Eqs. (16) with the corresponding inhomogeneous term. The second order estimate $[{}^{J}\mathcal{L}_{LL'}^{SS'}]$ is obtained from the first order solution using Eq. (14). The calculations have been performed by applying Eq. (14) to the S–matrix.

The extension of the method to energies above the three-body breakup deserves special interest. In this case the internal part asymptotically describes the breakup configuration. This behavior is obtained from the asymptotic form of the differential equation for the hyperradial functions $u_K^{\alpha}(\rho)$ [12]

$$\sum_{\alpha', k'} \left[\delta_{\alpha, \alpha'} \delta_{k, k'} \left(\frac{d^2}{d\rho^2} - \frac{\ell(\ell + 1)}{\rho^2} + Q^2 \right) - \frac{2Q\chi_{k,k'}^{\alpha',\alpha}}{\rho} + \frac{h_{k,k'}^{\alpha,\alpha'}}{\rho^3} \right] u_{\alpha'k'}(\rho) = 0 \tag{21}$$

where the χ term originates from the Coulomb potential matrix and shows the expected $1/\rho$ behavior. The kinetic operator and the nuclear potential contribute both to the h term. Here $Q^2 = mE/\hbar^2$ and $\ell = \ell_\alpha + L_\alpha + 2k + 3/2$. The solutions of the above equations are obtained numerically with outgoing boundary conditions at a given value of the hyperradius $\rho = \rho_0$. The value of the matching radius ρ_0 is not relevant provided that the asymptotic form is reached. For $\rho \to \infty$ such solutions evolve as

$$u_{\alpha k}(\rho) \to -\sum_{\alpha' k'} (e^{-i\chi \log 2Q\rho})_{\alpha\alpha'}^{kk'} S_{\alpha'k'} e^{iQ\rho} \tag{22}$$

corresponding to the asymptotic behavior of three outgoing particles interacting through the long-range Coulomb potential [15]. With the above considerations, for energies above the three-body breakup, the hyperradial functions are written as follows

$$u_K^\alpha(\rho) = \sum_m A_{K,m}^\alpha L_m^{(5)}(z) \exp(-z) + u_{\alpha,k}^0 \qquad (23)$$

where the functions $u_{\alpha,k}^0(\rho)$ have the appropriate asymptotic behavior for $\rho > \rho_0$. For smaller values of the hyperradius they are obtained by solving a single channel equation for each value of the indices α, k.

The present method has been applied to calculate the differential cross section and vector and tensor polarization observables in n-d and p-d elastic scattering. Applications to four nucleon scattering are given in ref. [8]. The interaction used was the new potential of Argonne, the AV18 model. The results are compared to experimental data and important conclusions about the capability of the interaction to reproduce the dynamics can be extracted. The comparison starts at a proton incident energy of $E_p = 650$ keV where a complete set of p-d observables has been measured recently [16]. Incident energies of $E_p = 2.5, 5.0, 10.0$ MeV are also considered. The experimental data are from ref. [17]. At these four energies the differential cross section (Fig.1) and the vector analyzing power A_y (Fig.2) are shown. For the differential cross section a good agreement is observed at all energies (the data at $E_p = 650$ keV are preliminary [18]), while for A_y the well known theoretical underprediction is observed. At $E_N = 10$ MeV theoretical curves for both n-d and p-d are given in order to evaluate Coulomb effects. The energies considered correspond to incident deuteron energies of $E_D = 1.3, 5.0, 10.0, 20.0$ MeV. At these energies the deuteron analyzing power iT_{11} (Fig. 3) and the three tensor analyzing powers T_{20} (Fig. 4), T_{21} (Fig. 5) and T_{22} (Fig. 6) are given. Again the theoretical underprediction for iT_{11} is observed in this energy range. Conversely, the tensor analyzing powers are well reproduced even though small differences are observed. For example, the p-d data at the minimum of T_{20} and T_{22} at $E_D = 20$ MeV are better reproduced by the n-d calculations than by the p-d ones.

3 Summary

In the present work the CHH (or PHH) basis method is discussed. It is shown that a correlated basis can be used to describe accurately few-body systems through variational principles. The Rayleigh-Ritz principle is applied for bound states while for scattering states the complex form of the Kohn variational principle is used. The extension of the method to processes above the three-body breakup threshold is outlined. Applications are presented to describe strongly interacting systems, for example, nuclei with $A = 3, 4$ and small helium clusters that were observed recently [19]. Furthermore, N-d scattering at several energies is discussed. From these studies important conclusions about the underlying dynamics of these systems as well as the evaluation of Coulomb effects and the importance of three-body forces can be deduced.

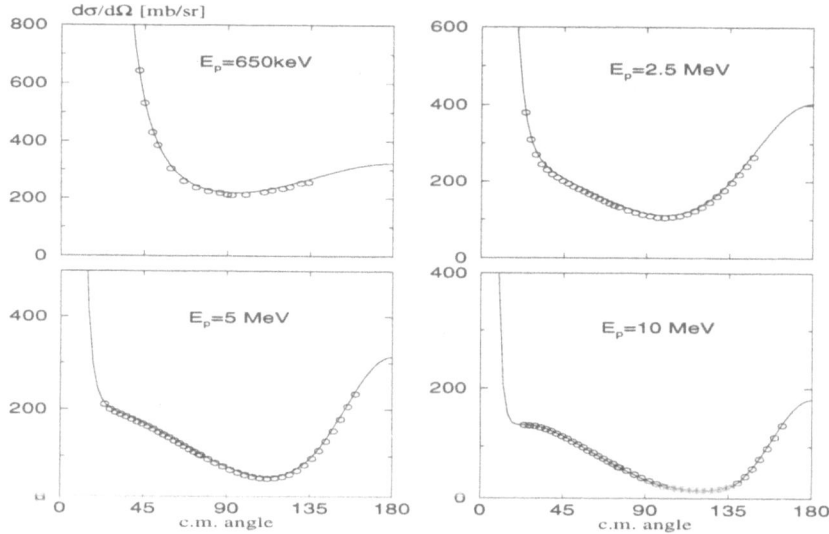

Figure 1. Differential cross section for p-d scattering

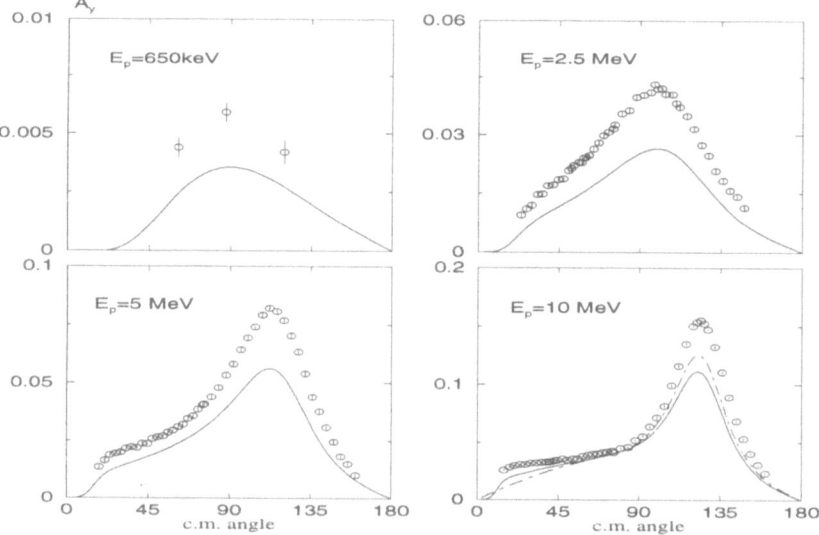

Figure 2. Vector analyzing power A_y for p-d (solid curve) and n-d (dashed curve) scattering

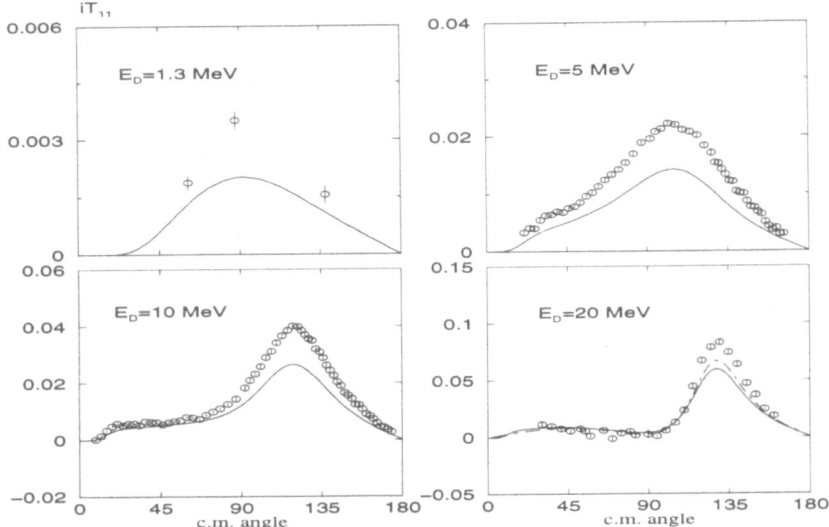

Figure 3. Deuteron analyzing power iT_{11} for p-d (solid curve) and n-d (dashed curve) scattering

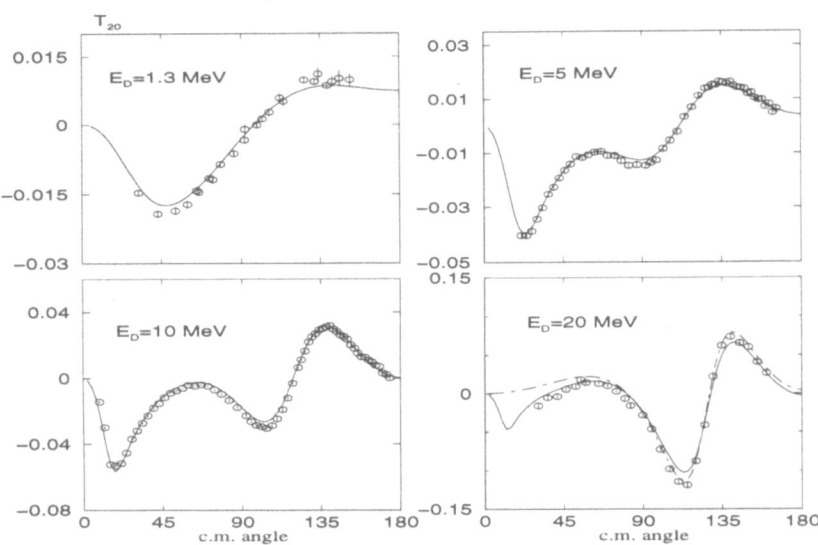

Figure 4. Tensor analyzing power T_{20} for p-d (solid curve) and n-d (dashed curve) scattering

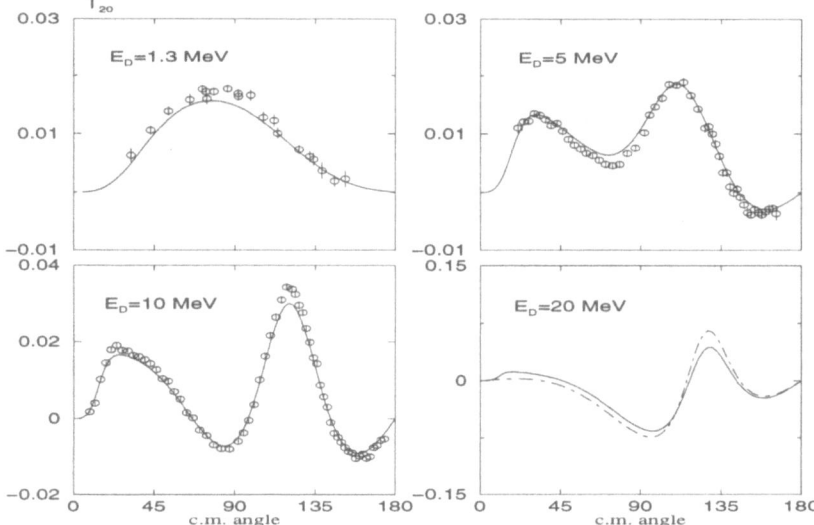

Figure 5. Tensor analyzing power T_{21} for p-d (solid curve) and n-d (dashed curve) scattering

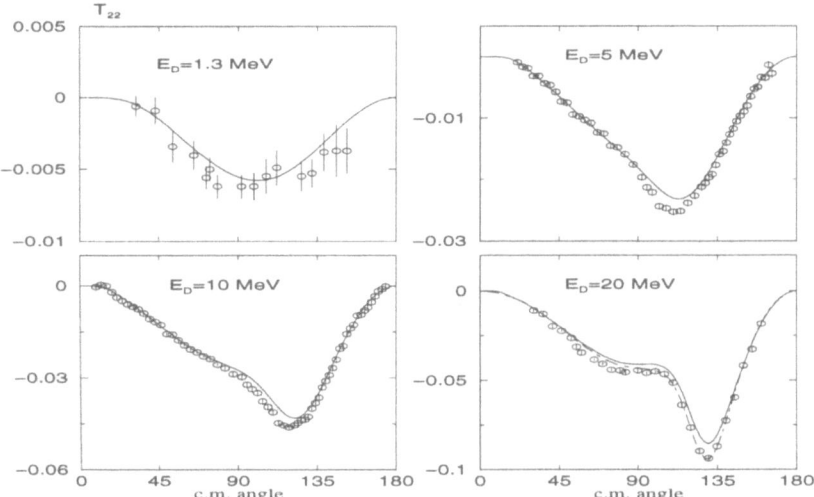

Figure 6. Tensor analyzing power T_{22} for p-d (solid curve) and n-d (dashed curve) scattering

Acknowledgement. I would like to thank S. Rosati, M. Viviani and W. Tornow for many fruitful discussions.

References

1. A. Kievsky, M. Viviani, S. Rosati: Nucl. Phys. **A551**, 241 (1993)

2. M. Viviani, A. Kievsky, S. Rosati: Few Body Syst. **18**, 25 (1995)

3. A. Kievsky, M. Viviani, S. Rosati: Nucl. Phys. **A577**, 511 (1994)

4. M. Viviani, S. Rosati, A. Kievsky: Phys. Rev. Lett. (in print)

5. A. Kievsky: Nucl. Phys. **A624**, 125 (1997)

6. R.A. Aziz, M.J. Slaman: J. Chem. Phys. **94**, 8047 (1991)

7. K.T. Tang, J.P. Toennies, C.L. Yiu: Phys. Rev. Lett. **74**, 1546 (1995)

8. M. Viviani: Contribution to this Conference

9. H. Kameyama, M. Kamimura, Y. Fukushima: Phys. Rev. **C40**, 974 (1989)

10. A. Nogga et al.: Phys. Lett. **B409**, 19 (1997)

11. E. Nielsen, D.V. Fedorov, A.S. Jenses: Preprint physics/9806020

12. A. Kievsky, M. Viviani, S. Rosati: Phys. Rev. **C56**, 2987 (1997)

13. R.R. Lucchese: Phys. Rev. **A40**, 6879 (1989)

14. A. Kievsky et al.: Nucl. Phys. **A607**, 402 (1997)

15. P. Merkuriev: Ann. Phys. (N.Y.) **130**, 395 (1980)

16. A. Kievsky et al.: Phys. Lett. **B406**, 292 (1997); C.R. Brune et al.: Phys. Lett. **B428**, 13 (1998)

17. S. Shimizu et al.: Phys. Rev. **C52**, 1193 (1995); J. Sowinski et al.: Nucl. Phys. **A464**, 223 (1987); F. Sperisen et al.: Nucl. Phys. **A422**, 81 (1984)

18. C.R. Brune: Private Communication

19. W. Schöllkopf, J.P. Toennies: J. Chem. Phys. **104**, 1155 (1996)

Few-Body Systems Suppl. 10, 37–40 (1999)

Few-
Body
Systems
© by Springer-Verlag 1999

Low-energy scattering in four nucleon systems. Method of Cluster Reduction

S.L. Yakovlev*, I.N. Filikhin

Department of Mathematical and Computational Physics, St. Petersburg State University, 198904 St. Petersburg, Petrodvoretz, Ulyanovskaya Str. 1, Russia

1   Cluster reduction of YDE

The elastic and rearrangement processes in the four-particle system with two clusters in the initial and final states can be treated adequately on the basis of Yakubovsky differential equations (YDE) [1, 2, 3]

$$(H_0 + V_{a_3} - E)\Psi_{a_3 a_2} + V_{a_3} \sum_{(c_3 \neq a_3) \subset a_2} \Psi_{c_3 a_2} = -V_{a_3} \sum_{d_2 \neq a_2} \sum_{(d_3 \neq a_3) \subset a_2} \Psi_{d_3 d_2}.$$

For the two cluster collisions the YDE admit a further reduction. Let $H_0 = T_{a_2} + T^{a_2}$ be the separation of the kinetic energy operator into the intrinsic one, T_{a_2}, with respect to the clusters of a_2 part and the kinetic energy, T^{a_2}, of the relative motion of a_2 clusters. The cluster reduction procedure consists in expanding the components $\Psi_{a_3 a_2}$ along the basis of the solutions to the Faddeev equations (FE) for subsystems of partition a_2

$$(T_{a_2} + V_{a_3})\psi_{a_2,k}^{a_3} + V_{a_3} \sum_{(c_3 \neq a_3) \subset a_2} \psi_{a_2,k}^{c_3} = \varepsilon_{a_2}^k \psi_{a_2,k}^{a_3}.$$

The expansion has the form

$$\Psi_{a_3 a_2}(\mathbf{X}) = \sum_{k=0}^{\infty} \psi_{a_2,k}^{a_3}(\mathbf{x}_{a_2}) F_{a_2}^k(\mathbf{z}_{a_2}). \tag{1}$$

Here, the unknown amplitudes $F_{a_2}^k(\mathbf{z}_{a_2})$ depend only on the relative position vector $\mathbf{z}_{a_2}$ between the clusters of the partition a_2 while $\mathbf{x}_{a_2}$ denote intrinsic

---

*E-mail address: yakovlev@mph.phys.spbu.ru

coordinates with respect to clusters of partition a_2. The basis of the solutions of FE is complete but not the orthogonal one [4, 5] due to not Hermiteness of FE. The biorthogonal basis is formed by the solutions conjugate to FE equations

$$(T^{a_2} + V_{a_3})\phi^{a_3}_{a_2,k} + \sum_{(c_3 \neq a_3) \subset a_2} V_{c_3}\phi^{c_3}_{a_2,k} = \varepsilon^k_{a_2}\phi^{a_3}_{a_2,k}.$$

Introducing the expansion of $\Psi_{a_3 a_2}(\mathbf{X})$ into YDE and projecting onto the elements of the biorthogonal basis $\{\phi^{a_3}_{a_2,k}(\mathbf{x}_{a_2})\}$ lead to resulting reduced YDE (RYDE)[6], [7] for $F^k_{a_2}(\mathbf{z}_{a_2})$

$$(T^{a_2} - E + \varepsilon^k_{a_2})F^k_{a_2} = -\sum_{a_3 \subset a_2} \langle \phi^{a_3}_{a_2,k} | V_{a_3} \sum_{d_2 \neq a_2} \sum_{(d_3 \neq a_3) \subset a_2} \sum_{l \geq 0} \psi^{d_3}_{d_2,l} F^l_{d_2} \rangle, \quad (2)$$

where the brackets $\langle . | . \rangle$ mean the integration over $\mathbf{x}_{a_2}$. The boundary conditions for $F^k_{a_2}(\mathbf{z}_{a_2})$ have the following *two body* form as $|\mathbf{z}_{a_2}| \to \infty$

$$F^k_{a_2}(\mathbf{z}_{a_2}) \sim \delta_{k0}[\delta_{a_2 b_2} \exp i(\mathbf{p}_{a_2}, \mathbf{z}_{a_2}) + \mathcal{A}_{a_2 b_2}\frac{\exp i\sqrt{E - \varepsilon^0_{a_2}}|\mathbf{z}_{a_2}|}{|\mathbf{z}_{a_2}|}],$$

where the index b_2 corresponds to the initial state, and $\mathbf{p}_{a_2}$ is the conjugate of the $\mathbf{z}_{a_2}$ momentum. The charged particles case can be treated in the framework of YDE formalism by adjusting the Coulomb potentials to the kinetic energy operator H_0 and by replacing the plane and spherical waves in the asymptotics by respective Coulomb modifications [3].

2   Application to low-energy scattering in the four nucleon system

RYDE (2) after suitable partial wave decomposition become one dimensional in variable $|\mathbf{z}_{a_2}|$. We solve numerically these equations by means of finite-difference approximation in $|\mathbf{z}_{a_2}|$ variable, spline expansion of the integrand in the right hand side and truncation of the summation over l by a finite number N. In all the cases, satisfactory convergence was observed with the parameter N not exceeding 20, what supports the efficiency of the expansion (1). The maximal size of the linear system to solve was of the order 10^5, so that the calculations were performed on a standard workstation. We have used the MT I-III model with parameters from [8] for NN forces.

The first group of results we are presenting concerns the isospin approximation (*i. e.* neglecting Coulomb interaction). The values of channel scattering lengths for nucleon scattered off three-nucleon cluster presented in Table 1 are in agreement with results of the Grenoble group obtained by a direct discretization of YDE. Note, that the $T = 1$ channels correspond to singlet and triplet $n - {}^3\text{H}$ scattering.

The second group of results is more realistic in view of taking account of the Coulomb interaction between protons. In Table 2 we collect our results for $p - {}^3\text{H}$ $({}^{2S+1}\text{A}_{pt})$ and $p - {}^3\text{He}$ $({}^{2S+1}\text{A}_{ph})$ elastic scattering lengths with results

Table 1. $S - T$ channel $N - NNN$ scattering lengths (in fm)

S	T	[9], [10]	our
0	0	14.75	14.7
1	0	3.25	3.2
0	1	4.13	4.0
1	1	3.73	3.6

obtained on the basis of Resonating Group Method (RGM) calculations [11] and experimental values from [12]. Last two columns of Table 2 show the position, E_r, and width, Γ, in MeV of the ^{4}He nucleus 0^+ resonance (measured relatively to $n - {}^3$He threshold) extracted from low energy behavior of calculated $p - {}^3$H phase-shifts. Table 3 contains results of calculations of scattering lengths for $n - {}^3$He ($^{2S+1}A_{nh}$) and ^{2}H$-^2$H ($^{2S+1}A_{dd}$) scattering. Due to open rearrangement channels, the scattering lengths in these cases have a non trivial imaginary part.

Table 2. Singlet and triplet $p - {}^3$H and $p - {}^3$He scattering lengths (in fm)

	$^1A_{pt}$	$^3A_{pt}$	$^1A_{ph}$	$^3A_{ph}$	E_r	Γ
our	-22.6	4.6	8.2	7.7	0.15	0.3
[11]	-21.46				0.12	0.26
[12]			10.8± 2.6	8.1±0.5		
[13]					0.3±0.05	0.27±0.05

Table 3. Singlet and triplet $p - {}^3$H and singlet and pentaplet ^{2}H$-^2$H scattering lengths (in fm)

	$^1A_{nt}$	$^3A_{nt}$	$^1A_{dd}$	$^5A_{dd}$
our	7.5-4.2i	3.0+0.0i	10.2-0.2i	7.5
[11]	7.25-3.92i			
[14]				6.68-0.135i

Acknowledgement. This work was partially supported by the Russian Foundation for Basic Research grant No. 98-02-18190. Authors would like to thank G.M. Halle for providing his preliminary result of R-matrix analysis of experimental data for pentaplet ^{2}H$-^2$H scattering. One of the authors (S.L. Ya.) is thankful to the Organizing Committee of the 16th European Conference on Few-Body Problem in Physics for financial support of his participation in the Conference.

References

1. S.P. Merkuriev, S.L. Yakovlev: Dans. Akad. Nauk (SSSR) **262**, 591 (1982)

2. S.P. Merkuriev, S.L. Yakovlev: Theor. Math. Phys. **56**, 673 (1983)

3. S.P. Merkuriev, S.L. Yakovlev, C. Gignoux: Nucl. Phys. **A431**, 125 (1984)

4. S.L. Yakovlev: Theor. Math. Phys. **102**, 323 (1995)

5. S.L. Yakovlev: Theor. Math. Phys. **107**, 513 (1996)

6. S.L. Yakovlev, I.N. Filikhin: Yad. Fiz. **58**, 817 (1995)

7. S.L. Yakovlev, I.N. Filikhin: Yad. Fiz. **60**, 1962 (1997)

8. G.L. Payne, J.L. Friar, B.F. Gibson: Phys. Rev. **C26**, 1385 (1982)

9. F. Ciesielski, J. Carbonell, C. Gignoux: Nucl. Phys. **A631**, 653 (1998)

10. F. Ciesielski, J. Carbonell, C. Gignoux, A. Fonseca: Contribution to this Conference

11. V.S. Vasilevsky et. al.: Yad. Fiz. **48**, 346 (1987)

12. M.T. Alley, L.D. Knutson: Phys. Rev. **C48**, 1901 (1993)

13. H. Kaiser et. al.: Phys. Lett. **B71** 321, (1977)

14. G.M. Halle: private communication

Few-Body Systems Suppl. 10, 41–44 (1999)

Few-
Body
Systems
© by Springer-Verlag 1999

An Investigation of Four Body Bound States

A. Nogga*, H. Kamada and W. Glöckle

Institute for Theoretical Physics II, Ruhr-Universität Bochum,
44780 Bochum, Germany

Abstract. We present solutions of the Yakubovsky equations for a four-boson-system interacting with a Malfliet-Tjon V potential.

1 Introduction

In this investigation we would like to take up again the method of solving the Faddeev-Yakubovsky equations in momentum space for the four-body bound state [1]. As an example we present the solution for four bosons interacting with the Malfliet-Tjon V potential. This system does not only possess a ground but also an excited state [2] both with $J = 0^+$.

Our intention is to show the efficiency of our algorithm on a parallel computer and test the feasibility in view of more complex physical systems like the four nucleon bound state and the $^4_\Lambda$H and $^4_\Lambda$He systems. The second aim is to demonstrate our numerical accuracy.

2 Yakubovsky Equations in Partial Wave Decomposition

In a system of four identical bosons there are only two independent Yakubovsky components (YC) ψ_1 and ψ_2 [1]. The two YC's fulfil the two Yakubovsky equations (YE)

$$
\begin{aligned}
\psi_1 &= G_0 \, t_{12} \, P \, ((1 + P_{34})\psi_1 + \psi_2) \\
\psi_2 &= G_0 \, t_{12} \, \tilde{P} \, ((1 + P_{34})\psi_1 + \psi_2)
\end{aligned}
\tag{1}
$$

Here G_0 and t_{12} denote the free propagator and the two-body T-Matrix, respectively. P and $\tilde{P}$ are permutation operators. They are composed from transpositions P_{ij}

$$
P = P_{12} \, P_{23} + P_{13} \, P_{23}
$$

---

*E-mail address: Andreas.Nogga@ruhr-uni-bochum.de

$$\tilde{P} = P_{13} P_{24} \tag{2}$$

According to the two partitions we choose two different sets of basis states in a partial wave decomposition (for the notation see [1]). The crucial point of our algorithm is to split permutations into two steps. Each step leaves one of the momentum variables unchanged and, therefore, is a three-body-like permutation operator.

3   Method of Iterated Orthogonal Vectors

A converged solution of the YE needs about 100 different partial waves. We need 20-30 mesh points for the continuous variables. In this case Eq. (1) turns into an energy dependent eigenvalue equation of dimension $\approx 10^6$

$$\psi = K(E) \ \psi \tag{3}$$

The binding energy E_0 of a bound state is found if the eigenvalue= 1 appears in the spectrum of K. We apply a Lanzcos-like method to solve Eq. (3).

By iteration we construct from an arbitrary starting vector ψ_0 a set of basis states $\{\psi_n\}$

$$\psi_n = K^n \ \psi_0 \tag{4}$$

Unfortunately these states are nearly linearly dependent. Therefore we create by Schmidt ortho normalisation a second basis $\{\tilde{\psi}_n\}$. This basis is linearly connected to the original one. The matrix elements $< \tilde{\psi}_n|K|\tilde{\psi}_m >$ can be calculated without any additional iterations of the kernel K. In this basis the eigenvalue equation reduces to a small dimension. With standard numerical packages one can obtain not only the largest eigenvalue in magnitude but also the next few eigenvalues and vectors. Typical iteration numbers are 5 (20) for the ground (excited) state. The method turned out to be a factor 4-5 faster than the simple power method and needs typically 10 min. on a Cray T3E with 512 Processors to find one eigenvalue and eigenvector of K.

4   Results

The convergence of the binding energies relative to the four particle threshold is shown in Table 1. One can see that an angular momentum of 6 is sufficient to get the binding energy of the ground state with four significant figures. With the same number of partial waves the binding energy of the excited state is converged within three figures.

The wavefunction Ψ is connected to the YC by some permutation operators [1]. This implies that the norm of Ψ fulfils the following relation

$$< \Psi|\Psi >= 12 \ < \psi_1|\Psi > +6 \ < \psi_2|\Psi > \tag{5}$$

The numerical accuracy of this relation is a good test for the symmetry of Ψ. If we normalise $|\Psi >$ by the right hand side of Eq. (5) we find $< \Psi|\Psi >= 0.9999$

Table 1. Convergence of the binding energy with increasing number of partial waves. All energies are given in MeV.

l_1,λ_1,λ_3	l_2	l_3	λ_2	E_{ground}	$E_{excited}$
0	0	0	0	-31.07	-8.24
2	2	0	0	-31.11	-8.47
4	4	0	0	-31.22	-8.50
6	6	0	0	-31.23	-8.50
4	6	2	0	-31.28	-8.50
4	6	4	0	-31.31	-8.50
4	6	6	0	-31.31	-8.50
4	6	4	2	-31.34	-8.51
4	6	4	4	-31.35	-8.51
4	6	4	6	-31.35	-
6	6	4	4	-31.36	-8.52
8	6	6	6	-31.36	-

Table 2. Expectation values of T, V and $T + V$ compared to the binding energies from Table 1

	$< T >$ [MeV]	$< V >$ [MeV]	$< T + V >$ [MeV]	E_0
ground state	69.73	-101.08	-31.35	-31.36
excited state	31.41	-39.81	-8.40	-8.52

Table 3. The quantum numbers and relative weights of the five most important partial waves of ground and excited state

a) ground state

l_1	l_2	j_3	l_3	Weight
0	0	0	0	97.8 %
2	2	0	0	0.6 %
0	2	2	2	0.3 %
2	0	2	2	0.2 %
0	1	1	1	0.2 %

b) excited state

l_1	l_2	j_3	l_3	Weight
0	0	0	0	64.3 %
0	1	1	1	7.7 %
2	2	0	0	5.0 %
0	2	2	2	4.1 %
2	1	1	1	3.8 %

for the ground state and $< \Psi|\Psi >= 0.9908$ for the excited state if one keeps all partial waves up to $l_i = 8$ and $\lambda_i = 8$.

A second test of the accuracy of Ψ is the calculation of the expectation values of the kinetic energies T, the potential V and the sum of both. The results are shown in Table 2 and compared to the binding energies. The achieved accuracy corresponds to the one found in the normalisation test.

To illustrate some of the properties of the wave function, Table 3 shows the five most important partial waves of the wave functions with their relative weight. In contrast to the excited state the ground state is nearly a pure s-wave state.

In Fig. 1 the s-wave correlation functions $C(r)$ in the form of [3] for the

44

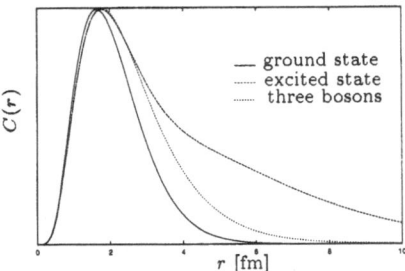

Figure 1. S-wave correlation functions of the three-boson wavefunction, the ground and the excited state of the four-boson system.

three and the ground and excited state of the four boson system are compared. That choice emphasises the long distance behaviour around 4 fm. The three functions are normalised to have the same peak heights. The similar behaviour for small distances is well known. We would like to point to the amazing agreement between the $C(r)$'s of the excited state and the three-boson ground state. Together with the long range shoulder of C(r) of the excited state this suggests the picture for the excited state of an nearly unperturbed three-boson ground state surrounded by the fourth particle. This structure was already suggested in [3] using a similar interaction.

5 Conclusion

We presented fully converged solutions for the wavefunctions of the ground and the exited state of four interacting bosons. We showed some features of the ground and the exited state of this system. The algorithm is reliable and powerful enough to be extended to more complex systems.

Acknowledgement. The work of A. Nogga was supported by the Deutsche Forschungsgemeinschaft under Grant No. GL-8727-1. The numerical calculations have been performed on the Cray T3E of the Höchstleistungsrechenzentrum in Jülich, Germany.

References

1. H. Kamada, W. Glöckle: Nucl. Phys. **A548**, 205 (1992)

2. H. Kameyama et al.: Phys. Rev. **C 40**, 974 (1989)

3. F. Cielsielski, J. Carbonell: Phys. Rev. **C 58**, 58 (1998)

Few-Body Systems Suppl. 10, 45–48 (1999)

Few-
Body
Systems
© by Springer-Verlag 1999

Extension of the Faddeev equation to Nuclei

M. Fabre de la Ripelle

Institut de Physique Nucléaire, 91406 Orsay Cedex, France

The Faddeev equation was derived for solving the three body problem. The wave function Ψ is written as the sum of three amplitudes. Each one $\psi_{ij}(\boldsymbol{r}_{ij}, \boldsymbol{r}_k; \alpha)$ is described in one of the three available Jacobi coordinate systems : $\boldsymbol{r}_{ij} = \boldsymbol{x}_i - \boldsymbol{x}_j$, $\boldsymbol{r}_k = \sqrt{3}(\boldsymbol{x}_k - \boldsymbol{X})$, $\boldsymbol{X} = \frac{1}{3}(\boldsymbol{x}_1 + \boldsymbol{x}_2 + \boldsymbol{x}_3)$ in terms of the coordinates $\boldsymbol{x}_i$, (i = 1, 2, 3), of equal mass particles. The other degrees of freedom α can be spin, isospin etc... Each amplitude is the solution of one Faddeev equation

$$(T - E)\psi_{ij} = -V(\boldsymbol{r}_{ij})\Psi \qquad (1)$$

for pairwise potentials $V(\boldsymbol{r}_{ij})$. Then each equation is projected on bipolar harmonics generating coupled two variables integro-differential equations in coordinate space.

Bosons in S-state - For central potentials and three body in S-state the amplitude $\psi_{ij} = P(z, r, \cos\theta)/r^{5/2}$ is a function of $z = \cos 2\phi$ and $r = (r_{ij}^2 + r_k^2)^{1/2}$ in the Jacobi coordinate system (ij, k) where $r_{ij} = r \cos\phi$ and $\boldsymbol{r}_{ij} \cdot \boldsymbol{r}_k = r_{ij} r_k \cos\theta$.

S-state projected potential - When the potential operates only on pairs in S-state the amplitude in the partition (ij, k) becomes a function $P(2r_{ij}^2/r^2 - 1, r)$ only. By integrating Eq. (1) over all variables except z and r one obtains the Faddeev equation for S-state projected potentials

$$\left\{ \frac{\hbar^2}{m} \left[-\frac{\partial^2}{\partial r^2} + \frac{\mathcal{L}_0(\mathcal{L}_0 + 1)}{r^2} - \frac{4}{r^2} \frac{1}{W_0} \frac{\partial}{\partial z}(1 - z^2)W_0 \frac{\partial}{\partial z} \right] - E \right\} P(z, r)$$
$$= -V(r\sqrt{\frac{1+z}{2}}) \left[P(z, r) + \int_{-1}^{1} f_0(z, z')P(z', r)dz' \right] \qquad (2)$$

The weight function $W_0 = (1 - z)^{\frac{D-5}{2}}\sqrt{1 + z}$, $\mathcal{L}_0 = (D - 3)/2$, in the D dimensional space spanned by the Jacobi coordinates where $D = 3A - 3$ i.e.

$D = 6$ for three body in the c.m. system. The projection function $f_0(z, z') = 2/\sqrt{3(1 - z^2)}$ for $Z_- \leq z' \leq Z_+$ where $Z_{\pm} = (-z \pm \sqrt{3(1 - z^2)})/2$ and zero otherwise in the 6-dimensional space. Equation (2) provides an exact solution Ψ of the Schrödinger equation for $D = 6$ and for a S-projected potential. Equation (2) can also be obtained by using an amplitude expansion

$$P(z, r) = \sum_{k=0}^{\infty} \mathcal{P}_{2k}^0(\Omega_{ij}, r) \int_{-1}^{1} P(z', r) \mathcal{P}_{2k}^0(\Omega') W_0(z') dz' \tag{3}$$

in terms of the 6-dimensional potential harmonics (PH)

$$\mathcal{P}_{2k}^0(\Omega_{ij}) = C_k^1(z)/\pi^{3/2}, \quad z = 2r_{ij}^2/r^2 - 1$$

where $C_k^1(z)$ is a Gegenbauer polynomial. When a projection of the amplitudes (3) for the connected pairs (jk) and (ki) is performed by projecting each PH occurring in the expansion on the PH for the pair (ij) in S-state Eq. (2) is recovered [2]. Since PH exist for any number of bosons in S-state the same procedure can be applied to obtain Eq. (2) for any boson system in S-state. For A-bosons in S-state the PH are the polynomials associated with the weight function W_0 i.e. properly normalised Jacobi polynomials $P_k^{\frac{D-5}{2}, \frac{1}{2}}$, and the projection function is given by [1]

$$f_0(z, z') = W_0(z') \sum_{k=0}^{\infty} (f_k^2 - 1) P_k^{\frac{D-5}{2}, \frac{1}{2}}(z) P_k^{\frac{D-5}{2}, \frac{1}{2}}(z')/h_k$$

$$f_k^2 - 1 = \left[2(A - 2) P_k^{\frac{D-5}{2}, \frac{1}{2}}\left(-\frac{1}{2}\right) + (A - 2)(A - 3) P_k^{\frac{D-5}{2}, \frac{1}{2}}(-1) \right] / P_k^{\frac{D-5}{2}, \frac{1}{2}}(1)$$

$$\tag{4}$$

where the sum in (4) can be calculated analytically [3].

Local central potentials - For solving the many-boson problem for local central potentials operating on all orbitals one uses the property of soft potentials to include a hypercentral component, invariant by rotation in the D-dimensional space, which contains most of the potential energy. The residual interaction is usually a strong repulsive core. it is fitted in nuclear potentials to the S-phase shifts vanishing around 300 MeV. It operates essentially on pairs of nucleons in S-state at the rather weak kinetic energy prevailing in nuclei. If we assume that we are in a situation where the residual interaction operates only on pairs in S-state the potential can be written as

$$V(r_{ij}) = V_0(r) + V_R(r_{ij}, r) P_{ij}^S \tag{5}$$

where $P_{ij}^S = 1$ for the pair (ij) in S-state and zero otherwise. For this potential Eq. (1) becomes

$$(T + A(A - 1)/2 \, V_0(r) - E)\psi_{ij} = -V_R(r_{ij}, r) P_{ij}^S \Psi. \tag{6}$$

The hypercentral potential $V_0(r) = \int V(r_{ij})d\Omega / \int d\Omega$ is the potential averaged over all angular coordinates Ω in the D-dimensional space. The equation which determine $P(z,r)$ is obtained by introducing $A(A-1)V_0(r)/2$ inside the parenthesis in the l.h.s. and the potential V_R instead of V in the r.h.s. of Eq. (2).

Fermions in bound states   - Any regular function $\Psi(x)$, $x(x_1, x_2, \cdots, x_A)$, can be expanded in harmonic polynomials

$$\Psi(x) = \sum_{[L]} H_{[L]}(x)u_{[L]}(r)/r^{L_m + \frac{D-3}{2}} \tag{7}$$

where $[L]$ is a set of $3A-4$ space quantum numbers and the spin and isospin states for nucleons characterising the harmonic polynomials of degree L. The sum starts from a minimal degree L_m and extends over all suitable quantum numbers. The $H_{[L_m]}(x)$ defines the state. The other polynomials are generated by the product of $H_{[L_m]}$ and V_R since $H_{[L_m]}$ is an eigenfunction of the kinetic energy operator. When one has to deal with two body potentials and V_R is used once only the wave function exhibits the structure

$$\Psi = H_{[L_m]}(x) \sum_{i<j} P(2r_{ij}^2/r^2 - 1, r)/r^{L_m + \frac{D-1}{2}} \tag{8}$$

The amplitude $P(z,r)$ is a solution of Eq. (6) projected on the pair (ij) in S-state :

$$\left\{ \frac{\hbar^2}{m} \left[-\frac{\partial^2}{\partial r^2} + \frac{\mathcal{L}_m(\mathcal{L}_m + 1)}{r^2} - \frac{4}{r^2} \frac{1}{W_{[L_m]}} \frac{\partial}{\partial z}(1 - z^2)W_{[L_m]}\frac{\partial}{\partial z} \right] \right.$$
$$\left. + \frac{A(A-1)}{2}V_{[L_m]}(r) - E \right\} P(z,r) =$$
$$-V_R(r\sqrt{\frac{1+z}{2}}, r)\left[P(z,r) + \int_{-1}^{1} f_0(z, z')P(z', r)dz' \right] \tag{9}$$

The effective hypercentral potential

$$V_{[L_m]}(r) = \int H_{[L_m]}^*(x)V(r_{ij})H_{[L_m]}(x)d\Omega / \int |H_{[L_m]}|^2 d\Omega$$

is an integral over all angular coordinates Ω in the D-dimensional space, and the weight function

$$W_{[L_m]}(z) = \int |H_{[L_m]}(x)|^2 d\Omega_1 / \int |H_{[L_m]}|^2 d\Omega$$

is an integral where Ω_1 is for all the variables Ω excluding z. For potentials including exchange operators details are given in ref. [4].

Table 1. Binding energies of 4 and 16 bosons in ground state (upper part) and ^{16}O and ^{40}Ca (lower part).

	4 bosons - ^4He				16 bosons			
	Ba	V	S3	MTV	Ba	V	S3	MTV
HCA	39.12	28.6	7.2	7.4	2969	1559	798	738
IDEA	40.05	30.4	27.3	31.0	2970	1630	1235	1363
OM	40.05	30.4	27.4	31.4			1403	1605

	^{16}O				^{40}Ca	
	B1	V	MS3	MTV	B1	MS3
HCA	106	1101	12.6	506	223	47.8
IDEA	165	1150	103	1027	483	273
FAHT	162		108	1103	478	335

A few numerical tests were performed to check the validity of our assumptions. The binding energies of 4 and 16 bosons in ground state and ^{16}O and ^{40}Ca are exhibited in Table 1 **a)** in the Hypercentral Approximation (HCA), where $V_R = 0$, which excludes correlations, **b)** by solving the Integro differential equation approach (IDEA) for local potentials with increasing strength of the repulsive core : the Bakker, a single attractive gaussian which does not generate correlation, the Volkov, Afnan-Tang S3 and MS3, Malfleit-Tjon MTV and Brink B1 potentials. These energies are compared to those obtained by the non variational coupled Cluster (FAHT) [5] or other methods (OM).

References

1. M. Fabre de la Ripelle: C.R. Acad. Sc. Paris. **299** Serie **II**, 839 (1984)

2. M. Fabre de la Ripelle: In *"Model and Methods in Few-Body Physics"* (Lecture Notes in Physics, vol. 283), p. 273. Berlin: Springer 1987

3. M. Fabre de la Ripelle, H. Fiedeldey, S. Sofianos: Phys. Rev. **C38**, 449 (1988)

4. R. Brizzi, M. Fabre de la Ripelle, M. Lassaut: Nucl. Phys. **A596**, 199 (1996)

5. R. Guardiola et al.: Nucl. Phys. **A371**, 79 (1981)

Few-Body Systems Suppl. 10, 49–52 (1999)

Few-
Body
Systems
© by Springer-Verlag 1999

Adiabaticity in Faddeev Equations

B.G. Giraud[1], J. Carbonell[2], N. Takigawa[3],

[1] Service Physique Théorique, DSM, C.E.Saclay, 91191 Gif/Yvette, France
[2] Institut des Sciences Nucléaires, 53 Av. des Martyrs, F-38026 Grenoble
[3] Physics Department, Tohoku University, Sendai, Japan

Abstract. A multicomponent Born-Oppenheimer approximation for Faddeev equations is introduced. It may provide a better understanding of non adiabatic effects.

Many efficient methods and numerical codes are available [1, 2, 3, 4, 5] for numerically precise solutions of Faddeev equations. However, even in the case of simple, short range interactions, a physical, intuitive understanding of such solutions often remains difficult. Furthermore, when connected kernel equations (CKE's) are used for the few-body problem, the curse of high dimensional integration grids soon reappears; as soon, actually, as the particle number N equals or exceeds 4.

This note is motivated by the following question: for reliable CKE estimates of binding energies or transition rates, can one use Born-Oppenheimer (BO) approximations when mass ratios suggest that a certain amount of adiabaticity is present? A positive answer is of some importance, because of the dimensional reductions brought by the BO approximation, and, moreover, because of the physical intuition which goes along.

This question was raised by Fonseca *et al.* [6] and already received a positive answer for several models or realistic cases. The same problem is visited again here, in a context where we attempt to simultaneously minimize *i)* the role of non localities, *ii)* the inconsistencies [7] they may raise into the formalism, and *iii)* the number of adiabatic approximations which must be accepted in such a CKE-BO procedure.

It's worth mentioning that other approaches to the adiabaticity exist in the literature, mainly those based in the hyperspherical coordinates [8]. However this formalism, although leading to relatively simple and accurate results, doesn't allow a direct comparison with the standard BO methods.

A special difficulty in our approach is that CKE's involve permutation operators which transform various Jacobi coordinate representations into each other. The traditional BO approach, which consists in temporarily disregarding the non locality linked to a Laplacian for a "slow" coordinate, must now face additional sources of non locality. This note explains a possible method to cope with such additional non localities when freezing slow degrees of freedom. For pedagogical reasons, this is shown through an outrightly simplified model. Generalizations, straightforward but slightly tedious, are left to the reader and will anyhow be the subject of future work by the present authors.

Our model uses the words "deuteron", "triton" and "muon" for illustrative purposes. But it is actually much more schematic, because, for the sake of simplicity, it assumes that the deuteron-triton interaction vanishes. Hence the number of Faddeev equations reduces to two. Furthermore, again for the sake of simplicity, the model assumes both remaining interactions $v_{\mu d}$ and $v_{\mu t}$ to be local, short range and scalar. Despite such simplifications, the model remains realistic in the sense that a light particle is exchanged between two heavy ones. All the necessary ingredients for an adiabatic and even a non adiabatic interplay between the "nuclei" via the screening by the "muon" are thus present in such a model.

Consider therefore a bare deuteron d, approaching a triton t, surrounded by a muon μ, in its lowest "atomic" orbital. In the following we select as "channel" Jacobi coordinates the relative coordinate $x = r_\mu - r_t$ between the triton and the muon, and the associated coordinate $y = r_d - (m_\mu r_\mu + M_t r_t)/(m_\mu + M_t)$ relating the deuteron to the center of mass of the muonic atom. The corresponding reduced masses

$$\frac{1}{m} = \frac{1}{m_\mu} + \frac{1}{M_t}$$
$$\frac{1}{M} = \frac{1}{m_\mu + M_t} + \frac{1}{M_d}$$

clearly indicate the "fast" nature of x and the "slow" nature of y. A similar statement holds for the set $x' = r_\mu - r_d$, $y' = (m_\mu r_\mu + M_d r_d)/(m_\mu + M_d) - r_t$ of Jacobi coordinates corresponding to the t(μd) channel, with the relevant masses $m' = m_\mu M_d/(m_\mu + M_d)$ and $M' = M_t(m_\mu + M_d)/(m_\mu + M_d + M_t)$.

With obvious notations, the kinetic energy operator is $\check{K} = K_x + K_y = K'_{x'} + K'_{y'}$. With our choice of coordinates, this means $K_x = -\Delta_x/(2m)$, $K_y = -\Delta_y/(2M)$, and the same for primed quantities. Then the interactions read $v(x) \equiv v_{\mu t}$ and $v'(x') \equiv v_{\mu d}$. The adiabaticity parameter is the ratio ε of the muon mass to the nucleon mass. Consider the expansions of x' and y' in terms of x and y, and the converse expansions,

$$\begin{aligned} x' &= \alpha x + \beta y & x &= \alpha' x' + \beta' y' \\ y' &= \gamma x + \delta y & y &= \gamma' x' + \delta' y' \end{aligned} \tag{1}$$

Here $\alpha = M_t/(m_\mu + M_t)$, $\beta = 1$, $\gamma = 1 - M_d M_t/[(m_\mu + M_d)(m_\mu + M_t)]$, $\delta = M_d/(m_\mu + M_d)$ and the inverse coefficients $\alpha' = \delta$, $\beta' = -\beta$, $\gamma' = -\gamma$ and $\delta' = \alpha$. With corrections of first order with respect to ε we find that $\alpha = \alpha' = \delta = \delta' = 1$ and $\gamma = \gamma' = 0$.

The two Faddeev equations read, in a differential form,

$$[E - K_X - \mathcal{K}_Y - v(X)]\Psi(X,Y) = v(X)\Psi'(\alpha X - Y, \gamma X + \delta Y) \quad (2)$$
$$[E - K'_X - \mathcal{K}'_Y - v'(X)]\Psi'(X,Y) = v'(X)\Psi(\alpha' X + Y, \gamma' X + \delta' Y) \quad (3)$$

We used here a unified notation X, Y for both pairs x, y and x', y' of Jacobi coordinates. In such equations, Eqs. (2-3), the sources of non locality with respect to the slow coordinate Y are not just the usual kinetic operators $\mathcal{K}, \mathcal{K}'$. Additional non localities are present because γ, γ' slightly differ from 0 and δ, δ' slightly differ from 1. As stated already, however, such effects are of first order with respect to ε. It is thus reasonable to approximate Eqs. (2-3) by

$$[E - K_X - \mathcal{K}_Y - v]\Psi(X,Y) = v(X)\Psi'(\alpha X - Y, Y) \quad (4)$$
$$[E - K'_X - \mathcal{K}'_Y - v']\Psi'(X,Y) = v'(X)\Psi(\alpha' X + Y, Y) \quad (5)$$

Within this approximation coordinate y' in (1) is identified to y. The resulting equations are still non local with respect to X but now contain only $\mathcal{K}, \mathcal{K}'$ as causes of non locality for Y. The usual freezing of Y and the BO ansatz:

$$\begin{aligned} \Psi(X,Y) &= \phi_Y(X)\,\chi(Y) \\ \Psi'(X,Y) &= \phi'_Y(X)\,\chi'(Y) \end{aligned} \quad (6)$$

generate the "fast" equations,

$$[\eta(Y) - K_X - v(X)]\,\phi_Y(X) = v(X)\phi'_Y(\alpha X - Y) \quad (7)$$
$$[\eta(Y) - K'_X - v'(X)]\,\phi'_Y(X) = v'(X)\phi_Y(\alpha' X + Y) \quad (8)$$

The eigenvalue η is chosen to be that of a "ground state". This maximizes the final binding (if one calculates a bound state), or the energy $E - \eta$ available for Y (if one calculates a collision, for which this is likely to be the best choice for the description of screening effects).

The "slow" BO equations can be found by inserting (6) into (4-5) and integrating over the fast degree of freedom $\phi_Y(X)$. This gives the coupled equations

$$[E - \mathcal{K}_Y - \eta(Y)]\,\chi(Y) = W(Y)\,[\chi'(Y) - \chi(Y)] \quad (9)$$
$$[E - \mathcal{K}'_Y - \eta(Y)]\,\chi'(Y) = W'(Y)\,[\chi(Y) - \chi'(Y)] \quad (10)$$

with

$$W(Y) = \int dX\,\phi_Y(X)V(X)\phi'_Y(\alpha X - Y)$$

$$W'(Y) = \int dX\,\phi'_Y(X)V'(X)\phi_Y(\alpha X + Y)$$

This way of doing disregards, as usual, terms related to $\nabla_Y \phi_Y$, $\nabla_Y \phi'_Y, \ldots \Delta_Y \phi'_Y$.

If the heavy particles have equal mass, $\chi = \chi'$ and the slow equations (9-10) are reduced to

$$[E - \mathcal{K}_Y - \eta(Y)]\,\chi(Y) = 0 \quad (11)$$

We remark also that even when the heavy masses are different the coupling of the slow equations is of first order with respect to ε and can be disregarded in a first approximation.

To summarize this note, a minimal sequence of approximations, based on adiabaticity, breaks the solution of Faddeev equations into two steps. The first step freezes a slow degree of freedom $\boldsymbol{Y}$, reducing the dynamics to a fast degree $\boldsymbol{X}$ only. The problem boils down to two coupled one-body equations. It is stressed, see Eqs. (4-8), that we set the limit $\varepsilon \to 0$ at those places only where it is strictly necessary. The non vanishing value of ε is retained otherwise. The second BO step reinstates the motion of $\boldsymbol{Y}$, as usual. The procedure respects the basic structure of Faddeev equations, in particular their boundary conditions. The multicomponent nature of the Faddeev formalism may still be interpreted as a configuration mixing theory.

The present model therefore, slightly paradoxically, retains both the advantage of adiabatic factorization and the possibility of non adiabatic coupling between channels. We are running numerical tests of this. Naturally, a further expansion, in terms of several solutions of the fast equations rather than just their ground state, is possible.

References

1. C.R. Chen et al. : Phys. Rev. **C31**, 2266 (1985)

2. N.W. Schellingerout, J.J. Schut, L.P. Kok: Phys. Rev. **C46**, 1192 (1992)

3. W. Glockle, H. Kamada: Phys. Rev. Lett. **71**, 971 (1993)

4. M. Viviani, A. Kievsky, S. Rosati: Few-Body Systems **18**, 25 (1995)

5. K. Varga : contribution to this Conference

6. A.C. Fonseca et al. : Ann. Phys. **117**, 268 (1979); Phys. Rev. **A35**, 4585 (1987); Nucl. Phys. **A487**, 92 (1988)

7. B.G. Giraud and Y. Hahn: Nucl. Phys. **A588**, 653 (1995); Y. Abe et al. : Phys. Rev. **C56** 2557 (1997)

8. D.V. Fedorov, A.S. Jensen: Phys. Rev. Lett. **71**, 4103 (1993); A.S. Jensen et al.: contribution to this Conference

Few-Body Systems Suppl. 10, 53–56 (1999)

Few-
Body
Systems
© by Springer-Verlag 1999

Jastrow-Correlated Configuration-Interaction Description of Light Nuclei

M. Portesi[1], J. Navarro[1], R. Guardiola[2], I. Moliner[2], R.F. Bishop[3], N.R. Walet[3]

[1] IFIC (Centre Mixt CSIC – Universitat de València), Avda. Dr. Moliner 50, E-46100 Burjassot, Spain
[2] Dept. de Física Atómica, Molecular i Nuclear, Universitat de València, Avda. Dr. Moliner 50, E-46100 Burjassot, Spain
[3] Department of Physics, UMIST, P.O. Box 88, Manchester M60 1QD, U.K.

Abstract. This work describes recent progress of the UMIST–VALENCIA collaboration on the *ab initio* study of ground states of light nuclei using realistic forces. The method presented here constructs trial variational wave functions by superimposing a central Jastrow correlation on a state-dependent translationally invariant linearly correlated state, with very promising results.

When dealing with finite systems of mutually interacting particles, one of the important difficulties found is the proper treatment of the centre-of-mass (CM) motion. One of the techniques to deal with this problem is the translationally invariant configuration-interaction (TICI) method [1]. This is a variational methodology inspired by a linear version of the coupled-cluster method (CCM), and it differs from the conventional CI formalism in the means of selecting the interacting configurations. The method has mainly been used in a two-body implementation (TICI2), where all correlations are truncated at a two-body level; the interacting configurations correspond to the most general single pair excitations which leave the CM unaffected. This kind of configuration is very similar to those considered in the so-called *potential basis* in the hyperspherical harmonics formalism. Recently [2, 3], TICI2 has been applied to saturated nuclei within the $0p$ shell, using state-dependent (SD) correlations between nucleons that interact via a realistic or semirealistic force. The method has also been extended for bosonic systems [4] to the so-called TICC2 approach, which is the full second-order CCM with translational invariance and contains the excitation of an arbitrary number of *independent* pairs. In the case of ^{4}He there is not a significant improvement when going from TICI2 to TICC2, indicating the lack of importance for this nucleus of two-pair excitations. In this work we report on an attempt to go beyond our previous TICI2 calculations. A study of the ^{4}He ground state (g.s.) with tensor forces shows that in order to get a good description we need to mix the operatorial structure of the TICI2 wavefunction with a scalar Jastrow-like factor. We hope to use this methodology as a means of shedding light on how to improve the prediction of the g.s. energy of few-body systems.

In the two-body approximation of CCM, the g.s. wavefunction for an N-particle system is written as $\exp(S_1 + S_2)|\Phi\rangle$, where $|\Phi\rangle$ is an appropriate independent-particle state, and the cluster operators S_n promote n particles out of it. The uncorrelated reference state is taken to be a Slater determinant of single-particle harmonic oscillators (HO), and the CM dependence is thereby factorised exactly. To preserve this translational invariance, the excitations produced by S_1 and S_2 can not be independent; instead, one has to construct a new operator, $S_{(1,2)}$, which mixes one- and two-body correlations properly. The TICC2 method is based on an expansion of the exponentiated form of $S_{(1,2)}$, at the same time imposing that triplets or higher excitations do not appear; the g.s. energy is obtained as the expectation value of the intrinsic Hamiltonian. TICI2 functions similarly, but one keeps only the linear term in the expansion of $\exp(S_{(1,2)})$. In coordinate representation, the TICI2 wavefunction reads

$$\Psi(\boldsymbol{r}_1,\ldots,\boldsymbol{r}_N) = \sum_{i<j} f(r_{ij})\,\Phi(\boldsymbol{r}_1,\ldots,\boldsymbol{r}_N).$$

In this form, it is quite straightforward to introduce state dependence into the pair correlation function by expanding it in an appropriate operatorial basis, $f(r_{ij}) = \sum_p f_p(r_{ij})\Theta_p(ij)$. The basis is generally chosen to be consistent with that of the internucleon potential. For interactions with V4 structure the operators used are $\{\Theta_p\} = \{\mathbb{1}, P_\sigma, P_\tau, P_\sigma P_\tau\}$, with P_σ and P_τ being the usual spin- and isospin-exchange operators, respectively, and for V6-like interactions, the basis also includes the tensor operators $\mathcal{S}_{ij}$ and $\mathcal{S}_{ij}P_\tau$ (we note that for ^{4}He the number of relevant operators is actually reduced due to spatial symmetry). The correlation functions f_p are taken as combinations of Gaussians, a common technique which gives a very accurate description [1]. Results obtained for ^{4}He [3] show a strong dependence on the shape of the interaction, specially for V6 forces, from what one can conclude that the method suffers a lack of correlations when dealing with realistic potentials.

It is known that Jastrow correlations are very well suited to treat the effects of strong short-range repulsions. This motivated us to introduce into the previous *ansatz* a scalar Jastrow factor. Simultaneously, we maintain the linear operatorial structure of the TICI2 state, leading to the following J–TICI2 wavefunction:

$$\Psi(\boldsymbol{r}_1,\ldots,\boldsymbol{r}_N) = \prod_{i<j} g(r_{ij}) \sum_{k<l}\sum_p f_p(r_{kl})\Theta_p(kl)\,\Phi(\boldsymbol{r}_1,\ldots,\boldsymbol{r}_N).$$

For the sake of simplicity, we use a single Gaussian for the Jastrow factor, $g(r) = 1 + ae^{-br^2}$. The trial wavefunction depends on the amplitude a and depth b of the Jastrow function, the HO length and the functions f_p. We must perform a variational search of the parameters to minimize $\langle\Psi|H|\Psi\rangle/\langle\Psi|\Psi\rangle - E_{CM}$, where E_{CM} is the energy related to the CM motion.

We have computed the g.s. energy of the ^{4}He nucleus using several interactions with V4 and V6 structure, the latter coming from truncating various realistic forces to their V6 parts. The calculated energies are given in Tables 1 and 2, respectively, together with a comparison with other methods.

Table 1. Comparison between different results for the ^{4}He g.s. energy (in MeV) using several V4 potentials. S3$_W$ is the Wigner part of the S3 potential.

Potential	TICI2 [2]	J–TICI2	SD-Jastrow [5]	DMC [6]
B1	−37.86	−38.28	−38.27 [6]	−38.32 ± 0.01
S3	−28.19	−30.16	−29.94	
S3$_W$	−25.41	−27.20	−27.21	−27.35 ± 0.02
MS3	−27.99	−29.97	−29.70	
MT V	−29.45	−31.21	−30.88	−31.32 ± 0.02
MT I/III	−30.81	−32.70	−32.01	

Table 2. Same as Table 1, but using the V6 part of several realistic potentials.

Potential	TICI2 [3]	J–TICI2	VMC [3]	GFMC [7]
GPDT	−27.37	−27.58	−27.71 ± 0.06	
SSC	−24.12	−26.74	−29.20 ± 0.12	
AV14	−14.77	−20.37	−23.24 ± 0.08	−24.79 ± 0.20
AV18	−15.40	−21.08	−24.80 ± 0.09	
REID	−5.67	−22.70	−27.82 ± 0.12	−28.30 ± 0.12

In Table 1, we compare our J–TICI2 results with previous TICI2 ones, and with a recent calculation using SD-Jastrow correlations parametrised by two Gaussians for each spin–isospin channel; for Wigner-like interactions, we also quote diffusion Monte Carlo (DMC) results. The main difference between J–TICI2 and SD-Jastrow lies in the way of describing the state dependence of the correlation: in the former, the Jastrow factor has a very simple scalar form and all the operatorial dependence is contained in the linear terms; in the latter, the correlation factor has the same operatorial structure as the potential (the product of correlated pairs must be symmetrised to maintain the Fermi statistics). It can be seen that: (i) the inclusion of the Jastrow factor lowers the energy of the pure TICI2 case by ~ 2 MeV for most interactions; (ii) the energies obtained for the Wigner potentials B1 and S3$_W$ from the Jastrow-times-linear and only-Jastrow correlations practically coincide, but for more realistic interactions with full V4 structure the mixed *ansatz* works better; (iii) for the MT V potential the one-Gaussian J–TICI2 value is lower than the two-Gaussian Jastrow result (this improvement can be explained by realizing that the MT forces contain a very short-range Yukawa repulsion and thus the Jastrow correlation centres in this region, treating poorly the longer distances); (iv) the proximity of the J–TICI2 and DMC results allows us to conclude that, for V4 interactions, the present approach gives a good way to treat the ^{4}He g.s.

In Table 2, in addition to the TICI2 and J–TICI2 descriptions, we include calculations with a V6 SD-Jastrow correlation computed with the variational Monte Carlo (VMC) method, and also the V6 contribution to the energy computed with the Green function Monte Carlo (GFMC) method. The pattern of results is now rather different. The TICI2 results are not good at all for strongly repulsive realistic forces; the effect of including the Jastrow correlation is impressive in some cases, like the Reid one with a gain in binding energy of 17 MeV. Nevertheless, with the exception of the very soft GPDT potential, we are still rather far from the MC results. The hybrid treatment seems however to be a promising method to deal with realistic forces.

We have analysed the optimal values of the variational parameters of the J–TICI2 wavefunction. The HO length does not change very much for different interactions, ranging between 0.71 and 0.78 fm^{-1}, and thus gives similar overall sizes for the system. By contrast, the Jastrow parameters present a wider range of variation, mainly due to the different short-range behaviour of the potentials.

In these model calculations we tried to ascertain the importance of superimposing a Jastrow correlation on linear SD pair correlations. The results appear to indicate that this might be a rather good way of describing light systems with realistic interactions. We still obtain several MeV less binding than the much more precise MC calculations; our belief is that the central part of the correlation used here is too simple and definite conclusions must be delayed until we are able to deal with a more complex scalar Jastrow correlation. Finally, we note the conceptual similarity between our approach and the correlated basis function (CBF) description of many-body systems.

Acknowledgement. This work is supported by the EEC network No. ER-BCHRXCT940456, DGICyT (Spain) contract No. PB92–0820, and EPSRC (UK) research grant GR/L22331.

References

1. R.F. Bishop et al.: Phys. Rev. **C42**, 1341 (1990); J. Phys. **G16**, L61 (1990); *Cond. Matt. Theo.* **5**, p. 253, ed. V.C. Aguilera-Navarro. New York: Plenum Press 1990
   R.F. Bishop et al.: *Cond. Matt. Theo.* **6**, p. 405, eds. S. Fantoni and S. Rosati New York: Plenum Press 1991; J. Phys. **G17**, 857 (1991); *ibid.* **18**, 1157 (1992); *ibid.* **19**, 1163 (1993)
2. R. Guardiola et al.: Nucl. Phys. **A609**, 218 (1996)
3. I. Moliner et al.: *Cond. Matt. Theo.* **13**, eds. J. da Providencia and R.F. Bishop. New York: Nova Science 1998, in press
4. R. Guardiola et al.: Nucl. Phys. **A628**, 187 (1998)
5. R. Guardiola and M. Portesi: J. Phys. **G24**, L37 (1998)
6. R.F. Bishop et al.: J. Phys. **G18**, L21 (1992)
7. J. Carlson: Phys. Rev. **C36**, 2026 (1987); *ibid.* **38**, 1879 (1988)

Few-Body Systems Suppl. 10, 57–60 (1999)

Few-
Body
Systems
© by Springer-Verlag 1999

Uniform Description of the Bound, Quasistationary and Scattering States in the Coulomb three-Body Systems Using Adiabatic Hyperspherical Basis

L.I. Ponomarev[1], D.I. Abramov[2], V.V. Gusev[3]

[1] Russian Research Center "Kurchatov Institute", Moscow, 123182, Russia
[2] St. Petersburg State University, St. Petersburg, 198904, Russia
[3] Institute for High Energy Physics, Protvino, 142284, Russia

Abstract. The method, its numerical realization and applications for the calculations of the different characteristics of the Coulomb three-body (CTB) problem are presented. The method was successfully applied for the description of the three-body bound states (including the local characteristics of the wave function) as well as for the scattering processes $2 \to 2$ (including the characteristics of the resonance states).

Theory. The backbone of the method is the reduced adiabatic hyperspherical approach (RAHSA) which was previously described in papers [1,2]. In this approach the basis functions have the form

$$\Phi_{jm}^{JK\lambda}(\rho|\chi,\vartheta,\Phi,\Theta,\varphi) = \varphi_{jm}(\rho|\chi,\vartheta)D_{Km}^{J\lambda}(\Phi,\Theta,\varphi), \qquad (1)$$

where ρ and (χ,ϑ) are the hyperradius and the hyperangles, (Φ,Θ,φ) are the Euler angles, $D_{Km}^{J\lambda}$ are the symmetrized Wigner D-functions, J and K are the total angular moment and its z-projection, λ is the total parity. Functions $\varphi_{jm}(\rho|\chi,\vartheta)$ in Eq. (1) are the eigenfunctions of the RAHS Hamiltonian acting in (χ,ϑ)-plane at fixed ρ [2]. The special feature of the RAHS basis (1) is the factorization of $\Phi_{jm}^{JK\lambda}$. The representation (1) in the form of the product of two functions depending on internal $(\rho|\chi,\vartheta)$ and external (Φ,Θ,φ) variables gives the significant advantage in the numerical calculations keeping the advantages of the original version of AHSA [3,4]. In particular every basis function $\varphi_{jm}(\rho|\chi,\vartheta)$ represents at large ρ the eigenfunction of the bound pair with correct reduced mass. That is especially important for the descriptions of the scattering processes $2 \to 2$. The CTB wave function $\Psi(\mathbf{r}, \mathbf{R})$ (R is the internuclear distance and r the muon coordinate) is decomposed over basis functions (1) with coefficients depending on ρ which satisfy N coupled differential equations [1,2] which approximately describe the relative nuclear motion for the bound and scattering states as well.

Bound states. Results of the calculations of energies and local characteristics of the bound states (Jv) of mesic molecules are presented in Tables 1, 2 [1,2,5].

Table 1. Binding energies (eV) of the states (Jv) of muonic molecules (the numbers of RAHS-harmonics used with m=0 and m=1 are equal to $N_0 = 15$ and $N_1 = 2$ correspondingly).

	Jv	ref. [6]	This work		Jv	ref. [6]	This work
$pp\mu$	00	253.1523	253.14	$dt\mu$	00	319.1396	319.09
	10	107.2658	107.21		01	34.8340	34.81
$pd\mu$	00	221.5494	221.51		10	232.4714	232.26
	10	97.4980	97.20		11	0.6600	0.65
$pt\mu$	00	213.8402	213.79		20	102.6486	102.15
	10	99.1262	98.76	$tt\mu$	00	362.9097	362.89
$dd\mu$	00	325.0735	325.05		01	83.7711	83.56
	01	35.8436	35.77		10	289.1419	289.07
	10	226.6815	226.60		11	45.2057	44.64
	11	1.97475	1.70		20	172.7022	172.50
	20	86.4936	86.18		30	48.8376	48.35

Table 2. The binding energies and local characteristics of $J = 0$ states of all mesic molecules.

v=0	E_0, eV	γ_1	γ_2	G, fm^{-3}	ρ, fm^{-6}	ω_0
$pp\mu$	253.14	0.568	——	0.231E-11	0.204E-18	——
$pd\mu$	221.51	0.641	0.508	0.868E-12	0.811E-19	0.990
$pt\mu$	213.79	0.663	0.490	0.533E-12	0.517E-19	0.948
$dd\mu$	325.05	0.587	——	0.1443E-12	0.153E-19	0.1376
$dt\mu$	319.09	0.610	0.570	0.5444E-13	0.582E-20	0.00874
$tt\mu$	362.89	0.596	——	0.126E-13	0.150E-20	——
v=1	E_1, eV	γ_1	γ_2	G, fm^{-3}	ρ, fm^{-6}	ω_0
$dd\mu$	35.89	0.506	——	0.986E-13	0.105E-19	0.1375
$dt\mu$	34.81	0.624	0.396	0.439E-13	0.486E-20	0.00867
$tt\mu$	83.65	0.518	——	0.143E-13	0.169E-20	——

The local characteristics are defined as follows:

$$G = \int d\mathbf{r} |\Psi_{Jv}(\mathbf{r}, \mathbf{R} = 0)|^2, \qquad (2)$$

$$\gamma_i = \frac{\pi}{\mu_i^3} \int d\mathbf{R} \, |\Psi_{Jv}(\mathbf{r} = \mathbf{R}_i, \mathbf{R})|^2, \qquad \rho_0 = |\Psi_{Jv}(\mathbf{r} = \mathbf{R} = 0)|^2,$$

where $\Psi_{Jv}(\mathbf{r}, \mathbf{R})$ is the normalized three-body function of the bound state with total moment J and vibration number v, $\mathbf{R}$ is the internuclear vector, $\mathbf{r}$ is the muon coordinate, $\mu_i^{-1} = m_\mu^{-1} + m_i^{-1}$, m_i is the mass of nucleus i, $\mathbf{R}_i$ is the coordinate of a nucleus i with respect to the c.m. of nuclei.

Probability ω_0 of muon sticking to the ground state of muonic helium created in the processes of muon catalyzed fusion from the state (Jv) of mesic molecule

$$\omega_0 = G^{-1} \left| \int d\mathbf{r} \varphi_0^*(\mathbf{r}) e^{-i\mathbf{q}\mathbf{r}} \Psi_{Jv}(\mathbf{r}, \mathbf{R} = 0) \right|^2 ,$$

where $\varphi_0(\mathbf{r})$ is the wave function of the muonic helium in the ground state, $q = m_\mu V$, V is the velocity of the muonic helium in c.m.s. of the reaction [5].
Scattering. The algorithm of the cross section calculation for the processes $2 \to 2$ in the CTB systems with two open channels is developed. The preliminary results are presented in [7]. Results for elastic scattering $t\mu + d$ and $d\mu + t$ are presented in Figs. 1,2.

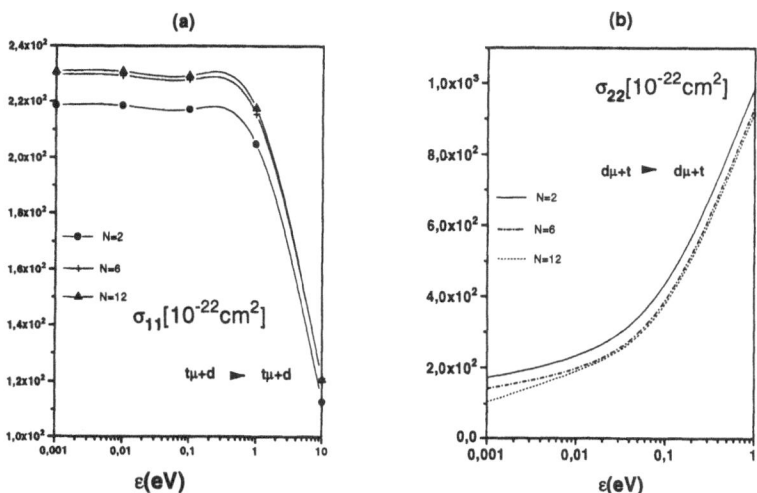

Figure 1. Cross sections of elastic scattering $t\mu + d$ (a) and $d\mu + t$ (b)

Resonances. The effective algorithm is developed for the investigation of resonances in the CTB systems. Position E_0 and width Γ of resonances are determined as the parameters of the Breit-Wigner formula which approximates the numerical results. The characteristics of the resonant state $J = 0$ of the system $d\mu^3 He$ are presented in Figs. 3,4 [8]. Here $G(E)$ is given by (2) and $A(E)$ is the amplitude of the outgoing wave in the open channel $d + \mu^3 He$ for the normalized three-particle wave function of the resonance state. The characteristics of the $d\mu^3 He$ resonance obtained at N=15 are the following [8]: $E_0 = -70.823$ eV, $\Gamma = 0.38 \times 10^{-3}$ eV, $G_0 = G(E_0) = 0.295 \times 10^{-12}$ m.a.u.,

$A_0 = A(E_0) = 0.167 \times 10^{-3}$ m.a.u, $\gamma_1 = 0.093$ m.a.u., $\gamma_2 = 0.894$ m.a.u..

Acknowledgement. This work was performed under support of the grant RFFR 096-02-17279 and INTAS - grant 96-041.

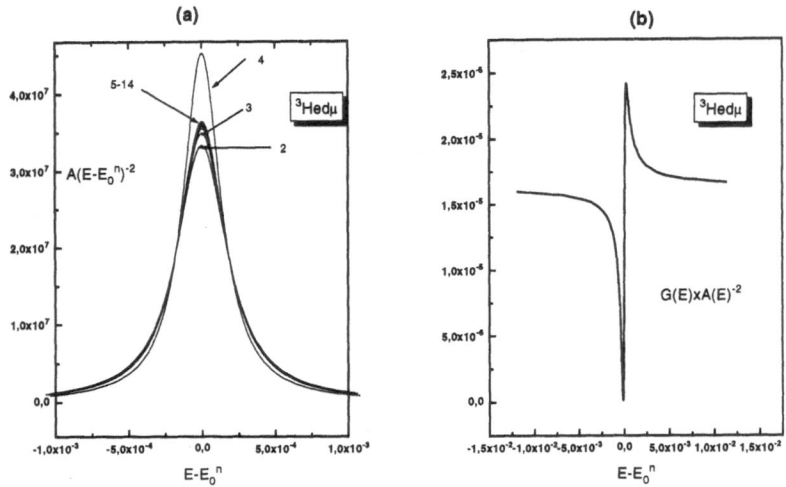

Figure 2. (a) Convergency of $A(E)$ with the number of basis functions, (b) Energy dependence of $G(E)$ in the resonance region

References

1. V.V. Gusev et al.: Few Body Systems **9**, 137 (1990)

2. D.I. Abramov, V.V. Gusev, L.I. Ponomarev: Contributed papers to 14th Int. Conf. on Few Body Problems in Physics, Williamsburg, pp. 745, 749, (1994); Hyp. Inter. **101/102**, 375 (1996); Yadernaya Fizika **60**, 1259 (1997)

3. J. Macek: J.Phys. **B1**, 831 (1968)

4. C.D. Lin: Phys.Rep. **257**, 1 (1995)

5. D.I. Abramov et al.: Hyp. Inter. **101/102**, 301 (1996); Yadernaya Fizika **61**, No 3 (1998)

6. V.I. Korobov, I.V. Puzynin, S.I. Vinitsky: Muon Cat. Fus. **7**, 63 (1992)

7. V.V. Gusev, L.I. Ponomarev, E.A. Solov'ev: Hyp. Inter. **82**, 53 (1993)

8. D.I. Abramov, V.V. Gusev: submitted to J. Phys. **B**

Few-Body Systems Suppl. 10, 61–64 (1999)

Few-
Body
Systems
© by Springer-Verlag 1999

The Muonic Helium Atom in CFHHM

R. Krivec[1], V.B. Mandelzweig[2]

[1] Department of Theoretical Physics, J. Stefan Institute, PO Box 3000, 1001 Ljubljana, Slovenia

[2] Racah Institute of Physics, Hebrew University, Jerusalem 91904, Israel

Abstract. The properties of the ground states of the muonic Helium atoms $e\mu\,^4\mathrm{He}^{2+}$ and $e\mu\,^3\mathrm{He}^{2+}$ have been calculated nonvariationally. In particular, the hyperfine splitting has been evaluated, including all relativistic and other corrections. The obtained theoretical precision is much larger than the discrepancies resulting from different versions of expressions for corrections, implying that the latter have to be reexamined before further comparison is possible.

To obtain precisely the wave function Ψ in both singular and nonsingular regions, we used the correlation function hyperspherical harmonic method (CFHHM), separating $\Psi = e^f \phi$ where e^f contains the cusps and the factor ϕ is expanded in the hyperspherical harmonic (HH) basis up to the maximum global angular momentum K_m, containing $N = (K_m/2 + 1)(K_m/2 + 2)/2$ functions.

Due to the two-scale nature of the muonic Helium the main difficulty was to find a nonlinear correlation function f [1] yielding stable results for a range of parameters: in the odd-man-out notation, $f = \sum_{i=1}^{3}\left[a_i + (b_i - a_i)e^{-r_i/(n_i \overline{r_i})}\right]r_i$,

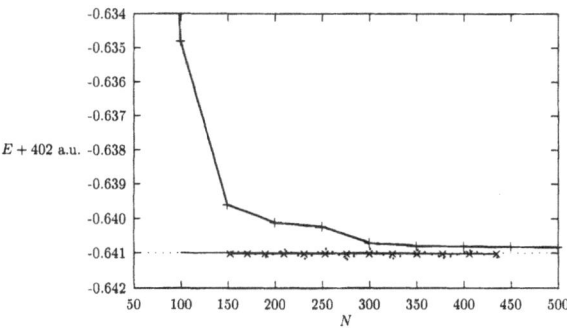

Figure 1. Convergence of $e\mu\,^4\mathrm{He}^{2+}$ ground state energy with basis size N for CFHHM [2] ("×": $\langle H \rangle$; "...": eigenvalue) and stochastic variational method (SVM) ("+")

Table 1. Comparison of HFS values for the ground state of muonic ^{4}He atom (MHz). The lowest order HFS is [5] $\nu_{\mathrm{HF}} = (8/3)\pi\alpha^2 (m_e/m_\mu)\langle\delta(\mathbf{r}_{e-\mu})\rangle$. Corrections are given as $\nu = \nu_{\mathrm{HF}} f_{\mathrm{corr}} = \nu_{\mathrm{HF}} + \nu_{\mathrm{corr}}$. CFHHM $\langle\delta(\mathbf{r}_{e-\mu})\rangle$ values are from ref. [2]

Source	$\langle\delta(\mathbf{r}_{e-\mu})\rangle$	ν_{HF}	ν	f_{corr}	ν_{corr}
Exp. [7]			4464.95(6)		
Exp. [8]			4465.004(29)		
Chen [6]		4454.181(1)	4464.907(1)	1.0024081	10.726[a]
CFHHM		4454.204(3)	4464.930(3)	1.0024081	10.726[b]
CFHHM	.3137622(2)	4454.206(3)	4464.982(3)	1.0024193	10.776[c]
CFHHM	.3137622(2)	4454.207(3)	4464.**983**(3)	1.0024193	10.776[d]
Smith [4]	.3137630(1)		4464.559(1)		[e]
	.3137630(1)	4454.226(1)			[f]
CFHHM	.3137621(2)	4454.213(3)			[g]
CFHHM	.3137621(2)	4454.213(3)	4464.**989**(3)	1.0024193	10.776[h]
CFHHM	.3137621(2)	4454.213(3)	4464.939(3)	1.0024081	10.726[i]
Ref. [9]			4464.8		

[a] f_{corr} derived from ref. [6]. Correct recoil (0.8004 MHz) is used (see [9, 6])

[b] Masses and α from ref. [6]

[c] f_{corr} via formulas of ref. [5], but the recoil correction by refs. [6, 9] (0.800 MHz). Masses: refs. [10, 5]; α: ref. [2] (differs from Huang's); ν_{HF} by CFHHM

[d] As [c], but using Huang's value of α [5]. This multiplies ν_{HF} of [c] by 1.00000011

[e] $\langle\delta(\mathbf{r}_{e-\mu})\rangle$ of ref. [4] times 14229.08 (ref. [4], containing all corrections?)

[f] $\langle\delta(\mathbf{r}_{e-\mu})\rangle$ of ref. [4] times 14196.1472 (masses of ref. [4], our α value)

[g] CFHHM $\langle\delta(\mathbf{r}_{e-\mu})\rangle$, using masses of ref. [4] (α not quoted), times 14196.1472, as in [f]

[h] As in [g], but $f_{\mathrm{corr}} = 1.0024193$ (not sensitive to masses of refs. [10, 5] or [4])

[i] As in [h], but using f_{corr} of ref. [6]

where particles $\{1, 2, 3\}$ correspond to the electron, the muon and He^{2+}; e.g., r_3 is the distance between the electron and the muon. b_i are the cusp factors; the constants $\bar{r}_i$ represent the equilibrium distances of the particles in the i-th pair: $\bar{r}_1 = 0.0037$ a.u., $\bar{r}_2 = \bar{r}_3 = 1.5$ a.u.. The optimized values of the free parameters are $a_3 = -4$, $n_3 = 0.5$ in both $e\mu\,^4$He^{2+} and $e\mu\,^3$He^{2+} [2, 3].

The convergence acceleration with respect to the linear cusp f ($a_i = b_i$) is as follows: at $K_m = 56$ the error of energy is less than 1×10^{-7} a.u. instead of 0.002 a.u.; the error of ν_{HF} is reduced from 440 MHz to 1 MHz.

Interpolating the doubly-convergent expectation values of operators as functions of K_m decreases the errors by additional 2–3 orders of magnitude.

With $N = 435$ we obtain the ground state energy -402.64101534 a.u., which is lower than most variational values, but may depend slightly on the roundoff errors in mass value manipulations. The rate of convergence is faster than in the SVM method (Fig. 1).

Table 2. Lowest order and total HFS of the ground state of $e\mu\,^3\text{He}^{2+}$ (MHz). $\nu_{\text{HF}} = \omega^{(e\mu)}\langle\delta(\mathbf{r}_{e-\mu})\rangle + \omega^{(e)}\langle\delta(\mathbf{r}_{3\text{He}-e})\rangle$

Source	$\nu_{\text{HF}}^{(e\mu)}$	$\nu_{\text{HF}}^{(e)}$	ν_{HF}	ν
Exp. (see [13])				4166.3(2)
CFHHM [3]	3339.830(3)	817.861(1)	4157.691(3)	4166.571(3)[a]
CFHHM [3]	3339.82(1)	817.859(5)	4157.68(1)	4166.56(1)[b]
Ref. [4]	3347.585	767.2	4166.34	[c,d]
	3339.837(1)	817.849(0)	4157.685(1)	[e]
Ref. [11]			4164.9(3.0)	[f]
Ref. [12]	3339.78(5)	817.85(5)	4157.623	[g]
Ref. [6]	3339.8037(5)	817.8558(5)	4157.659(1)	4166.620(1)[h]
Ref. [13]	3339.8037	817.8558	4157.6595	4166.540(5)[i]

[a] Mass set I (ref. [4]). Error estimate based on $n_3 = 0.5, 0.7$; $a_3 = $ -4, -5. $\alpha = 1/137.0359895$, other constants from ref. [12], see below. $\omega^{(e\mu)}$, $\omega^{(e)}$: 10647.110, 2550.926 MHz/a.u.. The correction factors of ref. [13] are used (see [i])

[b] As [a], but mass set II (close to [12]) and a less precise calculation. $\omega^{(e\mu)}$, $\omega^{(e)}$: see [g]

[c] $\omega^{(e\mu)}$, $\omega^{(e)}$ quoted [4]: 10671.81, 0.2393 MHz/a.u. (see also ref. [14])

[d] Quoted $\langle\delta(\mathbf{r}_k)\rangle$, $\omega^{(e\mu)}$, $\omega^{(e)}$ lead to 4114.801 MHz

[e] $\langle\delta(\mathbf{r}_k)\rangle$ of ref. [4], $\omega^{(e\mu)}$, $\omega^{(e)}$ of the present work. Errors inferred from [4]

[f] $m_\mu = 206.7686$ a.u., 1 a.u. $= 6.579684 \times 10^9$ MHz, $\alpha = 1/137.0360$, etc.

[g] Masses as in set II (m_μ, a.m.u.: ref. [10]), except $m_{3\text{He}} = 3.01602970$ a.m.u.; $m_\mu = 105.65946$ MeV (Erratum, ref. [15]); $\alpha = 1/137.03604$, 1 a.u. $= 6.57968413 \times 10^9$ MHz, $\mu_n = 0.00115873\mu_B$, $\omega^{(e\mu)}$, $\omega^{(e)}$: 10647.085, 2550.924 MHz/a.u.

[h] As in [f]. Correction factors are 1.002408, 1.001123

[i] As in [f]. Correction factors are 1.002408, 1.001025 (different from ref. [6])

The f described enabled us to obtain expectation values of singular operators better than in the literature. In particular the difference $\langle\delta(\mathbf{r}_2)\rangle - \langle\delta(\mathbf{r}_3)\rangle$ is smaller by about 6×10^{-6} a.u. in ref. [4] than in our work, while the effects of mass choice and of our computational error on $\langle\delta(\mathbf{r}_2)\rangle$, $\langle\delta(\mathbf{r}_3)\rangle$ are both smaller, i.e., of the order of 2×10^{-7} a.u. [2].

Tables 1 and 2 and Fig. 2 show the discrepancies in the values of the lowest-order hyperfine splitting (HFS) as well as the correction factors in the literature. The CFHHM values using Huang et al. [5] values for the corrections, except their recoil term which is claimed to be wrong [6] (see Table 1), agree with the more recent $e\mu\,^4\text{He}^{2+}$ experiment. If the more recent Chen et al. [6] values of corrections are used, the CFHHM values agree only with the less precise experiment. Overall, the CFHHM values agree better with experiment than any other in the literature, in particular by Chen et al. [6]. The agreement in the $e\mu\,^3\text{He}^{2+}$ case is worse but the same trend towards smaller values when using masses and corrections according to ref. [6] is observed.

The theoretical errors are larger than the dependence on fundamental constants but much smaller than the discrepancies between results using different

64

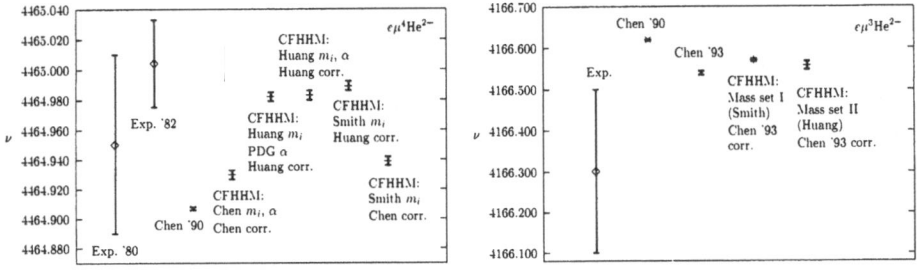

Figure 2. Comparison of theoretical and experimental results for muonic ^{4}He and ^{3}He; from Tables 1 and 2, respectively

expressions for correction terms. We conclude that the corrections need to be reevaluated before further comparison with experiment will be possible.

The research of one of the authors (VBM) was supported by the Israeli Science Foundation founded by the Israeli Academy of Sciences and Humanities. We thank N. Barnea for the SVM results and K. Varga for the SVM code.

References

1. M.I. Haftel, R. Krivec, V.B. Mandelzweig: J. Comp. Phys. **123**, 149 (1996)

2. R. Krivec, V.B. Mandelzweig: Phys. Rev. **A56**, 3614 (1997)

3. R. Krivec, V.B. Mandelzweig: Phys. Rev. **A57**, 4976 (1998)

4. V.H. Smith, A.M. Frolov: J. Phys. B: At. Mol. Phys. **28**, 1357 (1995)

5. K.-N. Huang, V.W. Hughes: Phys. Rev. **A20**, 706 (1979); Phys. Rev. **A21**, 1071 (1980)

6. M.-K. Chen: J. Phys. B: At. Mol. Phys. **23**, 4041 (1990)

7. H. Orth et al.: Phys. Rev. Lett. **45**, 1483 (1980)

8. C.J. Gardner et al.: Phys. Rev. Lett. **48**, 1168 (1982)

9. M.Ya. Amusia, M.Ju. Kuchiev, V.L. Yakhontov: Yad. Fiz. **47**, 877 (1988)

10. K.-N. Huang: Phys. Rev. **A15**, 1832 (1977)

11. S.D. Lakdawala, P.J. Mohr: Phys. Rev. **A24**, 2224 (1981)

12. M.-K. Chen, C.-S. Hsue: J. Phys. B: At. Mol. Phys. **22**, 3951 (1989)

13. M.-K. Chen: J. Phys. B: At. Mol. Phys. **26**, 2263 (1993)

14. R.J. Drachman: J. Phys. B: At. Mol. Phys. **16**, L749 (1983)

15. M.-K. Chen, C.-S. Hsue: Phys. Rev. **A40**, 5520 (1989); Phys. Rev. **A42**, 1830E (1990)

Few-Body Systems Suppl. 10, 65–74 (1999)

Few-
Body
Systems
© by Springer-Verlag 1999

The Few-Body Coulombian Problem

E.O. Alt*

Institut für Physik, Universität Mainz, D-55099 Mainz, Germany

Abstract. Recent advances in the treatment of scattering of charged composite particles are reviewed. In a first part I report on developments of the theory. Specifically I describe the recent completion of the derivation of the co-ordinate space asymptotic behaviour of the wave function for three charged particles in the continuum. This knowledge is increasingly being made use of in attempts to 'derive' three-Coulomb particle wave functions to be used in all of configuration space which are solutions of the Schrödinger equation, though not everywhere but at least in one or preferably all of the asymptotic regions. Their practical application in approximate calculations of ionisation and breakup processes is pointed out. The asymptotic three-charged particle wave functions find further use in investigations of asymptotic and analytic properties of matrix elements of the three-body Coulomb resolvent. An important example is the *nonperturbative* derivation, valid for all energies, of the large-distance behaviour of the optical potential. In the second part I describe a renewed attempt to establish the few-body approach as a valuable tool for calculating (energetic) atomic collision processes. At the end I briefly touch upon the recent successful above-breakup-threshold calculation of proton-deuteron elastic scattering for a realistic potential.

1 Theoretical Developments

1.1 Asymptotic Wave Function for Three Charged Particles in the Continuum

1.1.1 Introduction

Knowledge of the asymptotic form of the wave function is required, e.g., as boundary condition for selecting the physical solution of the Schrödinger equation (SE) for energies above the ionisation threshold, and for deriving asymptotic and/or analytic properties of quantities which are matrix elements of the three-body Coulomb resolvent, such as the optical potential or the kernel of momentum space integral equations for charged particles (for a discussion of

---

*E-mail address: alt@dipmza.physik.uni-mainz.de

the latter topic, see [1]). And, last but not least, it avails also for approximate calculations of breakup or ionisation in peripheral collisions.

I start by introducing some notation. Consider three distinguishable particles with masses m_ν and charges e_ν, $\nu = \alpha, \beta, \gamma$. In the c.m. system Jacobi co-ordinates are used: $\boldsymbol{k}_\alpha$ $(\boldsymbol{r}_\alpha)$ is the relative momentum (co-ordinate) between particles β and γ, and $\boldsymbol{q}_\alpha$ $(\boldsymbol{\rho}_\alpha)$ the momentum (co-ordinate) of particle α relative to the center of mass of the pair (β, γ). Frequently, two Jacobi three-vectors are combined into a sixdimensional vector: $\boldsymbol{P} := (\boldsymbol{k}_\nu, \boldsymbol{q}_\nu)$ and $\boldsymbol{X} := (\boldsymbol{r}_\nu, \boldsymbol{\rho}_\nu)$, $\nu = \alpha, \beta, \gamma$.

For three particles in the continuum the energy shell condition reads $E = q_\alpha^2/2M_\alpha + k_\alpha^2/2\mu_\alpha$, with the standard definitions $\mu_\alpha^{-1} = m_\beta^{-1} + m_\gamma^{-1}$ and $M_\alpha^{-1} = m_\alpha^{-1} + (m_\beta + m_\gamma)^{-1}$ of the appropriate reduced masses. The interaction between particles β and γ is assumed to consist of a short-range and a Coulomb part

$$V_\alpha = V_\alpha^S + V_\alpha^C, \quad \text{with} \quad V_\alpha^C(r_\alpha) = \frac{e_\beta e_\gamma}{r_\alpha}. \tag{1}$$

For the following it proves convenient to eliminate the three-body plane from the wave function by introducing a 'reduced' wave function $\tilde{\Psi}_{\boldsymbol{P}}^{(+)}(\boldsymbol{X})$ via

$$\Psi_{\boldsymbol{P}}^{(+)}(\boldsymbol{X}) := e^{i\boldsymbol{P}\cdot\boldsymbol{X}} \, \tilde{\Psi}_{\boldsymbol{P}}^{(+)}(\boldsymbol{X}) \equiv e^{i(\boldsymbol{k}_\alpha \cdot \boldsymbol{r}_\alpha + \boldsymbol{q}_\alpha \cdot \boldsymbol{\rho}_\alpha)} \, \tilde{\Psi}_{\boldsymbol{k}_\alpha \boldsymbol{q}_\alpha}^{(+)}(\boldsymbol{r}_\alpha, \boldsymbol{\rho}_\alpha). \tag{2}$$

1.1.2 Asymptotic Solution of the SE in Ω_0: 'Redmond Asymptotics'.

This region of configuration space is reached by letting r_α, r_β, and r_γ approach infinity, but such that *not* $r_\nu/\rho_\nu \to 0$ for $\nu = \alpha, \beta$, or γ; in other words, asymptotically all inter-particle distances grow uniformly. Introduce the Coulomb parameter, e.g., for the subsystem (β, γ) as $\eta_\alpha = e_\beta e_\gamma \mu_\alpha / k_\alpha$. Then, as stated by Redmond [2], in leading order one has

$$\tilde{\Psi}_{\boldsymbol{P}}^{(+)}(\boldsymbol{X}) \overset{\Omega_0}{\approx} \prod_{\nu=1}^{3} e^{i\eta_\nu \ln(k_\nu r_\nu - \boldsymbol{k}_\nu \cdot \boldsymbol{r}_\nu)} \tag{3}$$

in the so-called 'non-singular regions' (i.e., for $k_\nu r_\nu \neq \boldsymbol{k}_\nu \cdot \boldsymbol{r}_\nu$). An asymptotically equivalent form is

$$\tilde{\Psi}_{\boldsymbol{P}}^{(+)}(\boldsymbol{X}) \overset{\Omega_0}{\approx} \prod_{\nu=1}^{3} \tilde{\psi}_{\boldsymbol{k}_\nu}(\boldsymbol{r}_\nu), \tag{4}$$

where $\tilde{\psi}_{\boldsymbol{k}_\alpha}(\boldsymbol{r}_\alpha)$, when multiplied with the appropriate plane wave, is the continuum solution of two-body SE

$$\left\{ \frac{k_\alpha^2}{2\mu_\alpha} + \frac{\Delta_{r_\alpha}}{2\mu_\alpha} - V_\alpha^C(\boldsymbol{r}_\alpha) - V_\alpha^S(\boldsymbol{r}_\alpha) \right\} e^{i\boldsymbol{k}_\alpha \cdot \boldsymbol{r}_\alpha} \tilde{\psi}_{\boldsymbol{k}_\alpha}^{(+)}(\boldsymbol{r}_\alpha) = 0, \tag{5}$$

with leading asymptotic behaviour

$$\tilde{\psi}_{\boldsymbol{k}_\alpha}^{(+)}(\boldsymbol{r}_\alpha) \overset{r_\alpha \to \infty}{\longrightarrow} e^{i\eta_\alpha \ln(k_\alpha r_\alpha - \boldsymbol{k}_\alpha \cdot \boldsymbol{r}_\alpha)}. \tag{6}$$

For purely Coulombic interactions this is just the known asymptotic form of the explicit solution of (5),

$$\tilde{\psi}_{\boldsymbol{k}_\alpha}^{(+)}(\boldsymbol{r}_\alpha) \equiv \tilde{\psi}_{\boldsymbol{k}_\alpha}^{C(+)}(\boldsymbol{r}_\alpha) = N_\alpha\, F(-\mathrm{i}\eta_\alpha, 1; \mathrm{i}(k_\alpha r_\alpha - \boldsymbol{k}_\alpha \cdot \boldsymbol{r}_\alpha)) \quad (V_\alpha^S \equiv 0). \quad (7)$$

Here, $N_\alpha = e^{-\pi\eta_\alpha/2}\,\Gamma(1 + \mathrm{i}\eta_\alpha)$; $F(a, b; x)$ is the confluent hypergeometric, and $\Gamma(z)$ the Gamma function. In the general case the behaviour (6) follows from the fact that for sufficiently large r_α, $V_\alpha^S(\boldsymbol{r}_\alpha)$ can be neglected in (5) yielding $\tilde{\psi}_{\boldsymbol{k}_\alpha}^{(+)}(\boldsymbol{r}_\alpha) \overset{r_\alpha \to \infty}{\longrightarrow} \tilde{\psi}_{\boldsymbol{k}_\alpha}^{C(+)}(\boldsymbol{r}_\alpha)$. I, finally, mention that the form (4) with purely Coulombic wave functions (7) has been proposed in [3, 4].

1.1.3 Asymptotic Solution of SE in Ω_ν, $\nu = \alpha, \beta$ or γ: 'AM Asymptotics'.

Besides Ω_0, there exist three more asymptotic regions in which the boundary condition on the wave function must be prescribed. They are characterised by

$$\Omega_\nu: \quad \rho_\nu \to \infty, \; r_\nu/\rho_\nu \to 0, \quad \text{for} \quad \nu = \alpha, \beta, \text{ or } \gamma. \quad (8)$$

That is, e.g. in Ω_α, the distance ρ_α of particle α from the center of mass of the pair (β, γ) grows faster than the relative separation r_α between the particles of this pair. Recall the asymptotic behaviour (3) in Ω_0 which we rewrite as

$$\tilde{\psi}_{\boldsymbol{P}}^{(+)}(\boldsymbol{X}) \overset{\Omega_0}{\approx} e^{\mathrm{i}\eta_\alpha \ln(k_\alpha r_\alpha - \boldsymbol{k}_\alpha \cdot \boldsymbol{r}_\alpha)} \prod_{\nu \neq \alpha} e^{\mathrm{i}\eta_\nu \ln(k_\nu r_\nu - \boldsymbol{k}_\nu \cdot \boldsymbol{r}_\nu)}. \quad (9)$$

This suggests that to go from Ω_0 to Ω_α, which includes all values $r_\alpha \in [0, \infty)$ while still $r_\beta, r_\gamma \to \infty$, it might suffice to replace in (9) the asymptotic wave function (6) for the pair (β, γ) by the full solution $\tilde{\psi}_{\boldsymbol{k}_\alpha}(\boldsymbol{r}_\alpha)$ of the SE (5).

However, this simple idea turns out to be not correct. Instead, as has been shown by Alt and Mukhamedzhanov [5] (AM) the leading term is given by

$$\tilde{\Psi}_{\boldsymbol{k}_\alpha \boldsymbol{q}_\alpha}^{(+)}(\boldsymbol{r}_\alpha, \boldsymbol{\rho}_\alpha) \overset{\Omega_\alpha}{\approx} \tilde{\psi}_{\boldsymbol{k}_\alpha(\boldsymbol{\rho}_\alpha)}^{(+)}(\boldsymbol{r}_\alpha) \prod_{\nu \neq \alpha} e^{\mathrm{i}\eta_\nu \ln(k_\nu r_\nu - \boldsymbol{k}_\nu \cdot \boldsymbol{r}_\nu)}, \quad (10)$$

(for further and next-to-leading order contributions see [6, 7, 8, 9]). What appears is the exact continuum solution of two-body-like SE (for all $\boldsymbol{r}_\alpha \in \Omega_\alpha$)

$$\left\{ \frac{k_\alpha^2(\boldsymbol{\rho}_\alpha)}{2\mu_\alpha} + \frac{\Delta_{\boldsymbol{r}_\alpha}}{2\mu_\alpha} - V_\alpha^C(\boldsymbol{r}_\alpha) - V_\alpha^S(\boldsymbol{r}_\alpha) \right\} e^{\mathrm{i}\boldsymbol{k}_\alpha(\boldsymbol{\rho}_\alpha)\cdot\boldsymbol{r}_\alpha}\, \tilde{\psi}_{\boldsymbol{k}_\alpha(\boldsymbol{\rho}_\alpha)}^{(+)}(\boldsymbol{r}_\alpha) = 0. \quad (11)$$

It describes the relative motion of β and γ at an energy which depends parametrically on the distance from particle α, as determined by the *local* momentum

$$\boldsymbol{k}_\alpha(\boldsymbol{\rho}_\alpha) = \boldsymbol{k}_\alpha + \frac{\boldsymbol{a}_\alpha(\hat{\boldsymbol{\rho}}_\alpha)}{\rho_\alpha}, \quad \boldsymbol{a}_\alpha(\hat{\boldsymbol{\rho}}_\alpha) = -\sum_{\nu \neq \nu' = \beta, \gamma} \eta_\nu \lambda_{\nu\nu'} \frac{\epsilon_{\alpha\nu}\hat{\boldsymbol{\rho}}_\alpha - \hat{\boldsymbol{k}}_\nu}{1 - \epsilon_{\alpha\nu}\hat{\boldsymbol{\rho}}_\alpha \cdot \hat{\boldsymbol{k}}_\nu}. \quad (12)$$

Here, $\lambda_{\nu\nu'} = m_\nu/(m_\nu + m_{\nu'})$, $\hat{\boldsymbol{k}}_\nu := \boldsymbol{k}_\nu/k_\nu$, $\epsilon_{\alpha\beta} = -\epsilon_{\beta\alpha} = +1$ if (α, β) is a cyclic ordering of (1,2,3). If $V_\alpha^S \equiv 0$, the explicit solution of (11) is

$$\tilde{\psi}_{\boldsymbol{k}_\alpha(\boldsymbol{\rho}_\alpha)}^{C(+)}(\boldsymbol{r}_\alpha) = N_\alpha(\boldsymbol{\rho}_\alpha)\, F(-\mathrm{i}\eta_\alpha(\boldsymbol{\rho}_\alpha), 1; \mathrm{i}(k_\alpha(\boldsymbol{\rho}_\alpha)r_\alpha - \boldsymbol{k}_\alpha(\boldsymbol{\rho}_\alpha) \cdot \boldsymbol{r}_\alpha), \quad (13)$$

which is of the same form as (7) but with the asymptotic momentum $\boldsymbol{k}_\alpha$ replaced everywhere by the local momentum $\boldsymbol{k}_\alpha(\boldsymbol{\rho}_\alpha)$.

The physically decisive point is that the parametric dependence of the wave function $\tilde{\psi}^{(+)}_{\boldsymbol{k}_\alpha(\boldsymbol{\rho}_\alpha)}$ for the pair (β, γ) on the relative co-ordinate $\boldsymbol{\rho}_\alpha$ between its center of mass and particle α is a manifestation of a *long-ranged, noncentral, velocity-dependent, dynamic three-body correlation* which is a new genuine three-body effect. This is evident from rewriting the bracket in (11) as

$$\left\{ \cdots \right\} = \left\{ \frac{k_\alpha^2}{2\mu_\alpha} + \frac{\Delta_{\boldsymbol{r}_\alpha}}{2\mu_\alpha} - V_\alpha(\boldsymbol{r}_\alpha) - \frac{1}{\rho_\alpha} \cdot \frac{\boldsymbol{a}_\alpha(\hat{\boldsymbol{\rho}}_\alpha) \cdot \boldsymbol{k}_\alpha}{\mu_\alpha} \right\} + O\left(\rho_\alpha^{-2}\right). \tag{14}$$

Comparison with the bracket in the ordinary two-body SE (5) reveals that the additional, potential-like term in (14) is indeed of long range ($\sim \rho_\alpha^{-1}$), and depends on the relative velocity $\boldsymbol{k}_\alpha/\mu_\alpha$ within the pair (β, γ) and on the orientation of $\boldsymbol{k}_\alpha$ relative to $\boldsymbol{\rho}_\alpha$ (cf. the definition of $\boldsymbol{a}_\alpha$ in (12)). The latter properties can intuitively be understood as a result of the 'dynamic screening' of the charges when the three particles are in typical configurations pertaining to the asymptotic regions Ω_ν, $\nu = \alpha, \beta, \gamma$.

When can one expect to observe effects of these three-body correlations? As follows from (12), local ($\boldsymbol{k}_\alpha(\boldsymbol{\rho}_\alpha)$) and asymptotic momentum ($\boldsymbol{k}_\alpha$) will differ appreciably only if k_α is small (recall that $\rho_\alpha \to \infty$ in Ω_α). As a measure of the 'size' of their difference we can take $|\boldsymbol{a}_\alpha(\hat{\boldsymbol{\rho}}_\alpha)|$ which for $k_\alpha \to 0$ behaves as

$$|\boldsymbol{a}_\alpha(\hat{\boldsymbol{\rho}}_\alpha)| \overset{k_\alpha \to 0}{\sim} \frac{M_\alpha}{q_\alpha} \left| \left[\frac{e_\beta}{m_\beta} - \frac{e_\gamma}{m_\gamma} \right] \right|. \tag{15}$$

That is, it is inversely proportional to the relative velocity of particle α and the center of mass of the pair (β, γ), and proportional to the famous charge-over-mass ratio difference of the pair (β, γ) ('Post-acceleration effect').

1.1.4 Asymptotically Correct Three-body Coulomb Wave Functions

Recently many attempts at 'deriving' three-body Coulomb wave functions, which extrapolate the known asymptotic wave functions into the non-asymptotic region, have appeared. Obviously, such are highly non-unique and, thus, open to further optimisation. For instance, a wave function which is asymptotically correct in all regions $\Omega_1 \cup \Omega_2 \cup \Omega_3 \cup \Omega_0$ has been proposed in [5]

$$\tilde{\Phi}^{(+)}_{\boldsymbol{P}}(\boldsymbol{X}) = \prod_{\nu=1}^{3} \tilde{N}_\nu F(-\mathrm{i}\tilde{\eta}_\nu, 1; \mathrm{i}(\tilde{k}_\nu r_\nu - \tilde{\boldsymbol{k}}_\nu \cdot \boldsymbol{r}_\nu)), \tag{16}$$

with $\tilde{\boldsymbol{k}}_\alpha = \boldsymbol{k}_\alpha + 2\boldsymbol{a}_\alpha(\hat{\boldsymbol{\rho}}_\alpha)/R$, $R = \sum_{\nu=1}^{3} r_\nu$, $\tilde{\eta}_\nu := \eta_\nu(\tilde{k}_\nu)$, $\tilde{N}_\nu := N_\nu(\tilde{\eta}_\nu)$. A structurally identical wave function but with a different definition of the local momenta has recently been proposed in [10, 11]. For a wave function which contains - again differently defined - local momenta in only two hypergeometric functions and hence is asymptotically correct in three regions, say $\Omega_\alpha \cup \Omega_\beta \cup \Omega_0$, only see [12]. It is to be noted that all these admissible choices of local momenta differ from the original one [5] in next-to-leading order in $1/\rho_\alpha$ only.

Most effort has, however, been directed towards devising wave functions which are asymptotically correct in Ω_0 only. The prototype is (4) (with $\tilde{\psi}_{k_\alpha}^{(+)} \to \tilde{\psi}_{k_\alpha}^{C(+)}$) which, though it has the unphysical feature that the motion of the three particle pairs is completely *uncorrelated in all space*, is nevertheless widely used. To overcome this deficiency correlations are introduced, by modifying the Coulomb parameter which occur in each of the hypergeometric functions (7): $\eta_\alpha \to \mu_\alpha[e_\beta e_\gamma + \chi_\alpha]/k_\alpha$ where either $\chi_\alpha = \chi_\alpha(\mathbf{k}_\alpha, \mathbf{q}_\alpha)$ depends on the momenta ('velocity-dependent charges' [13, 14]) or on the positions $\chi_\alpha = \chi_\alpha(\mathbf{r}_\alpha, \boldsymbol{\rho}_\alpha)$ ('position-dependent charges' [15]). The fairly arbitrary functions χ_α are chosen such that the resulting three-body wave functions satisfy additional criteria like the cusp condition. For very general wave functions of this type see [11].

1.1.5 Applications

These - at least asymptotically correct - wave functions are increasingly being applied to approximate calculations of ionisation and breakup processes. In its simplest form (with incoming plane wave $|\mathbf{q}_\alpha\rangle$) the corresponding amplitude is

$$\mathcal{T}_{\alpha m,0}(\mathbf{q}_\alpha; \mathbf{k}'_\beta, \mathbf{q}'_\beta) = \langle \Psi_{\mathbf{k}'_\beta, \mathbf{q}'_\beta}^{(-)} | \bar{V}_\alpha | \psi_{\alpha m} \rangle | \mathbf{q}_\alpha \rangle, \tag{17}$$

with $\bar{V}_\alpha = \sum_{\nu \neq \alpha} V_\nu$ being the (entrance) channel interaction and $\psi_{\alpha m} = \psi_{\alpha m}(\mathbf{r}_\alpha)$ denoting the incoming bound state wave function. Expression (17) contains for the final state the full solution $\langle \Psi_{\mathbf{k}'_\beta, \mathbf{q}'_\beta}^{(-)} |$ of the three-charged particle SE which is being approximated by one of the aforementioned ansätze.

Many calculations have appeared for (e,2e) reactions on hydrogen and other atoms; for a comparative study of various types of asymptotically correct wave functions see [16]. A nuclear physics example is the Coulomb dissociation of a projectile in the field of a fully stripped nucleus which is an important tool for obtaining astrophysical S-factors [17]. Quite generally one expects the better results with such an approximation the more peripheral the collision is (the impact parameter should be larger than the range of the nuclear interaction).

1.2 Long-range Behaviour of the Optical Potential (OP)

1.2.1 Introduction

As is well known, elastic scattering of the type $\alpha + (\beta, \gamma)_m \to \alpha + (\beta, \gamma)_m$ can formally be described by means of a single, one-channel LS equation of the type

$$\mathcal{T}_{\alpha m, \alpha m}(z) = \mathcal{V}_{\alpha m, \alpha m}^{\text{opt}}(z) + \mathcal{V}_{\alpha m, \alpha m}^{\text{opt}}(z) \mathcal{G}_{0;\alpha m}(z) \mathcal{T}_{\alpha m, \alpha m}(z). \tag{18}$$

Though not useful for practical calculations, the very existence of an equation like (18) serves as justification for use of phenomenological OP's.

The solvability of (18) depends on the properties of the plane wave matrix elements $\mathcal{V}_{\alpha m, \alpha m}^{\text{opt}}(\mathbf{q}'_\alpha, \mathbf{q}_\alpha; z) = \langle \mathbf{q}'_\alpha | \mathcal{V}_{\alpha m, \alpha m}^{\text{opt}}(z) | \mathbf{q}_\alpha \rangle$ in the limit $\boldsymbol{\Delta}_\alpha = \mathbf{q}'_\alpha - \mathbf{q}_\alpha \to$

0 which reflects itself in a corresponding behaviour in co-ordinate space for large intercluster separation ρ_α. The exact expression for the OP comes as a sum of two terms $V^{\text{opt}}_{\alpha m, \alpha m} = V^{\text{stat}}_{\alpha m, \alpha m} + \tilde{V}^{\text{opt}}_{\alpha m, \alpha m}$. The first term, so-called static potential, has only the trivial Coulomb-type behaviour which for a spherically symmetric target reads as ($e_{\beta\gamma} = e_\beta + e_\gamma$)

$$V^{\text{stat}}_{\alpha m, \alpha m}(\boldsymbol{q}'_\alpha, \boldsymbol{q}_\alpha) \stackrel{\Delta_\alpha \to 0}{=} \frac{4\pi e_\alpha e_{\beta\gamma}}{\Delta_\alpha^2} \quad \Longleftrightarrow \quad V^{\text{stat}}_{\alpha m, \alpha m}(\rho_\alpha) \stackrel{\rho_\alpha \to \infty}{=} \frac{e_\alpha e_{\beta\gamma}}{\rho_\alpha}. \quad (19)$$

Thus it is the 'nonstatic' part $\tilde{V}^{\text{opt}}_{\alpha m, \alpha m}$ which is of interest. For energies $E < 0$, i.e. below the three-body threshold, its behaviour is known from 2nd order perturbation theory, and has been corroborated for the exact OP in [18]:

$$\tilde{V}^{\text{opt}}_{\alpha m, \alpha m}(\boldsymbol{q}'_\alpha, \boldsymbol{q}_\alpha; E) \stackrel{\Delta_\alpha \to 0}{\sim} \Delta_\alpha \quad \Longleftrightarrow \quad \tilde{V}^{\text{opt}}_{\alpha m, \alpha m}(\rho_\alpha) \stackrel{\rho_\alpha \to \infty}{\approx} -\frac{a}{2\rho_\alpha^4}. \quad (20)$$

Here, a is the static dipole polarisability of the composite particle.

But does this result hold also above the three-body threshold? In fact, the two *perturbative* answers are in support of it, but ref. [19] finds an energy-dependent "a" while in [20] the standard expression for a is obtained. And what happens to the *exact* nonstatic OP for $E > 0$? Answering requires the investigation of the zero-momentum transfer limit of $\tilde{V}^{\text{opt}}_{\alpha m, \alpha m}(\boldsymbol{q}'_\alpha, \boldsymbol{q}_\alpha; z)$ which is fully determined by the Coulomb part of the channel interaction $\bar{V}^C_\alpha = \sum_{\nu \neq \alpha} V^C_\nu$ and by the three-body Coulomb resolvent $G^C(z) = (z - H_0 - \sum_\nu V^C_\nu)^{-1}$,

$$\tilde{V}^{\text{opt}}_{\alpha m, \alpha m}(\boldsymbol{q}'_\alpha, \boldsymbol{q}_\alpha; z) \stackrel{\Delta_\alpha \to 0}{\approx} \tilde{V}^{\text{opt (as)}}_{\alpha m, \alpha m}(\boldsymbol{q}'_\alpha, \boldsymbol{q}_\alpha; z)$$
$$:= \langle \boldsymbol{q}'_\alpha | \langle \psi_{\alpha m} | \bar{V}^C_\alpha Q_{\alpha m} G^C(z) Q_{\alpha m} \bar{V}^C_\alpha | \psi_{\alpha m} \rangle | \boldsymbol{q}_\alpha \rangle. \quad (21)$$

Here, $Q_{\alpha m} = 1 - |\psi_{\alpha m}\rangle\langle\psi_{\alpha m}|$ is the projector onto all (discrete and continuum) target states except the incoming (= outgoing) bound state.

The procedure goes as follows: insert the spectral representation of G^C, with both two- and three-body scattering states; for the latter use the asymptotic three-charged particle wave function in Ω_α. Then extract the leading singularity in the limit $\Delta_\alpha \to 0$. (Note: 2nd-order perturbation theory is equivalent to replacing G^C by the Coulomb channel resolvent $G^C_\alpha = (z - H_0 - V^C_\alpha)^{-1}$.)

1.2.2 Nonperturbative Result Valid for All Energies

Let me just state the result without giving any details. For the latter I refer to [21]. Off the energy shell one finds

$$\tilde{V}^{\text{opt (as)}}_{\alpha m, \alpha m}(\boldsymbol{q}'_\alpha, \boldsymbol{q}_\alpha; E + i0) \stackrel{\Delta_\alpha \to 0}{=} C\Delta_\alpha + o(\Delta_\alpha), \quad (22)$$

with an energy- and momentum-dependent 'strength factor' $C = C(E, q_\alpha)$. On the energy shell, i.e. for $E = q_\alpha^2/2M_\alpha + \hat{E}_{\alpha m}$, $q'_\alpha = q_\alpha$, ($\hat{E}_{\alpha m}$ is the bound state energy), $\tilde{V}^{\text{opt (as)}}_{\alpha m, \alpha m}(\boldsymbol{\Delta}_\alpha)$ is a function of the momentum transfer only. Hence, in co-ordinate space one indeed ends up with a local, energy-independent potential

$$\tilde{V}^{\text{opt (as)}}_{\alpha m, \alpha m}(\boldsymbol{\rho}_\alpha) \overset{\rho_\alpha \rightarrow \infty}{\cong} -\frac{a}{2\rho_\alpha^4} + o\left(\frac{1}{\rho_\alpha^4}\right). \tag{23}$$

A further important point is that the exact 'strength factor' a coincides with the polarisability as found in perturbative approaches,

$$a = 2 \sum_{n \neq m} \frac{|\hat{\boldsymbol{\rho}}_\alpha \cdot \boldsymbol{D}_{nm}|^2}{\left[|\hat{E}_{\alpha m}| - |\hat{E}_{\alpha n}|\right]} + 2 \int \frac{d\boldsymbol{k}_\alpha^0}{(2\pi)^3} \frac{|\hat{\boldsymbol{\rho}}_\alpha \cdot \boldsymbol{D}_{\boldsymbol{k}_\alpha^0 m}|^2}{\left[|\hat{E}_{\alpha m}| + k_\alpha^{0\,2}/2\mu_\alpha\right]}, \tag{24}$$

$$\boldsymbol{D}_{nm} = \epsilon_{\alpha\beta} e_\alpha \mu_\alpha \left(e_\gamma/m_\gamma - e_\beta/m_\beta\right) \int d\boldsymbol{r}_\alpha \psi^*_{\alpha n}(\boldsymbol{r}_\alpha) \boldsymbol{r}_\alpha \psi_{\alpha m}(\boldsymbol{r}_\alpha) \tag{25}$$

$$\boldsymbol{D}_{\boldsymbol{k}_\alpha^0 m} = \epsilon_{\alpha\beta} e_\alpha \mu_\alpha \left(e_\gamma/m_\gamma - e_\beta/m_\beta\right) \int d\boldsymbol{r}_\alpha \psi^{(+)*}_{\boldsymbol{k}_\alpha^0}(\boldsymbol{r}_\alpha) \boldsymbol{r}_\alpha \psi_{\alpha m}(\boldsymbol{r}_\alpha) \tag{26}$$

being the matrix elements of the dipole operator between the incoming and all other (bound and continuum) target states. That is, no "renormalisation" of a arises from summing up all higher order terms in the perturbation expansion of G^C, and the E- and q_α-dependence of the 'off-shell strength' C has disappeared.

Concluding I mention that a new nonrelativistic contribution to the asymptotic part of the OP $\sim 1/\rho_\alpha^5$, which results from the long-range three-body correlations described in Sect. (1.1.3), has been derived in [22].

2   Calculational Developments

Considerable progress has been made in recent years also in applying few-body theory to the practical computation of atomic and nuclear reactions. Here, I will concentrate on describing an attempt at re-establishing this method as a powerful tool for atomic physics calculations. Advances in the calculation of proton-deuteron scattering with realistic nuclear potentials will only briefly be touched upon (cf. the separate Contribution to this Conference [23]).

2.1   Few-Body Approach to the Energetic Atomic Collision $\text{H}^+ + \text{H}$

Scattering of protons off hydrogen has attracted the interest of physicists since the '30. Because direct scattering, electron exchange (both including excitation) and ionisation are coupled by unitarity they should in principle constitute an 'ideal' field of application of three-body theory. However, except for a few isolated, and not so successful, attempts little has been accomplished in this respect.

Instead, most efforts have concentrated on developing, and improving, standard models based on the SE. For instance, at low and medium energies the expansion into target states ('close coupling') appears to be the method of choice. At higher energies (from keV to a few hundreds of keV), which is the regime I will concentrate on, DWBA-type models or models based on the multiple scattering expansion of Lippmann-Schwinger equation for wave functions are rather successful. However, most of these approaches treat only one channel explicitly, i.e., intermediate-state rearrangement and/or excitation is ignored. One consequence is the violation of (three-body) unitarity constraints. Hence, though they generally provide a good description of either electron exchange or (in)elastic cross sections, they are frequently unable to describe the other reaction. For recent reviews I refer to [24, 25].

Two-fragment processes of the type $\alpha + (\beta, \gamma)_m \to \beta + (\alpha, \gamma)_n$ are conveniently described by means of the effective-two-body formulation of the three-body theory [26]. There, the amplitudes for the various two-fragment processes satisfy multichannel Lippmann-Schwinger-type equations

$$\mathcal{T}_{\beta n, \alpha m} = \mathcal{V}_{\beta n, \alpha m} + \sum_{\nu, r} \mathcal{V}_{\beta n, \nu r} \, \mathcal{G}_{0;\nu r} \, \mathcal{T}_{\nu r, \alpha m} \tag{27}$$

$$\mathcal{V}_{\beta n, \alpha m} = \bar{\delta}_{\beta \alpha} \langle \chi_{\beta n} \mid G_0 \mid \chi_{\alpha m} \rangle + \sum_{\nu \neq \beta, \alpha} \langle \chi_{\beta n} \mid G_0 T'_\nu G_0 \mid \chi_{\alpha m} \rangle + \cdots, \tag{28}$$

with $\bar{\delta}_{\beta \alpha} = 1 - \delta_{\beta \alpha}$. The 'form factor' $\mid \chi_{\alpha m} \rangle$ is a certain off-shell extrapolation of the bound state wave function $\mid \psi_{\alpha m} \rangle$. Already the first-order terms ('triangle amplitudes') in the expansion of $\mathcal{V}$ contain matrix elements of an operator T'_ν, the most awkward part of which is the T-matrix T^C_ν describing Coulomb rescattering of the projectile off each of the target particles. Due to the singular nature of the latter, calculation of such terms is very difficult (this is particularly true if the rescattering particles have charges of opposite sign, because of the infinity of bound states). Indeed, the lack of applications of Faddeev-type approaches to atomic reactions is generally attributed to this fact [24].

Hence, the so-called Coulomb-Born Approximation (CBA) is usually made which consists in approximating T^C_ν by the Coulomb potential V^C_ν, $\nu = \alpha, \beta, \gamma$. Recently, however, we have succeeded in calculating exactly all (on-shell) triangle amplitudes with the full (attractive and repulsive) Coulomb T-matrix, for all energies and scattering angles, for arbitrary bound state wave functions. In this way we could show that the CBA almost always fails badly for atomic triangle amplitudes [27, 28, 29]. (I mention that we have also derived (semi-) analytical approximations for these triangle amplitudes, valid for the higher energies considered here, which greatly simplify practical calculations but are much more accurate than the CBA [29, 30]).

The exact triangle amplitudes have been employed in an investigation of direct and exchange scattering of H^+ off hydrogen atoms. For this purpose we have written Eqs. (27) in K-matrix approximation. When the resulting two-dimensional integral equations are transformed to the impact parameter representation they lead to a system of coupled algebraic equations [31]. The K-matrix has been approximated by the effective potential (28) were we keep the 0th and the full 1st order terms. For the latter, to get numerically calculated quantities, the transformation to the impact parameter representation has been performed by quadrature. Our results provide a simultaneous, satisfactory description of differential cross sections for both elastic scattering and electron exchange; likewise the total exchange cross section is well-reproduced [32].

2.2 Proton-Deuteron Scattering in the Integral Equations Approach

The motivation behind 30 years of investigation of nucleon-deuteron (Nd) scattering rests on the hope that a detailed understanding may lead to deeper insights into the properties of $2N$ and $3N$ forces etc. Experimentally the proton-induced reaction (pd) is preferred; but reliable and sophisticated calculations have become available only recently, for energies below the deuteron breakup threshold [33, 34]. Some of the underlying reasons are pointed out in [23]. Of course, the common practice of comparing nd calculations with pd data, though *in praxi* rather successful, is clearly unsatisfactory from principle point of view.

We [23] have obtained the first reliable results for pd scattering above the deuteron breakup threshold, for the Paris potential. We used the screening and renormalisation approach [35]. Comparison with elastic scattering cross section and polarisation data shows a reasonably good absolute agreement; but a much better reproduction is achieved of the difference between experimental pd and nd observables where available. Further progress requires, in particular, excellent, low-rank, separable representations of more modern potentials.

Acknowledgement. Most of the results presented here have been worked out in collaboration with G.V. Avakov, B.F. Irgaziev, A.S. Kadyrov, A.M. Mukhamedzhanov, and A.I. Sattarov.

References

1. A.M. Mukhamedzhanov, E.O. Alt, G.V. Avakov: Contribution to this Conference

2. P.J. Redmond, cited in L. Rosenberg: Phys. Rev. **D8**, 1833 (1972)

3. S.P. Merkuriev: Theor. Math. Phys. **32**, 680 (1977)

4. M. Brauner, J.S. Briggs, H.J. Klar: J. Phys. **B22**, 2265 (1989)

5. E.O. Alt and A.M. Mukhamedzhanov: JETP Lett. **56**, 435 (1992); Phys. Rev. **A47**, 2004 (1993)

6. A.M. Mukhamedzhanov, M. Lieber: Phys. Rev. **A54**, 3078 (1996)

7. Sh.D. Kunikeev and V.S. Senashenko: JETP **82**, 839 (1996)

8. Y.E. Kim and A.L. Zubarev: Phys. Rev. **A56**, 521 (1997)

9. A.M. Mukhamedzhanov, E.O. Alt, B.F. Irgaziev: Poster at this Conference

10. P.A. Macri et al.: Phys. Rev. **A55**, 3518 (1997)

11. F.D. Colavecchia, G. Gaseano, C.R. Garibotti: Phys. Rev. **A57**, 1018 (1998)

12. A. Engelns, H. Klar, A.W. Malcherek: J. Phys. **B30**, L811 (1997)

13. S. Jetzke and F.H.M. Faisal: Phys. Rev. **B25**, 1543 (1992)

14. J. Berakdar and J.S. Briggs: Phys. Rev. Lett. **72**, 3799 (1994)

15. J. Berakdar: Phys. Rev. **A53**, 2314 (1996) and references therein

16. S. Jones, D.H. Madison, D.A. Konovalov: Phys. Rev. **A55**, 444 (1997)

17. E.O. Alt, B.F. Irgaziev, A.M. Mukhamedzhanov: Contribution to this Conference and references therein

18. A.A. Kvitsinskii and S.P. Merkuriev: Sov. J. Nucl. Phys. **48**, 79 (1988)

19. I.E. McCarthy, B.C. Saha, A.T. Stelbovics: Phys. Rev. **A22**, 502 (1980)

20. K. Unnikrishnan and J. Callaway: Phys. Lett. **A138**, 285 (1989)

21. E.O. Alt and A.M. Mukhamedzhanov: Phys. Rev. **A51**, 3852 (1995)

22. A.M. Mukhamedzhanov: Phys. Rev. **A56**, 473 (1997)

23. E.O. Alt, A.M. Mukhamedzhanov, A.I. Sattarov: Contribution to this Conference and Preprint MZ-TH/98-28 University of Mainz 1998

24. D.P. Dewangan and J. Eichler: Phys. Rep. **247**, 59 (1994)

25. B.H. Bransden and M.C.R. McDowell: In: *Charge Exchange and the Theory of Ion-Atom Collisions*. Oxford: Clarendon Press 1992

26. E.O. Alt, P. Grassberger, W. Sandhas: Nucl. Phys. **B2**, 167 (1967)

27. E.O. Alt et al.: J. Phys. **B28**, 5137 (1995)

28. E.O. Alt, A.S. Kadyrov, A.M. Mukhamedzhanov: Phys. Rev. **A54**, 4091 (1996)

29. E.O. Alt, A.S. Kadyrov, A.M. Mukhamedzhanov: J. Phys. **B30**, 3659 (1997)

30. E.O. Alt, A.S. Kadyrov, A.M. Mukhamedzhanov: Phys. Rev. **A53**, 2438 (1996)

31. E.O. Alt et al.: J. Phys. **B27**, 4653 (1994) and references therein

32. E.O. Alt, A.S. Kadyrov, A.M. Mukhamedzhanov: Preprint MZ-TH/98-32 University of Mainz 1998

33. G.H. Berthold, A. Stadler, H. Zankel: Phys. Rev. **C41**, 1365 (1990)

34. A. Kievski et al.: Nucl. Phys. **A607**, 402 (1996)

35. E.O. Alt and W. Sandhas: In: *Coulomb Interactions in Nuclear and Atomic Few-Body Collisions*, eds. F.S. Levin, D.A. Micha. New York : Plenum 1996

Few-Body Systems Suppl. 10, 75–84 (1999)

Few-
Body
Systems
© by Springer-Verlag 1999

Structure of T– and S–Matrices in Unphysical Sheets and Resonances in Three–Body Systems

A.K. Motovilov[1,2], E.A. Kolganova[2]

[1] Physikalishes Institut der Universität Bonn, Endenicher Allee 11–13, D-53115 Bonn, Germany
[2] Joint Institute for Nuclear Research, 141980 Dubna, Moscow region, Russia

Abstract. Algorithm, based on explicit representations for the analytic continuation of Faddeev components of the three-body T-matrix in unphysical energy sheets, is employed to study mechanism of disappearance and formation of the Efimov levels of the helium ^{4}He trimer.

1 Introduction

Explicit representations for the Faddeev components of the three-body T-matrix continued analytically into unphysical sheets of the energy Riemann surface have been formulated and proved recently in ref. [1]. According to the representations, the T-matrix in unphysical sheets is explicitly expressed in terms of its components only taken in the physical sheet. Analogous explicit representations were also found for the analytic continuation of the three-body scattering matrices. These representations disclose the structure of kernels of the T- and S-matrices after continuation and give new capacities for analytical and numerical studies of the three-body resonances. In particular the representations imply that the resonance poles of the S-matrix as well as T-matrix in an unphysical sheet correspond merely to the zeros of the suitably truncated three-body scattering matrix taken in the physical sheet. Therefore, one can search for resonances in a certain unphysical sheet staying always, nevertheless, in the physical sheet and only calculating the position of zeros of the appropriate truncation of the total three-body scattering matrix. This statement holds true not only for the case of the conventional smooth quickly decreasing interactions but also for the case of the singular interactions described by different variants of the Boundary Condition Model, in particular for the inter–particle interactions of a hard-core nature like in most molecular systems.

As a concrete application of the method, we present here the results of our numerical study of the simplest truncation of the scattering matrix in the

^{4}He three-atomic system, namely of the $(2 + 1 \to 2 + 1)$ S-matrix component corresponding to the scattering of a ^{4}He atom off a ^{4}He dimer. The point is that there is already a series of works [2-4] (also see [5-7]) showing that the excited state of the ^{4}He trimer is initiated by the Efimov effect [8]. In these works, various versions of the ^{4}He–^{4}He potential were employed. However, the basic result of refs. [2-4] on the excited state of the helium trimer is the same: this state disappears after the interatomic potential is multiplied by the increasing factor λ when it approaches the value about 1.2. It is just such a nonstandard behavior of the excited-state energy as the coupling between helium atoms becomes more and more strengthening, points to the Efimov nature of the trimer excited state. The present work is aimed at elucidating the fate of the trimer excited state upon its disappearance in the physical sheet when $\lambda > 1$ and at studying the mechanism of arising of new excited states when $\lambda < 1$. As the interatomic He–He potential, we use the potential HFD-B [9]. We have established that for such He–He-interactions the trimer excited-state energy merges with the two-body threshold ϵ_d at $\lambda \approx 1.18$ and with further decreasing λ it transforms into a virtual level of the first order (a simple real pole of the analytic continuation of the $(2+1 \to 2+1)$ scattering matrix component) lying in the unphysical energy sheet adjoining the physical sheet along the spectral interval between ϵ_d and the three–body threshold. We trace the position of this level for λ increasing up to 1.5. Besides, we have found that the excited (Efimov) levels for $\lambda < 1$ also originate from virtual levels of the first order that are formed in pairs. Before a pair of virtual levels appears, there occurs a fusion of a pair of conjugate resonances of the first order (simple complex poles of the analytic continuation of the scattering matrix in the unphysical sheet) resulting in the virtual level of the second order.

2   Representations for three-body T– and S–matrices in unphysical energy sheets

The method used for calculation of resonances in the present work, is based on the explicit representations [1] for analytic continuation of the T- and scattering matrices in unphysical sheets which hold true at least for a part of the three-body Riemann surface. To describe this part we introduce the auxiliary vector-function $\mathsf{f}(z) = (\mathsf{f}_0(z), \mathsf{f}_{1,1}(z), ..., \mathsf{f}_{1,n_1}(z), \mathsf{f}_{2,1}(z), ..., \mathsf{f}_{2,n_2}(z), \mathsf{f}_{3,1}(z), ..., \mathsf{f}_{3,n_3}(z))$ with $\mathsf{f}_0(z) = \ln z$ and $\mathsf{f}_{\alpha,j}(z) = (z - \lambda_{\alpha,j})^{1/2}$. Here, by z we understand the total three-body energy in the c. m. system and by $\lambda_{\alpha,j}$, the respective binding energies of the two-body subsystems α, $\alpha = 1, 2, 3, j = 1, 2, ..., n_\alpha$, $n_\alpha < \infty$. The sheets Π_l of the Riemann surface of the vector-function $\mathsf{f}(z)$ are numerated by the multi-index $l = (l_0, l_{1,1}, ..., l_{1,n_1}, l_{2,1}, ..., l_{2,n_2}, l_{3,1}, ..., l_{3,n_3})$, where $l_{\alpha,j} = 0$ if the sheet Π_l corresponds to the main (arithmetic) branch of the square root $(z - \lambda_\alpha)^{1/2}$. Otherwise, $l_{\alpha,j} = 1$ is assumed. Value of l_0 coincides with the number of branches of the function $\ln z$, $\ln z = \ln |z| + i\, 2\pi l_0 + i\phi$ where $\phi = \arg z$. For the physical sheet identified by $l_0 = l_{\alpha,j} = 0$, $\alpha = 1, 2, 3$, $j = 1, 2, ..., n_\alpha$, we use the notation Π_0.

Surely, the structure of the total three-body Riemann surface is essentially more complicated than that of the auxiliary function $\mathfrak{f}$. For instance, the sheets Π_l with $l_0 = \pm 1$ have additional branching points corresponding to resonances of the two-body subsystems. The part of the total three-body Riemann surface where the representations of ref. [1] are valid, consists of the sheets Π_l of the Riemann surface of the function $\mathfrak{f}$ identified by $l_0 = 0$ (such unphysical sheets are called *two-body* sheets) and two *three-body* sheets identified by $l_0 = \pm 1$ and $l_{\alpha,j} = 1$, $\alpha = 1, 2, 3$, $j = 1, 2, ..., n_\alpha$.

In what follows by k_α, p_α ($k_\alpha, p_\alpha \in \mathbb{R}^3$ or $k_\alpha, p_\alpha \in \mathbb{C}^3$) we understand the standard reduced relative momenta of the three-body system while $P = \{k_\alpha, p_\alpha\}$ ($P \in \mathbb{R}^6$ or $P \in \mathbb{C}^6$) stands for the total relative momentum.

The representations [1] for the analytic continuation of the matrix $\mathbf{M}(z) = \{M_{\alpha\beta}(z)\}$, $\alpha, \beta = 1, 2, 3$, of the Faddeev components $M_{\alpha\beta}(z)$ (see [10]) of the three-body T-operator, into the sheet Π_l read as follows[1]:

$$\mathbf{M}(z)\Big|_{\Pi_l} = \mathbf{M}(z) - \mathbf{B}^\dagger(z)A(z)LS_l^{-1}(z)\widetilde{L}\mathbf{B}(z). \qquad (1)$$

Here, the factor $A(z)$ is the diagonal matrix, $A(z) = \mathrm{diag}\{A_0(z), A_{1,1}(z), ..., A_{1,n_3}(z)\}$ with $A_0(z) = -\pi i z^2$ and $A_{\alpha,j} = -\pi i \sqrt{z - \lambda_{\alpha,j}}$, $j = 1, 2, ..., n_\alpha$. Notations L and $\widetilde{L}$ stand for the diagonal number matrices combined of the indices of the sheet Π_l: $L = \mathrm{diag}\{l_0, l_{1,1}, ..., l_{3,n_3}\}$ and $\widetilde{L} = \mathrm{diag}\{|l_0|, l_{1,1}, ..., l_{3,n_3}\}$. By $S_l(z)$ we understand a truncation of the three-body scattering matrix $S(z)$ defined in $\widehat{\mathcal{G}} = L_2(S^5) \bigoplus_{\alpha=1}^{3} \bigoplus_{j=1}^{n_\alpha} L_2(S^2)$ by the equation

$$S_l(z) = \widehat{I} + \widetilde{L}\big[S(z) - \widehat{I}\big]L$$

where $\widehat{I}$ is the identity operator in $\widehat{\mathcal{G}}$. Also, we use the notations

$$\mathbf{B}(z) = \begin{pmatrix} J_0 \Omega \mathbf{M} \\ J_1 \boldsymbol{\Psi}^*[\boldsymbol{\Upsilon}\mathbf{M} + \mathbf{v}] \end{pmatrix} \text{ and } \mathbf{B}^\dagger(z) = \begin{pmatrix} \mathbf{M}(z)\Omega^\dagger J_0^\dagger, & [\mathbf{v} + \mathbf{M}\boldsymbol{\Upsilon}]\boldsymbol{\Psi}J_1^\dagger \end{pmatrix}.$$

Here, $\mathbf{v} = \mathrm{diag}\{v_1, v_2, v_3\}$ with v_α, the pair potentials, $\alpha = 1, 2, 3$. At the same time, $\Omega = (1, 1, 1)$, $\boldsymbol{\Upsilon} = \{\Upsilon_{\alpha\beta}\}$ with $\Upsilon_{\alpha\beta} = 1 - \delta_{\alpha\beta}$, $\alpha, \beta = 1, 2, 3$, and $\boldsymbol{\Psi} = \mathrm{diag}\{\boldsymbol{\Psi}_1, \boldsymbol{\Psi}_2, \boldsymbol{\Psi}_3\}$ where $\boldsymbol{\Psi}_\alpha$, $\alpha = 1, 2, 3$, are operators acting on $f = (f_1, f_2, ..., f_{n_\alpha}) \in \bigoplus_{j=1}^{n_\alpha} L_2(\mathbb{R}^3)$ as $(\boldsymbol{\Psi}_\alpha f)(P) = \sum_{j=1}^{n_\alpha} \psi_{\alpha,j}(k_\alpha)f_j(p_\alpha)$ where, in turn, $\psi_{\alpha,j}$ is the bound-state wave function of the pair subsystem α corresponding to the binding energy $\lambda_{\alpha,j}$. By $\boldsymbol{\Psi}^*$ we denote the operator adjoint to $\boldsymbol{\Psi}$. Notation $J_0(z)$ is used for the operator restricting a function on the energy-shell $|P|^2 = z$. The diagonal matrix-valued function $J_1(z) = \mathrm{diag}\{J_{1,1}(z), ..., J_{3,n_3}(z)\}$ consists of the operators $J_{\alpha,j}(z)$ with restriction on the energy surfaces $|p_\alpha|^2 = z - \lambda_{\alpha,j}$.

---

[1] One assumes that all the pair interactions fall off in the coordinate space not slower than exponentially and, thus, their Fourier transforms $v_\alpha(k)$, $k \in \mathbb{C}^3$, are holomorphic functions of the relative momenta k in a stripe $|\operatorname{Im} k| < b$ for some $b > 0$.

The operators $\Omega^\dagger$, $J_0^\dagger(z)$ and $J_1^\dagger(z)$ represent the "transposed" matrices Ω, $J_0(z)$ and $J_1(z)$, respectively. Operators $J_0^\dagger(z)$ and $J_1^\dagger(z)$ act in the expression for $\mathbf{B}^\dagger$ (as if) to the left.

With some stipulations (see [1]) the representations for the scattering matrix read

$$S(z)\Big|_{\Pi_l} = \mathcal{E}(l)\left\{\widehat{I} + S_l^{-1}(z)[S(z) - \widehat{I}]e(l)\right\}\mathcal{E}(l). \tag{2}$$

Here, $\mathcal{E} = \mathrm{diag}\{\mathcal{E}_0, \mathcal{E}_{1,1}, ..., \mathcal{E}_{3,n_3}\}$ where $\mathcal{E}_0$ is the identity operator in $L_2(S^5)$ if $l_0 = 0$ and inversion, $(\mathcal{E}_0 f)(\widehat{P}) = f(-\widehat{P})$, if $l_0 = \pm 1$. Analogously, $\mathcal{E}_{\alpha,j}$ is the identity operator in $L_2(S^2)$ for $l_{\alpha,j} = 0$ and inversion for $l_{\alpha,j} = 1$. Notation $e(l)$ is used for the diagonal number matrix $e(l) = \mathrm{diag}\{e_0, e_{1,1}, ..., e_{3,n_3}\}$ with nontrivial elements $e_{\alpha,j} = 1$ if $l_{\alpha,j} = 0$ and $e_{\alpha,j} = -1$ if $l_{\alpha,j} = 1$; for all the cases $e_0 = 1$.

It follows from the representations (1) and (2) that the resonances (the nontrivial poles of $\mathbf{M}(z)\Big|_{\Pi_l}$ and $S(z)\Big|_{\Pi_l}$) situated in the unphysical sheet Π_l are those points $z = z_{\mathrm{res}}$ in the physical sheet where the matrix $S_l(z)$ has zero as eigenvalue. Therefore, *calculation of resonances in the unphysical sheet Π_l is reduced to a search for zeros of the respective truncation $S_l(z)$ of the scattering matrix $S(z)$ in the physical sheet.*

3 Method for search of resonances in a three–body system on the basis of the Faddeev differential equations

In this work we discuss the example of the three-atomic ^{4}He system at the total angular momentum $L = 0$. We consider the case where the interatomic interactions include a hard core component and, outside the hard core domain, are described by conventional smooth potentials. In this case, the angular partial analysis reduces the initial Faddeev equation for three identical bosons to a system of coupled two-dimensional integro-differential equations (see ref. [4] and references therein)

$$\left[-\frac{\partial^2}{\partial x^2} - \frac{\partial^2}{\partial y^2} + l(l+1)\left(\frac{1}{x^2} + \frac{1}{y^2}\right) - E\right] F_l(x, y) \tag{3}$$

$$= \begin{cases} -V(x)\Psi_l(x, y), & x > c \\ 0, & x < c. \end{cases}$$

Here, x, y stand for the standard Jacobi variables and c, for the core range. At $L = 0$ the partial angular momentum l corresponds both to the dimer subsystem and a complementary atom. For the S-state three-boson system l is even, $l = 0, 2, 4, \ldots$. In our work, the energy z can get both real and complex values. The He–He potential $V(x)$ acting outside the core domain is assumed to be central. The partial wave function $\Psi_l(x, y)$ is related to the Faddeev components $F_l(x, y)$ by $\Psi_l(x, y) = F_l(x, y) + \sum_{l'}\int_{-1}^{+1} d\eta\, h_{ll'}(x, y, \eta)\, F_{l'}(x', y')$ where $x' = (\frac{1}{4}x^2 + \frac{3}{4}y^2 - \frac{\sqrt{3}}{2}xy\eta)^{1/2}$, $y' = (\frac{3}{4}x^2 + \frac{1}{4}y^2 + \frac{\sqrt{3}}{2}xy\eta)^{1/2}$ and

$-1 \leq \eta \leq 1$. The explicit form of the function $h_{ll'}$ can be found in refs. [10, 11]. The functions $F_l(x, y)$ satisfy the boundary conditions

$$F_l(x, y)\,|_{x=0} = F_l(x, y)\,|_{y=0} = 0\,, \qquad \Psi_l(c, y) = 0 \tag{4}$$

Note that the last of these conditions is a specific condition corresponding to the hard-core model (see ref. [4]).

Here we only deal with a finite number of equations (3), assuming that $l \leq l_{\max}$ where $l_{\max}$ is a certain fixed even number. The condition $0 \leq l \leq l_{\max}$ is equivalent to the supposition that the potential $V(x)$ only acts in the two-body states with $l = 0, 2, \ldots, l_{\max}$. The spectrum of the Schrödinger operator for a system of three identical bosons with such a potential is denoted by σ_{3B}. We assume that the potential $V(x)$ falls off exponentially and, thus, $|V(x)| \leq C\exp(-\mu x)$ with some positive C and μ. For the sake of simplicity we even assume sometimes that $V(x)$ is finite, i.e., $V(x) = 0$ for $x > r_0$, $r_0 > 0$. Looking ahead, we note that, in fact, in our numerical computations of the $^4\mathrm{He}_3$ system at complex energies we make a "cutoff" of the interatomic $\mathrm{He}-\mathrm{He}$-potential at a sufficiently large r_0. The asymptotic conditions as $\rho \to \infty$ and/or $y \to \infty$ for the partial Faddeev components of the $(2+1 \to 2+1; 1+1+1)$ scattering wave functions for $z = E + \mathrm{i}0$, $E > 0$, read (see, e. g., ref. [10])

$$\begin{aligned}
F_l(x, y; z) \;=\; & \delta_{l0}\psi_d(x)\left\{\sin(\sqrt{z - \epsilon_d}\,y) + \exp(\mathrm{i}\sqrt{z - \epsilon_d}\,y)\,[\mathrm{a}_0(z) + o\,(1)]\right\} \\
& + \frac{\exp(\mathrm{i}\sqrt{z}\rho)}{\sqrt{\rho}}\,[A_l(z, \theta) + o\,(1)]\,.
\end{aligned} \tag{5}$$

We assume that the $^4\mathrm{He}$ dimer has only one bound state with an energy ϵ_d, $\epsilon_d < 0$, and wave function $\psi_d(x)$, $\psi_d(x) = 0$ for $0 \leq x \leq c$. The notations ρ, $\rho = \sqrt{x^2 + y^2}$, and θ, $\theta = \arctan(y/x)$, are used for the hyperradius and hyperangle. The coefficient $\mathrm{a}_0(z)$, $z = E + \mathrm{i}0$, for $E > \epsilon_d$ is the elastic scattering amplitude. The functions $A_l(E + \mathrm{i}0, \theta)$ provide us, at $E > 0$, the corresponding partial Faddeev breakup amplitudes. For real $z = E + \mathrm{i}0$, $E > \epsilon_d$, the $(2 + 1 \to 2 + 1)$ component of the s-wave partial scattering matrix for a system of three helium atoms is given by the expression

$$S_0(z) = 1 + 2\mathrm{i}\mathrm{a}_0(z)\,.$$

Our goal is to study the analytic continuation of the function $S_0(z)$ into the physical sheet. As it follows from the results of ref. [1], the function $S_0(z)$ is just that truncation of the total scattering matrix whose roots in the physical sheet of the energy z plane correspond to the location of resonances situated in the unphysical sheet adjoining the physical one along the spectral interval $(\epsilon_d, 0)$.

There are the following three important domains in the physical sheet.

$1°$. The domain $\Pi^{(\Psi)}$ where the Faddeev components $F_l(x, y; z)$ (and, hence, the wave functions $\Psi_l(x, y; z)$) can be analytically continued in z so that the differences $\Phi_l(x, y; z) = F_l(x, y; z) - \delta_{l0}\psi_d(x)\sin(\sqrt{z - \epsilon_d}\,y)$ at $z \in \Pi^{(\Psi)} \setminus \sigma_{3B}$

are square integrable. This domain is described by the inequality

$$\mathrm{Im} \sqrt{z - \epsilon_d} < \min \left\{ \frac{\sqrt{3}}{2} \mu, \ \sqrt{3} \sqrt{|\epsilon_d|} \right\} .$$

2°. The domain $\Pi^{(A)}$ where both the elastic scattering amplitude $a_0(z)$ and the Faddeev breakup amplitudes $A_l(z, \theta)$ can be analytically continued in z, $z \notin \sigma_{3B}$, and where the continued functions $F_l(x, y; z)$ still obey the asymptotic formulas (5). This domain is described by the inequalities

$$\mathrm{Im} \sqrt{z} + \frac{1}{2} \mathrm{Im} \sqrt{z - \epsilon_d} < \frac{\sqrt{3}}{2} \sqrt{|\epsilon_d|}, \quad \mathrm{Im} \sqrt{z} + \mathrm{Im} \sqrt{z - \epsilon_d} < \frac{\sqrt{3}}{2} \mu .$$

3°. And finally, we distinguish the domain $\Pi^{(S)}$, most interesting for us, where the analytic continuation in z, $z \notin \sigma_{3B}$, can be only done for the amplitude $a_0(z)$ (and consequently, for the scattering matrix $S_0(z)$); the analytic continuability of the amplitudes $A_l(z, \theta)$ in the whole domain $\Pi^{(S)}$ is not required. The set $\Pi^{(S)}$ is a geometric locus of points obeying the inequality

$$\mathrm{Im} \sqrt{z - \epsilon_d} < \min \left\{ \frac{1}{\sqrt{3}} \sqrt{|\epsilon_d|}, \ \frac{\sqrt{3}}{2} \mu \right\} .$$

Since the spherical wave $\exp(i\sqrt{z}\,\rho)/\sqrt{\rho}$ in Eq. (5) is a function rapidly decreasing in all the directions, the use of the asymptotic condition (5) is justified even if $z \in \Pi^{(S)} \setminus \Pi^{(A)}$. Outside of the domain $\Pi^{(S)}$ the numerical construction of $S_0(z)$ by solving the Faddeev differential equations is, in general, impossible.

4 Numerical results

In the present work we search for the resonances of the ^{4}He trimer including the virtual levels as roots of $S_0(z)$ and for the bound-state energies as positions of poles of $S_0(z)$. All the results presented below are obtained for the case $l_{\max} = 0$. In all our calculations, $\hbar^2/m = 12.12 \, \mathrm{K\, \AA}^2$. As the interatomic He – He - interaction we employed the HFD-B potential constructed by Aziz and co-workers [9].

The value of the core range c is chosen to be so small that its further decrease does not appreciably influence the dimer binding energy ϵ_d and the trimer ground-state energy $E_t^{(0)}$. Unlike the paper [4], where c was taken to be equal $0.7 \, \text{Å}$, now we take $c = 1.3 \, \text{Å}$. We have found that such a value of c provides at least six reliable figures of ϵ_d and three figures of $E_t^{(0)}$.

Since the statements of Sect. 3 are valid, generally speaking, only for the potentials decreasing not slower than exponentially, we cut off the potential HFD-B setting $V(x) = 0$ for $x > r_0$. We have established that this cutoff for $r_0 \gtrsim 95 \, \text{Å}$ provides the same values of ϵ_d ($\epsilon_d = -1.68541 \, \text{mK}$), $E_t^{(0)}$ ($E_t^{(0)} = -0.096 \, \text{K}$) and scattering phases which were obtained in earlier calculations [4] performed with the potential HFD-B. Also, we have found that the trimer

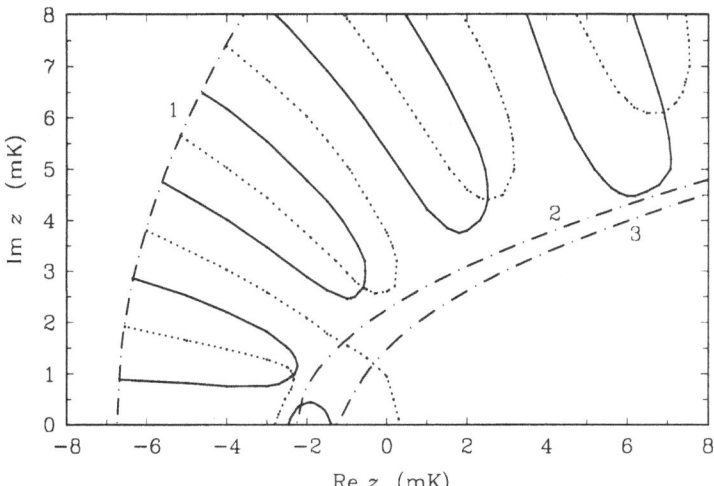

Figure 1. Root locus curves of the real and imaginary parts of the scattering matrix $S_0(z)$. The solid lines correspond to $\mathrm{Re}\,S_0(z) = 0$, while the tiny dashed lines, to $\mathrm{Im}\,S_0(z) = 0$. The numbers 1, 2 and 3 denote the boundaries of the domains $\Pi^{(\Psi)}$, $\Pi^{(S)}$ and $\Pi^{(A)}$, respectively. Complex roots of the function $S_0(z)$ are represented by the intersection points of the curves $\mathrm{Re}\,S_0(z) = 0$ and $\mathrm{Im}\,S_0(z) = 0$ and are located at $(-2.34 + \mathrm{i}\,0.96)\,\mathrm{mK}$, $(-0.59 + \mathrm{i}\,2.67)\,\mathrm{mK}$, $(2.51 + \mathrm{i}\,4.34)\,\mathrm{mK}$ and $(6.92 + \mathrm{i}\,6.10)\,\mathrm{mK}$.

excited-state energy $E_t^{(1)} = -2.46\,\mathrm{mK}$. Comparison of these results with results of other researchers can be found in ref. [4]. In all the calculations of the present work we take $r_0 = 100\,\text{Å}$. Note that if the formulas from Sect. 3 describing the holomorphy domains $\Pi^{(\Psi)}$, $\Pi^{(A)}$ and $\Pi^{(S)}$ are used for finite potentials, one should set in them $\mu = +\infty$.

A detailed description of the numerical method we use is presented in ref. [4]. When solving the boundary-value problem (3–5) we carry out its finite-difference approximation in polar coordinates ρ and θ. In this work, we used the grids of dimension $N_\theta = N_\rho = 600 - 1000$. Essentially, we chose the values of the cutoff hyperradius $\rho_{\max} = \rho_{N_\rho}$ from the scaling considerations (see [4]). We solve the resultant block-three-diagonal algebraic system on the basis of the matrix sweep method. This allows us to dispense with writing the system matrix on the hard drive and to carry out all the operations related to its inversion immediately in RAM. Besides, the matrix sweep method reduces almost by one order the computer time required for computations on the grids of the same dimensions as in [4].

Because of the symmetry relationship $\overline{S_0(z)} = S_0(\overline{z})$ we performed all the

Table 1. Dependence of the dimer binding energy ϵ_d and the differences $\epsilon_d - E_t^{(1)}$, $\epsilon_d - E_t^{(2)}$, $\epsilon_d - E_t^{(2)*}$ and $\epsilon_d - E_t^{(2)**}$ (all in mK) between this energy and the trimer exited-state energies $E_t^{(1)}$, $E_t^{(2)}$ and the virtual-state energies $E_t^{(2)*}$, $E_t^{(2)**}$ on the factor λ.

λ	ϵ_d	$\epsilon_d - E_t^{(1)}$	$\epsilon_d - E_t^{(2)*}$	$\epsilon_d - E_t^{(2)**}$	$\epsilon_d - E_t^{(2)}$
0.995	−1.160	0.710	−	−	−
0.990	−0.732	0.622	−	−	−
0.9875	−0.555	0.573	0.473	0.222	−
0.985	−0.402	0.518	0.4925	0.097	−
0.980	−0.170	0.39616	0.39562	0.009435	−
0.975	−0.036	0.2593674545	0.2593674502	−	0.00156

calculations for $S_0(z)$ only at $\mathrm{Im}\, z \geq 0$. First, we calculated the root lines of the functions $\mathrm{Re}\, S_0(z)$ and $\mathrm{Im}\, S_0(z)$. For the case of the grid parameters $N_\theta = N_\rho = 600$ and $\rho_{max} = 600\,\text{Å}$ these lines are depicted in Fig. 1. Both resonances (roots of $S_0(z)$) and bound-state energies (poles of $S_0(z)$) of the ^{4}He trimer are associated with the intersection points of the curves $\mathrm{Re}\, S_0(z) = 0$ and $\mathrm{Im}\, S_0(z) = 0$. In Fig. 1, along with the root lines we also plot the boundaries of the domains $\Pi^{(S)}$, $\Pi^{(A)}$ and $\Pi^{(\Psi)}$. One can observe that the "good" domain $\Pi^{(S)}$ includes none of the points of intersection of the root lines $\mathrm{Re}\, S_0(z) = 0$ and $\mathrm{Im}\, S_0(z) = 0$. The caption for Fig. 1 points out positions of four "resonances", the roots of $S_0(z)$, found immediately beyond the boundary of the domain $\Pi^{(S)}$. It is remarkable that the "true" (i.e., getting inside $\Pi^{(S)}$) virtual levels and then the energies of the excited (Efimov) states appear just due to these (quasi)resonances when the potential $V(x)$ is weakened.

Following [2-4] instead of the initial potential $V(x) = V_{\mathrm{HFD-B}}(x)$, we, further, consider the potentials $V(x) = \lambda \cdot V_{\mathrm{HFD-B}}(x)$. To establish the mechanism of formation of new excited states in the ^{4}He trimer, we have first calculate the scattering matrix $S_0(z)$ for $\lambda < 1$. In Table 1 for some values of λ from the interval between 0.995 and 0.975, we present the positions of roots and poles of $S_0(z)$, we have obtained at real $z < \epsilon_d(\lambda)$. We have found that for a value of λ slightly smaller than 0.9885, the (quasi)resonance closest to the real axis (see Fig. 1) gets on it and transforms into a virtual level (the root of $S_0(z)$) of the second order corresponding to the energy value where the graph of $S_0(z)$, $z \in \mathbb{R}$, $z < \epsilon_d$, is tangent to the axis z. This virtual level is preceded by the (quasi)resonances $z = (-1.04 + i\,0.11)\,\text{mK}$ for $\lambda = 0.989$ and $z = (-0.99 + i\,0.04)\,\text{mK}$ for $\lambda = 0.9885$. With a subsequent decrease of λ the virtual level of the second order splits into a pair of the first order virtual levels $E_t^{(2)*}$ and $E_t^{(2)**}$, $E_t^{(2)*} < E_t^{(2)**}$ which move in opposite directions. A characteristic behavior of the scattering matrix $S_0(z)$ when the resonances transform into virtual levels is shown in Fig. 2. The virtual level $E_t^{(2)**}$ moves towards the threshold ϵ_d and "collides" with it at $\lambda < 0.98$. For $\lambda = 0.975$ the function

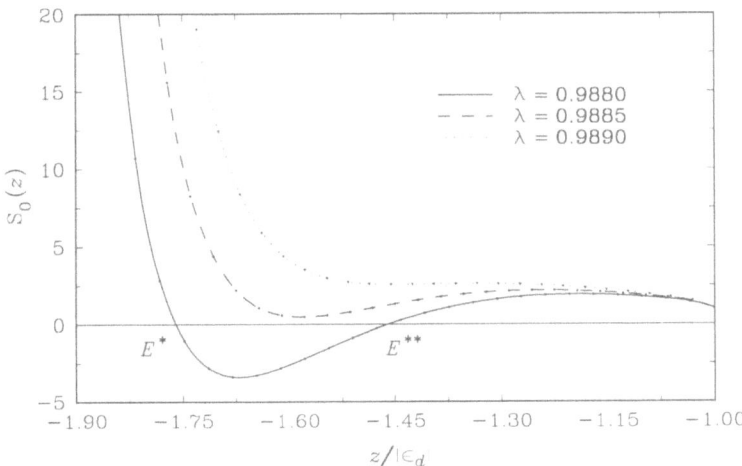

Figure 2. Graphs of the function $S_0(z)$ at real $z \leq \epsilon_d$ for three values of $\lambda < 1$. The notations used: $E^* = E_t^{(2)*}/|\epsilon_d|$, $E^{**} = E_t^{(2)**}/|\epsilon_d|$.

$S_0(z)$ has no longer the root corresponding to $E_t^{(2)**}$. Instead of the root, it acquires a new pole corresponding to the second excited state of the trimer with the energy $E_t^{(2)}$. We expect that the subsequent Efimov levels originate from the virtual levels just according to the same scheme as the level $E_t^{(2)}$ does.

The other purpose of the present investigation is to determine the mechanism of disappearance of the excited state of the helium trimer when the two-body interactions become stronger owing to the increasing $\lambda > 1$. It turned out that this disappearance proceeds just according to the scheme of the formation of new excited states; only the order of occurring events is inverse. The results of our computations of the energy $E_t^{(1)}$ when λ changes from 1.05 to 1.17 are given in Table 2. In the interval between $\lambda = 1.17$ and $\lambda = 1.18$ there occurs a "jump" of the level $E_t^{(1)}$ on the unphysical sheet and it transforms from the pole of the function $S_0(z)$ into its root, $E_t^{(1)*}$, corresponding to the trimer virtual level. The results of the calculation of this virtual level where λ changes from 1.18 to 1.5 are also presented in Table 2.

More details about our techniques and material presented here will be given in an extended article [12].

Acknowledgement. The authors are grateful to Prof. V.B. Belyaev and Prof. H. Toki for help and assistance in calculations at the supercomputer of the Re-

Table 2. Dependence of the dimer energy ϵ_d, the difference $\epsilon_d - E_t^{(1)}$ between ϵ_d and the trimer exited-state energy $E_t^{(1)}$ and the difference $\epsilon_d - E_t^{(1)*}$ between ϵ_d and the trimer virtual-state energy $E_t^{(1)*}$ (all in mK) on the factor λ.

λ	ϵ_d	$\epsilon_d - E_t^{(1)}$	λ	ϵ_d	$\epsilon_d - E_t^{(1)*}$
1.05	-12.244	0.873	1.18	-82.927	0.001
1.10	-32.222	0.450	1.20	-99.068	0.057
1.15	-61.280	0.078	1.25	-145.240	0.588
1.16	-68.150	0.028	1.35	-261.393	3.602
1.17	-75.367	0.006	1.50	-490.479	12.276

search Center for Nuclear Physics of the Osaka University, Japan. One of the authors (A. K. M.) is much indebted to Prof. W. Sandhas for his hospitality at the Universität Bonn, Germany. The support of this work by the Deutsche Forschungsgemeinschaft and Russian Foundation for Basic Research is gratefully acknowledged.

References

1. A.K. Motovilov: Math. Nachr. **187**, 147 (1997)

2. T. Cornelius, W. Glöckle: J. Chem. Phys. **85**, 3906 (1986)

3. B.D. Esry, C.D. Lin, C.H. Greene: Phys. Rev. **A54**, 394 (1996)

4. E.A. Kolganova, A.K. Motovilov, S.A. Sofianos: J. Phys. **B31**, 1279 (1998)

5. E. Nielsen, D.V. Fedorov, A.S. Jensen: LANL E-print physics/9806020

6. O.I. Kartavtsev, F.M. Penkov: In: *16th European Conference on Few-Body Problems in Physics (Autrans, 1–6 June 1998), Abstract Booklet*, p. 137. Grenoble 1998

7. L. Tomio, T. Frederico, A. Delfino, A.E.A. Amorim: *Ibid.*, p. 150.

8. V. Efimov: Nucl. Phys. **A210**, 157 (1973)

9. R.A. Aziz, F.R.W. McCourt, C.C.K. Wong: Mol. Phys. **61**, 1487 (1987)

10. L.D. Faddeev, S.P. Merkuriev: In: *Quantum Scattering Theory for Several Particle Systems*. Doderecht: Kluwer Academic Publishers 1993

11. S.P. Merkuriev, C. Gignoux, A. Laverne: Ann. Phys. (N.Y.) **99**, 30 (1976)

12. E.A. Kolganova, A.K. Motovilov: LANL E-print physics/9808027

Few-Body Systems Suppl. 10, 85–92 (1999)

Few-
Body
Systems
© by Springer-Verlag 1999

Spectral Properties of Faddeev Equations in Differential Form

S.L. Yakovlev*

Department of Mathematical and Computational Physics, St. Petersburg State University, 198904 St. Petersburg, Petrodvoretz, Ulyanovskaya Str. 1, Russia

1   Introduction

Faddeev equations in differential form were introduced by Noyes and Fiedeldey in 1968 [1]

$$(H_0 - E)\varphi_\alpha + V_\alpha \sum_{\beta=1}^{3} \varphi_\beta = 0, \tag{1}$$

and since that time are used extensively for investigating theoretical aspects of the three-body problem as well as for getting numerical solutions of three-body bound-state and scattering state problems. The simple formula

$$\sum_{\beta=1}^{3} \varphi_\beta = \Psi$$

allows one to obtain the solution to the three-body Schrödinger equation

$$(H_0 + \sum_{\beta=1}^{3} V_\beta - E)\Psi = 0$$

in the case when

$$\sum_{\beta=1}^{3} \varphi_\beta \neq 0. \tag{2}$$

Such solutions of (1) can be called **physical**. The proper asymptotic boundary conditions should be added to Eqs. (1) in order to guarantee (2). These conditions were studied by many authors and are well known [2], so that I will not discuss them here.

On the other hand, Eqs. (1) themselves allow solutions of the different type (with respect to physical ones) with the property

* *E-mail address:* yakovlev@mph.phys.spbu.ru

$$\sum_{\beta=1}^{3} \varphi_\beta = 0.$$

These solutions can be constructed explicitly and have the form

$$\varphi_\alpha = \sigma_\alpha \phi^0,$$

where ϕ^0 is an eigenfunction of operator H_0:

$$H_0\phi^0 = E^0\phi^0$$

and σ_α, $\alpha = 1, 2, 3$ are numbers such that $\sum_{\alpha=1}^{3} \sigma_\alpha = 0$. The solutions of this type can be called **spurious** or **ghost**, because they do not correspond to any three-body system and do not contain any information about interactions between particles. First observation of the existence of spurious solutions was made in ref. [3]. Some spurious solutions corresponding to particular values of the total angular momentum were found in refs. [4, 5]. All the spurious solutions on subspaces with fixed total angular momentum were constructed in ref. [6].

So that, there exist at least two types of solutions to Eqs. (1) corresponding to real energy:

physical ones with the property $\sum_{\beta=1}^{3} \varphi_\beta \neq 0$,

spurious ones with the property $\sum_{\beta=1}^{3} \varphi_\beta = 0$.

The QUESTION is whether these solutions form the complete set or there could exist solutions of a different type which do not belong to physical and spurious classes. The ANSWER is not so evident because the operator corresponding to Eqs. (1) is not self-adjoint and moreover even symmetrical:

$$\mathbf{H} = \begin{pmatrix} H_0 & 0 & 0 \\ 0 & H_0 & 0 \\ 0 & 0 & H_0 \end{pmatrix} + \begin{pmatrix} V_1 & 0 & 0 \\ 0 & V_2 & 0 \\ 0 & 0 & V_3 \end{pmatrix} \begin{pmatrix} 1 & 1 & 1 \\ 1 & 1 & 1 \\ 1 & 1 & 1 \end{pmatrix} = \mathbf{H}_0 + \mathbf{VX}, \quad (3)$$

and, in principle, this operator can not have real eigenvalues even if the ingredients H_0, V_α and the three-body Hamiltonian $H = H_0 + \sum_{\beta=1}^{3} V_\beta$ are self-adjoint operators.

In this report I will answer the QUESTION and will give a classification of eigenfunctions of the operator $\mathbf{H}$ and its adjoint. This report is based on refs. [7, 8].

2 Faddeev operator and its adjoint

Let us consider the Hilbert space $\mathcal{H}$ of three component vectors $F = \{f_1, f_2, f_3\}$. The operator $\mathbf{H}$ acts in $\mathcal{H}$ according to the formula

$$(\mathbf{H}F)_\alpha = H_0 f_\alpha + V_\alpha \sum_\beta f_\beta. \quad (4)$$

The adjoint $\mathbf{H}^*$ is defined as

$$\mathbf{H}^* = \mathbf{H}_0 + \mathbf{XV} = \begin{pmatrix} H_0 & 0 & 0 \\ 0 & H_0 & 0 \\ 0 & 0 & H_0 \end{pmatrix} + \begin{pmatrix} 1 & 1 & 1 \\ 1 & 1 & 1 \\ 1 & 1 & 1 \end{pmatrix} \begin{pmatrix} V_1 & 0 & 0 \\ 0 & V_2 & 0 \\ 0 & 0 & V_3 \end{pmatrix}$$

and acts as follows

$$(\mathbf{H}^* G)_\alpha = H_0 g_\alpha + \sum_\beta V_\beta g_\beta. \tag{5}$$

The equations for the eigenvectors of operators $\mathbf{H}$ and $\mathbf{H}^*$

$$\mathbf{H}\Phi = E\Phi, \qquad \mathbf{H}^*\Psi = E\Psi$$

decompose as:

$$H_0 \varphi_\alpha + V_\alpha \sum_{\beta=1}^3 \varphi_\beta = E\varphi_\alpha,$$

$$H_0 \psi_\alpha + \sum_{\beta=1}^3 V_\beta \psi_\beta = E\psi_\alpha.$$

The first ones coincide with the Faddeev equations (1) and the second ones have a direct connection to the so called triad of Lippmann-Schwinger equations [9].

It follows directly from the definitions (4) and (5) that operators $\mathbf{H}$ and $\mathbf{H}^*$ have the following invariant subspaces:
for $\mathbf{H}$

$$\mathcal{H}_s = \{F \in \mathcal{H}_s : \sum_\alpha f_\alpha = 0\},$$

for $\mathbf{H}^*$

$$\mathcal{H}_p^* = \{G \in \mathcal{H}_p^* : g_1 = g_2 = g_3 = g\}.$$

It is worth to notice that operators $\mathbf{H}$ and $\mathbf{H}^*$ on the subspaces $\mathcal{H}_s$ and $\mathcal{H}_p^*$ act as the free Hamiltonian H_0 and the three-body Hamiltonian H, respectively:

$$(\mathbf{H}F)_\alpha = H_0 f_\alpha \ , \text{if} \ F \in \mathcal{H}_s,$$

$$(\mathbf{H}^* G)_\alpha = Hg = H_0 g + \sum_\beta V_\beta g \ , \text{if} \ G \in \mathcal{H}_p^*.$$

As a consequence the spectrum of $\mathbf{H}$ on $\mathcal{H}_s$ coincides with the spectrum of H_0 and the spectrum of $\mathbf{H}^*$ on $\mathcal{H}^*$ does with the spectrum of three-body Hamiltonian H.

In order to describe the eigenfunctions of operators $\mathbf{H}$ and $\mathbf{H}^*$ let us introduce the resolvents

$$\mathbf{R}(z) = (\mathbf{H} - z)^{-1},$$

$$\mathbf{R}^*(z) = (\mathbf{H}^* - z)^{-1}.$$

The components of these resolvents can be expressed through the resolvent of three-body Hamiltonian and the free Hamiltonian as follows

$$R_{\alpha\beta}(z) = R_0(z)\delta_{\alpha\beta} - R_0(z)V_\alpha R(z), \tag{6}$$

$$R^*_{\alpha\beta}(z) = R_0(z)\delta_{\alpha\beta} - R(z)V_\beta R_0(z). \tag{7}$$

Here

$$R(z) = (H - z)^{-1} = (H_0 + \sum_\beta V_\beta - z)^{-1}, \quad R_0(z) = (H_0 - z)^{-1}.$$

It is worth to note that the components of the resolvents obey the following Faddeev equations

$$R_{\alpha\beta}(z) = R_\alpha(z)\delta_{\alpha\beta} - R_\alpha(z)V_\alpha \sum_{\gamma\neq\alpha} R_{\gamma\beta}(z), \tag{8}$$

$$R^*_{\alpha\beta}(z) = R_\alpha(z)\delta_{\alpha\beta} - R_\alpha(z) \sum_{\gamma\neq\alpha} V_\gamma R^*_{\gamma\beta}(z). \tag{9}$$

Here $R_\alpha(z) = (H_0 + V_\alpha - z)^{-1}$ is the two-body resolvent for the pair α in the three-body space.

In order to proceed it is convenient to introduce the spectral representation for the resolvent of the three-body Hamiltonian

$$R(z) = \sum_{E_i} \frac{|\psi^i\rangle\langle\psi^i|}{E_i - z} + \sum_\gamma \int dp_\gamma \frac{|\psi^\gamma(p_\gamma)\rangle\langle\psi^\gamma(p_\gamma)|}{p_\gamma^2 - z} + \int dP \frac{|\psi^0(P)\rangle\langle\psi^0(P)|}{P^2 - z}.$$

It is implied here that the system of eigenfunctions of the operator H is complete $i.e.$,

$$I = \sum_i |\psi^i\rangle\langle\psi^i| + \sum_\gamma \int dp_\gamma |\psi^\gamma(p_\gamma)\rangle\langle\psi^\gamma(p_\gamma)| + \int dP |\psi^0(P)\rangle\langle\psi^0(P)|.$$

Introducing this representation into (6) and (7) one arrives to the spectral representations for components $R_{\alpha\beta}(z)$:

$$R_{\alpha\beta}(z) = \sum_{E_i} \frac{|\psi_\alpha^i\rangle\langle\psi^i|}{E_i - z} + \sum_\gamma \int dp_\gamma \frac{|\psi_\alpha^\gamma(p_\gamma)\rangle\langle\psi^\gamma(p_\gamma)|}{p_\gamma^2 - z} +$$

$$\int dP \frac{|\psi_\alpha^{10}(P)\rangle\langle\psi^0(P)|}{P^2 - z} + \sum_{k=1}^2 \int dP \frac{|u_\alpha^k(P)\rangle\langle w_\beta^k(P)|}{P^2 - z}. \tag{10}$$

Here ψ_α^i, $\psi_\alpha^\gamma(p_\gamma)$ and $\psi_\alpha^{10}(P)$ are the Faddeev components of the eigenfunctions of the three-body Hamiltonian:

$$\psi_\alpha^i = -R_0(E_i)V_\alpha\psi^i,$$

$$\psi_\alpha^\gamma(p_\gamma) = -R_0(\varepsilon_\gamma + p_\gamma^2 + i0)V_\alpha\psi^\gamma(p_\gamma),$$

$$\psi_\alpha^{10}(P) = \delta_{\alpha 1}\phi^0(P) - R_0(P^2 + i0)V_\alpha\psi^0(P),$$

where $\phi^0(P)$ is an eigenfunction of the free Hamiltonian:

$$H_0\phi^0(P) = P^2\phi^0(P).$$

A new feature in (10) is the appearance of the last term related to the spurious solutions of Faddeev equations and its adjoint. The explicit formulas for the spurious eigenfunctions $u_\alpha^k(P)$ of **H** are of the form

$$u_\alpha^k(P) = \sigma_\alpha^k \phi^0(P),$$

where σ_α^k, $k = 1, 2$, are the components of two noncollinear vectors from $\mathbf{R}^3$ lying on the plane $\sum_\alpha \sigma_\alpha = 0$. The spurious eigenfunctions $w_\beta^k(P)$ of $\mathbf{H}^*$ can be expressed by the formula

$$w_\beta^k(P) = \theta_\beta^k \phi^0(P) - \sum_\alpha [\mathcal{P}_p^*]_{\beta\alpha} \theta_\alpha^k \phi^0(P),$$

where

$$[\mathcal{P}_p^*]_{\beta\alpha} = \sum_i |\psi^i\rangle\langle\psi_\alpha^i| + \sum_\gamma \int dp_\gamma' |\psi^\gamma(p_\gamma')\rangle\langle\psi_\alpha^\gamma(p_\gamma')| + \int dP' |\psi^0(P')\rangle\langle\psi_\alpha^{01}(P')|.$$

Here the vectors $\theta^k \in \mathbf{R}^3$ are defined the by following biorthogonality conditions

$$\sum_\alpha \theta_\alpha^i \sigma_\alpha^j = \delta_{ij}, \quad i, j = 0, 1, 2,$$

with $\sigma_\alpha^0 = \delta_{\alpha 1}$ and $\theta_\alpha^0 = 1$.

For the components of the resolvent $R_{\alpha\beta}^*(z)$ one can obtain a formula similar to (10)

$$R_{\alpha\beta}^*(z) = \sum_{E_i} \frac{|\psi^i\rangle\langle\psi_\beta^i|}{E_i - z} + \sum_\gamma \int dp_\gamma \frac{|\psi^\gamma(p_\gamma)\rangle\langle\psi_\beta^\gamma(p_\gamma)|}{p_\gamma^2 - z} + \int dP \frac{|\psi^0(P)\rangle\langle\psi_\beta^{10}(P)|}{P^2 - z} +$$

$$+ \sum_{k=1}^2 \int dP \frac{|w_\alpha^k(P)\rangle\langle u_\beta^k(P)|}{P^2 - z}. \tag{11}$$

It follows from (10) and (11) that operators **H** and $\mathbf{H}^*$ have the following system of eigenfunctions:

$$\{ \Phi^i, \Phi^\gamma(p_\gamma), \Phi^{10}(P) \text{ and } U^k(P) \}$$

$$\mathbf{H}\Phi^i = E_i\Phi^i,$$

$$\mathbf{H}\Phi^\gamma(p_\gamma) = (\varepsilon_\gamma + p_\gamma^2)\Phi^\gamma(p_\gamma),$$

$$\mathbf{H}\Phi^{10}(P) = P^2\Phi^{10}(P),$$

$$\mathbf{H}U^k(P) = P^2 U^k, \quad k = 1, 2;$$

$$\{ \Psi^i, \Psi^\gamma(p_\gamma), \Psi^{10}(P) \text{ and } W^k(P) \}$$

$$\mathbf{H}^*\Psi^i = E_i\Psi^i,$$

$$\mathbf{H}^*\Psi^\gamma(p_\gamma) = (\varepsilon_\gamma + p_\gamma^2)\Psi^\gamma(p_\gamma),$$

$$\mathbf{H}^* \Psi^{10}(P) = P^2 \Psi^{10}(P),$$

$$\mathbf{H}^* W^k(P) = P^2 W^k, \quad k = 1, 2,$$

with components of physical eigenfunctions:
for $\mathbf{H}$

$$\phi_\alpha^i = -R_0(E_i) V_\alpha \psi^i,$$

$$\phi_\alpha^\gamma(p_\gamma) = -R_0(\varepsilon_\gamma + p_\gamma^2 + i0) V_\alpha \psi^\gamma(p_\gamma),$$

$$\phi_\alpha^{10}(P) = \delta_{\alpha 1} \phi^0(P) - R_0(P^2 + i0) V_\alpha \psi^0(P),$$

for $\mathbf{H}^*$

$$\psi_\alpha^i = \psi^i,$$

$$\psi_\alpha^\gamma(p_\gamma) = \psi^\gamma(p_\gamma),$$

$$\psi_\alpha^{10}(P) = \psi^0(P).$$

Physical eigenfunctions span the physical subspace of $\mathcal{H}$. This subspace can be defined as

$$\mathcal{H}_p = \mathcal{P}_p \mathcal{H},$$

where the projection $\mathcal{P}_p$ is defined by the formula

$$\mathcal{P}_p = \sum_i |\Phi^i\rangle\langle\Psi^i| + \sum_\gamma \int dp_\gamma |\Phi^\gamma(p_\gamma)\rangle\langle\Psi^\gamma(p_\gamma)| + \int dP |\Phi^{10}(P)\rangle\langle\Psi^{10}(P)|.$$

Spurious solutions span the spurious subspace of $\mathcal{H}$:

$$\mathcal{H}_s = \mathcal{P}_s \mathcal{H}.$$

where

$$\mathcal{P}_s = \sum_{k=1}^{2} \int dP |U^k(P)\rangle\langle W^k(P)|.$$

It follows from the construction and the completeness of eigenfunctions of the three-body Hamiltonian, that the physical and spurious subspaces are complete in $\mathcal{H}$:

$$\mathcal{H} = \mathcal{H}_p + \mathcal{H}_s.$$

The same is valid for the physical and spurious subspaces of operator $\mathbf{H}^*$:

$$\mathcal{H} = \mathcal{H}_p^* + \mathcal{H}_s^*,$$

where the subspaces $\mathcal{H}_p^*$ and $\mathcal{H}_s^*$ are defined as

$$\mathcal{H}_p^* = \mathcal{P}_p^* \mathcal{H}, \quad \mathcal{H}_s^* = \mathcal{P}_s^* \mathcal{H}.$$

Here the operators $\mathcal{P}_p^*$ and $\mathcal{P}_s^*$ are the Hilbert space adjoints for $\mathcal{P}_p$ and $\mathcal{P}_s$.

The results described above can be summarized as follows

Theorem: The *Faddeev operator* **H**

$$\mathbf{H} = \begin{pmatrix} H_0 & 0 & 0 \\ 0 & H_0 & 0 \\ 0 & 0 & H_0 \end{pmatrix} + \begin{pmatrix} V_1 & 0 & 0 \\ 0 & V_2 & 0 \\ 0 & 0 & V_3 \end{pmatrix} \begin{pmatrix} 1 & 1 & 1 \\ 1 & 1 & 1 \\ 1 & 1 & 1 \end{pmatrix}$$

and its adjoint **H***

$$\mathbf{H}^* = \begin{pmatrix} H_0 & 0 & 0 \\ 0 & H_0 & 0 \\ 0 & 0 & H_0 \end{pmatrix} + \begin{pmatrix} 1 & 1 & 1 \\ 1 & 1 & 1 \\ 1 & 1 & 1 \end{pmatrix} \begin{pmatrix} V_1 & 0 & 0 \\ 0 & V_2 & 0 \\ 0 & 0 & V_3 \end{pmatrix}$$

have coinciding spectra of real eigenvalues

$$\sigma(\mathbf{H}) = \sigma(\mathbf{H}^*) = \sigma(H) \cup \sigma(H_0),$$

where the physical part of the spectrum $\sigma(H)$ is the spectrum of the three-body Hamiltonian $H = H_0 + \sum_\alpha V_\alpha$ and the spurious part $\sigma(H_0)$ is the spectrum of the free Hamiltonian H_0. The sets of physical and spurious eigenfunctions are complete and biorthogonal in the sense:

$$\mathcal{P}_p + \mathcal{P}_s = \mathcal{P}_p^* + \mathcal{P}_s^* = I,$$

$$\mathcal{P}_{p(s)}^2 = \mathcal{P}_{p(s)}, \quad \mathcal{P}^*{}_{p(s)}^2 = \mathcal{P}^*{}_{p(s)}, \quad \mathcal{P}_p \mathcal{P}_s^* = 0, \quad \mathcal{P}_s \mathcal{P}_p^* = 0.$$

3 Extension on CCA equations

It is shown that the matrix operator generated by Faddeev equations in differential form has, in addition to the physical ones, a spurious spectrum. The existence of this spectrum strongly relates to the invariant spurious subspace formed by components of which sum is equal to zero. The theorem formulated in the preceding section can be extended to any matrix operator corresponding to few-body equations for components of wave functions obtained in the framework of the so called coupled channel array (CCA) method [10] as follows. CCA equations can be written in the matrix form as

$$\mathbf{H}\Phi = E\Phi, \tag{12}$$

where **H** is a $n \times n$ matrix operator acting in the Hilbert space $\mathcal{H}$ of vector-functions Φ with components $\phi_1, \phi_2, ..., \phi_n$ each belonging to few-body system Hilbert space h. The equivalence of Eq. (12) with the Schrödinger equation $H\psi = (H_0 + \sum_\beta V_\beta)\psi = E\psi$ by requiring $\sum_\alpha \phi_\alpha = \psi$ can be reformulated as the following intertwining property for operators **H** and H

$$\mathcal{S}\mathbf{H} = H\mathcal{S}. \tag{13}$$

Here $\mathcal{S}$ is the summation operator

$$\mathcal{S}\Phi = \sum_\alpha \phi_\alpha$$

acting from $\mathcal{H}$ to h. Due to (13) the subspace $\mathcal{H}_s$ formed by spurious vectors such that $\mathcal{S}\Phi = 0$ is invariant with respect to $\mathbf{H}$ and as a consequence the operator $\mathbf{H}$ has the spurious spectrum σ_s. Clearly, that the concrete form of σ_s and of corresponding eigenfunctions depends on the particular form of the matrix operator $\mathbf{H}$ is the subject of special investigation.

The physical part σ_p of the spectrum of $\mathbf{H}$ can be found with adjoint variant of (13)

$$\mathbf{H}^* \mathcal{S}^* = \mathcal{S}^* H, \tag{14}$$

where adjoint $\mathcal{S}^*$ acts from h to $\mathcal{H}$ according to the formula

$$[\mathcal{S}^* \phi]_\alpha = \phi.$$

It follows from Eq. (14) that the range $\mathcal{H}_p^*$ of operator $\mathcal{S}^*$ consisting of vector-functions with the same components is invariant with respect to $\mathbf{H}^*$ and the restriction of $\mathbf{H}^*$ on $\mathcal{H}_p^*$ is reduced to few-body Hamiltonian H. So that, $\sigma_p = \sigma(H)$ and, similarly to the case of the Faddeev operator, the same formula for the spectra of operators $\mathbf{H}$ and $\mathbf{H}^*$ is valid

$$\sigma(\mathbf{H}) = \sigma(\mathbf{H}^*) = \sigma(H) \cup \sigma_s,$$

where $\sigma(H)$ is the spectrum of the few-body Hamiltonian $H = H_0 + \sum_\alpha V_\alpha$.

Acknowledgement. This work was partially supported by the Russian Foundation for Basic Research grant No. 98-02-18190. The author is grateful to the Organizing Committee of the 16th European Conference on Few-Body Problem in Physics for financial support of his participation in the Conference.

References

1. H.P. Noyes, H. Fiedeldey: In: *Three-particle scattering in quantum mechanics*, p. 195. New-York-Amsterdam 1968

2. L.D. Faddeev, S.P. Merkuriev: In: *Quantum Scattering Theory for several particle systems*. Doderecht: Kluwer Academic Publishers 1993

3. J.L. Friar, B.F. Gibson, G.L. Paine: Phys. Rev. **C22**, 284 (1980)

4. V.V. Pupyshev: Theor. Math. Phys. **81**, 86 (1989)

5. V.V. Pupyshev: Phys. Lett. **A140**, 284 (1989)

6. V.A. Roudnev, S.L. Yakovlev: Phys. of Atom. Nuclei **58**, 1662 (1995)

7. S.L. Yakovlev: Theor. Math. Phys. **102**, 323 (1995)

8. S.L. Yakovlev: Theor. Math. Phys. **107**, 513 (1996)

9. W. Glöckle: Nucl. Phys. **A141**, 620 (1970)

10. F.S. Levin: Ann. Phys. (N.Y.) **130**, 139 (1980)

Few-Body Systems Suppl. 10, 93–96 (1999)

Few-
Body
Systems
© by Springer-Verlag 1999

Jost function for coupled channels

S.A. Rakityansky * and S.A. Sofianos †

Department of Physics, University of South Africa, P.O.Box 392, Pretoria 0003, South Africa

Almost any textbook on the scattering theory has a chapter devoted to the Jost function, but none of them gives a practical recipe for its calculation. Thus one usually gets a feeling that the Jost function is a pure mathematical object, elegant and useful in formal theory, but impractical in computations. This is even more so in the case of multi-channel problems, where the Jost function is, in fact, a matrix. Recently, we proposed an exact method for direct calculation of the Jost function for central potentials (which may have Coulombic tails) [1, 2] and the Jost matrix for non-central short range potentials [3]. This method works for all real or complex momenta of physical interest, including the spectral points corresponding to bound and Siegert states. In this work we extend it for potentials which couple channels with different thresholds.

There are three different types of physical problems associated with the Schrödinger equation: bound, scattering, and resonance state problems. They differ in the boundary conditions imposed on the wave function at large distances. Alternatively, a solution can be specified by the boundary conditions at the origin. For example, in the single–channel case the solution which vanishes as $r \to 0$ exactly like the Riccati-Bessel function, is called the *regular* solution. It is well known that if the potential couples N channels, the Schrödinger equation

$$\left[\partial_r^2 + 2\mu_n(E - E_n) - \frac{\ell(\ell + 1)}{r^2}\right] u_n(E, r) = \sum_{n'=1}^{N} V_{nn'}(r) u_{n'}(E, r) \qquad (1)$$

has $2N$ linearly independent column–solutions

$$\begin{pmatrix} u_1 \\ u_2 \\ \vdots \\ u_N \end{pmatrix}_1, \begin{pmatrix} u_1 \\ u_2 \\ \vdots \\ u_N \end{pmatrix}_2, \begin{pmatrix} u_1 \\ u_2 \\ \vdots \\ u_N \end{pmatrix}_3, \cdots \begin{pmatrix} u_1 \\ u_2 \\ \vdots \\ u_N \end{pmatrix}_{2N}$$

*E-mail address: rakitsa@kiaat.unisa.ac.za
†E-mail address: sofiasa@kiaat.unisa.ac.za

and only half of them are regular at the point $r = 0$. We can therefore combine all the regular solutions (to distinguish them we use the notation ϕ) in a square matrix,

$$\begin{pmatrix} \phi_1 \\ \phi_2 \\ \vdots \\ \phi_N \end{pmatrix}_1 \oplus \begin{pmatrix} \phi_1 \\ \phi_2 \\ \vdots \\ \phi_N \end{pmatrix}_2 \oplus \cdots \Longrightarrow \begin{pmatrix} \phi_{11} & \phi_{12} & \cdots & \phi_{1N} \\ \phi_{21} & \phi_{22} & \cdots & \phi_{2N} \\ \vdots & \vdots & \ddots & \vdots \\ \phi_{N1} & \phi_{N2} & \cdots & \phi_{NN} \end{pmatrix} \equiv \|\phi(E, r)\| \, ,$$

which obeys the (matrix) Schrödinger equation with the boundary condition

$$\lim_{r \to 0} \phi_{nn'}(E, r)/j_\ell(k_{n'} r) = \delta_{nn'} \tag{2}$$

where $j_\ell(z)$ is the Riccati–Bessel function and $k_n = \sqrt{2\mu_n(E - E_n)}$ is the channel momentum. Any physical solution must be regular at the origin and therefore it is a linear combination of the regular columns. In this respect the matrix $\|\phi\|$ can be called the *regular basis*.

In order to obtain this basis in the form which is most suitable for describing its long–range asymptotics, we replace the unknown matrix–function $\|\phi(E, r)\|$ by two others, $\|\mathcal{F}^{(-)}(E, r)\|$ and $\|\mathcal{F}^{(+)}(E, r)\|$, via the following ansatz

$$\phi_{nn'}(E, r) = \frac{1}{2} \left[h_\ell^{(-)}(k_n r) \mathcal{F}_{nn'}^{(-)}(E, r) + h_\ell^{(+)}(k_n r) \mathcal{F}_{nn'}^{(+)}(E, r) \right] \, , \tag{3}$$

with the additional constraint conditions

$$h_\ell^{(-)}(k_n r) \partial_r \mathcal{F}_{nn'}^{(-)}(E, r) + h_\ell^{(+)}(k_n r) \partial_r \mathcal{F}_{nn'}^{(+)}(E, r) = 0 \tag{4}$$

which make the derivative of the wave functions continuous even if the potential has a sharp cut–off. Substituting this ansatz into the Schrödinger equation we derive the following system of first order differential equations for these new unknown matrices,

$$\partial_r \mathcal{F}_{nn'}^{(\mp)} = \mp \frac{1}{2ik_n} h_\ell^{(\pm)}(k_n r) \sum_{n''} V_{nn''}(r) \left[h_\ell^{(+)}(k_{n''} r) \mathcal{F}_{n''n'}^{(+)} + h_\ell^{(-)}(k_{n''} r) \mathcal{F}_{n''n'}^{(-)} \right] \, . \tag{5}$$

The boundary conditions corresponding to (2) are

$$\lim_{r \to 0} \mathcal{F}_{nn'}^{(\pm)}(E, r) = \delta_{nn'} \, . \tag{6}$$

Explicit inclusion of the Riccati–Hankel functions into the basis (3) guarantees the correct asymptotic behaviour of the physical solution.

In constructing a physical wave function $\varphi_n(E, r)$ one requires the appropriate coefficients c_n in the linear combination of the regular columns,

$$\varphi_n(E, r) = \sum_{n'} \phi_{nn'}(E, r) c_{n'} \, .$$

For bound and resonant states, for example, in the physical wave function only the term proportional to $h^{(+)}$ survives at large distances. This gives the homogeneous system of equations

$$\sum_{n'} \mathcal{F}_{nn'}^{(-)}(E, \infty) c_{n'} = 0$$

which has a non-trivial solution if and only if its determinant is zero, i. e.,

$$\det \|\mathcal{F}^{(-)}(E, \infty)\| = 0 . \tag{7}$$

It is easily seen that this is the determinant of the Jost matrix $\|F(E)\|$ which is defined by the asymptotic behaviour of the regular basis [4]

$$\phi_{nn'}(E, r) \xrightarrow[r \to \infty]{} \frac{1}{2} \left[h_\ell^{(-)}(k_n r) F_{nn'}(E) + h_\ell^{(+)}(k_n r) \tilde{F}_{nn'}(E) \right] .$$

We can calculate this matrix by solving the differential Eqs. (5) from $r = 0$ up to a large radius where the potential is insignificant. Since the potential vanishes, the right hand sides of Eqs. (5) disappear at large distances. Hence the derivatives $\partial_r \|\mathcal{F}^{(\mp)}(E, r)\|$ become zero which means that asymptotically the matrix–functions $\|\mathcal{F}^{(\mp)}(E, r)\|$ become constants coinciding with $\|F(E)\|$ and $\|\tilde{F}(E)\|$ respectively.

Similarly to the single–channel case (see refs. [1, 2, 3]) the above scheme can be easily implemented only for bound and scattering states. In the resonance domain of the complex energy plane the Riccati–Hankel functions $h_\ell^{(+)}(k_n r)$, and therefore the derivatives of $\|\mathcal{F}^{(-)}(E, r)\|$ at large distances, become infinitely large which means that the limit $\|\mathcal{F}^{(-)}(E, \infty)\|$ does not exist. Thus in order to calculate the Jost matrix in this domain we need to perform a complex rotation of the independent variable, $r = x \exp(i\theta)$, in Eqs. (5), where $x \geq 0$ and $0 \leq \theta < \pi/2$. This results in a movement of the unitary cuts at each threshold downwards as is shown in Fig. 1. After such a rotation $h_\ell^{(+)}(k_n r)$ vanishes

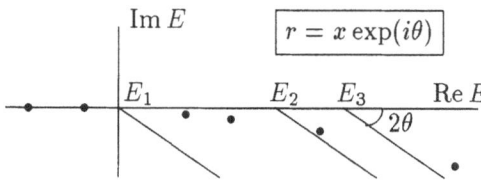

Figure 1. Rotation of r by an angle θ results in a rotation of the cuts by 2θ.

in all areas above the turned cuts, which guarantees the existence of the limit $\|\mathcal{F}^{(-)}(E, \infty)\|$ there.

By locating the zeros of the Jost–matrix determinant we can find the resonance energies and the total widths. The determination of the partial widths, though it is a difficult task for most methods used in the past, in our approach is not a problem at all. Indeed, knowing the Jost matrix, we can easily obtain the S-matrix at a given complex energy as

$$\|S(E)\| = \|\mathcal{F}^{(+)}(E,\infty)\| \cdot \|\mathcal{F}^{(-)}(E,\infty)\|^{-1}$$

which, generally, consists of two terms, viz.

$$S_{nn'}(E) = S_{nn'}^{\text{bg}} - i\frac{\sqrt{\Gamma_n \Gamma_{n'}}}{E - E_r + i\Gamma/2}.$$

One of them is the smooth background term $S_{nn'}^{\text{bg}}$, and the other a singular resonance term. Thus a partial width is the limit

$$\Gamma_n = \left| \lim_{E \to (E_r - i\Gamma/2)} (E - E_r + i\Gamma/2) S_{nn}(E) \right|.$$

To demonstrate the ability of the method we consider a numerical example proposed by Noro and Taylor in ref. [5] where they used the S-wave potential

$$\|V(r)\| = \begin{pmatrix} -1.0 & -7.5 \\ -7.5 & 7.5 \end{pmatrix} r^2 e^{-r}$$

which describes a two-channel system with the thresholds $E_1 = 0$ and $E_2 = 0.1$ (atomic units). By expanding the resonance wave function in a series of square-integrable functions, Noro and Taylor located the first resonance generated by this potential. We reproduce their result and found also three bound states and several broad resonances. For comparison we put the parameters of the first resonance in Table 1. The digits shown do not change when the rotation angle is changed. This demonstrates the accuracy of our method.

Table 1. Parameters of the first resonance given in atomic units

	our results	Noro & Taylor
E_r	4.7681968188	4.7682
Γ	0.0014201920	0.001420
Γ_1	0.000051103	0.000059
Γ_2	0.001368733	0.001361

References

1. S.A. Rakityansky, S.A. Sofianos, K. Amos: Nuovo Cim., **B111**, 363 (1996)

2. S.A. Sofianos, S.A. Rakityansky: Journal of Phys. **A30**, 3725 (1997)

3. S.A. Rakityansky, S.A. Sofianos: Journal of Phys. **A31**, 5149 (1998)

4. J.R. Taylor: In: *Scattering Theory*. New York: John Wiley & Sons

5. T. Noro, H.S. Taylor: J. Phys. **B13**, L377 (1980)

Few-Body Systems Suppl. 10, 97–100 (1999)

Few-
Body
Systems
© by Springer-Verlag 1999

About Compactness of Faddeev Integral Equations for Three Charged Particles

A.M. Mukhamedzhanov[1][*][†], E.O. Alt[2][**], and G.V. Avakov[3][††]

[1] Cyclotron Institute, Texas A&M University, College Station, TX, 77843, USA

[2] Institut für Physik, Universität Mainz, D-55099 Mainz, Germany

[3] Institute for Nuclear Physics, Moscow State University, Moscow, 119899, Russia

Abstract. Momentum space three-body integral equations of the Faddeev type can not be used for Coulomb-like potentials, for energies above the breakup threshold. The reason is the occurrence of singularities in their kernels which destroy the compactness properties known to exist for purely short-range interactions. Using the rigorously equivalent formulation in terms of an effective-two-body theory, we prove that the nondiagonal kernels occurring therein possess on and off the energy shell only integrable singularities, provided all three particles have charges of the same sign (i.e., only repulsive Coulomb interactions). In contrast, if some of the charges have opposite signs the nondiagonal kernels develop nonintegrable singularities which destroy the compactness properties.

1 Introduction

Since pioneering work of Faddeev [1], the three-body quantum scattering theory has became a powerful tool for the investigation of processes in many different areas of physics. However, one major obstacle which has impeded its widerspread application, in particular to atomic reactions, has remained, *viz.*, the question of how to incorporate the long-ranged Coulomb forces into the integral equations formalism. The main problem to be solved is whether the kernels of Faddeev-type equations with Coulomb interactions are compact or not.

Instead of the Faddeev equations, it proves convenient to use the rigorously equivalent effective-two-body Alt-Grassberger-Sandhas (AGS) equations

[*] Supported by DOE Grant DE-FG05-93ER40773

[†] *E-mail address:* akram@comp.tamu.edu

[**] *E-mail address:* alt@dipmza.physik.uni-mainz.de

[††] *E-mail address:* avakov@srdlan.npi.msu.su

[2] which have the - in the present context important - advantage that the reaction amplitudes for $2 \to 2$ processes are decoupled from those with three particles in the initial and/or final state. Quite generally, the question of compactness of the effective-two-body kernels depends on their analytical properties. However, up to now no thorough investigation of the singularities of these kernels in the presence of (unscreened) Coulomb potentials has been performed. This deficiency has been overcome now. The present paper deals specifically with the nondiagonal kernels. In fact, it will show that, if all Coulomb potentials are repulsive, off the energy shell the singularity of the effective potential in the momentum-transfer plane which is located nearest to the physical region and, therefore, is the most dangerous one, is a branch point. An additional pole singularity arises from the effective free Green's function. The important fact is that these two singularities can never coincide for values of the momenta in the integration region, and are thus integrable. For attractive Coulomb potentials, however, nonintegrable singularities occur in the nondiagonal kernels.

A forthcoming paper will deal with the singularity structure of the diagonal kernels. There, it will be shown that, if the charges of all three particles are of the same sign, nonintegrable singularities appear only on the energy shell, and coincide with those considered by Veselova [3] below the breakup threshold. These singularities can, however, be explicitly singled out and inverted as has been done by Alt and Sandhas [4] (see also [5]). The off-shell singularities turn out to be integrable. Hence, the effective-two-body AGS equations can be converted by standard techniques into integral equations with compact kernels.

2 Effective-two-body Alt-Grassberger-Sandhas (AGS) equations and Analytical Behaviour of the Nondiagonal Kernels

Consider the collision of particle α with the bound state of particles β and γ (channel α). We use the standard notation: m_ν is the mass and e_ν the charge of particle $\nu = 1, 2, 3$. On a one-body quantity an index α characterises the particle α, on a two-body quantity the particle pair (β, γ), with $\beta, \gamma \neq \alpha$, and on a three-body quantity the two-fragment partition $\alpha + (\beta, \gamma)$. Jacobi co-ordinates are introduced as follows: $\boldsymbol{k}_\alpha (\boldsymbol{r}_\alpha)$ is the relative momentum (co-ordinate) between particles β and γ, and $\mu_\alpha = m_\beta m_\gamma / (m_\beta + m_\gamma)$ their reduced mass; $\boldsymbol{q}_\alpha (\boldsymbol{\rho}_\alpha)$ denotes the relative momentum (co-ordinate) between particle α and the center of mass of the pair (β, γ), the corresponding reduced mass being $M_\alpha = m_\alpha (m_\beta + m_\gamma) / (m_\alpha + m_\beta + m_\gamma)$. The AGS equations, which have the structure of coupled Lippmann-Schwinger-type equations, are given by [2]

$$\mathcal{T}_{\beta\alpha}(\boldsymbol{q}'_\beta, \boldsymbol{q}_\alpha; z) = \mathcal{V}_{\beta\alpha}(\boldsymbol{q}'_\beta, \boldsymbol{q}_\alpha; z) + \sum_{\nu=1}^{3} \int d\boldsymbol{q}''_\nu \, \mathcal{K}_{\beta\nu}(\boldsymbol{q}'_\beta, \boldsymbol{q}''_\nu; z) \, \mathcal{T}_{\nu\alpha}(\boldsymbol{q}''_\nu, \boldsymbol{q}_\alpha; z), \quad (1)$$

where for simplicity we have assumed that in each channel only one bound state exists. $\mathcal{T}_{\beta\alpha}(\boldsymbol{q}'_\beta, \boldsymbol{q}_\alpha; z)$ is the off-shell reaction amplitude corresponding to the transition from channel α to channel β. The kernel is defined as

$$\mathcal{K}_{\beta\alpha}(\boldsymbol{q}'_\beta, \boldsymbol{q}_\alpha; z) := \mathcal{V}_{\beta\alpha}(\boldsymbol{q}'_\beta, \boldsymbol{q}_\alpha; z) \mathcal{G}_{0;\alpha}(\boldsymbol{q}_\alpha; z), \quad (2)$$

where $\mathcal{G}_{0;\alpha}(\boldsymbol{q}_\alpha; z)$ is the effective Green function describing the propagation of the noninteracting particles α and (β, γ). Furthermore, $\bar{V}_\alpha^C = V_\beta^C + V_\gamma^C$, and

$$\mathcal{V}_{\beta\alpha}(\boldsymbol{q}'_\beta, \boldsymbol{q}_\alpha; z) = \langle \boldsymbol{q}'_\beta, \phi_\beta \mid [G_0^{-1} + V_\gamma^C + \bar{V}_\beta^C G^C \bar{V}_\alpha^C] \mid \phi_\alpha, \boldsymbol{q}_\alpha \rangle \ (\gamma \neq \alpha, \beta). \quad (3)$$

$\phi_\alpha(\boldsymbol{k}_\alpha; \hat{z}_\alpha)$ is the off-shell form factor in the vertex $\alpha \to \beta + \gamma$, $\hat{z}_\alpha = z - q_\alpha^2/2M_\alpha$. $G_0(z)$ and $G^C(z)$ are the three-body free and Coulomb resolvents.

We investigated the analytical properties of $\mathcal{V}_{\beta\alpha}(\boldsymbol{q}'_\beta, \boldsymbol{q}_\alpha; z)$ near its leading singularity and have found the following results.

1) The behaviour of the first term in (3) near the leading singularity which is located at $\sigma_\beta(q'_\beta; \hat{z}_\beta) = k'^2_\beta + 2\mu_\beta(-\hat{z}_\beta) = 0$, $\hat{z}_\beta = z - q'^2_\beta/2M_\beta \neq 0$, is given by

$$\mathcal{V}_{\beta\alpha}^{(a)}(\boldsymbol{q}'_\beta, \boldsymbol{q}_\alpha; z) \overset{\sigma_\beta(q'_\beta; z) \to 0}{\sim} \frac{1}{\sigma_\beta(q'_\beta; z)^{1 - \mathrm{i}(\hat{\eta}_\alpha - \hat{\eta}_\beta)}}, \quad (4)$$

where $\hat{\eta}_\beta$ is the Coulomb parameter of particles α and γ moving with relative kinetic energy $\hat{z}_\beta$, and analogously for $\hat{\eta}_\alpha$.

2) We proved the theorem: Each term of Eq. (3) has the same leading singularity as the first term $\mathcal{V}_{\beta\alpha}^{(a)}$. The most important part of this theorem is that it holds for the last term containing the three-body Coulomb resolvent G^C, too.

3) Given the leading singularity of the nondiagonal effective potential $\mathcal{V}_{\beta\alpha}(\boldsymbol{q}'_\beta, \boldsymbol{q}_\alpha; z)$ we investigated the singularity structure of the kernel $\mathcal{K}_{\beta\alpha}(\boldsymbol{q}'_\beta, \boldsymbol{q}_\alpha; z)$. We found that the leading singularity of $\mathcal{V}_{\beta\alpha}(\boldsymbol{q}'_\beta, \boldsymbol{q}_\alpha; z)$ at $\sigma_\beta(q'_\beta; \hat{z}_\beta) = 0$ ($\hat{z}_\beta \neq 0$) does not coincide with the pole of $\mathcal{G}_{0;\alpha}(\boldsymbol{q}_\alpha; z)$ for real values of the (integration) momenta $\boldsymbol{q}_\alpha$.

4) We investigated the leading singular behaviour of $\mathcal{V}_{\beta\alpha}(\boldsymbol{q}'_\beta, \boldsymbol{q}_\alpha; z)$ for momenta $q'_\beta = \tilde{q}_\beta \equiv \sqrt{2M_\beta E}$, where $\hat{z}_\beta = 0$ so that $\sigma_\beta(q'_\beta; \hat{z}_\beta = 0) = 0$ if $k'_\beta = 0$. We show that for repulsive Coulomb potentials, $\mathcal{V}_{\beta\alpha}(\boldsymbol{q}'_\beta, \boldsymbol{q}_\alpha; z)$ does not develop any dangerous singularity at $k'_\beta = 0$. However, the singularity at $k'_\beta = 0$ becomes nonintegrable for an attractive Coulomb potential V_β^C. In that case the behaviour of $\mathcal{V}_{\beta\alpha}(\boldsymbol{q}'_\beta, \boldsymbol{q}_\alpha; z)$ is given by $(\hat{\boldsymbol{q}}'_\beta := \boldsymbol{q}'_\beta / q'_\beta)$

$$\mathcal{V}_{\beta\alpha}(\hat{\boldsymbol{q}}'_\beta \tilde{q}_\beta, \boldsymbol{q}_\alpha; E + i0) \overset{k'_\beta \to 0}{=} O\left(\frac{1}{k'^{4 + 2\mathrm{i}\hat{\eta}_\beta - 2\mathrm{i}\hat{\eta}_\alpha + \mathrm{i}\eta_\beta + \mathrm{i}\eta_\gamma}_\beta} \right). \quad (5)$$

Here, η_β is the Coulomb parameter of particles α and γ moving with the relative momentum k'_β, and analogously for η_γ. Since the point $k'_\beta = 0$ lies on the integration contour, such a singular behaviour is not integrable, thereby destroying the compactness properties of the kernel of the AGS equations.

5) The singular behaviour of the on-shell nondiagonal effective potential for $\bar{\sigma}_\beta \to 0$ is given by $\mathcal{V}_{\beta\alpha}(\bar{\boldsymbol{q}}'_\beta, \bar{\boldsymbol{q}}_\alpha; z) \sim \bar{\sigma}_\beta^{-1 + \eta_\alpha^{(bs)} + \eta_\beta^{(bs)}}$, where $\bar{\sigma}_\beta = \bar{k}'^2_\beta + \kappa_\beta^2$, κ_β is the wave number of the bound state (α, γ), $\eta_\alpha^{(bs)}$ the Coulomb parameter of the bound state (β, γ), and $\bar{k}'_\beta$ the usual linear combination of the external on-shell momenta $\bar{\boldsymbol{q}}'_\beta$ and $\bar{\boldsymbol{q}}_\alpha$. Since $\bar{\sigma}_\beta > 0$, the singular point $\bar{\sigma}_\beta = 0$ is located outside of the integration region.

3 Conclusions

Let us summarise the results obtained.

(i) We have shown that repulsive Coulomb interactions change, but do not amplify, the singular behaviour of the nondiagonal kernels of the effective-two-body AGS equations as compared to the case of purely short-range interactions. Hence, all considerations performed for the nondiagonal kernels for short-range potentials are valid also for the nondiagonal kernels for short-range plus unscreened Coulomb interactions. However, if some of the Coulomb potentials are attractive, the effective potentials develop nonintegrable singularities which impede the applicability of standard integral equations methods to Eqs. (1). This deficiency is shared by the integrodifferential equations approach [6].

(ii) Even in the presence of only repulsive Coulomb interactions there exists - from the practical point of view possibly important - a complication: according to the theorem proved by us, all the terms in the Quasi-Born series of the effective nondiagonal potentials, which is obtained by a series expansion of G^C in Eq. (3), have the same branch point singularity. This could imply that, in principle, all terms of this series should be taken into account unless, of course, their contribution to the singularity strength is found to decrease with increase of the number of intermediate-state Coulomb scatterings, i.e., with the order of iteration. Note that this remark does not apply when only two of the three particles are charged because in that case the Quasi-Born series rigorously collapses to its first two terms [2, 4].

Let us add a final remark. The singularity (4) is the leading one, i.e., it is the singularity which is strongest and closest to the integration region. In addition, each term of the Quasi-Born series has its own singularities which are, however, more distant than this singularity. These latter singularities are not caused by the Coulomb interactions and occur even if all or some of them switched off. That is, they are present already for purely short-range interactions and are, therefore, known to be not dangerous.

References

1. L.D. Faddeev: Sov. Phys. JETP **12**, 1014 (1961); *Mathematical Aspects of the Three-Body Problem in the Quantum Scattering Theory*. Jerusalem: Israel Program for Scientific Translations 1965

2. E.O. Alt, P. Grassberger, W. Sandhas: Nucl. Phys. **B2**, 167 (1967)

3. A.M. Veselova: Theor. Math. Phys. **3**, 542 (1970); **13**, 369 (1972)

4. E.O. Alt and W. Sandhas: Phys. Rev. **C21**, 1733 (1980)

5. E.O. Alt, W. Sandhas, H. Ziegelmann: Phys. Rev. **C17**, 1981 (1978)

6. S.P. Merkuriev and L.D. Faddeev: In: *Quantum Scattering Theory for Systems of Few Particles*. Moscow: Nauka 1985 (in Russian)

Few-Body Systems Suppl. 10, 101–104 (1999)

© by Springer-Verlag 1999

Poles near the Thresholds in the Coupled $\Lambda N - \Sigma N$ System

H. Yamamura* and K. Miyagawa†

Department of Applied Physics, Okayama University of Science, 1-1 Ridai-cho, Okayama 700, Japan

Abstract. We find t-matrix poles near the ΣN threshold for the meson-theoretical Nijmegen YN interactions including hard-core models. These poles are connected with the strength of the $\Lambda N - \Sigma N$ coupling. We also observe antibound-state poles below the ΛN threshold which correlate with scattering lengths.

1 Introduction

Recent studies of few-baryon systems with strangeness have made fairly clear the nature of S-wave YN interaction. The analysis of the hypertriton [1] not only constrained the strengths of $^1S_0$ and $^3S_1$ YN force components, but showed that the effect of $\Lambda - \Sigma$ conversion is crucial for its binding. It dimonstrated that the expectation values of the interactions $V_{\Lambda N, \Sigma N}$ and $V_{\Sigma N, \Lambda N}$ amount to 8% of the total potential energy. To clarify the feature of this coupling interaction, the analysis should be extended to the region near the Σ threshold where $\Lambda - \Sigma$ conversion effects will emerge sharply.

This paper thus investigates t-matrix poles for various meson-theoretical $\Lambda N - \Sigma N$ interactions around the ΣN threshold. Since these t-matrices show enhancements caused by the poles around the threshold, the knowledge of the poles is indispensable for few-body analyses in future to incorporate this coupling interaction precisely.

2 Analytic Continuation of the T-matrix in Momentum Space

The coupled t-matrices for the $\Lambda N - \Sigma N$ system are defined by the integral equations

---

*E-mail address: yamamu-s@dap.ous.ac.jp

†E-mail address: miyagawa@dap.ous.ac.jp

$$T_{ij}(E) = V_{ij} + \sum_k V_{ik} \, G_0^{(k)}(E) \, T_{kj}(E), \qquad i,j,k = 1,2 \qquad (1)$$

with

$$G_0^{(k)}(E) = \left(E + i\varepsilon - H_0^{(k)} \right)^{-1}, \quad H_0^{(k)} = \frac{p^2}{2\mu_k} + m_N + m_Y^{(k)} \qquad (2)$$

where the numbers 1 and 2 are assigned to the ΛN and ΣN channels, and the masses $m_Y^{(k)}(k = 1,2)$ indicate m_Λ and m_Σ, respectively. After partial-wave decomposition in momentum space, we express projected t-matrix elements again simply by T, which read

$$<p\,|\,T_{ij}(E)\,|\,p'> \; = \; <p\,|\,V_{ij}\,|\,p'>$$
$$+ \sum_k 2\mu_k \int_0^\infty dp'' p''^2 <p\,|\,V_{ik}\,|\,p''> \frac{1}{q_k^2 - p''^2 + i\varepsilon} <p''\,|\,T_{kj}(E)\,|\,p'> \qquad (3)$$

with

$$\frac{q_k^2}{2\mu_k} = E - m_N - m_Y^{(k)}. \qquad (4)$$

We analytically continue these t-matrices into the complex plane of q_k in a usual manner, and solve the resulting equations for various q_k. We search the values of q_k for which the t-matrix elements take infinity.

3 T-matrix for a hard core potential

Although it has become old-fashioned to represent the short-range repulsion of NN interaction by a hard core, the Nijmegen D and F models of YN interaction with hard cores have still influence on hypernucler physics. We therefore explore a method to obtain the off-shell t-matrix for a hard core potential in momentum space. This is done based on the two-potential formula.

Following this formula, we express the off-shell t-matrix as

$$<p\,|\,T(E)\,|\,k> \; = \; <p\,|\,U\,|\,\Phi_{q,k}^{(+)}> + <\Phi_{q,p}^{(-)}\,|\,\hat{V}\,|\,\Psi_{q,k}^{(+)}> \qquad (5)$$

where U indicates the pure hard core part of a potential, and $\hat{V}$ the remainder. $\Phi^{(\pm)}$ are the scattering states by U while Ψ means scattering state by the full potential. As shown in ref. [2], the first term of the right-hand side is expressed analytically, and the second term satisfies an integral equation similar to the Lippmann-Schwinger equation, which can be solved in an ordinary way.

To obtain the analytic expression of the first term, we follow Takemiya [3], who proposed a method to evaluate off-shell t-matrices for a hard-core potential in coordinate space. We only use this method to treat the pure hard-core part in the formula (5). Partial wave projected elements of the first term, expressed here by $\tilde{t}_{l's'ls}^J(p,k;q)$, become

$$
\tilde{t}^J_{l's'ls}(p,k;q) = \delta_{l'l}\delta_{s's}\frac{\hbar^2}{2\mu}\frac{2}{\pi}i^{-l'+l}
$$

$$
\times \left[-c\,j_l(pc)\frac{d}{dr}r\,h_l^{(+)}(qr)\bigg|_{r=c}\frac{j_l(kc)}{h_l^{(+)}(qc)} \right.
$$

$$
+\frac{d}{dr}r\,j_l(pr)\bigg|_{r=c}c\,j_l(kc)
$$

$$
\left. -(q^2-p^2)\int_0^c dr\,r^2\,j_l(pr)\,j_l(kr) \right] \tag{6}
$$

where c indicates the hard-core radius.

On the other hand, the second term of Eq. (5), expressed by $< p\,|\,\hat{t}(E)\,|\,k >$ below, satisfies the integral equation

$$
< p\,|\,\hat{t}(E)\,|\,k > = < \Phi_{q,p}^{(-)}\,|\,\hat{V}\,|\,\Phi_{q,k}^{(+)} >
$$
$$
+\int d\boldsymbol{p}' < \Phi_{q,p}^{(-)}\,|\,\hat{V}\,|\,\boldsymbol{p}' >\frac{1}{E-{p'}^2/2\mu+i\varepsilon}< \boldsymbol{p}'\,|\,\hat{t}(E)\,|\,\boldsymbol{k} > . \tag{7}
$$

This equation can be solved in a similar way to Eq. (3). Thus, one can easily obtain the off-shell t-matrix in momentum space. For the details, we refer the readers to ref. [2].

4   Results and Discussions

We search t-matrix poles for various meson theoretical YN interactions, the soft core model (NSC89) and the hard core models D and F (abbreviated as ND and NF respectively) of the Nijmegen group. This group has recently proposed a new soft core model (NSC97) [4], which includes five different versions named a,b,c,d and e. Among them, we analyze NSC97d and NSC97e. All three soft core models that we use reproduce the correct binding energy of the hypertriton [1].

Poles around the ΣN threshold are shown in Table 1. For the interactions, NSC97e, NSC97d and NF, the poles locate in the second quadrant of $q_{\Sigma N}$, while for the NSC89 and ND, the poles lie in the third quadrant. These positions of the poles are related to the shapes of the ΛN elastic total cross sections. In general, a pole in the second quadrant causes a round resonance peak, while a pole in the third quadrant near the threshold produces a cusp just at the threshold [2]. Since these poles are closely connected with the strength of the $\Lambda N - \Sigma N$ coupling, experimental studies around the threshold are highly important.

We also find antibound-state poles below the ΛN threshold which are displayed in Table 2. These poles are correlated to the ΛN scattering lengths, which are fairly constrained by both of the hypertriton binding energy and the ΛN cross section data [1], but are still not determined. To fix the scattering lengths is also left as a challenge in hypernuclear physics. .

Table 1. The poles near the ΣN threshold for the force component $^3S_1 - {}^3D_1$. The relative momenta in the ΛN and ΣN channels are indicated by $q_{\Lambda N}$ and $q_{\Sigma N}$, respectively. The center-of-mass energy is shown by E.

model	$q_{\Lambda N}$ [fm^{-1}]	$q_{\Sigma N}$ [fm^{-1}]	E [MeV]
NSC97d	(1.46,-0.04)	(-0.38,0.16)	(2136.3,-4.51)
NSC97e	(1.47,-0.04)	(-0.40,0.16)	(2136.9,-4.74)
NSC89	(1.37,0.01)	(-0.04,-0.39)	(2126.3,1.07)
D	(1.43,0.01)	(-0.18,-0.08)	(2132.8,1.07)
F	(1.44,0.02)	(-0.28,0.12)	(2134.2,-2.49)

Table 2. The poles below the ΛN threshold. The scattering lengths indicated by a are also shown. See the caption of Table 1 for other details.

model	partial wave	$q_{\Lambda N}$ [fm^{-1}]	$q_{\Sigma N}$ [fm^{-1}]	E [MeV]	a [fm]
NSC97d	$^1S_0$	(0,-0.27)	(0,1.47)	(2051.8,0)	-2.63
	$^3S_1 - {}^3D_1$	(0,-0.38)	(0,1.50)	(2049.0,0)	-1.62
NSC97e	$^1S_0$	(0,-0.27)	(0,1.47)	(2051.8,0)	-2.69
	$^3S_1 - {}^3D_1$	(0,-0.39)	(0,1.50)	(2048.7,0)	-1.60
NSC89	$^1S_0$	(0,-0.28)	(0,1.47)	(2051.5,0)	-2.48
	$^3S_1 - {}^3D_1$	(0,-0.45)	(0,1.52)	(2046.8,0)	-1.32
D	$^1S_0$	(0,-0.35)	(0,1.49)	(2050.0,0)	-1.83
	$^3S_1 - {}^3D_1$	(0,-0.35)	(0,1.49)	(2050.0,0)	-1.89
F	$^1S_0$	(0,-0.31)	(0,1.48)	(2050.8,0)	-2.19
	$^3S_1 - {}^3D_1$	(0,-0.36)	(0,1.49)	(2049.7,0)	-1.83

References

1. K. Miyagawaet al.: Phys. Rev. **C51**, 2905 (1995)

2. K. Miyagawa, H.Yamamura: submitted for publication

3. T. Takemiya: Prog. Theor. Phys. **48**, 1547 (1972)

4. Th.A. Rijken: In: *Proc.of the Int. Conf. on Hypernuclear and Strange Particle Physics (HYP97)*, Brookhaven National Laboratory, USA, October, 1997

Few-Body Systems Suppl. 10, 105–114 (1999)

Few-
Body
Systems
© by Springer-Verlag 1999

Lorentz covariance of three-dimensional equations

V. Pascalutsa and J.A. Tjon

Institute for Theoretical Physics, University of Utrecht, Princetonplein 5,
3584 CC Utrecht, The Netherlands

Abstract. We show how the invariance under the charge conjugation and CPT symmetry, present in the Bethe-Salpeter equation, is lost in the reduction to certain relativistic three-dimensional equations. This in particular leads to the breakdown of the standard Lorentz structure and renormalization procedures for the resulting single-particle propagators. We formulate the equal-time approximation of the Bethe-Salpeter equation in the form which manifestly satisfies the above symmetries, and apply it to the description of the pion-nucleon interaction in a dynamical model based on hadron exchanges. We also consider the one-body limit of various three-dimensional equations for the case of t- and u-channel one-particle-exchange potential.

1 Introduction

Quantum field theory (QFT) provides us with a suitable framework unifying principles of relativistic covariance and quantum mechanics. Unfortunately, any systematic calculation beyond perturbation theory is extremely complicated within QFT. On the other hand, in the ordinary quantum mechanics the scattering and bound state problems are well understood in terms of the Schrödinger or Lippmann-Schwinger equation. Its relativistic generalizations, referred to as *quasipotential* (QP) approximations to field theory, can therefore be very useful for practical studies of relativistic effects in the strongly interacting systems.

The QP equations can conveniently be obtained from the manifestly covariant four-dimensional Bethe-Salpeter (BS) equation by approximating the kernel in some way. This approximation involves an assumption about the singularities of the BS kernel, after which the integration over the 0-th component (time or energy) can easily be done explicitly, leading to the three-dimensional (3-D) equation. One of the first such reductions of the BS equation was studied by Salpeter [1], using the *instantaneous* approximation.

Since the covariant reductions can be done in infinitely many different ways it is desirable to establish certain criteria which would constrain the choice. For instance, an important property one would like to have for a relativistic two-body equation is the *correct one-body limit*, which means that in the limit when one of the particles becomes infinitely heavy the two-body equation must reduce to the corresponding equation of motion of the light particle (e.g., the Klein-Gordon equation) in an external potential. The fact that the ladder BS equation does not have the correct one-body limit, created further motivation for the QP approach since the one-body limit can relatively simply be incorporated into a 3-D equation for the one-particle-exchange (OPE) potential. Some of the first equations of this type were suggested by Gross [2] and by Todorov [3]. Since then the one-body limit is regarded as an important criterion, even though not a very restrictive one, as many of the equations can be adjusted to satisfy it. In particular, Mandelzweig and Wallace [4] incorporated the one-body limit in Salpeter's instantaneous equation.

In this contribution we would like to demonstrate the importance of the constraint put by *charge conjugation* symmetry. Also, the one-body limit constraint will be examined at the one loop level for various OPE potentials. As an application, results of a quasipotential modelling of the pion-nucleon system will be presented as well.

2 The role of charge conjugation symmetry

To begin with, let us recall several basic definitions concerning the Lorentz group. A general Lorentz transformation L of a four-momentum is given by real (pseudo-)orthogonal tensor $\Lambda_{\mu\nu}$, and may belong to one of the following four domains:

$$
\begin{aligned}
L_+^\uparrow &: \quad \det \Lambda = +1, \ \Lambda_{00} \geq +1, \\
L_-^\uparrow &: \quad \det \Lambda = -1, \ \Lambda_{00} \geq +1, \\
L_+^\downarrow &: \quad \det \Lambda = +1, \ \Lambda_{00} \leq -1, \\
L_-^\downarrow &: \quad \det \Lambda = -1, \ \Lambda_{00} \leq -1.
\end{aligned}
\tag{1}
$$

From these, only transformations $L_+^\uparrow$ form a group by themselves, called the proper orthochronous Lorentz group. The other domains do not contain the unity element, however their multiplication with $L_+^\uparrow$ may form a group. Fields and corresponding Green functions are transformed according to unitary representations of the Lorentz group. Let us remark, that the full Lorentz group transformations [which include the proper continuous transformations, and the (anti-)unitary transformations of parity and time reversal] do not connect Green functions defined for positive energy (upper light-cone) with those for negative energy (lower light-cone). In other words, applying such transformations to the momentum-space Green functions can induce only $L_+^\uparrow$ and $L_-^\uparrow$ transformations of the relevant four-momenta. To be able to induce $L_+^\downarrow$ and $L_-^\downarrow$ transformations of the four-momenta, one needs to include charge conjugation in addition to the above mentioned Lorentz transformations.

In considering some existing relativistic 3-D equations, we find that they in general do not yield the correct Lorentz structure. For example, the calculated self-energy of a spin-1/2 particle *does not* have the following form,

$$\Sigma(\slashed{P}) = \slashed{P}A(P^2) + B(P^2),\tag{2}$$

where A and B are scalar functions of the invariant P^2 only. At first this is surprising, naively we would expect form (2) to come out in any covariant formalism. However, Eq. (2) holds only if there is a symmetry under all Lorentz transformations of the four-momenta. Therefore, in order to obtain the self-energy consistent with Eq. (2), the relativistic equation in question should be covariant under the Lorentz group and charge conjugation.

The four-dimensional BS equation, of course, preserves the standard structure, such as Eq. (2). The symmetry can obviously be lost in doing the QP reduction. To illustrate this, consider the example of a scalar self-energy, given by

$$\Sigma(P^2) = i \int \frac{\mathrm{d}^4 q}{(2\pi)^4} \frac{\Phi(q^2, P^2, P \cdot q)}{[(\frac{1}{2}P - q)^2 - m^2 + i\varepsilon][(\frac{1}{2}P + q)^2 - m^2 + i\varepsilon]},\tag{3}$$

where P is the relevant four-vector, Φ is an "interaction function" which corresponds to the product of the two vertex functions, and which may also have some particle propagation poles.

We can immediately see that Eq. (3) is a function of P^2 only: a sign change of P can be absorbed by a change of the loop variable q to $-q$. In a QP description this substitution in general cannot be applied in view of the constraint in q_0. To see what happens then, consider the poles of the integrand of Eq. (3) in the complex q_0 plane.

There are four poles (two in the upper and two in the lower half-plane) coming from the propagators in the two-particle Green function:

$$q_0 = \pm\tfrac{1}{2}P_0 - \sqrt{m^2 + (\tfrac{1}{2}\boldsymbol{P} \mp \boldsymbol{q})^2} + i\varepsilon, \text{ and } q_0 = \pm\tfrac{1}{2}P_0 + \sqrt{m^2 + (\tfrac{1}{2}\boldsymbol{P} \mp \boldsymbol{q})^2} - i\varepsilon.$$

We can see that a simultaneous sign reflection of P_0 and q_0 interchanges the poles of the upper half-plane with the poles of the lower half-plane. The same symmetry exists for the singularities of Φ. Therefore, in order for Σ to be even in P_0 the integration over q_0 must be independent of the choice of the half-plane where we close the contour. In performing a 3-D reduction, however, one usually neglects the contribution of certain poles, hence the result becomes dependent on the contour. In that case Σ is not anymore an even function of P_0, consequently it cannot be a function of P^2 only. In this sense the standard Lorentz structure of the self-energy is violated. (An example of the QP prescription, which is covariant under the Lorentz group but violates the charge conjugation symmetry, is the spectator approximation of Gross [2]. In this approximation, one of the particles inside the loop is restricted to its mass shell, therefore only a single pole is taken in calculating the q_0 integral.)

Similar arguments apply for the spin-1/2 particle self-energies. Consider the dressed fermion propagator given by

$$S(\slashed{P}) = [\slashed{P} - m - \Sigma(\slashed{P}) + i\epsilon]^{-1},\tag{4}$$

where $\Sigma(P)$ is the self-energy. For simplicity we work in the c.m. frame, where $P = (P_0, \mathbf{0})$. In this frame the Dirac structure of the self-energy can be represented as

$$\Sigma(P_0) = \Sigma_+(P_0)\gamma_+ + \Sigma_-(P_0)\gamma_-, \tag{5}$$

where $\gamma_\pm = \frac{1}{2}(I \pm \gamma_0)$. A similar decomposition holds for the propagator:

$$S(P_0) = S^{(+)}(P_0)\gamma_+ + S^{(-)}(P_0)\gamma_-, \tag{6}$$

with $S^{(\pm)}(P_0) = \pm[P_0 \pm (-m - \Sigma_\pm(P_0) + i\epsilon)]^{-1}$. Obviously, $S^{(+)}$ corresponds to the positive and $S^{(-)}$ to the negative energy-state propagation.

It is easy to see that, if the self-energy can be written in the general covariant form (2), then the following identity holds,

$$\Sigma_r(P_0) = \Sigma_{-r}(-P_0), \quad r = \pm 1, \tag{7}$$

and vise versa (in the c.m. frame). This identity is particularly useful to test numerically Eq. (2) in models based on QP equations which are usually solved for partial waves in the c.m. system.

Performing the standard renormalization procedure by subtracting the counter-term: $Z_2(m_0 - m) + (1 - Z_2)(P - m)$, where m_0 is the bare mass, and Z_2 is the field renormalization constant, we find that the on-shell renormalization scheme requires

$$
\begin{aligned}
Z_2(m_0 - m) &= \Sigma_+(m) = \Sigma_-(-m), \\
1 - Z_2 &= \left.\frac{\partial \Sigma_+(P_0)}{\partial P_0}\right|_{P_0 = m} = -\left.\frac{\partial \Sigma_-(P_0)}{\partial P_0}\right|_{P_0 = -m}.
\end{aligned}
\tag{8}
$$

Obviously, it is not possible to satisfy these relations if Eq. (7) is violated. In other words, the violation of the extended Lorentz symmetry leads to the different renormalization of the positive and negative energy states. This can be understood, as the violation of the charge conjugation symmetry in a Lorentz-covariant framework implies violation of CPT symmetry.

To recapitulate, relativistic equations obtained from the BS equation via the 3-D reduction which discriminates between the positive and negative energy poles (e.g. by putting particles on-shell, or using the positive energy projection operators) lead to results which do not have the standard Lorentz structure, even if the symmetry under the full Lorentz group remains intact. Such equations necessarily violate the charge conjugation and CPT symmetries, and thus lead to the breakdown of the usual renormalization procedures which rely on constructing the counter-terms from a CPT invariant Lagrangian. Also, one then cannot use the standard covariant arguments to construct the transformation properties of the calculated amplitudes (as well as any other functions involving loop corrections), which for instance is needed to incorporate the basic interaction in more particle systems.

3 Manifestly covariant three-dimensional equation

One of the ways to perform a 3-D reduction consistent with charge conjugation and unitarity is by removing the poles of the interaction in q_0 complex plane, while treating exactly the poles of the two-particle propagator. This procedure is realized in the equal-time (ET) approximation [4, 5]. In this approximation the poles are removed from the interaction piece by fixing the relative-energy variable q_0 in some way. Most frequently the constraint $q_0 = 0$, or its Lorentz-invariant generalization, $P \cdot q = 0$, is used. Moreover, the two-particle propagator is sometimes modified to include approximately the crossed graphs [4, 6].

On the other hand, it is well known that the $P \cdot q = 0$ constraint is troublesome in the inelastic or more-particle problems, see, e.g., the introductory remarks in refs. [7, 8]. The weak point resides in the fact that the constraint is embedded through a δ-function. In the following we formulate a 3-D formulation which exhibits manifest Lorentz covariance, does not make use of δ-functions, and for the elastic two-body problem is equivalent to the ET approximation.

Recall the two-particle Bethe-Salpeter equation in the momentum-space:

$$T(p', p) = V(p', p) + i \int \frac{\mathrm{d}^4 q}{(2\pi)^4} \, V(p', q) \, G(q) \, T(q, p), \qquad (9)$$

where we assume p', p and q are the relative four-momenta of the final, initial and intermediate state, respectively. To transit to the 3-D formulation we impose the condition that the interaction is insensitive to the off-shellness along the direction defined by unit four-vector n_μ. For the two-body case this means that V and T entering the scattering equation depend on the projections of the relative four-vectors onto a 3-D hyperplane orthogonal to n_μ. Defining the projection operator: $O_{\mu\nu} = \mathrm{g}_{\mu\nu} - n_\mu n_\nu$, we write the corresponding equation as follows:

$$T(\tilde{p}', \tilde{p}) = V(\tilde{p}', \tilde{p}) + i \int \frac{\mathrm{d}^4 q}{(2\pi)^4} \, V(\tilde{p}', \tilde{q}) \, G(q) \, T(\tilde{q}, \tilde{p}), \qquad (10)$$

where $\tilde{p}_\mu = O_{\mu\nu} p^\nu$, and similarly for $\tilde{p}'$, $\tilde{q}$. Equation (10) is manifestly covariant, and on the other hand it can easily be reduced to the 3-D form. For example, let us choose the frame where $n = (1, 0, 0, 0)$, and therefore V and T are independent of the 0-th component of relative momenta (since any scalar product will depend only on the spatial components, e.g., $\tilde{q} \cdot \gamma = -\boldsymbol{q} \cdot \boldsymbol{\gamma}$). The integration over q_0 in Eq. (10) can now be readily done leading to the 3-D equation.

Obviously, the newly introduced four-vector n will enter the final covariant forms. To prevent this dependence one may choose it along some physical four-momentum, for instance the total momentum of the system, i.e.,

$$n_\mu = P_\mu / \sqrt{P^2}. \qquad (11)$$

It is then easy to see that for the two-body elastic scattering Eq. (10) in the c.m. system becomes equivalent to the usual ET approximation.

4 Box graphs and the one-body limit

Analyzing the box and the crossed-box graphs in QFT, for a neutral particle exchange, Gross revealed a cancellation among various pole contributions, and proved that the only pole which survives in the one-body limit is that of the heavy particle in the intermediate state of the box graph [2, 9]. This led him to formulate the spectator equation where the heavy particle is on the mass-shell. Obviously, the heavier is the spectator, the closer should the Gross (spectator) equation be to the QFT result. Therefore, for instance the πN system would seem to be a particularly good application for this equation, since the nucleon is much heavier than the pion. Recently, however, Gross and Surya, applying the spectator equation to πN system, have argued that the light particle (the pion) must be taken as the spectator [10, Sec. II.A]. Studying the box and the crossed-box graphs at threshold, they have conjectured that "the essential difference is the mass of the exchanged particle". We have examined their conjecture, studying the graphs for more general situations, and find that the argument should be related more to the type of the OPE potential, rather than to the mass of the exchanged particle.

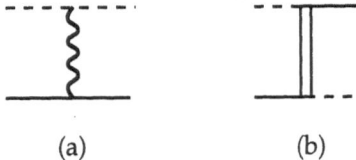

(a)   (b)

Figure 1. The t-channel (a) and u-channel (b) exchange potentials.

Namely, we consider two types of the potentials, see Fig. 1: (a) t-exchange potential, and (b) u-exchange potential. (We shall refer to the dashed line particle as to pion and the solid line as to nucleon with corresponding masses m_π and m_N, the exchange particle mass is denoted as μ.) Substituting these

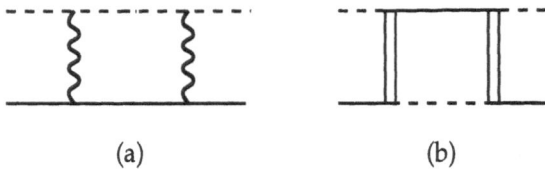

(a)   (b)

Figure 2. The box graphs obtained by iterating once the potentials of Fig. 1.

potentials into the scattering equation, Fig. 4, and iterating once, we obtain the box graphs depicted in Fig. 2 (a) and (b), respectively. Note that in QFT, due to the crossing symmetry, one in addition has the corresponding crossed-box graphs.

We have calculated such box and crossed-box graphs in 4-D field theory numerically and compared with the box graph calculation within various quasipo-

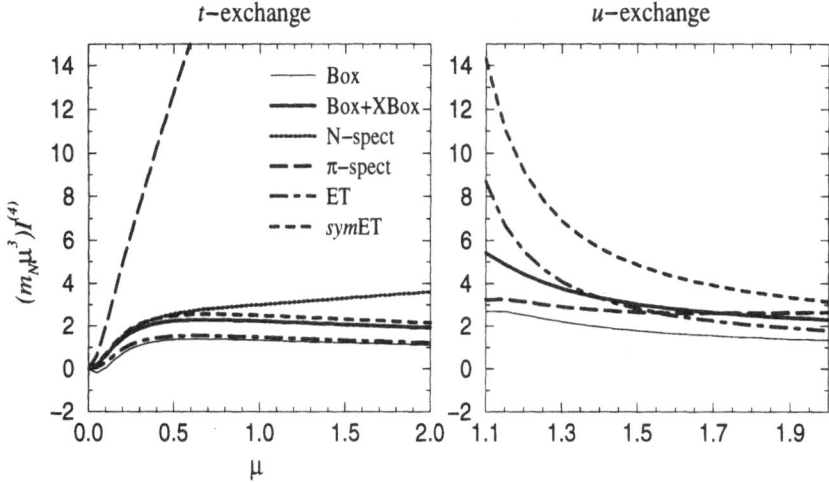

Figure 3. Results for $m_N = 1$, $m_\pi = 0.01$, $\sqrt{s} = 1.1$ as the function of the exchange particle mass.

tential formulations. Namely, the nucleon and the pion *spectator* [2, 9, 10], the *equal-time* [5], and the *symmetrized equal-time* [4, 6].

The results of these calculations for the case close to the one-body limit (the nucleon is much heavier that the pion) is plotted in Fig. 3, as a function of the exchange particle mass[1]. The energy is fixed slightly above the threshold, $\sqrt{s} = 1.1 m_N$, and $t = 0$. One can see that for the t-exchange potential the one-body limit is achieved in the symmetrized ET formulation independently of the mass of the exchanged particle. The nucleon spectator indeed deviates from the limit for large μ, however the pion spectator does not produce a better result in this situation.

On the other hand, in the u-exchange case, both the nucleon spectator and symmetrized ET disagree substantially with the QFT result (the spectator calculation is an order of magnitude larger and hence beyond the scale of the figure). The pion spectator is in a much better agreement. Thus, we conclude that the difference between the NN and πN situation encountered by Gross and Surya [10] appears due to the different *type of the potential*.

It would be interesting to see if there is a possibility to develop a prescription which would give the proper limit in both the t- and u-exchange cases. It should be emphasized though, that the one-body limit situation is physically very different for the two cases: for the t-exchange potential it corresponds to the light particle moving in an external potential of the heavy particle, while in the u-exchange case the heavy particle obviously does not act as a static external source, and therefore there seems to be no correspondence to any one-body situation.

---

[1]Note that we multiply the results by $\mu^3 m_N$ in order to obtain reasonable values for various limiting cases.

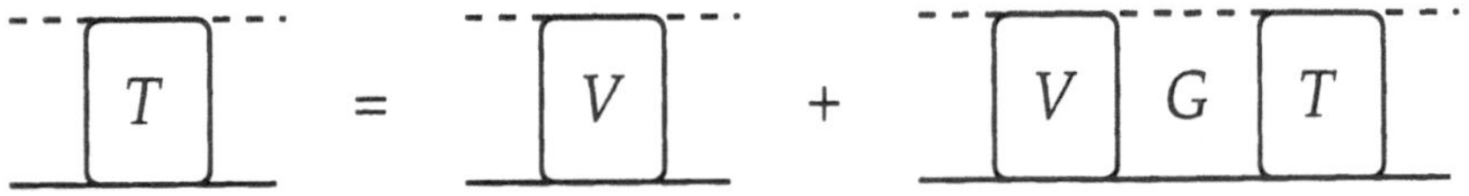

Figure 4. Diagrammatic form of a relativistic two-body scattering equation.

5 πN scattering

We have studied the ET approximation of the BS equation in a dynamical model for πN scattering. The corresponding equation, Fig. 4, is solved for the πN partial-wave amplitudes with the OPE potential represented by $N(938)$, $N^*(1450)$, $\Delta(1232)$, $D_{13}(1525)$, $S_{11}(1555)$, $\rho(770)$ and $\sigma(550)$ exchanges, see Fig. 5. The model is very close to the one presented earlier [11], even though presently we have used a different form of the $\pi N \Delta$ coupling [12], and D_{13} and S_{11} exchanges are included in addition. The latter has a considerable effect only in the S_{11} partial wave.

The model parameters (coupling constants, resonance and cutoff masses) were adjusted to reproduce the low-energy quantities, such as scattering lengths, volumes and ranges, and the energy behavior of the phase-shifts. The resulting description of the phase-shifts up to 600 MeV pion kinetic lab energy is depicted in Fig. 6.

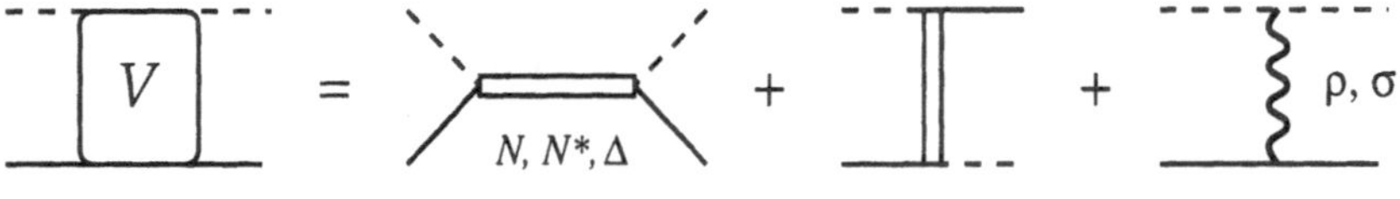

Figure 5. The tree-level πN potential, the driving force of the scattering equation.

6 Conclusions

The relativistic scattering and bound state problems are often formulated in terms of a 3-D (or quasipotential) relativistic equation of the Lippmann-Schwinger type. Such equations can be obtained from the manifestly covariant (3+1)-dimensional Bethe-Salpeter equation by integrating out the time variable in some approximate way, thus performing the so called 3-D reduction. Adopting the 3-D formulation in favor of the 4-D one leads to major technical simplifications, since the field-theoretical BS kernel may, in principle, contain many singularities in the time variable. However, the charge conjugation symmetry can easily be violated in performing such a reduction. On the other hand, it plays an important role in obtaining the standard Lorentz structure

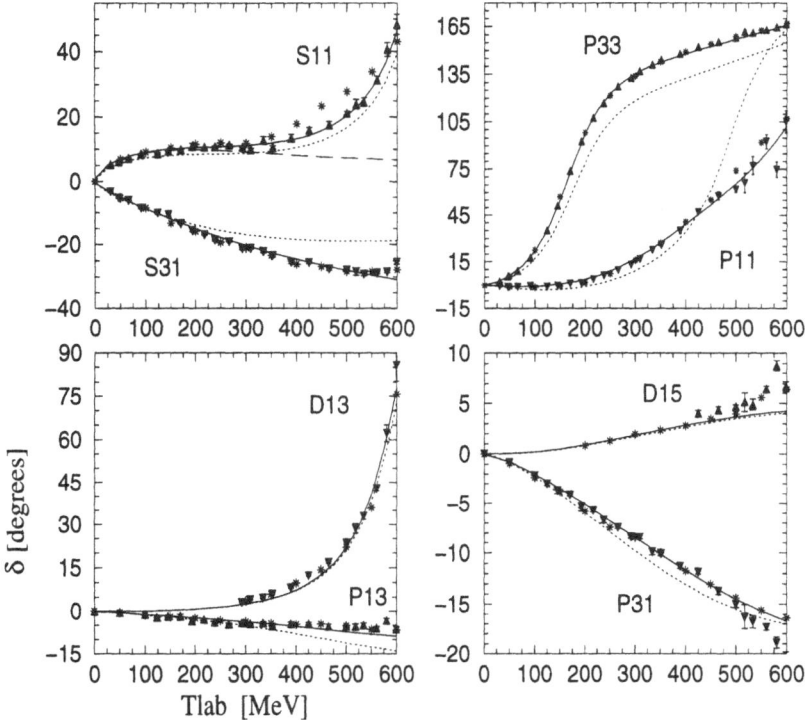

Figure 6. The description of S-, P- and D-wave πN phase-shifts. Data points are from the SM95 [13] (triangles) and KH80 [14] (stars) partial-wave analyses. Solid lines represent the model solution. Dotted lines represent the calculation where the principal value part of the rescattering integrals is switched off (i.e.,the K-matrix approximation with the same set of parameters). Dashed line for the S_{11} shows the calculation when the pole contribution of the S_{11} resonance is switched off.

of the loop corrections. Therefore, the equations which respect charge conjugation symmetry are preferable, and thus the choice among the infinite number of possible relativistic 3-D equations is somewhat restricted in this way.

We have presented a 3-D reduction which is manifestly covariant under the complete set of Lorentz transformations as well as charge conjugation. The two-body equation, obtained by using this reduction, in the c.m. system is equivalent to the Salpeter equation.

We have studied the one-body limit for several 3-D equations with the OPE potential. In the limit, a large qualitative difference is observed between the situation when the potential in question has the form of t- or u-channel exchange. The 3-D equations, such as the nucleon spectator [2, 9] and the symmetrized ET [4, 6], developed to satisfy the one-body limit for the t-type exchange potential, have a poor agreement with the exact calculation if the u-type exchange potential is used.

The pion spectator approximation describes the u-exchange case better, but fails in the other case. Therefore, in the situation where both types of the potential are present, either of the spectator equations cannot be justified. Analyzing the πN situation with realistic parameters we find that the ET type of prescriptions can be fairly close to the QFT answer for both types of the potential, and, hopefully, is a reasonable dynamical framework in this case. We have therefore applied the ET approximation of the BS equation to the description of the πN scattering.

Acknowledgement. We thank F. Coester, R. Timmermans and B. de Wit for illuminating discussions on Lorentz covariance. One of us (V.P.) had the pleasure to discuss with V. Mandelzweig during the conference.

References

1. E. Salpeter: Phys. Rev. **87**, 328 (1952)

2. F. Gross: Phys. Rev. **186**, 1448 (1969)

3. I.T. Todorov: In: *Properties of the Fundamental Interactions*, vol. 9, part C, p. 953; ed. A. Zichichi. Bologna: 1973, and references therein

4. V.B. Mandelzweig, S.J. Wallace: Phys. Lett. **B197**, 469 (1987); S.J. Wallace, V.B. Mandelzweig: Nucl. Phys. **A503**, 673 (1989)

5. J.A. Tjon: In: *Hadronic Physics with multi-GeV Electrons*, Les Houches Series, p. 89. New York: New Science Publishers 1990

6. P.C. Tiemeijer, J.A. Tjon: Phys. Rev. **C48**, 896 (1993); *ibid.* **C49**, 494 (1994); P.C. Tiemeijer: Ph.D. thesis. Univ. Utrecht 1993 (unpublished)

7. N.K. Devine, S.J. Wallace: Phys. Rev. **C48**, R973 (1993)

8. D.R. Phillips, S.J. Wallace: Few-Body Systems **24**, 175 (1998)

9. F. Gross: In: *Relativistic Quantum Mechanics and Field Theory*, Chapter 12. New York: John Wiley & Sons 1993

10. F. Gross, Y. Surya: Phys. Rev. **C47**, 703 (1993)

11. V. Pascalutsa, J.A. Tjon: Nucl. Phys. **A631**, 534c (1998); Phys. Lett. **B435**, 245 (1998)

12. V. Pascalutsa: Phys. Rev. **D58**, 076003 (1998)

13. R.A. Arndt, I. Strakovsky, R.L. Workman: Phys. Rev. **C52**, 2120 (1995)

14. R. Koch, E. Pietarinen: Nucl. Phys. **A336**, 331 (1980)

Few-Body Systems Suppl. 10, 115–118 (1999)

Few-
Body
Systems
© by Springer-Verlag 1999

Null-Plane Invariance of Hamiltonian Null-Plane Dynamics

F. Coester[1], W.H. Klink[2] and W.N. Polyzou[2]

[1] Physics Division, Argonne National Laboratory, Argonne IL 60439, USA
[2] Department of Physics, University of Iowa, Iowa City IA 52242, USA

Relativistic Hamiltonian few-body dynamics [1, 2] involves two unitary representations of the Poincaré group on the Hilbert space $\mathcal{H}$ of physical states, with and without interactions. These two representations, $U(\Lambda, a)$ and $U_0(\Lambda, a)$, coincide for a kinematic subgroup H. The "Hamiltonians" are the generators not in the Lie algebra of the kinematic subgroup. The kinematic subgroup of null-plane dynamics leaves the null-plane $z \cdot x \equiv x^0 + x_3 = 0$ invariant.

Few-body Hamiltonians satisfying the required commutation relations can be constructed as functions of a mass operator and kinematic quantities. For more than two particles there are nontrivial problems in satisfying cluster separability [3]. Consistency of electro-weak interactions with strong interactions also involves significant problems: Poincaré covariance of current operators requires the construction of appropriate interaction currents.

In formal Fock-space representations of local field theories, the Poincaré generators are obtained by integrating the energy momentum tensor over the hyper-surface that is invariant under the kinematic subgroup [4]. The elements of the Lie algebra so constructed are bilinear functionals over the Fock space, but they are not self-adjoint Fock-space operators. Thus there is no unitary Poincaré representation, $U(\Lambda, a)$, on the Fock space.

In the context sketched above it is useful to consider the manifold of null planes $n \cdot x \equiv x^0 + \hat{n} \cdot \boldsymbol{x} = 0$ parameterized by the unit vectors $\hat{n}$, and treat the orientations $\hat{n}$ of the null planes as an additional unphysical degree of freedom [5, 6].

The essential mathematical feature is the coset decomposition $\Lambda = \mathcal{R}_H \Lambda_H$, $\Lambda_H \in H$ for all Lorentz transformations Λ, where Λ_H leaves the null-plane $z \cdot x = 0$ invariant and the rotations $\mathcal{R}_H$ are parameterized by the components of the spatial unit vector $\hat{n}$, $n := \{-1, \hat{n}\} = \mathcal{R}(n)\{-1, 0, 0, 1\}$. The Poincaré transformations Λ_{H_n}, a_{H_n} that leave the null plane $n \cdot x = 0$ invariant are related to Λ_H, a_H by $\Lambda_{H_n} = \mathcal{R}(n)\Lambda_H \mathcal{R}^{-1}(n)$ and $a_{H_n} = \mathcal{R}(n)a_H$. The corresponding coset representatives are $\mathcal{R}_{H_n}(\Lambda) = \Lambda \Lambda_{H_n}^{-1}$. It follows that

$$\mathcal{R}(\Lambda n) = \Lambda \mathcal{R}(n) \Lambda_H^{-1} = \mathcal{R}_H(\Lambda) \Lambda_H \mathcal{R}(n) \Lambda_H^{-1} \ .$$

It is easy to verify that the unitary operators $C(n) := U_0[\mathcal{R}(n)]U^\dagger[(\mathcal{R}(n)]$ satisfy the intertwining relations $U_0(\Lambda)C(n) = C(\Lambda n)U(\Lambda)$ [7]. The representations $U_n(\Lambda, a) \equiv \exp(iP_n \cdot a)U_n(\Lambda) := C(n)U(\Lambda, a)C^\dagger(n)$ are identical to $U_0(\Lambda, a)$ for $\Lambda, a \in H_n$. The fact that the coset representatives form a compact manifold is a special feature of null-plane dynamics.

The treatment of the null-plane orientation $\hat{n}$ as a degree of freedom depends on the existence of an irreducible unitary representation, $\tilde{U}(\Lambda)$, on a Hilbert space $\tilde{\mathcal{H}}$ of functions $f(\hat{n})$, with the inner product $(f, g) = \int d^2 \hat{n} f^*(\hat{n}) g(\hat{n})$. Translations are represented by the identity. The extended Hilbert space of states is the tensor product $\bar{\mathcal{H}} := \mathcal{H} \otimes \tilde{\mathcal{H}}$. On the extended Hilbert space we have the Poincaré representation

$$\bar{U}(\Lambda, a) = \exp(i\bar{P} \cdot a)[U_0(\Lambda) \otimes \tilde{U}(\Lambda)]$$

where only the momentum operator, $\bar{P}^\mu := \int dE_n P_n^\mu$, depends on the dynamics.

From properties of the representation $U(\Lambda, a)$ follows the existence of generalized wave operators, $\Omega_\pm$, which map the Hilbert space $\mathcal{H}$ onto the Hilbert space $\mathcal{H}_f$ of free physical particles [3]. The corresponding extended wave operators, $\bar{\Omega}_\pm := \int dE_n C(n) \Omega_\pm$ map $\bar{\mathcal{H}}$ onto $\bar{\mathcal{H}}_f := \mathcal{H}_f \otimes \tilde{\mathcal{H}}$ with the Poincaré representation

$$\bar{U}_f(\Lambda, a) := \exp(i\bar{P}_f \cdot a)[U_f(\Lambda) \otimes \tilde{U}(\Lambda)] \ .$$

The operators $\bar{P}_f$ and $\bar{S} := \bar{\Omega}_+^\dagger \bar{\Omega}_-$ commute with $\tilde{U}(\Lambda)$. Assuming existence and completeness of the wave operators, null-plane independence of the observables, $[\tilde{U}(\Lambda), \bar{\Omega}^\dagger \bar{P}^\mu \Omega] = 0$ and $[\tilde{U}(\Lambda), \bar{S}] = 0$, is both necessary and sufficient for the existence of the unitary operator $C(n)$,

$$C(n)dE_n = \bar{\Omega}_+ dE_n U_f^\dagger[\mathcal{R}(n)]\bar{\Omega}_+^\dagger U_0[\mathcal{R}(n)] = \bar{\Omega}_- dE_n U_f^\dagger[\mathcal{R}(n)]\bar{\Omega}_-^\dagger U_0[\mathcal{R}(n)].$$

Attempts to construct Poincaré covariant Hamiltonian dynamics from the Fock-space representation of a local Lagrangian requires additional considerations [8]. Since for a local interaction Lagrangian $\int d^3\mathbf{x} \mathcal{L}(\mathbf{x})\Psi$ is not in the Fock space, $\mathcal{F}$, for any $\Psi \in \mathcal{F}$, there is no unitary Poincaré representation on the Fock space. There is, of course, the free-field unitary representation $U_0(\Lambda)$ and it is possible to regularize the interaction in such a manner that the four-momentum $\bar{P}$ is a covariant self-adjoint operator with commuting components. That much is sufficient for the existence of a unitary representation $\bar{U}(\Lambda, a)$ on $\mathcal{F} \otimes \tilde{\mathcal{H}}$. For a renormalizable field theory, it should be straightforward to verify the null-plane independence of perturbative results in the limit of infinite cutoff. The framework we have outlined may provide a systematic approach to non-perturbative approximations in suitably defined subspaces of the Fock space.

Null-plane dynamics has the unique feature that, for space-like momentum transfer the impulse approximation is kinematically consistent, provided the null-plane orientations are restricted to $n \cdot Q = 0$. The invariant $Q^2 = Q_\perp^2$ depends only on the kinematic components of the momentum operator, and the momenta of all contributing constituents are related by kinematic Lorentz transformations. The treatment of the null-plane orientation so restricted as an additional degree of freedom provides a convenient framework for the determination of "minimal" interaction currents required for full Lorentz covariance. The operator $N^\mu := n^\mu/(n \cdot P)$ transforms as a four vector and commutes with the Fourier transform $\tilde{I}^\mu(Q)$ of the current density operator. Since $N^2 = 0$, $N \cdot Q = 0$ and $N \cdot P = 1$ there are no $\hat{n}$-dependent invariants. The projection, $\tilde{P}(Q)$, of the four-momentum operator into the hyperplane perpendicular to Q also commutes with $\tilde{I}^\mu(Q)$. Let the boosts $B(\tilde{P}, Q)$ be defined by $Q = B(\tilde{P}, Q)\{0, 0, 0, \sqrt{Q^2}\}$ and $\tilde{P}(Q) = B(\tilde{P}, Q)\{M(Q^2), 0, 0, 0\}$. Define $\mathcal{N} := B^{-1}(\tilde{P}, Q)N\tilde{M}(Q^2) = \{-1, \cos\varphi, \sin\varphi, 0\}$ and $\mathcal{I} := B^{-1}(\tilde{P}, Q)\tilde{I}(Q)$. Under Lorentz transformations the operator $\mathcal{I}(Q^2, \mathcal{N})$ undergoes $SO(2)$ Wigner rotations, $\mathcal{R}_W(\Lambda; \tilde{P}, Q) := B^{-1}(\Lambda\tilde{P}, \Lambda Q)\Lambda B(\tilde{P}, Q)$. The operators $\mathcal{I}^\beta$, $\beta = \{0, \pm1\}$ transform respectively as an $O(2)$ scalar and vector and their matrix elements are $O(2)$ Clebsch-Gordan coefficients multiplied by invariant form factors, $\langle\sigma', j'|\mathcal{I}^\beta(Q^2, \mathcal{N})|j, \sigma\rangle = e^{im\varphi}\delta(m + \beta + \sigma - \sigma')F(Q^2; |\beta|, m, |\sigma|))$ The null-plane dependent part of the current, $m \neq 0$ must be cancelled by "minimal" interaction currents. This construction is equally applicable to impulse currents and "model dependent" meson exchange currents.

Equivalent results obtain by consideration of $O(1, 3)$ covariant current kernels $(\rho', j', p'|I^\mu(0, N^\nu)|p, j, \rho)$ where ρ denotes generalized spinor indices. Such kernels are linear combinations of invariants multiplied by irreducible tensors which may involve monomials of N. Since there are no N-dependent invariants the null-plane independent part of the current is well defined [6].

Acknowledgement. Supported in part by the U.S. Department of Energy, Nuclear Physics Division, contracts W-31-109-ENG-38 and DE-FG02-86ER40286.

Appendix A. Unitary Representation of $SO(1,3)$ and Degenerate Representation of the Poincaré Group

Let $\hat{n}$ be a unit vector parameterized by an azimuthal angle φ and $n_z = \cos\theta$.

$$\hat{n}(n_z, \varphi) := \{\sqrt{1 - n_z^2}\cos\varphi, \sqrt{1 - n_z^2}\sin\varphi, n_z\}$$

The Hilbert space $\tilde{\mathcal{H}}$ of square integrable functions has the inner product

$$(f, g) := \int_0^{2\pi} d\varphi \int_{-1}^{+1} dn_z f^*(n_z, \varphi)g(n_z, \varphi) = \int d^2\hat{n} f^*(\hat{n})g(\hat{n}), \quad \forall f, g \in \tilde{\mathcal{H}}.$$

The Lorentz transformation $\Lambda \in O(1, 3)$ acts on $\tilde{H}$ according to

$$[\tilde{U}(\Lambda)f](\hat{n}) = s(\Lambda; \hat{n})f(\hat{n}'), \quad \text{where} \quad s(\Lambda, \hat{n})\{-1, \hat{n}'\} := \Lambda^{-1}\{-1, \hat{n}\}$$

118

defines how the Lorentz transformation $\Lambda \in SO(1,3)$ changes $\hat{n} \to \hat{n}'$ and induces a scale factor $s(\Lambda, \hat{n})$. For space reflection and time reversal we have $[\mathcal{P}f](\hat{n}) = f(-\hat{n})$ and $[\mathcal{PT}f](\hat{n}) = f^*(\hat{n})$

The group property of the scale factors,

$$s(\Lambda_1 \Lambda_2, \hat{n}) = s(\Lambda_1, \hat{n}) s(\Lambda_2, \hat{n}')$$

follows from their definition.

$$s(\Lambda_1 \Lambda_2, \hat{n})\{-1, \hat{n}''\} = \Lambda_2^{-1} s(\Lambda_1, \hat{n})\{-1, \hat{n}'\} = s(\Lambda_1, \hat{n}) s(\Lambda_2, \hat{n}')\{-1, \hat{n}''\}$$

The Jacobians $\mathcal{J}(\Lambda, \hat{n})$,

$$\mathcal{J}(\Lambda, \hat{n}) := \frac{\partial(u', \varphi')}{\partial(u, \varphi)} \equiv \frac{d^2 \hat{n}'}{d^2 \hat{n}}.$$

have the same group property,

$$\mathcal{J}(\Lambda_1 \Lambda_2, \hat{n}) = \mathcal{J}(\Lambda_1, \hat{n}) \mathcal{J}(\Lambda_2, \hat{n}')$$

Thus

$$(\tilde{U}(\Lambda)f)(\hat{n}) = \sqrt{\mathcal{J}(\Lambda, \hat{n})} f(\hat{n}')$$

defines a unitary, irreducible representation of the Lorentz group $\forall f \in \tilde{\mathcal{H}}$.

$$\|\tilde{U}(\Lambda)f\|^2 = \int d^2 \hat{n} \mathcal{J}(\Lambda, \hat{n}) |f(\hat{n}')|^2 = \int d^2 \hat{n}' |f(\hat{n}')|^2 = \|f\|^2$$

Improper basis states $|\hat{n}\rangle$, transform according to

$$\tilde{U}(\Lambda)|\hat{n}\rangle = |\hat{n}'\rangle \, s(\Lambda, \hat{n})$$

so that the spectral projector $dE_n := d^2 \hat{n} |\hat{n}\rangle\langle\hat{n}|$ is invariant.

References

1. P. A. M. Dirac: Reviews of Modern Physics **21**, 392 (1949)

2. B.D. Keister and W. N. Polyzou: Adv. in Nucl. Phys. **20**, 225 (1991)

3. F. Coester and W.N. Polyzou: Phys. Rev. **D26**, 1348 (1982)

4. J. Schwinger: Phys. Rev. **127**, 324 (1962)

5. V.A. Karmanov: ZhETF **71**, 399 (1976) [transl.: JETP **44**, 210 (1976)]

6. J. Carbonell et al.: Phys. Rep. **300**, 215 (1998)

7. M. Fuda: Ann. Phys. **197**, 265 (1990)

8. F. Coester: Prog. Part. Nuc. Phys. **29**, 1 (1992)

Few-Body Systems Suppl. 10, 119–122 (1999)

Few-
Body
Systems
© by Springer-Verlag 1999

Entanglement of Fock-Space Expansion and Covariance in Light-Front Dynamics

Ben Bakker and Nico Schoonderwoerd

Vrije Universiteit, Amsterdam, The Netherlands

Abstract. The entanglement of the expansion of operators in Fock space with covariance is discussed in light-front (LF) dynamics. Numerical examples are given for a scalar Yukawa model.

1 Advantages and disadvantages of Light-Front Dynamics

Two options are available to treat relativistic systems: Lagrangian or Hamiltonian methods. The Lagrangian method may be preferred if only S-matrix elements are needed. For bound states the Hamiltonian methods are intuitively more appealing. Light-front dynamics (LFD) is a Hamiltonian description of relativistic systems. It makes use of a Hamiltonian to describe states, both scattering states and bound states. In this form of dynamics, already discussed by Dirac[1] one quantizes the theory on planes of equal LF time x^+. LF coordinates are defined by:

$$x^+ = (x^0 + x^3)/\sqrt{2}, \quad x^- = (x^0 - x^3)/\sqrt{2}, \quad x^\perp = (x^1, x^2).$$

Using the plane $x^+ = 0$ as the quantization plane one finds that the LF energy operator is

$$p^- = \frac{p_\perp^2 + m^2}{2p^+}.$$

Positive LF energy is kinematically distinguished from negative LF energy by the sign of p^+. Therefore one can limit all states in LFD to those where all particles have non-negative p^+. This so-called spectrum condition limits the possible time-orderings and is believed to play an important role in making contributions from higher Fock sectors numerically smaller in LFD than in the ordinary, instant form of dynamics (IFD). In LFD the momentum components p^+ and $p^\perp$ are conserved, so any state must obey

$$p_i^+ > 0, \quad \sum_i p_i^+ = P^+.$$

LFD has some drawbacks: covariance is not *manifest*, a feature it shares with all Hamiltonian formalisms, in particular rotational invariance. The reason for this lack of manifest rotational invariance is clear. While in IFD J is *kinematical*, i.e., it is free of interactions, in LF dynamics J^3 is the only *kinematical* component of J. On the other hand, in LFD K^3 the boost in the z-direction is also free of interaction, while in IFD K is dynamical. This means that one has to be more careful when using LFD to maintain rotational invariance. It can only be done when the interaction is treated correctly. It is the purpose of this work to show numerically the importance of the connection between the dynamical input and covariance.

2 Boson Exchange as Dynamical Model

We studied the boson-exchange model that is used extensively in nuclear and particle physics. Our focal point here is meson exchange in nuclear physics.

Boson exchange potentials are usually used in the ladder approximation. In that case corrections are needed for restoration of covariance. These are related to the expansion in Fock space.

In covariant perturbation theory, using Feynman diagrams, the S-matrix can be expanded in powers of the coupling constant. This expansion is covariant order by order. The Fock-state expansion, i.e., an expansion in the number of particles, is not, as the interaction is not diagonal in Fock space. The Fock-space expansion may depend on the quantization procedure, so it makes sense to study it specifically in LFD.

We studied numerically the Yukawa model with the interaction Lagrangian

$$\mathcal{L}_{\text{int}} = g\,\Phi(x)\Phi(x)\phi(x).$$

We did not include intrinsic spin to keep the discussion as simple as possible. The particles involved are Φ, "nucleons", mass $m = 940$ MeV/c^2 and ϕ, "mesons", mass $\mu = 140$ MeV/c^2. The covariant one-boson exchange diagram is split into two LF time-ordered diagrams: ‾| ‾ = ‾\‾ + ‾/‾ . Formally this means that Feynman propagators are replaced by LF energy denominators:

$$\frac{1}{q^2 - \mu^2 + i\epsilon} \rightarrow \frac{1}{2P^+}\left(\frac{1}{P^- - H_0 + i\epsilon}\right), \quad H_0 = \sum_i \frac{p_{i\perp}^2 + m_i^2}{2p_i^+}.$$

Upon iteration one obtains ladder diagrams. Next in line is the covariant box diagram, which is the sum of six LF time-ordered diagrams (upper left panel of Fig. 1). Iterated LF time-ordered OBEP diagrams occur, for obvious reasons denoted as trapezia and diamonds. The other two, the stretched boxes, are not obtained by iterating LF time-ordered diagrams, although they are obtained upon splitting the covariant box diagram in time-ordered ones.

If one would use the OBEP as a driving term in a Lippmann-Schwinger equation, one would have to make a Fock-space expansion of the interaction $V = V_1 + V_2 + \dots$ such that V_1 consists of all contributions with at most one meson exchanged at a time, and V_2 can have two mesons in the air, et cetera.

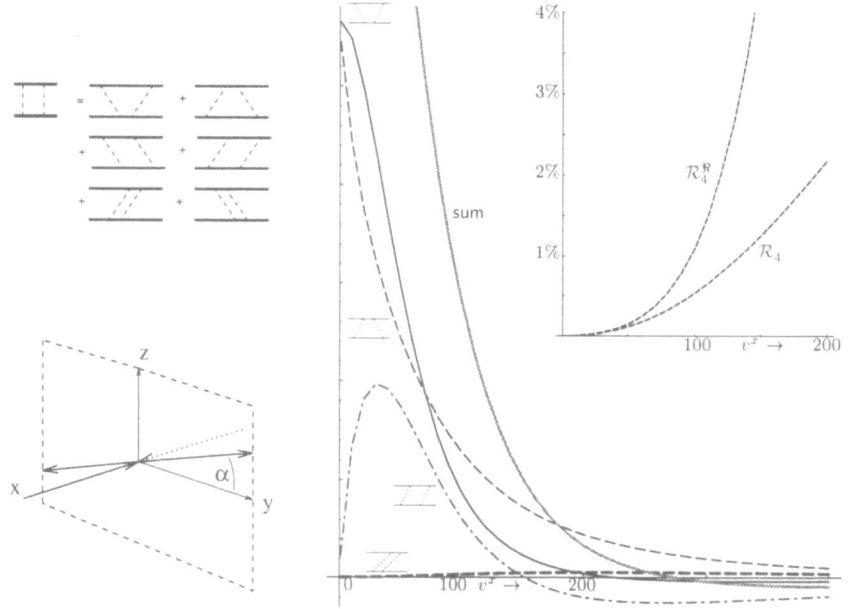

Figure 1. Upper left: expansion of the covariant box diagram. Lower left: kinematics of two-particle scattering. α is the azimuthal angle and z the quantization axis. Right: the ratios $\mathcal{R}_4$ and $\mathcal{R}_4^{\text{Re}}$ with azimuthal angle $\alpha = 90°$.

3 Numerical results and conclusions

Using this model we calculated all box diagrams and compared them to the covariant box. Above threshold the covariant diagram is an S-matrix element. It is complex. We found that the sum of LF time-ordered diagrams is equal to the covariant diagram. This is in agreement with the result proven by Ligterink[2].

Now we show, that the different time-ordered contributions individually violate rotational invariance, because they depend on the orientation of the LF. We studied the situation where the nucleons come in along the x-axis, and scatter by 90° with an azimuthal angle α (lower left panel of Fig. 1).

In Fig. 2 we show the contributions of the time-ordered diagrams, normalized to the covariant amplitude. We see that the LF time-ordered diagrams depend on α, while the covariant diagram does not, as it should not.

The energy dependence of the different diagrams is shown in the right panel of Fig. 1. The quantities $\mathcal{R}_4$ and $\mathcal{R}_4^{\text{Re}}$ are the ratios of the stretched boxes to the magnitude of the covariant amplitude or its real part, respectively. The amplitudes are all calculated for $\alpha = 90°$ for which angle the stretched boxes become maximal.

From the results of our computations we draw the following conclusions. LFD gives covariant results order by order, as the LF time-ordered amplitudes

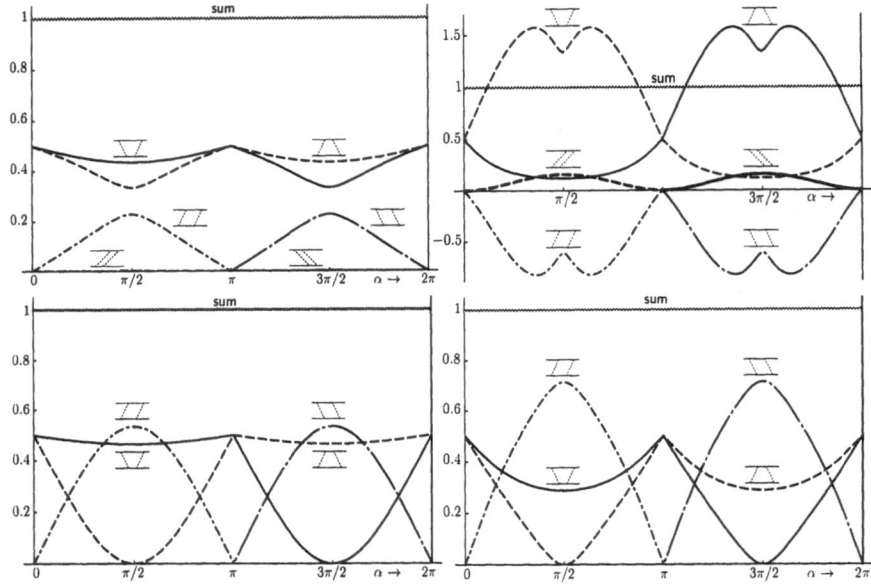

Figure 2. Real (up) and imaginary (down) part of the LF time-ordered diagrams as a function of α: (left) $p_{CMS} = 40$ MeV/c^2, (right) $p_{CMS} = 200$ MeV/c^2.

add up to the covariant one. The Fock-space expansion is entangled with covariance, because in no way can we obtain an invariant amplitude from diagrams that are restricted in Fock space. If we do, then we find that limiting the Fock space sectors included to two-nucleon and two-nucleon–one-meson one finds that the violation of rotational symmetry is small for p_{CMS} below 200 MeV.

In order to do practical calculations one has to truncate Fock space somewhere. The outlook of our work is to describe bound states in LFD. Then one needs to check the Fock-state expansion for off-shell states. Based on the present work [3] we expect violation of rotational invariance to be small.

References

1. P.A.M. Dirac: Rev. Mod. Phys. **21**, 392 (1949)

2. N.E. Ligterink and B.L.G. Bakker: Phys. Rev. **D 52**, 5954 (1995)

3. N.C.J. Schoonderwoerd, B.L.G. Bakker and V.A. Karmanov: submitted to Phys. Rev. **C**, **hep-ph/9806365** (1998)

Few-Body Systems Suppl. 10, 123–126 (1999)

Few-
Body
Systems
© by Springer-Verlag 1999

Ladder Bethe-Salpeter equation on the Light-Front

T. Frederico[1], J.H.O. Sales[1], B.V. Carlson[1], P.U. Sauer[2]

[1] Dep. de Física, Instituto Tecnológico de Aeronáutica, Centro Técnico Aeroespacial, 12.228-900 São José dos Campos, São Paulo, Brazil
[2] Institute for Theoretical Physics, University Hannover, D-30167 Hannover, Germany

Abstract. We study the ladder Bethe-Salpeter (BS) equation on the light-front for two-bosons exchanging a third one. The kernel is constructed from the perturbative light-front propagator of the two-boson system up to $O(g_S^4)$. Intermediate states with up to four particles are considered. We compare the numerical solutions of the light-front and four-dimensional BS equations for the two-particle bound state mass and wave-function.

1 Introduction

It is well known that light-front perturbation theory is equivalent to a covariant perturbative expansion in field theory [1, 2]. Thus, in principle, it is possible to find the exact representation of the four-dimensional BS equation in the light-front. This is achieved when all the intermediate states with any number of exchanged bosons are allowed in the kernel, which corresponds to consider the light-front propagator to all orders in g_S^2 in the ladder. The formal proof of the above statement demands the construction of the light-front two-body Green's function to any order in the ladder [3]. The truncation of the Fock-state breaks the covariance of the light-front BS equation. Covariance is approached by increasing the number of particles in the intermediate states.

For our purpose we use a bosonic model for which two bosons of mass m exchange an intermediate boson of mass μ. The coupling constant is g_S. We solve numerically the two-particle bound state BS equation, considering up to four particles in lowest order in the intermediate states and compare to the four-dimensional solution in the ladder approximation. The numerical solution of the light-front BS equation with truncated kernel containing up to three-particles has been obtained already [4].

2 Light-Front BS Equation

The irreducible two-body Green's function that has intermediate propagation in the light-front time of states with three and four particles is found from the perturbative ladder corrections up to $\mathcal{O}(g_S^4)$ of the two-body propagator between two light-front hypersurfaces with $x^+ = 0$ and $x^+ > 0$. In this approximation, the kernel of the light-front BS equation has two parts, one corresponds to the propagation of three virtual particles, the other to the propagation of four virtual particles. Consequently, the bound state vertex as a function of the light-front variables obeys the integral equation,

$$F_{LF}(\mathbf{q}_\perp, y) = \int \frac{d\mathbf{k}_\perp dx}{x(1-x)} \frac{K^{(3)}(\mathbf{q}_\perp, y; \mathbf{k}_\perp, x) + K^{(4)}(\mathbf{q}_\perp, y; \mathbf{k}_\perp, x)}{M_2^2 - M_0^2} F_{LF}(\mathbf{k}_\perp, x),$$

(1)

where M_2 is the bound state mass, $0 \le x \le 1$ and the free two body mass is $M_0^2 = (\mathbf{k}_\perp^2 + m^2)/x(1-x)$.

The part of the kernel containing three virtual particles propagating forward in the light-front time, is given by:

$$K^{(3)}(\mathbf{q}_\perp, y; \mathbf{k}_\perp, x) = \frac{g_S^2}{16\pi^3} \frac{\theta(x-y)}{(x-y)\left(M_2^2 - \frac{\mathbf{q}_\perp^2 + m^2}{y} - \frac{\mathbf{k}_\perp^2 + m^2}{1-x} - \frac{(\mathbf{q}_\perp - \mathbf{k}_\perp)^2 + \mu^2}{x-y}\right)}$$
$$+ \; [x \leftrightarrow y, \mathbf{k}_\perp \leftrightarrow \mathbf{q}_\perp].$$

(2)

The contribution to the kernel from the four-body virtual propagation is given by:

$$K^{(4)}(\mathbf{q}_\perp, y; \mathbf{k}_\perp, x) = \left(\frac{g_S^2}{16\pi^3}\right)^2$$

$$\times \int \frac{d\mathbf{p}_\perp dz}{z(1-z)(z-x)(y-z)} \frac{\theta(z-y)\theta(x-z)}{\left(M_2^2 - \frac{\mathbf{q}_\perp^2 + m^2}{y} - \frac{\mathbf{p}_\perp^2 + m^2}{1-z} - \frac{(\mathbf{q}_\perp - \mathbf{p}_\perp)^2 + \mu^2}{z-y}\right)}$$

$$\times \frac{1}{\left(M_2^2 - \frac{\mathbf{q}_\perp^2 + m^2}{y} - \frac{\mathbf{k}_\perp^2 + m^2}{1-x} - \frac{(\mathbf{q}_\perp - \mathbf{p}_\perp)^2 + \mu^2}{z-y} - \frac{(\mathbf{p}_\perp - \mathbf{k}_\perp)^2 + \mu^2}{x-z}\right)}$$

$$\times \frac{1}{\left(M_2^2 - \frac{\mathbf{p}_\perp^2 + m^2}{z} - \frac{\mathbf{k}_\perp^2 + m^2}{1-x} - \frac{(\mathbf{p}_\perp - \mathbf{k}_\perp)^2 + \mu^2}{x-z}\right)} + [x \leftrightarrow y, \mathbf{k}_\perp \leftrightarrow \mathbf{q}_\perp].$$

(3)

The covariant Bethe-Salpeter equation in the ladder approximation is:

$$F(q^\mu) = ig_S^2 \int \frac{d^4k}{(2\pi)^4} \frac{F(k^\mu)}{((q-k)^2 - \mu^2 + i\varepsilon)(k^2 - m^2 + i\varepsilon)((P-k)^2 - m^2 + i\varepsilon)},$$

(4)

where $P^\nu = (M_2, \mathbf{0})$ is the total momentum. We solve the four-dimensional equation in Euclidean space. To avoid the nontrivial analytical continuation from the Euclidean to Minkowski space, we define a transverse distribution by

$$f(q) = \int_0^1 \frac{dx}{x(1-x)} \frac{F_{LF}(\mathbf{q}_\perp, x)}{M_2^2 - \frac{\mathbf{q}_\perp^2 + m^2}{x(1-x)}}$$

$$= \int dq^0 dq^3 \frac{F(q^\mu)}{(q^2 - m^2 + i\varepsilon)((P - q)^2 - m^2 + i\varepsilon)} , \tag{5}$$

which can be easily calculated from the Euclidean solution, $q = \mathbf{q}_\perp$.

3 Numerical Results

We found numerically the S-wave solution of Eqs. (4) and the light-front equation, (1), with *i)* three and *ii)* up to four particles in the intermediate state. We use $m = 1$ and $\mu = 0.5$. The coupling constant g_S as a function of M_2 is shown in Fig. 1. In the strong-binding limit, the coupling constants obtained solving the case *i)* is few-percent greater that the solution *ii)*. The higher Fock-state components contribute to the kernel of Eq. (1) attractively. Including up to four-particles in the intermediate state the covariant result is practically obtained. The coupling constants of our four-dimensional calculations agree with the results of ref. [5, 6]. Observe that if one fixes the coupling constant and looks for the mass of the bound state, the effect of the higher Fock-state components is noticeable in the strong-binding limit. In the weak-binding limit, the presence of higher Fock-state components are unimportant for the coupling constant or for the two-body binding. In Fig. 2, the results for the transverse distribution

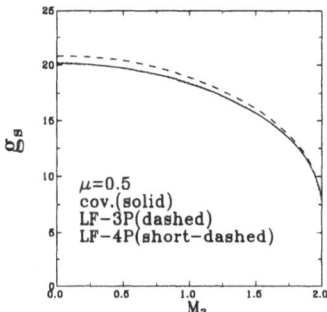

Figure 1. g_S as a function of the bound state mass M_2 for $\mu = 0.5$. Covariant BS (cov.); light-front BS with intermediate three-particle states (LF-3P) and with four-particle states (LF-4P).

are shown; $f(0)$ is normalized to 1. For small transverse momentum all results are in agreement, because in this region the wave-functions are dominated by the value of the mass of the bound-state, which is the same in all calculations. With the increase of q, deviations are present but small. The effect of higher Fock-state intermediate states is to approach the four-dimensional results. In the weak binding limit, we verified that the four-dimensional result is well approximated by the light-front BS solution with the kernel given by Eq. (2). Thus, the higher Fock-states are unimportant in the weak binding limit.

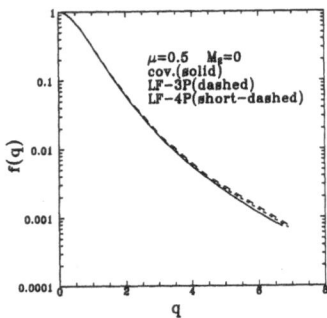

Figure 2. Transverse distribution as a function of q for $M_2 = 0$ and $\mu = 0.5$. Explanation of the curves as in Fig. 1.

4 Conclusion

We compared the numerical solutions of the S-wave light-front and four-dimensional BS equation for the two-body bound-state in the ladder aproximation. We show that the four-dimensional results are well approximated by the light-front solution of the BS equation with up to four particles in the intermediate state in lowest order. Moreover, the higher Fock states are irrelevant in the weak binding limit for the bound state mass and transverse distribution.

Acknowledgement. This project was supported by CAPES/DAAD/Probral (project 015/95) and the Brazilian agencies FAPESP and CNPq. T.F. thanks the hospitality of the Institute for Theoretical Physics, University of Hannover, where part of this work was done.

References

1. S.-J. Chang and T.-M. Yan: Phys. Rev. **D7**, 1147 (1973); T.-M. Yan: *ibid.* **D7**, 1780 (973)

2. N.E.Ligterink and B.L.G. Bakker: Phys. Rev. **D52**, 5917 (1995); Phys. Rev. **D52**, 5954 (.1995); N.C.J. Schoonderwoerd and B.L.G. Bakker: Phys. Rev. **D57**, 4965 (1998); hep-ph/9801433 to appear in Phys. Rev. **D**

3. J.H.O. Sales, T.Frederico, B.V. Carlson and P.U.Sauer: in preparation

4. C.-R. Ji and R.J. Furnstahl: Phys. Lett. **B167**, 11 (1986); C.-R. Ji: Phys. Lett. **B167**, 16 (1986); C.-R. Ji: Phys. Lett. **B322**, 389 (1994)

5. T. Nieuwenhuis and J.A. Tjon: Few-Body Syst. **21**, 167 (1996); T. Nieuwenhuis and J.A. Tjon: Phys. Rev. Lett. **77**, 814 (1996)

6. K.Kusaka, K.Simpson, A.G. Willians: Phys. Rev. **D56**, 5071 (1997)

Few-Body Systems Suppl. 10, 127–130 (1999)

Few-
Body
Systems
© by Springer-Verlag 1999

3D reduction of the three-fermion Bethe-Salpeter equation

J. Bijtebier*†

Theoretische Natuurkunde, Vrije Universiteit Brussel, Pleinlaan 2 B-1050 Brussel, Belgium

Abstract. We present a 3D approximation of the three-fermion Bethe-Salpeter equation. Our 3D equation is covariantly cluster separable and the two-fermion cluster separated limits are exact equivalents of the corresponding two-fermion Bethe-Salpeter equations. The potentials include positive free energy projectors in order to avoid continuum dissolution.

1 Introduction

The elimination of the relative times in the three-fermion Bethe-Salpeter equation can be performed in many ways. This equation can for example be approximated by 3D Schrödinger-Pauli or Faddeev equations. In principle a lot of higher-order correction terms of various origins, often neglected, should restore the equivalence with the initial Bethe-Salpeter equation. We are searching for a 3D equation which would be an element in a chain of approximations transforming the original Bethe-Salpeter equation into a manageable equation and would also satisfy at best the following list of requirements: Lorentz invariance, cluster separability, hermiticity and slow energy dependence of the potentials, correct heavy mass limits, absence of continuum dissolution. The solutions of the corresponding two-fermion problem will provide the building blocks of our three-fermion equation.

2 The two-fermion problem

The Bethe-Salpeter equation for the bound states of two fermions is

$$\Phi = G_0 K \Phi$$

*Senior Research Associate at the Fund for Scientific Research (Belgium).
† *E-mail address:* jbijtebi@vub.ac.be

where Φ is the Bethe-Salpeter amplitude, K the Bethe-Salpeter kernel (sum of the irreducible Feynman graphs) and

$$G_0 = G_{01}G_{02}, \qquad G_{0i} = \frac{1}{p_{i0} - h_i + i\epsilon h_i}\,\beta_i$$

the free propagator. The h_i are the free Dirac hamiltonians

$$h_i = \boldsymbol{\alpha}_i \cdot \boldsymbol{p}_i + \beta_i\, m_i \qquad (i = 1, 2).$$

We shall denote the total and relative momenta, and the corresponding combinations of the free hamiltonians by

$$P = p_1 + p_2 \ , \quad p = \frac{1}{2}(p_1 - p_2), \quad S = h_1 + h_2 \ , \quad s = \frac{1}{2}(h_1 - h_2).$$

We shall also need the positive and negative free energy projectors:

$$\Lambda^{\pm\pm} = \Lambda_1^\pm \Lambda_2^\pm, \quad \Lambda_i^\pm = \frac{E_i \pm h_i}{2E_i}, \qquad E_i = \sqrt{h_i^2} = (\boldsymbol{p}_i^2 + m_i^2)^{\frac{1}{2}}.$$

The free propagator G_0 will be written as the sum of an approached propagator G_δ (combining a constraint fixing the relative energy, and a global 3D propagator) and a rest G_R. Salpeter's 3D propagator, which appears automatically in case of an "instantaneous kernel" is

$$\int dp_0 G_0(p_0) = -2i\pi\,\frac{\Lambda^{++} - \Lambda^{--}}{P_0 - S}\,\beta_1\beta_2.$$

We shall skip the Λ^{--} projector and write the free propagator as

$$G_0 = G_\delta + G_R, \qquad G_\delta(p_0) = -2i\pi\,\delta(p_0 - s)\,\frac{\Lambda^{++}}{P_0 - S}\,\beta_1\beta_2.$$

The Bethe-Salpeter equation becomes then the inhomogeneous equation

$$\Phi = \Psi + G_R K \Phi, \qquad \Psi = G_\delta K \Phi.$$

Eliminating Φ :

$$\Psi = G_\delta K(1 - G_R K)^{-1}\Psi = G_\delta K_T \Psi, \qquad K_T = K + K G_R K + \ldots$$

The reduction series K_T re-introduces in fact the reducible graphs into the Bethe-Salpeter kernel, but with G_0 replaced by G_R. The equation becomes

$$\Psi = G_\delta K_T \Psi = -2i\pi\,\delta(p_0 - s)\,\frac{\Lambda^{++}}{P_0 - S}\,\beta_1\beta_2\,K_T\,\Psi.$$

Eliminating the relative energy dependence gives a single 3D equation:

$$\Psi = \delta(p_0 - s)\,\psi, \qquad \psi = \frac{\Lambda^{++}}{P_0 - E_1 - E_2}\,V(P_0)\,\psi,$$

$$V(P_0) = -2i\pi\int dp_0' dp_0\,\delta(p_0' - s)\,\beta_1\beta_2 K_T(p_0', p_0, P_0)\,\delta(p_0 - s).$$

We explicitated the dependence of the operator K_T in the conserved total momentum P_0 and in the relative momentum p_0 (this last dependence being non-local).

Similar results are obtained with other constraints and 3D propagators.

3 The three-fermion problem

The Bethe-Salpeter equation is now:

$$\Phi = [G_{01}G_{02}K_{12} + G_{02}G_{03}K_{23} + G_{03}G_{01}K_{31} + G_{01}G_{02}G_{03}K_{123}]\,\Phi$$

where K_{123} is the sum of the purely three-body irreducible contributions. We shall neglect it and replace the two-body kernels by instantaneous ones, equivalent at the cluster-separated limits. Among an infinity of choices, we shall use directly the two-body potentials of the previous section, putting the spectator fermion on the mass shell:

$$K_{12}(p'_{120}, p_{120}, P_{120}) \approx \beta_1\beta_2\,\Lambda_{12}^{++}$$

$$\times \int dp'_{120}dp_{120}\,\delta(p'_{120} - s_{12})\beta_1\beta_2 K_{T12}(p'_{120}, p_{120}, P_0 - h_3)\,\delta(p_{120} - s_{12})\Lambda_{12}^{++}$$

$$= \frac{-1}{2i\pi}\,\beta_1\beta_2\,\Lambda_{12}^{++}\,V_{12}(P_0 - h_3)\,\Lambda_{12}^{++},\ldots$$

where P is now the total energy-momentum of the three-fermion system. The Bethe-Salpeter equation becomes

$$\Phi = \frac{-1}{2i\pi}\,G_{01}G_{02}G_{03}\,\beta_1\beta_2\beta_3\left[\Lambda_{12}^{++}\,V_{12}\,\Lambda_{12}^{++}\,\psi_{12} + \cdots + \cdots\right],$$

$$\psi_{ij}(p_{k0}) = \beta_k\,G_{0k}^{-1}\int dp_{ij0}\,\Phi.$$

This leads to a set of three coupled integral equations in the ψ_{ij}. We shall search for solutions analytical in the $\mathrm{Im}(p_{k0}) < 0$ half planes and close the integration paths clockwise in these planes. The only singularities will then be the poles of the free propagators. Performing the integrations with respect to the p_{ij0} gives then

$$\begin{aligned}\psi_{12}(p_{30}) &= \frac{\Lambda_{12}^{++}}{(P_0 - S) - (p_{30} - h_3) + i\epsilon}\left[\Lambda_{12}^{++}\,V_{12}\,\Lambda_{12}^{++}\,\psi_{12}(p_{30})\right.\\ &\quad + \left.\Lambda_{23}^{++}\,V_{23}\,\Lambda_{23}^{++}\,\psi_{23}(h_1) + \Lambda_{31}^{++}\,V_{31}\,\Lambda_{31}^{++}\,\psi_{31}(h_2)\right]\end{aligned}\quad(1)$$

and similarly for ψ_{23} and ψ_{31}. Solving (1) with respect to $\psi_{12}(p_{120})$ confirms its analyticity in the $\mathrm{Im}(p_{k0}) < 0$ half plane. Furthermore, Eq. (1) shows that the three projections $\Lambda_k^+\psi_{ij}(h_k)$ are equal (let us call them ψ) and satisfy the 3D equation

$$\psi = \frac{\Lambda^{+++}}{P_0 - E_1 - E_2 - E_3}\left[V_{12}(P_0 - E_3) + V_{23}(P_0 - E_1) + V_{31}(P_0 - E_2)\right]\psi.$$

Moreover, it can be shown that ψ is the integral of the Bethe-Salpeter amplitude with respect to the relative times:

$$\psi = \frac{-1}{2i\pi}\int dp_{10}dp_{20}dp_{30}\,\delta(p_{10} + p_{20} + p_{30} - P_0)\,\Phi.$$

4 Conclusions: pro's and con's of our three-cluster equation

— The positive-energy projectors included in the equation forbid the mixing of the physical bound states with a continuum combining positive and negative energy free states (equations suffering of this "continuum dissolution" disease have no normalisable solutions).

— When the two potentials acting on one of the fermions are "switched off", one gets a free Dirac equation for this fermion and a correct two-fermion equation for the two other fermions (cluster separability). Furthermore, this last equation is an exact 3D equivalent of the two-fermion Bethe-Salpeter equation. Our approximations concern thus only three-body terms (primary ones at the Bethe-Salpeter level (in K_{123}) or generated by the 3D reduction).

— We did not specify our reference frame until now. Our equations are not explicitly covariant, but if we assume that they are written in the three-fermion rest frame, we can always render them covariant by using the conserved total energy-momentum vector P. At the cluster-separated limits, however, the cluster separability requirement forbids the use of this vector. The equations for the two-fermion clusters must then be covariant. The fact that these equations are exact equivalents of the covariant two-fermion Bethe-Salpeter equations insures an implicit covariance without introducing Lorentz boosts by hand.

— The 3D reduction of the two-fermion Bethe-Salpeter equation of Sec. 2 is only an example. The requirement of preserving the equivalence with the original equation leaves a large freedom which could be used to suit the needs of the three-fermion phenomenology.

— The potentials are hermitian and their dependence in the total energy is an higher-order effect.

— When the mass of one of the fermions becomes infinite, its presence should be translated in the equations by a potential (Coulombian in QED) acting on the other fermions. This requirement is only approximately satisfied. Satisfying it exactly would demand the reintroduction of some of the neglected three-body terms.

— Our two-body potentials are the sum of an infinity of contributions symbolized by Feynman graphs. Keeping only the first one (Born approximation) or a finite number of them renders the Lorentz covariance of the two-fermion clusters only approximate. One can use another 3D reduction based on a covariant second-order two-body propagator of Sazdjian, combined with a covariant substitute of Λ^{++}. This leads to a 3D three-cluster equation which is covariantly Born approximable, but more complicated.

— Our equation can also be written as a set of three Faddeev equations. These Faddeev equations can also be obtained as an approximation of Gross' spectator model equations. This approximation being of the same order than these already made in Gross' model, further investigations would be needed to decide which model is closer to the exact Bethe-Salpeter equation.

— The higher-order three-body contributions we neglected in our approximation are explicitly given at the Bethe-Salpeter level. We are presently trying to transform them into a correcting three-body potential in our 3D equation.

Few-Body Systems Suppl. 10, 131–134 (1999)

Few-
Body
Systems
© by Springer-Verlag 1999

Semileptonic Decays of Pseudoscalar Mesons in the Light-Front Quark Model

Chueng-Ryong Ji and Ho-Meoyng Choi

Department of Physics, North Carolina State University, Raleigh, N.C. 27695-8202, U.S.A

Abstract. We investigate the form factors and decay rates for the semileptonic decays of pseudoscalar mesons using the light-front quark model. The zero-mode problems in calculating the form factor $f_-(q^2)$ at $q^+ = 0$ frame are circumvented by using the transverse component of the weak current. Our numerical results are in a good agreement with the available experimental data.

The interest in exclusive semileptonic decays of heavy mesons as well as light mesons lies in a possibility to obtain the most accurate values of the Cabibbo-Kobayashi-Maskawa (CKM) matrix element and test various approaches to the description of the internal hadron structure. In this work, we apply our recent light-front quark model [1] to the form factors and decay rates for the semileptonic decays of a pseudoscalar meson into another pseudoscalar meson.

In the light-front approach, a hadron is characterized by a set of Fock state wave functions, the probability amplitudes for finding different combinations of bare quarks and gluons in the hadron at a given light-front time $\tau = t + z/c$. These wave functions provide the essential link between hadronic phenomena at short distances (perturbative) and at long distances (non-perturbative). The distinguished features in the light-front approach are the simplicity of the vacuum except the zero-modes and the dynamical property of rotation operators. The vacuum at equal τ presents a dramatic difference compared to the vacuum at equal t. For a particle which has the mass m and the four-momentum $k = (k^0, k^1, k^2, k^3)$, the relativistic energy-momentum relation of the particle at equal-τ is given by

$$k^- = \frac{\boldsymbol{k}_\perp^2 + m^2}{k^+},\tag{1}$$

where the light-front energy conjugate to τ is given by $k^- = k^0 - k^3$ and the light-front momenta $k^+ = k^0 + k^3$ and $\boldsymbol{k}_\perp = (k^1, k^2)$ are orthogonal to k^- and form the light-front three-momentum $\underline{k} = (k^+, \boldsymbol{k}_\perp)$. The rational relation given by Eq. (1) is drastically different from the irrational energy-momentum relation at equal t, $k^0 = \sqrt{\boldsymbol{k}^2 + m^2}$, where the energy k^0 is conjugate to t and the

132

three-momentum vector k is given by $k = (k^1, k^2, k^3)$. The important point is that the signs of k^+ and k^- are correlated while at equal t the signs of k^0 and k are not correlated. Thus the momentum k^+ is always positive because only the positive energy k^- makes the system evolve to the future direction (i.e. positive τ), while the momentum k^3 can be either positive or negative even though k^0 is positive to evolve the system in the future direction (i.e. positive t). This provides a remarkable feature to the light-front vacuum; i.e., the Fock state vacuum is an eigenstate of the full Hamiltonian. Consequently, all bare quanta in an hadronic Fock state are associated with the hadron and none are disconnected elements of the vacuum. Furthermore, the problem of boost operators at equal t changing particle numbers can then be cured since the quantization surface $\tau = 0$ is invariant under both longitudinal and transverse boosts defined at equal τ. However, the quantization surface $\tau = 0$ is not invariant under the transverse rotation whose direction is perpendicular to the direction of the quantization axis z at equal τ. Thus, the transverse angular momentum operator involves the interaction that changes the particle number and it is not easy to specify the total angular momentum of a particular hadronic state. Also τ is not invariant under parity. We avoid these problems by using the Melosh transformation of each constituent from equal t to equal τ. This transformation uniquely determines the assignment of angular momentum, parity and charge conjugation for the light-front wave function of hadrons.

The key idea in our light-front quark model for mesons is to saturate the Fock state expansion by the constituent quark and anti-quark and use the variational principle for a QCD motivated Hamiltonian to fix all the model parameters by comparing with the meson mass spectra. The QCD motivated effective Hamiltonian for the description of the meson mass spectra is given by

$$H_{q\bar{q}} = H_0 + V_{q\bar{q}} = \sqrt{m_q^2 + k^2} + \sqrt{m_{\bar{q}}^2 + k^2} + V_{q\bar{q}}. \tag{2}$$

We use the usual confining interaction potential $V_{q\bar{q}} = V_0(r) + V_{\text{hyp}}(r)$ given by

$$V_{q\bar{q}} = a + bV_{\text{conf.}}(r) - \frac{4\kappa}{3r} + \frac{2S_q \cdot S_{\bar{q}}}{3m_q m_{\bar{q}}} \nabla^2 V_{\text{Coul}}, \tag{3}$$

where $V_{\text{conf.}}(r) = r[r^2]$ is linear [harmonic oscillator] type potential. Our variational method first evaluates $< \phi|[H_0 + V_0]|\phi >$ with a trial function $\phi(k^2) = Ne^{-k^2/2\beta^2}$ that depends on the parameters (m, β) and varies these parameters until the expectation value of $H_0 + V_0$ is a minimum. Once these model parameters are fixed, then, the mass eigenvalue of each meson is obtained by $M_{q\bar{q}} = < \phi|[H_0 + V_0]|\phi > + < \phi|H_{\text{hyp}}|\phi >$. Our model parameters are summarized in Table 1. Both the mass spectra and the wave functions of the light pseudoscalar (π, K, η, η') and vector (ρ, K^*, ω, ϕ) mesons were analyzed using our light-front constituent quark model [1]. The mixing angles of $\omega - \phi$ and $\eta - \eta'$ were predicted and various observables such as decay constants, charge radii, and radiative decay rates etc. were calculated [1]. Our numerical results [1] had a good agreement with the available experimental data. In Table 2, the decay constants and charge radii of heavy mesons are summarized.

Table 1. Optimized quark masses m_q [GeV] and the gaussian parameters β [GeV] for both harmonic oscillator and linear potentials obtained from the variational principle. $q=u$ and d.

Potential	m_q	m_s	m_c	m_b	$\beta_{q\bar{q}}$	$\beta_{s\bar{s}}$	$\beta_{q\bar{s}}$	$\beta_{q\bar{c}}$	$\beta_{q\bar{b}}$
H.O.	0.25	0.48	1.8	5.2	0.32	0.37	0.34	0.42	0.50
Linear	0.22	0.45	1.8	5.2	0.37	0.41	0.39	0.47	0.53

Table 2. Decay constants [MeV] and charge radii [fm^2] for various heavy pseudoscalar and vector mesons.

Potential	f_D	f_{D^*}	f_B	f_{B^*}	$r^2_{D^+}$	$r^2_{D^0}$	$r^2_{B^+}$	$r^2_{B^0}$
H.O.	127.1	149.6	113.8	122.3	0.182	- 0.309	0.420	- 0.208
Linear	139.2	168.9	121.2	131.4	0.176	- 0.301	0.438	- 0.217

The semileptonic decays of a pseudoscalar meson $Q_1\bar{q}$ into another pseudoscalar meson $Q_2\bar{q}$ are governed by the weak vector current as follows

$$< P_2|\bar{Q}_2\gamma^\mu Q_1|P_1 > = f_+(q^2)(P_1 + P_2)^\mu + f_-(q^2)(P_1 - P_2)^\mu, \qquad (4)$$

where $q^\mu = (P_1 - P_2)^\mu$ is the four-momentum transfer to the leptons. In the LFQM calculations presented in ref. [2], $q^+ \neq 0$ frame has been used to calculate the weak decays in the time-like region $m_l^2 \leq q^2 \leq (M_1 - M_2)^2$, with $M_{1[2]}$ and m_l being the initial [final] meson mass and the lepton (l) mass, respectively. However, when the $q^+ \neq 0$ frame is used, the inclusion of the non-valence contributions arising from quark-antiquark pair creation ("Z-graph") is inevitable and this inclusion may be very important for heavy-to-light and light-to-light decays. Nevertheless, the previous analyses [2] in $q^+ \neq 0$ frame considered only valence contributions neglecting non-valence contributions. In this work, we circumvent this problem by calculating the processes in $q^+ = 0$ frame and analytically continuing to the time-like region. The $q^+ = 0$ frame is useful because only valence contributions are needed. However, one needs to calculate the component of the current other than J^+ to obtain the form factor $f_-(q^2)$. Since J^- is not free from the zero-mode contributions even in $q^+ = 0$ frame [3], we use $J_\perp$ instead of J^- to obtain f_-. In the standard $q^+ = 0$ frame, we obain the form factors $f_+(q^2)$ and $f_-(q^2)$ using the matrix element of the "+" and "⊥"-components of the current, J^μ, respectively, and then continued to the time-like $q^2 > 0$ region by changing $q_\perp$ to $iq_\perp$ in the form facors. We note that our analytic continuation method is equivalent to that of ref. [4] where the form factors are obtained by the dispersion representations through the (gaussian) wave functions of the initial and final mesons. The more detailed formulations were presented in ref. [5]. The rates for the semileptonic decays

Table 3. Branching ratio $Br(A \to Xl\nu_l)$ for various semileptonic decays.

Process		$Br^{\text{Th.}}$	$Br^{\text{exp.}}$
$K \to \pi$	H.O.	$(37.60 \pm 0.60)\%$	$(38.78 \pm 0.27)\%$
	Linear	$(37.90 \pm 0.61)\%$	
$B \to \pi$	H.O.	$(1.20 \pm 0.14^{+0.24+0.63}_{-0.25-0.36}) \times 10^{-4}$	$(1.8 \pm 0.4 \pm 0.3 \pm 0.2)$
	Linear	$(1.33 \pm 0.16^{+0.27+0.71}_{-0.28-0.40}) \times 10^{-4}$	$\times 10^{-4}$
$D \to K$	H.O.	$(3.43 \pm 1.33)\%$	$(3.64 \pm 0.20)\%$
	Linear	$(3.46 \pm 1.34)\%$	
$D \to \pi$	H.O.	$(2.24 \pm 0.33) \times 10^{-3}$	$(3.8^{+1.2}_{-1.0}) \times 10^{-3}$
	Linear	$(2.28 \pm 0.34) \times 10^{-3}$	
$B \to D$	H.O.	$(2.36 \pm 0.36)\%$	$(2.35 \pm 0.2 \pm 0.44)\%$
	Linear	$(2.47 \pm 0.37)\%$	

$K \to \pi$, $B \to \pi(D)$, and $D \to \pi(K)$ are presented in Table 3. Our numerical results are in a good agreement with the available experimental data [6].

In summary, we investigated the semileptonic decays of a pseudoscalar meson into another pseudoscalar meson using the light-front quark model. The form factors $f_\pm$ are obtained in $q^+ = 0$ frame and then analytically continued to the time-like region by changing $q_\perp$ to $iq_\perp$ in the form factors. Our numerical results are in a good agreement with the available experimental data. The zero-mode contribution seems highly suppressed as the quark mass increases. More detailed analysis on the zero-mode contribution is currently underway.

Acknowledgement. This work was supported by the U.S. Department of Energy (DE-FG-02-96ER40947). The North Carolina Supercomputing Center and the National Energy Research Scientific Computing Center are also acknowledged for the grant of computing time allocation.

References

1. H.-M. Choi and C.-R. Ji: hep-ph/9711450

2. N.B. Demchuk et al.: Phys. At. Nuclei **59**, 2152 (1996);
   I. L. Grach et al.: Phys. Lett. **B385**, 317(1996);
   H.-Y. Cheng et al.: Phys. Rev. **D55**, 1559 (1997)

3. H.-M. Choi and C.-R. Ji: to appear in Phys. Rev. **D** (hep-ph/9805438)

4. D. Melikhov: Phys. Lett. **B394**, 385 (1997); Phys. Rev. **D53**, 2460 (1996)

5. H.-M. Choi and C.-R. Ji: hep-ph/9807500

6. Particle Data Group, R. M. Barnett et al.: Phys. Rev. **D54**, 1 (1996)

Few-Body Systems Suppl. 10, 135–138 (1999)

Few-
Body
Systems
© by Springer-Verlag 1999

A Poincaré Covariant Current Operator for Low and High Energy Phenomena

F.M. Lev [1], E. Pace [2], G. Salmè [3]

[1] Laboratory of Nuclear Problems, Joint Institute for Nuclear Research, Dubna, Moscow region 141980, Russia

[2] Dipartimento di Fisica, Università di Roma "Tor Vergata", and Istituto Nazionale di Fisica Nucleare, Sezione Tor Vergata, Via della Ricerca Scientifica 1, I-00133, Rome, Italy

[3] Istituto Nazionale di Fisica Nucleare, Sezione di Roma, P.le A. Moro 2, I-00185 Rome, Italy

Abstract. Within front-form dynamics and in the Breit frame where initial and final three-momenta of the system are directed along the z axis, Poincaré covariance constrains the current operator only through kinematical rotations around the z axis. Therefore, in this frame the current can be taken in the one-body form. Applications to deep inelastic structure functions in an exactly solvable model and to the deuteron magnetic form factor are presented.

The electromagnetic (em) and weak current operators should properly commute with the Poincaré generators and satisfy Hermiticity. The em current should also satisfy parity and time reversal covariance, as well as continuity equation. For instance, the tensor $\langle P', \chi' | J^\mu(x) J^\nu(0) | P', \chi' \rangle$ in deep inelastic scattering will have correct transformation properties relative to the Poincaré group only if both the nucleon state $|P', \chi'\rangle$, with four-momentum P' and internal wave function χ', and the operator $J^\mu(x)$ have correct transformation properties with respect to the *same* representation of the Poincaré group.

In ref. [1] we investigated within the front-form dynamics [2] the constraints imposed on the current operator $J^\mu(x)$ by extended Poincaré covariance (continuous + discrete transformations), Hermiticity and current conservation using a spectral decomposition of the current operator and we showed that Poincaré covariance of $J^\mu(x)$ can take place if $J^\mu(0)$ satisfies Lorentz covariance [1].

Let $\mathcal{H}$ be the space of states for an N particle system and let Π_i be the projector onto the subspace $\mathcal{H}_i \equiv \Pi_i \mathcal{H}$ corresponding to the mass M_i and the spin S_i. Since $J^\mu(0) = \sum_{ij} \Pi_i J^\mu(0) \Pi_j$ the operator $J^\mu(0)$ is fully defined by the operators $J^\mu(P_i, P'_j)$, acting in the internal space and corresponding to definite values of the masses: $J^\mu(P_i; P_j') \equiv \langle \boldsymbol{P}_\perp, P^+ | \Pi_i J^\mu(0) \Pi_j | \boldsymbol{P}_\perp', P^{'+} \rangle$, with $\boldsymbol{P}_\perp \equiv (P_x, P_y)$ and $P^\pm = (P^0 \pm P^z)/\sqrt{2}$). In the front form rotations around the z axis are kinematical, while the ones around x and y axes are dynamical. To take advantage of this fact, we use the Breit frame with the initial and final three-momenta of the system directed along the z axis. In order to satisfy Poincaré covariance, the operator $j^\nu(K\boldsymbol{e}_z; M_i, M_j)$ (which is the current operator $J^\mu(K_i, K_j')$ in this particular Breit frame, with $K\boldsymbol{e}_z = \boldsymbol{K}_i$) has to be covariant with respect to rotations around the z axis [1]

$$ j^\mu(K\boldsymbol{e}_z; M_i, M_j) = L(u_z)^\mu_\nu D^{S_i}(u_z) j^\nu(K\boldsymbol{e}_z; M_i, M_j) D^{S_j}(u_z)^{-1} \tag{1} $$

where $D^S(u)$ is the rank S representation of the rotation u_z around the z axis.

Therefore for a non-interacting system the continuous Lorentz transformations constrain the current $j^\mu(K\boldsymbol{e}_z; M_i, M_j)$ in the same way as in the interacting case. The same property holds for the covariance with respect to the reflection of the y axis, $\mathcal{P}_y$, and to the product of parity and time reversal, θ, which leave the light cone $x^+ = 0$ invariant and are kinematical. Hence the constraints imposed on the current for an interacting system by extended Lorentz covariance can be fulfilled by a current composed in our frame by the sum of one-body currents (i.e., by $J^\mu_{free}(0) = \sum_{n=1}^N j^\mu_{free,n}$, where N is the number of constituents).

The Hermiticity, $j^\mu(\boldsymbol{K}; M_i, M_j)^* = j^\mu(-\boldsymbol{K}; M_j, M_i)$, is satisfied if

$$ j^\mu(K\boldsymbol{e}_z; M_i, M_j)^* = L[r_x(\pi)]^\mu_\nu D^{S_j}[r_x(\pi)] j^\nu(K\boldsymbol{e}_z; M_j, M_i) D^{S_i}[r_x(\pi)])^{-1} \tag{2} $$

where the symbol $*$ means Hermitian conjugation in internal space and $r_x(\pi)$ represents a rotation by π around the x axis. Equation (2) is a non trivial constraint when $M_i = M_j$ (i.e., for elastic form factors), because in this case the rhs and the lhs contain the same operator [1].

As a first application, we will study the deep inelastic scattering in an exactly solvable model. Because of Hermiticity, the hadronic tensor for a system of mass m, ground state χ_o and initial momentum P, in the Breit reference frame where $\boldsymbol{P}_\perp = \boldsymbol{q}_\perp = 0$, $P_z = -P_z' = K > 0$, can be written as follows

$$ W^{\mu\nu} = \frac{1}{4\pi} \sum (2\pi)^4 \delta^{(4)}(P + q - P') \langle \chi_o | j^\mu(K\boldsymbol{e}_z; m, M') | \chi' \rangle \cdot $$
$$ \overline{\langle \chi_o | j^\nu(K\boldsymbol{e}_z; m, M') | \chi' \rangle} \tag{3} $$

where the sum is taken over all final states $|P', \chi'\rangle$ of mass M'.

As we have seen, the free current fulfills the extended Lorentz covariance in our Breit frame. Furthermore in the calculation of the structure functions only three components of the current are needed and can be chosen unconstrained by the current conservation, while the fourth component can be determined through the continuity equation. Therefore the structure functions can be calculated by using the $+$ and $\perp$ components of the free current in our Breit frame, even in the case where the final state interaction is present (cf. [1], [3]).

Let us consider a system of two particles, each one of mass m_o and spin $(1/2, \sigma_n)$, interacting through a relativistic harmonic oscillator potential, with ground state $\chi_o(\boldsymbol{k}_\perp, \xi, \sigma_1, \sigma_2)$ $(\xi = p^+/P^+)$, and mass eigenvalues $M_n = 2\{m_o^2 + a^2[2(n_x + n_y + n_z) + 3]\}^{1/2}$ [3]. If the exact harmonic oscillator eigenstates are used, the hadronic tensor and then the structure functions become sums of δ functions. In order to obtain continuous functions, one can consider average values, $\overline{F}_{1(2)}(x, Q)$, of the structure functions over small intervals of x, which resemble the finite experimental resolution. Taking exactly into account the interaction, both in the initial and in the final states, in the Bjorken limit $(Q^2 \to \infty,\ x = Q^2/(2Pq))$ the averaged structure functions yield exactly the parton model results [3]:

$$x = \xi, \quad \overline{F}_1(x) = \overline{W}^{11} = \sum_{\sigma_1 \sigma_2} \int |\chi_o(\boldsymbol{k}_\perp, x, \sigma_1, \sigma_2)|^2 d\boldsymbol{k}_\perp / \{4(2\pi)^3 x(1-x)\}$$

$$W^{+\nu} = 0, \quad \overline{F}_2(x, Q) = 2x\overline{F}_1(x, Q). \tag{4}$$

As a second application, we calculate elastic deuteron form factors. Let Π be the projector onto the subspace of bound states $|m, S, S_z\rangle$. The following current:

$$j^\mu(K\boldsymbol{e}_z; m, m) = \frac{1}{2}\{\mathcal{J}^\mu(K\boldsymbol{e}_z; m, m) + \mathcal{J}^\mu(-K\boldsymbol{e}_z; m, m)^*\}$$

$$\mathcal{J}^\mu(-K\boldsymbol{e}_z; m, m) = L[r_x(-\pi)]^\mu_\nu exp(\imath\pi S_x)\mathcal{J}^\nu(K\boldsymbol{e}_z; m, m)exp(-\imath\pi S_x). \tag{5}$$

(with $\mathcal{J}^\mu(K\boldsymbol{e}_z; m, m) = \langle 0, P^+ | \Pi J^\mu_{free}(0)\Pi | 0, P'^+\rangle$) is compatible with extended Lorentz covariance and Hermiticity, Eq. (2), and fulfills current conservation [1].

In the elastic case one has only $2S + 1$ independent matrix elements for the em current defined in Eq. (5), corresponding to the $2S + 1$ elastic form factors [1]. Therefore, the extraction of em form factors is no more plagued by the ambiguities which are present when the reference frame $q^+ = 0$ is used (see, e.g., [4]). Only three matrix elements $\langle m_d S S_z | j^\mu(K\boldsymbol{e}_z; m_d, m_d)|m_d S S'_z\rangle$ are independent for the deuteron (e.g., $\langle m_d 10|j^+|m_d 10\rangle$, $\langle m_d 11|j^+|m_d 11\rangle$, $\langle m_d 11|j_x|m_d 10\rangle$), corresponding to the three em elastic form factors.

In Fig. 1 we report our result for the deuteron magnetic form factor

$$B(Q^2) = 2\langle m_d 11|j_x(K\boldsymbol{e}_z; m_d, m_d)|m_d 10\rangle^2/(3m_d^2) \tag{6}$$

corresponding to different $N - N$ interactions and the Gari-Krümpelmann nucleon form factors. A reasonable agreement with the available experimental data is achieved. A more stringent comparison with the new TJNAF data will be possible in the near future. In Table 1 our results for the deuteron magnetic moment

$$\mu_d = \lim_{Q \to 0} 2^{1/2}m_p\langle m_d 11|j_x(K\boldsymbol{e}_z; m_d, m_d)|m_d 10\rangle/(Qm_d) \tag{7}$$

are compared with the results of ref. [5], obtained using the free current in the $q^+ = 0$ reference frame. While the $q^+ = 0$ approach points to a low P_D, our covariant approach prefers higher P_D values.

Calculations for charge and quadrupole form factors are in progress.

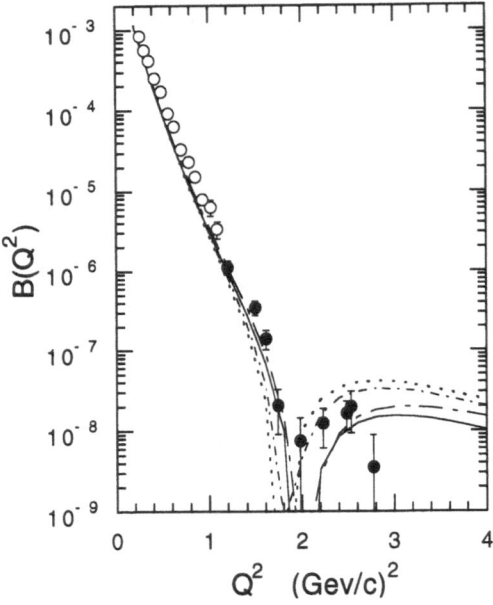

Figure 1. The deuteron magnetic form factor $B(Q^2)$ obtained with the Gari-Krümpelmann nucleon form factors and different $N - N$ interactions: RSC (dotted line), Av14 (solid line), Av18 (dashed line), Paris (dot-dashed line). Experimental data are from refs. [6a] (open dots) and [6b] (full dots).

Table 1. μ_d^{th} for different D-state percentages, P_D ($\mu_d^{exp} = 0.8574$).

Interaction	P_D	μ_d (ref.[5])	μ_d (this paper)
RSC	6.47	0.8500	0.8611
Av14	6.08	0.8516	0.8608
Paris	5.77	0.8531	0.8632
Av18	5.76		0.8635

References

1. F. Lev, E. Pace, G. Salmè: Nucl. Phys. **A**, in press

2. P.A.M. Dirac: Rev. Mod. Phys. **21**, 392 (1949)

3. E. Pace, G. Salmè, F. Lev: Phys. Rev. **C 58**, 2655 (1998)

4. I.L. Grach and L.A. Kondratyuk: Yad. Fiz. **39**, 316 (1984)

5. P.L. Chung et al.: Phys. Rev. **C37**, 2000 (1988)

6. a) S. Platchkov et al.: Nucl. Phys. **A510**, 740 (1990);
   b) P.E. Bosted et al.: Phys. Rev. **C 42**, 1 (1990)

Few-Body Systems Suppl. 10, 139–142 (1999)

Few-
Body
Systems
© by Springer-Verlag 1999

The nucleon Drell-Hearn-Gerasimov sum rule within a relativistic constituent quark model

B. Pasquini[1]*, F. Cardarelli[2] and S. Simula[2]

[1] Institut für Kernphysik, Universität Mainz, Germany
[2] Istituto Nazionale di Fisica Nucleare, Sezione Roma III, Italy

Abstract. We investigate the Drell-Hearn-Gerasimov sum rule within a relativistic constituent quark model based on the light-front formalism. The contribution of the $N - \Delta(1232)$ transition is explicitly evaluated using different forms of the baryon wave functions and adopting a one-body relativistic current which includes Dirac and Pauli form factors for the constitutent quarks.

1 Introduction

The Drell-Hearn-Gerasimov (DHG) sum rule [1] provides a very interesting relation between a ground state property of the nucleon and an integral property of its whole excitation spectrum. It connects the squared anomalous magnetic moment of the nucleon (κ_N) with an energy-weighted integral of the difference of the polarized photoabsorption cross sections for parallel $(\sigma_{3/2})$ and anti-parallel $(\sigma_{1/2})$ alignements of the photon and nucleon helicities. Specifically, the DHG sum rule reads

$$I_N = \int_{\omega_{th}}^{\infty} d\omega \frac{\sigma_{1/2}(\omega) - \sigma_{3/2}(\omega)}{\omega} = -2\pi^2 \alpha \frac{\kappa_N^2}{m_N^2}, \tag{1}$$

where m_N is the nucleon mass, $\alpha \simeq 1/137$ is the fine structure constant and ω is the photon energy running from the pion production threshold to infinity.

The DHG sum rule has been investigsted within non-relativistic [2] and relativized versions [2, 3] of the constituent quark (CQ) model. In spite of their good overall description of the excitation spectrum of the nucleon, these versions of the CQ model fail in describing the DHG sum rule. The purpose of this communication is to reconsider the DHG integral within the relativistic

*Fellow of the Marie Curie-TMR Project

CQ model of refs. [4]. In particular, this model includes: i) hadron eigenfunctions of an effective Hamiltonian able to reproduce the mass spectroscopy; ii) the relativistic composition of the CQ spins obtained via the introduction of the Melosh rotations; iii) a relativistic one-body electromagnetic (e.m.) current which contains Dirac and Pauli form factors for the CQ's.

2 Results and discussion

Within the so-called zero-width approximation, the contribution to the DHG integral from the photo-excitation of baryon resonances is given by

$$I_N^{res} = \sum_R I_N^R = \sum_R 4\pi \frac{m_N}{m_R^2 - m_N^2} \left(|A_{1/2}^R|^2 - |A_{3/2}^R|^2 \right), \tag{2}$$

where A_λ^R is the helicity amplitude describing the e.m. excitation of the nucleon to a resonance with mass m_R and spin projection $\lambda = 1/2, 3/2$. In the following discussion we only consider the contribution of the $N - \Delta(1232)$ transition, corresponding to the leading term in the sum of Eq. (2). To investigate the different ingredients of our calculation, three different forms of the N and $\Delta(1232)$ wave functions are adopted. In model A, the N and $\Delta(1232)$ wave functions are derived from the full interacting Hamiltonian proposed by Capstick and Isgur (CI) [5], while in model B the effects of the hyperfine terms of the CI interaction are switched off and in model C only the linear confining part of the CI potential is retained. In all the models the mass of u and d quark is $m_u = m_d = m = 0.22$ GeV. Within the light-front formalism, the helicity amplitudes of the $\Delta(1232)$ resonance can be calculated from the matrix elements of the plus component of the e.m. current, $\mathcal{I}^+ \equiv \mathcal{I}^0 + \hat{n} \cdot \mathcal{I}$, with $\hat{n}$ corresponding to the spin quantization axis. The explicit calculation has been performed in ref. [4], where the following relativistic one-body e.m. current has been used

$$\mathcal{I}^+(0) \simeq \sum_{q=1}^3 \mathcal{I}_q^+(0) = \sum_{q=1}^3 \left(e_q \gamma^+ f_1^{(q)}(Q^2) + i\kappa_q \frac{\sigma^{+\rho} q_\rho}{2m_q} f_2^{(q)}(Q^2) \right). \tag{3}$$

In Eq. (3), e_q (κ_q) is the CQ charge (anomalous magnetic moment) and $f_{1(2)}^{(q)}$ is the corresponding Dirac (Pauli) form factor.

First of all, we have calculated the nucleon magnetic moments, $\mu_{p[n]}$. It turns out that both the values of μ_p and $|\mu_n|$ calculated with $\kappa_q = 0$ in Eq. (3) significantly underestimate the experimental values ($\mu_{p[n]}^{exp} = 2.793 [-1.913]$). Then, non-vanishing κ_q are introduced, fixing their values by the request of reproducing the experimental nucleon magnetic moments. In Table 1, we show the results obtained with the three adopted forms of the nucleon wave function.

The results for the $N - \Delta$ transition contribution to the DHG integral (I_N^Δ) are listed in Table 2 and compared with the predictions of the sum rule (Eq. (1)) corresponding to the nucleon magnetic moments in the different models. As expected, a large fraction of the sum rule is saturated by the $N - \Delta(1232)$

contribution, both in the case of $\kappa_q = 0$ and $\kappa_q \neq 0$. The introduction of the CQ anomalous magnetic moments sharply affects our results and leads the calculated values of I_N^A to reproduce almost totally the experimental value expected from the sum rule.

Table 1. Values of the nucleon magnetic moments calculated within our models $A - C$ of the baryon wave functions. The rows labelled κ_u and κ_d give the values of the CQ anomalous magnetic moments needed in Eq. (3) to reproduce the experimental nucleon magnetic moments. (See also the discussion in ref. [6]).

Model	A	B	C
$\mu_p(\kappa_q = 0)$	2.28	2.44	2.74
$\mu_n(\kappa_q = 0)$	-1.18	-1.30	-1.60
κ_u	0.087	0.051	-0.0057
κ_d	-0.157	-0.129	-0.0700

Table 2. Δ-resonance contribution to the nucleon DHG integral (given in μbarn), calculated within the $A - C$ models. The entries with $\kappa_q \neq 0$ and $\kappa_q = 0$ are the results with and without the effects of the CQ anomalous magnetic moments, respectively. The columns labelled $(p + n)/2$ contain the values of $(I_p + I_n)/2$ predicted by the DHG sum rule using the nucleon magnetic moments calculated within each model.

Model	$\kappa_q = 0$		$\kappa_q \neq 0$	
	$N - \Delta$	$(p+n)/2$	$N - \Delta$	$(p+n)/2$
A	-107	-96	-204	-219
B	-114	-120	-193	-219
C	-165	-183	-197	-219

Anomalous magnetic moments $\kappa_q \neq 0$ are indication of an internal structure in the quarks which may contribute to the nucleon DHG sum rule. As shown in [2, 7] with a non-relativistic e.m. current, the resonance contribution I_N^{res} (2) satisfies the nucleon DHG sum rule only if $\kappa_q = 0$. When $\kappa_q \neq 0$, the sum rule contains an additional term which should be compensated by requiring a subtraction at infinity of the dispersion integral. Instead of a modification of the sum rule, we suggest to interpret this additional term as the contribution to the DHG integral from non-resonant processes due to the occurence of inelastic channels at level of finite-size quarks. Neglecting any interference between elastic and inelastic channels, the DHG integral may be written as

$$I_N \simeq I_N^{res} + I_N^{bkg}. \tag{4}$$

Using the results of ref. [7], we may approximate the contribution I_N^{bkg} as

$$I_N^{bkg} \simeq -2\pi^2\alpha \, \langle \psi_N, \tfrac{1}{2} | \sum_{q=1}^{3} \frac{\kappa_q^2}{m_q^2} \, \sigma_3^{(q)} \, | \psi_N, \tfrac{1}{2} \rangle$$

$$= -\frac{2\pi^2\alpha}{m^2} \, \langle \gamma_M \rangle \left[\frac{\kappa_u^2 + \kappa_d^2}{2} + \tau_3 \, \frac{5}{3} \frac{\kappa_u^2 - \kappa_d^2}{2} \right], \tag{5}$$

Table 3. Values of the nucleon DHG integrals $(I_p + I_n)/2$ and $(I_p - I_n)/2$ (given in μbarn), calculated within the $A - C$ model using Eq. (4). The entries with $\langle \gamma_M \rangle \neq 1$ and $\langle \gamma_M \rangle = 1$ correspond to the calculation of the background contribution with and without the relativistic dilution factor, respectively. The row labelled DHG contains the values predicted from the DHG sum rule (Eq. (1)).

Model	$(I_p + I_n)/2$		$(I_p - I_n)/2$	
	$\langle \gamma_M \rangle = 1$	$\langle \gamma_M \rangle \neq 1$	$\langle \gamma_M \rangle = 1$	$\langle \gamma_M \rangle \neq 1$
A	-223	-213	16.5	8.3
B	-204	-198	13.6	6.9
C	-200	-199	4.7	2.9
DHG	-219		14.7	

where we have considered a $SU(6)$-symmetric nucleon wave function, while $\langle \gamma_M \rangle$ is a dilution factor resulting from the Melosh rotations of the CQ spins [6]. In Table 3 are listed our final results for DHG integral of Eq. (4), where the resonant part is calculated by retaining only the contribution of the $N - \Delta(1232)$ transition. As a consequence, the $(I_p - I_n)/2$ combination of the nucleon sum rule receives contribution only from the background term and it results proportional to $(\kappa_u^2 - \kappa_d^2)$. Given the relation $|\kappa_u| < |\kappa_d|$ in all our models, the sign of such contribution is always positive, according to the predictions of the sum rule. However, such results need to be confirmed once included contributions from resonances other than the $\Delta(1232)$. Moreover, the relevant role played by the quark anomalous magnetic moment in our relativistic calculations motivates further theoretical investigation into the problems related with the physical meaning of the CQ anomalous magnetic moment and its consequences in the nucleon DHG sum rule.

References

1. S.D. Drell, A.C. Hearn: Phys. Rev. Lett. **16**, 908 (1966);
   S.B. Gerasimov: Sov. J. Nucl. Phys. **2**, 430 (1966)

2. D. Drechsel, M.M. Giannini: Few-Body Syst. **15**, 99 (1993);
   M. De Sanctis, D. Drechsel, M.M. Giannini: Few-Body Syst. **16**, 143 (1994)

3. Z. Li: Phys. Rev. **D47**, 1854 (1993)

4. F. Cardarelli et al.: Phys. Lett. **B357**, 267 (1995); Phys. Lett. **B371**, 7 (1996); Phys. Lett. **B397**, 13 (1997)

5. S. Capstick, N. Isgur: Phys. Rev. **D34**, 2809 (1986)

6. F. Cardarelli, B. Pasquini, S. Simula: Phys. Lett. **B418**, 237 (1998)

7. S.J. Brodsky, J.R. Primack: Ann. of Phys. **52**, 315 (1969);
   J.L. Friar: Phys. Rev. **C16**, 1504 (1977)

Few-Body Systems Suppl. 10, 143–146 (1999)

Few-Body Systems
© by Springer-Verlag 1999

Relativistic corrections to the electric polarizability of the neutron

M. Bawin[1], S. A. Coon[2]

[1] Université de Liège, Institut de Physique B.5, Sart Tilman, B-4000 Liège 1, Belgium

[2] Physics Department, New Mexico State University, Las Cruces, NM 88003, U.S.A.

Abstract. We demonstrate in a solvable model the connection between the intrinsic electric polarizability $\bar{\alpha}$ and the value α_{Sch} obtained from neutron-atom scattering.

1 Introduction

According to the Low Energy Theorems of Compton scattering, the electric $\bar{\alpha}$ and magnetic $\bar{\beta}$ polarizabilities are fundamental quantities that characterize the neutron. Given the lack of free neutron targets and the much smaller Compton cross-section (compared to the charged proton), the two current experimental approaches are low energy neutron-atom scattering [1, 2] and quasi-free Compton scattering from the neutron bound in a deuteron [3, 4]. The tool used to analyze the data in neutron-atom scattering is the Foldy-Wouthuysen reduction of the Dirac equation [5]. For a neutron of magnetic moment μ and mass m, it has been shown [6] that $\bar{\alpha}$ is *not* the coefficient (labeled α_{Sch}) of $-\frac{1}{2}\mathbf{E}^2$ in this wave equation. Instead, $\bar{\alpha} = \alpha_{Sch} + \mu^2/m$.

The purpose of this communication is to illustrate these results by solving exactly the Foldy-Wouthuysen-Dirac equation for a neutron in a constant $\mathbf{E}$ field. Actually, this is how electric polarizability is defined [7]. Our main result is the following: the energy eigenvalues are independent of μ^2 in the neutron rest frame defined by

$$\langle m\mathbf{v} \rangle = \langle \mathbf{p} - (\mathbf{E} \times \boldsymbol{\mu}) \rangle = 0 ,$$

where $\mathbf{p}$ is the canonical momentum, $\mathbf{v}$ the velocity operator, and $\boldsymbol{\mu} = \mu\boldsymbol{\sigma}$. Since the components of $\mathbf{v}$ implicitly include non-commuting spin operators, the rest frame can only be defined in an average sense. Given this definition, the rest frame energy eigenvalues do not include the $-\mu^2/m$ coefficient multiplying $-\frac{1}{2}\mathbf{E}^2$ present in the Foldy-Wouthuysen-Dirac equation. Thus we confirm by an explicit calculation our earlier identification [6] $\bar{\alpha} = \alpha_{Sch} + \mu^2/m$ which was of a somewhat formal nature.

2 Calculation

The Foldy-Wouthuysen-Dirac equation for a neutron of magnetic moment $\boldsymbol{\mu} = \mu\boldsymbol{\sigma}$ and mass m in a constant external $\mathbf{E}$ field is [6]:

$$H\psi = \left(\frac{\mathbf{p}^2}{2m} - \frac{\mathbf{p}}{m}\cdot(\mathbf{E}\times\boldsymbol{\mu}) + \frac{\mu^2\mathbf{E}^2}{2m} - \tfrac{1}{2}\bar{\alpha}\mathbf{E}^2 \right)\psi = \varepsilon\psi \tag{1}$$

where $\bar{\alpha}$ is the intrinsic neutron electric polarizability and ε the energy eigenvalue. Equation (1) can be solved exactly to give:

$$\left[\varepsilon - \frac{p^2}{2m} - \frac{\mu^2 E^2}{2m} + \tfrac{1}{2}\bar{\alpha}E^2 \right] = \frac{E\mu}{m}\lambda, \tag{2}$$

where λ is given by:

$$\mathbf{p}\cdot(\boldsymbol{\sigma}\times\mathbf{E})\psi = E\lambda\psi, \tag{3}$$

and we have taken $\mathbf{E}$ to be along the x-axis, ie. :

$$\mathbf{E} = (E, 0, 0).$$

Writing

$$\mathbf{p} = (p_1, \mathbf{p}_\perp)$$

one finds for (3):

$$\lambda = \pm p_\perp, \qquad \text{with} \qquad p_\perp = |\mathbf{p}_\perp|. \tag{4}$$

With the aid of the explicit eigenstates of Eq. (3), one finds that the *average* value of the particle velocity operator

$$\mathbf{v} \equiv \frac{\partial H}{\partial\mathbf{p}} = \frac{1}{m}\mathbf{p} - (\mathbf{E}\times\boldsymbol{\mu})$$

for $\lambda = -p_\perp$ is

$$\langle\mathbf{v}_\perp\rangle = \frac{\mathbf{p}_\perp}{m}\left(1 - \frac{\mu E}{p_\perp} \right) \tag{5}$$

and

$$\langle v_1\rangle = v_1 = \frac{p_1}{m}. \tag{6}$$

Formulae (5) and (6) imply that

$$\langle\mathbf{v}\rangle = 0 \qquad \text{for} \qquad \mu E = p_\perp \text{ and } p_1 = 0. \tag{7}$$

This finally leads, from Eqs. (2), (7), and $\lambda = -p_\perp$ to:

$$\varepsilon = -\tfrac{1}{2}\bar{\alpha}\mathbf{E}^2$$

Thus, in the neutron rest frame ($\langle\mathbf{v}\rangle = 0$), all terms quadratic in μ cancel exactly. This implies, as stated in the Introduction, that the rest frame energy eigenvalues do not include the $-\mu^2/m$ coefficient multiplying $-\tfrac{1}{2}\mathbf{E}^2$ present in Eq. (1). It is this Eq. (1), however, which is used to analyze neutron-atom scattering, so that the quantity μ^2/m should be *added* to the quoted polarizability result (α_{Sch}) of the experiment, thus leading to an *increase* of the electric polarizability $\bar{\alpha}$ (in the Compton sense), in complete agreement with [6].

3 Discussion

The explicit calculation just described sharpens the argument of ref. [6]. In that earlier paper we asserted "that the rest frame of a neutron in an external electric field is defined by a vanishing value of the velocity operator, as confirmed by experimental measurements of the Aharonov-Casher effect." As we have seen in Sect. 2, the velocity *operator* cannot vanish in a three dimensional geometry because it contains non-commuting spin operators, and the neutron rest frame is defined by the *expectation value* of the velocity operator. The observation [8] of the Aharonov-Casher effect was made in a two dimensional geometry (a neutron diffracting around a line of electric charge) where the velocity *operator* does indeed vanish.

Finally, our result that the rest frame energy eigenvalues do not include the $-\mu^2/m$ coefficient multiplying $-\tfrac{1}{2}\mathbf{E}^2$ is also consistent with Foldy's well known result. Foldy solved exactly the problem of a structureless neutral Dirac particle with an anomalous magnetic moment μ in a homogeneous static electric field $\mathbf{E}$ [9]. He did find a term quadratic in μ^2 in the non-relativistic limit of the energy eigenvalues (implying a *negative* polarizability of magnitude μ^2/m from this Dirac-Foldy term). Foldy's solution was obtained in the frame $\mathbf{p} = 0$ which is not the rest frame of the neutron.

Acknowledgement. The work of M.B. was supported by the National Fund for Scientific Research, Belgium and that of S.A.C. by NSF grant PHY-9722122.

References

1. J. Schmiedmayer et al.: Phys. Rev. Lett. **66**, 1015 (1991)

2. L. Koester et al.: Phys. Rev. C **51**, 3363 (1995)

3. K.W. Rose et al.: Nucl. Phys. **A514**, 621 (1990)

4. R. Wissmann, M. I. Levchuk, M. Schumacher: Eur. Phys. J. A **1**, 193 (1998)

5. H. Leeb, G. Eder, and H. Rauch: J. de Physique **C45**, 3 (1984)

6. M. Bawin and S.A. Coon: Phys. Rev. **C55**, 419 (1997)

7. B.R. Holstein: Comments Nucl. Part. Phys. **20**, 301 (1992)

8. A. Cimmino et al.: In J.C. Zorn and R.R. Lewis (eds.): *Atomic Physics 12*, AIP Conf. Proc. no. 233, p.247. New York: AIP 1991

9. L.L. Foldy: Phys. Rev. Lett. **3**, 105 (1959)

Few-Body Systems Suppl. 10, 147–150 (1999)

Few-
Body
Systems
© by Springer-Verlag 1999

Covariant formulation of pion-nucleon scattering

A.D. Lahiff, I.R. Afnan

Department of Physics, The Flinders University of South Australia, GPO Box 2100, Adelaide 5001, Australia

Abstract. A covariant model of elastic pion-nucleon scattering based on the Bethe-Salpeter equation is presented. The kernel consists of s- and u-channel N and $\Delta(1232)$ exchanges, along with ρ and σ exchange in the t-channel. A good fit is obtained to the s- and p-wave phase shifts up to the two-pion production threshold.

Pion-nucleon (πN) scattering is an important example of a strong interaction, and as such plays a significant role in many other nuclear reactions involving pions, for example pion photoproduction. Ideally a theory describing πN scattering should be derived from Quantum Chromodynamics (QCD), but since QCD is not at present solvable for low energies, it is necessary to use a chirally invariant effective Lagrangian, where the degrees of freedom are baryons and mesons, rather than quarks and gluons. Following the success of meson-exchange models in describing the nucleon-nucleon interaction, a number of meson-exchange models for πN scattering have been developed over the last few years [1]. These models invariably begin with an effective Lagrangian which describes the couplings between the various mesons and baryons. The tree-level diagrams obtained from this Lagrangian are then unitarized in a 3-dimensional approximation to the Bethe-Salpeter (BS) equation [2]. Convergence is guaranteed by the introduction of phenomenological form factors at each vertex. There are an infinite number of 3-dimensional reductions to the BS equation, and there is no overwhelming reason to choose one particular approximation over any other. Here we describe a covariant model of elastic πN scattering in which the BS equation is solved without any reduction to 3-dimensions.

In principle, the exact $\pi N \leftarrow \pi N$ amplitude for a given Lagrangian can be obtained from the BS equation, with a potential consisting of all 1- and 2-particle irreducible diagrams, and dressed propagators in the πN intermediate state. Since it is impossible to construct such a potential, as it would contain an infinite number of diagrams, it is common practice to truncate this kernel and include only the tree-level diagrams. Furthermore, if only two-body unitarity is required to be maintained, then the dressed nucleon propagator is replaced by a bare propagator with a pole at the physical mass.

Our kernel consists of s- and u-channel N and $\Delta(1232)$ exchanges, along with ρ and σ exchange in the t-channel. We do not include any higher baryon resonances because these contributions are not expected to be significant for πN scattering below the two-pion production threshold. The couplings to the pion field are always through derivative couplings, as required by chiral symmetry.

There is an ambiguity as to the choice of propagator for a particle with spin-3/2. The most commonly used propagator is the Rarita-Schwinger propagator, which is known to have both spin-3/2 as well as background spin-1/2 components [3]. Other forms have been introduced by Williams [4] and Pascalutsa [5], which each have only a spin-3/2 component. In the present paper we use the Rarita-Schwinger propagator.

To guarantee the convergence of all integrals, we need to associate with each vertex a cut-off function. We take this cut-off function to be the product of form factors that depend on the 4-momentum squared of each particle present at the vertex [1]. Each form factor is chosen to be of the form

$$f(q^2) = \left(\frac{\Lambda^2 - m^2}{\Lambda^2 - q^2} \right),$$

where q^2 is the 4-momentum squared of the particle and m is the mass. A different cutoff mass Λ is used for each particle.

The s-channel pole terms present in the potential become dressed when the BS equation is iterated. Therefore bare coupling constants and masses are used in the N and Δ s-channel pole diagrams in the potential. The bare nucleon parameters are determined by requiring that in the P_{11} partial wave, there is a pole at the physical nucleon mass with a residue related to the physical πNN coupling constant. Since the dressed Δ has a width, it would be necessary to analytically continue the BS equation into the complex s-plane in order to carry out the renormalization for the Δ. Rather than doing this we treat the bare Δ mass and the bare $\pi N\Delta$ coupling constant as free parameters. Since the P_{33} partial wave is dominated by the s-channel Δ pole diagram, the bare Δ parameters are essentially fixed by the P_{33} phase shifts. While only the s-channel diagrams in the potential are dressed by the ladder BS equation, Pearce and Afnan [6] have shown that if 3-body unitarity is satisfied, then the u-channel diagrams also become dressed. We approximate this by using physical masses and coupling constants in the u-channel N and Δ diagrams.

We solve the BS equation by first expanding the nucleon propagator in the πN intermediate state into positive and negative energy components, and then sandwiching the resulting equation between Dirac spinors. This gives two coupled 4-dimensional integral equations which are reduced to 2-dimensional integral equations after a partial wave decomposition. A Wick rotation [7] is carried out in order to obtain equations suitable for numerical solution. With our choice of form factors, there is no interference from the form factors when carrying out the Wick rotation, provided the cutoff masses are large enough. The cutoff masses also have to be chosen large enough so that unphysical thresholds, which are generated by form factor singularities and propagator

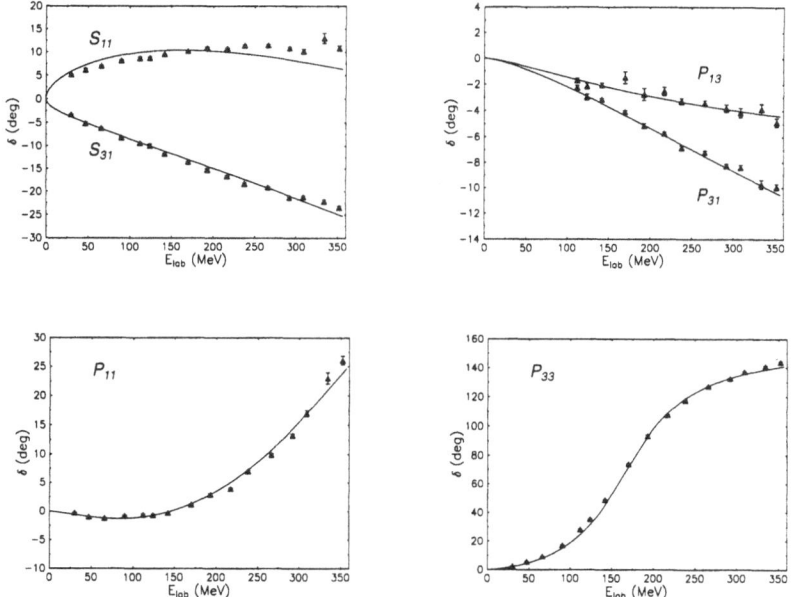

Figure 1. The phase shifts obtained from the Bethe-Salpeter equation are shown versus the pion laboratory energy, compared to the VPI SM95 partial wave analysis.

singularities pinching the integration contour, are far above the two-pion production threshold. More details are given in [8].

The renormalized πNN coupling constant is fixed at its physical value of $g^2_{\pi NN}/4\pi = 13.5$. The nucleon renormalization procedure fixes the bare N parameters. The remaining parameters are determined in a fit to the phase shifts and scattering lengths from the SM95 partial wave analysis of Arndt et al. [9]. For the cutoff masses we obtain $\Lambda_N = 3.12$, $\Lambda_\pi = 1.73$, $\Lambda_\rho = 4.67$, $\Lambda_\Delta = 4.86$, and $\Lambda_\sigma = 1.4$ (all in GeV). The coupling constants are $g_{\rho\pi\pi}g_{\rho NN}/4\pi = 3.2$, $\kappa_\rho = 2.57$, $g_{\sigma\pi\pi}g_{\sigma NN}/4\pi = -0.3$, $f^2_{\pi N\Delta}/4\pi = 0.44$, and $f^{(0)2}_{\pi N\Delta}/4\pi = 0.35$. The remaining masses are $m^{(0)}_\Delta = 2.13$ GeV and $m_\sigma = 700$ MeV. With these parameters, the renormalization procedure gives $m^{(0)}_N = 1.37$ GeV and $g^{(0)2}_{\pi NN}/4\pi = 3.8$.

We obtain a good fit to the s- and p-wave phase shifts. The resulting phase-shifts are shown in Fig. 1, and the scattering lengths and volumes are shown in Table 1. In general our coupling constants are consistent with those used in the other πN models. Assuming universality ($g_\rho \equiv g_{\rho\pi\pi} = g_{\rho NN}$), our value of $g^2_\rho/4\pi$ is close to that obtained from the decay $\rho \to 2\pi$, i.e. $g^2_\rho/4\pi = 2.8$, and κ_ρ is close to the value $\kappa_\rho = 3.7$ arising from vector meson dominance. Our physical $\pi N\Delta$ coupling constant is slightly larger than the value $f^2_{\pi N\Delta}/4\pi = 0.36$ obtained from calculations of the width of the Δ (the use of the larger coupling was necessary in order to obtain a good fit to the P_{13} and P_{31} phase shifts). The contribution of σ-exchange is very small, while the u-channel Δ

Table 1. Scattering lengths and volumes obtained from the Bethe-Salpeter equation (BSE). Units are $m_\pi^{-(2\ell+1)}$.

	BSE	SM95		BSE	SM95
S_{11}	0.184	0.175	S_{31}	−0.101	−0.087
P_{11}	−0.079	−0.068	P_{31}	−0.045	−0.039
P_{13}	−0.037	−0.022	P_{33}	0.181	0.209

diagram provides a large amount of attraction in all partial waves except S_{31} and P_{33}. The attraction in the P_{11} partial wave is dominated by ρ-exchange and the u-channel Δ. Notice that some additional attraction is required for high energies in the S_{11} partial wave.

The cutoff masses turn out to be quite large, with the result that the dressing is significant, as is evident from the large size of the bare N and Δ masses. The baryon self energies are dominated by the one-pion loop diagrams. In view of the significance of the dressing, it is interesting to examine the effect of the dressing on the πNN form factor. We can calculate a renormalized cutoff mass by comparing the dressed πNN vertex, with the nucleons on-mass-shell and the pion off-shell, to a monopole form factor. We find $\Lambda_\pi^R = 1.17$ GeV (recall that the bare cutoff mass is 1.73 GeV). Therefore the vertex dressing softens the πNN form factor.

References

1. B.C. Pearce and B.K. Jennings: Nucl. Phys. **A528**, 655 (1991);
   F. Gross and Y. Surya: Phys. Rev. **C47**, 703 (1993);
   C. Schütz et al.: Phys. Rev. **C49**, 2671 (1994);
   C.T. Hung, S.N. Yang, T.-S.H. Lee: J. Phys. **G20**, 1531 (1994);
   V. Pascalutsa and J.A. Tjon: Nucl. Phys. **A631**, 534c (1998)

2. E.E. Salpeter and H.A. Bethe: Phys. Rev. **84**, 1232 (1951)

3. M. Benmerrouche, R.M. Davidson, N.C. Mukhopadhyay: Phys. Rev. **C39**, 2339 (1989)

4. H.T. Williams: Phys. Rev. **C31**, 2297 (1985)

5. V. Pascalutsa: hep-ph/9802288

6. B.C. Pearce and I.R. Afnan: Phys. Rev. **C34**, 991 (1986)

7. G.C. Wick: Phys. Rev. **96**, 1124 (1954)

8. A.D. Lahiff and I.R. Afnan: in preparation

9. R.A. Arndt et al.: Phys. Rev. **C52**, 2120 (1995)

Few-Body Systems Suppl. 10, 151–160 (1999)

An Exotic Three-Body System
- Antiprotonic Helium Atomcules

Toshimitsu Yamazaki *

Institute of Particle and Nuclear Studies, High Energy Accelerator Research Organization, 3-2-1 Midori-cho, Tanashi, Tokyo, 188 Japan and
Japan Society for the Promotion of Science, 5-3-1 Koji-machi, Chiyoda-ku, Tokyo, 102 Japan

Abstract. Recent development of high precision spectroscopy of metastable antiprotonic helium atomcules, which were discovered in 1991 as anomalous longevity of antiprotons in matter, is described.

1 Discovery of Antiprotonic Helium Atomcules

The metastable antiprotonic helium atom-molecule (in short, *atomcule*), discovered in 1991 [1, 2], is an unusual exotic three-body system composed of $e^- + \bar{p} + He^{2+}$, interfacing between matter and anti-matter. In the initial phase the longevity of antiprotons in various helium media was studied from the shape of delayed annihilation time spectra (DATS) [3 − 5]. The observed DATS showed little change with the helium density in pure helium media [3], but striking quenching effects when foreign atoms/molecules were added [2, 4]. Subsequently, laser spectroscopy of the $\bar{p}He^+$ atomcule emerged [6 − 8], and abundant experimental data on the structure and reactions have been obtained to date [8 − 21].

2 Basic Properties of the Atomcule

The longevity was ascribed to the population of series of metastable states with large principal (n) and orbital (l) quantum numbers, as predicted earlier by Condo [22] and Russell [23]. The structure was further studied theoretically [24 − 34]. The level diagram in Fig. 1 shows long-lived radiation-dominated states and short-lived Auger-dominated states calculated by Ohtsuki [2, 6, 25].

---

*E-mail address: yamazaki@nucl.phys.s.u-tokyo.ac.jp

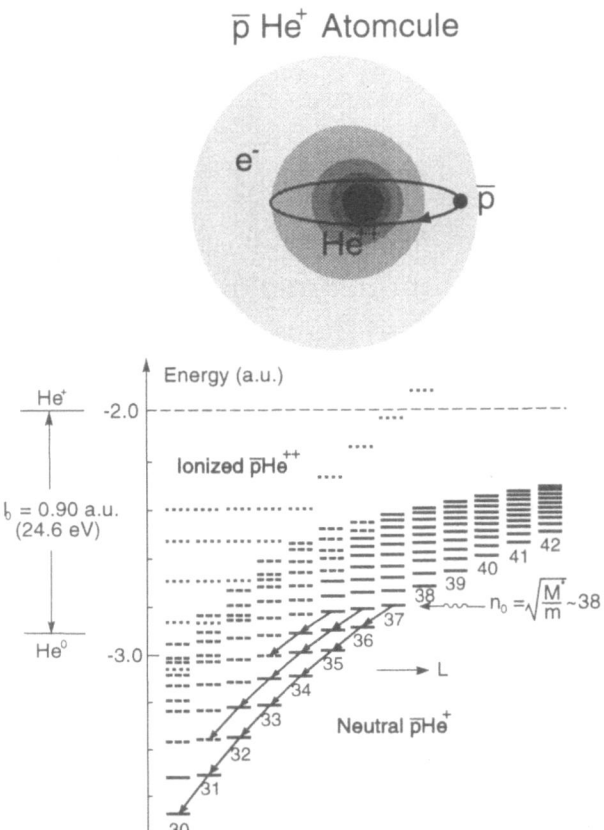

Figure 1. (a) The structure of the $\bar{p}He^+$ atomcule, where the $\bar{p}$ with a large (n, l) quantum numbers circulates in a localized orbit around the He^{2+} nucleus, while the electron occupies the distributed 1s state. (b) The level scheme of large-(n, l) states of the $\bar{p}He^+$ atomcule. The solid bars indicate radiation-dominated metastable states, while the broken lines are for Auger-dominated short-lived states. The ionized states are also shown by dotted lines. From ref. [2].

The division into the two domains is in accordance with the rule

$$\Delta L > 3 : \quad \text{Radiation dominated,}$$
$$\Delta L \leq 3 : \quad \text{Auger dominated,}$$

where ΔL is a minimum change of the angular momenta between a $\bar{p}He^+$ state and its daughter $\bar{p}He^{2+}$ ion. The Auger transition rates have recently been calculated extensively [35, 36].

This three-body system is called *atom-molecule*, or, in short *atomcule*, because it shows a unique dual nature as an atomic system and a molecular

ATOMCULE: MOLECULAR ASPECT OF YRAST ATOMS

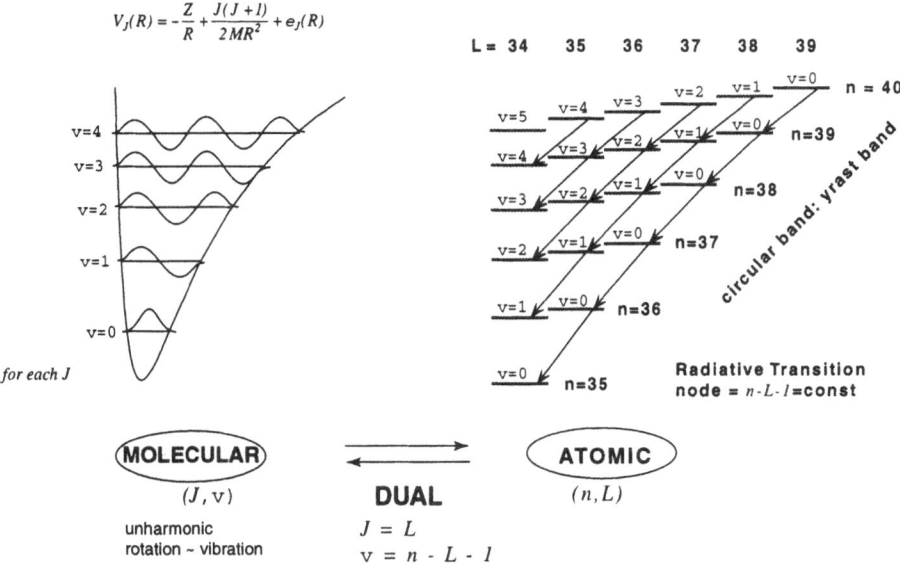

Figure 2. Atomic and molecular views of the antiprotonic helium atomcule system. The large (n, l) states in the atomic yrast region in the atomic model are also assigned as the molecular states of corresponding rotational and vibrational quantum numbers $(J, v) = (l, n - l - 1)$ in the one-dimensional potential for each J. The radiative transitions with $\Delta v = 0$, as shown by arrows, are favoured because of the maximum overlapping of the radial densities. In this sense, the atomcule system has a dual character by itself.

system, as schematically shown in Fig. 2. Each state possesses simultaneously two kinds of quantum numbers, *atomic* (n, l) and *molecular* (J, v), where $J = l$ and the vibrational quantum number is related to the radial node number as $v = n - l - 1$. Since the $\bar{p}$ in a large (n, l) orbit in this atomic yrast region is localized, the radiative decay process follows a propensity rule [25, 26]

$$\Delta v = \Delta(n - l - 1) = 0.$$

The atomic configuration mixing theory shows a systematic reduction of the radiative rates over the single-particle estimates due to the repulsive correlation between the $\bar{p}$ and the e^- (called *atomic core polarization* effect [25], similar to the well known effects in nuclear physics). This effect is readily taken into account in the molecular model, in which the electron motion is calculated by the Born-Oppenheimer approximation on the $\bar{p}$-He^{2+} *molecule*.

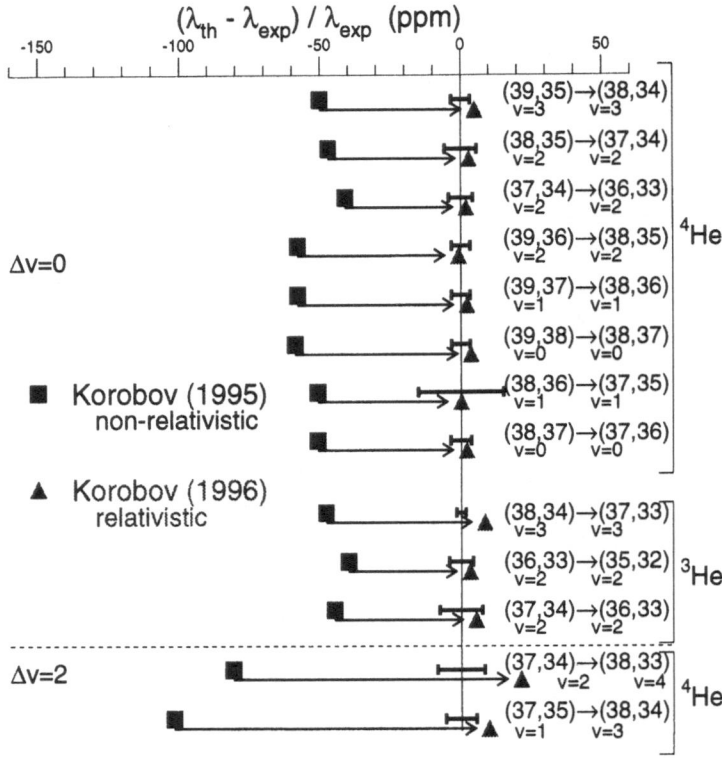

Figure 3. Comparison between experiment and theory of the observed resonance wavelengths including unfavoured transitions [13, 17].

3   Advent of Laser Spectroscopy

As the radiative transition energies of the atomcule are all in the range of visible light, its laser spectroscopy was invented [8, 9]. In the first phase a favoured transition from a metastable to its daughter short-lived state such as $(39, 35) \rightarrow (38, 34)$ was chosen to detect a forced annihilation spike in DATS. The observed transition energies showed good agreement with the then available theoretical calculations [25, 26, 28] to the 10^{-3} level. The lifetimes of the metastable states were also determined, and turned out to be in excellent agreement with the theoretical radiative rates that include the atomic core polarization effect [25, 26].

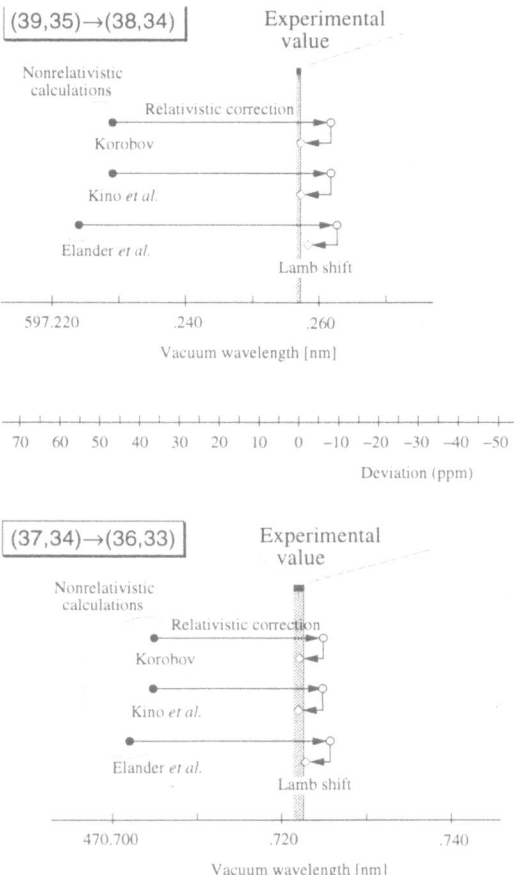

Figure 4. The precisely determined transition wavelengths after the correction for pressure shifts are compared with the recent theoretical values (non-relativistic values, after the relativistic corrections and also the Lamb shift corrections). From Torii et al. [14].

4   High-Precision Frontier Both Experimental and Theoretical

In the second phase, a new calculation of Korobov [30] based on the molecular expansion variational method emerged. The theoretical values show an excellent agreement with the experimental values to the level of 50 ppm. This theory with a high-precision predictability helped observing new resonance transitions greatly. Thus, not only favoured $\Delta v = 0$ transitions but also unfavoured $\Delta v = 2$ transitions were found [13], and a systematic deviation as much as 50 ppm for the $\Delta v = 0$ transitions and 100 ppm for the $\Delta v = 2$ transitions was revealed,

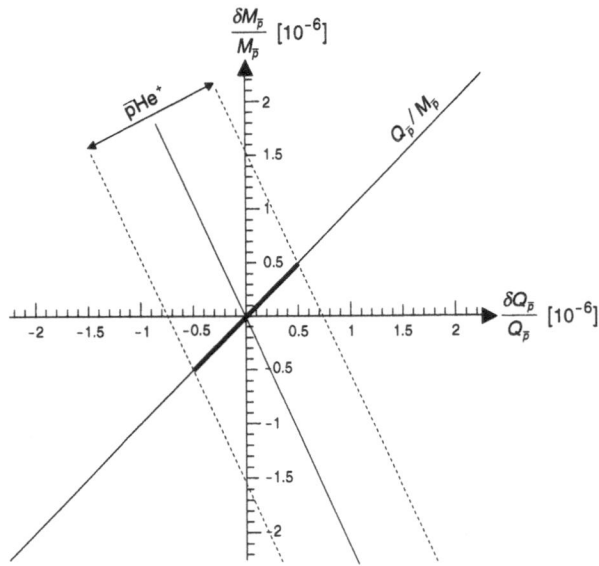

Figure 5. Two-dimensional constraint on $\Delta M_{\bar{p}}/M_{\bar{p}} = (M_{\bar{p}} - M_p)/M_{\bar{p}}$ and $\Delta Q_{\bar{p}}/Q_{\bar{p}} = (Q_{\bar{p}} - Q_p)/Q_{\bar{p}}$ obtained from the cyclotron frequency of $\bar{p}$ [37] and from the present spectroscopic studies of $\bar{p}He^+$ [14].

as shown in Fig. 3. This discrepancy was immediately explained by taking into account the relativistic corrections on the motion of the electron [34], and the remaining difference between theory and experiment became around several ppm. Two other groups obtained similar theoretical values [32, 33].

To further clarify the situation more precise determination of the resonance wavelength taking into account the possible pressure shifts of the wavelength was carried out [14]. The most precise experimental values to date are compared with the most recent theoretical values in Fig. 4. The comparison showed a furthermore excellent agreement to the 1 ppm level, when the Lamb shift is taken into account. This provides a surprisingly high precision frontier for the three-body Coulomb system.

Thus, the present state of art is its capability of imposing experimental constraints on the fundamental constants of the antiproton (mass $M_{\bar{p}}$ and charge $-Q_{\bar{p}}$) which are assumed to be equal to those of the proton in the theoretical calculations. If we allow $\Delta M_{\bar{p}}/M_{\bar{p}} = (M_{\bar{p}} - M_p)/M_{\bar{p}}$ and $\Delta Q_{\bar{p}}/Q_{\bar{p}} = (Q_{\bar{p}} - Q_p)/Q_{\bar{p}}$ to be free, the present agreement between theory and experiment yields a constraint on these two values, since the transition wavelengths depend on the antiproton Rydberg constant which is proportional to $M_{\bar{p}} \times Q_{\bar{p}}^2$. The constraints imposed are a linear relation, as shown in Fig. 5.

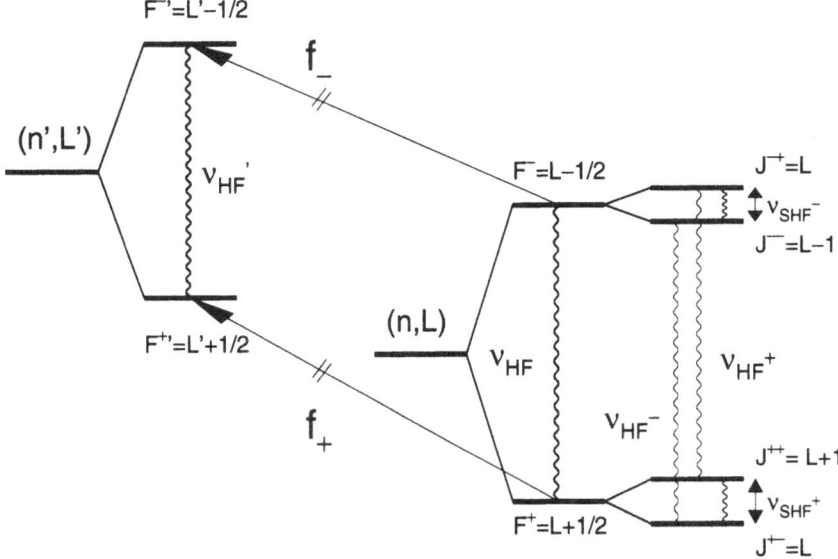

Figure 6. Hyperfine quadruplet structure of the p̄He$^+$ atomcule predicted by Bakalov and Korobov [39]. The dominant doublet structure was recently revealed as a doublet in the $(37, 35) \rightarrow (38, 34)$ unfavoured resonance transition by Widmann et al. [15].

On the other hand, the cyclotron frequency of p̄, which was very precisely measured by Gabrielse et al. [37], gives a strong constraint which crosses the line set by the p̄He$^+$ atomcule, as shown. Thus, we obtain

$$\Delta M_{\bar{p}}/M_{\bar{p}} < 5 \times 10^{-7}, \tag{1}$$

$$\Delta Q_{\bar{p}}/Q_{\bar{p}} < 5 \times 10^{-7}. \tag{2}$$

In the past only a poor constraint ($\sim 5 \times 10^{-5}$) was known by using antiprotonic x-ray data [38].

5 Hyperfine Structure

Each state of (n, l) has finer structure resulting from the coupling of the p̄ angular momentum ($\boldsymbol{L}$), the electron spin ($\boldsymbol{S_e}$) and the p̄ spin ($\boldsymbol{S_{\bar{p}}}$). Figure 6 shows a quadruplet structure, as predicted by Bakalov and Korobov [39]. The dominant hyperfine doublet with $F^{+-} = L \pm S_e$ is further split into superhyperfine doublets with $J = F^{+-} \pm S_{\bar{p}}$. Recently, hyperfine splitting of a resonance line was observed [15] and confirmed the theoretical expectation. Further experiments for direct observation of the transitions between these substates are being planned for the near future by employing a new method of laser-microwave triple resonances.

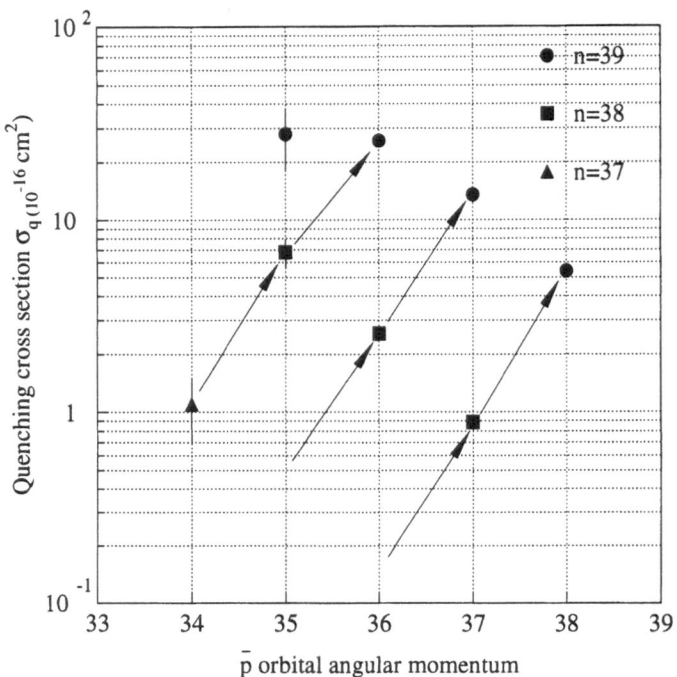

Figure 7. The (n, l) dependence of the quenching cross sections by H_2 molecules observed by using laser resonance tagging method. The arrows indicate hydrogen-assisted inverse resonance transitions. From Ketzer et al. [18].

6   Chemical Physics Aspects

Very interesting chemico-physical aspects of this atomcule have also been revealed by using laser-tagged individual states, such as lifetime shortening by collision with H_2 molecules [16 − 18]. The laser tagging method was powerful in yielding surprisingly large state-dependent quenching cross sections, namely, the higher-n states are more strongly destroyed by H_2. This selective quenching behavior was used to purposely shorten the lifetimes of higher-lying states to induce detectable laser resonances between two normally metastable states (hydrogen-assisted inverse resonances) [17]. The state-dependent quenching cross sections deduced from the experiment [18] is shown in Fig. 7.

A strong density dependence of the lifetimes of individual states in pure helium has also been found [19]. Here, the lower-lying (37,34) state is easily destroyed by increasing the He density, while the higher-lying (39,35) state is not. A different quenching behavior in the case of O_2 was also revealed [20]. All these phenomena are still unexplained theoretically.

7 Future Scopes

Further experiments are planned for a new antiproton beam facility (Antiproton Decelerator) at CERN, which will provide a pulsed beam of $\bar{p}$ of 100 MeV/c. A new group named ASACUSA was formed, one of the purposes being more extensive studies of this atomcule [40].

Acknowledgement. The author would like to thank all the members of the CERN PS205 experimental group for the collaborative work since 1991 and also Drs. K. Ohtsuki, I. Shimamura, V.I. Korobov and D. Bakalov for the theoretical discussions. The present work is supported by the Grant-in-Aid for the Specially Promoted Research of the Japanese Ministry of Education, Science, Culture and Sport (Monbusho).

References

1. M. Iwasaki et al.: Phys. Rev. Lett. **67**, 1246 (1991)

2. T. Yamazaki et al.: Nature **361** (1993) 238;
   S.N. Nakamura et al.: Phys. Rev. **A49**, 4457 (1994)

3. E. Widmann et al.: Phys. Rev. **A51**, 2870 (1995)

4. E. Widmann et al.: Phys. Rev. **A53**, 3129 (1996)

5. B. Ketzer et al.: Phys. Rev. **A53**, 2108 (1996)

6. N. Morita, K. Ohtsuki and T. Yamazaki: Nucl. Instr. and Meth. **A330**, 439 (1993)

7. H.A. Torii et al.: Nucl. Instr. and Meth. **A396**, 257 (1997)

8. N. Morita et al.: Phys. Rev. Lett. **72**, 1180 (1994)

9. R.S. Hayano et al.: Phys. Rev. Lett. **73**, 1485 (1994); **73**, 3181 (1994) (E)

10. F.E. Maas et al.: Phys. Rev. **A52**, 4266 (1995)

11. H.A.Torii et al.: Phys. Rev. **A53**, R1931 (1996)

12. R.S. Hayano et al.: Phys. Rev. **A55**, 1 (1997)

13. T. Yamazaki et al.: Phys. Rev. **A55**, R3295 (1997)

14. H.A. Torii et al.: Phys. Rev. **A** (1998), (in print)

15. E. Widmann et al.: Phys. Lett. **B404**, 15 (1997)

16. T. Yamazaki et al.: Chem. Phys. Lett. **265**, 137 (1997)

17. B. Ketzer et al.: Phys. Rev. Lett. **78**, 1671 (1997)

18. B. Ketzer et al.: J. Chem. Phys. *accepted* (1998)

19. M. Hori et al.: Phys. Rev. **A57**, 1698 (1998)

20. R. Pohl et al.: Phys. Rev. **A** (1998), (in print)

21. J. Hartmann et al.: Phys. Rev. **A** (submitted)

22. G.T. Condo: Phys. Lett. **9**, 65 (1964)

23. J.E. Russell: Phys. Rev. **A1**, 721 (1970); **A1**, 735 (1970); **A1**, 742 (1970)

24. R. Ahlrichs et al.: Z. Phys. **A306**, 297 (1982)

25. T. Yamazaki and K. Ohtsuki: Phys. Rev. **A45**, 7782 (1992);
    K. Ohtsuki: Private Communication (1993)

26. I. Shimamura: Phys. Rev. **A46**, 3776 (1992);
    Private Communication (1993)

27. P.T. Greenland and R. Thürwächter: Hyperfine Interactions **76**, 355 (1993)

28. O.I. Kartavtsev: Private Communication (1994);
    Phys. At. Nucl. **59**, 1483 (1996)

29. V.I. Korobov: Phys. Rev. **A54**, R1749 (1996)

30. V.I. Korobov and D.D. Bakalov: Phys. Rev. Lett. **79**, 3379 (1997)

31. E. Yarevsky and N. Elander: Europhys. Letters **37**, 453 (1997)

32. N. Elander and E. Yarevsky: Phys. Rev. **A56**, 1855 (1997); **58**, 2256 (1998) (E)

33. Y. Kino, M. Kamimura and H. Kudo: In: *Proc. XV. Int. Conf. on Few-Body Problems in Physics*, Groningen, The Nederlands (1997)

34. V.I. Korobov: (1998) (in print)

35. V.I. Korobov and I. Shimamura: Phys. Rev. **A56**, 4587 (1997)

36. J. Revai and A.T. Kruppa: Phys. Rev. **A57**, 174 (1998)

37. G. Gabrielse et al.: Phys. Rev. Lett. **65**, 1317 (1990)

38. R.J. Hughes and B.I. Deutch: Phys. Rev. Lett **69**, 578 (1992)

39. D. Bakalov and V.I. Korobov: Phys. Rev. **A57**, 1662 (1998)

40. ASACUSA Proposal (1997)

Few-Body Systems Suppl. 10, 161–168 (1999)

Few-
Body
Systems
© by Springer-Verlag 1999

Positrons in atomic physics

J.W. Humberston

Department of Physics and Astronomy, University College London,
Gower Street, London WC1E 6BT, UK

Abstract. Comparisons of positron and electron scattering from atoms provide important insights into collision processes, particularly at low energies where the differences between the results for the two projectiles are most pronounced. Also, there are two processes in positron collisions that have no counterpart in electron collisions, namely positronium formation and electron-positron annihilation.

Recent advances in experimental positron physics now enable detailed comparisons to be made between experimental measurements and theoretical predictions for a variety of scattering and annihilation parameters, providing important tests of the accuracy of the experimental measurements. After discussing the basic features of positron-atom scattering, an account is given of some accurate theoretical results which have recently been obtained for annihilation and positronium formation in hydrogen and helium. Comparisons with available experimental data are made where possible.

1 Introduction

Since the prediction by Dirac in 1930 of the existence of the positron, the anti-particle of the electron, and its subsequent discovery by Anderson in 1932, considerable theoretical attention has been given to atomic systems involving positrons. Some of the earliest investigations were concerned with the possible binding of positrons to atomic systems, but from the mid 1950s studies were also made of positron-atom scattering, even though no experimental results were then available with which to compare theoretical predictions. This situation changed dramatically in the early 1970s with the development of monoenergetic positron beams. Initially these were of such low intensity that only total scattering cross sections could be measured, but subsequent developments have resulted in beams with intensities many orders of magnitude greater and of much narrower energy spread, enabling a variety of partial and differential scattering cross sections to be measured. These experimental developments have in turn stimulated further theoretical investigations which have enabled increasingly detailed comparisons to be made between experimental and theoretical results. Such comparisons have played an important role in helping to reveal systematic errors, and errors in interpretation, in some experimental data.

The positron has several properties which make it a particularly interesting projectile for scattering by atoms and molecules. It is distinguishable from the electrons in the target, and therefore there are no exchange effects with the target system. Its mass and intrinsic spin are the same as those of the electron, but its charge is of opposite sign so that its static interaction with an atomic system is repulsive, whereas that for electrons is attractive. However, the polarization interaction with the target is attractive for both projectiles, and therefore two major components of the overall projectile-target interaction, which are both attractive for electrons, are of opposite sign for positrons. Consequently, total scattering cross sections are usually significantly smaller for positrons than for electrons, particularly at low impact energies where polarization effects are most important. Exceptions to this pattern arise with the alkali atoms because the positronium formation channel is open even at zero incident positron energy. In all cases, however, the total cross sections for the two projectiles merge at higher energies (approximately 200 eV for helium), but it is interesting to note that the energy at which this merging occurs is significantly lower than the energies at which individual partial cross sections merge.

Because of the repulsive character of the static interaction, a positron is much less likely than an electron to bind to an atom. A positron can bind to positronium to form the anti-system of the positronium negative ion, Ps^-, but it has been rigorously proved that it cannot bind to atomic hydrogen or helium. It might be expected to bind to a highly polarizable atom such as an alkali atom, and states of the positron-alkali atom system do indeed exist below the positron-atom elastic scattering threshold, but a true bound state must also lie below the threshold for positronium-ion elastic scattering. No such states were definitely known to exist until Ryzhik and Mitroy [1] rigorously proved that a positron can bind to lithium, although the configuration is more appropriately described as positronium weakly bound to Li^+. It is probable that a positron can also bind to Be, Na and Mg.

Positronium, the bound state of a positron and an electron, has a zero static interaction with any other system because its centre of mass is mid-way between the two equal but opposite charges. Nevertheless, it is known to bind to itself to form the positronium molecule, Ps_2, and also to atomic hydrogen to form positronium hydride, PsH, and there is fairly convincing evidence that it can bind to alkali atoms.

Processes which may occur in positron-atom scattering are as follows:

$$e^+ + X \;\rightarrow\; e^+ + X \qquad \text{(elastic scattering)}$$
$$\rightarrow\; Ps + X^+ \qquad \text{(positronium formation)}$$
$$\rightarrow\; e^+ + X^* \qquad \text{(excitation of the atom)}$$
$$\rightarrow\; e^+ + X^+ + e^- \qquad \text{(ionization)}$$
$$\rightarrow\; 2\gamma \text{ or } 3\gamma + X^+ \qquad \text{(positron - electron annihilation)}$$

Positronium formation and positron-electron annihilation have no counterparts in electron scattering, and we give particular attention to these processes in this article. Annihilation is always possible, but the annihilation cross section is very much less than the elastic scattering cross section (typically, $\sigma_a \approx 10^{-6} \times \sigma_{el}$), although annihilation is, of course, the ultimate fate of all positrons.

Of the two target systems which we consider here, atomic hydrogen is the simplest to treat theoretically, but is difficult to investigate experimentally, whereas helium is simple to investigate experimentally but is much more difficult theoretically. Nevertheless, very accurate theoretical scattering and annihilation results have been obtained for both systems, particularly at low energies where the differences between electron and positron collisions are most pronounced.

2   Elastic scattering

Low energy elastic positron scattering is strongly influenced by the balance between the attractive polarization interaction and the repulsive static interaction and this fact, together with the strong positron-electron correlations, makes the accurate determination of the scattering parameters a particularly challenging task. At very low energies (< 2 eV) the overall interaction is attractive and the s-wave phase shift is therefore positive. As the positron energy is increased, however, the polarization interaction becomes less attractive and the repulsive static interaction becomes dominant, resulting in the s-wave phase shift turning negative. The zero in the s-wave phase shift manifests itself as a Ramsauer minimum in the elastic scattering cross section, and this has been observed experimentally for helium.

Some of the most accurate theoretical results, particularly at low energies, have been obtained using the Kohn variational method with very flexible trial functions to represent all the interparticle correlations [2]. Among several other approximation methods which have been used to obtain accurate results, mention should also be made of the coupled-state method in which the total wave function is expanded in terms of eigenstates of the target atom and of positronium. This expansion is doubly complete, and in principle an expansion in terms of the states of target system alone should be sufficient. However, the convergence of the phase shifts with respect to increasing the number of terms in the expansion is very slow unless both types of term are included. This approximation method has been used particularly effectively by Walters and his collaborators.[3, 4] to calculate elastic and various inelastic scattering cross sections for positron scattering by atomic hydrogen, helium and the alkali atoms over quite a wide energy range.

3   Positron-electron annihilation

If the positron and the electron in the target with which it is about to annihilate are in a singlet spin state (S = 0) annihilation is into two γ-rays, each with an energy of approximately $m_o c^2 = 511$ keV. If, however, the two particles are in a triplet spin state (S = 1) annihilation is into three γ-rays, but at a much slower rate than that for a singlet pair. When a beam of unpolarized positrons passes through a gas target, annihilation is therefore predominantly into two γ-rays, the cross section for which is

$$\sigma_a = \pi r_o^2 c Z_{\text{eff}}/v, \tag{1}$$

where r_o is the classical radius of the electron, v is the speed of the positron relative to the atom, c is the speed of light and Z_{eff} is an effective number of electrons in each atom. This parameter is a measure of the electron density in the immediate vicinity of the positron at the time of annihilation and its value can be calculated from the elastic scattering wave function, Ψ, as follows;

$$Z_{\text{eff}} = \sum_{i=2}^{Z+1} \int |\Psi(r_1, r_2, r_3, \dots, r_{Z+1})|^2 \delta(r_1 - r_i) dr_1 dr_2 dr_3 \dots dr_{Z+1}, \tag{2}$$

where r_1 is the coordinate of the positron and $r_2, \dots, r_{Z+1}$ are the coordinates of the Z electrons. In the Born approximation, in which there is no distortion of the wave functions of either the target or the positron, $Z_{\text{eff}} = Z$, independent of the energy of the positron beam. However, the exact value of Z_{eff} is greater than Z for almost all target systems, and is also energy dependent. Using very accurate scattering wave functions at a mean positron energy corresponding to room temperature (40 meV), the following values of Z_{eff} have been obtained: 8.9 for atomic hydrogen [5] and 3.88 for helium [6]. The latter value is in good agreement with the experimental value of 3.9 [7]. As the positron energy is

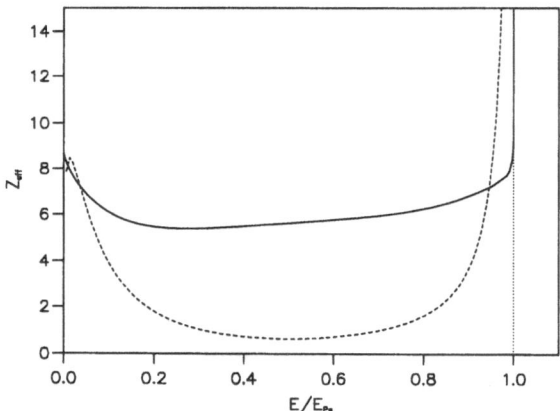

Figure 1. The energy dependence of Z_{eff} for positron-hydrogen scattering compared with the predictions of the model of Laricchia and Wilkin (broken curve).

increased from zero the target electrons have progressively less time in which to adjust to the attractive influence of the positron, and the value of Z_{eff} falls. Thereafter, Z_{eff} increases slowly until just below the positronium formation threshold when it starts to increase very rapidly, tending to infinity at the threshold itself. The rapid rise is attributed to the increasing significance of the configuration of virtual positronium bound to the residual positive ion. The positron is then trapped in the vicinity of the electron for longer than the usual collision time, resulting in a much larger probability of annihilation. Calculations of Z_{eff} at energies above the positronium formation threshold also yield very large values (infinite if a time-independent scattering formalism is used) because the positron remains close to the electron in the positronium which is formed until annihilation certainly occurs. The positronium formation cross section is therefore, in a sense, also the annihilation cross section, and because it is several orders of magnitude larger than the typical magnitude of the annihilation cross section below the threshold, so too is the equivalent value of Z_{eff} obtained using Eq. 1. The energy dependence of Z_{eff} for hydrogen is shown in Fig. 1 [8] where it is compared with the predictions of a rather simple model [9] which attempts to explain the enormously wide range of values of Z_{eff} for different atoms and molecules in terms of virtual positronium formation. Quite large quantitative differences exist between the two sets of results, but the accurate results nevertheless provide qualitative confirmation of the main features of the predictions of the model.

In the centre of mass coordinate system of an annihilating electron-positron pair in a singlet spin state, the angle between the two emerging γ-rays is π, and the energy of each one is $E_o = 511$ keV. However, after transforming to the laboratory frame of reference, in which the atomic nucleus is at rest, the angle between the two γ-rays becomes $(\pi - \theta)$ and their energies are Doppler shifted to $E = E_o \pm \Delta E$. The values of θ and ΔE depend on the momentum, p, of the annihilating electron-positron pair in the laboratory frame of reference, the relationship being $\Delta E = m_o c^2 \theta / 2 = cp/2$. Measurement of the angular correlation of the two γ-rays, or the Doppler broadening of the energy of either one of them, therefore provides information on the momentum distribution of the electron-positron pair at the moment of annihilation. This distribution function, $\Gamma(\boldsymbol{p})$, may also be calculated from the wave function for elastic positron-atom scattering; thus, for positron-helium scattering

$$\Gamma(\boldsymbol{p}) = \int d\boldsymbol{r_3} \left| \int \exp(-i\boldsymbol{p}.\boldsymbol{r_2})\Psi(\boldsymbol{r_1} = \boldsymbol{r_2}, \boldsymbol{r_2}, \boldsymbol{r_3})d\boldsymbol{r_2} \right|^2 . \qquad (3)$$

Van Reeth et al. [6] have obtained excellent agreement between theoretical and experimental energy distributions for very low energy positrons annihilating in helium gas, as shown in Fig. 2.

4 Positronium formation

The threshold energy for positronium formation is $E_{Ps} = (E_i - 6.8)$ eV, where E_i is the ionization energy of the target atom. For atoms with ionization energies less than 6.8 eV, such as the alkali atoms, the positronium formation channel is open at zero incident positron energy, where the formation cross section is infinite. For hydrogen and helium, however, the threshold energies

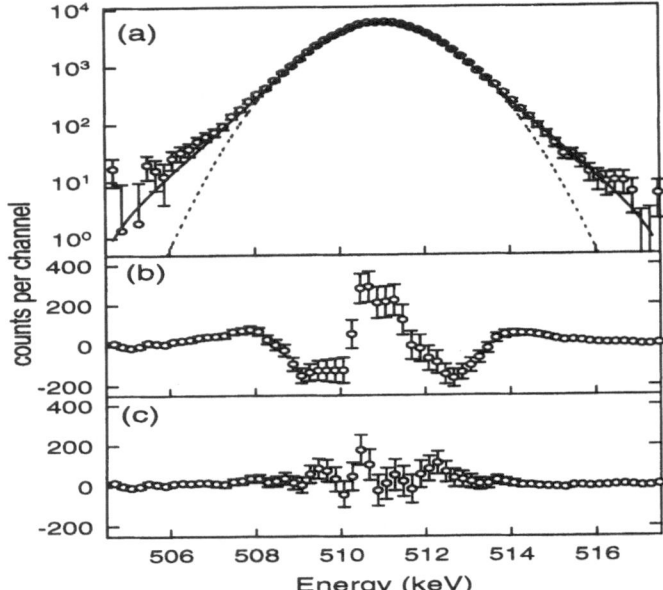

Figure 2. (a) Annihilation γ-ray spectrum for positrons interacting with helium atoms. Solid line: theoretical prediction convolved with the response of the Ge detector; dashed line: Gaussian function fitted to the experimental data; o, experimental measurements. (b) Residuals from the Gaussian fit. (c) Residuals from the theoretical calculation.

are 6.8 eV and 17.8 eV respectively. Because the positronium formation threshold has a lower energy than that for any other inelastic process there is an energy interval between this threshold and the lowest positron impact excitation threshold, known as the Ore gap, within which the only two processes are elastic scattering and positronium formation. The most detailed and accurate theoretical studies of positronium formation have been made within the Ore gap using similar variational methods to those used for elastic scattering. Particular attention has recently been given to investigating structures in the elastic scattering and positronium formation cross sections in the vicinity of the positronium formation threshold. The lth partial wave contribution to the positronium formation cross section should exhibit an energy dependence of the form [10]

$$\sigma_{Ps}(l) \propto (E - E_{Ps})^{l+1/2}, \tag{4}$$

and therefore the s-wave contribution ($l = 0$) should have an infinite slope at the threshold itself. Accurate results do indeed reveal this threshold structure, although the energy range over which the threshold law is valid is very narrow [11], as may be seen for hydrogen in Fig. 3. The opening of the positronium formation channel influences the elastic scattering cross section on both sides of the threshold, with the s-wave contribution displaying a Wigner 'cusp' at the threshold itself ([12] for hydrogen; [13] for helium). Among several interesting features of these results, which are discussed in more detail by Van Reeth and

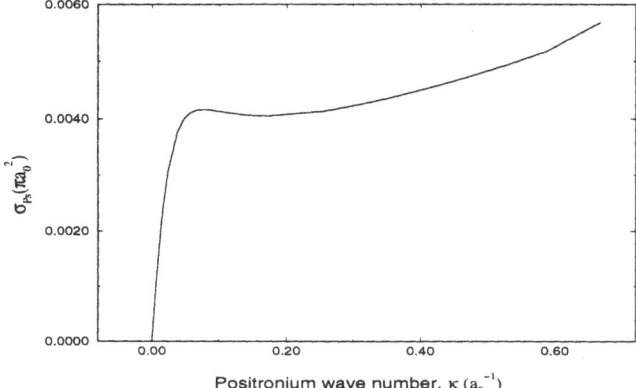

Figure 3. Variation of $\sigma_{Ps}(0)$ with positronium wavenumber for positron-hydrogen scattering.

Humberston in this volume, is the fact that the s-wave positronium formation cross sections for hydrogen and helium are very similar in magnitude and energy dependence. Also, the sum of the s-wave elastic scattering and positronium formation cross sections within the Ore gap matches well with a smooth extrapolation of the s-wave elastic scattering cross section from some way below the positronium formation threshold. Thus, apart from a rather narrow feature just below the threshold itself, the s-wave total scattering cross section continues rather smoothly across the threshold. This structure has been observed in both hydrogen and helium.

Accurate values of the higher partial wave contributions to the positronium formation cross section have also been calculated for hydrogen [12] and helium [14], enabling comparisons to be made between the experimental and theoretical values of the total positronium formation cross section. The experimental results for hydrogen have rather large statistical errors associated with them, but they are broadly in agreement with the theoretical results. The experimental results for helium [15] which are expected to be reasonably accurate, display an energy dependence which is quite accurately proportional to $(E - E_{Ps})^{1.5}$ all the way up to the ionization threshold, apparently implying that the p-wave contribution to the positronium formation cross section is dominant throughout this energy range. However, the theoretical results for helium reveal that the p-wave contribution is not dominant. Admittedly it rapidly exceeds the s-wave contribution as the positron energy is increased, but it is in turn exceeded by the d-wave contribution, which is the largest single contribution throughout most of the Ore gap. Nevertheless, the sum of these and higher partial wave contributions add up to give an energy dependence of the total positronium formation cross section which is reasonably similar to that of the experimental results, as shown in Fig. 4.

In this brief review some of the interesting aspects of low energy positron-atom systems have been discussed and comparisons have been made between theoretical and experimental results in order to show the level of precision and

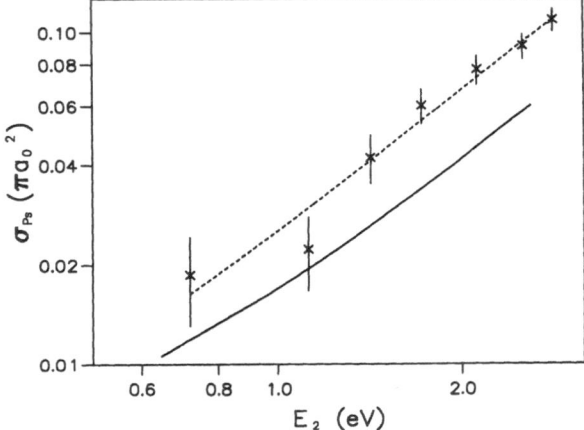

Figure 4. Theoretical and experimental total positronium formation cross sections for positron-helium scattering, where $E_2 = E - E_{Ps}$. The experimental data ($\times$) are from [15]

detail that can now be obtained. The availabilty of reliable theoretical data against which to check the accuracy of experimental results is an important factor in the continuing development of this exciting field.

Acknowledgement. I wish to thank P. Van Reeth, with whom I have had a close and productive collaboration in these investigations during the past few years. This work was supported by the United Kingdom EPSRC on grant number GR/L38431.

References

1. G.G. Ryzhikh and J. Mitroy: Phys. Rev. Lett. **79**, 4214 (1997)
2. E.A.G. Armour and J.W. Humberston: Phys. Reports **204**, 165 (1991)
3. A.A. Kernoghan et al.: J. Phys. **B29**, 2089 (1996)
4. C.P. Campbell et al.: Nucl. Instrum. Methods **B** (1998) (to be published)
5. J.W. Humberston and J.B.G. Wallace: J. Phys. **B5**, 1138 (1972)
6. P. Van Reeth et al.: J. Phys. **B29**, L465 (1996)
7. P.G. Coleman: J. Phys. **B8**, 1734 (1975)
8. P. Van Reeth and J.W. Humberston: J. Phys. **B31**, L231 (1998)
9. G. Laricchia and C. Wilkin: Phys. Rev. Lett. **77**, 2241 (1997)
10. E.P. Wigner: Phys. Rev. **73**, 1002 (1948)
11. P. Van Reeth and J.W. Humberston: J. Phys. **B31**, L621 (1998)
12. J.W. Humberston et al.: J. Phys. **B30**, 2477 (1997)
13. P. Van Reeth and J.W. Humberston: J. Phys. **B28**, L511 (1995)
14. P. Van Reeth and J.W. Humberston: J. Phys. **B30**, L95 (1997)
15. J. Moxom et al.: Phys. Rev. **A50**, 3129 (1994)

Few-Body Systems Suppl. 10, 169–178 (1999)

Few-
Body
Systems
© by Springer-Verlag 1999

Few Body Problems in Muon Catalyzed Fusion

V.E. Markushin*

Paul Scherrer Institute, CH-5232 Villigen, Switzerland

Abstract. Recent theoretical and experimental results in muon catalyzed fusion are reviewed from the viewpoint of intersections with few-body problems. Nuclear reactions in muonic molecules are considered as a selective probe of low energy nuclear interactions. Precision calculations of the bound and scattering states in the Coulomb three-body system are discussed.

1 Introduction

Muon catalyzed fusion (μCF) is a nuclear fusion reaction in which the Coulomb repulsion between the nuclei in the initial state is partially screened by a muon in a bound state with the initial nuclei (a muonic molecule) or with one of them (a muonic atom). In a broad sense, the term μCF includes various reactions with muonic atoms and molecules, which lead to or compete with the processes resulting in the fusion reaction. Several μCF reactions have been observed experimentally in muonic molecules of hydrogen isotopes (see [1, 2, 3] and references therein):

$$pd\mu \rightarrow \gamma + \mu^3\text{He}, \ \mu + {}^3\text{He} \tag{1}$$

$$pt\mu \rightarrow \gamma + \mu^4\text{He}, \ \mu + {}^4\text{He} \tag{2}$$

$$dd\mu \rightarrow p + t + \mu, \ n + {}^3\text{He} + \mu, \ n + \mu^3\text{He} \tag{3}$$

$$dt\mu \rightarrow n + \mu^4\text{He}, \ n + {}^4\text{He} + \mu \tag{4}$$

$$tt\mu \rightarrow 2n + \mu^4\text{He}, \ 2n + {}^4\text{He} + \mu \tag{5}$$

The characteristic scales of muonic atoms and molecules are determined by the muon mass m_μ: the space scale $a_\mu = (m_\mu\alpha)^{-1} = 256$ fm and the energy scale $E_\mu = m_\mu\alpha^2 = 5.63$ keV (here and below we use $\hbar = c = 1$).

The muonic molecules of the hydrogen isotopes are analogous to the hydrogen molecular ion H_2^+, but they are much smaller than the ordinary molecules (by about a factor of $m_\mu/m_e \approx 200$ which makes the muon screening much

---

*E-mail address: markushin@psi.ch

Table 1. The energies E_{Jv} (eV) of the rotational-vibrational states (Jv) of muonic molecules associated with the muon configuration $1s\sigma$ (see [4] and references therein). The energy is counted from the lowest threshold $a + (\mu b)_{1S}$ (b is the heaviest nucleus), real negative values correspond to bound states, complex values to resonances.

(Jv)	(00)	(10)	(20)	(30)	(01)	(11)
$pp\mu$	-253.15	-107.27	—	—	—	—
$pd\mu$	-221.55	-97.50	—	—	—	—
$pt\mu$	-213.84	-99.13	—	—	—	—
$dd\mu$	-325.07	-226.68	-86.49	$+34.7 + i4.5$	-35.84	-1.97
$dt\mu$	-319.14	-232.47	-102.65	$+22.2 + i0.7$	-34.83	-0.66
$tt\mu$	-362.91	-289.14	-172.65	-48.70	-83.77	-45.21

more efficient than the electron one in assisting the nuclear fusion) and have only a few rotational-vibrational levels as shown in Table 1.

The average distance between nuclei in muonic molecules is much larger than the range of the nuclear interaction, therefore the structure of muonic molecules is determined mainly by the electromagnetic interaction. Neglecting the nuclear interaction, the muonic molecules are few-body systems with exactly known Hamiltonian, and numerical calculations of their properties (spectrum, wave functions) can be pushed to the limits of modern computational methods. It is remarkable that some of these precision calculations can be tested experimentally.

Muon catalyzed fusion has many features attracting both theoretical and experimental interest; this paper will focus on those which are directly related to the physics of few body systems. The nuclear reactions in muonic molecules are considered in Sect.2. The resonances in molecular formation are described in Sect. 3. The precision calculations of muonic molecules are discussed in Sect. 4 and the scattering problem in Sect. 5.

2 Nuclear Reactions in Muonic Molecules

In muonic molecules, the width of the Coulomb barrier between the nuclei is about a_μ which is approximately equivalent to the conventional fusion in-flight ("hot" fusion) at energy $E = \alpha/a_\mu = E_\mu \approx 6$ keV. The nuclear interaction in muonic molecules can be treated as a perturbation using the smallness of the nuclear scattering amplitude in comparison with the molecular scale a_μ. In the leading order, the nuclear fusion rate for molecular states $(ab\mu)$ with orbital momentum $J = 0$ is given by the factorization formula:

$$\lambda^f = A_{ab}\,\rho_{ab\mu} \tag{6}$$

$$A_{ab} = \lim_{E \to 0} \frac{v^2}{2\pi\alpha}(e^{\frac{2\pi\alpha}{v}} - 1)\,\sigma(E) = \frac{S(E = 0)}{\pi\alpha m_{ab}}, \quad E = m_{ab}v^2/2 \tag{7}$$

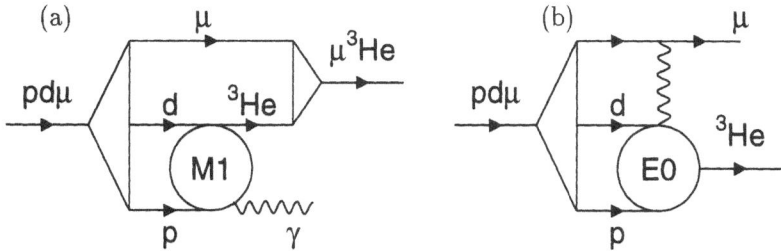

Figure 1. Nuclear reaction amplitudes for the $pd\mu$ molecule: (a) the radiative pd capture, (b) the nonradiative pd capture (muon conversion).

$$\rho_{ab\mu} = \int |\Psi_{ab\mu}(\boldsymbol{R} = 0, \boldsymbol{r})|^2 d^3\boldsymbol{r}. \tag{8}$$

Here A_{ab} is the reaction constant determined by the low energy behaviour of the reaction cross section $\sigma(E)$, $S(E)$ is the astrophysical factor, m_{ab} is the reduced mass of the nuclei. The nuclear density at zero internuclear distance, $\rho_{ab\mu}$, is obtained from the molecular wave function $\Psi_{ab\mu}(\boldsymbol{R}, \boldsymbol{r})$ where $\boldsymbol{R}$ is the internuclear separation and $\boldsymbol{r}$ is the muon coordinate.

2.1 Nuclear Reactions in the $pd\mu$ Molecule

The $pd\mu$ molecule is a good example of using μCF as a probe of nuclear reactions in few-body systems [5, 6]. The amplitudes for the two reaction channels (1) are given by the diagrams in Fig. 1. The $pd\mu$ molecule works as a partial-wave filter: the fusion always takes place in the rotational-vibrational ground state of the $pd\mu$ (as soon as the molecule is formed, it deexcites to the state $(Jv) = (00)$ much faster than the fusion occurs), so that the pd system has the relative angular momentum $L = 0$. The fusion rates depend on the total nuclear spin $\boldsymbol{S}_{pd} = \boldsymbol{S}_p + \boldsymbol{S}_d$. The radiative capture proceeds via $M1$ transitions and the fusion rates for the spin states $S_{pd} = \frac{1}{2}, \frac{3}{2}$ are given according to (6,8) by the formulas [7]:

$$\lambda^{\gamma}_{S_{pd}} = \frac{S_{S_{pd}}}{\pi \alpha m_{pd}} \cdot \rho_{pd\mu} \tag{9}$$

$$S_{S_{pd}} = \frac{\pi \alpha^2 m_{pd} \omega^3}{3(2S_{pd} + 1)m_p^2} |\langle {}^3\text{He}||M1||pd; L = 0, S_{pd}\rangle|^2. \tag{10}$$

Here $S_{S_{pd}}$ is the S-wave astrophysical factor for the spin state S_{pd}, $\langle {}^3\text{He}||M1||pd; L = 0, S_{pd}\rangle$ is the corresponding reduced matrix element, m_{pd} is the reduced mass of the pd system, and ω is the photon energy. The $M1$-transitions are very sensitive to the meson exchange currents [7, 8], thus the $pd\mu$ molecule can be used as a very selective probe (only the S-wave contributes to the pd capture) of non-nucleonic degrees of freedom in $A = 3$ systems.

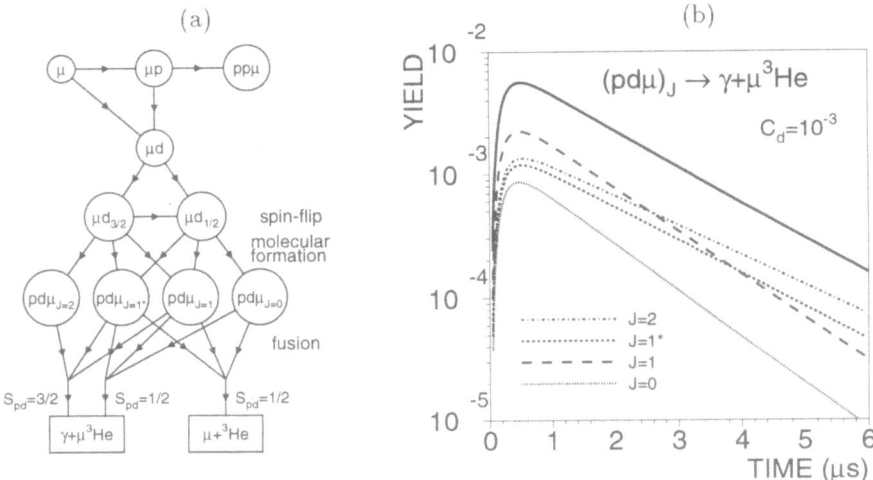

Figure 2. (a) The scheme of the $pd\mu$ cycle. (b) The time distribution of the nuclear fusion $pd\mu \to \gamma + {}^3\mathrm{He}$ (solid line). The muons are stopped in the H+D mixture at $t = 0$. The time dependence at large t is determined by the disappearance rates λ_J of the individual states $(pd\mu)_J$ (their contributions are shown separately with dashed and dotted lines, the rate λ_J is the sum of the muon decay rate and the corresponding total fusion rate). The build-up at small time is determined by the $pd\mu$ formation.

The nonradiative fusion $pd\mu \to \mu + {}^3\mathrm{He}$ is dominated by the $E0$-transition, and its rate is given by formula [5]:

$$\lambda^\mu_{S_{pd}=1/2} \;=\; \frac{2\pi\alpha^2}{9}\, m_{\mu^3\mathrm{He}} q_\mu \, |\langle ||E0|| \rangle|^2\, |\Psi_{pd\mu}(\boldsymbol{R}_{pd} = 0, \boldsymbol{r} = 0)|^2. \quad (11)$$

Here q_μ is the muon momentum, $m_{\mu^3\mathrm{He}}$ is the reduced mass of the final state, and the reduced matrix element $\langle ||E0|| \rangle = \langle {}^3\mathrm{He}||E0||pd; L = 0, S_{pd} = \frac{1}{2}\rangle$ is calculated at the recoil momentum q_μ.

The fusion rates $\lambda^\gamma_{1/2}$ and $\lambda^\gamma_{3/2}$ can be determined using the hyperfine structure (h.f.s.) of the $pd\mu$ molecule (see Fig. 2). The h.f.s. states with total spin $J = |\boldsymbol{S}_p + \boldsymbol{S}_d + \boldsymbol{S}_\mu| = 0$ and $J = 2$ have definite nuclear spin $S_{pd} = \frac{1}{2}$ and $S_{pd} = \frac{3}{2}$ correspondingly, and the other two h.f.s. states with $J = 1$ are superpositions of $S_{pd} = \frac{1}{2}, \frac{3}{2}$. The population of the $pd\mu$ h.f.s. states can be controlled by changing the composition of the hydrogen-deuterium mixture where the $pd\mu$ molecules are formed in $\mu d + \mathrm{H}_2$ collisions. At low deuterium concentration C_d the h.f.s. states of μd atom ($F = |\boldsymbol{S}_\mu + \boldsymbol{S}_d| = \frac{1}{2}, \frac{3}{2}$) are populated statistically that leads to the statistical population of the $pd\mu$ states. With increasing C_d the relative population of the lowest μd state with $F = \frac{1}{2}$ increases due to the irreversible spin flip in the collisions $\mu d_{3/2} + d \to \mu d_{1/2} + d$ (the h.f.s. splitting $\Delta E = E_{\mu d_{3/2}} - E_{\mu d_{1/2}} = 0.0485\,\mathrm{eV}$ is larger than the energy of thermal motion at temperature $T < 300\,\mathrm{K}$). The resulting change in the relative population of the $pd\mu$ states enhances the contribution of the nuclear spin state $S_{pd} = \frac{1}{2}$

Table 2. The $pd\mu$ fusion rates λ^{γ} and λ^{μ}, the astrophysical factor $S = (S_{1/2} + 2S_{3/2})/3$ for the S-wave radiative pd capture at $E \to 0$, and the quartet-to-doublet ratio of the $M1$ strengths $M1_{3/2}/M1_{1/2} = \sqrt{S_{3/2}/S_{1/2}}$: (a) analysis of the $pd\mu$ data, (b) theoretical calculations including MEC, (c) the impulse approximation, (d) measurements of the reaction $p + d \to \gamma + {}^3\mathrm{He}$.

	$pd\mu$ fusion rates (μs^{-1})			S-factors (eV·b)	
	$\lambda^{\gamma}_{1/2}$	$\lambda^{\gamma}_{3/2}$	$\lambda^{\mu}_{1/2}$	S	$(\frac{S_{3/2}}{S_{1/2}})^{\frac{1}{2}}$
PSI-90[a] [6]	0.35(2)	0.11(1)	—	0.105(6)	0.56(4)
TRIUMF-98[a] [9]	0.40(2)	0.10(1)	—	0.110(6)	0.50(4)
Bogdanova et al.[a] [5]	—	—	0.056(6)	—	—
Friar et al.[b] [7]	0.37(1)	0.107(6)	0.062(2)	0.108(4)	0.54(2)
Viviani et al.[b] [8]	—	—	—	0.105	0.49
IA[c] [8]	—	—	—	0.065	1.0
Griffiths et al.[d] [10]	—	—	—	0.12(3)	—
TUNL[d] [11, 12]	—	—	—	0.109(10)	0.50(15)

to the fusion reaction. By analysing the fusion time distributions (Fig. 2b) at different C_d one can determine the fusion rates.

The summary of the theoretical and experimental results related to the S-wave pd radiative capture at low energy is given in Table 2[1]. The very good agreement between the results obtained from the analysis of $pd\mu$ data [6, 9], the recent measurements of the pd capture [11, 12] and the theoretical calculations using realistic nuclear forces and meson-exchange currents (MEC) [7, 8] is remarkable. The meson-exchange currents are most important for the doublet ($S_{pd} = \frac{1}{2}$) $M1$ matrix element which is enhanced by a factor of about 2 in comparison with the impulse approximation.

After catalyzing a fusion reaction the muon either forms an atomic state with the charged nucleus or ends up in a continuum state ready for the next μCF cycle. In the sudden approximation, the probability of the muon sticking to helium in the reaction $ab\mu \to (\mu He)_{nl} + X$ is given by the formula:

$$\omega_{nl}(ab\mu) = \rho_{ab\mu}^{-1} \left| \int \phi_{nl}^{*}(\boldsymbol{r}) \, e^{-im_{\mu}\boldsymbol{v}\boldsymbol{r}} \, \Psi_{ab\mu}(\boldsymbol{R} = 0, \boldsymbol{r}) d^3 r \right|^2 \qquad (12)$$

where $\phi_{nl}(\boldsymbol{r})$ is the $(\mu He)_{nl}$ wave function, $\boldsymbol{v}$ is the velocity of the muonic helium in the final state. The sticking probabilities are very sensitive to the molecular wave function at small internuclear separations [13, 14]. For the radiative $pd\mu$ fusion $\omega_{1S}(pd\mu) = 0.96$, $\omega_{n=2}(pd\mu) = 0.028$ and the total sticking probability is $\omega(pd\mu) = 0.99$ [14, 15].

---

[1]For comparison, the muon decay rate is $\lambda_0 = 0.455 \ \mu s^{-1}$ and the $pd\mu$ formation rate in liquid hydrogen is $\lambda_{pd\mu} = 5.6 \ \mu s^{-1}$.

2.2 Nuclear Reactions in the $pt\mu$ Molecule

The radiative capture and the muon conversion in the $pt\mu$ molecule are similar to the $pd\mu$ case. The $pt\mu$ state $(Jv) = (00)$ has three h.f.s. states $S_{pt\mu} = \frac{1}{2}, \frac{1}{2}', \frac{3}{2}$. The radiative capture is possible only from the configurations with nuclear spin $S_{pt} = 1$ via $M1$-transition and the muon conversion from the configurations $S_{pt} = 0$ via $E0$-transition. The measured radiative fusion rate $\lambda^\gamma_{S_{pt}=1} = 0.067(5)$ μs^{-1} [16] corresponds to a reaction constant $A_{pt} = 3.3(3)$ nb. This value is much larger than the result $A_{n^3\text{He}} = 0.39(4)$ nb [17] for the mirror reaction $n + ^3\text{He} \to \gamma + ^4\text{He}$, and this difference points to a problem waiting for a solution.

3 Resonance Formation of Muonic Molecules

When a muonic molecule is formed in a collision of a muonic atom with a molecule of hydrogen isotopes, the molecular binding energy is transferred either to the electron (Auger process) or to the rotational-vibrational excitations of the resulting molecular complex. The Auger process dominates for the molecular states with binding energy larger than the ionization potential (the molecules $pd\mu$ and $pt\mu$ are formed in this way), the molecular formation rate being weakly dependent on the collision energy. The resonance formation [1, 2] has a strong energy dependence, and in case of weakly bound state $(dt\mu)_{Jv=11}$ the molecular formation rates exceed the muon decay rate by orders of magnitude[2]. The best studied case of the resonant formation is the $dd\mu$ cycle that we discuss below.

The weakly bound state $(dd\mu)_{Jv=11}$ is formed in the resonance reaction:

$$\mu d_F + (D_2)_{\nu_i K_i} \quad \to \quad [(dd\mu)^+_{(Jv)=(11)} d^+ e^- e^-]_{\nu_f K_f} \tag{13}$$

where the μd atom is in the h.f.s. state $F = \frac{1}{2}, \frac{3}{2}$, and the initial molecule D_2 and the final complex $[(dd\mu)^+_{(Jv)=(11)} d^+ e^- e^-]$ are in the rotational-vibrational states $(\nu_i K_i)$ and $(\nu_f K_f)$ with the energies $E_{\nu_i K_i}$ and $E_{\nu_f K_f}$ correspondingly. The resonances occur at the collision energy $E = E_{\nu_f K_f} - E_{\nu_i K_i} + E_{(dd\mu)_{11}}$ with the strength dependent on the form factors for the transitions $(\nu_i K_i) \to (\nu_f K_f)$.

The energy dependence of the resonance formation can be studied experimentally by changing the temperature of the deuterium target where the $dd\mu$ molecules are formed. The kinetics scheme of the $dd\mu$ μCF cycle is shown in Fig. 3a; important processes are also the nonresonant $dd\mu$ formation and the spin-flip $\mu d_{3/2} + d \leftrightarrow \mu d_{1/2} + d$. The $dd\mu$ formation rates have been measured in many experiments, the most accurate results obtained recently by the Gatchina-PSI collaboration [18] are shown in Fig. 3b (see [19] and references therein for earlier results).

The strong temperature dependence of the $dd\mu$ formation rates allows one to determine the binding energy of the $(Jv) = (11)$ state with an accuracy of

---

[2]An important consequence for the $dt\mu$ fusion cycle is the possibility to catalyze more than 10^2 dt fusions during the muon life-time [1, 2].

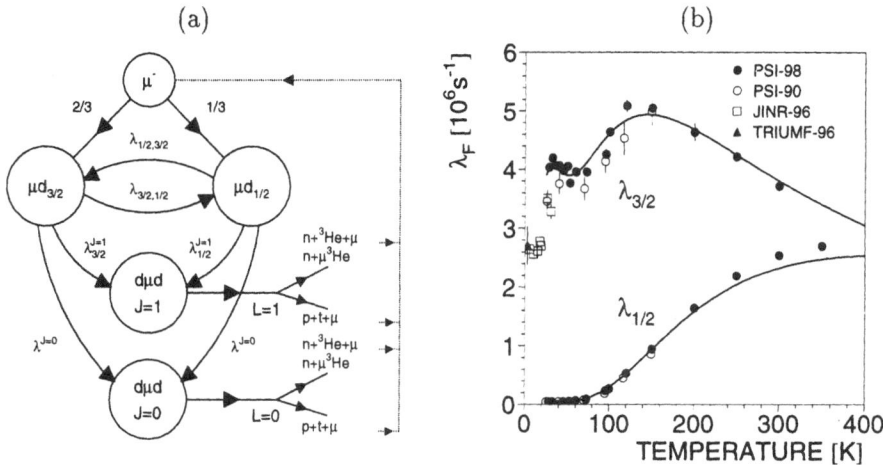

Figure 3. (a) The scheme of the $dd\mu$ cycle. (b) The temperature dependence of the $dd\mu$ formation rates from the μd h.f.s. states $F = \frac{1}{2}, \frac{3}{2}$. The theoretical curves are from [20].

about 10^{-4}eV ~ 1 K (see [20] and references therein). In comparison with the μ-atomic scale $E_\mu = 5.6$ keV this corresponds to a relative accuracy of about 10^{-8}, and the theory of muonic molecules stands up to this challenge.

4 Precision Calculations of the Bound States

While the adiabatic picture can be used for classification of molecular states (Table 1), the nonadiabatic effects in muonic molecules are very important [1], and that requires new methods beyond the Born-Oppenheimer approximation. A brief illustration of high-accuracy calculations of $dd\mu$ and $dt\mu$ molecules in the framework of the nonrelativistic Coulomb three-body problem is given in Table 3. The best results for the binding energies have been obtained with the variational methods (the level of precision reached is comparable with the current uncertainty in the muon mass). The corrections to the nonrelativistic result (relativistic terms, vacuum polarization, finite nuclear size, nuclear polarization, molecular environment) becomes important at the level 10^{-2}eV (see [24] and references therein).

Recently a significant progress was made in using the adiabatic hyperspherical expansion (AHSA) for solving the Coulomb 3-body problem both for bound and scattering states [15, 25, 26, 27]. In particular, the local characteristics like nuclear densities (8) and the sticking fractions (12) can be calculated with relative precision better than 10^{-2} using only 15 terms of the AHSA expansion [15] (about 10^3 variational parameters are usually needed [28] to reach the same accuracy).

Table 3. The binding energies B_{Jv} (eV) of the $dd\mu$ and $dt\mu$ states $(Jv) = (00), (11)$ calculated with different variational methods: (RTS) the random-tempered Slater-type basis $(N = 500, 1200)$ [21], (CRA) coupled-rearrangement with Gaussian basis $(N = 1988)$ [22], (ESC) expansion in spheroidal coordinates $(N = 2000)$ [23]. The masses and physical constants correspond to set 2 from [21].

Method	$dd\mu$		$dt\mu$	
	B_{00}	B_{11}	B_{00}	B_{11}
RTS [21]	325.0735402	1.9748717	319.1397226	0.6601721
CRA [22]			319.139606	0.660104
SCE [23]	325.0735	1.97475	319.1396	0.6600

5 Scattering Problem

Extensive calculations of the collisions of muonic atoms in the $1S$ state with hydrogen atoms and molecules have been done using various methods [29, 30, 31, 32], and the results are available in detailed tables of cross sections [29, 33].

One interesting result concerns the Ramsauer-Townsend (RT) effect in the $\mu d + p$ and $\mu t + p$ scattering (Fig. 4a): the elastic cross section has a deep minimum corresponding to a zero of the S-wave scattering amplitude. As a result, hydrogen becomes nearly transparent for μd and μt atoms in the corresponding energy range. This effect is used in TRIUMF experiment [34] where μd and μt atoms are produced following μ transfer from muonic protium in a layer of solid hydrogen with a small fraction of deuterium or tritium. Emitted as atomic beams in vacuum (Fig. 4b), the μ-atoms are used for measurements of the energy dependence of various reactions by time-of-flight method.

6 Conclusion

The field of μCF spans the gap between the atomic and nuclear scales, the muonic molecules playing a role complementary to that of exotic atoms. In particular, μCF allows one to study nuclear reactions in well defined partial waves at very low energies. Precision calculations have been done for many molecular states resulting in a set of very useful benchmarks for few-body calculations. The size of this paper is too small to discuss all interesting intersections of μCF with few-body systems. Among the missing, but not less important topics, are the following: scattering of muonic atoms in excited states and Feshbach resonances [35]; Coulomb transitions and muon transfer in the atomic cascade [36, 37]; kinetics of $dt\mu$ and $dd\mu$ cycles [1, 2]; muonic molecules with hydrogen and helium isotopes (d^3Heμ, etc.) [38].

Therefore, the interplay of nuclear and electromagnetic interactions in μCF processes leads to various interesting and even unique phenomena the study of which can be beneficial for few-body physics.

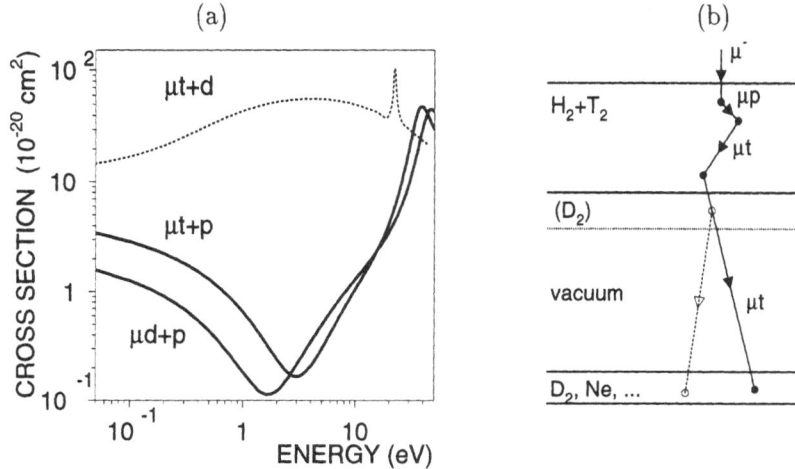

Figure 4. (a) The energy dependence of the elastic cross sections for the $\mu d + p$, $\mu t + p$, and $\mu t + d$ collisions [29]. (b) The μt emission from a multilayered solid target based on the RT effect [34]: (1) μt production layer, (2) moderator, (3) reaction layer.

Acknowledgement. The author thanks his colleagues and collaborators at MU-CATEX, PNPI, PSI, RRCKI, TRIUMF for fruitful discussions and the Organising Committee of FB-XVI for partial support of his participation in the FB-XVI Conference.

References

1. L.I. Ponomarev: Contemp. Phys. **31**, 218 (1991)

2. J.S. Cohen: In: *Review of Fundamental Processes and Applications of Atoms and Ions, Ch. 2*, ed. C.D. Lin. Singapore: World Scientific 1994

3. L.N. Bogdanova: Surveys in High Energy Physics **6**, 177 (1992)

4. V.I. Korobov: Hyperfine Interactions **101/102**, 307 (1996)

5. L.N. Bogdanova, V.E. Markushin: Muon Catal. Fusion **5/6**, 189 (1990/91)

6. C. Petitjean et al.: Muon Catal. Fusion **5/6**, 199 (1990/91)

7. J.L. Friar, B.F. Gibson, H.C. Jean, G.L. Payne: Phys. Rev. **66**, 1827 (1991)

8. M. Viviani, R. Schiavilla, A. Kievsky: Phys. Rev. **C54**, 534 (1996)

9. A. Adamczak et al.: Hyperfine Interactions (to be published)

10. G.M. Griffiths, M. Lal, C.D. Scarfe: Can. J. Phys. **41**, 724 (1963)

11. G.J. Schmid, M. Viviani, B.J. Rice et al.: Phys. Rev. Lett. **76**, 3088 (1996)

12. H.R. Weller: Nucl. Phys. **A631**, 627c (1998)

13. L.N. Bogdanova et al.: Nucl. Phys. **A454**, 653 (1986); L.N. Bogdanova et al.: Sov. J. Nucl. Phys. **50**, 848 (1989)

178

14. V.E. Markushin: Muon Catal. Fusion **3**, 395 (1988)

15. D.I. Abramov et al.: Phys. At. Nucl. **61**, 457 (1998)

16. P. Baumann et al.: Phys. Rev. Lett. **70**, 3720 (1993)

17. F.L. Wolfs et al.: Phys. Rev. Lett. **63**, 2721 (1989)

18. C. Petitjean et al.: Hyperfine Interactions (to be published)

19. J. Zmeskal et al.: Phys. Rev. **A42**, 1165 (1990); C. Petitjean et al.: Hyperfine Interactions **101/102**, 1 (1996); D.L. Demin et al.: Hyperfine Interactions **101/102**, 13 (1996)

20. M.P. Faifman: Muon Catal. Fusion **2**, 247 (1988); A. Scrinzi et al.: Phys. Rev. **A47**, 4691 (1993); M.P. Faifman (to be published)

21. S.A. Alexander, H.J. Monkhorst: Phys. Rev. **A38**, 26 (1988)

22. M. Kamimura: Phys. Rev. **A38**, 621 (1988)

23. V.I. Korobov, I.V. Puzynin, S.I. Vinitsky: Muon Catal. Fusion **7**, 63 (1992)

24. G. Aissing, D. Bakalov, H.J. Monkhorst: Phys. Rev. **A42**, 116 (1990); M. Kamimura et al.: Phys. Rev. **A45**, 94 (1992)

25. D.I. Abramov, V.V. Gusev, L.I. Ponomarev: Phys. At. Nucl. **60**, 1133 (1997)

26. V.V. Gusev, L.I. Ponomarev, E.A. Solov'ev: Hyperfine Interactions **82**, 53 (1993)

27. L.I. Ponomarev: talk at this Conference

28. S.A. Alexander, P. Froelich, H.J. Monkhorst: Phys. Rev. **A41**, 2854 (1990)

29. L. Bracci et al.: Muon Catal. Fusion **4**, 247 (1989); **5/6**, 21 (1990/91); C. Chiccoli et al.: Muon Catal. Fusion **7**, 87 (1992)

30. J.S. Cohen, M.C. Struensee: Phys. Rev. **A43**, 3460 (1991); J.S. Cohen: Phys. Rev. **A43**, 4668 (1991)

31. Y. Kino, M. Kamimura: Hyperfine Interactions **82**, 45 (1993)

32. A. Adamczak: Hyperfine Interactions **101/102**, 113 (1996)

33. V.S. Melezhik, J. Woźniak: Muon Catal. Fusion **7**, 203 (1992); Hyperfine Interactions (in press)

34. G.M. Marshall et al.: Hyperfine Interactions **101/102**, 47 (1996)

35. J. Wallenius, M. Kamimura: Hyperfine Interactions **101/102**, 319 (1996)

36. L.I. Ponomarev, E.A. Solov'ev: JETP Lett. **64**, 135 (1996); JETP Lett. (in press); V.E. Markushin: Phys. Rev. **A50**, 1137 (1994)

37. E.C. Aschenauer, V.E. Markushin: Z. Phys. **D39**, 165 (1997); B. Lauss et al.: Phys. Rev. Lett. **76**, 4693 (1996)

38. L.N. Bogdanova, S.S. Gerstein, L.I. Ponomarev: Pisma ZhETF **67**, 89 (1998)

Few-Body Systems Suppl. 10, 179–188 (1999)

Few-
Body
Systems
© by Springer-Verlag 1999

Few nucleon dynamics in a nuclear medium

Michael Beyer*

Fachbereich Physik, Universität Rostock, D-18051 Rostock, Germany

Abstract. Few body methods are used in many particle physics to describe correlations, bound states, and reactions in strongly correlated quantum systems. Although this has already been recognized earlier, rigorous attempts to treat three-body collisions have only been done recently. In this talk I shall give examples and areas where few-body methods have been and might be of use in the future.

1 Introduction

Describing an ensemble of many particles (fermions/bosons) becomes challenging and interesting as soon as interactions (e.g. Coulomb or strong interaction) are considered. Examples for Coulombic systems are *ionic plasmas* as they occur in the sun and stars, *liquid metals* and *electron-hole plasmas*. *Nuclear matter* and the *quark gluon plasma* are examples of strongly interacting systems. Because of the interaction it is not possible to treat even single particle dynamics without regarding effects of the other particles. Further on the system may be in equilibrium or out of equilibrium, depending on the boundary conditions imposed.

The density temperature planes of matter and nuclear matter are shown in Figs. 1 and 2. The phase diagram of nuclear matter turns out to be very rich. In particular the superfluid phase reflecting strong pairing is relevant for the structure of neutron stars [1]. At lower densities bound states occur. This part of the phase diagram may be accessed in the laboratory through heavy ion collisions at intermediate energies. The conditions for the formation of bound states are reached in particular during the final stage, where the nuclear density drops below the Mott density.

---

*E-mail address: beyer@darss.mpg.uni-rostock.de

180

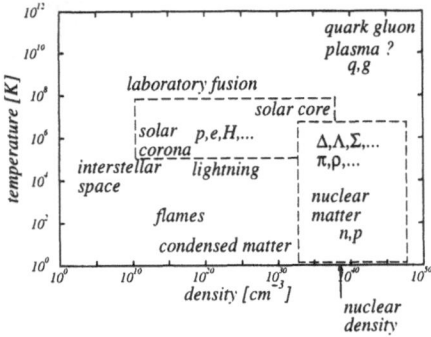

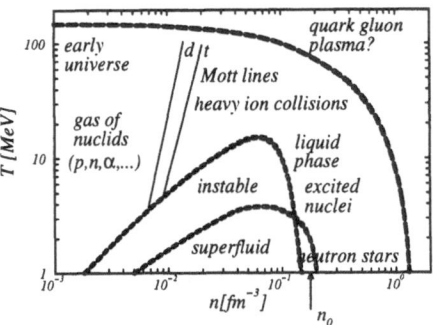

Figure 1. Density temperature phase diagram of matter.

Figure 2. Density temperature plane of nuclear matter.

2 Theory

Quantum statistics provides powerful methods to tackle the many particle systems. Here I follow the Green function formalism [2], which is convenient to introduce few-body methods. Let the Hamiltonian of the system be given by

$$H(t) = \sum_1 H_0(1)\psi_1^\dagger(t)\psi_1(t) + \sum_{12} V_2(12, 1'2')\psi_1^\dagger(t)\psi_2^\dagger(t)\psi_{2'}(t)\psi_{1'}(t), \quad (1)$$

where $\psi_1(t)$ denotes the Heisenberg operator of the particle with quantum numbers s_1, k_1, etc. for spin, momentum etc. The one particle Green function is defined by

$$i\mathcal{G}_1(1, 1') = \langle T\psi_1(t)\psi_{1'}^\dagger(t')\rangle \equiv \mathrm{Tr}\{\rho_0 T\psi_1(t)\psi_{1'}^\dagger(t')\}, \quad (2)$$

where averaging is due to the density operator ρ_0. For an open system in thermodynamical equilibrium the extremum condition for the entropy leads to the following expression for the quantum grand canonical density operator,

$$\rho_0 = \frac{e^{-\beta(H-\mu N)}}{\mathrm{Tr}\{e^{-\beta(H-\mu N)}\}}. \quad (3)$$

The temperature $(1/\beta = T)$ and the chemical potential μ are the corresponding Lagrange parameters. Using the Heisenberg equation for ψ_1 results in the following equation for $\mathcal{G}_1$ [2]

$$\mathcal{G}_1(1, 1') = \mathcal{G}_1^{(0)}(1, 1') - \sum_{\tilde{1}2\tilde{2}} \mathcal{G}_1^{(0)}(1, \tilde{1})iV_2(\tilde{1}2, \tilde{1}\bar{2}) \mathcal{G}_2(\tilde{1}\bar{2}, 1'2^+)\big|_{t_1=t_2}. \quad (4)$$

The argument 2^+ means that $t_{2+} = t_2 + 0^+$ and

$$(i\partial_{t_1} - H_0(1))\mathcal{G}_1^{(0)}(1, 1') = \delta_{11'}. \quad (5)$$

Equation (4) shows already the basic problem of many particle physics, the hierarchy. To find a useful truncation of the $n+1$ particle Green function from the n particle one, some notion of the system is needed. Usually the hierarchy is truncated at the two particle level, assuming binary collisions only. Three-particle collisions have been treated at most approximately using Born (for Coulombic systems) or impulse approximation (for nuclear matter). This may not be sufficient, in particular, if explicit three-particle processes are considered (e.g. such as cluster formation, where a third particle is needed to achieve momentum conservation, etc.).

Equation (4) may be formally decoupled by introducing the self energy $\Sigma(1,\bar{1})$.

$$\sum_{\bar{1}} \Sigma(1,\bar{1})\,\mathcal{G}_1(\bar{1},1') = -\sum_{\bar{1}\bar{2}2} iV_2(12,\bar{1}\bar{2})\,\mathcal{G}_2(\bar{1}\bar{2},1'2^+). \tag{6}$$

In the simplest case the self energy may then be treated in mean field (e.g. Hartree-Fock) approximation, i.e. $\mathcal{G}_2 \rightarrow \mathcal{G}_2^{(0)} = \mathcal{G}_1(1,1')\,\mathcal{G}_1(2,2') - \mathcal{G}_1(1,2')\,\mathcal{G}_1(2,1')$ viz. no two particle correlations (leading to an ideal gas of quasi particles). Using $V_2(12,\bar{1}\bar{2}) = -V_2(12,\bar{2}\bar{1})$ the self energy $\Sigma^{HF}(1,1')$ is given by

$$\Sigma^{HF}(1,1') = i\sum_{2\bar{2}} V_2(12,1'\bar{2})\mathcal{G}_1(\bar{2},2^+), \tag{7}$$

which reduces to the standard expression, if a static potential is used, $V_2(12,1'2') = \delta_{11'}\delta_{22'}V_2(12)$, since $\mathcal{G}_1(2,2^+) = \langle\psi_2^\dagger\psi_2\rangle = f_2$ is the one particle distribution function for fermions. For a given potential Eq. (7) constitutes a self consistent problem to determine μ and β, solved by iteration

$$f_1 \equiv f(\varepsilon) = \frac{1}{e^{\beta(\varepsilon-\mu)}+1}, \qquad \varepsilon = \frac{k^2}{2m} + \Sigma^{HF}(k). \tag{8}$$

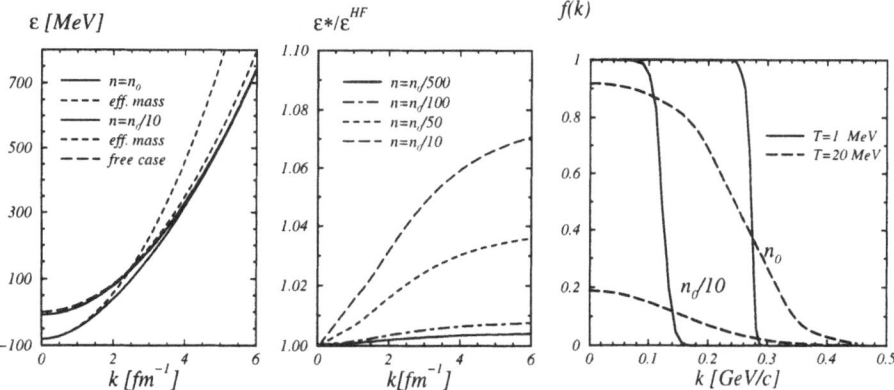

Figure 3. The nucleon self energy (left side). Deviation of the effective mass approximation from the exact model result for small densities (right side).

Figure 4. One particle distribution function as a function of momentum k, $n_0 = 0.17$ fm^{-3}.

As a further simplification useful in later applications effective masses may be introduced. The self energy for the nucleon in a nuclear medium of $T = 10$ MeV is shown in Fig. 3 along with its effective mass approximation $\varepsilon^* = k^2/(2m^*) + \Sigma^{HF}(k = 0)$. The distribution function $f(k)$ is shown in Fig. 4. Calculations are done using a separable Yamguchi type potential (that will later be used for the three body calculations). Finally, analytic continuation of the Green function defined in Eq. (2) leads to the Kubo-Martin-Schwinger boundary condition as $e^{itH} = e^{\beta H}$, i.e.

$$\mathcal{G}_1(1,1')|_{t_1=0} = -e^{\beta\mu}\,\mathcal{G}_1(1,1')|_{t_1=-i\beta}. \tag{9}$$

As a consequence the time like component of the Fourier transform is restricted to certain values only (Matsubara frequencies), $\mathcal{G}_1^{t_1-t_1'} \to G_1(z_\nu)$, where $z_\nu = i\pi\nu/\beta + \mu$ and $\nu = \pm 1, \pm 3, \ldots$ for fermions.

3 Correlations

A better treatment that goes beyond the quasi particle approximation is provided, e.g. through the cluster approximation that includes correlations [3]. As a consequence the one particle spectral function $A_1(\omega)$, defined through

$$G_1(z_\nu) = \int \frac{d\omega}{2\pi} \frac{A_1(\omega)}{z_\nu - \omega} \tag{10}$$

which is given by $A_1(\omega) = 2\pi\,\delta(\omega - \epsilon)$ for the quasi particle approximation, is more complicated, viz.

$$A_1(\omega) = \frac{2\,\mathrm{Im}\Sigma(\omega)}{[\omega - E - \mathrm{Re}\Sigma(\omega)]^2 + [\mathrm{Im}\Sigma(\omega)]^2}. \tag{11}$$

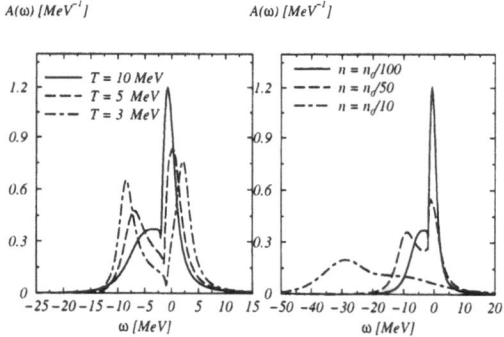
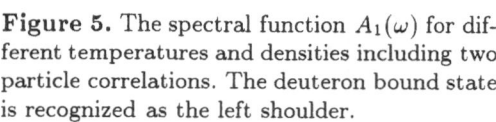
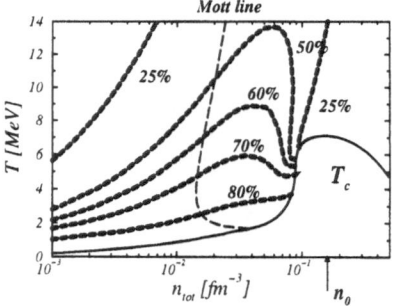

Figure 5. The spectral function $A_1(\omega)$ for different temperatures and densities including two particle correlations. The deuteron bound state is recognized as the left shoulder.

Figure 6. Temperature density plane for nuclear matter as a function of the total density. The amount of correlates density is given in per cent of the total density.

As an example Fig. 5 shows the spectral function using the full two body t-matrix to describe the correlations [4]. As a consequence nuclear matter cannot be considered as a system of independent quasi particles but for a large part it is correlated up to full pairing in the superfluid phase. This is depicted in Fig. 6, where the dashed lines show equal contribution of the correlated density to the total density [5]. The basis to treat correlated densities is provided by a generalization of the Bethe-Uhlenbeck approach [6]. The nuclear density $n = n(\mu, T)$ is given by

$$n = n_{\text{free}} + n_{\text{corr}}, \qquad n_{\text{corr}} = 2n_2 + 3n_3 + \dots \qquad (12)$$

where $n_{2,3}$ denotes the two, three-particle correlations, present as bound/scattering states. In first iteration these correlations may be treated on the basis of residual interactions between the quasi particles. The exact two particle equations to be solved are known as Bethe-Goldstone or Feynman-Galitski equations depending on some details. In ladder approximation the equation for the two body Green function reads

$$G_2(z) = \frac{\bar{f}_1 \bar{f}_2 - f_1 f_2}{z - \varepsilon_1 - \varepsilon_2} + \frac{\bar{f}_1 \bar{f}_2 - f_1 f_2}{z - \varepsilon_1 - \varepsilon_2} V_2 \, G_2(z), \qquad (13)$$

where $\bar{f} = 1 - f$. Introducing the two-body t-matrix in a standard fashion both bound and scattering states have been solved [6].

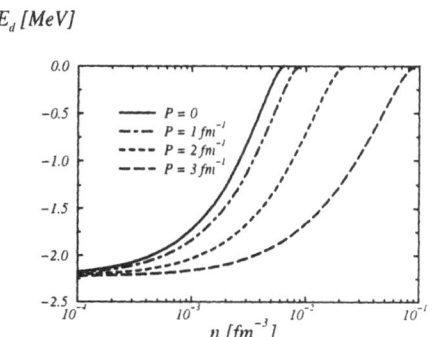

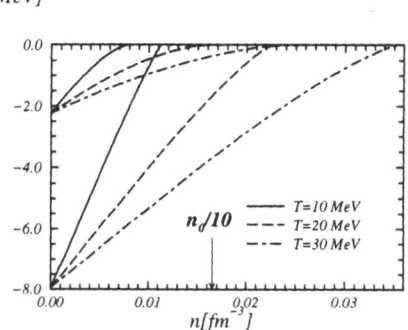

Figure 7. The deuteron binding energy as a function of the nuclear density for $T = 10$ MeV. P denotes the relative momentum between the deuteron and the medium.

Figure 8. The triton binding energy as a function of nuclear density for different temperatures T. The triton rests in the medium. The respective Nd continua are also indicated.

As an example the deuteron energy is shown in Fig. 7. The Mott density is defined through the condition $E_d = 0$. Note that the Fermi functions f_1 for particle 1 etc. depends on the relative momentum P between the deuteron and the medium. The respective three-particle Faddeev type equation for the Green function has been given in ref. [7],

$$G_3(z) = \frac{\bar{f}_1 \bar{f}_2 \bar{f}_3 + f_1 f_2 f_3}{z - \varepsilon_1 - \varepsilon_2 - \varepsilon_3} + \frac{(\bar{f}_1 \bar{f}_2 - f_1 f_2) V_2(12) + \text{perm.}}{z - \varepsilon_1 - \varepsilon_2 - \varepsilon_3} G_3(z). \qquad (14)$$

Equation (14) has been rewritten using the AGS approach [8]. The resulting in-medium AGS equations are

$$U_{\alpha\beta} = (1 - \delta_{\alpha\beta}) \left(\frac{\bar{f}_1 \bar{f}_2 \bar{f}_3 + f_1 f_2 f_3}{z - \varepsilon_1 - \varepsilon_2 - \varepsilon_3} \right)^{-1} + \sum_{\gamma \neq \alpha} T_3^{(\gamma)} \frac{\bar{f}_1 \bar{f}_2 \bar{f}_3 + f_1 f_2 f_3}{z - \varepsilon_1 - \varepsilon_2 - \varepsilon_3} U_{\gamma\beta}, \quad (15)$$

where $T_3^{(\gamma)}$ is the solution of the in-medium two-body problem, e.g. for $\gamma = 3$

$$T_3^{(3)} = (1 - f_3 + g(\varepsilon_1 + \varepsilon_2))^{-1} V_2 + V_2 \frac{\bar{f}_1 \bar{f}_2 - f_1 f_2}{z - \varepsilon_1 - \varepsilon_2 - \varepsilon_3} T_3^{(3)}, \qquad (16)$$

and the Bose function is given by $g(\omega) = 1/(e^{\beta(\omega - 2\mu)} - 1)$. This equation has been solved for the Nd reaction relevant for deuteron formation [7, 9] (see below), assuming $f(\varepsilon)^2 \ll f(\varepsilon)$, compare Fig. 4. Recently, the triton bound state equation has also been solved [10]. The triton binding energy is shown in Fig. 8. The deuteron as well as the triton binding energies weaken if the nuclear density is increased until the Mott density is reached. This tendency is dominated by the Pauli blocking of the surrounding medium.

The four-nucleon correlation is believed to play a significant role for lower densities and temperatures. Exploratory calculations using a simple variational ansatz for the ^{4}He wave function predict an α condensate/quartetting on top of the deuteron condensate/triplet pairing that leads to superfluidity [11].

Bose systems behave quite differently with respect to the occurrence of bound states. The bose functions enhance the effective residual interaction that might lead to an "opposite Mott effect", i.e. existence of bound states and also pairing (condensate) even if no bound state exist for the isolated case. An example is provided by a pion gas, were a pionic condensate may occur [12] that are discussed, e.g. in the context of neutron stars [1].

4 Reactions

Nuclear reaction rates play an essential role in the formation of stars like the sun. The standard solar model is based on binary collisions. Recently, triple reactions, e.g. $e + ^3\text{He} + \alpha \to ^7\text{Be} + e$ to be compared to $^3\text{He} + \alpha \to ^7\text{Be} + \gamma$ have been investigated and found to be rather small in plasmas at solar conditions [13]. However, note that triple collisions are mostly non-radiative and that they may be more important for other stars than the sun or at the early universe [13].

Another example are dense ionic plasmas, where the ionisation rate depends on three-particle reactions that are presently treated in Born approximation. Since the residual interaction is Coulombic this may be considered a good approximation and it reproduces the experimental results for hydrogen like plasmas [14]. For higher ionized plasmas this might not be the case and the application of Faddeev like methods may be in order. These will be sketched in the following for nuclear matter.

The generalized quantum kinetic Boltzmann equations for the nucleon $f_N(p, t)$ (momentum p) and deuteron $f_d(P, t)$ (momentum P) distribution functions [15]

$$
\begin{aligned}
f_N(p, t) &= \langle a^\dagger_{Np} a_{Np} \rangle \equiv \mathrm{Tr}\{\rho(t) a^\dagger_{Np} a_{Np}\} \\
f_d(P, t) &= \langle b^\dagger_{dP} b_{dP} \rangle
\end{aligned}
\tag{17}
$$

are coupled and read

$$
\begin{aligned}
f_N(p, t) &= -\mathcal{D}_N(p, t) + \mathcal{I}_N(p, t) \\
f_d(P, t) &= -\mathcal{D}_d(P, t) + \mathcal{I}_d(P, t).
\end{aligned}
\tag{18}
$$

The first term reflects the so called Vlasov term and is related to the mean field. The second term is the collision term that is responsible for equilibration of the system. The explicit form of the integral $\mathcal{I}_N(p, t)$ is

$$
\mathcal{I}_N(p, t) = \mathcal{I}^>_N(p, t) f_N(p, t) - \mathcal{I}^<_N(p, t) \bar{f}_N(p, t),
\tag{19}
$$

where, e.g. $\mathcal{I}^>_N(p, t)$ is given by

$$
\begin{aligned}
\mathcal{I}^>_N(p, t) &= \int dk\, dk_1\, dk_2 \left| \langle kp | T_{NN \to NN} | k_1 k_2 \rangle \right|^2 \bar{f}_N(k_1, t) \bar{f}_N(k_2, t) f_N(k, t) \\
&+ \int dk\, dk_1\, dk_2\, dk_3 \left| \langle kp | U_{Nd \to NNN} | k_1 k_2 k_3 \rangle \right|^2 \\
&\quad \times \bar{f}_N(k_1, t) \bar{f}_N(k_2, t) \bar{f}_N(k_3, t) f_d(k, t) + \dots
\end{aligned}
\tag{20}
$$

The solution of this equation are the distribution functions f_N and f_d, that however also appear in the transition matrices $T_{NN \to NN}$, $U_{Nd \to NNN}$, etc. Therefore a full solution of this equation is a difficult problem. For small fluctuations from the equilibrium distributions the equations may be linearized in the framework of linear response theory. The binary collision approximation may the transition matrix elements depend on the equilibrium distribution only and the results of the previous sections can be utilized. Even this in-medium dependence has hardly been considered in modeling of heavy ion collision [16]. Also, for three-nucleon collisions so far only the impulse approximation has been used [17]. Here we solve the in-medium Faddeev type equation that includes the Hartree-Fock self energy shift and the Pauli blocking in a consistent way, Eq. (14). The resulting break-up cross section for a typical temperature of $T = 10$ MeV and densities below the deuteron Mott density is shown in Fig. 9. For a comparison of the quality of the model the isolated cross section along with the experimental data [18] are shown in Fig. 10. From inspection of Fig. 9 we see that the in-medium cross section is significantly enhanced compared to the isolated

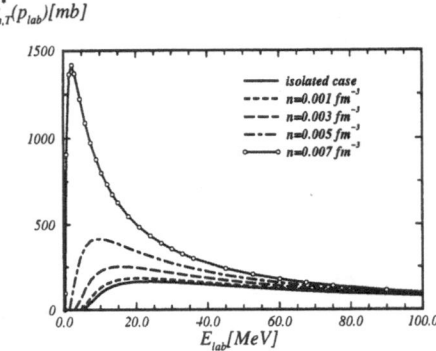

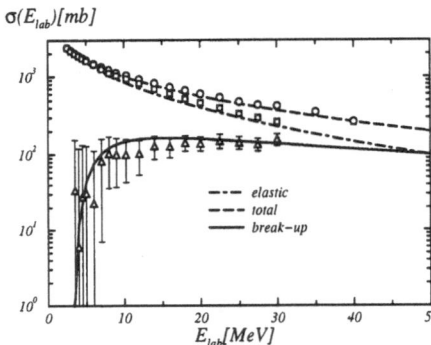

Figure 9. In-medium break-up cross section at $T = 10$ MeV. Isolated cross section is shown as solid line. Other lines are due to different nuclear densities.

Figure 10. A comparison of the total, elastic, and break-up cross sections $nd \to nd$, $nd \to nnp$ with the experimental data of ref. [18].

one. The threshold is shifted to smaller energies, which is because the binding energy of the deuteron becomes smaller. We observe that for higher energies the medium dependence of the cross section becomes much weaker.

Though the change of the cross section looks dramatic, the quantum Boltzmann equation still has some additional medium dependence that may change this effect in the observables f_N and f_d. Within the linear response theory it is possible to calculate the chemical reaction time due to the break-up process. For small fluctuations $\delta f(t)$ linear response leads to

$$\partial_t \delta f_d^{\text{reaction}}(P, t) = \frac{1}{\tau_P} \delta f_d^{\text{reaction}}(P, t) \tag{21}$$

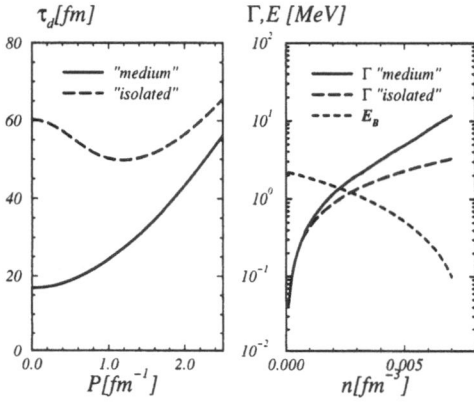

Figure 11. Fluctuation time of the deuteron distribution as a function of the deuteron momentum P. As input the in-medium cross section is compared to the isolated one, nuclear density $n = 0.007$ fm^{-3}, temperature $T = 10$ MeV.

where the "life time" of deuteron fluctuations has been introduced,

$$\tau_d^{-1} = \frac{4}{3!} \int dk_N dk_1 dk_2 dk_3 \ |\langle kp|U_{Nd \to NNN}|k_1 k_2 k_3 \rangle|^2 \ \bar{f}_1 \bar{f}_2 \bar{f}_3 f_\varepsilon \ 2\pi\delta(E - E_0).$$
(22)

which can be related to the break-up cross section given in Eq. (9). For low densities the life time (as a function of the deuteron momentum P) and the inverse life time, i.e. the width, at $P = 0$ along with the deuteron binding energy for comparison is shown in Fig. 11 [9]. These times have to be compared to the approximate duration of the heavy ion collision of about 200 fm.

5 Conclusion

The treatment of correlations and triple collisions in non-ideal many particle quantum systems opens up a new field for few-body methods. The examples shown have been mostly from nuclear physics. However, applications are possible for stellar plasmas to improve the description of the basic quantity, which is the spectral function and to include correlations into the equation of state. In the laboratory the ionisation rate of dense ionic plasmas is determined by three-particle collisions. A description is presently restricted to hydrogen like plasmas, where the Born approximation for the three-particle reaction is sufficient. Typical applications in nuclear physics are related to heavy ion collisions, here in particular the formation of light clusters such as deuterons, helium-3, tritium, and alpha particles. The conditions are satisfied during the final stage of the heavy ion collision, where a temperature of $T \simeq 5 \ldots 10$ MeV may be meaningful and the densities are below the Mott densities of cluster formation. The results are also relevant for the equation of states of neutron stars.

The approach given here follows the quantum statistical description as it provides a rigorous, systematic treatment of many particle systems. The major approximation utilized is the cluster expansion to decouple the infinite hierarchy of equations (Green functions or kinetic equations). This approximation clearly goes beyond the quasi particle picture as it includes the residual interactions in a systematic way. This is done rigorously using few-body methods. To this end few-body equations have to be substantially generalized. Presently, these equations resemble an RPA structure [19], however extended to finite temperatures.

The validity of this already ambitious approach has to be checked by facing experimental results. This seems easier for ionic plasmas, e.g. to calculate the ionisation rate, or for electron-hole plasmas in the context of exciton formation. Testing the validity of the approach for nuclear physics needs a handle of heavy ion collisions. Presently one relies on numerical simulations of the complicated dynamics of a heavy ion collision, which is subject to discussions by its own. Typical heavy ion simulation codes that require microscopic input are based on e.g. a Boltzmann-Uehling-Uhlenbeck treatment or on quantum molecular dynamics. The results presented here may however also be relevant for standard nuclear physics, e.g. electron scattering off heavy nuclei, when correlations are considered, and one therefore needs to go beyond the quasi particle picture.

Acknowledgement. It is a pleasure to thank my colleagues in Rostock who provided me with some material presented during the talk.

References

1. D. Pines, R. Tamagaki, S. Tsuruta (editors): In: *The structure and evolution of neutron stars.* Reading: Addison-Wesley 1992

2. L.P. Kadanoff, G. Baym: In: *Quantum Theory of Many-Particle Systems.* New York: Mc Graw-Hill 1962; A.L. Fetter, J.D. Walecka: In: *Quantum Theory of Many-Particle Systems.* New York: McGraw-Hill 1971

3. G. Röpke et al.: Nucl. Phys. **A339**, 587 (1983)

4. A. Schnell, T. Alm, G. Röpke: Phys. Lett. **B387**, 443 (1996)

5. H. Stein, A. Schnell, T.Alm, G. Röpke: Z. Phys. **A351**, 295 (1995)

6. M. Schmidt, G. Röpke, H. Schulz: Ann. Phys. **202**, 57 (1990)

7. M. Beyer, G. Röpke, A. Sedrakian: Phys. Lett. **B376**, 7 (1996)

8. E.O. Alt, P. Grassberger, W. Sandhas: Nucl. Phys. **B2**, 167 (1967)

9. M. Beyer, G. Röpke: Phys. Rev. **C56**, 2636 (1997)

10. M. Beyer et al.: in preparation

11. G. Röpke et al.: Phys. Rev. Lett. **80**, 3177 (1998)

12. T. Alm et al.: Nucl. Phys. **A612**, 472 (1997)

13. D.E. Monakhov et al.: Nucl.Phys. **A635**, 257 (1998); N.V. Shevchenko et al.: nucl-th/9804020; S.A. Rakityansky et al.: Nucl. Phys. **A613**, 132 (1997)

14. T. Bornath et al.: J. Quant. Radiat. Transfer **58** (1997); T. Bornath, M. Schlanges, R. Prenzel: Phys. Rev. **E5**, 1485 (1998)

15. D.N. Zubarev, V.G. Morozow, G. Röpke: In: *Statistical Mechanics of Nonequlibrium Processes Volume 1+2* . Berlin: Akademie-Verlag 1996

16. T. Alm et al.: Nucl. Phys. **A587**, 815 (1995)

17. P. Danielewicz, G.F. Bertsch: Nucl. Phys. **A533**, 712 (1991)

18. P. Schwarz et al.: Nucl. Phys. **A398**, 1 (1983)

19. P. Schuck: Z. Phys. **241**, 395 (1971); P. Schuck, F. Villars, P. Ring: Nucl. Phys. **A208**, 302 (1973); J. Dukelsky, P. Schuck: Nucl. Phys. **A512**, 466 (1990); P. Schuck: Nucl. Phys. **A567**, 78 (1994)

Few-Body Systems Suppl. 10, 189–198 (1999)

Few-
Body
Systems
© by Springer-Verlag 1999

Few-Electron Artificial Atoms

J. Adamowski[1], B. Szafran[1], S. Bednarek[1], and B. Stébé[2]

[1] Faculty of Physics & Nuclear Techniques, Technical University (AGH), Kraków, Poland
[2] Institut de Physique et d'Electronique, Université de Metz, Metz, France

Abstract. Artificial atoms, i.e., bound systems of excess electrons confined in semiconductor quantum dots, are studied by the variational and Hartree-Fock methods. The confinement potential is assumed to have the form of a spherical potential well of finite depth, which provides a theoretical model for electron states in a spherical semiconductor nanocrystal embedded in an insulating matrix. For the two- and three-electron artificial atoms, we have applied the variational method and obtained the binding of both the ground states and excited states. The Hartree-Fock method has been applied to the N-electron artificial atoms with $N = 1, \ldots, 20$. It is shown that the shells of the artificial atoms are filled by electrons in the same manner like those of the natural atoms. In particular, Hund's rule is fulfilled. The radial probability density calculated for artificial atoms is different from that for natural atoms.

1 Introduction

Recently fabricated semiconductor nanostructures, called quantum dots (QDs), can confine electrons in all three dimensions [1]. A strength and range of the confinement potential can be changed intentionally, which leads to the designed properties of the electron energy spectrum. The excess electrons confined in the QDs can form bound systems called artificial atoms. The QDs, besides promising applications in modern electronics, provide a unique laboratory for studying new physical properties of interacting few-electron systems with a controllable number of electrons. The most common are the QDs of cylindrical shape, which are fabricated on semiconductor substrates by a combination of electron or ion lithography and chemical etching [2], and of spherical shape, which are grown as nearly spherical nanocrystals in an insulating matrix [3, 4]. Electronic properties of QDs are studied by various methods, e.g., far-infrared magnetoabsorption [5], capacitance spectroscopy [6], and transport spectroscopy [7]. All these methods allow us to study the atomic-like properties of the confined electron systems. Recently, the filling of shells of artificial atoms formed in cylindrical quantum dots has been observed [8] and theoretically described by the present authors [9]. The present paper is devoted to a theoretical study of the artificial atoms formed in spherical QDs [3].

In most of the theoretical papers on the electronic properties of the QDs, the confinement potential for electrons is assumed to possess the infinite depth. This is either the infinitely deep potential well [10] or the harmonic (parabolic) potential [11]. For both these potentials, the continuum-energy threshold does not exist, which leads to the fundamental physical deficiency of these model potentials: the unbound (free) states of electrons do not exist. Therefore, the binding and dissociation processes, that play an important role in experiments, cannot be described with the use of these model confinement potentials.

In order to solve this problem, we apply in the present paper the confinement potential of finite depth and range. We consider the few-electron artificial atoms, which are formed in the system of excess electrons confined in the spherical QD. We have applied the direct variational approach to the two- and three-electron systems, i.e., artificial helium and lithium atoms, and the Hartree-Fock method for more complex artificial atoms with a larger number of electrons. We have studied the artificial atoms with the following occupied shells: $1s$, $1p$, $1d$, and $2s$, i.e., consisting of up to 20 electrons. The calculated energy levels allow us to determine the conditions of binding for the artificial atoms.

The paper is organized as follows: the theoretical model applied is presented in Sect. 2, the computational method and results for the two- and three-electron artificial atoms are given in Sect. 3, the Hartree-Fock method and results for the more complex artificial atoms are contained in Sect. 4. Section 5 consists of a discussion of the results.

2 Theoretical model

We consider a single spherical QD, i.e., a semiconductor nanocrystal of almost spherical shape, which forms a potential-well region, embedded in an insulating matrix, which forms a potential-barrier region. In the effective-mass approximation, the Hamiltonian of the system of N excess electrons confined in the spherical QD has the form

$$H = \sum_{i=1}^{N} h(\mathbf{r}_i) + \sum_{i=1}^{N} \sum_{j>i}^{N} U(r_{ij}) , \tag{1}$$

where the one-electron Hamiltonian is given by

$$h(\mathbf{r}) = -\frac{\hbar^2}{2m_e}\nabla^2 + V_{conf}(\mathbf{r}) , \tag{2}$$

m_e is the electron effective mass, and the confinement potential energy is assumed to be

$$V_{conf}(\mathbf{r}) = \begin{cases} -V_0 & \text{if } r < R , \\ 0 & \text{otherwise} . \end{cases} \tag{3}$$

Here, V_0 is the depth of the potential well ($V_0 > 0$) and R is the radius of the QD. The energy of the conduction-band minimum of the barrier material is taken on equal to zero. This is the reference energy in the present work. We assume that the excess electrons interact with themselves via the screened Coulomb interaction, i.e.,

$$U(r_{ij}) = \frac{\kappa e^2}{\varepsilon_s r_{ij}} , \qquad (4)$$

where r_{ij} is the electron-electron distance, ε_s is the static dielectric constant, $\kappa = 1/4\pi\varepsilon_0$, and ε_0 is the permittivity of vacuum.

The condition of binding for the N-electron system has the form

$$E_\nu^{(N)} < E_0^{(N-1)} , \qquad (5)$$

where $E_\nu^{(N)}$ is the energy of N-electron state ν and $E_0^{(N-1)}$ is the ground-state energy of the $(N-1)$-electron system. For the assumed reference energy, the one-electron state is bound if

$$E_\nu^{(1)} < 0 . \qquad (6)$$

The binding energy is defined as $W_\nu^{(N)} = E_0^{(N-1)} - E_\nu^{(N)}$ for the N-electron system in state ν. For one electron $W_\nu^{(1)} = -E_\nu^{(1)}$. Binding conditions (5) and (6) are equivalent to the requirement for the binding energy to be positive. This amount of energy is released when the electron goes over from the minimum-energy conduction-band state of the barrier region into state ν of the QD. The same value of energy should be supplied to the N-electron QD in order to liberate one electron transferring it into the barrier to the conduction-band state of the lowest energy. This second process corresponds to the dissociation of the artificial atom. Let us notice that this description is valid for the single QD embedded in the barrier material. The application of the confinement potential of the finite depth allows us to describe these binding and dissociation processes.

Throughout the present paper, we use the following units: the donor rydberg R_D is the unit of energy ($R_D = m_e \kappa^2 e^4 / 2\hbar^2 \varepsilon_s^2$) and the donor Bohr radius a_D is the unit of length ($a_D = \hbar^2 \varepsilon_s / m_e \kappa e^2$). For semiconductors, these quantities take on the following typical values: $R_D = (\sim 10 \div \sim 100)$ meV and $a_D = (\sim 1 \div \sim 10)$ nm.

3 Two- and three-electron artificial atoms

In order to describe the artificial atoms that consist of the two and three electrons we apply the variational method and assume that the total spin and total angular momentum are well defined (Russell-Saunders coupling). The variational trial wave function is proposed as a linear combination of Slater determinants, which are constructed from the $1s$, $1p$, $1d$, and $2s$ one-electron wave functions for the spherical potential well. These one-electron wave functions are taken on as linear combinations of Slater-type orbitals. We have checked [12] that such an approach allows us to reproduce the subsequent energy levels for an electron in the spherical potential well with a large accuracy. In particular, it properly accounts for a change of curvature of the wave function at the QD boundary, which results from a jump of the confinement potential (3).

Using the four s, p, and d orbitals, we have constructed the following two-electron states: $(1s^2)\,^1S$, $(1s1p)\,^3P$, $(1s1p)\,^1P$, $(1p^2)\,^3P$, $(1p^2)\,^1D$, $(1p^2)\,^1S$, $(1s1d)\,^3D$, $(1s1d)\,^1D$, $(1s2s)\,^3S$, and $(1s2s)\,^1S$, and three-electron states: $(1s^21p)\,^2P$, $(1s1p^2)\,^4P$, $(1s^21d)\,^2D$, and $(1s^22s)\,^2S$. We use the following notation for the quantum states: S, P, and D correspond to the total angular-momentum quantum numbers $L = 0$, 1, and 2, respectively, the term inside the brackets, e.g., $(1s^2)$ denotes the occupied one-electron orbitals, and the superscript outside the brackets denotes the total-spin multiplet.

The few-electron trial wave function for the state ν can be written in the form

$$\Psi_\nu(\{\xi_i\}) = \sum_{\{n_i\}} c_{\nu\{n_i\}} \psi^\nu_{\{n_i\}}(\{\xi_i\}) , \qquad (7)$$

where $\xi_i = (\mathbf{r}_i, \sigma_i)$, $\{\xi_i\} = (\xi_1, \ldots, \xi_N)$ denotes the set of the position vectors $\mathbf{r}_i$ and spin variables σ_i, and $c_{\nu\{n_i\}}$ are the linear variational parameters $(i = 1, \ldots, N)$. The one-electron basis wave functions $\psi_{\nu\{n_i\}}$ are proposed as properly symmetrized linear combinations of the Slater determinants

$$\mathcal{A}[\varphi^\nu_{n_1 l_1 m_1}(\mathbf{r}_1)\chi_{s_1}(\sigma_1)\varphi^\nu_{n_2 l_2 m_2}(\mathbf{r}_2)\chi_{s_2}(\sigma_2)\ldots] , \qquad (8)$$

where $\mathcal{A}$ denotes the antisymmetrization operator, $\chi_{+1/2}(\sigma) = \alpha(\sigma)$ and $\chi_{-1/2}(\sigma) = \beta(\sigma)$ are the spinors. The spatial wave functions have been assumed in the form of the Slater orbitals, i.e.,

$$\varphi^\nu_{n,lm}(\mathbf{r}) = r^{n_i} \exp(-\gamma_\nu r) Y^l_m(\theta, \phi) , \qquad (9)$$

where γ_ν are the nonlinear variational parameters, $Y^l_m(\theta, \phi)$ are the spherical harmonics, the parameters $\{n_i\} = (n_1, n_2, n_3)$ take on the values $n_i = 0, \ldots, 12$.

Slater determinants (8) allow us to build basis wave functions $\psi^\nu_{\{n_i\}}$ [Eq. (7)]. Below, we give the two exemplary basis functions:

$$\begin{aligned}
\psi^\nu_{n_1 n_2}(\xi_1, \xi_2) &= \mathcal{A}[\varphi^\nu_{n_1 00}(\mathbf{r}_1)\alpha(\sigma_1)\varphi^\nu_{n_2 00}(\mathbf{r}_2)\beta(\sigma_2)] \\
&- \mathcal{A}[\varphi^\nu_{n_1 00}(\mathbf{r}_1)\beta(\sigma_1)\varphi^\nu_{n_2 00}(\mathbf{r}_2)\alpha(\sigma_2)]
\end{aligned} \qquad (10)$$

for the two-electron state $\nu = (1s^2)\,^1S$ and

$$\begin{aligned}
\psi^\nu_{n_1 n_2 n_3}(\xi_1, \xi_2, \xi_3) &= \\
\mathcal{A}[\varphi^\nu_{n_1 00}(\mathbf{r}_1)&\alpha(\sigma_1)\varphi^\nu_{n_2 00}(\mathbf{r}_2)\beta(\sigma_2)\varphi^\nu_{n_3 10}(\mathbf{r}_3)\alpha(\sigma_3)] \\
- \mathcal{A}[\varphi^\nu_{n_1 00}(\mathbf{r}_1)&\beta(\sigma_1)\varphi^\nu_{n_2 00}(\mathbf{r}_2)\alpha(\sigma_2)\varphi^\nu_{n_3 10}(\mathbf{r}_3)\alpha(\sigma_3)]
\end{aligned} \qquad (11)$$

for the three-electron state $\nu = (1s^21p)\,^2P$. The basis wave functions for the other considered states have been constructed in a similar manner. When using expansion (7) in the present calculations, we have included 91 terms of form (10) for two-electron state $(1s^2)\,^1S$ and 364 terms of form (11) for three-electron state $(1s^2 1p)$ ^2P. The numbers of terms in expansion (7) for the other considered states were between these numbers.

We comment on the properties of the trial wave functions applied for the two- and three-electron systems. These variational wave functions are similar to those used in a configuration-interaction method for natural atoms and partially take into account the electron-electron correlation. The correlation would be fully included if all the few-electron states of the required symmetry were taken into account. In order to check the quality of the trial wave function (7), we have performed the variational calculations for the two-electron states with the more flexible trial wave function of the form: $\Psi_{cor} = (1 + cr_{12})\Psi$, that explicitly includes the electron-electron correlation. We have found that the relative improvement of the variational estimates is small if the radius of the QD is not very large. This means that the correlation can be neglected for the strong- and intermediate-confinement regimes [13]. Moreover, we have also checked the variational wave function (7) by applying it to the natural helium atom and to the H^- ion both in an infinite space and confined in a spherical microcavity [12]. The estimates so obtained [12] agree very well with the exact results for the He atom and H^- ion in the infinite space and the results obtained [14] by the quantum Monte Carlo method for the confined H^- ion.

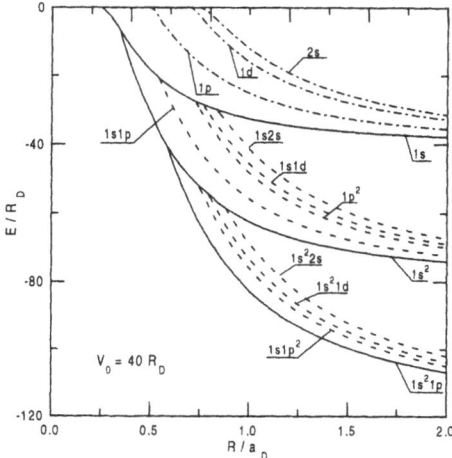

Figure 1. Energy levels of artificial atoms consisting of one electron in states $1s$, $1p$, $1d$, and $2s$, two electrons in states $(1s^2)^1S$, $(1s1p)^3P$, $(1p^2)^3P$, $(1s1d)^1D$, and $(1s2s)^1S$, and three electrons in states $(1s^2 1p)^2P$, $(1s1p^2)^4P$, $(1s^2 1d)^2D$, and $(1s^2 2s)^2S$ for $V_0 = 40R_D$ as functions of QD radius R.

The results for the electron states in spherical QDs are displayed in Figs. 1 and 2. Figure 1 shows the energy levels of one-, two-, and three-electron artificial atoms calculated as functions of the QD radius for $V_0 = 40R_D$. For QDs of very small radius, the electrons do not form bound states. If the radius of the QD increases, the one, two, and three electrons become bound in the ground states for $R \simeq 0.25, 0.3$, and 0.6 a_D, respectively. Several excited states are also

194

bound for sufficiently large R. The results in Fig. 1 correspond to the strong-
and intermediate-confinement regimes ($R \leq 2a_D$).

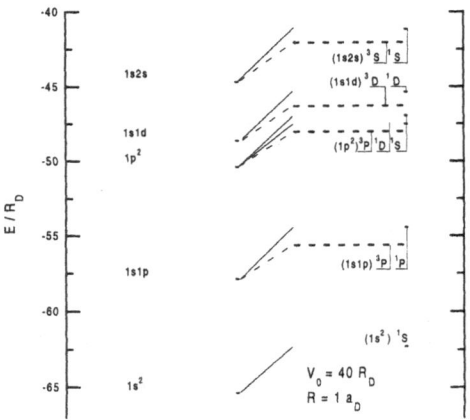

Figure 2. Energy levels of two-electron systems in the QD. The left part corresponds
to the noninteracting electrons and the right part to interacting electrons.

The left (right) part of Fig. 2 shows the energy levels of the noninteracting
(interacting) electron pair in the spherical QD with $V_0 = 40\,R_D$ and the radius
$R = a_D$. Switching on the electron-electron interaction shifts all the energy levels upwards. Nevertheless, the two-electron states are bound since the binding
condition (5) is fulfilled for this QD.

4 N-electron artificial atoms

Let us consider the N-electron artificial atoms with $N = 1, \ldots, 20$. We can
say that now we deal with the following systems: from the artificial hydrogen
atom to artificial calcium atom. In order to calculate the ground-state energy
of the system of N electrons confined in the spherical QD, we have applied the
unrestricted Hartree-Fock method. According to this approach, the N-electron
wave function is the Slater determinant, which is built from the one-electron
spinorbitals

$$\psi_i(\xi_j) = \varphi_{\mu s}(\mathbf{r}_j)\chi_s(\sigma_j) \,, \tag{12}$$

where $\mu = (nlm)$, $s = \pm 1/2$, and $i, j = 1, \ldots, N$. In the present calculations,
the one-electron orbitals are expanded in the Gaussian basis

$$\varphi_{\mu s}(\mathbf{r}) = \sum_{k n_1 n_2 n_3} c^{\mu s}_{k n_1 n_2 n_3} g_{k n_1 n_2 n_3}(\mathbf{r}) \,, \tag{13}$$

where

$$g_{kn_1n_2n_3}(\mathbf{r}) = x^{n_1} y^{n_2} z^{n_3} \exp(-\gamma_k r^2) \,, \tag{14}$$

γ_k $(k = 1, \ldots, 4)$ are the nonlinear variational parameters, and the parameters n_1, n_2, and n_3 take on the values $n_i = 0, 1, 2$. The Hartree-Fock equations have been solved for each one-electron state (μ, s) by the iterative procedure with the minimization performed over the variational parameters γ_k.

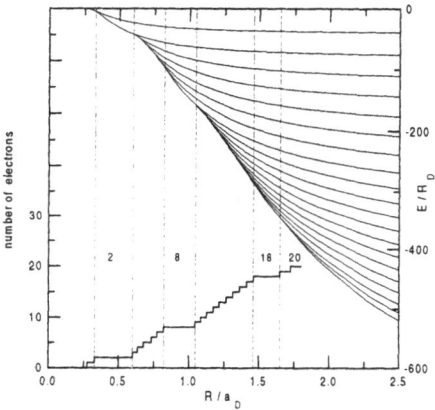

Figure 3. Ground-state energy (upper right part) and number of electrons (lower left part) associated with stable N-electron artificial atoms for $V_0 = 40R_D$ as functions of the QD radius R. Vertical thin lines correspond to the filled shells.

The results of the Hartree-Fock calculations are shown in Fig. 3. The right upper part of Fig. 3 shows the ground-state energy levels of N-electron artificial atoms for $N = 1, \ldots, 20$. The results for 1, 2, and 3 electrons agree with those in Fig. 1 with a very good accuracy. We have found that this good agreement between the Hartree-Fock and configuration-interaction results takes place for QDs of small and intermediate size, for which the correlation plays a minor role. In this case, the results obtained by the Hartree-Fock method can be regarded as accurate. The application of condition (5) tells us that the binding of each additional electron occurs if the QD radius increases by some amount. Therefore, the artificial atoms with the increasing number of electrons become stable if the size of the QD increases. The left lower part of Fig. 3 shows the numbers of electrons, that correspond to the stable artificial atoms. The obtained staircase structure consists of two types of stairs: (i) the stairs of the small "width" (measured as the change of the QD radius), which correspond to the filling by electrons of the states belonging to the same shell, (ii) the stairs of the large "width", which correspond to the change of shell after the full filling by electrons of the inner shell. The subsequent shells are fully filled by 2, 8, and 18 electrons, which is in agreement with the filling by electrons of the shells in the natural atoms. This property of the artificial and natural atoms results from the spherical symmetry of both the systems. We have found that the subshells $1p$ and $1d$ are filled by electrons according to Hund's rule: the minimum ground-state energy has been obtained for the states with the maximum spin for the half-filled subshells, i.e., for the artificial atoms with 4 and 12 electrons, respectively. In order to obtain the stability of the artificial atoms consisting of 5 and 13 electrons, for which the one-electron states with the opposite spin are filled, we have to increase the QD radius by slightly larger values than those for the states with the same spin. This slight increase of the "width" of the stairs is also visible in Fig. 3.

5 Discussion

It is well known that one electron forms bound states in a spherical potential well of finite depth if the effective quantum "capacity" of the dot, defined as $\Omega = V_0 R^2$, is sufficiently large. In particular, the one-electron states $1s$, $1p$, $1d$, and $2s$ become bound if $\Omega/R_D a_D^2$ exceeds the values $\pi^2/4$, π^2, $2\pi^2$, and $9\pi^2/4$, respectively. We have found that the subsequent two- and three-electron states are bound if Ω exceeds certain critical values. For example, these values are 5.1 and 17.8 (in units $R_D a_D^2$) for the ground states $(1s^2)^1S$ and $(1s^2 1p)^2P$ of the two-electron and three-electron QDs, respectively. For the N-electron artificial atoms with $N > 3$, the results of Fig. 3 allow us to determine the critical values of the quantum "capacity", above which the binding occurs.

It is interesting to compare electron probability distributions for the artificial and natural atoms with the same number of electrons. The total radial probability density is calculated as the sum over the one-electron states of the one-electron radial probability densities, which are obtained from the Hartree-Fock solutions. The results for the ground states of the artificial atoms with $N = 1, \ldots, 8$ electrons are shown in Fig. 4 (a). These results can be compared with those depicted in Fig. 4 (b) for the oxygen ions (from O^{+7} to O^{+1}) and oxygen atom (O). Figure 4 (a) shows that the bound N-electron states are strongly localized within the QD and the electron penetration into the barrier region is small. Moreover, the shell structure of the artificial atoms is not visible in the total radial probability density. On the contrary, the shell structure is very pronounced in the oxygen atom and ions [Fig. 4 (b)]. The probability distribution of Fig. 4 (a) also allows us to point out the origin of binding of the artificial atoms: this binding results from the confinement of electrons in the QD potential well.

In summary, we have shown that the spherically symmetric N-electron artificial atoms with $N = 1, \ldots, 20$ electrons can form bound states if the effective quantum "capacity" of the QD is sufficiently large. The fundamental properties of the artificial atoms are the same as those of the natural atoms, i.e., the atomic shells are filled according to the Pauli principle and Hund's rule. The differences are visible in the radial probability density.

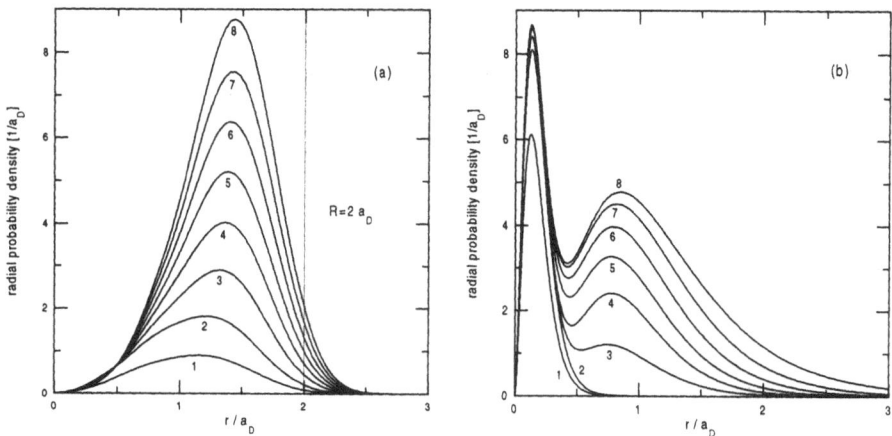

Figure 4. Total radial probability density for (a) the artificial atoms with $N = 1, \ldots, 8$ electrons ($V_0 = 20R_D, R = 2a_D$) and (b) the corresponding oxygen atoms and ions with the given number of valence electrons as functions of electron-dot center (nucleus) distance. In part (b), the unit of length (a_D) is equal to the hydrogen Bohr radius.

Acknowledgement. This work is performed in the frame of the French-Polish scientific cooperation agreement and partially supported by the Scientific Research Committee (KBN) under grant No. 2 P03B 5613. The computations have been performed at the Academic Computer Center CYFRONET in Kraków (grant KBN/S2000/AGH/050/1998).

References

1. For review articles, see: U. Merkt: Phys, Bl. **47**, 509 (1991); M.A. Kastner: Phys. Today **46**, 24 (1993); A.D. Yoffe: Adv. Phys. **42**, 173 (1993); N.F. Johnson: J. Phys. Condens. Matter **7**, 965 (1995); M. A. Kastner: Comments Cond. Mat. Phys. **17**, 349 (1996)

2. M.A. Reed et al.: J. Vac. Sci. Technol. **B4**, 358 (1986)

3. Q. Ye, R. Tsu. and E.H. Nicollian: Phys. Rev. **B44**, 1806 (1991)

4. A.P. Alivisatos: J. Phys. Chem. **100**, 13226 (1996)

5. Ch. Sikorski and U. Merkt: Phys. Rev. Lett. **62**, 2164 (1989)

6. R.C. Ashoori et al.: Phys. Rev. Lett. **68**, 3088 (1992)

7. U. Meirav et al.: Z. Phys. **B85**, 357 (1991); J. Weis et al: Phys. Rev. Lett. **71**, 4019 (1993)

8. S. Tarucha et al.: Phys. Rev. Lett. **77**, 3613 (1996)

9. B. Szafran, S. Bednarek and J. Adamowski: Proc. XXVII Int. School on Physics of Semiconducting Compounds, Jaszowiec, Poland, June 7-12, 1998 – in print

10. G.W. Bryant: Phys. Rev. Lett. **59**, 1140 (1987); D. Babić, R. Tsu and R.F. Greene: Phys. Rev. **B45**, 14150 (1992); M. Iwamatsu et al.: J. Phys.: Condens. Matter **9**, 9881 (1997)

11. U. Merkt, J. Huser, and M. Wagner: Phys. Rev. **B43**, 7320 (1991); D. Pfannkuche, V. Gudmundsson and P.A. Maksym: Phys. Rev. **B47**, 2244 (1993); P. Hawrylak: Phys. Rev. Lett. **71**, 3347 (1993); A. Matulis and F.M. Peeters: J. Phys.: Condens. Matter **6**, 7751 (1994); F.M. Peeters and V.A. Schweigert: Phys. Rev. **B53**, 1468 (1996); M. Fujito, A. Natori and H. Yasunaga: Phys. Rev. **B53**, 9952 (1996); L. Jacak, J. Krasnyj and A. Wójs: Physica **B229**, 279 (1997)

12. B. Szafran, J. Adamowski and B. Stébé: J. Phys.: Condens. Matter (1998) – in print

13. L. Bányai and S.W. Koch: In: *Semiconductor Quantum Dots*. Singapore: World Scientific 1993

14. C. Joslin and S. Goldman: J. Phys. B: At. Mol. Opt. Phys. **25**, 1965 (1992)

Few-Body Systems Suppl. 10, 199–202 (1999)

Few-
Body
Systems
© by Springer-Verlag 1999

Universal description of the He3 system at low energy

O.I. Kartavtsev

Bogoliubov Laboratory of Theoretical Physics
Joint Institute for Nuclear Research, 141980, Dubna, Russia

Abstract. Universal laws have been considered for the near-threshold three-body scattering in the three-body system with short-range interactions.
    To solve the scattering and bound state problems all the terms of the system of hyperradial equations have been obtained analytically in the asymptotic domain of large hyperradius ρ. The solution reveals universal description for the He2 + He elastic scattering below the three-body threshold and the uppermost He3 bound state regardless of the exact interaction of helium atoms.

Near-threshold phenomena attract a growing attention due to recent studies of the loosely bound He2 and He3 molecules [1, 2, 3] , Bose-Einstein condensation in a dilute vapor of alkali atoms [4, 5, 6] and related problem of the three-body recombination in these systems. Observations of the helium dimer and trimer [1, 2, 3] suppose a progress in studying the He2 + He collisions. Nevertheless, in spite of the thorough theoretical investigations of He3 [7, 8, 9, 10, 11], the scattering problem is in the primary stage [10, 11].
    The exceptionally small energies of the He2 bound state and He3 excited state challenge to a search for the universal picture. Also, the universal law is wanted to describe the system for all the variety of inter-atomic interactions. Indeed, the uncertainty in the He – He interaction potential is large enough to produce the third He3 bound state [7]. Besides, the three-body forces give rise to the undefined contribution in the properties of He3.
    The method of hyperspherical surface functions is appropriate for solving the problem posed. In the low-energy limit the total angular momentum $L = 0$ is taken into account. The particle mass m and two-body scattering length a are chosen as mass and length units. Using the scaled Jacobi coordinates $\mathbf{x}_i = \mathbf{r}_j - \mathbf{r}_k$, $\mathbf{y}_i = (2\mathbf{r}_i - \mathbf{r}_j - \mathbf{r}_k)/\sqrt{3}$, hyperspherical coordinates ρ, α_i, θ_i are defined by $x_i = \rho \cos \alpha_i$, $y_i = \rho \sin \alpha_i$, $\cos \theta_i = (\mathbf{x}_i \cdot \mathbf{y}_i)(x_i y_i)^{-1}$. Particle coordinates are $\mathbf{r}_i$ and $V(x_i)$ is the interaction potential of pair j, k.
    The expansion of the wave function Ψ in a set

$$\Psi = \rho^{-5/2} \sum_n u_n(\rho) \sum_{i=1}^{3} \chi_i^n(\alpha_i, \theta_i; \rho), \tag{1}$$

where the χ_i^n are solutions of the Faddeev equations on a hypersphere

$$\left[\Delta^* - \rho^2 V_i(\rho\cos\alpha_i) + \gamma_n^2\right]\chi_i^n - \rho^2 V_i(\rho\cos\alpha_i)\sum_{j\neq i}\chi_j^n = 0,$$

$$\Delta^* = \frac{1}{\sin^2 2\alpha_i}\frac{\partial}{\partial\alpha_i}\left(\sin^2 2\alpha_i\frac{\partial}{\partial\alpha_i}\right) + \frac{1}{\sin\theta_i}\frac{\partial}{\partial\theta_i}(\sin\theta_i\frac{\partial}{\partial\theta_i}),$$

(2)

normalized by the condition $\sum_{ij}\langle\chi_j^n|\chi_i^m\rangle = \delta_{nm}$, results in the system of ordinary differential equations

$$\left[-\frac{d^2}{d\rho^2} + \frac{\gamma_n^2 - 1/4}{\rho^2} - E\right]u_n(\rho) +$$

$$\sum_m\left[P_{nm}(\rho) - Q_{nm}(\rho)\frac{d}{d\rho} - \frac{d}{d\rho}Q_{nm}(\rho)\right]u_m(\rho) = 0.$$

(3)

Coupling terms in (3) are defined by the integrals on the hypersphere over $\sin^2 2\alpha_i d\alpha_i d\cos\theta_i$

$$Q_{nm}(\rho) = \sum_{i,j=1}^{3}\langle\chi_i^n|\frac{\partial}{\partial\rho}\chi_j^m\rangle, \quad P_{nm}(\rho) = \sum_{i,j=1}^{3}\langle\frac{\partial}{\partial\rho}\chi_i^n|\frac{\partial}{\partial\rho}\chi_j^m\rangle.$$

(4)

In the low-energy limit, inter-particle distances are large in comparison with the potential range b and one should consider the leading order terms in b/ρ. The interaction is negligible in the domain defined by conditions $\cos\alpha_i \gg b/\rho$, $i = \overline{1,3}$ and solution of Eq. (2) takes the form

$$\chi_i^n(\alpha_i) = B_n(\rho)\frac{\sin\gamma_n(\rho)\alpha_i}{\sin 2\alpha_i}.$$

(5)

Matching the expression (5) and the solution in small domains of interactions $\cos\alpha_i < b/\rho$, one comes to the equation for eigenvalues $\gamma_n(\rho)$

$$\gamma_n\cos\gamma_n\frac{\pi}{2} = \rho\sin\gamma_n\frac{\pi}{2} + \frac{8}{\sqrt{3}}\sin\gamma_n\frac{\pi}{6}.$$

(6)

The matching condition is assumed independent of energy and determined by the two-body scattering length $a \gg b$. Similar derivation of the equation for $\gamma_n(0)$ and $\gamma_n(\rho)$ was given in [12, 13, 14].

Roots of (6), $\gamma_n(\rho)$, are enumerated in accordance with increasing $\gamma_n^2(\rho)$. The first one $\gamma_1^2(\rho) \to -\rho^2$ ($\gamma_1 \to i\rho$) as $\rho \to \infty$ and the attractive potential in the first channel of (3) has a natural limit $\gamma_1^2(\rho)/\rho^2 \to -1$ (two-body bound state energy is -1 for $a \gg b$). Other $\gamma_n(\rho)$ are real-valued and $\gamma_n(\rho) \to 2n$, $n \geq 2$, as $\rho \to \infty$. A straightforward calculation, taking into account Eqs. (5,6), gives rise to a simple form for the coupling terms and normalization factor via corresponding eigenvalues γ_n and their derivatives

$$Q_{nm}(\rho) = 6B_n(\rho)B_m(\rho)(\gamma_n^2 - \gamma_m^2)^{-1}\sin\gamma_n\frac{\pi}{2}\sin\gamma_m\frac{\pi}{2},$$

(7)

$$P_{nm}(\rho) = Q_{nm}(\rho)\left[\frac{d}{d\rho}\ln\frac{B_n\sin\gamma_m\frac{\pi}{2}}{B_m\sin\gamma_n\frac{\pi}{2}} + \frac{\frac{d}{d\rho}(\gamma_n^2 + \gamma_m^2)}{(\gamma_m^2 - \gamma_n^2)}\right], \quad n \neq m, \qquad (8)$$

$$P_{nn}(\rho) = \left(\frac{d\gamma_n^2}{d\rho}\right)^2\left[\pi^3 B_n^2(\rho)\left(\frac{1}{8} + \frac{8\sin\gamma_n\frac{\pi}{3} - \sin\gamma_n\frac{2\pi}{3}}{27\sqrt{3}\gamma_n}\right) - \frac{1}{2}\frac{d^2\ln B_n(\rho)}{d\gamma_n^2}\right], \qquad (9)$$

$$B_n^2(\rho) = -\left(6\sin^2\gamma_n\frac{\pi}{2}\right)^{-1}\frac{d\gamma_n^2}{d\rho}. \qquad (10)$$

The exact analytical expressions (7–10) seem to be unknown in the literature. So far coefficients of the system (3) were calculated numerically [9, 14, 15]. The exact asymptotic in the method of hyperspherical surface functions are widely applicable in different problems. Among them, one should mention the inter-atomic scattering and three-body recombination of Bose-condensed atoms. After evident modification, the exact asymptotic is applicable to study the halo nuclei.

In the asymptotic region $\rho \gg b$ the system (3) is parameterless due to universality of functions (6), (7), (8), (9). Nevertheless, the solution of Eq. (3) in the asymptotic region should match the solution in the region of small ρ, i. e. the boundary conditions have to be posed on u_n in the region of small ρ. The boundary condition is regarded as energy independent for the low-energy processes. As far as channel functions $u_n(\rho)$, $n \geq 2$ decrease rapidly at $\rho \to 0$ due to repulsive terms γ_n^2/ρ^2, only the boundary condition for $u_1(\rho)$ is significant. Running this boundary condition, one obtains a set of one-parameter solutions of Eq. (3). Note an analogy with the one-parameter limit for the three-body Hamiltonian found in the investigation of the Efimov effect [16].

Among different, though equivalent, opportunities to look through all the solutions of the truncated system (3), the convenient way is to impose the boundary conditions $u_n(\rho_0) = 0$, $n = 1, 2$ and run the parameter ρ_0. For $\rho \to 0$ the solution takes the form $u_1(\rho) = \rho^{1/2}(C_1\sin|\gamma_1(0)|\ln\rho + C_2\cos|\gamma_1(0)|\ln\rho)$, $u_2(\rho) \to 0$. Clear, the replacement ρ_0 to $\rho_0\exp(-\pi/|\gamma_1(0)|)$ leaves any solution unchanged and produce one more bound state. Thus it is sufficient to run the parameter ρ_0 in any interval $\rho_1 < \rho_0 < \rho_2$ provided $\rho_1/\rho_2 = \exp(-\pi/|\gamma_1(0)|)$ (recall that $|\gamma_1(0)| = 1.006237825$). The artificial parameter ρ_0 has no exact meaning and it seems reasonable to use the uppermost trimer state $\varepsilon = -1 - E_b$ as a parameter in the universal law. The next trimer state with zero energy arises at $\varepsilon = \varepsilon_c$, therefore, ε runs the interval $[0, \varepsilon_c)$.

In spite of a simple analytical form for the universal coefficients (6), (7), (8), (9), the three-body characteristics (bound state energy, $2 + 1$ scattering phase and length) should be obtained by a numerical solution of Eq. (3).

Asymptotical form of the system (3) at $\rho \to \infty$ follows from the analytical expressions (6), (7), (9). For example, the effective potential in the first channel tends exponentially to the two-body threshold $(\gamma_1^2(\rho) - 1/4)/\rho^2 + P_{11} \longrightarrow -1 + O(e^{-c\rho})$, $c > 0$, as $\rho \to \infty$. At last, the asymptotic boundary conditions at $\rho \to \infty$ are $u_n \to 0$, $n = 1, 2$ for the bound state problem ($E_b < -1$) and $u_1 \to \sin(k\rho + \delta_k)$, $u_2 \to 0$ for the scattering problem, where the wavenumber k is defined by $k^2 = E + 1$ ($-1 < E < 0$) and δ_k is the phase shift of the $2 + 1$

elastic scattering. Numerical solution of the truncated system (3) provides the elastic $2+1$ scattering phase shifts δ_k and cross sections $\sigma(k^2)$ below the three-body threshold $(0 < k < 1)$ for different binding energies of the uppermost trimer state ε.

Generally, the universal law depends on two parameters that should be fixed to complete a description of the system. The parameters describe the wave function in small regions, where either two and all three particles interact with each other. One parameter (two-body scattering length) is used as a scale, thus producing one-parameter universal picture. The second one is fixed by the measurement or calculation of an arbitrary three-body (e. g. the uppermost trimer binding energy or $2 + 1$ scattering length) characteristic. In principle, more experimental (or computational) efforts are superfluous, as far as all the low-energy properties follow from the universal law. On the other hand, any set of measurements (or realistic calculations) may be tested by a comparison with the universal law.

References

1. F. Luo et al.: J. Chem. Phys. **98**, 3564 (1996)

2. F. Luo, C.F. Giese and W.R. Gentry: J. Chem. Phys. **104**, 1151 (1996)

3. W. Schöllkopf and J.P. Toennies: J. Chem. Phys. **104**, 1155 (1996)

4. M.H. Anderson et al.: Science **269**, 198 (1995)

5. K.B. Davis et al.: Phys. Rev. Lett. **75**, 3969 (1995)

6. C.C. Bradley et al.: Phys. Rev. Lett. **75**, 1687 (1995)

7. Th. Cornelius and W. Glöckle: J. Chem. Phys. **85**, 3906 (1986)

8. J. Carbonell, C. Gignoux and S.P. Merkuriev: Few-Body Syst. **15**, 15 (1993)

9. B.D. Esry, C.D. Lin, and Ch.H. Greene: Phys. Rev. **A54**, 394 (1996)

10. S. Nakaichi-Maeda and T.K. Lim: Phys. Rev. **A28**, 692 (1983)

11. E.A. Kolganova, A.K. Motovilov and S.A. Sofianos: Phys. Rev. **A56**, R1686 (1997)

12. V.N. Efimov: Nucl. Phys. **A210**, 157 (1973)

13. A.C. Phillips: Rep. Progr. Phys. **40**, 905 (1977)

14. D.V. Fedorov, A.S. Jensen: Phys. Rev. Lett. **71**, 4103 (1993)

15. Z. Zhen and J. Macek: Phys. Rev. **A38**, 1193 (1988)

16. S.A. Albeverio, R. Hoegh-Kron and T.T. Wu: Phys. Lett. **83A**, 105 (1981)

Few-Body Systems Suppl. 10, 203–206 (1999)

Few-
Body
Systems
© by Springer-Verlag 1999

Three Helium Atoms and the Scaling limit

L. Tomio[1], T. Frederico[2], A. Delfino[1] and A.E.A. Amorim[1]

[1] Instituto de Física Teórica, UNESP, São Paulo, Brasil
[2] Instituto Tecnológico de Aeronáutica, CTA, São José dos Campos, Brasil
[3] Instituto de Física, Universidade Federal Fluminense, Niterói, Brasil
[4] Faculdade de Tecnologia de Jahu, CEETEPS, Jahu, Brasil

Abstract. We show that a scaling limit approach, previously applied in three-body low-energy nuclear physics, is realized for the first excited state of ^4He trimer. The present result suggests that such approach has a wider application.

The ^{4}He$-^4$He dimer (B_{He}) has a measured size of about 50 Å [1], being the best known illustration of a loosely bound system. Qualitatively, the interaction has roughly a depth of 11 K with a scattering length (a_{He}) around 100 Å. In addition, it is well known that, approaching the limit where the two boson energy, E_2, is zero, the three-boson Faddeev equation presents the Efimov effect, which is characterized by an increasing number of three-body bound states [2]. Therefore, considering the large scattering length compared to the range of the potential, it is natural the search for Efimov states in the helium-trimer system. Several theoretical studies, using realistic interactions, have already clearly identified one excited state for this system [3, 4, 5, 6, 7], suggesting its identification with the first excited Efimov state. Another well known effect also occurs in the zero range limit of the interaction, when the two-boson scattering length (a_2) is kept fixed. In this case, the three-boson ground state ($B_3^{(0)}$) collapses and an infinite number of strongly bound states appears. This is known as the Thomas effect [8]. The Efimov and Thomas states are related by a scale transformation [9, 10], which can be easily visualized when a zero range interaction it is used. The presence of such states is governed by the ratio between the scattering length and the range of the force.

A scaling limit approach for three-body systems is defined when the range of two body interaction goes to zero, and it was applied in three-body low-energy nuclear physics in ref. [10]. The existence of such a scaling is a consequence of the renormalizabity of quantum mechanics with zero range interactions [11]. In ref. [10], the scaling limit approach was applied to a three-body nuclear system consisting of a massive core and two neutrons. The conclusion reached about the existence of one Efimov state in Carbon-20 ($(^{20})$C), for example, confirmed a suggestion made in ref. [12]. In this example, the scaling limit was also used to estimate quantitatively the energy of the first excited Efimov state. In the present case of ^4He-trimer, the scaling limit formalism is simplified, considering that we have three identical particles.

By using the scaling limit approach, we consider only the essential ingredients in the three-body equation, which leads to the Efimov limit, as it was done in ref. [10] for the case of halo-nuclei. So, in this respect the approach differs fundamentally from the previous ones (that consider realistic interactions) [3, 4, 5, 6, 7], as it will become clearer the identification of the first excited state. Different parametrizations in the interactions can disturb the analysis of the results obtained for the excited states, considering their sensibility to the range of the interaction or other parameters. By keeping the two-body energy fixed, as well as the three-body ground state, as done in case of the scaling limit approach, we can show that almost all such sensibility disappears.

We show that the scaling limit is in fact approached in the particular system of three ^{4}He atoms for which several calculations with realistic potentials exist in the literature [5, 6, 7]. The scaling limit of weakly bound tri-atomic systems (exploring the possibility of excited states) is found from the renormalization procedure of the zero range (RZR) three-body model [11]. The renormalization procedure keeps the general features of the two-body and three-body physics, that are given by the physical two-body inputs (like two-body bound state, scattering lengths and masses) and one three-body physical parameter, which can be the ground state energy [11]. The three-body ground state energy, if not available from experiments, can be determined using a realistic two-body interaction, with finite range [5]. In the latter case (when a realistic calculation has already been used for the ground state), one should still think about the advantages of using the RZR model to obtain other low-energy three body observables in the scaling limit, considering the extensive numerical task involved in a realistic calculation (see the appendix of ref. [5]).

In ref. [10], coupled Faddeev integral equations were solved for a three body system, in the zero-range limit of two-body interaction, considering that only two particles are identical. Here, with three identical particles, we can simplify the zero-range formalism and obtain a single integral equation. It is used a regularization parameter Λ, in the momentum integration, which represents the inverse of the interaction range [2, 10]. Later on, the limit of Λ going to infinity is considered while the ratio between the two-body energy, E_2, and the three-body ground state energy, $B_3^{(0)}$, is kept fixed. For bound two-body systems, $E_2(= -|E_2|)$ is the separation energy and $a_2 = 1/\sqrt{|E_2|}$ (in units such that the mass $= 1$ and $\hbar = 1$) is positive. In the following we also consider the case where the two-boson system has a virtual state, which implies that $E_2 = -|E_2|$, with a negative two-body scattering length, $a_2 = -1/\sqrt{|E_2|}$.

The $s-$wave partial wave equation we solve is given by

$$\chi(q) = \frac{2/\pi}{\sqrt{B_3^{(N)} + \frac{3}{4}q^2 \mp \frac{1}{a_2}}} \int_0^\Lambda dk \log \frac{B_3^{(N)} + qk + q^2 + k^2}{B_3^{(N)} - qk + q^2 + k^2} \chi(k), \quad (1)$$

where $B_3^{(0)}$ is the three-body ground state and, for $N > 1$, $B_3^{(N)}$ is the energy of the $N-$th Efimov state.

We solve the above equation in units such that $\Lambda = 1$. The corresponding dimensionless quantities are: $\epsilon_3^{(N)} \equiv B_3^{(N)}/\Lambda^2$, $\kappa_2 \equiv 1/(a_2\Lambda)$. The two-body observables can be written in terms of the three-body binding energy $B_3^{(N)}$, by replacing Λ, such that $\kappa_2/\sqrt{\epsilon_3^{(N)}} = 1/\left(a_2\sqrt{B_3^{(N)}}\right)$. The Thomas effect occurs for $\Lambda \to \infty$ with the energies of the two-body systems kept fixed, whereas the Efimov states arise when $a_2 \to \infty$ with Λ kept fixed.

In the strict Efimov limit (when $E_2 = 0$), the ratio between two close Efimov states, given by $B_3^{(N+1)}/B_3^{(N)}$, is practically constant. For equal masses, such a ratio is close to $1/500$. In the general case, for a finite and large two-body scattering length a_2, such a ratio can define a *scaling function* f, which will depend on a single dimensionless observable z, given by the inverse of the product of a_2 with the square-root of the three-body Efimov state energy, $B_3^{(N)}$:

$$\frac{B_3^{(N+1)}}{B_3^{(N)}} = f(z), \quad \text{where} \quad z \equiv \frac{1}{\left(a_2\sqrt{B_3^{(N)}}\right)} \tag{2}$$

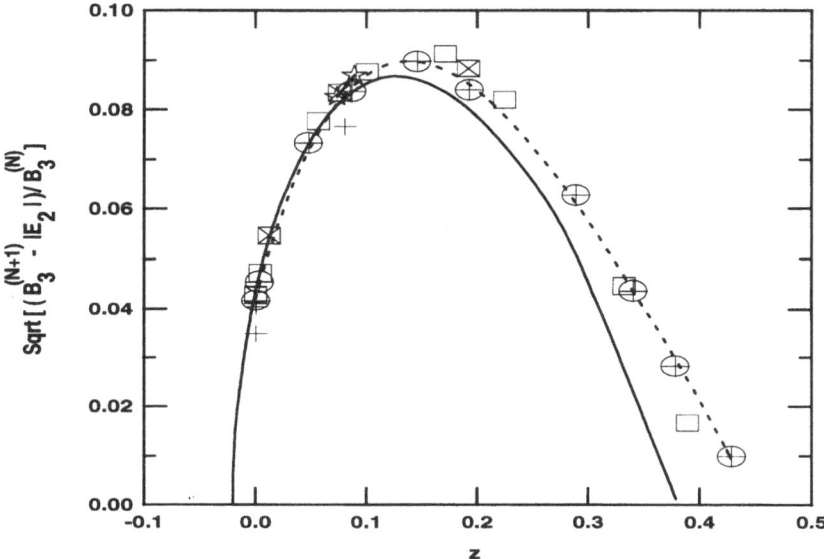

Figure. 1 Our scaling limit (solid line) is compared with realistic model results, for the helium-trimer. The dashed-line guides the eyes through realistic model results.

The results presented in Fig. 1 lead to our scaling function $f(z)$, which is compared to results from realistic interactions, taken from ref. [5] (empty boxes for $s-$wave and $N = 0$; crossed circles for $s + d$ waves and $N = 0$; crosses for $s + d$ waves with $N = 1$), from ref. [4] (stars), and from ref. [6] (crossed boxes).

Recent results from ref. [7] are also consistent with this plot. As we can see, the results from realistic interactions approach the scaling limit.

The universality of the scaling function is a characteristic of the Efimov regime, which does not depend on the detailed nature of the short-range two-body interaction. As the two-body binding energy is increased, a small deviation between the scaling limit and realistic models can be observed. By including the effective range effect, one could improve the scaling function without requiring further details about the two-body interaction.

In summary, we have shown that the ^{4}He$-$trimer ground and excited states approach the scaling limit, obtained from a calculation with the renormalized zero-range model. We have also extended the study of the critical conditions for the appearance of one Efimov state to other three atomic systems, with two like atoms plus a third one, for a wide range of atom-atom scattering lengths, in a work to be published.

Acknowledgement. This work was partially supported by the Brazilian agencies Fundação de Amparo à Pesquisa do Estado de São Paulo and Conselho Nacional de Desenvolvimento Científico e Tecnológico.

References

1. W. Schöllkopf, J.P. Toennies: Science **266**, 1345 (1994)

2. V. Efimov: Phys. Lett. **B33**, 563 (1970)

3. T.K. Lim, K. Duffy, W.C. Damert: Phys. Rev. Lett. **38**, 341 (1977); H.S. Huber, T.K. Lim, D.H. Feng: Phys. Rev. **C 18**, 1534 (1978); H.S. Huber, T.K. Lim: J. Chem. Phys. **68**, 1006 (1978)

4. S. Huber: Phys. Rev. **A31**, 3981 (1985)

5. Th. Cornelius, W. Glöckle: J. Chem. Phys. **85**, 1 (1986)

6. B.D. Esry, C.D. Lin, C.H. Greene: Phys. Rev. **A54**, 394 (1996)

7. E.A. Kolganova, A.K. Motovilov, S.A. Sofianos: Phys. Rev. **A56**, R1686 (1997)

8. L.H. Thomas: Phys. Rev. **47**, 903 (1935)

9. S.K. Adhikari et al.: Phys. Rev. **A37**, 3666 (1988); Phys. Rev. **A 47**, 1093 (1993)

10. A.E.A. Amorim, L. Tomio, T. Frederico: Phys. Rev. **C56**, R2378 (1997)

11. S.K. Adhikari, T. Frederico, I.D. Goldman: Phys. Rev. Lett. **74**, 487 (1995); S.K. Adhikari, T. Frederico: Phys. Rev. Lett. **74**, 4572(1995)

12. D.V. Fedorov, A.S. Jensen, K. Riisager: Phys. Rev. Lett. **73**, 2817 (1994)

Few-Body Systems Suppl. 10, 207–210 (1999)

Few-
Body
Systems
© by Springer-Verlag 1999

Antiproton-Hydrogen and Antihydrogen-Hydrogen annihilation at sub-kelvin temperatures

A.Yu. Voronin[1], J. Carbonell[2]

[1] P. N. Lebedev Physical Institute, 53 Leninsky pr., 117924 Moscow, Russia
[2] Institut des Sciences Nucléaires, 53 Av. des Martyrs, F-38026 Grenoble

Abstract. Cross sections of ultra low-energy antiprotons ($E \leq 10^{-6}$ a.u.) and antihydrogen with atomic hydrogen are calculated within a quantum mechanical coupled channels approach. The results differs from the extrapolations of semiclassical models. The main features of the observables are found to be determined by a family of $\bar{p}H$, $\bar{H}H$ nearthreshold metastable states.

1 Introduction

The project of obtaining ultracold antimatter in traps [1] require theoretical calculations of the low energy matter-antimatter inelastic collisions rates. We present here a quantum mechanical model of antiproton ($\bar{p}$) and atomic hydrogen (H) collisions, describing the elastic ($\bar{p} + H \rightarrow \bar{p} + H$) and Protonium (Pn) formation ($\bar{p} + H \rightarrow Pn^* + e$) channels at energies less than 10^{-6}a.u.

A direct solution of the Faddeev three-body equations is made extremely difficult by the large number of Pn open channels which contain in addition fast oscillating asymptotics. As a first step towards a full solution of this problem we developed [2] a "Faddeev inspired" unitary coupled channel model which incorporate the main physical inputs and provide, we hope, reliable estimations for the desired cross sections. Our results will be comprared to those provided by the semiclassical methods [3, 4, 5, 6]. We will show that in our approach the low energy energy properties are determined by a rich spectrum of nearthreshold S-matrix singularities generated by the long range charge-dipole $H\bar{p}$ interaction. The model has been extended to the case of $H + \bar{H}$ interaction.

2 The formalism

The details of the formalism can be found in [2]. Their final aim was to obtain a $\bar{p}$-H optical potential V_e to be inserted in the Schrodinger equation governing the p̄-H wavefunction $\chi(R)$

$$\left[-\frac{1}{M}\partial_R^2 + \frac{L(L+1)}{MR^2} + \hat{V}_e + V_p(R) - E + \varepsilon_B\right]\chi(R) = 0 \tag{1}$$

where M is the p̄ mass, L the p̄-H angular momentum, E the p̄ center of mass energy, ε_B the H-atom Bohr energy and V_p the long range polarization forces which are treated separatedly. The effective interaction $\hat{V}_e$, complex energy-dependent and nonlocal, account for the coupling with the Pn formation channels. In practical calculations only the channels with Pn principal quantum number $10 < n < 40$ and Pn angular momentum $l \leq 1$ were included. The effect of the infinite number of closed Pn states ($n > 30$) is known to be responsible for long-range polarization terms in the interaction and have been taken into account by explictly introducing the polarization potential in the form $V_p(R) = \frac{1}{2}\frac{\alpha(R)}{R^4}$ with $\alpha(R)$ such that for $R >> r_B$ (H-Bohr radius) $\alpha(R) \to -\alpha_d$ the H dipole polarizability. Numerical checks showed that a 10% accuracy is achieved with the restrictions done in the the the Pn channels.

3   Results

The $H\bar{p}$ scattering length is found to be $a = (-7.8 - i11.5)r_B$ and the corresponding zero energy elastic cross-section $\sigma_{el}(0) = 2426.4\,r_B^2$. Such a big value is a consequence of the long range polarization forces. The $\bar{p}H$ annihilation rate in the low energy limit is $v\sigma_{ann} = 1.5\,10^{-9}cm^3/s$. The S-wave dominance and the scattering length approximation are valid for $E \leq 10^{-8}a.u.$

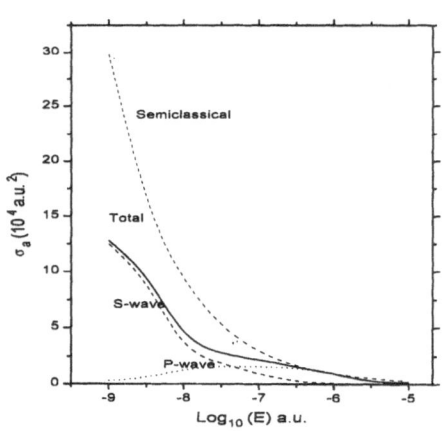

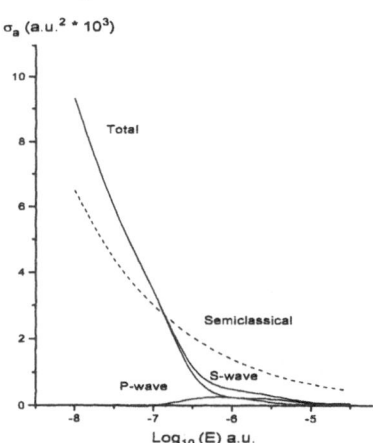

Figure 1. Annihilation cross section for $H + \bar{p} \to Pn^* + e$ reaction

Figure 2. Annihilation cross section for $H + \bar{H} \to Pn^* + (e^+e^-)$ reaction

The total annihilation cross-section is shown on Fig. 1. It follows the $1/v$-law for $E_{\bar{p}} < 10^{-8}$ a.u. and decreases nonmonotonously for $E_{\bar{p}} > 10^{-8}$ a.u. It is compared with the semiclassical results of [5] and turns to be ~ 2.5 times bigger than the one we got for $E_{\bar{p}} < 10^{-8}$ a.u.

The scattering cross sections significantly change their behavior at $E_{\bar{p}} \sim 10^{-8}$ a.u. . This is due to the presence of nearthreshold $H\bar{p}$ bound and virtual states generated by the polarization potential [7]. Such states, being nearthreshold S-matrix singularities, determine the energy dependence of the $H\bar{p}$ scattering cross-section. The energies, inelastic widths and mean radius of such states are shown in Table 1. In the threshold vicinity the elastic S-matrix for L=0 is dominated by its singularities. The position of the nearest to the origin S-matrix zero (and pole) corresponds to an energy of $E_c \sim 10^{-8}$ a.u. and plays a role of characteristic energy for the reaction. We notice however, that this nearest to the threshold S-matrix pole lies on the non-physical list, i.e. $Im(k) < 0$ and $Re(k) < 0$, and corresponds to a virtual state.

Table 1. Energies, Auger widths and mean radii (a.u) of L=0 $H\bar{p}$ states. We denote by index I the results in V_p alone and by index II those obtained with $(V_p + V_e)$

E_I	E_{II}	$\bar{x}_{II}$
	$-5.1\ 10^{-8} + i\ 7\ 10^{-9}$	
$-4.2\ 10^{-7}$	$-2.5\ 10^{-6} - i\ .2\ 10^{-7}$	27.0
$-3.6\ 10^{-5}$	$-7.0\ 10^{-5} - i\ 8.4\ 10^{-6}$	11.3
$-2.6\ 10^{-4}$	$-4.1\ 10^{-4} - i\ 3.2\ 10^{-5}$	7.3
$-9.2\ 10^{-4}$	$-1.5\ 10^{-3} - i\ 8.6\ 10^{-5}$	5.3
$-2.3\ 10^{-3}$	$-4.2\ 10^{-3} - i\ 2.0\ 10^{-4}$	4.2

One can see from Table 1 that the position of nearthreshold states is mainly determined by the polarization potential and only modified by nonlocal complex inelastic interaction $\hat{V}_e$. This suggests the possibility to obtain a local complex potential which would be equivalent to the full $H\bar{p}$ interaction in the energy range of interest. We search for such equivalent local complex potential as a sum of three terms $V_e^l(R) = V_s(R) + V_{cs}(R) + V_p(R)$ in which V_{cs} is a screened Coulomb potential and V_s is a local short rang part to be determined. It was assumed to have the form:

$$V_s(R) = \begin{cases} -V_1\, e^{-\alpha_1\left(\frac{R}{r_B}\right)} - i\, W_1\, e^{-\beta_1\left(\frac{R}{r_B}\right)} & \text{if } R < R_c \\ -i\, W_2\, e^{-\beta_2\left(\frac{R}{r_B}\right)} & \text{if } R \geq R_c \end{cases} \tag{2}$$

and a satisfactory fit is obtained with the values: $V_1 = 0.572$, $W_1 = W_2 = 0.040$, $\alpha_1 = 1.20$, $\beta_1 = \beta_2 = 3.20$ and $R_c = 2r_B$. The agreement between the results of the nonlocal effective potential and its local approximation are within few percent in the energy range $0.5\ 10^{-9} - 0.5\ 10^{-6}$ a.u.

The results obtained for $H\bar{p}$ interaction can be used for a qualitative treatment of $H\bar{H}$ low-energy collisions $H + \bar{H} \rightarrow Pn^* + (e^+e^-)$. The $H\bar{H}$ system interacts at long distances via a dipole-dipole potential of the form $V_{dd} \sim -6.5/R^6$. An estimation of the $H\bar{H}$ potential can be obtained by adding to the same short-range part as in $H\bar{p}$ case the dipole-dipole long-range tail V_{dd}. The reaction cross-section calculated in such a way is shown on Fig. 2. The characteristic energy for this reaction was found to be $\sim 10^{-5}$ a.u., corresponding to the position of the nearest to the threshold S-matrix singularity (virtual state with energy $-7.8\ 10^{-6}$ a.u.). The scattering length was found to be $a = (-3.1 - i3.7)r_B$ and the $H\bar{H}$ annihilation rate in the low energy limit $v\sigma_{ann} = 3\ 10^{-10}$ cm^3/s. The annihilation cross-section, obtained in the classical approach [5] is also plotted on Fig.2. The energy dependence of such cross-section in the low energy limit turns to be $v^{-2/3}$ instead of v^{-1} in a quantum mechanical treatment.

4 Conclusion

A coupled channels model describing the $H\bar{p}$ system at energies less than 10^{-6} a.u. has been developed. The results thus obtained substantially differ from the low energy extrapolations of the black sphere model and other classical or semiclassical approaches.

The reaction dynamics is found to be determined by the existence of several nearthreshold states. Such states are produced by the long-range polarization potential and are shifted in the complex momentum plane by the coupling with Protonium formation channels.

A local approximation of the effective potential has been proposed for further applications. It reproduces the scattering observables in the considered energy range and has the same nearthreshold spectral properties.

A qualitative extension of this approach to $H\bar{H}$ system has been discussed. We have estimated the rate of $H\bar{H}$ annihilation to be $v\sigma_{ann} = 3\ 10^{-10} cm^3/s$.

References

1. Hyp. Int. **76** (1993); G. Gabrielse et al.: Hyp. Int. **89**, 371 (1994); M. Charlton et al.: Phys. Rep. **241** 65 (1994); Nucl. Phys. B **56A** (1997)

2. A.Yu. Voronin, J. Carbonell: Phys. Rev. **A57**, 4335 (1998)

3. E. Fermi, E. Teller, Phys. Rev. **72**, 399 (1947)

4. D.L. Morgan Jr. and V.W. Hughes: Phys. Rev. **D2**, 1389 (1970); W. Kolos et al.: Phys. Rev. **A11**, 1792 (1975)

5. D.L. Morgan Jr. : Hyp. Int. **44**, 399 (1988)

6. G.V. Shlyapnikov et al.: Hyp. Int. **76**, 31 (1993)

7. J. Carbonell et al.: Few-Body Systems Suppl. **8**, 428 (1995)

Few-Body Systems Suppl. 10, 211–214 (1999)

Few-
Body
Systems
© by Springer-Verlag 1999

Collisional survival of antiprotonic atomcules

S. Sauge[1]*, P. Valiron[1], J. Carbonell[2]

[1] Observatoire de Grenoble, Laboratoire d'Astrophysique, U. Joseph Fourier,
BP 53X, 38041 Grenoble Cedex, France
[2] Institut des Sciences Nucléaires, Av. des Martyrs, 38026 Grenoble, France

Abstract. The collisional metastability of antiprotonic atomcules $He^+ - \bar{p}$ in pure helium is discussed in the framework of a classical trajectory approach for three slow nuclei ($\alpha\bar{p} - \alpha$) moving in an ab initio Born-Oppenheimer electronic potential. Our results support the destruction of the $n > 41$ states during the thermalisation stage, as well as the strong depopulation of the elliptic states adjacent to Auger-dominated short-lived levels.

1 Introduction

When massive negatively charged particles such as antiprotons ($\bar{p}$) are stopped in ordinary matter, they slow down as they excite and ionize atoms and some are eventually captured, forming bound states of exotic atoms. In most materials, antiprotons annihilate in a few picoseconds, to the notable exception of helium, since 3% of the $\bar{p}$ survive several μs, for helium densities varying over several orders of magnitude [1]. While the exotic atoms may resist to millions of collisions in pure helium, molecular contaminants such as H_2 are likely to destroy them in a single collision [2].

The metastability of the isolated $He^+\bar{p}$ system is well understood from an atomic or molecular point of view. The atomic description is supported by the Condo model [3], in which metastable $\bar{p}$ states are identified to quasi-circular Rydberg orbits $l \simeq 40$, with radiative lifetimes of a few μs, in agreement with experiments. The $He^+\bar{p}$ system can also be considered as a special diatomic molecule where metastability of the "anti-bounding" is ensured by the centrifugal force from the high rotational excitation $J \simeq 40$. The atomcule denomination [1] was introduced to account for this duality of description, which was confirmed by the remarkable agreement between the theoretical predictions of transition wavelengths [4] and the laser spectroscopy measurements performed at LEAR in CERN [5].

---

*E-mail addresses: sauge and valiron@obs.ujf-grenoble.fr, carbonel@isn.in2p3.fr

Nevertheless, the collisional stability of atomcules remains mysterious. Korenmann [6], assuming that Stark mixing with Auger nonstable states is dominant in pure helium because of the weak splitting of energy levels, applied the Rosen-Zener-Demkov model and concluded that the $n > 40$ states would be quenched in times shorter than a few ns, compatible with experimental results. Voronin [7] used quantum coupled channels model and confirmed Korenman's mechanism for $n > 42$ states only, lower states being primarily quenched by Auger decay enhancement due to collisions or via the formation of an $(He^+ - \bar{p} - He^+)$ molecular ion. We present here an alternate *physicochemistry* approach to account for collisional effects arising from the detailed features of the intermolecular interaction.

2 Classical trajectory approach

Since metastable states are Rydberg ones, a classical treatment of the three slow nuclei $(\alpha\bar{p} - \alpha)$ in the adiabatic electronic potential provides a valid first step towards a full quantum mechanical study, at least as far as finer quantum effects such as tunneling or electronic nonadiabaticity are not considered.

In our molecular picture, Rydberg metastable states for the $He^+\bar{p}$ subsystem were described by enforcing a semi-classical quantization (Fig. 1).

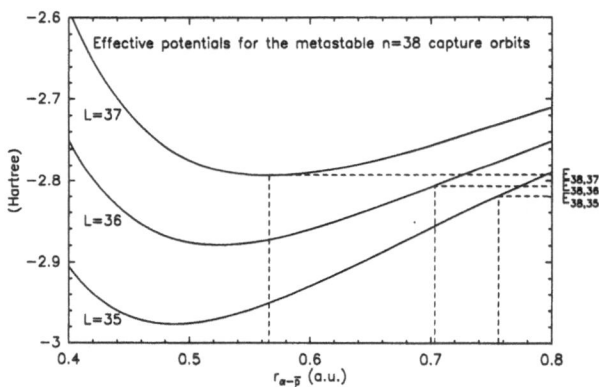

Figure 1. Semi-classical quantization of the adiabatic $He^+\bar{p}$ potential. Reproducing theoretical energies [4] of given n states constrains the classical orbit of the circular state and semi-major axis of the elliptic ones. Alternatively, a quantization in the spirit of Bohr's model predicts the energies of the circular states with a relative accuracy of a few 10^{-3}, validating the consistency of the approach.

3 Atomcule-He interaction

We solved the Schrödinger equation for three electrons in the field of three fixed $(\alpha-\bar{p}-\alpha)$ nuclei, using fourth order perturbation treatment (UMP4-SDTQ) [8] for electronic correlation and extended polarized basis sets. Basis set superposition error was corrected using a generalized counterpoise (CP) scheme [9]. Calculations were repeated over a fine 3-D mesh and a spline interpolation of the hypersurface was performed. The potential corresponding to the circular $n = 38$ capture orbit is shown in Fig. 2.

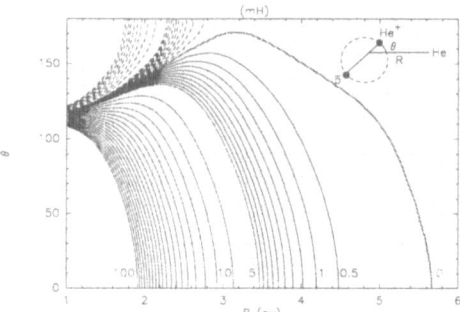

Figure 2. Atomcule-He interaction potential for the $(n, l) = (38, 37)$ circular capture orbit. The instantaneous potential is strongly modulated along the $\bar{p}$ trajectory (i.e. along θ). This modulation might induce transitions from circular to elliptic states.

Averaging this potential over $\bar{p}$ fast rotating orbit allows a preliminary interpretation in terms of classical reactive channels for frozen geometries. As seen in Fig. 3, the potential (averaged over θ) presents an activation barrier, which qualitatively accounts for the collisional survival of thermalized atomcules in pure He. The kinetic energy of freshly-formed atomcules should be sufficient to overcome the activation barrier in the initial thermalisation stage.

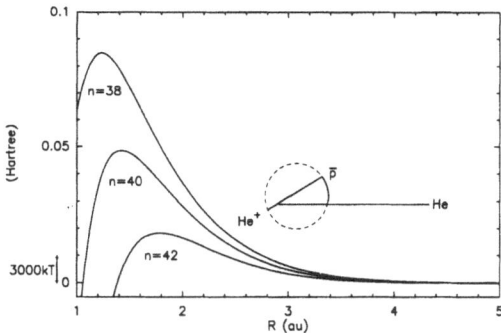

Figure 3. Atomcule-He interaction potential for different circular capture orbits. Activation barrier presents a net reduction for orbits with $n > 42$, accounting for the spectroscopic observation that the corresponding states are not populated [5]. This can be attributed to a reduction of Pauli shielding, for high n states.

214

4 Conclusions and perspectives

Our results support the destruction of the $n > 41$ states during the thermalisation stage, as well as the strong depopulation of the elliptic states adjacent to Auger-dominated short-lived levels [1]. The classical trajectory approach can be easily generalized to study the collisional destruction of atomcules by impurities such as noble gases and H_2. Strong quenching by H_2 is supported by the existence of reactive channels with low activation barriers (Fig. 4). It may explain the laser induced transitions observed between two normally metastable states, using resonant depopulation of the states quenched by H_2 [2].

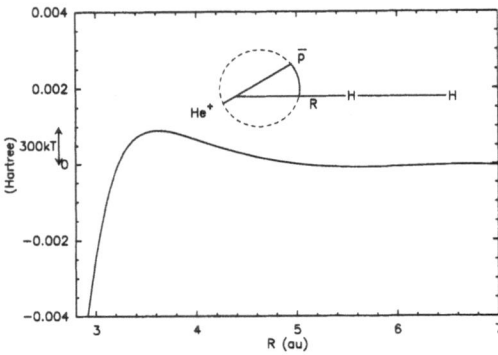

Figure 4. Atomcule-He interaction for the $(n, l) = (38, 37)$ circular capture orbit in the (coplanar) colinear approach.

References

1. T. Yamazaki et al.: Nature **361**, 238 (1993)

2. B. Ketzer: Phys. Rev. Lett. **78**, 1671 (1997)

3. G.T. Condo: Phys. Lett. **9**, 65 (1964)

4. V. Korobov: Hyp. Int. **101/102**, 479 (1996)

5. T. Yamazaki: Proceedings of the LEAP94 Conference, p. 553 (1994)

6. G.Ya. Korenman: Hyp. Int. **103**, 341 (1996)

7. A. Voronin, O. Dalkarov: Private Communication

8. W.J. Hehre et al.: *Ab initio molecular orbital theory.* New York: Wiley 1986

9. P. Valiron, I. Mayer: Chem. Phys. Lett. **275**, 46 (1997) *and Refs. therein*

---

[1]Detailed destruction probabilities and mean lifetimes of capture states as well as repopulating of adjacent levels will be published elsewhere.

Few-Body Systems Suppl. 10, 215–218 (1999)

Few-
Body
Systems
© by Springer-Verlag 1999

The fate of ^{7}Be in the Sun. Triple collisions

V.B. Belyaev[1,2], D.E. Monakhov[1], D.V. Naumov[1,3] and F.M. Pen'kov[1]

[1] Joint Institute for Nuclear Research, Dubna, 141980, Russia
[2] Research Center for Nuclear Physics, Osaka University, Japan
[3] Physics Department, Irutsk State University, Irkutsk, 664003, Russia

Abstract. We evaluate the effect of screening by bound electron in ^{7}Be(p,γ)^{8}B reaction, where ^{7}Be target contains bound electron, in the framework of the adiabatic representation of the three particle problem. A comparison with two other approximations (united atom and folding) is presented. A good agreement between the "united atom" approximation and the exact solution is found. We also discuss the screening corrections induced by two K-shell electrons on a ^{7}Be target. The bound electron screening effect consequences for ^{7}Be and ^{8}B solar neutrino fluxes are discussed.

1 Introduction

In recent years, an increasing attention has been devoted to an accurate estimation of electron screening effect for nuclear fusion reactions in stellar plasma and for the interactions of low-energy ion beams with atomic or molecular targets in laboratory experiments (see refs. [1, 2, 3] and references therein).

In this letter we present the first quantum mechanical calculation of screening effect by bound electron in ^{7}Be + p $\longrightarrow$ ^{8}B + γ nuclear fusion from the pp-cycle in the sun. Contribution of this reaction to the total luminosity of the sun is negligibly small, but it is directly related to the long-standing "Solar Neutrino Problem", – one of the most intriguing issue in the present-day neutrino astrophysics.

We consider the phenomenon within the framework of the adiabatic approximation, which comes from the well known Born-Oppenheimer (BO) approximation (see for example ref. [4]). Corrections to the BO approximation are expected to be negligible at solar conditions. At low and moderate energies, the fusion cross section of "bare" charged nuclei colliding with the relative momenta p in the center-of-mass frame is expressed as (see ref. [5]):

$$\sigma_b(E) = \frac{S(E)}{E} e^{-2\pi\eta}, \tag{1}$$

where $S(E)$ is the so-called astrophysical factor which incorporates all nuclear features of the process, E is the collision energy of the nuclei, $\eta = MZ_1Z_2/(m_e a_0 p)$ is the usual Coulomb parameter, m_e and M are the electron and reduced nuclear masses, respectively, and a_0 is the hydrogen Bohr radius. The exponential factor originates from the Coulomb wave function of the internuclear motion $\psi_E^C(R)$ at $R = 0$. The screened cross section $\sigma_s(E)$ differs by the enhancement factor

$$\gamma(E) \equiv \frac{\sigma_s}{\sigma_b} = \frac{|\psi_E(0)|^2}{|\psi_E^C(0)|^2}, \qquad (2)$$

where $\psi_E(R)$ is the wave function of the internuclear motion which accounts for the bound electron.

Our calculation of the screening effect is compared with two relevant approximations, "united atom" (UA) and folding approximations which, as we will demonstrate below, give respectively upper and lower estimates for the problem. In the framework of the UA approximation we estimate also the screening effect for the ^{7}Be nucleus with two K-shell electrons.

2   Method of Calculation

The three particle Schrödinger equation solution is represented as $\Psi(\mathbf{r}, \mathbf{R}) = \sum_k \psi_k(\mathbf{R})\phi_k(\mathbf{r}; \mathbf{R})$. Here, $\mathbf{R}$ is the internuclear radius-vector, and $\mathbf{r}$ is the electron radius-vector from the center-of-mass of the nuclei. The two-center eigenfunctions $\phi_k(\mathbf{r}; \mathbf{R})$ are derived from the Schrödinger equation for three particles, when two nuclei are fixed at a distance R, and the eigenfunctions have a dependence on R as a parameter. The nuclei with electric charges Z_1 and Z_2 interact with the effective potential $V_{nlm}(R) = Z_1 Z_2/R + U_{nlm}(R)$, where $U_{nlm}(R)$ is the electron energy in the field of the nuclei, and $\psi_k(\mathbf{R})$ are the eigenfunctions of nuclear motion in this potential. We shall write the parameters in atomic units, except that we shall display the units. Thus, the energy unit is $m_e e^4/\hbar^2 \approx 27.21$ eV, the lenght unit is a_0, and the electric charge unit is e.

To compute the screening effect we used only one two-center eigenfunction corresponding to the ground state of the system, since only the ground state energy has the correct asymptotic behaviour [6] (see below) and the high energy states corrections to the screening are negligible. The fusing nuclei are considered in S-wave. Thus, the total three particle wave function reads: $\Psi(\mathbf{r}, \mathbf{R}) = \psi(\mathbf{R})\phi(\mathbf{r}; \mathbf{R})$. The tabulated values of $U_{nlm}(R)$ [7] are used for our purposes. The ground state eigenvalue energy $U(R)$ ($\equiv U_{000}(R)$), with zero quantum numbers, has the correct asymptotic behaviour: the electron eigenvalue energy reaches the energy of the united ion as R approaches zero: $U(0) = (Z_1 + Z_2)^2/2$, and $U(\infty) = Z_1^2/2$ (that is the energy of the isolated ion eZ_1). In Fig. 1 a) the values for the ground state are plotted. We solved numerically the Schrödinger equation for two scattering nuclei with $Z_1 = 4$ and $Z_2 = 1$ at center-of-mass kinetic energy E within the potential V_{000}. The values

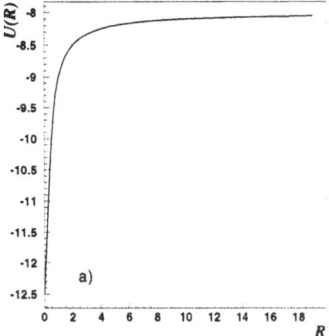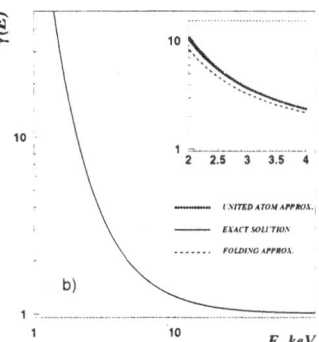

Figure 1. a) Electron energy in the field of the two nuclei ^7Be and p. b) Enhancement of nuclear fusion rate due to the electron screening

of the squared wave function of relative motion of the two nuclei $|\psi_E(0)|^2$ are calculated and compared with $|\psi_E^C(0)|^2 = 2\pi\eta/(e^{2\pi\eta} - 1)$.

The essence of the UA approximation, serving as an upper limit to the screening corrections, consists in replacing the kinetic energy E by the shifted energy $E + \Delta E$, where ΔE is the energy difference of the bound electron between the final and initial states. For one bound electron this energy difference is given by $\Delta E = (Z_1 + Z_2)^2/2 - Z_1^2/2$. Thus, the UA approximation changes $|\psi_E^C(0)|^2$ to $|\psi_{E+\Delta E}^C(0)|^2$. It is easy to ensure that the UA approximation applicability at solar conditions is well satisfied.

The folding approximation can be used as a lower limit to the screening corrections. This approximation embodies the static wave function of the bound electron, which does not depend on the internuclear distance. The total three particle wave function in this approximation is presented as a product: $\Psi_f(\mathbf{r}, \mathbf{R}) = \psi_f(\mathbf{R})\phi_f(\mathbf{r})$, where the wave function for the electron bound in the nucleus with electric charge Z_1 reads: $\phi_f(\mathbf{r}) = \sqrt{Z_1^3/\pi}e^{-Z_1 r}$, and $\psi_f(\mathbf{R})$ is the wave function of the two nuclei. Averaging the electron coordinates, $\psi_f(\mathbf{R})$ can be derived from the Schrödinger equation with the potential: $V_f(R) = Z_1 Z_2/R - Z_2 \cdot \left(1 - e^{-2RZ_1}/R - Z_1 e^{-2RZ_1}\right)$. The attractive part of the potential V_f reaches the value: $Z_1 Z_2$ as $R \longrightarrow 0$. Since $Z_1 Z_2 \leq (Z_1 + Z_2)^2/2$ that is the value of the electron energy in united ion, the folding approximation gives a lower estimate for the enhancement factor. The area of the folding approximation applicability is the "impact" collision of the nuclei.

3 Summary and discussion

The enhancement factor (2) is plotted in Fig. 1 b) for all three approximations. It is seen that the UA approximation always overestimates the exact solution, while the folding approximation underestimates it. Nevertheless, it is easy to see that simple UA prescription gives very close values to the exact solution at kinetic energies above 2 keV. Therefore, the latter can be used not only as a qualitative, but as a good quantitative approximation to the electron screening by bound electrons. In a plasma, most of the nuclear fusions come at the Gamow peak energy ($E_0 \approx 18$ keV in the considered reaction), that is defined by the strong dependence of the nuclear cross section on energy (1) and the fast decrease of the exponential particle distribution. Then $\gamma(E_0) = 1.1$, that is, there is 10% enhancement by one bound electron in the boron production rate. Simple computations within the UA approximation give the screening effect as: $\gamma(E_0) = e^{\Delta E/kT}$. One can apply this formula also for a ^{7}Be nucleus with two bound electrons. Then, $\Delta E = 227.98$ eV. Thus, two bound electrons enhancement factor is given by $\gamma(E_0) = 1.196$, i.e. roughly 20%.

The probabilites that one or two K-shell electrons are associated with any given ^{7}Be nucleus in the solar interior are $f_1 = 30\%$, and $f_2 = 3\%$ [8]. Using the calculated abundances of ^{7}Be ions, one can estimate the thermal averaged screening effect induced by both one and two bound electrons on a ^{7}Be nucleus: $\langle \gamma \rangle - 1 = 0.30 \times 0.1 + 0.03 \times 0.2 \approx 0.04$.

In summary, in the solar interior K-shell bound electrons enhance ^{7}Be(p,γ)^{8}B rate and increase ^{8}B neutrino production rate by of about 4%. Therefore, bound electrons screening have an effect on the solar ^{8}B neutrinos, and acts with the opposite effect to the berrylium neutrinos. The main essence of the electron screening in nuclear fusions is the change in electron density on a nucleus during the collision of the nuclei. This effect can be treated only in a dynamical calculation like the present three particle calculation or the discussed UA approximation.

We thank J. N. Bahcall, A. V. Gruzinov and V. A. Naumov for useful discussions.

References

1. A.V. Gruzinov and J.N. Bahcall: astro-ph/9801028

2. L.S. Brown and R.F. Sawyer: Rev. Mod. Phys. **69**, 411 (1997)

3. C.W. Johnson, E. Kolbe, S.E. Koonin and K. Langanke: Ap. J. **392**, 320 (1992)

4. L. Bracci, G. Innocenti, V.S. Melezhik, G. Mezzorani and P. Pasini: Phys. Lett. **A153**, 456 (1991)

5. Astrophysical formulae (Springer, Berlin, 1974)

6. S.S. Gershtein and V.D. Krivchenkov: Sov. Phys. JETP, 1491 (1961)

7. L.I. Ponomarev and T.P. Puzynina: Dubna preprint R4–3175 (1967), in Russian

8. I. Iben, K. Kalata and J. Schwartz: Ap. J. **150**, 1001 (1967)

Few-Body Systems Suppl. 10, 219–222 (1999)

Few-Body Systems
© by Springer-Verlag 1999

Coulomb Corrections to the Rate of the Astrophysical Reaction $p + p \to d + e^+ + \nu_e$

L.D. Blokhintsev[1][*][†], G.V. Avakov[1][*][**], A.M. Mukhamedzhanov[2][††][***],
E.N. Voronina[1][*][**]

[1] Institute of Nuclear Physics, Moscow State University, Moscow 119899,
Russia
[2] Cyclotron Institute, Texas A&M University, College Station, TX 77843,
USA

Abstract. The effect of the $e^+ p$ interaction in the intermediate state of the astrophysical reaction $p + p \to d + e^+ + \nu_e$ has been considered. The corresponding correction to the astrophysical S factor of this reaction at near-zero energy turned out to be about 1%.

1 Introduction

The solar neutrino puzzle is one of the most pressing problems in physics of today. The major part of the solar neutrino flux is supplied by the reaction

$$p + p \to d + e^+ + \nu_e, \tag{1}$$

which is the primary process of the astrophysical hydrogen cycle. As is known, it is conventional to describe the intensity of nuclear reactions between charged particles at very low energies by the astrophysical S factor $S(E)$

$$S(E) = E e^{2\pi\eta} \sigma(E), \tag{2}$$

where E is the energy in c.m. system, $\sigma(E)$ is the cross section of the reaction and $\eta = Z_1 Z_2 e^2 / \hbar v$ is the Sommerfeld parameter (Z_1 and Z_2 are the charges of colliding nuclei and v is their relative velocity).

[*]Supported by Russian Foundation for Basic Research, Grant 98-02-16275
[†]*E-mail address:* blokh@srdlan.npi.msu.su
[**]*E-mail address:* avakov@srdlan.npi.msu.su
[††]Supported by DOE Grant DE-FG05- 93ER40773
[***]*E-mail address:* akram@comp.tamu.edu

For those reactions of astrophysical interest which are induced by strong or electromagnetic interactions, the information on cross sections (or astrophysical S factors) at astrophysical energies can typically be obtained by extrapolation of laboratory data measured at higher energies down to stellar energies. For the reaction under consideration, which is due to weak interaction, this procedure is not realizable since the cross section is too small. Reaction (1) was studied theoretically beginning from 1938 [1]. The most comprehensive analysis of this reaction has been done in a series of papers by J.N. Bahcall and his collaborators (see [2, 3] and references therein). In these papers, besides the main mechanism of reaction (1), various corrections have been considered (in particular, the meson corrections, the deuteron D state contribution, vacuum polarization, etc.) and uncertainties of $S(E \approx 0)$ have been evaluated. The subject of our work is a correction term to $S(E \approx 0)$, which has not been considered before.

2   Calculation of the Correction Term

The correction under consideration is related to the interaction in the intermediate state of reaction (1). The matrix element of this reaction can be schematitcally written as

$$M = \langle \Psi_f \mid \Gamma_\beta \mid \Psi_i \rangle, \tag{3}$$

where Γ_β is the β transition operator and Ψ_i (Ψ_f) is the initial (final) state wave function. This expression is exact if $\Psi_{i,f}$ are the exact solutions for the corresponding Hamiltonians. There are no problems with the calculation of Ψ_i. It is naturally obtained as a solution, Ψ_{pp}, of the Schrödinger equation for two protons interacting via the sum of Coulomb and nuclear potentials. In doing so, one can consider two protons at astrophysical energies being in the singlet s state. The situation with Ψ_f is more problematic. Of course, the neutrino is described by a plane wave which can be incorporated into Γ_β. As to the other final state particles, it is a common practice in β decay theory to allow for the Coulomb interaction of the electron or the positron in the final state by introducing the wave function which describes the Coulomb scattering of the electron (positron) off a residual nucleus A (in our case, off the deuteron). In actual practice, it is done by including the corresponding correction factor into the final state phase space volume. This factor is determined essentially by the value of the $e^\pm A$ scattering wave function taken at the point where the nucleus is situated. Within this approach, the matrix element M can be displayed graphically as is shown in Fig. 1 where the ovals indicate that the corresponding particles interact with each other.

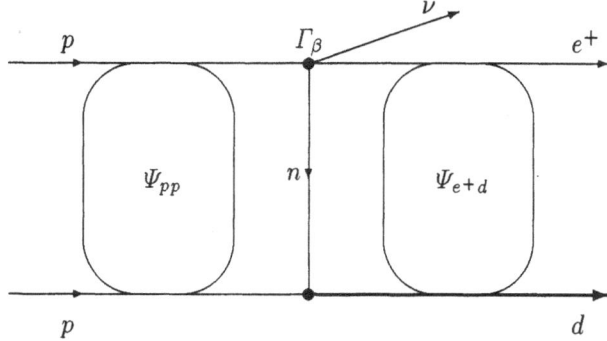

Figure 1. Diagram corresponding to the dominant mechanism of reaction (1)

In Fig. 1 the final state is treated as a two-body state $(e^+ + d)$. But actually there are three interacting final state particles, namely, e^+, n and p. Hence Ψ_f should be the solution of the three-particle Faddeev-type equation. Within the three-particle approach, the diagram of Fig. 1 results from the diagram, in which first the neutron interacts with the proton and then the positron scatters off the proton. But there is a possibility that the positron first scatters off the proton and only thereafter the proton interacts with the neutron forming the deuteron (see Fig. 2).

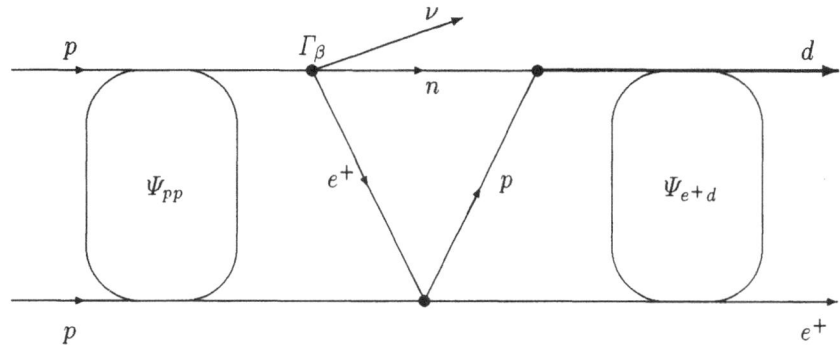

Figure 2. Diagram corresponding to the Coulomb correction term

Denote the amplitudes of the diagrams of Fig. 1 and Fig. 2 by M_0 and M_1, respectively. The amplitudes M_0 and M_1 correspond to two different Faddeev components of the full three-body wave function Ψ_f. One of these components "begins" with the np interaction (M_0), the other "begins" with the e^+p interaction (M_1). Therefore the amplitudes M_0 and M_1 can be summed up without double counting. Note that the aforementioned two Faddeev components add up to the full wave function Ψ_f if the positron-neutron interaction is neglected.

The function Ψ_{e+d}, describing the Coulomb interaction between the positron and the center of mass of the deuteron, allows approximately for the higher-order terms of the rescattering series in the $(e^+ + p + n)$ system.

The mechanism corresponding to M_1 can be called the intermediate state Coulomb correction. The evaluation of this correction to the S factor for reaction (1) is the object of our studies. We do not pursue the goal of obtaining a very accurate absolute value of S; rather, we estimate the value of the relative correction to S due to M_1 and check the sensitivity of this correction to the nuclear interaction. Therefore we use the rather simple model. The nuclear pp interaction in the singlet S state is described by the one-term separable potential with the Yamaguchi form factor which describes the low-energy pp scattering parameters. For the deuteron wave function we use the Hulthen form with the correct asymptotic normalization. We also use the allowed transition approximation, as our predecessors did. The Coulomb e^+p amplitude is taken in the first Born approximation. In this approach we were able to get the closed expressions in terms of hypergeometric functions both for M_0 and M_1.

We have calculated $S_0(E)$ which is proportional to $\mid M_0 \mid^2$, $S(E)$ proportional to $\mid M_0 + M_1 \mid^2$ and $\gamma(E) = \{S_0(E) - S(E)\}/S_0(E)$ in the range $E = 0 \div 100$ keV. The relative correction factor $\gamma(E)$, for all practical purposes, does not depend on energy within that range:

$$\gamma(E) = 1.069\% \div 1.071\%, \qquad E = 1 \div 100\,\text{keV}. \tag{4}$$

On the other hand, $S(E)$ changes by about a factor of 3 within that interval. Moreover, $\gamma(E)$ is insensitive to the details of the nuclear pp interaction in the initial state. If one completely neglects that interaction, then $S(E)$ decreases by about a factor of 4 whereas $\gamma(E)$ is practically unchanged ($\gamma = 1.069\% \rightarrow \gamma = 1.061\%$).

In conclusion, it can be said that the considered correction due to the intermediate state e^+p interaction is small but not negligible. It is comparable to other corrections and uncertainties considered earlier. It acts in the right direction, decreasing the S factor and narrowing the gap between the theory and the experiment. The possible ways to improve our evaluation are as follows:
i) the inclusion of the full Coulomb e^+p amplitude instead of the Born term;
ii) solving the Faddeev equations for Ψ_f, that is for the (e^+np) system;
iii) allowing for the relativistic corrections.

The authors believe though that such improvements will not seriously influence the results obtained.

References

1. H. Bethe, C.L. Critchfield: Phys. Rev. **54**, 248 (1938)

2. J.N. Bahcall, M. Kamionkowski: Nucl. Phys. **A625**, 893 (1997)

3. E.G. Adelberger et al.: Rev. Mod. Phys., to be published (1998)

Few-Body Systems Suppl. 10, 223–226 (1999)

Few-
Body
Systems
© by Springer-Verlag 1999

Novel approach to the three-body Coulomb problem

Zoltán Papp

Institute of Nuclear Research of the Hungarian Academy of Sciences,
P.O. Box 51, H–4001 Debrecen, Hungary

Abstract. I give an overview on a recently developed approach to the three-body Coulomb problem. I argue that the set of Faddeev–Noble and Lippmann–Schwinger integral equations provide unique solution. The integral equations are solved in Coulomb–Sturmian-space representation. The power of the method is demonstrated by numerical illustrations, and the results indicate that this is a viable method.

The first approach to solving the nuclear three-body Coulomb problem is due to Noble [1]. In his procedure the interaction has to be split into short-range and long-range terms as

$$v_\alpha(\xi_\alpha) = v_\alpha^{(s)}(\xi_\alpha) + v_\alpha^{(l)}(\xi_\alpha), \tag{1}$$

where α denotes the subsystem and superscripts s and l indicate the short- and long-range terms, respectively, with ξ_α and η_α being the usual configuration space Jacobi coordinates. The short-range and the long-range parts have to be treated in different ways: the Faddeev procedure has to be applied only to the short-range potentials. The resulting Faddeev-Noble integral equations read

$$|\psi_\beta\rangle = \delta_{\beta\alpha}|\Phi_\alpha^{(l)}\rangle + G_\beta^{(l)}(E_\alpha)v_\beta^{(s)} \sum_{\gamma \neq \beta} |\psi_\gamma\rangle, \tag{2}$$

with a cyclic permutation in α, β, γ. Here $G_\alpha^{(l)}(z) = (z - H_\alpha^{(l)})^{-1}$, where $H_\alpha^{(l)} = H^0 - v_\alpha^{(s)} - v_\alpha^{(l)} - v_\beta^{(l)} - v_\gamma^{(l)}$ and $H^0 = h_{\xi_\alpha}^0 + h_{\eta_\alpha}^0$ with h^0 being the two-body kinetic energy operator, and $|\Phi_\alpha^{(l)}\rangle$ is an eigenstate of $H_\alpha^{(l)}$. This framework is formally exact. However, $G_\alpha^{(l)}$ and $|\Phi_\alpha^{(l)}\rangle$ are not known; they belong to the auxiliary three-body Coulomb Hamiltonian $H_\alpha^{(l)}$.

It is generally assumed that Noble's procedure is of little help because the solution of the auxiliary Hamiltonian is as complicated as the solution of the original problem. This is, however, not the case. Let us introduce the channel-distorted long-range Hamiltonian $\widetilde{H}_\alpha = H^0 + v_\alpha + u_\alpha^{(l)}$, together with its resolvent $\widetilde{G}_\alpha(z) = (z - \widetilde{H}_\alpha)^{-1}$. The auxiliary potential $u_\alpha^{(l)} = u_\alpha^{(l)}(\eta_\alpha)$ should have the asymptotic form $u_\alpha^{(l)} \sim e_\alpha(e_\beta + e_\gamma)/\eta_\alpha$ as $\eta_\alpha \to \infty$. In fact, $u_\alpha^{(l)}$ is an effective Coulomb interaction between the center of mass of subsystem α (with charge $e_\beta + e_\gamma$) and the third particle (with charge e_α). Formally, one can write down Lippmann-Schwinger equations for the wave function

$$|\Phi_\alpha^{(l)}\rangle = |\tilde{\Phi}_\alpha\rangle + \tilde{G}_\alpha(E_\alpha)U^\alpha|\Phi_\alpha^{(l)}\rangle, \tag{3}$$

where $|\tilde{\Phi}_\alpha^{(\pm)}\rangle$ is an eigenstate of $\tilde{H}_\alpha$ and $U^\alpha = v_\beta^{(l)} + v_\gamma^{(l)} - u_\alpha^{(l)}$, and a similar equation holds for the Green's operator

$$G_\alpha^{(l)} = \tilde{G}_\alpha + \tilde{G}_\alpha U^\alpha G_\alpha^{(l)}. \tag{4}$$

The potential $u_\alpha^{(l)}$ has been chosen such that in the two-body channels α it compensates the long-range Coulomb part of $v_\beta^{(l)} + v_\gamma^{(l)}$. If in Eq. (1) $v^{(l)}$ is chosen such that it does not support any bound state, the system described by the Hamiltonian $H_\alpha^{(l)}$ has only one kind of two-body channels, the channels of α fragmentation. Thus there are no rearrangement channels, the non-uniqueness problem of the Lippmann-Schwinger equation [2] does not come up and a single Lippmann-Schwinger equation guarantees unique solution [3]. Although this uniqueness has only been proven for short-range interactions, there seems to be no reason to rule out its validity in the Coulomb case, so the Faddeev-Noble equations (2), together with the Lippmann-Schwinger equations (3) and (4), provide unique solutions.

The integral equations can be solved in Coulomb–Sturmian-space (CS) representation. The CS functions form a biorthonormal discrete basis in the Hilbert space, so Eqs. (2), (3) and (4) become matrix equations. A crucial point is the representation of the Green's operator $\tilde{G}_\alpha$, which is a resolvent of the sum of two commuting Hamiltonians, $h_{\xi_\alpha} = h_{\xi_\alpha}^0 + v_\alpha$ and $h_{\eta_\alpha} = h_{\eta_\alpha}^0 + u_\alpha^{(l)}$, acting in different two-body Hilbert spaces. By using the convolution theorem [4], the three-body Green's operator $\tilde{G}_\alpha$ can be expressed as a convolution integral of two-body Green's operators, whose CS representations are known [5], i.e.,

$$\tilde{G}_\alpha(z) = (z - h_{\xi_\alpha} - h_{\eta_\alpha})^{-1} = \frac{1}{2\pi i}\oint_C dz'\, g_{\xi_\alpha}(z - z')\, g_{\eta_\alpha}(z'), \tag{5}$$

where $g_{\xi_\alpha}(z) = (z - h_{\xi_\alpha})^{-1}$ and $g_{\eta_\alpha}(z) = (z - h_{\eta_\alpha})^{-1}$. The contour C should be taken counterclockwise around the continuous spectrum of h_{η_α} so that g_{ξ_α} is analytic in the domain encircled by C. For bound-state energies the spectra of the two Green's operators are well separated, for below-breakup scattering states the bound state pole of g_{ξ_α} meets the continuous spectrum of g_{η_α}, for above-breakup scattering states even the two continua overlap and for resonant-state energies the two spectra penetrate into each others second sheets. All these cases are covered by the contour of Fig. 1.

The method has been tested so far in bound-state nuclear [6] and atomic [7] problems, as well as in low energy $p - d$ scattering [8] calculations. If in the

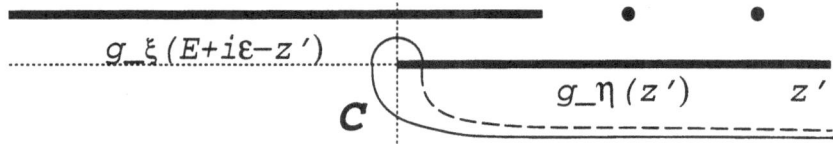

Figure 1. Analytic structure of $g_{\xi_\alpha}(z-z')\,g_{\eta_\alpha}(z')$ as a function of z' with $z = E+i\varepsilon$, $E > 0$, $\varepsilon > 0$. The contour C encircles the continuous spectrum of h_{η_α} and avoids the singularities of h_{ξ_α}. A part of it, which goes on the second Riemann sheet of h_{η_α}, is drawn by broken line.

three-body Coulomb problem all the Coulomb interactions are repulsive, the splitting (1) is straightforward: one should take $v_\alpha^{(l)}$ as the Coulomb interaction and $v_\alpha^{(s)}$ is the rest. If some of the Coulomb interactions are attractive one should introduce a cut-off function [7] and should split the Coulomb interaction into long-range and short-range terms is such a way that in the energy spectrum of physical interest $v_\alpha^{(l)}$ does not support bound states. As an illustration, in Table 1 I present the convergence of the binding energy of the helium atom and the positronium ion with respect to the angular momentum channels. The convergence is very fast, much faster than in previous Faddeev calculations.

Table 1. Convergence of the binding energy of the He atom and the positronium ion $(e^-e^+e^-)$ with increasing the number of angular momentum channels. The values are given in atomic units.

Channels:	$l = 0$	$l = 2$	$l = 4$	$l = 6$	$l = 8$
He atom	2.892977	2.903781	2.903736	2.903727	2.903725
$e^-e^+e^-$	0.250559	0.261760	0.261991	0.262003	0.262005

Table 2. $^4\delta_{pd}$ and $^2\delta_{pd}$ phase shifts for the three-nucleon system interacting via the MTI-III potential at various energies. The phase shifts are in degrees.

	$^4\delta_{pd}$		$^2\delta_{pd}$	
E	Ref. [9]	This method	Ref. [9]	This method
0.05	-2.69	-2.694	-0.113	-0.112
1.0	-46.5	-46.55	-16.2	-16.24
2.0	-63.8	-63.74	-28.8	-28.78

The real advantages of this method come up in scattering-state calculations because, as integral equations are solved, the boundary conditions are automatically incorporated. In Table 2 the $p-d$ scattering results obtained by this method are compared with the configuration-space Faddeev calculations of ref. [9]; the agreement is perfect in all cases. Moreover, in this method the phase shift contribution coming from the genuine short-range nuclear potential $v_\alpha^{(s)}$

and from the polarization potential U^α can be separated, and the Coulomb plus polarization potential modified scattering length $a_{pd}^{cp,s}$ can be calculated [see Table 3]. This is the quantity which really carries information on the nuclear potential.

Table 3. $p - d$ scattering lengths for the three-nucleon system interacting via the MTI-III potential.

	$^4a_{pd}^{c,ps}$	$^4a_{pd}^{cp,s}$	$^2a_{pd}^{c,ps}$	$^2a_{pd}^{cp,s}$
This method	13.76	13.79	0.161	0.195
Ref. [9]	13.8		0.17	

In this paper I suggested that the set of Faddeev-Noble and Lippmann-Schwinger integral equations provide unique solution for the three-body Coulomb problem. It is practicable to solve these equations in Coulomb–Sturmian-space representation, as then there is an analytic representation of the two-body Green's operator, thus making the practical evaluation of $\widetilde{G}_\alpha$, via the contour integral, possible for all physically relevant regions of the complex energy plane. The numerical results obtained so far indicate that this new approach is a viable, and likely that it will work above the breakup threshold.

Acknowledgement. This work has been supported by OTKA under Contracts No. T17298 and No. T026233. The author is indebted to W. Plessas for his contribution to the early stages of this work.

References

1. J. Noble: Phys. Rev. **161**, 945 (1967)

2. See. e.g. S.K. Adhikari and K.L. Kowalsky: In: *Dynamical Collision Theory.* San Diego: Academic Press 1991

3. W. Sandhas: In: *Few-Body Nuclear Physics*, (IAEA Vienna), 3 (1978)

4. N.M. Hugenholtz: Physica, **23**, 481 (1957); L. Bianchi and L. Favella: Nuovo Cim. **6**, 6873 (1964)

5. Z. Papp: J. Phys. **A20**, 153 (1987); Comp. Phys. Comm. **70**, 426, 435 (1992); B. Kónya, G. Lévai and Z. Papp: J. Math. Phys. **38**, 4832 (1997)

6. Z. Papp and W. Plessas: Phys. Rev. **C54**, 50 (1996)

7. Z. Papp: Few-Body Systems, (in print), (physics/9702001)

8. Z. Papp: Phys. Rev. **C55**, 1080 (1997)

9. C.R. Chen et al.: Phys. Rev. **C39**, 1261 (1989)

Few-Body Systems Suppl. 10, 227–230 (1999)

Few-
Body
Systems
© by Springer-Verlag 1999

Nuclear Fusion in $d\mu^3$He Mesic Molecule

L.N. Bogdanova[1], S.S. Gershtein[2], L.I. Ponomarev[3]

[1] Institute for Theoretical and Experimental Physics, Moscow, 117218, Russia

[2] Institute for High Energy Physics, Protvino, 142284, Russia

[3] Russian Research Center "Kurchatov Institute", Moscow, 123182, Russia

Abstract. A new scheme of physical processes leading to the nuclear fusion reaction $d(^3\text{He},^4\text{He})p$ catalyzed by a negatively charged muon (μ^-) is presented. It is shown that the observable rate and yield of the nuclear reaction depend on a chain of ion-molecular reactions in which the $d\mu^3$He-molecule participates. New calculations of the nuclear fusion rates in the $d\mu^3$He-molecule are presented.

The muon catalysis phenomenon gives the possibility to study the reaction

$$d +{}^3\text{He} \rightarrow {}^4\text{He} + p \tag{1}$$

(and other fusion reactions [1]) at practically zero collision energy from the mesic molecular state $(d\mu^3\text{He})^{++}$ (below $(d\mu^3\text{He})^{++} \equiv d\mu^3\text{He}$):

$$(d\mu^3\text{He})_J \xrightarrow{\lambda_f^J} \mu^4\text{He} + p, \quad \mu +{}^4\text{He} + p . \tag{2}$$

The rates λ_f^J of nuclear reactions (2) from the states $(d\mu^3\text{He})_J$ with total angular momentum J were calculated many times [2], however, results differ by several orders of magnitude. The most recent experimental upper limit for this rate is $\lambda_f < 1.3 \cdot 10^6 \ s^{-1}$ [3]. The scheme of the processes of $(d\mu^3\text{He})_J$ mesic molecule formation and decay is presented in Fig. 1. States $(d\mu^3\text{He})_J$ are quasistationary due to decay $(\mu^3\text{He})_{1s} + d$ by γ-emission (3), Auger transitions (4) and predissociation (5), with rates λ_γ^J, λ_A^J and λ_p^J, respectively:

$$(d\mu^3\text{He})_J \quad \rightarrow \quad (\mu^3\text{He})_{1s} + d + \gamma \ , \tag{3}$$

$$(d\mu^3\text{He})_J \quad \rightarrow \quad (\mu^3\text{He})_{1s} + d + e \ , \tag{4}$$

$$(d\mu^3\text{He})_J \quad \rightarrow \quad (\mu^3\text{He})_{1s} + d \ . \tag{5}$$

In collisions of $[(d\mu\text{He})e]^+$ with D_2 and He [5] transition

$$(d\mu^3\text{He})_{J=1} \rightarrow (d\mu^3\text{He})_{J=0} \tag{6}$$

228

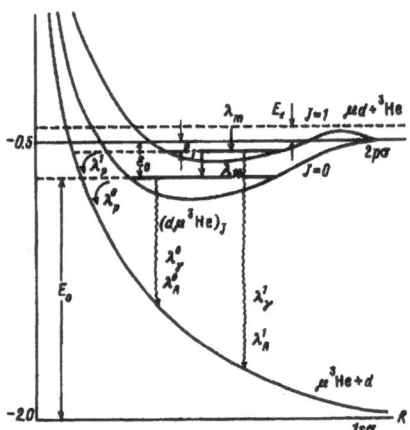

Figure 1. Scheme of formation and decay of the $d\mu^3$He molecule. The muonic molecule $d\mu^3$He is formed in $d\mu +^3$He collisions in the bound state $(d\mu^3\text{He})_{J=1}$ [4], which decays into system $\mu^3\text{He}+d$ with rates λ_γ^1, λ_p^1 and λ_A^1; transition $(d\mu^3\text{He})_{J=1} \rightarrow (d\mu^3\text{He})_{J=0}$ with rate λ_{10} competes with these decays. Binding energies of the states $(d\mu^3\text{He})_J$ equal ε_J; collision energies in the states $d\mu +^3$He and $\mu^3\text{He} + d$ equal E_1 and E_0, respectively; fusion rates from states $J = 0$ and $J = 1$ are λ_f^0 and λ_f^1.

can occur with the rate λ_{10}. The yield N_f of nuclear fusion per one stopped μ^- is determined by fusion rates λ_f^J and populations w_J of the states J, which depend on the rates of processes (3)-(6) and kinetics of ion-molecular reactions in which the $d\mu^3$He molecule participates. We present new results of λ_f^J calculations [6, 7] and a new scheme of kinetics of ion-molecular reactions preceding fusion in $d\mu^3$He mesic molecule [5].

The rates λ_f^J are determined by the relation [1]

$$\lambda_f^J = \sum_L A_L G_L^J \ . \tag{7}$$

Here, A_L are reaction constants for the nuclear states with orbital angular momenta L, determined by the extrapolation of reaction (1) partial cross section to zero collision energy. Since the s-wave $J^P = 3/2^+$ resonance dominates near the threshold, only values G_0^J are of interest:

$$G_0^J = \int d\mathbf{r} \, | \, \Psi^J(\mathbf{r}, \mathbf{R} = 0) \, |^2 \ , \tag{8}$$

where $\Psi^J(\mathbf{r}, \mathbf{R})$ is the mesic molecule wave function ($\mathbf{R}$ is the internuclear distance, $\mathbf{r}$ is the muon coordinate with respect to the nuclei center of mass).

States of the $(d\mu^3\text{He})_J$ molecule are localized in the potential $W_{2p\sigma}(R)$ with quantum numbers $(Nlm) = (210)$ of the system $(d\mu)_{1s} +^3$He. However, due to the strong coupling between channels $1s\sigma$ and $2p\sigma$, all components $\psi^L(\mathbf{R})$ representing the relative motion of nuclei with angular momenta $L = |J - l|, \ \dots \ |J + l|$ in the potential $W_j(\mathbf{R})$, given by the muon motion in the quantum state $j = (Nlm)$ at fixed distance R between nuclei, are essential in the wave function $\Psi^J(\mathbf{r}, \mathbf{R})$:

$$G_0^0 = \sum_N \mid \psi_{N00}^{L=0}(0) \mid^2, \quad G_0^1 = \sum_{N,m} \mid \psi_{N1m}^{L=0}(0) \mid^2 \quad . \tag{9}$$

($\phi_j(\mathbf{r}; R)$ are orthonormalized adiabatic basis functions [8]).

Functions $\psi_{Nlm}^L(\mathbf{R})$ have been recently calculated in [6] by complex coordinate rotation method, expanding variational function $\Psi^J(\mathbf{r}; \mathbf{R})$ over the adiabatic basis $\phi_j(\mathbf{r}; R)$. The calculated G_0^J-factors equal:

$$G_0^0 = 3.8 \cdot 10^{19} cm^{-3}, \quad G_0^1 = 5.1 \cdot 10^{16} cm^{-3} \tag{10}$$

In the other approach [7] exploiting expansion of the function $\Psi^J(\mathbf{r}, \mathbf{R})$ over the adiabatic hyperspherical basis [9] the finite width of the quasistationary $d\mu^3He$ state was explicitly taken into account. The preliminary result obtained

$$G_0^0 = 4.0 \cdot 10^{19} cm^{-3} \tag{11}$$

is in a reasonable agreement with result (10).

With the reaction constant for unpolarized nuclei $A_0 = 0.34 \cdot 10^{-14} cm^3 s^{-1}$ [10] the nuclear fusion rates λ_f^J are

$$\lambda_f^0 = 3/2 \cdot A_0 G_0^0 = 1.9 \cdot 10^5 s^{-1}, \quad \lambda_f^1 = 0.65 \cdot 10^3 s^{-1} \tag{12}$$

With G_0^0 from (11) the rate is

$$\lambda_f^0 = 2.1 \cdot 10^5 s^{-1} \quad . \tag{13}$$

Total rates $\lambda_{dec}^J = \lambda_\gamma^J + \lambda_p^J + \lambda_A^J$ of quasistationary states $(d\mu^3He)_J$ decay to channels (3)-(5) equal $\lambda_{dec}^0 \simeq 0.9 \cdot 10^{12} s^{-1}$, $\lambda_{dec}^1 \simeq 1 \cdot 10^{12} s^{-1}$ (see refs. [11]). The rate λ_f^1 of nuclear fusion from the state $J = 1$, in which mesic molecules $d\mu^3He$ are formed in the reaction $(d\mu)_{1s} + {}^3He \rightarrow (d\mu^3He)_J^+ + e$ by dipole $E1$-transition [12], is negligibly small compared to its decay rate λ_{dec}^1. Hence fusion in the $d\mu^3He$ mesic molecule can be observed if the rate λ_{10} of reaction (6) is comparable to the rate λ_{dec}^1. Transition (6) is possible only in collisions of $(d\mu^3He)_J$ with atoms of the medium (internal Auger transition $(d\mu^3He)_{J=1} \rightarrow (d\mu^3He)_{J=0}^+ + e$ is forbidden by energy considerations).

The whole set of ion-molecular reactions has been considered in [5]. It has been established that at mixture density $\varphi \sim 0.1$ and $C_{He} \leq 0.1$ the dominating channel leading to transition (6) is related with the formation of the complex $[(d\mu^3He)_{J=1} e D_2]^+$ and its subsequent decay with conversion of an electron of the D_2 molecule. In comparison with this process the external Auger transition is negligibly slow, in contrast to the statement of [13]. Calculations [5] show that at $\varphi \simeq 0.1$ and $C_{He} \lesssim 0.1$ a noticeable fraction (~ 0.2) of mesic molecules $(d\mu^3He)_{J=1}$ reaches the state $J = 0$, where fusion (2) can be observed.

Simple considerations allow to calculate the expected fusion yield N_f per muon stop [14]: at $\varphi = 0.075$ and $C_{He} = 0.05$

$$N_f \simeq 0.12 \cdot \lambda_f^0/\lambda_{dec}^0 \approx 3 \cdot 10^{-8}/\mu^- \quad . \tag{14}$$

The detailed analysis of the kinetics of processes in $D_2 + {}^3$He mixture is yet to be done. Experiment R-94-05.1 performed at PSI will probe the correctness and self-consistency of the theoretical scheme. In principle, observation of the φ-dependence of the yield N_f might be the test of the ion-molecular mechanism of the transition $(d\mu^3\text{He})_{J=1} \rightarrow (d\mu^3\text{He})_{J=0}$ via formation of clusters $[(d\mu^3\text{He})eD_2]^+$. Comparison of the fusion rate λ_f^0, extracted from measured yield N_f, with its theoretical value will check the sophisticated calculations of the Coulomb three-body problem.

Acknowledgement. The authors are grateful to D.I. Abramov, M.P. Faifman, V.V. Gusev, V.I. Korobov, L.I. Menshikov for the information about their results prior to publication, and to C. Petitjean and A.A. Vorobyov for stimulating and challenging discussions. This work has been supported (L.N.B. and L.I.P.) by grant 96-02-17279 of the Russian Foundation for Basic Research.

References

1. L.N. Bogdanova: Muon Cat. Fusion **2**, 359 (1988)

2. A.V. Kravtsov, N.P. Popov, G.E. Solyakin: Sov. Phys. JETP Lett. **40**, 875 (1984); D. Harley, B. Muller, J. Rafelsky: J. Phys. **616**, 281 (1990); Y. Kino and M. Kamimura: Hyperfine Int. **82**, 195 (1993); W. Czaplinski et al.: Phys. Lett. **A219**, 86 (1996); F.M. Penkov: Yad. Fiz. **60**, 1003 (1997)

3. D.V. Balin et al.: to appear in Hyperfine Int.

4. Yu.A. Aristov et al.: Sov. J. Nucl.Phys. **33**, 564 (1981)

5. L.I. Menshikov and M.P. Faifman: to appear in Hyperfine Int.

6. L.N. Bogdanova and V.I. Korobov: to appear in Hyperfine Int.

7. S.S. Gershtein, V.V. Gusev et al.: to appear in Hyperfine Int.

8. S.I. Vinitsky and L.I. Ponomarev: Sov. J. Part. Nucl. **13**, 557 (1982)

9. D.I. Abramov, V.V. Gusev, L.I. Ponomarev: Yad.Fiz. **60**, 1259 (1997)

10. H.S. Bosch and G.M. Hale: Nuclear Fusion **32**, 611 (1992)

11. S. Hara and T. Ishihara: Phys. Rev. **A39**, 5633 (1989); S.S. Gershtein and V.V. Gusev: Hyperfine Int. **82**, 205 (1993); A.V. Kravtsov, A.I. Mikhailov, V.I. Savichev: Z. Phys. **D29**, 49 (1994)

12. V.K. Ivanov et al.: Sov. Phys. JETP **64**, 210 (1986)

13. W. Czaplinski et al.: Z. Phys. **D37**, 283 (1996)

14. L.N. Bogdanova, S.S. Gershtein, L.I. Ponomarev: JETP Lett. **67**, 99 (1998)

Few-Body Systems Suppl. 10, 231–234 (1999)

Few-
Body
Systems
© by Springer-Verlag 1999

Theoretical Studies of Low Energy Positron Collisions with Hydrogen and Helium

P. Van Reeth, J.W. Humberston

Department of Physics and Astronomy, University College London,
Gower Street, London WC1E 6BT, UK

Abstract. Detailed theoretical investigations are made of elastic scattering and positronium formation in low energy collisions of positrons with hydrogen and helium, using a two channel version of the Kohn variational method with very flexible trial functions and the full non-relativistic Hamiltonian. Well converged results are obtained, enabling the behaviour of the cross sections in the vicinity of the positronium formation threshold to be analysed. The s-wave positronium formation cross sections for both hydrogen and helium are very similar in energy dependence and magnitude, but in both cases this contribution to the total positronium formation cross section is very small.

Theoretical studies of low energy positron-atom scattering have developed significantly in recent years, and improvements in experimental techniques have made it possible to obtain accurate measurements of various scattering parameters with which theory can be compared [1].

There are two additional processes which differentiate e^+-atom scattering from e^--atom scattering; positron-electron annihilation and positronium formation. Here we concentrate on the cross sections for elastic scattering, σ_{el}, and for positronium formation, σ_{Ps}, in the Ore gap, the energy region between the positronium formation threshold and the lowest threshold for excitation by positron impact. Furthermore, we only consider positron scattering by hydrogen and helium, for which it is possible to undertake detailed ab-initio calculations including all inter-particles coordinates and using the full non-relativistic Hamiltonian. For a more general overview of positron-atom scattering see the article by Humberston in these proceedings.

The two channel Kohn variational method [2] is used with the $\boldsymbol{K}$ matrix functional; $K_{pq}^{v} = K_{pq}^{t} - \langle \Psi_p | L | \Psi_q \rangle$ $(p, q = 1, 2)$, where the superscript v indicates the variational values, t the trial values, and $L = 2(H - E)$ with H and E the total Hamiltonian and total energy of the system. The requirement that the Kohn functional be stationary with respect to variations of all linear parameters in the trial functions gives rise to a set of linear simultaneous equations from which the variational K matrix can be found. The cross sections for the various processes are then given, in units of πa_0^2, by

$$\sigma_{pq} = \frac{4(2l+1)}{k_p^2} \left| \left(\frac{K}{1-iK} \right)_{pq} \right|^2 \qquad (p, q = 1, 2)$$

where $k_1 = k$ and $k_2 = \kappa$ are the positron and positronium wave numbers respectively, which are related through energy conservation by $\frac{1}{2}k^2 + E_A = \frac{1}{4}\kappa^2 + E_{Ps} + E_{A+}$, and $\sigma_{11} \equiv \sigma_{el}$, $\sigma_{12} \equiv \sigma_{Ps}$.

In the case of s-wave e^+-He scattering the two component trial function is

$$\Psi_1 = \frac{1}{\sqrt{4\pi}} \left\{ \Phi_{He}(\boldsymbol{r_2}, \boldsymbol{r_3})\sqrt{k} \left\{ j_0(kr_1) - K_{11}^t n_0(kr_1)f_{sh}(r_1) \right\} \right.$$

$$\left. - \frac{(1+P_{23})}{\sqrt{2}} \Phi_{Ps}(r_{12})\Phi_{He^+}(r_3)\sqrt{2\kappa}K_{21}^t n_0(\kappa\rho)f_{sh}(\rho) + \frac{(1+P_{23})}{\sqrt{2}} \sum_{i=1}^{N} c_i\phi_i \right\}$$

$$\Psi_2 = \frac{1}{\sqrt{4\pi}} \left\{ \frac{(1+P_{23})}{\sqrt{2}} \Phi_{Ps}(r_{12})\Phi_{He^+}(r_3)\sqrt{2\kappa}\{j_0(\kappa\rho) - K_{22}^t n_0(\kappa\rho)f_{sh}(\rho)\} \right.$$

$$\left. - \Phi_{He}(\boldsymbol{r_2}, \boldsymbol{r_3})\sqrt{k}K_{12}^t n_0(kr_1)f_{sh}(r_1) + \frac{(1+P_{23})}{\sqrt{2}} \sum_{j=1}^{N} d_j\phi_j \right\} \qquad (1)$$

where $f_{sh}(r_1)$ and $f_{sh}(\rho)$ shield the singularities in the Neuman functions, Φ_{He} is a very accurate approximation to the exact helium wave function [3], and Φ_{He^+} and Φ_{Ps} are the ground state wave functions of the helium positive ion and positronium. The position vector of the centre of mass of the positronium relative to the nucleus is $\boldsymbol{\rho} = (\boldsymbol{r_1} + \boldsymbol{r_2})/2$, the position vectors of the positron and the two electrons are respectively $\boldsymbol{r_1}$, $\boldsymbol{r_2}$ and $\boldsymbol{r_3}$, and $r_{ij} = |\boldsymbol{r_i} - \boldsymbol{r_j}|$. The short-range correlation terms, ϕ_i, are of the form

$$\phi_i = \exp\left[-(\alpha r_1 + \beta r_2 + \beta r_3)\right] r_1^{k_j} r_2^{l_j} r_{12}^{m_j} r_3^{n_j} r_{13}^{p_j} r_{23}^{q_j}, \qquad (2)$$

and all terms with $k_i + l_i + m_i + n_i + p_i + q_i \leq \omega$ are included. The flexibility of the trial function is improved by increasing ω, and the accuracy of the results can be estimated by considering the convergence of the K matrix elements with respect to ω. Similar trial functions have been used in the calculations of e^+-H scattering parameters [4].

Very well converged cross sections have been calculated for e^+-H scattering [4] which agree well with accurate results obtained by other methods [5, 6]. In the case of e^+-He scattering, the theoretical data are not yet quite as well converged as for e^+-H and there are no other detailed calculations in this energy region with which to compare. However, the results we have obtained for the total σ_{Ps} [7] show a similar energy dependence in the Ore gap as the experimental data [8] and we believe that the 30% disagreement in magnitude between the experimental and theoretical cross sections can be explained partly by the lack of convergence of the theoretical results and also the uncertainties in the experimental measurements.

The flexibility and accuracy of the Kohn method have enabled us to make detailed investigations of features in both σ_{el} and σ_{Ps} in the vicinity of the positronium formation threshold [9]. We have found remarkable similarities in $\sigma_{Ps}(l = 0)$ for the two systems, both in energy dependence and in magnitude, as seen in Fig. 1. The infinite slope at the threshold is a consequence of the energy dependence of the K_{12} matrix element, $K_{12} \propto \kappa^{l+1/2}$, as predicted by Wigner's threshold theory. The similarity in magnitude is even more remarkable when one considers that the ratio of the total positronium formation cross sections, $\sigma_{Ps}(H)/\sigma_{Ps}(He)$, is ≈ 30. Also, in both cases the $l = 0$ contribution to the total σ_{Ps} is found to be not very significant. We believe that the similarity in $\sigma_{Ps}(l = 0)$ may be due in part to the fact that the final states for the two systems are rather similar; a diffuse Ps atom interacting with a positive charge (either the proton or the tightly bound He$^+$ ion). This suggests that $\sigma_{Ps}(l = 0)$ is mainly dependent on the final state of the scattering process, but this needs to be confirmed by further detailed calculations of positronium formation in positron scattering by noble gas atoms.

It is of interest to consider the total s-wave scattering cross section at low energy, which is purely $\sigma_{el}(l = 0)$ below the positronium formation threshold and the sum of $\sigma_{el}(l = 0)$ and $\sigma_{Ps}(l = 0)$ above, as shown in Fig. 2. At the threshold itself, we have a Wigner cusp in $\sigma_{el}(l = 0)$ which is due to the opening of the positronium formation channel. There is also another wider feature below the threshold which is related to the influence of virtual positronium. Because the phase shift is negative in this energy region the increased attraction arising from the influence of virtual positronium makes the phase shifts less negative and the elastic scattering cross section decreases from the value it would have without virtual positronium. Therefore, if we extrapolate $\sigma_{el}(l = 0)$ from far enough below the threshold, where the influence of virtual positronium is insignificant, we would obtain $\sigma_{el}(l = 0)$ in the threshold region if no threshold were present. Such an extrapolation matches reasonably well with the accurate

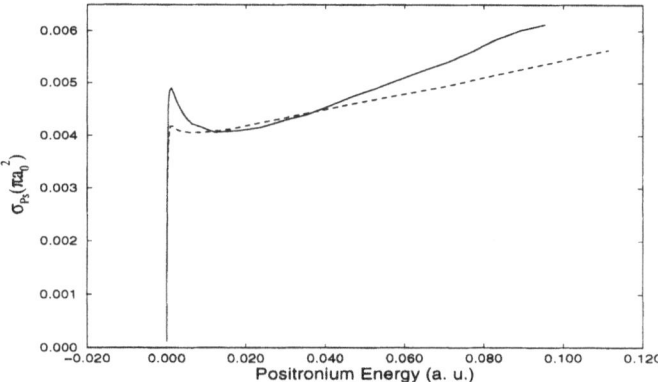

Figure 1. The variation of $\sigma_{Ps}(l = 0)$ with positronium energy. Full line; helium: dashed line; hydrogen.

total s-wave cross section above the threshold, as shown for e^+-He in figure 2, and it seems to provide a ceiling up to which $\sigma_{Ps}(l = 0)$ rises. It is not possible to determine from our results whether the magnitude of $\sigma_{Ps}(l = 0)$ is due to the effect of virtual positronium below the threshold or the strength of the influence of virtual positronium is given by the magnitude of $\sigma_{Ps}(l = 0)$. However, we have shown here that there is clearly a link between these processes which gives rise to interesting features in $\sigma_{Ps}(l = 0)$ although further insight may require the investigation of simpler model problems.

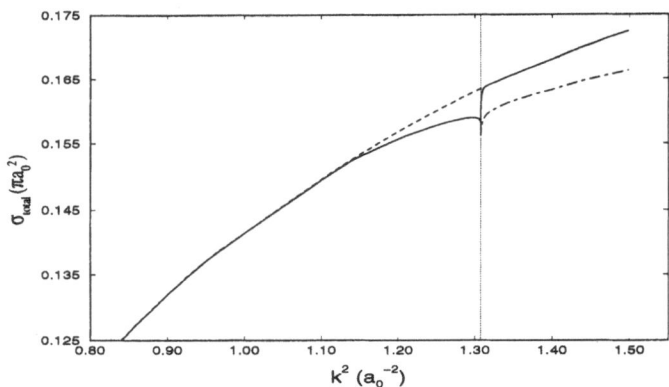

Figure 2. The total cross section for s-wave positron- helium scattering. The dashed line is an extrapolation of $\sigma_{el}(l = 0)$ from some way below the threshold, and the dashed-dotted line is the variational result for $\sigma_{el}(l = 0)$ above the threshold.

Acknowledgement. This work was supported by the United Kingdom EPSRC on grant number GR/L38431.

References

1. G. Laricchia and M. Charlton: J. Phys. B: At. Mol. Opt. Phys. **23**, 1045 (1990)

2. E. A. G. Armour and J. W. Humberston: Phys. Reports **204**, 165 (1991)

3. P. Van Reeth and J. W. Humberston: J. Phys. **B28**, L23 (1995)

4. J. W. Humberston et al.: J. Phys. **B30**, 2477 (1997)

5. Y. R. Kuang and T. T. Gien: Phys. Rev. **A55**, 256 (1997)

6. J. Mitroy and K. Ratnavelu: J. Phys. **B28**, 287 (1995)

7. P. Van Reeth and J. W. Humberston: J. Phys. **B30**, L95 (1997)

8. J. Moxom et al.: Phys. Rev. **A50**, 3129 (1994)

9. P. Van Reeth and J. W. Humberston: J. Phys. **B31**, L621 (1998)

Few-Body Systems Suppl. 10, 235–238 (1999)

Few-
Body
Systems
© by Springer-Verlag 1999

New Approach to Calculations of (e,2e) Reactions with Light Atoms

Yu.V. Popov[1]*, V.A. Knyr[2], V.V. Nasyrov[2], A. Lahmam-Bennani[3]†

[1] Nuclear Physics Institute, Moscow State University, Moscow 119899, Russia
[2] Departament of Physics, Khabarovsk State University of Technology, Khabarovsk 680035, Russia
[3] LCAM, bat. 351, Université Paris XI, 91405 Orsay Cedex, France

Abstract. The conclusion, that the traditional Hartree-Fock approach is not adequate for light atoms, is made from the studies of triple differential cross sections (TDCS) for the single ionization of atomic helium by electron impact at high incident energy, coplanar asymmetric geometry and small momentum transfers.

Dipolar high energy $He(e, 2e)He^+$ experiments carried out by the Lahmam-Bennani's scientific team [1,2] seem to contain a direct information on the asymptotics of the helium ground state. In these experiments where the projectile beam had the energy $E_0 \sim 4 \div 8KeV$, the scattered electron was practically of the same energy. Such high incident energies allow to use the first Born approximation (FBA) with high accuracy. In the experiments the residual ion He^+ was left both in the ground state (n=1) and in the excited state (n=2). Our special interest is focused on small momentum transfers $Q \sim 0.1 \div 0.2$ a.u. and small ejected electron energies $E_b \sim 4 \div 10$ eV.

This choice is not occasional. Two characteristic features pay one's attention (in the paper we shall mainly refer to the case n=1, i.e. the residual ion stays in the ground state). Firstly, most cross section calculations performed in this kinematical domain do coincide with each other. Secondly, the calculations in the vicinity of the backward peak are situated about $15 \div 20\%$ lower than the experimental points, which in turn reproduce almost near values of peaks along and opposite the momentum transfer direction.

The calculations have been presented both in the papers [1,2] and later by Kheifets et al. [3] (we shall call this model KBMS) and by Knyr et al. [4] (we

---

*E-mail address: popov@srdlan.npi.msu.su
†E-mail address:azzedine@lcam.u-psud.fr

shall call it KNSP). The theories can be conditionally divided in two groups. The first one usually operates with some simple or multiconfiguration Hartree-Fock wave function as a ground state and a distorted or a coulomb wave for the slow ejected electron. The models named OCW and V1/U/SEP in [1,2] and the KBMS model belong to this set. The second group is based on the preliminary reduction of the six-dimensional Schrödinger equation (SE) to an infinite system of one-dimensional coupled equations [4,5].

Thus, the aim of the analysis presented here is to answer the question: Would it be possible to obtain any new substantial information on a target or ionization mechanisms studying high incident energy dipolar (e,2e) experiments in the limit of vanishing momentum transfer and small ejected electron energy?

The first Born TDCS for a helium target is given by:

$$\frac{d^3\sigma_n}{d\Omega_a d\Omega_b dE_b} = \frac{16 p_a p_b}{(2\pi)^3 p_0 Q^4} \sum_{lm} | < \Phi^-_{(\mu)} | \hat{T}(\boldsymbol{Q}) | \Phi_0 > |^2, \tag{1}$$

where $(E_0, \boldsymbol{p}_0)$, $(E_a, \boldsymbol{p}_a)$, $(E_b, \boldsymbol{p}_b)$ are the energies and the momenta of the incident, scattered and ejected electrons respectively, $\boldsymbol{Q} = \boldsymbol{p}_0 - \boldsymbol{p}_a$ is the momentum transfer, $|\Phi^-_{(\mu)} >$ and $|\Phi_0 >$ are the singlet wave functions of the two-electron final and initial states, $\hat{T}(\boldsymbol{Q})$ is the transition operator, index $(\mu) = (nlm)$ denotes the reaction channel.

Different forms of the transition operator are acceptable: length form (L), velocity form (V), acceleration form etc. If the initial and the final state wave functions are exact then the transition operator form plays no role, but it is not so for models, especially if functions in the couple are obtained by different methods. The V-form as a rule gives better results for cruder final state wave functions than the L-form. The couple of wave functions yielding the closest results of the TDCS calculations for different forms of the transition operator can be called optimal.

At present the couple of wave functions computed by Knyr et al. seems to be the optimal one. Some of KNSP-calculations are presented in Fig. 1 both in the L- and in the V- forms. One easily sees that even for $n = 2$ the L- and the V-calculations almost coincide with each other. For instance, the L- and V-KBMS calculations perform even the different shapes of the TDCSs (see figures in [3]).

As we mentioned in the introduction, all couples of the wave functions presented now "on the market of calculations" in the kinematic region chosen and for the ionization leaving the residual ion in the ground state ($\mu = (100)$ and we omit it in this case) practically display results coincident with each other for all angles θ_b if Q and E_b are rather small. To understand the reason of this effect we write down:

$$< \Phi^- | \hat{T}(\boldsymbol{Q}) | \Phi_0 > = \int d\boldsymbol{r}_1 \exp(i\boldsymbol{Q}\boldsymbol{r}_1) I(\boldsymbol{r}_1, \boldsymbol{p}_b) \tag{2}$$

where

$$I(\boldsymbol{r}_1, \boldsymbol{p}_b) = \int d\boldsymbol{r}_2 \Phi^{-*}(\boldsymbol{p}_b; \boldsymbol{r}_1, \boldsymbol{r}_2) \Phi_0(\boldsymbol{r}_1, \boldsymbol{r}_2). \tag{3}$$

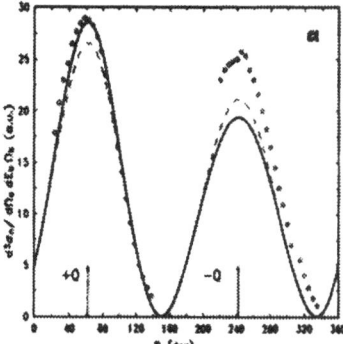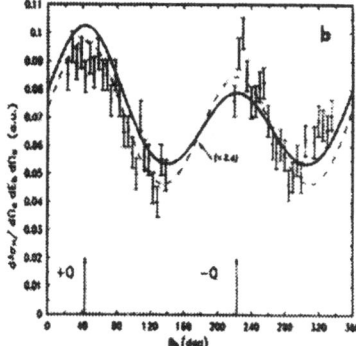

Figure 1. Triple differential cross section $d^3\sigma_n/d\Omega_a d\Omega_b dE_b$ (in atomic units) for the $He(e, 2e)He_n^+$ reaction as a function of the angle θ_b. The kinematic conditions are as follows: $E_a = 5500eV$; $E_b = 10eV$; a) $n = 1$; $Q=0.129$ a.u.; b) $n = 2$; $Q=0.177$ a.u.; The experimental points are taken from [2]. The L-form of the transition operator (solid curve) practically coincides with its V-form (solid-dotted curve).

The orthogonality condition gives $\int d\boldsymbol{r} I(\boldsymbol{r}, \boldsymbol{p}) = 0$.

Taking into account that Q is small we obtain from (1) and (2)

$$\frac{d^3\sigma}{d\Omega_a d\Omega_b dE_b} = \frac{2^5}{\pi} \frac{p_a p_b}{p_0 Q^2} |Q a_0 + a_1 \cos\theta + Q a_2 \cos^2\theta + O(Q^2)|^2. \qquad (4)$$

In (4), θ is the angle between the vectors $\boldsymbol{Q}$ and $\boldsymbol{p}_b$.

The calculations within the J–matrix formalism show that all coefficients a_i are of the same order of magnitude, therefore the coefficient a_1 determines the main $cos^2\theta$–behaviour of the TDCS, otherwise $Q(a_0+a_2)$ cause the small difference of the peaks' altitude. Theoretical and experimental quantities of the values $|a_1|^2$ and $\beta = Re(a_1^*(a_0 + a_2))$ for the kinematical conditions characterizing Fig. 1 are equal accordingly $|a_1|^2_{exp} = 0.11$, $\beta^{exp} = 0.037$ and $|a_1|^2_{theor} = 0.097$, $\beta^{theor} = 0.092$.

Let us try to understand what can stand behind an almost three times difference between β^{exp} and β^{theor} which are connected to the sum $(a_0 + a_2)$. We determine:

$$f(z) = 2\pi \int\limits_0^\infty \rho I(\rho, z; p_b)d\rho, \quad f_{\substack{even \\ odd}}(z) = \frac{1}{2}[f(z) \pm f(-z)].$$

As follows from (2), (3) and the orthogonality condition:

$$\int\limits_0^\infty f_{even}(z)dz = 0, \quad \int\limits_0^\infty z f_{odd}(z) = a_1, \quad \frac{i}{2}\int\limits_0^\infty z^2 f_{even}(z)dz = (a_0 + a_2). \qquad (5)$$

The most specific feature of the Hartree-Fock (HF) helium ground state wave functions is the dominant contribution of the $1s^2$ - state, on the level of $93 \div 98\%$. Physically the dominance of the $1s^2$-state means that both electrons occupy the same orbit $(r_1 = r_2)$, and the role of correlations reduces to some widening of its mean radius compared to the radius of an electron in a hydrogen-like ion with the charge $Z = 2$. For example, the only absolute maximum of the KNSP zero partial ground state component $|r_1 r_2 \Phi_{00}(r_1, r_2)|^2$ appears at $r_1 = r_2 = 0.6$ a.u., that corresponds to $Z_{eff} = 1.67$. This means that the mean radius of an "independent" electron in He is almost twice less than the Bohr radius.

According to (5) the factor z^2 cuts off effectively the innermost domain of the integral which is zero without this factor, therefore the value $(a_0 + a_2)$ is determined by the peripherical part of the function $I(\boldsymbol{r}, \boldsymbol{p}_b)$ in (3) which should be practically the same for all wave functions with the same asymptotics.

The presented simple arguments make clear why different couples of the standard helium wave functions give coincident results for the calculations at small Q. The matrix element (2) in this case is nothing more than the overlap between the OCW final state and a tail of the typical HF $1s^2$ initial wavefunction. To alter the value $Q(a_0 + a_2)$ we have to suppose the presence of some noticeable additional structure in $I(\boldsymbol{r}, \boldsymbol{p}_b)$ in the vicinity of the Bohr radius $(r_B = 1)$. Another word, we are obliged to suppose that the traditional simple asymptotics of the He ground state wave function $\Phi_0(\boldsymbol{r}_1, \boldsymbol{r}_2)$ obtained by the variational method are not correct, and we must perform the microscopical three-body calculations in this particular case. Unfortunately the space of this publication is not enough to present the theory for new calculations.

Acknowledgement. The work is partially supported by the grants of the Royal Society (UK) and Russian Ministry of Science and High Technologies.

References

1. A. Duguet et al.: J. Phys. **B20**, 6145 (1987)

2. C. Dupre et al.: J. Phys. **B25**, 259 (1992)

3. A.S. Kheifets et al.: Phys. Rev. **A50**, 4700 (1994)

4. V.A. Knyret al.: Rus. JETP **22**, 190 (1996)

5. F. Mota-Furtado, P.F. O'Mahony: J. Phys. **B21**, 137 (1988)

Few-Body Systems Suppl. 10, 239–242 (1999)

Few-
Body
Systems
© by Springer-Verlag 1999

Dynamics in Few Body Coulomb Problems

S.Yu. Ovchinnikov * †, J.H. Macek, R.S. Tantawi, A.S. Sabbah **

Physics Division, Oak Ridge National Laboratory, PO Box 2008, Oak Ridge, TN 37831-6373; Department of Physics, University of Tennessee, Knoxville, TN 37996-1501, USA

Abstract. We develop the "positive energy Sturmian technique" for the solution of time-dependent Schrödinger equations which describe few Coulomb centers with scattering initial conditions. The "positive energy Sturmian technique" is based on the following main steps: (i) time-dependent scaled transformation; (ii) Fourier transformation into the frequency domain; (iii) outgoing wave Sturmian expansions; and (iv) solution of coupled equations. The technique has been applied in electron-atom and ion-atom collisions for calculations of energy and angular distributions of emitted electrons and excitations of atoms.

1 Introduction

The calculation of accurate electron-impact excitation and ionization cross sections for atoms remains a long-standing problem in atomic collision physics. The time-dependent solution of a time-dependent Schrödinger equation is used in order to avoid problems with asymptotical conditions involving three or more charged particles. The time-dependent techniques obviate the need for answers to questions about the asymptotic form of the wave function. It is recognized that the electron wave-packet in the continuum region expand approximately according to classical relations $\mathbf{r} = \mathbf{k}t$. Basic states which expand with time such that $\mathbf{q}_s = \mathbf{r}_s/t$ are expected to represent time-dependent functions efficiently. In this note we formulate such a representation.

---

*_Permanent address:_ Ioffe Physical Technical Institute, St. Petersburg 194021, Russia
†_E-mail address:_ serge@mail.phy.ornl.gov
**_Permanent address:_ Department of Mathematics, Faculty of Science, Zagazig University, Zagazig, EGYPT.

2 Time-Dependent Schrödinger Equations

The time-dependent Schrödinger equation used in a few body Coulomb problem has the form

$$\left[i\frac{\partial}{\partial t} + \sum_{s=1}^{n} \frac{1}{2m_s} \nabla_{\mathbf{r}_s}^2 - \sum_{s>j}^{n} \frac{Z_s Z_j}{|\mathbf{r}_s - \mathbf{r}_j|} \right] \psi_{i,f}(\mathbf{r}_1, \cdots, \mathbf{r}_n, t) = 0, \qquad (1)$$

where Z_s and m_s are the charges and masses of the particles. Atomic units with $e = m_e = \bar{h} = 1$ are used throughout. The initial conditions are imposed at $t \to -\infty$ for ψ_i^- and at $t \to \infty$ for ψ_f^+, i.e

$$\begin{aligned}
\lim_{t \to -\infty} \psi_i^-(\mathbf{r}_1, \cdots, \mathbf{r}_n, t) &= \psi_i^0(\mathbf{r}_1, \cdots, \mathbf{r}_n, t), \\
\lim_{t \to \infty} \psi_f^+(\mathbf{r}_1, \cdots, \mathbf{r}_n, t) &= \psi_f^0(\mathbf{r}_1, \cdots, \mathbf{r}_n, t).
\end{aligned} \qquad (2)$$

The equality $\psi^+(\mathbf{r}_1, \cdots, \mathbf{r}_n, t) = \psi^-(-\mathbf{r}_1, \cdots, -\mathbf{r}_n, -t)^*$ allows us to define the initial conditions for both initial and final states at $t \to -\infty$ and describe collisional processes on semi-infinite time axis $-\infty < t \leq 0$. Transition amplitudes between the initial and final states are given by matrix elements

$$\begin{aligned}
T_{\mathbf{k},i} &= \int \psi_f^+(\mathbf{r}_1, \cdots, \mathbf{r}_n, t)^* \, \psi_i^-(\mathbf{r}_1, \cdots, \mathbf{r}_n, t) \, d^3\mathbf{r}_1 \cdots d^3\mathbf{r}_n \\
&= \int \psi_f^-(-\mathbf{r}_1, \cdots, -\mathbf{r}_n, 0) \, \psi_i^-(\mathbf{r}_1, \cdots, \mathbf{r}_n, 0) \, d^3\mathbf{r}_1 \cdots d^3\mathbf{r}_n. \qquad (3)
\end{aligned}$$

3 Schrödinger Equation in a Scaled Space

In order to transform plane waves to wave functions that are manifestly Galilean covariant, we introduce a time-dependent scaling transformation [1] with scaled coordinates $\mathbf{q}_s = -\mathbf{r}_s/t$, scaled time $\tau = -1/t$ and wave functions

$$\phi_{i,f}(\mathbf{q}_1, \cdots, \mathbf{q}_n, \tau) = \tau^{-3n/2} \prod_{s=1}^{n} \exp\left[i\frac{m_s q_s^2}{2\tau} \right] \psi_{i,f}(\mathbf{q}_1/\tau, \cdots, \mathbf{q}_n/\tau, -1/\tau), \quad (4)$$

The wave function $\varphi(\tau, \mathbf{q}_1, ..., \mathbf{q}_n)$ is the solution of a Schrödinger equation

$$\left[i\frac{\partial}{\partial \tau} - H_0(\mathbf{q}_1, \cdots, \mathbf{q}_n) + \frac{1}{\tau} V(\mathbf{q}_1, ..., \mathbf{q}_n) \right] \varphi_{i,f}(\mathbf{q}_1, \cdots, \mathbf{q}_n, \tau) = 0, \qquad (5)$$

where

$$\begin{aligned}
H_0(\mathbf{q}_1, \cdots, \mathbf{q}_n) &= -\sum_{s=1}^{n} \frac{1}{2m_s} \nabla_{\mathbf{q}_s}^2 \\
V(\mathbf{q}_1, \cdots, \mathbf{q}_n) &= \sum_{s>j}^{n} \frac{Z_s Z_j}{|\mathbf{q}_s - \mathbf{q}_j|} \qquad (6)
\end{aligned}$$

The Hamiltonian depends on the scaled time τ only through a factor $1/\tau$ multiplying the potential. The collisional processes are described on the semi-infinite axis $0 < \tau < \infty$ with the initial conditions at $\tau = 0$

$$\lim_{\tau \to 0} \varphi_{i,f}(\mathbf{q}_1, \cdots, \mathbf{q}_n, \tau) = \varphi^0_{i,f}(\mathbf{q}_1, \cdots, \mathbf{q}_n, \tau), \tag{7}$$

Note that the time-dependent scaled transformation transforms the plane-wave function $\psi_{\mathbf{k}}(t, \mathbf{r}) = (2\pi)^{-3/2} \exp\left(i\mathbf{r} \cdot \mathbf{k} - ik^2 t/2\right)$ to the free-particle propagator

$$\varphi_{\mathbf{k}}(\mathbf{q}, \tau) = (2\pi i \tau)^{-3/2} \, \exp\left[\frac{i}{2\tau} (\mathbf{q} - \mathbf{k})^2\right] \tag{8}$$

which is covariant (invariant in form) under the Galilean transformation $\mathbf{r}' = \mathbf{r} + \mathbf{v}t$ and $\mathbf{k}' = \mathbf{k} + \mathbf{v}$ in contrast to the plane-wave function.

4 Schrödinger Equation in Frequency Space

In order to use the Sturmian bases, we write the wave functions as Fourier transform

$$\varphi_{i,f}(\mathbf{q}_1, \cdots, \mathbf{q}_n, \tau) = \frac{1}{\sqrt{-2i\pi}} \int_{-\infty}^{\infty} d\omega \, \exp(-i\omega\tau) \, \chi_{i,f}(\mathbf{q}_1, \cdots, \mathbf{q}_n, \omega). \tag{9}$$

Then the Schrödinger equation in frequency space becomes

$$\left\{ i\frac{\partial}{\partial\omega} \left[H_0(\mathbf{q}_1, ..., \mathbf{q}_n) - \omega\right] - V(\mathbf{q}_1, \cdots, \mathbf{q}_n) \right\} \chi_{i,f}(\mathbf{q}_1, \cdots, \mathbf{q}_n, \omega) = 0. \tag{10}$$

Using the stationary phase method in the limit $t \to -\infty$ we obtain that the initial conditions are

$$\lim_{\omega \to -\infty} \chi_{i,f}(\mathbf{q}_1, \cdots, \mathbf{q}_n, \omega) = \chi^0_{i,f}(\mathbf{q}_1, \cdots, \mathbf{q}_n, \omega), \tag{11}$$

where

$$\begin{aligned}
\chi^0_{i,f}(\mathbf{q}_1, \cdots, \mathbf{q}_n, \omega) &= \frac{1}{\sqrt{2i\pi}} \int_0^{\infty} \frac{d\tau}{\tau^{3n/2}} \exp\left(i\omega\tau + i\frac{\sum_s m_s q_s^2}{2\tau}\right) \\
&\times \psi^0_{i,f}\left(\frac{\mathbf{q}_1}{\tau}, \cdots, \frac{\mathbf{q}_n}{\tau}, -\frac{1}{\tau}\right).
\end{aligned} \tag{12}$$

5 Sturmian Expansions and Coupled Equations

We expand the wave functions $\chi_{i,f}$ in terms of a discrete set of orthonormal Sturmian basis functions

$$\chi_{i,f}(\mathbf{q}_1, \cdots, \mathbf{q}_n, \omega) = \sum_{\nu} B^{i,f}_{\nu}(\omega) \, S_{\nu}(\mathbf{q}_1, \cdots, \mathbf{q}_n; \omega) + \chi^0_{i,f}(\mathbf{q}_1, \cdots, \mathbf{q}_n, \omega). \tag{13}$$

The Sturmian functions satisfy the following equation:

$$[H_0(\mathbf{q}_1, \cdots, \mathbf{q}_n) - \omega + \rho_\nu(\omega) V_0(\mathbf{q}_1, \cdots, \mathbf{q}_n)] S_\nu(\mathbf{q}_1, \cdots, \mathbf{q}_n; \omega) = 0 \qquad (14)$$

where

$$V_0(\mathbf{q}_1, \cdots, \mathbf{q}_n) = \sum_{s>1}^{n} \frac{Z_s Z_1}{|\mathbf{q}_s - \mathbf{q}_1|} \qquad (15)$$

and $\rho_\nu(\omega)$ are the Sturmian eigenvalues. The expansion coefficients $B_\nu(\omega)$ are solutions of coupled differential equations

$$\frac{d}{d\omega} \left[\rho_\nu(\omega) B_\nu^{i,f}(\omega) \right] + \sum_{\nu' \neq \nu} \frac{\langle S_\nu(\omega) | S_{\nu'}(\omega) \rangle}{\rho_\nu(\omega) - \rho_{\nu'}(\omega)} \rho_{\nu'}(\omega) B_{\nu'}^{i,f}(\omega)$$

$$- \frac{i}{v} \sum_{\nu'} [\delta_{\nu,\nu'} - \langle S_\nu(\omega) | V - V_0 | S_{\nu'}(\omega) \rangle] B_{\nu'}^{i,f}(\omega)$$

$$+ \frac{i}{v} \langle S_\nu(\omega) | V - V_{i,f} | \chi_{i,f}^0(\omega) \rangle = 0, \qquad (16)$$

with initial conditions $B_\nu^{i,f}(\omega \to -\infty) = 0$.

Acknowledgement. This work is supported by the Office of Basic Energy Sciences, Division of Chemical Sciences, USDOE, under contract DE-AC05-96OR22464 with Lockheed Martin Energy Research Corp. J. H. M. acknowledges support by NSF under Grant No. PHY-9600017 and NATO Grant No. CRG-950407.

References

1. E.A. Solov'ev and S.I. Vinitsky: J. Phys. **B18**, L557 (1985)

2. S.Yu. Ovchinnikov and J.H. Macek: Phys. Rev. Lett. **75**, 2474 (1995)

Few-Body Systems Suppl. 10, 243–252 (1999)

Few-
Body
Systems
© by Springer-Verlag 1999

Halo Nuclei

N.A. Orr

Laboratoire de Physique Corpusculaire, IN2P3-CNRS, ISMRA et Université de Caen, Boulevard du Maréchal Juin, 14050 Caen Cedex, France

Abstract. A brief review of some recent developments in the study of halo nuclei is presented. Particular emphasis is placed on high energy experimental probes and the relationship to the few-body character of such systems.

1 Introduction

The size and distribution of matter in the nucleus have long played a central role in nuclear physics. Indeed, such gross properties reflect the combined effects of many fundamental aspects of nuclear models. In stable nuclei the neutron and proton distributions exhibit essentially identical RMS radii. For some light nuclei far from stability, which combine large neutron (or proton) excesses with very weak binding, considerable differences have been found. Such systems are well represented by a core, resembling a normal nucleus, surrounded by an extended valence nucleon density distribution, the so-called "halo" [1].

In general terms the formation of such an extended spatial distribution is governed by the low binding energy ($\sim$0.1–1.0 MeV) and orbital angular momentum of the valence nucleons (typically $l \leq 1$) [2, 3]. Further constraints on the development of a halo arise from the mass of the system, due to the contribution of the core to the overall size, and the Coulomb barrier, which leads to significant suppression of proton haloes [2]. Consequently the known halo systems are confined primarily to the light, drip and near dripline, neutron-rich nuclei, ^{6}He, ^{11}Li, ^{14}Be (two-neutron haloes) and ^{11}Be (one-neutron halo), with evidence also suggesting the presence of a one-neutron halo in ^{19}C [4, 5, 6] and a two-neutron halo in ^{17}B [7].

Experimentally the investigation of halo nuclei is hampered by the secondary beam intensities available for such nuclei, typically many orders of magnitude below stable beam intensities, $\sim$1 pps for ^{17}B to $\sim 10^5$ pps for ^{6}He. The very short half-lives ($\lesssim$ 1s) of the species of interest further preclude the use of standard nuclear structure techniques. Consequently, high cross section processes, such as breakup reactions, have been employed as probes. The principal signatures of halo nuclei are large total reaction and halo removal cross sections (particularly for high-Z targets) and characteristic fragment momentum distributions, typically narrow and very forward peaked.

Despite the relative simplicity of such measurements, the interpretation has undergone some revision in recent years, as discussed in Sects. 2 and 3. With the advent of high acceptance devices and continuing progress in secondary beam intensities, exclusive measurements have also become possible, allowing more detailed investigations to be made (Sects. 4 and 5).

Given the interdisciplinary nature of this conference, the present paper is intended to provide a review to the non-specialist of some of the recent progress in the investigation of nuclear haloes. Particular emphasis is given to experimental probes arising from high energy reactions and their interpretation in the light of the few-body character of halo systems (many of the recent theoretically advances are treated in accompanying contributions). Finally, although excited state haloes must exist, very little is known about them, primarily due to the lack of a direct probe. The present discussion is thus confined to ground state neutron haloes.

2 Radii and Density Distributions

The determination of the radii and density distributions of stable nuclei are usually undertaken using high energy electron and hadron scattering [8]. As noted above, the very short half-lives and limited production rates of nuclei very far from stability, have, until very recently [9] precluded such measurements for halo nuclei.

Following the pioneering work of Tanihata and collaborators [10], measurements of total reaction cross sections have been employed to probe sizes and matter distributions of nuclei far from stability. At very high energies ($\sim$100-1000 MeV/nucleon), such measurements on a light target may be used, following simple geometrical considerations, to estimate an effective interaction radius. Moreover, by combining total reaction cross section measurements made over a range of energies with the Glauber model, it has proven possible to map matter distributions (see, for example, [11]).

Recently it has been pointed out that while such an approach is valid for normal nuclei it underestimates the sizes of halo nuclei [12]. In particular, in the standard Glauber approach (optical limit), nucleon-nucleon cross sections are combined with single-particle density distributions to derive the cross section. At high energies, however, the use of single-particle density distributions for a system displaying such a distinct clustering (core + halo nucleons) is no longer valid as the reaction time scale is very much shorter than the orbital time of a nucleon. The halo nucleons are thus essentially "frozen" during the collision. This adiabatic limit leads to an increased transparency, as illustrated schematically in Fig. 1, and thus an upward revision in the size of the system for a given cross section. Such a few-body treatment requires explicit three-body wave functions for which the cross section is then calculated based on all pos-

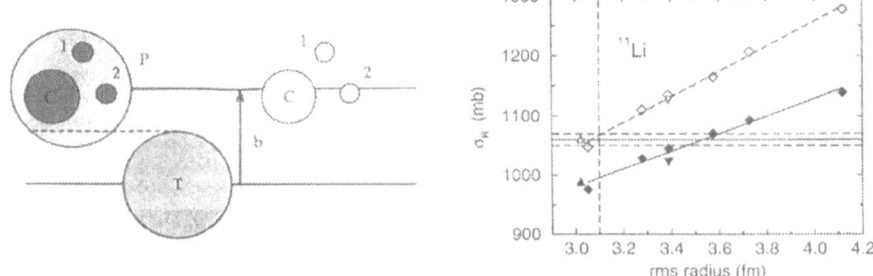

Figure 1. Left panel: schematic representation of the standard Glauber and few-body adiabatic limit treatments of a collision at impact parameter b (see text). Right panel: calculated total reaction cross sections for ^{11}Li (800 MeV/nucleon) + ^{12}C. The horizontal lines represent the measured value (and uncertainty) [10] and the vertical line the RMS radius obtained from a standard Glauber calculation employing Gaussian single-particle density distributions. The solid [open] symbols correspond to few-body adiabatic [standard] Glauber calculations based on three-body wavefunctions incorporating various (increasing with r_{RMS}) s-wave admixtures [12].

sible resulting reaction configurations. A systematic reanalysis of halo systems for which total reaction cross sections have been measured at high energies has been carried out by the Surrey group [12], resulting in upward revisions in the sizes, notably for ^{11}Li (Fig. 1) for which the revised r_{RMS} (3.54±0.10 fm) is consistent with a significant s-wave admixture in the wave function.

3 Fragment Momentum Distributions

The high cross sections for the dissociation of halo nuclei and the possibility of a relatively simple interpretation have lead to the examination of fragment (core and neutron) distributions as not only signatures, but also probes of the structure of the halo [13].

Based on the results of early measurements indicating a target independence of the distributions from the breakup of ^{11}Li [14], fragment momentum distributions were interpreted, in the context of the Serber or sudden approximation [15], as a direct measure of the intrinsic momentum distribution of the halo nucleons. Systematic studies of ^{11}Be [16, 17] and ^{6}He [18] have demonstrated, however, that this simple picture is modified by a number of factors. In particular, the perturbing effects of the reaction process and final-state interactions (FSI) in the exit channels play important roles in defining the final momentum distributions. Owing to the core–neutron mass difference, these effects are much more pronounced for the neutron distributions.

The reaction is governed by the nuclear interaction for light targets and a mixture of Coulomb and nuclear for heavy targets, with the former dominating only for the most well developed halo systems, such as ^{11}Li and ^{11}Be. Four reaction mechanisms may be identified as a function of impact parameter, b.

Taking the example of a two-neutron halo system and proceeding with decreasing b these are:

–Coulomb induced dissociation ($b > R_{target} + R_{proj}$) – resulting in two forward focussed, beam velocity halo neutrons.

–Diffractive dissociation ($R_{target} + R_{core} < b < R_{target} + R_{proj}$) – two beam velocity halo neutrons, one of which is forward focussed and the other, diffracted, forming a broad distribution.

–Absorption or "stripping" ($R_{target} + R_{core} < b < R_{target} + R_{proj}$) – one or fewer, depending on the degree of correlation (Sect. 4.2), forward focussed, beam velocity halo neutrons.

–Core-breakup ($b < R_{target} + R_{core}$) – two or fewer, depending on the degree of correlation (Sect. 4.2), forward focussed, beam velocity halo neutrons and a broad distribution of neutrons arising from the core–target interaction and a small contribution from halo neutrons involved in the collision [5, 17, 19].

In the first three cases a beam velocity, forward focussed core fragment is observed in coincidence with the neutrons, whilst core-breakup reactions result in a lighter charged fragment. In all cases FSI between the charged fragment and neutron have been found to play a role, typically resulting in a narrowing of the neutron momentum distribution. Of particular interest are the core–neutron FSI which, for two-neutron halo systems carry information concerning the resonances in the unbound A-1 systems (Sect. 4.1).

As indicated above, the reaction process also defines the impact parameter range for the collision, thus imposing an additional bias. Importantly, the core fragment distributions for breakup on a light target (long assumed for measurements made parallel to the beam direction to provide the most direct measure of the intrinsic distributions [13]) are limited to probing only radii greater than that of the core [20]. Consequently the observed momentum distributions are narrower than the intrinsic distributions, an effect that will in general be more pronounced for the less well developed haloes. Indeed, recognition of this effect was important in the case of ^{8}B in reconciling measurements of the ^{7}Be fragment momentum distribution and the total reaction cross section [21].

4 Three-Body Observables

1. *Core–neutron Subsystems* One of the ingredients of primary importance in constructing three-body models of two-neutron halo systems is the core–neutron interaction. Ideally the parameters for describing this interaction may be derived from the level schemes of the corresponding nucleus — ^{5}He, ^{10}Li, ^{13}Be and ^{16}B. Beyond ^{5}He, however, data on these unbound nuclei is scant and often contradictory. Indeed, until recently, only multinucleon transfer (or other exotic) reactions with very low yields and complex selectivities enabled these systems to be accessed using stable beams and targets [22].

As noted above, the imprint of the core–neutron FSI is observed in the neutron momentum distributions and can, in principal, be used to extract information about the core–neutron interaction [23]. More direct information may be

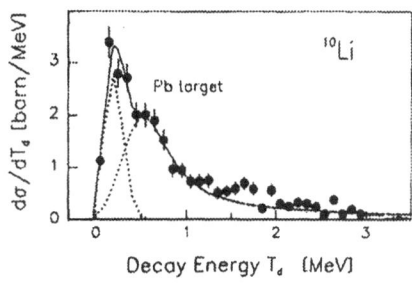 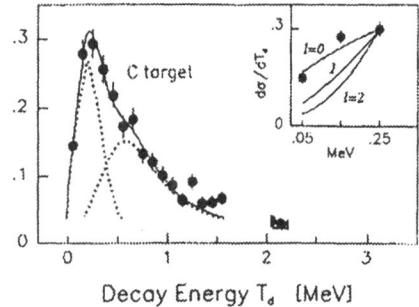

Figure 2. Invariant mass spectrum for ^{9}Li + neutron events from the reaction of a ^{11}Li beam (280 MeV/nucleon) on C and Pb targets [24].

obtained if the core–neutron relative energy (or invariant mass) spectrum can be constructed from precise measurements of the core and neutron momenta. In Fig. 2, the result is shown from a recent study of ^{10}Li populated via breakup of ^{11}Li [24]. Given the conjecture surrounding the low-lying level structure of ^{10}Li [1], the feature observed very close to threshold is of considerable interest and has been associated, following the well known N=7 inversion for ^{11}Be, with a $\pi p_{3/2} \nu s_{1/2}$ configuration. Further support for this identification may be derived from the core–neutron angular correlations [25].

2. *Neutron–Neutron Correlations* The Borromean character of two-neutron halo systems[1] is one of the most intrinsically intriguing features of halo nuclei. Of particular interest, then, in three-body descriptions is the degree of correlation between the halo neutrons. Considering only the intrinsic momentum distributions, a dineutron like correlation will be characterised by $p_{core} = 2p_n$, whereas for a system with no correlations $p_{core} = \sqrt{2}p_n$. Unfortunately, as described in Sect. 3, no unperturbed measure of the momentum distributions is accessible.

On geometrical grounds the degree of spatial correlation may, in principle, be derived from two-neutron removal reactions [26]. As detailed in Sect. 3, for reactions on a light target, dissociation resulting in a core fragment in the exit channel arises from diffraction and stripping of one neutron and, depending on the degree of correlation between the halo neutrons, stripping of two neutrons. The latter process is difficult, however, to identify experimentally as it corresponds to a neutron multiplicity of zero.

The tool employed to date to probe the neutron-neutron correlations has thus been measurements of the neutron-neutron relative momenta, p_{n-n}, following breakup reactions. Clearly processes must be sought in which the intrinsic neutron-neutron correlations are perturbed as little as possible. The avenue exploited to date has been dissociation reactions on high-Z targets. Here the relatively low-energy excitations (Sect. 5) involved in the breakup are hoped to provoke only minimal perturbations in the intrinsic neutron-neutron momenta.

---

[1] That is, none of the constituent subsytems, core-neutron or neutron-neutron, are bound.

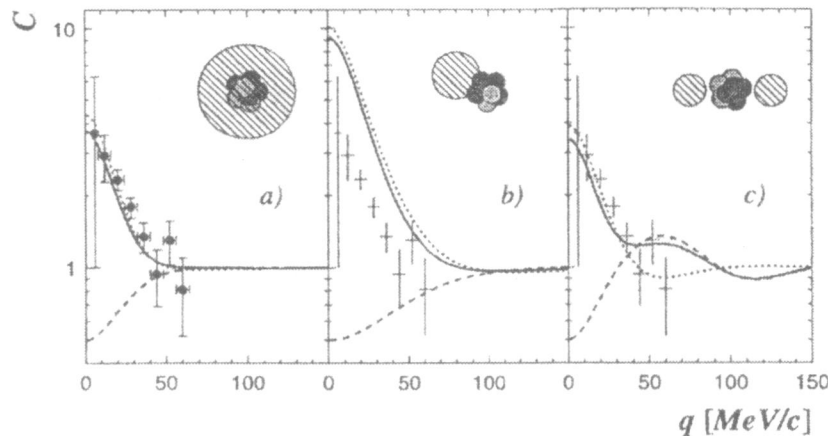

Figure 3. Neutron–neutron correlation functions, C, for different halo configurations. The calculations are based on Gaussian sources with sizes, σ, of (a) 6 fm, (b) 3 fm and (c) 2 fm separated by 10 fm. The contributions from Fermi–Dirac statistics and the neutron–neutron FSI are shown by the dashed and dotted lines respectively [32]. The data points are from ref. [27] (see text).

Such measurements are experimentally difficult, owing primarily to the interfering effects of cross-talk, whereby neutrons may be scattered from one detector (or passive support material) into another, thus mimicking a two-neutron event or deforming the momentum of one neutron and hence the p_{n-n}.

To date all three kinematically complete breakup experiments on ^{11}Li [24, 27, 28, 29] have been used to reconstruct the p_{n-n}. Attempts have been made to compare the results with respect to the very simple pictures of a strong dineutron correlation or none whatsoever, with the former appearing to be excluded [24, 28, 29]. The simplicity of the analyses, which do not include, for example, the neutron–neutron FSI, must be stressed. A method better adapted to probing correlations is intensity interferometry [30]. In this approach the relative motion of the two outgoing neutrons is governed, as illustrated in Fig. 3, by the neutron–neutron FSI and quantum statistics (here Fermi-Dirac), both of which are directly related to the spatial[2] characteristics of the source [31]. Providing then that any other effects on the neutron momenta may be neglected or eliminated in the construction of the correlation function, interferometry provides a means to probe the spatial correlations of the halo neutrons. Importantly, owing to the low momentum content of the halo neutrons, the standard approach to constructing the correlation functions (applied, for example, in the measurement of Ieki et al. [27]) is no longer valid. In particular, the narrow momentum distributions result in strong residual correlations and consequently a significant overestimate of the source size. As a result a new iterative technique has been developed and is currently being applied to ^{11}Li and ^{14}Be [32].

---

[2] Assuming direct breakup (Sect. 5), the source lifetime is negligible.

5 Soft-Dipole Excitations

The very weak binding of the valence neutrons and the separation of the system into a core and halo distribution has lead to the proposal of a new type of excitation in halo nuclei; namely the "Soft Dipole Resonance" or SDR [33]. In normal nuclei, oscillations (or resonances) with a multipolarity of $\lambda = 1$ can be induced between the *total* neutron and *total* proton distributions. Such "Giant (Electric) Dipole Resonances" (GDR) have long been known and occur, owing to the relatively strong restoring force between the neutron and proton distributions, at excitation energies of $\sim 15 - 25$ MeV. In a similar spirit, E1 oscillations have been postulated to occur between the *core* and the *neutron halo* [33]. Given the low binding and large core–halo separation, the correspondingly weak restoring force would be expected to drive such a resonance to low excitation energies (~ 1 MeV); hence the term "soft".

As with the GDR, any soft-dipole mode should be preferentially excited via Coulomb excitation, with considerable enhancement expected for the E1 strength function, $dB(E1)/dE_x$, at low excitation energies. Owing to the very low binding the nucleus will, however, undergo dissociation into the core and halo neutrons. The enhanced low-lying E1 strength will thus translate into a large EMD dissociation cross section. Indeed, large cross sections are observed for reactions with high-Z targets.

Experimentally the E1 strength function may be derived from a measurement of the excitation spectrum ($d\sigma/dE$) from reactions on a high-Z target. The EMD contribution to such reactions may be viewed as the absorption of a virtual photon corresponding to the Coulomb field of the target. Importantly the form with excitation energy of the virtual photon spectrum depends on the beam energy — at low energy the spectrum is dominated by low energy photons, while at higher beam energies the spectrum becomes flatter, and excitations to higher energies become possible.

As noted above, even relatively small excitations lead to the dissociation of halo nuclei. The experimental determination of $d\sigma/dE$ is thus challenging, requiring measurement of the momenta of the charged core and halo neutrons following the breakup of very low intensity secondary beams. From such "kinematically complete" measurements the excitation energy in the nucleus prior to breakup may be reconstructed.

In the case of the reaction of a ^{11}Be beam on a Pb target [34] the E1 strength was found to be concentrated at low energies, in line with expectations for an SDR. As the wave function is reasonably well known, the dipole strength function could be calculated. Interestingly, the data was well reproduced assuming direct breakup. Further support for such non-resonant breakup was found in the velocities of the core fragments (^{10}Be) and the neutrons, which were observed to vary with scattering angle (or b, assuming Coulomb trajectories) as expected for direct breakup within the Coulomb field of the target.

In the case of ^{11}Li a number of experiments have been performed [24, 27, 28]. Similar enhancements in the low-lying E1 strength function are observed in all measurements at $E_x \sim 1$ MeV (Fig. 4), however the observation of a core-

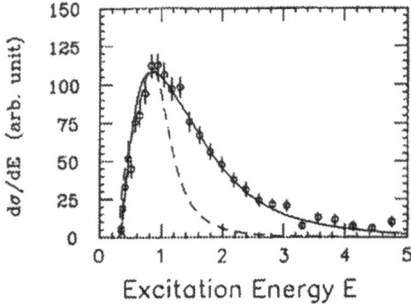 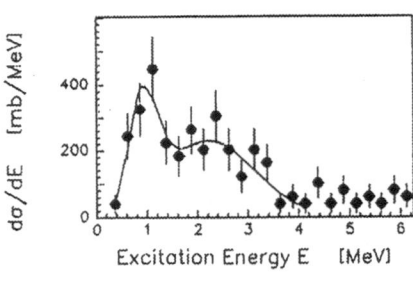

Figure 4. Invariant mass spectra for the breakup of [11]Li on Pb at 43 MeV/nucleon [28] (left panel) and at 280 MeV/nucleon [24] (right panel). The lineshape obtained at 28 MeV/nucleon [27] is included in the left panel (dashed line) for comparison; the normalisation is arbitrary and the uncertainties are not included. The solid line in this panel is a direct breakup prediction (see ref. [28] for details of the wave function used). The solid line in the right panel is a two component fit to the data.

neutron velocity difference[3] [27] and the form of the strength function [28] suggest, once again, that the breakup is direct.

Experimentally then, soft-dipole resonances in [11]Be and [11]Li seem all but ruled out by present breakup reaction studies. For completeness it should be noted that in a very recently reported measurement of the dissociation of [11]Be on Au, no velocity dependence on impact parameter was observed for [10]Be (unfortunately no corresponding measurement of the neutron was possible) [35]; a result consistent with breakup taking place on a very long timescale.

In comparing the various measurements and theoretical predictions for the low-energy excitations, two important constraints must be underlined. Firstly, as noted above, the virtual photon spectrum is heavily weighted towards low energies at the lower beam energies, thus biasing these experiments towards the population of low-energy excitations. This effect is further compounded by the acceptances of the experimental setups, which decrease with increasing decay energy, ($E_{decay} = E_x - S_{(2)n}$). It is important to note that the acceptance effects may only be calculated based on an assumed form for the excitation energy spectrum, thus precluding any model independant analysis. Both effects may, however, be reduced by employing very high beam energies (100 MeV/nucleon). At such energies, not only are higher excitations attainable through the increase in flux of high energy virtual photons, but the strong forward focussing of the reaction products leads to 100% geometrical efficiencies. The results from the first experiment of this kind, carried out at GSI, is shown in Fig. 4, where additional strength is observed in the region of 2 MeV excitation energy.

Finally, it should also be recognised that other means may be employed to explore the low-energy excitations. In particular, proton inelastic scattering

---

[3] In direct breakup the dissociation will on average occur at the distance of closest approach of the projectile and target. Consequently only the charged core is subject to reacceleration on the outgoing leg of the trajectory. In contrast, for a resonance, the dissociation would take place on average well after the point of closest approach.

on ^{11}Li has been recently measured [36]. Intriguingly, a state was observed at 1.3 MeV with an angular distribution identified as corresponding to an $L = 1$ excitation. This interpretation has, however, been subsequently questioned and the feature suggested as arising from "shakeoff", whereby the primary process in the reaction is proton elastic scattering on the ^{9}Li core[4][37].

6 Conclusions

A review of some selected advances in the study of halo nuclei has been presented. Clearly many open questions remain, some of which have been addressed here. In the context of this conference, it is of particular interest to note that the inherent few-body character of halo nuclei is present at some level in nearly all experimental observables.

While the present discussion has concentrated on high energy probes much valuable information can be derived from low energy studies. For example, β-decay spectroscopy offers a potentially rich source of information via a well understood mechanism. Additionally, current advances in secondary beam intensities should soon allow conventional spectroscopic tools, such as single-nucleon transfer reactions [39], to come into play.

Acknowledgement. My thanks are due to Jim Al-Khalili for discussions concerning the recent work of the Surrey theory group and to my colleague at LPC Miguel Marqués who has been primarily responsible for the neutron–neutron correlation work.

References

1. P.G. Hansen, A.S. Jensen, B. Jonson: Ann. Rev. Nucl. Part. Sci **45**, 591 (1995)

2. K. Riisager, A.S. Jensen, P. Moller: Nucl. Phys. **A548**, 393 (1992)

3. D.V. Fedorov, A.S. Jensen, K. Rissager: Phys. Lett. **B312**, 1 (1993); Phys. Rev. **C49**, 201 (1994)

4. D. Bazin et al.: Phys. Rev. Lett. **74**, 3569 (1995); Phys. Rev. **C57**, 2156 (1998)

5. F.M. Marqués et al.: Phys. Lett. **B381**, 407 (1996); E. Liegard, N.A. Orr, F.M. Marqués et al.: LPC Report, LPCC 98-03, to be published

6. A. Ozawa et al.: GSI Scientific Report 1997, p. 21; T. Baumann et al.: to be published; T. Nakamura et al.: In: *Proc. of ENAM98*, to be published

7. I. Tanihata: In: *Proc. 1st Int. Conf. on Radioactive Nuclear Beams*, Sept. 1989, Berkeley, USA, p. 429, eds. W.D. Myers, J.M. Nitscke, E.B. Norman. World Scientific 1990; M.G. St-Laurent et al.: Z. Phys. **A332**, 457 (1989); E. Liatard et al.: Europhys. Lett. **13**, 401 (1990)

8. C.J. Batty et al.: Advances Nucl. Phys. **19**, 1 (1989)

---

[4]This proposal appears difficult to reconcile with the evidence also seen for a state at $\sim$1.3 MeV in the ^{11}B$(\pi^-,\pi^+)^{11}$Li reaction [38].

9. G.D. Alkhazov et al.: Phys. Rev. Lett. **78**, 2313 (1997); J.S. Al-Khalili, J.A. Tostevin: Phys. Rev. **C57**, 1846 (1998)

10. I. Tanihata et al.: Phys. Rev. Lett. **55**, 676 (1985); Phys. Lett. **106B**, 380 (1985); Phys. Lett. **B206**, 592 (1988)

11. I. Tanihata et al.: Phys. Lett. **B287**, 307 (1992)

12. J.S. Al-Khalili, J.A. Tostevin: Phys. Rev. Lett. **76**, 3903 (1996); J.S. Al-Khalili, J.A. Tostevin, I.J. Thompson: Phys. Rev. **C54**, 1843 (1996)

13. N.A. Orr: Nucl. Phys. **A616**, 155 (1997) and references therein

14. R. Anne et al.: Phys. Lett. **B250**, 19 (1990); N.A. Orr et al.: Phys. Rev. Lett. **69**, 2050 (1992); Phys. Rev. **C51**, 3116 (1995)

15. R. Serber: Phys. Rev. **72**, 1008 (1947)

16. R. Anne et al.: Phys. Lett. **B304**, 55 (1993)

17. R. Anne et al.: Nucl. Phys. **A575**, 125 (1994)

18. A.A. Korsheninnikov, T. Kobayashi: Nucl. Phys. **A567**, 971 (1994)

19. T. Nilsson et al.: Europhys. Lett. **30**, 16 (1995)

20. P.G. Hansen: Phys. Rev. Lett. **77**, 1016 (1996); H. Esbensen: Phys. Rev. **C53**, 2007 (1996); K. Hencken et al.: Phys. Rev. **C54**, 3043 (1996)

21. W. Schwab et al.: Z. Phys. **A350**, 283 (1995); M.M. Obuti et al.: Nucl. Phys. **A609**, 74 (1996)

22. W. Mittig, A. Lépine-Szily, N.A. Orr: Ann. Rev. Nucl. Part. Sci. **47**, 27 (1997)

23. M. Zinser et al.: Phys. Rev. Lett. **75**, 1719 (1995)

24. M. Zinser et al.: Nucl. Phys. **A619**, 151 (1997)

25. L.V. Chulkov, G. Schrieder: Z. Phys. **A359**, 231 (1997)

26. G. Bertsch, K. Hencken, H. Esbensen: Phys. Rev. **C57**, 1366 (1998)

27. K. Ieki et al.: Phys. Rev. Lett. **70**, 730 (1993); D. Sackett et al.: Phys. Rev. **C48**, 118 (1993)

28. S. Shimora et al.: Phys. Lett. **B348**, 29 (1995)

29. K. Ieki et al.: Phys. Rev. **C54**, 1589 (1996)

30. D.H. Boal, C.K. Gelbke, B.K. Jennings: Rev. Mod. Phys. **62**, 553 (1990)

31. R. Lednicky, V.L. Lyuboshits: Sov. J. Nucl. Phys. **35**, 770 (82)

32. F.M. Marqués, M. Labiche, N.A. Orr et al.: in preparation

33. K. Ikeda: Nucl. Phys. **A538**, 355 (1992)

34. T. Nakamura et al.: Phys. Lett. **B331**, 296 (1994)

35. J.E. Bush et al.: Phys. Rev. Lett. **81**, 61 (1998)

36. A.A. Korsheninnikov et al.: Phys. Rev. Lett. **78**, 2317 (1997)

37. S. Karataglidis et al.: Phys. Rev. Lett. **79**, 1447 (1997)

38. T. Kobayashi et al.: Nucl. Phys. **A538**, 355 (1992); KEK-Report 91-22

39. J.S. Winfield, W.N. Catford, N.A. Orr: Nucl. Instr. Meth. **A396**, 147 (1997)

Few-Body Systems Suppl. 10, 253–262 (1999)

Few-
Body
Systems
© by Springer-Verlag 1999

Double Rydberg Atoms

Jan M. Rost*

Theoretical Quantum Dynamics.
Fakultät für Physik, Universität Freiburg, Hermann–Herder–Str. 3, D-79104
Freiburg, Germany

Abstract. Since the first attempts to calculate the helium ground state in the early days of Bohr-Sommerfeld quantization two-electron atoms have posed a series of unexpected challenges to theoretical physics. Despite the seemingly simple problem of three charged particles with known interactions it took more than half a century after quantum mechanics was established to describe spectra of two-electron atoms satisfactorily. The evolution of the understanding of *correlated* two-electron dynamics and its importance for doubly excited resonance states is sketched in this overview by contrasting a forgotten idea from the old days with recent theoretical concepts and new challenges.

1 Introduction– A brief history on the theory of the helium atom

The theory of two–electron atoms has played an important role for the development of theoretical physics in this century. Roughly, three periods may be distinguished.

1.1 The quest for quantum mechanics (1920–1960)

The failure of the old quantum theory to describe a stable two-electron atom triggered the development of quantum mechanics. Once the basic formalism had been established by Heisenberg and Schrödinger early variational calculations produced remarkably good results for the ground state of Helium and broke the ground for a the general acceptance of quantum mechanics.

In Sect. 2 we will briefly sketch the attempts of the old quantum theory to understand the helium ground state. In particular a forgotten idea by Heisenberg comes surprisingly close to recent findings of classical dynamics which turns out to be relevant for high lying resonances of helium.

---

*This contribution is based on an extended review written by G. Tanner. K. Richter and myself for Review of Modern Physics [1].

1.2 The need to go beyond the Hartree-Fock approximation (1960–1990)

With the seminal synchrotron absorption experiments detecting doubly-excited states in the early 60's it became clear that the effective single particle picture, familiar from the successful Hartree-Fock approximation, was inadequate to understand two-electron resonances. As a consequence, sophisticated alternative quantum approximations were developed over the next 30 years. The most important concepts were a group theoretical approach and two adiabatic approximations. These concepts successfully explained the high degree of nontrivial regularity in the spectra of two-electron resonances, i.e. features which could and can not be accounted for by an effective single particle picture. Rather, they are intrinsically related to the *correlated* dynamics of two electrons. Section 3 briefly summarizes the results achieved by the approximate quantum methods, in particular the existence of approximate collective quantum numbers and propensity rules for the decay of resonances.

1.3 The quest for new concepts in the extreme excitation regime (since 1990)

Over the last decade the regime of extremely high excitation, i.e. principal quantum numbers N, n for both electrons of $N \approx n \geq 10$, has become feasible, both, experimentally and computationally. For these high excitations the approximate quantum numbers begin to loose their meaning and the regularities in the two-electron resonance spectrum start to dissolve. Moreover, even if applicable, the spectroscopic concept of isolated resonances identified by a set of quantum numbers becomes very questionable if the density of resonances per unit energy tends to infinity which is the case towards the three body break up limit $E = 0$ with $N \approx n \to \infty$. Hence, one needs a new concept to understand two-electron dynamics in this regime of extreme double-excitation. The concept is provided by a modern semiclassical approach. Its development over the last few years reveals an impressive progress in the quantitative description of the resonances. The backbone of these semiclassical descriptions are the periodic orbits of the full classical two-electron system without approximations. Surprisingly, from the shortest and simplest periodic orbits and from their stability properties one can draw a picture of two-electron excitation dynamics which agrees perfectly well with the results of the quantum approximations explaining the regular spectrum of intermediate double excitation. However, the periodic orbits have one advantage that goes beyond the simple structural picture: A representation of the spectra in terms of periodic orbits does not rely on an explicit quantization scheme based on the quasi-separability of the problem in (collective) coordinates or the existence of approximate quantum numbers. This advantage might lead the way into the extreme excitation regime where the increase of resonances renders a description in terms of the assignment of quantum numbers meaningless. One tool, which has already been proven to be very useful to characterize dynamical features in the extreme excitation limit for hydrogen in the magnetic field, is scaled periodic orbit spectroscopy. As we will discuss in Sect. 4 an incredibly complex energy spectrum with many (overlapping) resonances can look quite ordered, if properly Fourier transformed into the time domain where peaks at certain times indicate the periods of relevant periodic orbits.

We will deal exclusively with the nonrelativistic two-electron Hamiltonian in atomic units,

$$H = \frac{\boldsymbol{p}_1^2 + \boldsymbol{p}_2^2}{2} - \frac{Z}{r_1} - \frac{Z}{r_2} + \frac{1}{r_{12}} , \tag{1}$$

where Z is the nuclear charge, r_k $(k = 1, 2)$ denote the electron-nucleus distances and the distance between the electrons is r_{12}.

2 Pre-quantum attempts to quantize helium and Heisenberg's forgotten idea

In 1913 Bohr succeeded to explain the energy levels of the hydrogen atom in terms of a quantization of the classical Kepler orbits. Numerous attempts to explain the ground state of helium by quantizing different two-electron periodic orbits in a similar manner failed[1]. All these attempts shared the unproven but common belief that (i) the ground state of helium is related to a single periodic orbit of the electron pair, (ii) the electrons move on symmetric orbits with $r_1 = r_2$ for all times, (iii) orbits where the electrons hit the nucleus ("Pendelbahnen" [3]) are excluded and (iv) the quantum number n in the quantization condition is an integer.

Heisenberg's interest in the helium problem was stimulated by Bohr in 1922. Sommerfeld and Heisenberg devised as a possible classical ground state configuration a model where both electrons move *asymmetrically* on two different Kepler ellipses of the same shape but oriented in opposite directions, sketched by Heisenberg in a letter to Sommerfeld [4] (see Fig. 1). Heisenberg outlines in the letter a calculation of the ground state energy based on this electron pair motion. As an important achievement from a today's point of view, he introduced a second quantization condition, $\oint p_\varphi \mathrm{d}\varphi = n_\varphi h$, for the motion of the angle φ between the major axes of the two orbits under the influence of the mutual electron interaction (see Fig. 1). Moreover, he allowed for n_φ to be *half-integer*. Including the electron repulsion in a perturbative manner, Heisenberg arrived at an ionization potential of 24.6 Volt, compared to the best experimental value of 24.5 Volt at that time! Discouraged by Bohr and Pauli who harshly criticized the concept of non-integer quantum numbers, Heisenberg never published his calculation. This may explain why the Heisenberg-Sommerfeld model, to the best of our knowledge, has never been mentioned in the atomic physics literature.[2]

In the meantime it has become clear that Heisenberg's success was accidental [1, 6]. Consequently, attempts by Born and Heisenberg to describe excited states of helium in a similar fashion failed [7]. The pessimistic mood in the

---

[1] A summary can be found in [2] .

[2] Sommerfeld, encouraged by the results obtained by Heisenberg, wrote a paper [5], refering to Heisenberg's perturbative results.

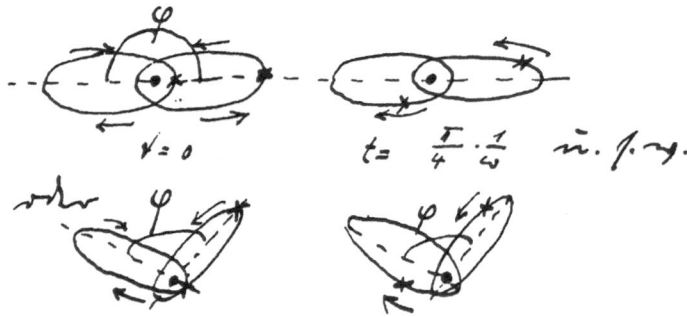

Figure 1. Sketch of the periodic electron pair motion proposed by Heisenberg in a letter to Sommerfeld in 1922 as a candidate for a classical ground state configuration of helium.

early twenties concerning the treatment of two-electron atoms becomes apparent in a conclusion given by van Vleck [8]: *"The conventional quantum theory of atomic structure does not appear to be able to account for the properties of even such a simple element as helium, and to escape from this dilemma some radical modification in the ordinary conceptions of the quantum theory or of the electron may be necessary."* History has shown that van Vleck was right. Quantum mechanics as we know it today took off quickly once the concepts developed by Schrödinger and Heisenberg had been proven to work, e.g. for the Helium ground state which Hylleraas was able to calculate accurately in 1929 using variational techniques [9]. The singly excited states of two electron atoms were also relatively simple to understand and with the mean field theory, i.e. the Hartree-Fock approach, the structure of almost the entire periodic table could be explained.

3 Quantum approximations for double Rydberg atoms

The surprise and a new – productive – crisis came with the first experiment on *doubly*–excited states of two–electron atoms by Madden and Coddling in 1963 [10]. The observed resonance spectrum looked quite different from what had been expected from a Hartree-Fock picture: Only two instead of three Rydberg series were observed below the second ionization threshold of Helium, $N = 2$. Moreover one series was very weak, and none of the energies of the resonances agreed with the mean field predictions. This was correctly interpreted as a signature of strong electron-electron correlation – and the tools to understand it were developed over the last 30 years. The adiabatic hyperspherical, molecular and a group theoretical approximation (for reviews see [11], [12], and [13]) have mainly shaped today's understanding of moderately excited double Rydberg atoms in terms of approximate quantum numbers and propensity rules for the decay and radiative transitions (dipole matrix elements) of the resonances. These properties come from a remarkably well defined nodal structure of resonant wave function near the maximum probability density exemplified in Fig. 2. These results are in remarkable agreement with the latest photoabsorption experiments, for a review see [16].

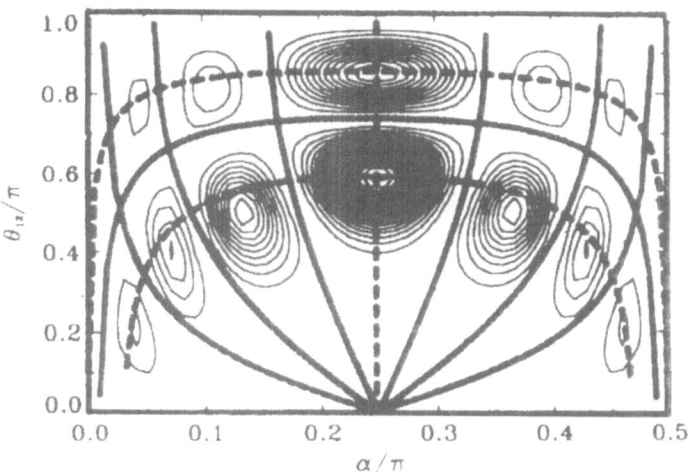

Figure 2. Two-electron density at $(r_1^2 + r_2^2)^{1/2} = 80$a.u. for a $^1P^o$ symmetry of H^- from an adiabatic hyperspherical calculation [14]. The coordinates are $\theta_{12} = \angle(\boldsymbol{r}_1, \boldsymbol{r}_2)$ and $\tan\alpha = r_1/r_2$. Overlaid are the spheroidal nodal lines $\lambda = (r_1 + r_2)/r_{12}$ and $\mu = (r_1 - r_2)/r_{12}$ predicted by the adiabatic molecular approximation [15].

4   Semi–classical dynamics of double Rydberg atoms

In contrast to the semiclassical methods of the old quantum theory discussed in Sect. 2, contemporary semiclassical theory has a sound mathematical foundation [17]. Based on this background it is indeed possible to understand the spectrum of double Rydberg atoms in terms of a few, more exactly three, periodic orbits only. This is very much in the spirit of the old quantum theory. However, the three orbits are not guessed or arbitrarily picked, rather they represent the first order approximation in a well defined expansion of the semiclassical Green function in terms of periodic orbits (PO).

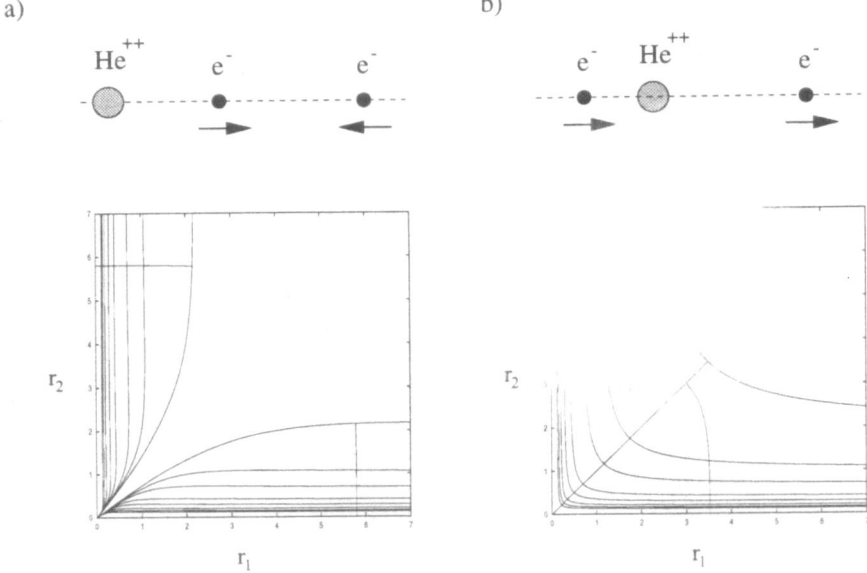

Figure 3. The potential $V(r_1, r_2) = -\frac{Z}{r_1} - \frac{Z}{r_2} + \frac{1}{|r_1 \pm r_2|}$ in the classically allowed regime $V \leq E$ together with the shortest periodic orbits; a) the Zee–configuration with both electrons on one side of the nucleus together with the stable frozen planet orbit (FP); b) the eZe–configuration with electrons on different sides of the nucleus together with the unstable asymmetric stretch orbit (AS) and the symmetric stretch orbit ($r_1 = r_2$).

4.1 The three fundamental periodic orbits and their role in two-electron dynamics

In Fig. 3 the three fundamental periodic two-electron orbits, which are all collinear, are sketched. The *frozen planet* orbit (Fig. 3a) has the two electron on the same side of the nucleus. This orbit is classically stable and obtained its name from the fact that the outer electron moves only very little and represents almost a "frozen planet". The *asymmetric stretch* orbit has the two electrons on opposite sides of the nucleus. The orbit is slightly unstable and exhibits an asymmetry in the electron distances, i.e. $r_1(t) \neq r_2(t)$. This is in contrast to the third orbit, the so called *symmetric stretch* or Wannier orbit with $r_1(t) = r_2(t)$, also drawn in Fig. 3b.

4.2 Scaling and Rydberg formulae

If a periodic orbit of a system with pure Coulomb forces is quantized it produces automatically a Rydberg series. The reason lies in the scaling property of the classical Hamiltonian which can be made energy independent in scaled coordinates $\boldsymbol{p}_i^s = \boldsymbol{p}_i(-E)^{-1/2}$ and $\boldsymbol{r}_i^s = \boldsymbol{r}_i(-E)$. This leads to the scaled action $S_s = S(-E)^{-1/2}$. Hence, quantization of the (unscaled) action produces a Rydberg series of the form $E_n = -(S_c/2\pi)^2/(n-\mu)^2$, where the semiclassical quantum defect μ consists of the stability properties of the quantized PO, i.e. the characteristics of the dynamics in phase space orthogonal to the PO.

The *frozen planet* orbit is stable. Hence, one can apply torus quantization according to Einstein-Brilloin-Keller (EBK), leading to a Rydberg formula [18]

$$E_{m,k,\bar{n}} = -\frac{(S_{\text{FP}}/2\pi)^2}{[(m+\frac{1}{2}) + 2(k+\frac{1}{2})\sigma_1 + (\bar{n}+\frac{1}{2})\sigma_2]^2}, \quad m,k,\bar{n} = 0,1,2,\ldots. \tag{2}$$

For Helium the scaled action of the orbit is $S_{\text{FP}}/2\pi = 1.4915$ and $\sigma_1 = 0.4616$, $\sigma_2 = 0.0677$ are the winding numbers for the dynamics in the degree of freedom perpendicular to the collinear subspace and for the dynamics in the collinear plane of the PO, respectively. The quantum number m corresponds to nodes along the PO, k and $\bar{n}$ to nodes in the degrees of freedom perpendicular to the PO.

The asymmetric stretch orbit on the other hand is unstable. Quantization of this orbit alone is in the strict sense not appropriate and may be viewed as a first order truncation of a series expansion of the semiclassical Green function in terms of POs [19], see also [20, 1],

$$E_{m,k} = -\frac{(S_{\text{AS}}/2\pi)^2}{[m+\frac{1}{2} + 2(k+\frac{1}{2})\sigma_{AS}]^2}, \quad m,k = 0,1,2,\ldots. \tag{3}$$

For Helium $S_{\text{AS}}/2\pi = 1.8290$ and $\sigma_{AS} = 0.5393$ is the winding number for the dynamics in bending degree of freedom. The quantum numbers m and k describe excitation along and perpendicular to the orbit which in turn corresponds to intra-shell excitation and vibrational excitation in the bending degree of freedom, respectively.

The energies of resonances characterized by excitation along these two orbits follow very closely the single orbit quantization from Eqs. (2, 3) as can be seen on Fig. 4. The symmetric stretch PO (Fig. 3b) is highly unstable [21] and incapable of supporting quantum resonances. However, this PO is the main decay path for resonances in two-electron systems. It can be easily reached by the asymmetric stretch type resonances but not by the frozen planet ones. Hence, the latter ones have much longer lifetimes on average.

In summary, the three simplest periodic orbits give an excellent qualitative and semi-quantitative understanding of the double-Rydberg dynamics for moderate excitation.

4.3 Scaled periodic orbit spectroscopy

With infinitely many, partially overlapping resonances in the spectrum when approaching the double ionization threshold a description of the spectrum in terms of these resonances becomes meaningless. The previous paragraph gives

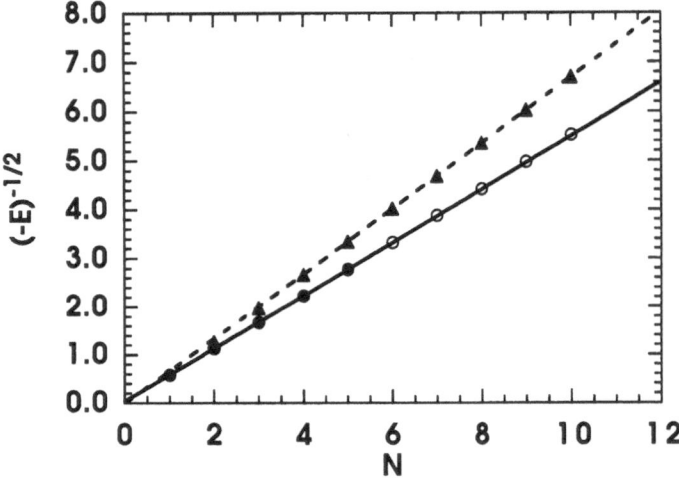

Figure 4. The energies of quantum resonances of asymmetric stretch (circles) and frozen planet (triangles) character. The lines are the semiclassical energies predicted from Eq. (3) (solid) and Eq. (2) (dashed) respectively. Open circles indicate states which are are not observable in the experiment due to strong interaction with an entire Rydberg series. In the quantum simulation they can be calculated by discarding the Hilbert space representing the Rydberg series, see [22].

already a hint for an alternative: Each of the discussed periodic orbits carries an infinite series of resonances. Hence, the question arises if there is a way to describe the spectrum directly in terms of the properties (stability and length, i.e. period or action) of the periodic orbits. This is indeed the case and the tool which has been developed is scaled periodic orbit spectroscopy [23]. It makes heavily use of the classical scaling property of the Hamiltonian, in particular that the action $S_j(E)$ of the periodic orbit j can be written as $S_j(E) = S_j(E = -1)|E|^{-1/2} \equiv s_j z$. In a conventional way we can express the density of resonant states with position E_n and widths Γ_n as $d(E) = \sum_n \frac{\Gamma_n/2}{(E-E_n)^2-(\Gamma/2)^2}$. As a function of $z = |E^{-1/2}|$ instead of E we can define the Fourier transform with respect to the scaled action space s, $f(s) = \int dz e^{-isz} d(z)$. This finally transforms the density of states $d(E)$ to the power spectrum of the periodic orbits where a peak appears at each action s which corresponds to a periodic orbit [24],

$$F(s) = |f(s)|^2 \approx \sum_{jk} |a_{jk}|^2 \delta(s_{jk} - s). \tag{4}$$

An actual action s_{jk} belongs to the kth repetition of the jth periodic orbit. For Helium resonances the impressive simplification of the spectrum going from the energy to the action domain is demonstrated in [24, 25]. This relatively new tool can either be used to construct semiclassical spectra directly from periodic orbits, or to interpret experimentally or numerically obtained quantum spectra by merely Fourier transform them after proper scaling.

5 Summary and Outlook

We have given a very brief overview over the research on double Rydberg atoms, covering the time since 1920. The astonishing order in the non-separable two-electron dynamics for moderate double-excitation energies is well understood in terms of quasi-separable approximations leading to approximate quantum numbers and propensity rules for autoionization and radiative transitions. The agreement of theoretical precision with recent high resolution experiments is very good [16].

In the extreme excitation regime ($N \approx n \geq 10$) the order in the resonance spectrum begins to dissolve [22] and the increasing density of resonances calls for new concepts to describe two-electron spectra in this energy regime. Periodic orbit spectroscopy provides such a tool. In the future enhanced experimental as well as numerical capabilities will permit to study the limit of double excitation $E \to 0$ towards the double ionization threshold. This limit is a very interesting singular point in the two-electron spectrum and connects extreme excitation ($E < 0$) with threshold double-ionization ($E > 0$) [26].

Acknowledgement. It is a pleasure to acknowledge the long and fruitful collaboration with John Briggs on two electron atoms, as well as with Andre Bürgers, the late Dieter Wintgen, Klaus Richter and Gregor Tanner. Financial support from the Deutsche Forschungsgemeinschaft (SFB 276 in Freiburg and Gerhard Hess-Programm) is also gratefully acknowledged.

References

1. G. Tanner, K. Richter and J.M. Rost: Rev. Mod. Phys., submitted (1998)

2. J.G. Leopold and I.C. Percival: J. Phys. B **13**, 1037 (1980)

3. M. Born: In: *Vorlesungen über Atommechanik.* Berlin: Springer 1925; english translation: *The Mechanics of the Atom.* New York: Ungar 1927

4. W. Heisenberg: letter to Sommerfeld, 28 October 1922; archive of the *Deutsches Museum*, Munich, 1922. The role of W. Heisenberg at this stage of the "old quantum theory" is also described in: D.C. Cassidy, *Werner Heisenberg*, Spektrum Akademischer Verlag Heidelberg 1995, Sec. II.8: "Der blonde Bauernjunge"

5. A. Sommerfeld: J. Opt. Soc. America, **7**, 509 (1923)

6. E.A. Solov'ev: Sov. Phys. JETP **62**, 1148 (1985); Zh. Eksp. Teor. Fiz. **89**, 1991 (1985)

7. M. Born and W. Heisenberg: Z. Phys. **16**, 229 (1923)

8. J.H. van Vleck: Phil. Mag. **44**, 842 (1922)

9. E.A. Hylleraas: Z. Phys. **54**, 347 (1929)

10. R.P. Madden and K. Codling: Phys. Rev. Lett. **10**, 516 (1963)

11. C.D. Lin: Adv. At. Mol. Phys. **22**, 77 (1986)

12. J.M. Rost and J.S. Briggs: J. Phys. **B24**, 4293 (1991)

13. D.R. Herrick: Adv. Chem. Phys. **52**, 1 (1983)

14. H.R. Sadeghpour and C.H. Greene: Phys. Rev. Lett. **65**, 313 (1990)

15. J.M. Rost, J.S. Briggs and J.M. Feagin: Phys. Rev. Lett. **66**, 1642 (1991)

16. J.M. Rost et al.: J. Phys. **B30**, 4663 (1997)

17. M. Gutzwiller: In: *Chaos in Classical and Quantum Mechanics*. New York: Springer 1990

18. K. Richter et al.: J. Phys. **B25**, 3929 (1992)

19. G.S. Ezra et al.: J. Phys. **B24**, L413 (1991)

20. G. Tanner and D. Wintgen: Phys. Rev. Lett. **75**, 2928 (1995).

21. K. Richter and D. Wintgen: J. Phys. **B23**, L197 (1990)

22. A. Bürgers, D. Wintgen and J.M. Rost: J. Phys. **B28**, 3163 (1995).

23. H. Friedrich and D. Wintgen: Phys. Reports **183**, 37 (1989)

24. Y. Qiu, J. Müller and J. Burgdörfer: Phys. Rev. **A54**, 1922 (1996)

25. B. Grémaud and P. Gaspard: J. Phys. **B31**, 1671 (1998)

26. J.M. Rost and G. Tanner: In: *Classical, Semiclassical and Quantum Dynamics in Atoms*, eds. H. Friedrich and B. Eckhardt (Lecture Notes in Physis 485, p. 274). Berlin: Springer 1997

Few-Body Systems Suppl. 10, 263–272 (1999)

Few-
Body
Systems
© by Springer-Verlag 1999

Few-Body Approach to Diffraction of Small Helium Clusters by Nanostructures

Gerhard C. Hegerfeldt[1][*], Thorsten Köhler[2][,1][†]

[1] Institut für Theoretische Physik, Universität Göttingen, Bunsenstr. 9, D-37073 Göttingen, Germany
[2] Max-Planck-Institut für Strömungsforschung, Bunsenstr. 10, D-37073 Göttingen, Germany

Abstract. We use few-body methods to investigate the diffraction of weakly bound systems by a transmission grating. For helium dimers, He_2, we obtain explicit expressions for the transition amplitude in the elastic channel.

1 Introduction

In recent years typical optics experiments have been carried over to atoms, among them diffraction by double slits [1] and by transmission gratings which were produced by nanostructure techniques [2, 3]. The usual wave-theoretical methods of classical optics (Huygens, Kirchhoff) seem to give excellent descriptions of these experiments [4] so that apparently not much more than a good textbook on classical optics seems to be needed.

This simple picture, however, has changed. Recently Schöllkopf and Toennies [2, 5, 6] have performed impressive diffraction experiments with helium dimers and helium clusters consisting of up to 26 atoms. A typical diffraction pattern is shown in Fig. 1. The helium dimer He_2, discovered a few years ago [7, 2], has an extremely low binding energy of -0.11 μeV [8], with no excited states. The binding energy is much smaller than the incident kinetic energy of typical experiments. Its diameter is estimated to be about 6 nm [9]. Present day slit widths are as low as 50 nm and a further reduction is expected.

When a system is observed at a nonzero diffraction angle it clearly has received a lateral moment transfer from the grating. For weakly bound systems like He_2 or higher clusters this might induce breakup processes which in turn might change the diffraction pattern. Similarly, the size of the bound system

---

[*]e-mail: hegerf@theorie.physik.uni-goettingen.de
[†]e-mail: koehler@theorie.physik.uni-goettingen.de

may have an influence. If so, what will be the effect? Will the diffraction pattern change drastically? Will the diffraction peaks still be at the same locations as for point particles? These questions will be investigated and answered in the following, with emphasis on He_2. More technical details can be found in ref. [10].

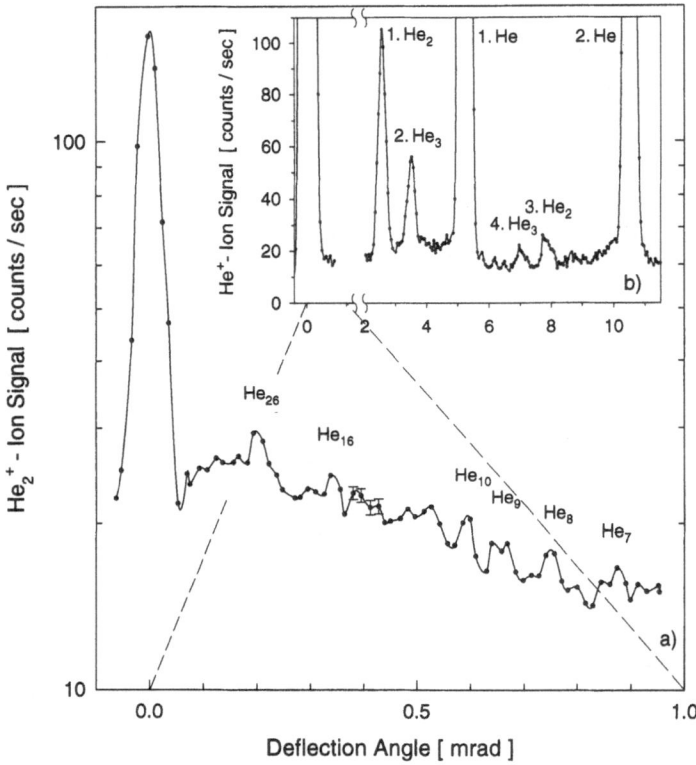

Figure 1. Diffraction of helium clusters for a source temperature of 6 K by a 100-nm-period transmission grating [6].

Diffraction by a transmission grating really means scattering by the bars of the grating, and to include breakup processes and finite-size effects we use multichannel scattering theory. We describe the effect of the grating bars by means of a short-range repulsive ('reflective') potential. The possible role of attractive parts to the potential will be discussed at the end. As usual we neglect the electronic degrees of freedom.

We thus consider a weakly bound system of *two* particles and its scattering by an *external potential*. Due to the presence of the external potential this problem cannot be reduced to a one-particle problem but has more resemblance, at least mathematically, to a three-body problem. In fact, we will show that the Faddeev approach to the three-body problem in its formulation by Alt, Grassberger, and Sandhas [11] can be carried over and adapted to the two-body problem with external potential.

2   AGS Equations for Two Particles with External Potential

In this section we consider a bound two particle system ('molecule') scattered by a quite general obstacle. The coordinates are $\mathbf{x}_1$ and $\mathbf{x}_2$, the binding potential is $V(\mathbf{x}_1 - \mathbf{x}_2)$, with negative binding energies E_γ and bound states $\phi_\gamma(\mathbf{x})$. The external (obstacle) potential is

$$W(\mathbf{x}_1, \mathbf{x}_2) = W_1(\mathbf{x}_1) + W_2(\mathbf{x}_2) \tag{1}$$

where W_1 and W_2 are short-range and strongly repulsive, representing Dirichlet boundary conditions in the limit. For dimers one has $W_1 = W_2$.

We first consider the process $|\mathbf{P}', \phi_{\gamma'}\rangle \to |\mathbf{P}, \phi_\gamma\rangle$ from an incoming molecule of momentum $\mathbf{P}'$ and internal state $\phi_{\gamma'}$ to an outgoing molecule of momentum $\mathbf{P}$ and internal state ϕ_γ, i.e. no breakup. We introduce the six-dimensional Green's operators $G_0(z) = (z - H_0)^{-1}$, $G(z) = (z - H_0 - V - W)^{-1}$, $G_V(z) = (z - H_0 - V)^{-1}$, and similarly $G_W(z)$, where H_0 is the free Hamiltonian (kinetic energy). One has the usual resolvent equations

$$G = G_V + G_V W G \tag{2}$$
$$G = G_W + G_W V G. \tag{3}$$

The T matrices for V and W are defined as

$$T_V(z) \equiv V + V G_V(z) V \tag{4}$$
$$T_W(z) \equiv W + W G_W(z) W .$$

Writing

$$T_V = V G_V (G_V^{-1} + V) = V G_V G_0^{-1} \tag{5}$$

and similarly for W gives

$$T_V G_0 = V G_V \tag{6}$$
$$T_W G_0 = W G_W .$$

We now introduce operators $U_{VV}(z)$ and $U_{WV}(z)$ through

$$G(z) = G_V(z) + G_V(z) U_{VV}(z) G_V(z) \tag{7}$$
$$G(z) = G_W(z) U_{WV}(z) G_V(z). \tag{8}$$

As in ref. [11] one easily shows that the transition amplitude for the above process is

$$t(\mathbf{P}, \phi_\gamma; \mathbf{P}', \phi_{\gamma'}) = \langle \mathbf{P}, \phi_\gamma | U_{VV}(E + \mathrm{i}0) | \mathbf{P}', \phi_{\gamma'}\rangle . \tag{9}$$

Equating Eqs. (2) and (7),

$$G_V + G_V W G = G_V + G_V U_{VV} G_V, \tag{10}$$

and inserting Eq. (8) gives

$$U_{VV} = W G_W U_{WV} = T_W G_0 U_{WV} \tag{11}$$

Similarly, equating Eqs. (3) and (8) and inserting Eq. (7) gives

$$U_{WV} = G_0^{-1} + T_V G_0 U_{VV}. \tag{12}$$

The last two equations are the AGS equations for two particles interacting through a potential V and with an additional external potential W. These AGS equations decouple here, as seen by insertion of Eq. (12) into Eq. (11), which gives

$$U_{VV} = T_W + T_W G_0 T_V G_0 U_{VV}. \tag{13}$$

This equation is exact and contains both breakup and finite-size effects. It lends itself to iteration, and to lowest order in T_V one has

$$U_{VV} \cong T_W . \tag{14}$$

By Eq. (9) this gives

$$t(\mathbf{P}, \phi_\gamma; \mathbf{P}', \phi_{\gamma'}) \cong \langle \mathbf{P}, \phi_\gamma | T_W(E + \mathrm{i}0) | \mathbf{P}', \phi_{\gamma'} \rangle. \tag{15}$$

Although T_W is the transition operator for scattering of two asymptotically free particles by the external potential W it is not on the energy shell in Eq. (15). Here the interaction V is taken into account through the wave-functions ϕ_γ and $\phi_{\gamma'}$.

Similarly to Eq. (9) one can show that

$$U_{0V} \equiv G_0^{-1} + T_V G_0 U_{VV} + T_W G_0 U_{WV} \tag{16}$$

is the transition operator for breakup processes, and for this one only has to know U_{VV} and U_{WV}.

3 Application to Diffraction of Helium Dimers

The scattering amplitude in Eq. (15) can be evaluated in case of small diffraction angles for a transmission grating of N bars of period d and slit width s [10]. We consider a bound system of two particles, each of mass m, and normal incidence. With P the total momentum of the system and P_2 the lateral momentum transfer, the elastic scattering amplitude becomes of the form

$$t(\mathbf{P}, \phi_\gamma; \mathbf{P}', \phi_\gamma) = t_{\mathrm{coh}}(P_2) + t_{\mathrm{incoh}}(P_2). \tag{17}$$

The first term is a superposition of amplitudes from individual bars, while the second results from interactions with separate bars and it can be shown to be *negligible* if $d - s$ and s are larger than the molecule. The coherent part is of the form

$$t_{\text{coh}}(P_2) = t_{\text{bar}}^{\text{mol}}(\gamma, P/M; P_2)\frac{\sin(P_2 N d/2\hbar)}{\sin(P_2 d/2\hbar)}\delta(P_3) \tag{18}$$

where the first factor is the molecular transition amplitude for a single bar of width $d - s$, for total momentum P and total mass M. The second factor is the usual sharply peaked grating function known from optics, and it gives the diffraction peaks at the same locations as for point-particles with corresponding lateral momentum transfer. The δ function is due to the assumed infinite extent of the grating in the 3 direction. The amplitude of the diffraction peaks is determined by $t_{\text{bar}}^{\text{mol}}$. The latter has been explicitly calculated by means of Eq. (15) in ref. [10] and it depends only on the absolute value of the ground-state wave-function. One obtains

$$t_{\text{bar}}^{\text{mol}}(\gamma, P/M; P_2) = -\frac{2iP}{(2\pi)^2 M}\frac{\sin[P_2(d-s)/2\hbar]}{P_2}\int d^3x\, 2\, e^{iP_2 x_2/2\hbar}|\phi_\gamma(\mathbf{x})|^2 \tag{19}$$

$$+ \frac{4iP}{(2\pi)^2 M}\int dx_1 dx_3 \int_0^{d-s} dx_2 |\phi_\gamma(\mathbf{x})|^2 \sin\left[\frac{P_2}{2\hbar}(d - s - x_2)\right]/P_2\,.$$

The expression preceding the first integral is the single-bar amplitude for a point particle of mass M and total momentum P. The complete expression on the r.h.s. reduces to this if $|\phi_\gamma(\mathbf{x})|^2$ contracts to a point.

In Fig. 2 we have plotted $|t_{\text{bar}}^{\text{mol}}|^2$ for He$_2$ and the corresponding quantity for a point particle. The bar width is 25 nm. For a symmetric grating of period 50 nm the vertical lines in the inset indicate where the peaks of the grating functions cut out the diffraction peaks of first order, second order and so on. For a symmetric grating and point particles there are no even-order diffraction peaks since the single-bar amplitude vanishes there, as seen in the inset. For the dimer this is not true, and it therefore has small even-order peaks which become more pronounced for smaller bar and slit widths.

Intuitively we expect that finite-size effects of He$_2$ might be partially taken into account by a larger bar and smaller slit width. Figure 3 shows that the single-bar amplitude of He$_2$, for a bar width of 25 nm, can be approximated for small lateral momentum transfer by that of a point particle for bar width of $(25 + 2.8)$ nm. For a grating this would mean a nonsymmetric grating of correspondingly smaller slit width. The expectation value of $|x_2|$ for the ground-state wave-function of He$_2$ we are using is just 2.8 nm. A cumulant expansion of $t_{\text{bar}}^{\text{mol}}$ shows that the equality of these two numbers is not a coincidence [12].

4 Location of the Diffraction Peaks

Can one understand more directly and intuitively why the diffraction peaks occur at the same locations as for point particles of the same lateral momentum transfer? This is easy to see if one accepts that t_{incoh} in Eq. (17) is negligible and that one just has to add the amplitudes from the individual bars. Indeed, the potential of the nth bar is the translate

$$\exp\{-in\hat{P}_2 d/\hbar\}W_{\text{bar}}\exp\{in\hat{P}_2 d/\hbar\} \tag{20}$$

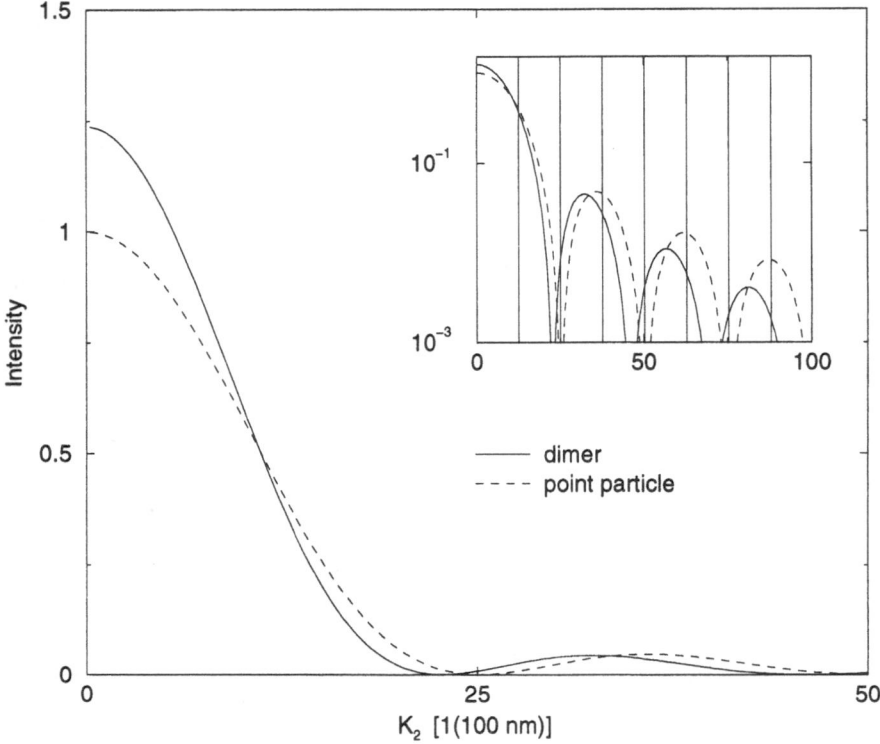

Figure 2. Diffraction by a single bar of width 25 nm for helium dimers and point particles. For a symmetric grating with this bar and slit width the vertical lines in the inset pick out the corresponding diffraction peaks. For dimers there are even order peaks which are absent for point particles. The lateral momentum transfer is $P_2 = \hbar K_2$.

of the first bar potential, and similarly for the nth bar transition amplitude $U_{VV}^{(n)}$. Since $|\mathbf{P}, \phi_\gamma\rangle$ is an eigenvector of the momentum operator $\hat{P}_2$, the associated amplitude differs just by the phase factor $\exp\{-inP_2d/\hbar\}$ from that of the first bar. Summation over n leads to a sum over the phase factors, resulting in the grating function and Eq. (18).

There is also a symmetry argument for an infinite grating, exact to all orders and holding for larger clusters, too. For an infinite grating the complete Hamiltonian has period d under translations in the 2 direction. Hence the Hamiltonian as well as the associated time-development operator $U(t)$ commute with $\exp\{-i\hat{P}_2d/\hbar\}$. As a consequence

$$\langle \mathbf{P}, \phi_\gamma | U(t) | \mathbf{P}', \phi_{\gamma'} \rangle$$

must vanish unless

$$P_2/\hbar = \frac{2\pi}{d}n \,, \qquad n = 0, \pm 1, \cdots \tag{21}$$

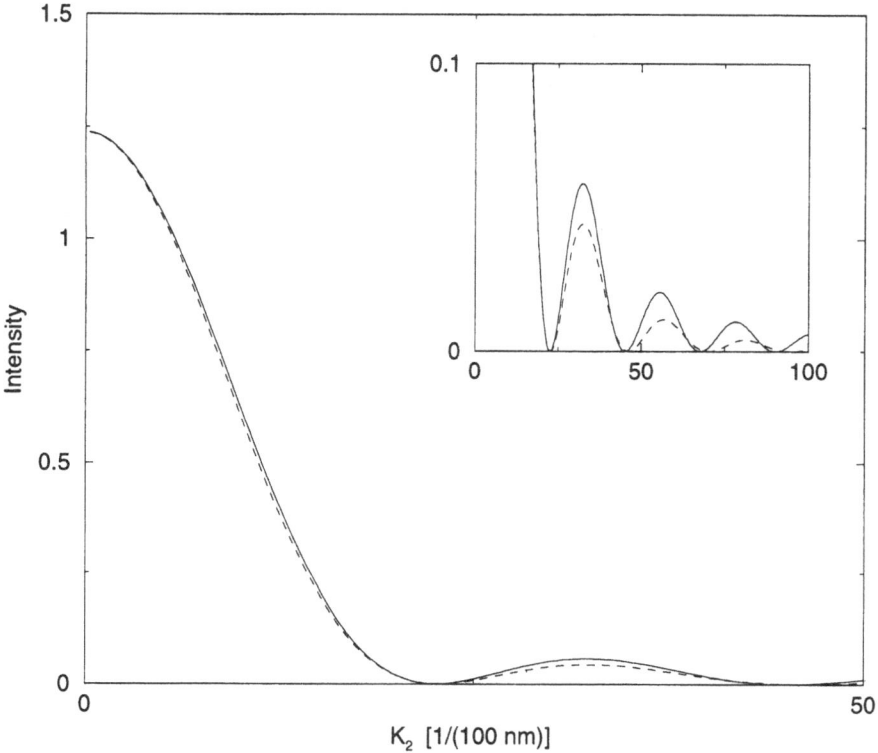

Figure 3. Diffraction of helium dimers (solid line) by a single bar of width 25 nm and diffraction of point particles (dashed line) by a single bar of width (25 + 2.8) nm. There is agreement for small lateral momentum transfer $P_2 = \hbar K_2$.

and thus there are only transitions in these directions.

We recall that experimentally one always has a certain momentum spread and that only a part of the grating is illuminated by the incoming beam. To take this into account one can modify the above plane-wave formulation and it allows one to let N become arbitrarily large [12]. The momentum spread then just leads to a broadening of the diffraction peaks. Incidentally, this procedure yields the correct height for the zeroth-order peak which otherwise would be misrepresented by a finite grating illuminated by an infinite plane wave. Figures 4 and 5 show some results.

5 Further Developments

Recent experiments by Schöllkopf and Toennies [13] have shown that an additional attractive potential may have an unexpectedly large influence on the diffraction patterns for higher noble atoms, increasing from neon to argon to krypton, in line with the increasing polarizability of these atoms. This attractive potential will also influence the diffraction pattern of bound systems to

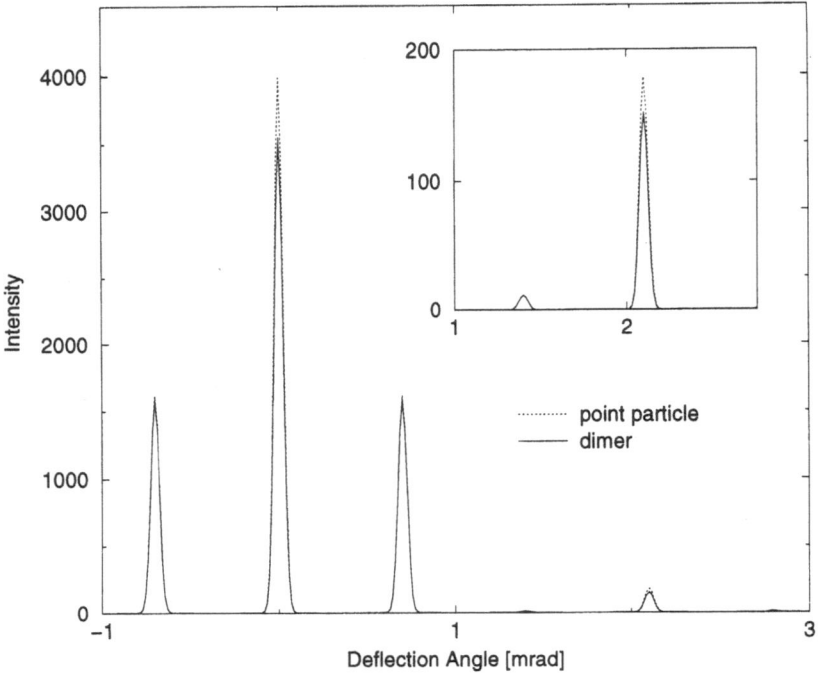

Figure 4. Diffraction by a symmetric grating with slit width 50 nm for helium dimers (solid line) and for corresponding point particles (dotted line). The differences are small, the second order peak barely visible.

some extent.

In a first step we have taken this additional surface potential into account for diffraction of atoms. This can be done by using similar resolvent methods as in Sect. 2, with the surface potential treated as a perturbation. In the course of this something interesting happened. At first our results did not even qualitatively agree with the experiments for higher noble atoms – in particular the hierarchy of the peaks did not come out correctly – until we learned that due to the etching process the cross section of the grating bars was not rectangular as assumed by us but rather trapeze-like, with the angle off the perpendicular by 8^0. Using this geometry agreement with the experiments was then obtained [12]. This, incidentally, shows the sensitivity of the theoretical methods.

The next step is to take the attractive surface potential into account for helium dimers. This work is presently in progress.

Another interesting topic is to use the cumulant expansion mentioned in the last section to obtain some measure of the 'size' of He_2 from diffraction data and to obtain consequences for the binding potential.

In principle it should be possible to carry the approach of Sect. 2 over

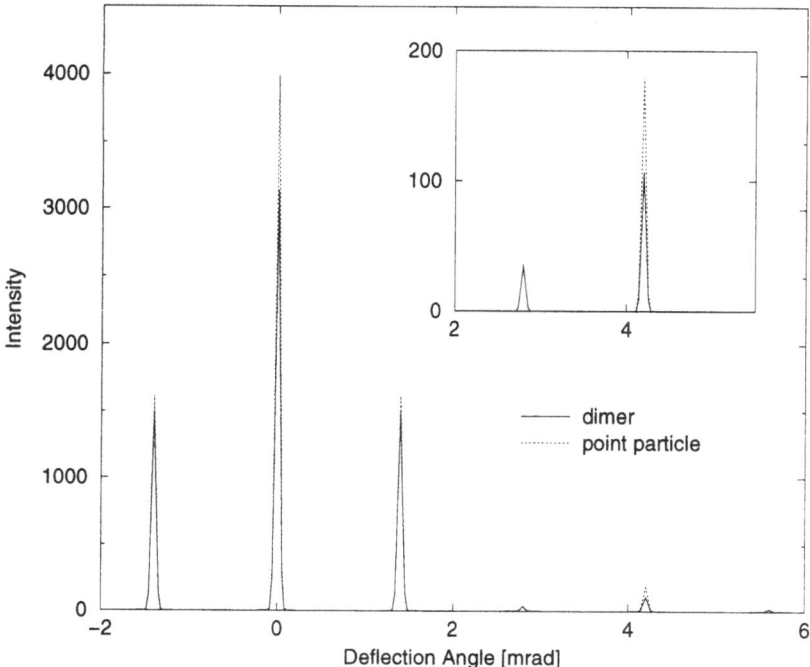

Figure 5. Diffraction by a symmetric grating with slit width 25 nm for helium dimers (solid line) and for point particles (dotted line). The differences are more pronounced, the second order peak is larger.

to three particles with external potential and apply it to the helium trimer, He$_3$. In this connection a very interesting question is whether one can draw any conclusions from He$_3$ diffraction patterns for the theoretically predicted Efimov state of He$_3$ [14, 15].

6 Conclusions

We have used few-body methods to investigate the diffraction of weakly bound two-particle systems by a transmission grating and have obtained explicit expressions for the transition amplitude in the elastic channel. The diffraction pattern is a product of the single-bar amplitude $t_{\text{bar}}^{\text{mol}}$ and the usual grating function. As for point particles our result depends only on $P_2/\hbar = (2\pi/\lambda)\sin\varphi$ and on the bar width. The zeros of $t_{\text{bar}}^{\text{mol}}$ may differ from those for point particles so that even-order diffraction peaks may appear for symmetric gratings.

For large bar and slit widths ($\sim 10 \times$ bound-state diameter) the diffraction pattern is still very close to that for point particles, although minor deviations exist (second order peak). For smaller bar and slit widths ($\sim 5 \times$ diameter) significant deviations occur. The third order peak can be up to 40% lower. This can be attributed to breakups which should increase for larger momentum transfer. There is also an appreciable second order peak which is absent for point particles. This is more difficult to understand by breakups and is rather a finite-size effect since the finite size can be partially taken into account by considering point particles and a nonsymmetric grating with smaller slit widths.

Additional attractive surface potentials which can, and do, modify the diffraction patterns can be taken into account.

Extension to helium trimers seems feasible, with a possible connection to Efimov states.

Acknowledgement. We would like to thank W. Schöllkopf and J.P. Toennies for letting us use some of their as yet unpublished results.

References

1. O. Carnal and J. Mlynek: Phys. Rev. Lett. **66**, 2689 (1991); D.W. Keith et al.: Phys. Rev. Lett. **66**, 2693 (1991)

2. W. Schöllkopf and J.P. Toennies: Science **266**, 1345 (1994)

3. M.S. Chapman et al.: Phys. Rev. Lett. **74**, 4783 (1995)

4. For an overview of atom optics see the special issue: Appl. Phys. **B54** (1992), eds. J. Mlynek, V. Balykin and P. Meystre or C.S. Adams, M. Sigel and J. Mlynek: Phys. Reports **240**, 143 (1994).

5. W. Schöllkopf and J.P. Toennies: J. Chem. Phys. **104**, 1155 (1996)

6. W. Schöllkopf and J.P. Toennies: to be published

7. F. Luo et al.: J. Chem. Phys. **98**, 3564 (1993)

8. K.T. Tang, J.P. Toennies and C.L. Yiu: Phys. Rev. Lett. **74**, 1546 (1995)

9. F. Luo, C.F. Giese and W.R. Gentry: J. Chem. Phys. **104**, 1151 (1995)

10. G.C. Hegerfeldt and T. Köhler: Phys. Rev. **A57**, 2021 (1998)

11. E.O. Alt, P. Grassberger and W. Sandhas: Nucl. Phys. **B2**, 167 (1967)

12. G.C. Hegerfeldt and T. Köhler: in preparation

13. W. Schöllkopf and J.P. Toennies: private communication

14. T. Cornelius and W. Glöckle: J. Chem. Phys. **85**, 3906 (1986)

15. B.D. Esry, C.D. Lin, and C.H. Green: Phys. Rev. **A54**, 394 (1996)

Few-Body Systems Suppl. 10, 273–276 (1999)

Few-
Body
Systems
© by Springer-Verlag 1999

The Borromean Halo Excitations

B. Danilin[1]*, I. Thompson[3], M. Zhukov[4], J. Vaagen[5], S. Ershov[6], J. Bang[2]

[1] RRC The Kurchatov Institute, Moscow, Russia
[2] Niels Bohr Institute, Copenhagen, Denmark
[3] Department of Physics, University of Surrey, Guildford , U.K.
[4] Chalmers University of Technology Göteborg, Sweden
[5] SENTEF, Department of Physics, University of Bergen, Norway
[6] JINR, Dubna, Russia

Abstract. The model, which describes the system as a three-body $core+N+N$ cluster structure and the method of hyperspherical harmonics (HH) both for bound state and continuum has been used to study the low-lying resonances and dipole soft mode in ^{6}He. The structure of the continuum is probed via numerous methods revealing the physical nature for three-body resonances.

1 Introduction

Halo nuclei are a new domain for the application of three-body methods, and show a considerably richer structure than that of the three-nucleon problem. These are nuclei in which there is very weak binding of the last pair of neutrons or protons. There is then a large probability for these neutrons to tunnel into the classically-forbidden regions outside a more tightly bound core. The overall r.m.s. matter radius of the nucleus becomes very large and the momentum distributions following fragmentation of the halo nucleus at high velocity are correspondingly very narrow, and also abnormally large E1 Coulomb breakup cross sections are found (see reviews [1, 2]).

Except for momentum distributions from fragmentation experiments , the only data are from charge-exchange reactions with ^{6}Li to the ^{6}He continuum . Recent experiments on electromagnetic dissociation and inelastic (p,p') reaction revealed the structure in the ^{11}Li excitation spectrum.

The properties of the ground state of the ^{6}He,^{11}Li and ^{14}Be halo nuclei have been examined by a variety of three-body methods [3, 4]. Different treatments

---

E-mail address: danilin@cerber.mbslab.kiae.ru

of the Pauli principle have also been employed, since the halo nucleons must not overlap with occupied states in the core.

A further progress in the theoretical and experimental studies have recently been made by exploring structure of the ^{6}He continua, by considering the full range of aspects of the three-body problem and its solution [5, 6] based on the coordinate-space hyperspherical representation. The scattering in this region is strongly influenced by the neutron-neutron correlation, but it does not appear that this correlation is sufficiently strong to constitute a soft dipole resonance. Compression of the spectrum of continuum resonances to the threshold is predicted, with a wide variety of structures, namely for ^{6}He [5, 6]: well-known 2_1^+ state , 2_2^+ resonance , 1^+ spin-flip resonance, 1^- soft dipole mode and 0_2^+ soft monopole breathing mode .

The structure of the continuum is probed via charge-exchange reactions and inelastic scattering using four-body DWIA theory [7] and electromagnetic dissociation (EMD) cross sections .

The search for persistence of halo structure related to the concept of halo analogue states is also the subject of theoretical [8] and experimental [2] interest.

2 Resonance Properties

The various properties of the ground [3] and three-body scattering states [5] of the halo nuclei were explored in an core + n + n three-body model with realistic interactions for A=6 nuclei and the different plausible structure scenarios for ^{11}Li. These were also successfully employed in our previous calculations for the dipole strength function [9, 10] and inelastic ^{6}He(n, n') as well as charge-exchange (n, p) reactions with ^{6}Li to the ^{6}He continuum [7].

Also the different methods for calculations of three-body continuum using the adiabatic hyperspherical (AH) method [11], Gamov states (complex energy) method [12, 11] and the coordinate complex rotation method [13] were applied to the ^{6}He nucleus. They show some common results such as the compression of continous spectra in comparison with shell-model -like expectations [6] and the existence of 2_2^+, 1^+ and 0^+ resonanses in the energy interval 1-5 MeV, which are not distinguished experimentally yet.

Let us remind in brief the different criteria of the resonance phenomena and the methods discussed in [6] to explore three-body continuum structure.

Since the three-body problem in the HH method has been reduced to the form of a single particle motion in the extremely deformed collective potential (which is projection of sum of binary interactions on the partial components), we have used the well-known methods of coupled channels theory and nuclear reactions to both investigate the exact solution, and to make a physical (qualitative) analysis of the continuum structure. The items B(III)–(VI) concern the intrinsic structure of the continuum, whereas items C(VII)–(VIII) deal with the response generated by transition operators from the ground state.

A. Coupling potentials and shell-like states:

(I) Exploration of the effective three-body interaction in the hyperradius: diagonal parts and adiabatically recoupled interactions;

(II) Dependence on λ of the energy of any bound states created by potential scaling factors $\lambda > 1$ (shell-model-like or Sturmian-like method, giving the relative positions of the possible continuum states);

B. Intrinsic structure of the continuum:

(III) Three-body phase shifts extracted from the diagonal elements of the S-matrix, and eigenphases of the whole S-matrix for $3 \to 3$ scattering;

(IV) Total cross sections for $3 \to 3$ continuum scattering - it could be realized in the very dense matter like supernovae or neutron stars;

(V) Interior norms of the scattering wave functions, reflecting the amplification of wave function in the interaction region in case of 'real' resonance (necessary condition of any resonance);

(VI) A quasi-bound treatment of resonant states, which is a good approximation for narrow resonances, enabling to calculate the widths with a high accuracy;

C. Transition properties:

(VII) Study of the continuum based on response (strength) functions for long range operators of electromagnetic nature and short range (zero range) operators of nuclear origin and different multipolarity.

The three-body problem is more complicated and includes the phenomenon of long-range effective three-body interaction with effective range about sum of scattering lengths in binary subsystems (Efimov effect [14]), which is responsible for the spectra compression near three-body threshold.

The most remarkable feature of the 'true' three-body resonance is that it exists in lowest configurations (with lowest hypermoment), which corresponds to three particles interacting closely to each other.

Comparison of effective three-body interactions for the lowest HH in case of true three-body resonances (2_1^+ , a second 2_2^+ and 1^+) and dipole partial component shows that the pocket in the dipole effective interaction is too small to support the resonance. It shows also three-body resonant structure at both - interior norms and 3-body phase shifts.

In view of the controversy surrounding the concept of a 'soft dipole mode' we wish to carefully distinguish the *continuum structure* from the *continuum response*.

This difference is most dramatic if the binary potentials are set to zero for the continuum wave functions: this is the often-treated no-final-interaction case. Even in this case, there might be relatively narrow peaks in the continuum response from the ground state, especially if a long range transition operator such as $\mathbf{r}$ or $\mathbf{r^2}$ is used, but it is clear that although these peaks reflect the time delay of excitation processes, they should never be identified as intrinsic 'states' with widths and lifetimes etc.

For example, in case of 1^- scattering in ^{6}He, we have to deal with a situation somewhat unfavourable for the HH method. The s-wave Pauli repulsion and

276

p-wave attraction approximately cancel each other, and only the *nn* interaction prevails. Thus we need rather many HH for describing the *nn* correlated pair at large distances.

Our calculations also produced 0^- and 2^- unnatural parity dipole modes, but the associated response functions are two orders of magnitude less than that for dipole excitation. The 3^+ mode is four orders of magnitude less than for the second quadrupole resonance.

Since this continuum structure is concentrated in the vicinity of the very dominant first 2^+ resonance, high resolution experiments with detailed energy - momenta and angular distributions are needed.

References

1. I. Tanihata: J. Phys. **G22**, 157 (1996)

2. P.G. Hansen, A.S. Jensen and B. Jonson: Ann. Rev. Nucl. Part. Sci. **45**, 591 (1995)

3. M.V. Zhukov, B.V. Danilin, D.V. Fedorov et al.: Phys. Reports **231**, 151 (1993)

4. I.J. Thompson, J.S. Al-Khalili, J.M. Bang et al.: Nucl. Phys. **A588**, 59c (1995)

5. B.V. Danilin and M.V. Zhukov: Phys. At. Nucl. **56**, 460 (1993)

6. B.V. Danilin et al.: Nucl. Phys. **A632**, 383 (1998)

7. S.N. Ershov, T. Rogde, B.V. Danilin et al.: Phys. Rev. **C56**, 1483 (1997)

8. M.V. Zhukov et al.: Phys. Rev. **C52**, 2461 (1995)

9. B.V. Danilin et al.: Phys. Lett. **B302**, 129 (1993)

10. B.V. Danilin, I.J. Thompson, M.V. Zhukov et al.: Phys. Lett. **B333**, 299 (1994)

11. A. Cobis, D.V. Fedorov and A.S. Jensen: Phys. Rev. Lett. **79**, 2411 (1997)

12. B.V. Danilin: Izv. AN USSR (ser. fiz.) **54(11)**, 2212 (1990)

13. A. Csoto: Phys. Rev. **C49**, 3035, 2244 (1994); S. Aoyama et al.: Progr. Theor. Phys. **94**, 343 (1995); K. Arai, Y. Suzuki and K. Varga: Phys. Rev. **C51**, 2488 (1995)

14. V.M. Efimov: Comments Nucl. Part. Phys. **19**, 271 (1990)

Few-Body Systems Suppl. 10, 277–280 (1999)

Structure of Atomic Helium Trimers

E. Nielsen, A.S. Jensen and D.V.Fedorov

Institute of Physics and Astronomy, Aarhus University, DK-8000 Aarhus C

Abstract. The structure of the bound states of the atomic helium trimers is calculated by use of hyperspherical coordinates and an adiabatic expansion of the Faddeev equations. The ground state of ^{4}He$_3$ has a triangular shape, the excited state is an Efimov state and the ^{3}He^{4}He$_2$ bound state is a halo state.

Introduction. Weakly bound three-body systems exhibit general properties related to the large-distance behavior of the corresponding wave function [1]. Halos are spatially extended bound systems where the density distributions reach far into classically forbidden regions. Halo structure arises when the binding energy and the orbital angular momentum are small. Efimov states are extreme examples of weakly bound and spatially extended states. They appear as infinitely many bound states in three-body systems where the scattering lengths for at least two of the two-body subsystems are infinitely large. When the condition is approximately fulfilled, only a finite number of these Efimov states would exist.

However, due to the extreme properties even one Efimov state would be interesting to observe in nature. We must then search for two-body systems with very large scattering lengths compared to the range of the potentials. This might occur for neutron dripline nuclei described as two neutrons and a core. The neutron-core scattering length must then be very large and the Efimov conditions are fulfilled. The search is at the moment more advanced in molecular physics where an excited state of the atomic ^{4}He-trimer reveal one of the characteristic feature of an Efimov state, i.e. it disappears both by decreasing and increasing the strength of the potential [2].

Only recently the bound state of the ^{4}He-dimer was experimentally observed although the binding energy is too low to be measured directly. Instead the dimer size is obtained to be $\langle r \rangle = 60 \pm 10$ Å [3, 4] indicating a very large scattering length of about twice the size. The realistic LM2M2 potential suggests a binding energy of 1.31 mK and a size (consistent with the measurement) of 51.9 Å[5]. The purpose of this work [6] is to describe the spatial structure of the bound states of the helium-trimers and in particular to search for halos and Efimov states.

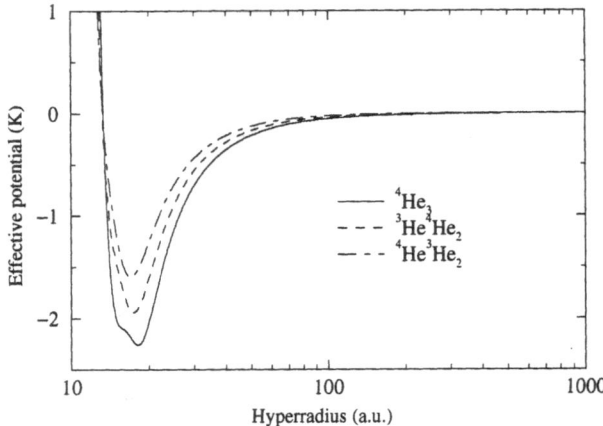

Figure 1. The lowest adiabatic potential as a function of hyperradius for the helium trimers. The potentials only support bound states for ^{4}He$_3$ and ^{3}He^{4}He$_2$ where the asymptotic value (-1.31 mK) is the binding energy of the ^{4}He-dimer. For ^{4}He^{3}He$_2$ the asymptotic value is 0.

Method and results. We solve the Scrödinger equation for three helium atoms interacting with the LM2M2 potential [5]. We use hyperspherical coordinates and an adiabatic expansion of the Faddeev equations. The advantage of using the Faddeev equations instead of the Schrödinger equation is that only components with low relative angular momentum are needed for each Faddeev component – especially at large hyperradius where only s-waves are needed. The advantage of the adiabatic expansion is that the angular variables are defined in finite intervals with a corresponding discrete spectrum. The angular part of the Faddeev equations are then relatively easily solved for each hyperradius.

For large distances only the s-waves are important. We therefore use the corresponding semi-analytical solutions in the validity range of distances larger than a few times the range of the potentials. For small distances we expand each of the Faddeev components on hyperspherical harmonics. The large repulsive core of the LM2M2 potential unfortunately requires a very large basis at small distances. The maximum hyperspherical quantum number in the basis is $K_{max} = 300$ whereas only low relative angular momenta ℓ are necessary. Sufficient accuracy is obtained when we include $\ell \leq 8$ for ^{4}He$_3$ and $\ell \leq 14$ for ^{3}He^{4}He$_2$ and ^{4}He^{3}He$_2$. The size of the angular basis is then in total 840 and 1500, respectively. Restriction to the correct symmetries reduces these basis sizes to 235 and 638. Finally, the coupled set of radial equations are solved with up to 10 adiabatic potentials [1, 6]. The energies are then obtained with a relative accuracy better than 10^{-3}.

The lowest adiabatic potential for each of the three trimers is shown in Fig. 1. The potentials for ^{4}He$_3$, ^{3}He^{4}He$_2$ and ^{4}He^{3}He$_2$ correspond to two, one and zero bound states, respectively. The computed energies of these three bound states are -125.2 mK, -2.269 mK and -13.66 mK, where zero energy

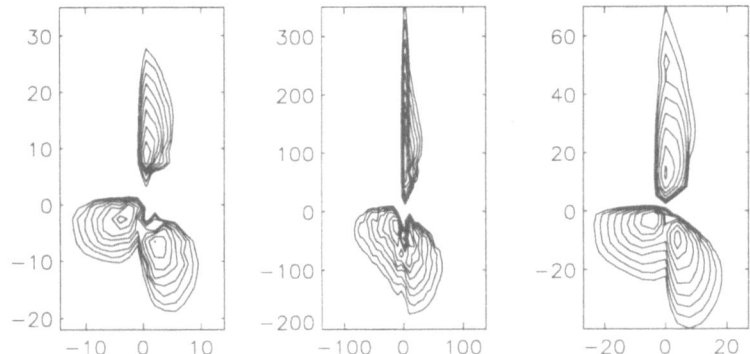

Figure 2. Particle contour diagrams in the internal rest frame for the $^4\text{He}_3$ ground (left) and excited state (middle) and $^3\text{He}^4\text{He}_2$ ground state (right). The distances are in atomic units. There is a factor of 1.5 between each contour.

in all cases corresponds to all three particles infinitely far apart. The related root mean square radii, $\sqrt{\langle r^2 \rangle}$, of the same three bound states are 11.8 a.u., 115 a.u. and 26 a.u. The corresponding value for the ^4He-dimer is 66.9 a.u.

The bound states all have zero total angular momentum and the density distributions are therefore spherically symmetric in the laboratory system. However, the distributions can be computed in the *internal rest frame* defined as the system where the moment of inertia is diagonal. The y-axis corresponds to the lowest and the z-axis to the highest moment of inertia implying that the center of mass of all three particles are in the (x, y)-plane. The signs of the axes are chosen to place one particle in the first quadrant and the other two particles in the lower half-plane. The distributions then become asymmetric.

The particle densities in the internal rest frame are shown in Fig. 2 for the three bound trimer states. The excited state of $^4\text{He}_3$ is 10 times as big as its ground state which essentially corresponds to an equal sided triangle. The excited state is much more elongated with one particle relatively far away from the other two. This structure is characteristic for an Efimov state. It is only at very large distance that the wave function describes a particle far away from a dimer. This configuration, arising due to the dimer bound state, occurs with very small probability. The bound state of $^3\text{He}^4\text{He}_2$ represents an intermediate structure between ground and excited states of $^3\text{He}_4$. This structure is characteristic for a halo state. Comparing the energies and sizes of these states emphasizes the extreme nature of the Efimov states even when only the lowest is present.

The constituent particles are different in the bound states of $^3\text{He}^4\text{He}_2$. The individual density distributions of ^4He and ^3He can then be distinguished and plotted as in Fig. 3. The two distributions are roughly equal except for an additional small component of ^3He at the large distances. In fact the substitution of one ^4He-atom by ^3He in $^4\text{He}_3$ only leads to a relatively small change of

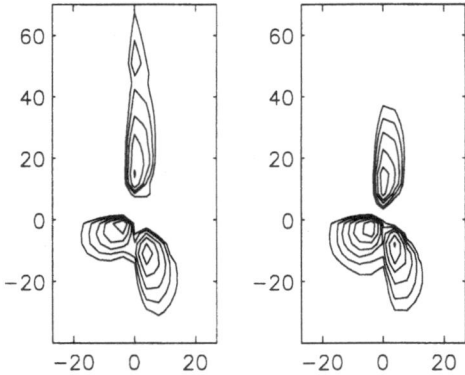

Figure 3. The ^{3}He (left) and half of the ^{4}He (right) density contour diagrams in the internal rest frame for the ^{3}He^{4}He$_2$ ground state. The distances are in atomic units. The contours are at the same densities as in the third plot (right) of Fig. 2.

structure, i.e. a little less binding and a little larger spatial extension. Only the additional small component deviates substantially corresponding to tunneling of the lighter ^{3}He further away from the two ^{4}He-atoms. Thus the structure of the trimer state is not a ^{4}He-dimer and an attached ^{3}He-atom.

Additional structure information is obtained from the expectation value of the ratio between the smallest and the largest interparticle distance $\left\langle \frac{\min r_{ij}}{\max r_{ij}} \right\rangle$. This number is 0.52 for a wave function independent of all angles, i.e. only depending on the hyperradius. For an Efimov state this number is 0.38. For the ground and excited states of ^{4}He$_3$ and ^{3}He^{4}He$_2$ we find 0.51, 0.34 and 0.46, respectively. Thus the excited state of ^{4}He$_3$ resembles an Efimov state in contrast to the two ground states. The marginally lower ratio than for an exact Efimov state is due to the admixture of the configuration with a ^{4}He-atom attached to a ^{4}He-dimer.

References

1. A.S. Jensen et al.: Supplements to Few-Body Systems, this volume

2. B.D. Esry, C.D. Lin and C.H. Greene: Phys. Rev. **A54**, 394 (1996)

3. W. Schöllkopf and J.P. Toennies: J. Chem. Phys. **104**, 1155 (1996)

4. F. Lou, C.F. Giese, and W.R. Gentry: J. Chem. Phys. **104**, 1151 (1996)

5. R.A. Aziz and M.J. Slaman: J. Chem. Phys. **94**, 8047 (1991)

6. E. Nielsen, D.V. Fedorov and A.S. Jensen: J. Phys. **B** in press

Few-Body Systems Suppl. 10, 281–284 (1999)

Few-
Body
Systems
© by Springer-Verlag 1999

A few-body treatment of low energy atomic collisions involving exotic particles

R.A. Sultanov*

Instituto de Física Teórica, UNESP, 01405-900 São Paulo, SP, Brazil

Abstract. Faddeev-type equations are applied to three-charged particle systems. The rather satisfactory results are obtained for low energy e^+H elastic scattering and muonic transfer reactions. The cross sections for antihydrogen formation from antiproton-positronium collisions are calculated using a six state model ($Ps[1s2s2p]$, $\bar{H}[1s2s2p]$).

1. The quantum-mechanical few-body problem plays an important role for adequate description of few-particle systems in nuclear, atomic-antiatomic and molecular physics. Methods developed in this field are based on detailed few-body equations for correct description of the few-particle dynamics. Recently the 3- and 4-body Coulombic systems have attracted considerable interest in relation to the problems of antihydrogen physics [1], muonic transfer reactions in muon catalyzed fusion cycle [2], etc. Therefore, the development of alternative dynamical few-body approaches to systems of 3- and 4-charged particles is actual.

For a 3-charged particle system (*123*) (two positive, *1+* and *2+*, and one negative) only two asymptotic configurations are possible below breakup threshold. This suggests to write down a set of two coupled equations for Faddeev-type components of the system wave function. In this work we shall consider a method based on such an integral differential equation approach.

2. Let r_ξ be a coordinate, and m_ξ be a mass of the ξ-th particle ($\xi = 1, 2, 3$). Taking the system of units to be $e = \hbar = m_3 = 1$, let us introduce Jacobi coordinates

$$r_{j3} = r_3 - r_j, \qquad \rho_k = (r_3 + m_j r_j)/(1 + m_j) - r_k, \qquad j \neq k = 1, 2. \quad (1)$$

Where energies are below the three-body breakup threshold, two cluster asymptotic configurations are possible in the system (*123*), that is (*23*) $-$ *1* and (*13*) $-$ *2*, each being determined by their own Jacobi coordinates (r_{j3}, ρ_k).

---

*E-mail address: sultanov@ift.unesp.br

Because of this, in the process of the Faddeev division for the three-body wave function Ψ, one can confine oneself only to two components rather than three ones ,

$$\Psi = \Psi_1(r_{23}, \rho_1) + \Psi_2(r_{13}, \rho_2), \qquad (2)$$

where $\Psi_1(r_{23}, \rho_1)$ is quadratically integrated over the variables r_{23}, whereas $\Psi_2(r_{13}, \rho_2)$ is over the variables r_{13}.

To define Ψ_l $(l = 1, 2)$, a set of two coupled equations is written down

$$(E - H_0 - V_{23} - U_1)\Psi_1(r_{23}, \rho_1) = (V_{23} + V_{12} - U_2)\Psi_2(r_{13}, \rho_2)$$
$$(E - H_0 - V_{13} - U_2)\Psi_2(r_{13}, \rho_2) = (V_{13} + V_{12} - U_1)\Psi_1(r_{23}, \rho_1) , \qquad (3)$$

where H_0 is the kinetic energy operator of the three-particle system; $V_{ij}(r_{ij})$ are pair interaction potentials $(i \neq j = 1, 2, 3)$; E is the total energy of the system; $U_i(\rho_i)$, $(i = 1, 2)$ are supplementary asymptotic potentials, for instance, polarization interaction in the asymptotic region. Equations constructed in this way satisfy the Schrödinger equation exactly. For the energies below the three-body breakup threshold, they possess the same advantages as the Faddeev equations, since they are formulated for the wave function components with correct physical asymptotics. In the following, these equations will be referred to as Faddeev-Hahn-type equations (FH) or Faddeev-type equations.

In the general case, a component of the three-body wave function has the asymptotic form which includes all open channels: elastic/inelastic, transfer and breakup [3]. In this work, we shall use an approximation, where each component of the total wave function corresponds just to one definite channel: for the elastic/inelastic channel

$$\Psi_1(r_{23}, \rho_1) \underset{\rho_1 \to +\infty}{\sim} e^{ik_1 z}\varphi_1(r_{23}) + \sum_n^\infty A_n^{el/ex}(\Omega_{\rho_1})\frac{e^{ik_n \rho_1}}{\rho_1}\varphi_n(r_{23}), \qquad (4)$$

and for the transfer channel

$$\Psi_2(r_{13}, \rho_2) \underset{\rho_2 \to +\infty}{\sim} \sum_m^\infty A_m^{tr}(\Omega_{\rho_2})\frac{e^{ik'_m \rho_2}}{\rho_2}\varphi_m(r_{13}), \qquad (5)$$

it is easy to see that the asymptotic behaviour of the total wave function (2) becomes similar to Merkuriev's asymptotic [3]. Such an approximation allows to simplify the solution procedure and simultaneously to provide a correct asymptotic behaviour of the solution below the 3-body breakup threshold.

Let us write down Eq. (3) in terms of the adopted notations

$$\left[E + \frac{\triangle_{\rho_k}}{2M_k} + \frac{\triangle_{r_{j3}}}{2\mu_j} - V_{j3} - U_i\right]\Psi_i(r_{j3}, \rho_k) =$$
$$(V_{j3} + V_{jk} - U_{i'})\Psi_{i'}(r_{k3}, \rho_j) , \qquad (6)$$

where $i \neq i' = 1, 2$ and $M_k^{-1} = m_k^{-1} + (1 + m_j)^{-1}$, $\mu_j^{-1} = 1 + m_j^{-1}$. Substituting expansion

$$\Psi_i(r_{j3}, \rho_k) = \sum_{LM\lambda l} \sum_n \frac{1}{\rho_k} f_{nl\lambda}^{(i)LM}(\rho_k) R_{nl}^{(i)}(r_{j3}) \{Y_\lambda(\hat{\rho}_k) \otimes Y_l(\hat{r}_{j3})\}_{LM} \qquad (7)$$

into (6), multiplying this by the appropriate biharmonic functions and the corresponding functions $R_{nl}^i(r_{j3})$, integrating over the corresponding angular coordinates of the vectors r_{j3} and ρ_k and $r_{j3}^2 dr_{j3}$ yield a set of integral differential equations for the unknown functions $f_{nl\lambda}^i(\rho_i)$ [4]

$$
\left[k_n^2 + \frac{\partial^2}{\partial \rho_i^2} - \frac{\lambda(\lambda+1)}{\rho_i^2} - U_i^p(\rho_i)\right] f_\alpha^i(\rho_i) = g \sum_{\alpha'} \frac{\sqrt{(2\lambda+1)(2\lambda'+1)}}{2L+1}
$$

$$
\int_0^\infty d\rho_{i'} f_{\alpha'}^{i'}(\rho_{i'}) \int_0^\pi d\omega \sin\omega R_{nl}^i(r_{i'3})(V_{i'3}(r_{i'3}) + V_{ii'}(r_{ii'}) - U_2(\rho_{i'}))
$$

$$
R_{n'l'}^{i'}(r_{i3})\rho_{i'}\rho_i \sum_{mm'} D_{mm'}^L(0,\omega,0) C_{\lambda 0 lm}^{Lm} C_{\lambda' 0 l'm'}^{Lm'} Y_{lm}(\nu_i, \pi) Y_{l'm'}^*(\nu_{i'}, \pi) \ . \quad (8)
$$

For the sake of simplicity $\alpha \equiv (nl\lambda)$ are quantum numbers of a three-body state and L is the total angular momentum of the three-body system; $g = 4\pi M_i/\gamma^3$; $k_n^2 = 2M_i(E - E_n^i)$, where E_n^i is the binding energy of the subsystem $(i'3)$ and $\gamma = 1 - m_i m_{i'}/((m_i+1)(m_{i'}+1))$; $D_{mm'}^L(0,\omega,0)$ are the Wigner functions, $C_{\lambda 0 lm}^{Lm}$ the Clebsh-Gordon coefficients, Y_{lm} the spherical functions; ω is the angle between the Jacobi coordinates ρ_i and $\rho_{i'}$, ν_i the angle between $r_{i'3}$ and ρ_i, $\nu_{i'}$ the angle between r_{i3} and $\rho_{i'}$. In addition, let us write down useful formulas for the triangle (123): $\sin\nu_i = (\rho_k/\gamma r_{kj})\sin\omega$ and $\cos\nu_i = (1/\gamma r_{kj})(\beta\rho_i + \rho_k\cos\omega)$. To find a unique solution, Eqs. (8) should be completed by appropriate boundary conditions.

3. The numerical procedure involves the following steps: first we separate off the incoming wave by representing the function of the open channel by $\tilde{f}_{1s}^1(\rho_1) = [f_{1s}^1(\rho_1) - \sin(k_1\rho_1)/k_1]$, the Eqs. (8) go over into inhomogeneous equations and integrals over ρ_i ($i = 1, 2$) are replaced by sums. A set of liner equations for the unknown coefficients $f_{\alpha'}^i(k)$ ($k = 1, N_{max}$) is solved by methods of linear algebra. The Gauss elimination method is employed.

The proposed method is used for calculations of (i) the scattering length and low energy phase shifts of the e^+H elastic scattering; (ii) the rates $\lambda_{tr} = \sigma_{tr} v N_0$ (with v being the relative velocity of the incident particles and N_0 the density of the hydrogen medium) of the low energy muonic transfer reactions $(h\mu)_{1s} + t \to (t\mu)_{1s} + h$, where h is a proton or a deuteron and t is a triton; (iii) the antihydrogen atom formation from antiproton-positronium collision $\bar{p} + (e^+e^-) \to \bar{H} + e^-$ using a six state model $(Ps[1s2s2p], \bar{H}[1s2s2p])$.

Our results concerning e^+H elastic scattering are presented in Table 1 together with results of variational calculations [5].

Table 1. e^+H $s-$wave elastic scattering length and low energy phases

k_1	δ, FH	δ [5]	k_1	δ, FH	δ [5]
0.0	-2.1 ± 0.1	-2.103	0.1	0.15	0.1483
0.2	0.19	0.1877	0.3	0.17	0.1677

In case of muonic transfer reactions, our results for the reaction $(d\mu)_{1s} + t \to (t\mu)_{1s} + d$: $\lambda_{dt}^{tr}(0.001eV) = (2.6 \pm 0.2) \times 10^8 s^{-1}$, and for the reaction

$(p\mu)_{1s} + t \rightarrow (t\mu)_{1s} + p$: $\lambda_{pt}^{tr}(0.001eV) = (0.6 \pm 0.2) \times 10^{10}s^{-1}$ are in a good agreement with calculations [6]. One can see from Table 2 that at energies $\sim (0.5eV - 2eV)$ our results for the antihydrogen formation agree fairly well

Table 2. Cross sections of the reaction $\bar{p} + (Ps)_{1s} \rightarrow \bar{H}_{1s} + e^-$

E, eV	$\sigma_{\bar{H}}, \pi a_0^2$	$\sigma_{\bar{H}}[7], \pi a_0^2$	$E(eV)$	$\sigma_{\bar{H}}, \pi a_0^2$	$\sigma_{\bar{H}}[7], \pi a_0^2$
0.05	4.0	2.5	0.5	2.0	3.0
1.0	3.5	3.5	2.0	1.7	3.8

with findings of hyperspherical coupled-channel calculations [7], and at very low energies our cross section $\sigma_{\bar{H}}$ increases in accordance with results of Mitroy (1995) and Humberston (1987) [1].

4. As a test of the method and numerical procedure, the cross section of the $(t\mu)_{1s} + d$ elastic scattering below the $(t\mu)_{n=2}$ threshold was calculated as far as the level spectra and wave functions of the corresponding muonic molecules $(dt\mu)$, $(dd\mu)$ and $(tt\mu)$ within the six state approximation of the close coupling expansion [4].

It is shown that within this formalism, the application of a close-coupling-type ansatz leads to surprisingly good results already in lowest order. The whole procedure leads to a reduction of the usual technical effort and is definitely worth to be used. It seems reasonable to suppose that the method of the integral differential equation, Eqs. (8), should be an effective tool for the description of other atomic-antiatomic and μ-atomic collisions.

Acknowledgement. The author acknowledges the support from the Organizing Committee of the 16th European Conference on Few-Body Problems in Physics (1-6 June, 1998, Autrans, France) and FAPESP, São Paulo, Brazil.

References

1. A. Yu. Voronin and J. Carbonell: Phys. Rev. **A57**, 4335 (1998); J. Mitroy: Phys. Rev. **A52**, 2859 (1995); J. W. Humberston et al.: J. Phys. **B20**, L25 (1987)

2. S. Tresch et al.: Phys. Rev **A57**, 2496 (1998); Y.-A. Thalman et al.: Phys. Rev. **A57**, 1713 (1998)

3. S. P. Merkuriev: Ann. Phys. NY **130**, 395 (1980)

4. R. A. Sultanov: In: *Proceedings of INNOCOM97*, Osaka, Japan (1997). World Scientific Publishing Co (in print)

5. A. K. Bhatia et al.: Phys. Rev. **A3**, 1328 (1971)

6. Y. Kino and M. Kamimura: Hyper. Inter. **82**, 45 (1993)

7. A. Igarashi, N. Toshima and T. Shirai: J. Phys. **B27**, L497 (1994)

Few-Body Systems Suppl. 10, 285–294 (1999)

Few-
Body
Systems
© by Springer-Verlag 1999

Hadronic Probes of Few-Nucleon Systems

W. Tornow[1], D.E. González Trotter[1], C.R. Howell[1], F. Salinas[1], H. Witała[2]

[1] Duke University and Triangle Universities Nuclear Laboratory, Durham, NC 27708-0308, USA
[2] Jagellonian University, Cracow, PL-30059, Poland

1 Introduction

The common theme of this talk is "probing the nuclear dynamics at low energies". This subtitle implies the study of nuclear systems at rather large internucleon distances. We will focus our attention on the three-nucleon (3N) system and our goal here is to explore the role of 3N forces in the 3N continuum. The result anticipated from such a study may turn out to be trivial, considering our restriction to low energies, i.e., sizeable internucleon distances. Traditionally, one expects 3N forces in the continuum to be important only at fairly high energies ($E_N > 100$ MeV) where the three nucleons can come closer together. However, experimental evidence, some of which may be controversial, makes 3N-force studies at low energies not only interesting, but also necessary.

2 Three-Nucleon Forces and NN $^1S_0$ Scattering Lengths

The availability of rigorous 3N continuum calculations, especially from the Bochum-Cracow group [1] and more recently also from the Pisa group [2], allows accurate conclusions to be drawn from the comparison of experimental data and 3N calculations. These calculations employ the modern high-accuracy nucleon-nucleon (NN) potential models [3, 4, 5] without resorting to finite-rank approximations. While the Bochum-Cracow group solves the 3N Faddeev equations in the continuum, the Pisa group's approach is based on the Correlated Hyperspherical Harmonic Method which, unlike the momentum-space approach of the Bochum-Cracow group, allows also to incorporate the long-range Coulomb force in 3N continuum calculations.

Experimentally, by far the best way to find out whether 3N forces influence the 3N continuum is to determine from 3N reactions two-nucleon properties, which are known accurately from free NN scattering. If it turns out that the results for certain NN parameters obtained from the analysis of 3N reactions are different from the known NN values, then there is convincing evidence for the action of 3N forces. In all other cases where discrepancies have been found between experimental 3N data and rigorous 3N calculations based on NN forces only, one always has to face the criticism that the discrepancy may be due to some shortcoming of the NN interaction used in the 3N calculations and not due to 3N forces.

The best candidates for such NN parameters are the $^1S_0$ NN scattering lengths. Especially, the neutron-proton (n-p) $^1S_0$ scattering length is extremely well known from free n-p scattering: $a_{np} = -23.749 \pm 0.008$ fm [6]. If the result for a_{np} obtained from the measurement of the n-p final-state interaction (FSI) cross section in the neutron-deuteron (n-d) breakup reaction $n+d \rightarrow n_1+p+n_2$ is different from the result given above, then one clearly can establish the role of 3N forces, i.e., the action of, for example, neutron n_2 on the interacting n_1-p pair. Similarly, the neutron-neutron (n-n) scattering length a_{nn} can be determined from the n-n FSI cross section in the n-d breakup reaction, and its results can be compared to the value $a_{nn} = -18.6 \pm 0.5$ fm [7] obtained from the $\pi^- + d \rightarrow n_1 + n_2 + \gamma$ capture reaction with only two strongly interacting particles in the exit channel. The world average for a_{nn} extracted from kinematically-complete n-d breakup data is $a_{nn} = -16.7 \pm 0.5$ fm [8], in clear disagreement with the value obtained from the π^--d reaction. Historically, this difference has been interpreted as the first evidence for 3N forces in the 3N continuum at low energies [8]. However, it is important to remember that the associated experimental data were analyzed using theoretical methods which in most cases turned out to be not as accurate as assumed at the time the analyses were performed. In addition, some of the experimentalists may have underestimated the size of the error bars associated with their data. Anyway, we tried to reanalyze published kinematically complete n-d breakup data using modern NN interactions in rigorous 3N calculations. However, it was impossible to obtain some of the important experimental details from the relevant publications or from the authors to make an accurate Monte-Carlo simulation of the experimental setups possible. In the case of kinematically-incomplete n-d breakup data most of the experimental details are still available and our reanalyses [9] have changed the average value obtained for a_{nn} from the magnitude and shape of the proton energy distribution from $a_{nn} = -19.0 \pm 0.9$ fm [8] to -15.6 ± 0.4 fm [10]. Again, the former result and its comparison to both the value -18.6 ± 0.5 fm obtained from π^--capture experiments and -16.7 ± 0.5 fm based on kinematically-complete n-d data has been interpreted [8] as additional evidence for the importance of 3N forces in the 3N continuum. However, our reanalyses clearly showed that the accuracy of the theoretical methods used in the past, especially at high neutron bombarding energies, was not adequate. In addition, the experimental accuracy of these, in principle straight-forward experiments, but in practice very difficult measurements, turned out to be not as high as assumed in the past, therefore, raising suspicions also about the accuracy of both experimental data and theoretical analyses of all the published kinematically-complete n-d breakup data. This is the second reason why we performed a new experiment and determined for the first time

a_{np} and a_{nn} simultaneously in a kinematically-complete n-d breakup experiment. The associated data were analyzed using rigorous 3N calculations and modern NN potential models. A measurement of the proton energy spectrum from the n-d breakup reaction is presently underway at TUNL. Hopefully, this experiment and its analysis will produce an accurate result for a_{nn} obtained in kinematically-incomplete geometry as well.

Similarly to a_{np}, the proton-proton (p-p) scattering length a_{pp} can be measured to extremely high precision ($a_{pp} = -7.813 \pm 0.004$ fm [11]), via free p-p scattering, but the influence of the Coulomb interaction is very large and its treatment is model-dependent, resulting in a value for the nuclear part of the p-p scattering length of $a_{pp}^N = -17.3 \pm 0.4$ fm [6]. Although in principle possible, the Pisa group has not yet performed proton-deuteron (p-d) breakup calculations using modern NN interactions. Therefore, the available high-accuracy p-p FSI data obtained from the p-d breakup reaction $p + d \to p_1 + p_2 + n$ have not been analyzed yet with modern theoretical techniques.

3 The TUNL n-p and n-n Scattering Length Experiments

The experimental setup used at TUNL for the simultaneous determination of the n-p and n-n scattering lengths a_{np} and a_{nn}, respectively, from kinematically-complete n-d breakup FSI cross-section data is shown in Fig. 1. Neutrons of mean energy $E_n = 13$ MeV and an energy spread of ± 200 keV are produced via the $^2\text{H}(d,n)^3\text{He}$ reaction in a deuterium gas cell. The neutrons are collimated by a double-truncated heavy metal collimator resulting in a well-defined neutron beam at $0°$. This neutron beam interacts with the liquid scintillator C_6D_{12} center detector (4 cm in diameter and 6 cm high), where recoil deuterons from n-d elastic scattering and protons from the n-d breakup reaction are detected. Elastically scattered neutrons and the lower energy neutrons from the breakup reaction are detected in 20 liquid scintillator detectors labeled #1 to #20 as double coincidences (between the center detector and one neutron detector) and triple coincidences (between the center detector and two appropriate neutron detectors), respectively. The energy of the neutrons is determined from their time-of-flight between the center detector and the neutron detectors of interest. The deuteron and proton energies in the center detector are obtained from the associated pulse heights measured in the center detector. The detectors #1 to #16, symmetrically positioned on the left and right side of the incident neutron beam, and very well shielded from the neutron production gas cell, are used for the n-n FSI cross-section measurements at four production angles ($20.5°$, $28.0°$, $35.5°$, and $43.0°$ in the laboratory system) of the n-n pair. For example, considering the n-n FSI at $43.0°$ on the right side, one of the two neutrons passes through the opening (7.3 cm diameter) of the ring detector #8 and is detected in detector #9 (12.6 cm diameter and 5.1 cm thick) at a distance of 250 cm from the center detector, while the companion neutron of the n-n FSI pair is detected in the active volume (cylinder with inner and outer diameter of 7.6 cm and 13.4 cm, respectively, and 4 cm deep) of the ring detector #8 at a distance of 150 cm from the center detector. The individual detectors #8 and #9 are also used to measure the yield for elastic n-d scattering events. The n-d elastic cross section is used together with the accurately determined neutron detector efficiencies to obtain absolute FSI cross-section data.

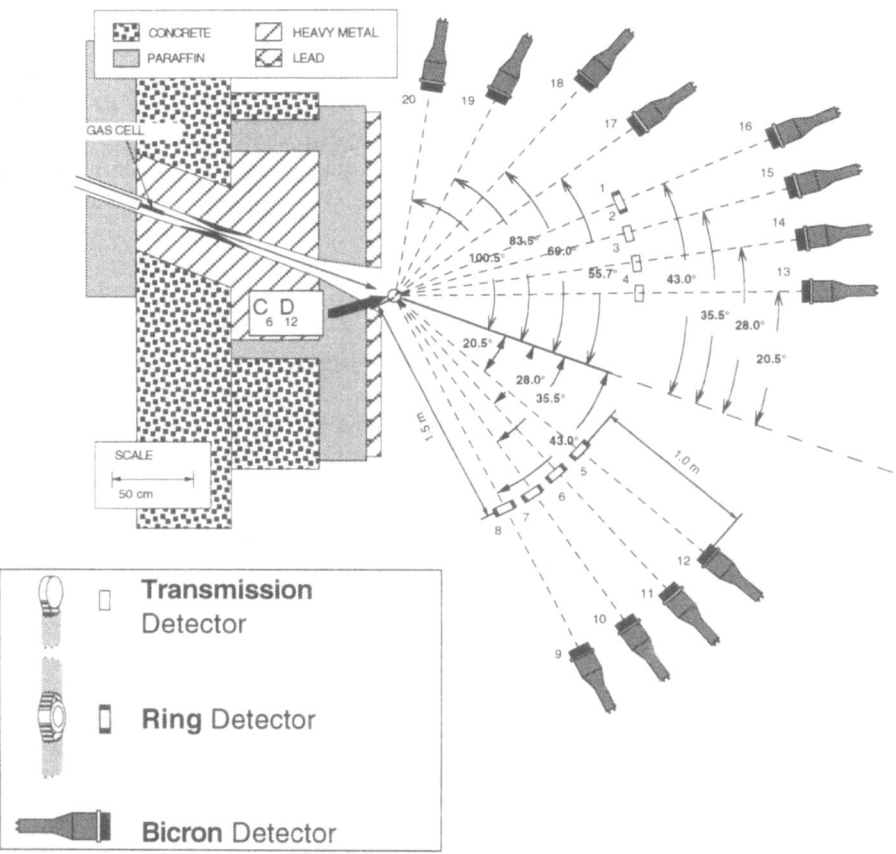

Figure 1. Experimental setup of the TUNL n-p and n-n scattering lengths determination using kinematically complete n-d breakup data.

The reason for measuring the n-n FSI cross section at four different production angles of the n-n pair is based on the theoretical predictions that the action of a possible 3N force must result in a slightly modified angular dependence of the FSI cross section and therefore of a_{nn}. More precisely, at 43° the effect of the Tucson-Melbourne (TM) 3N force [12] on the FSI cross section is predicted to be zero, while at smaller production angles of the n-n pair, especially at 20.5°, 3N force effects may be measurable. At the angle of 43° (for $E_n = 13$ MeV) the n-n FSI cross section is practically identical for any NN interaction used in the rigorous 3N calculations, thus making this angle ideal for obtaining a gauge for a_{nn} obtained from the n-d breakup reaction. The detectors #1 to #4 and #13 to #16 on the left side of the incident neutron beam are used to improve the statistical accuracy of the n-n FSI cross-section measurements.

In the case of a_{np} and the n-p FSI cross-section measurements, we detect the neutron of the n-p pair emitted at 43° in either the ring detector #8 or the detector #9, while the proton's energy is measured in the center detector, as in the case of a_{nn}. The associated second neutron is detected in detector #17 at 55.7° on the left side of the incident neutron beam. Like for a_{nn}, we are interested in a possible angular dependence of the extracted value for a_{np}. Therefore, we study the n-p FSI at four production angles of the n-p pair. Again, at 43° the influence of the TM 3N force is predicted to be zero and NN potential models give practically the same result for the FSI cross section. The predicted 3N-force effect is again largest at the smallest angle (20.5°) of our experimental setup. Contrary to the a_{nn} measurements, for the a_{np} determinations we have not duplicated neutron detectors also on the right side of the incident neutron beam (in this case, detectors #17 to #20), because the statistical accuracy is sufficient due to the fact that we are using already two neutron detectors for each individual n-p pair production angle on the right side. It is important to point out that the n-p and n-n FSI cross-section measurements share the neutron detectors #6 to #11. One of the neutron energies associated with the n-p pair emitted at 20.5° is very small and therefore close to the detection threshold of our neutron detectors, thus making a reliable determination of the n-p FSI cross section at this angle impossible.

If one represents the two neutron energies E_{n_1} and E_{n_2} in a two-dimensional plot, then ideally, all n-d breakup events fall on a well defined kinematical point-geometry locus, called S curve, as shown in Fig. 2 (left side) for the n-n FSI configuration at $\theta_{nn} = 28.0°$. However, due to the extended size of the center detector and the neutron detectors as well as due to the energy spread of the incident neutron beam, in practice all valid events are distributed around the ideal, point-geometry locus. Therefore, we project the measured events on the point-geometry locus. We do exactly the same in our sophisticated Monte-Carlo simulation of the entire experimental setup, where we use 3N cross sections obtained from rigorous 3N calculations using a modified version of the Bonn B NN potential [13] with properly adjusted values for the individual σ-meson coupling constants to provide a range of different values for both a_{np} and a_{nn}. Fig. 2 (right side) shows a 3N calculation for the n-n FSI cross section at $\theta = 28.0°$ along the ideal S curve (starting at $E_2 = 0$) for the n-d breakup reaction at an incident neutron energy of $E_n = 13$ MeV. The NN potential used in this calculation is Bonn B with $a_{nn} = -17.7$ fm. The dashed curve is the result obtained from the Monte-Carlo simulation of our experimental setup using a detailed, Bonn B based 3N cross-section library covering finite angular range and energy spread of the experimental arrangement.

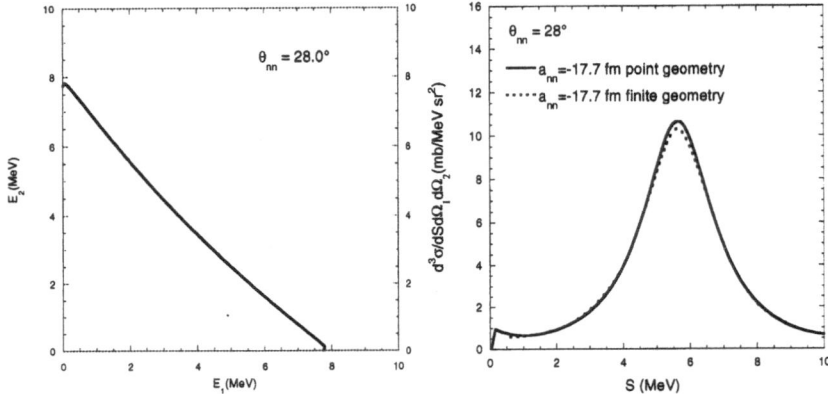

Figure 2. Left side: Kinematic locus (S curve) for n-n FSI events at $\theta_{nn} = 28°$ (lab) using the reaction $n + d \rightarrow n_1 + p + n_2$ at $E_n = 13.0$ MeV. Right side: Point geometry (solid curve) and finite geometry (dashed curve) cross section along the S curve for the n-d breakup reaction at $\theta_{nn} = 28°$ using 13 MeV incident neutrons.

4   Results for a_{np}

Fig. 3 represents the measured n-p FSI cross sections using the ring detectors at our three production angles of the n-p pair in comparison to finite-geometry, rigorous 3N calculations using different values for a_{np}. A value for a_{np} is determined at each angle from a single-parameter χ^2 analysis of the cross-section data and the Monte-Carlo simulations. Our results obtained for a_{np} from the 6 individual measurements are given in Table 1. As can be seen, there is no significant angle dependence in the value of a_{np} obtained from our data. The main sources of the ± 0.8 fm systematic uncertainty are the uncertainties in absolute neutron detection efficiency and the integrated target-beam luminosity. Combining the statistical and systematic uncertainties in quadrature, we obtain a final value for $a_{np} = -23.5 \pm 0.8$ fm. Our result is in agreement with the value of a_{np} (-23.749 ± 0.008 fm) obtained from 2N-scattering data and can be used to set an upper limit on the effective shift in the n-p $^1S_0$ potential strength due to 3N-force contributions.

Table 1. Results of TUNL a_{np} experiment.

Angle (°)	a_{np} (fm)	Error (fm)
43.0	-23.6	± 0.3
35.5	-23.2	± 0.3
28.0	-23.7	± 0.3
Average	-23.5	± 0.2 (*stat*)
		± 0.8 (*syst*)

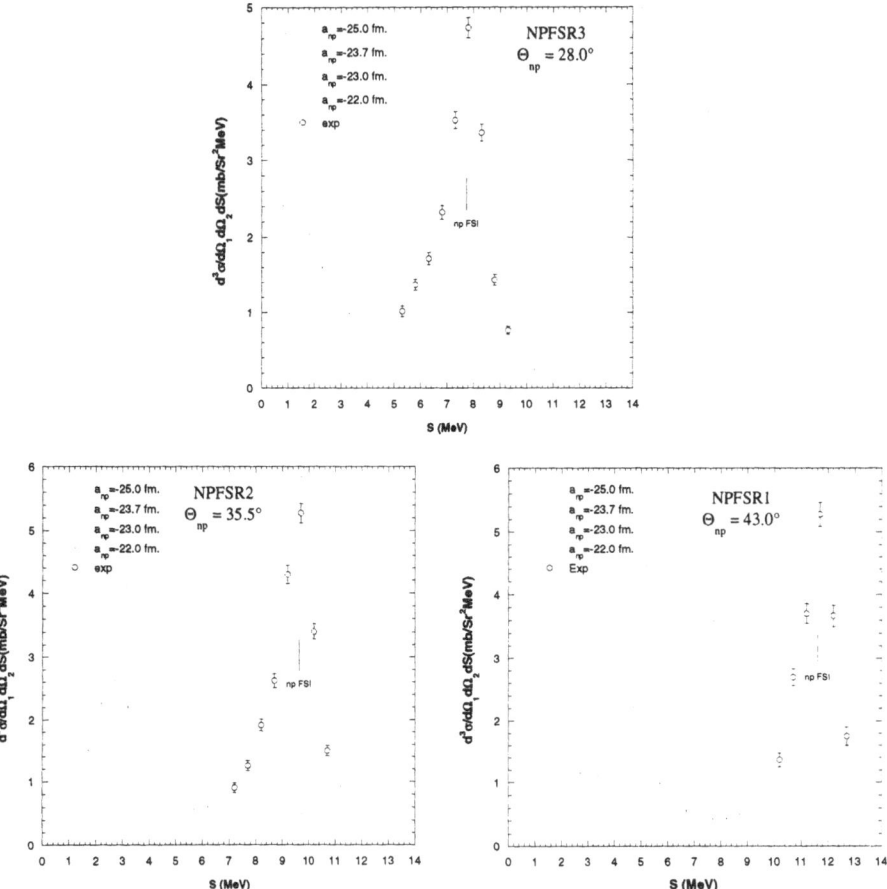

Figure 3. Measured and predicted (for the TUNL experimental setup using the ring detectors) n-p FSI cross section at $\theta_{np} = 28.0°, 35.5°$, and $43.0°$ (lab). The curves are based on different values for a_{np}.

The variation of the scattering length with the strength of the nuclear potential can be estimated using effective-range theory and low-energy approximations. The fractional change in the NN scattering length $\left(\frac{\delta a}{a}\right)$ due to a shift in the NN potential strength $\left(\frac{\delta V}{V}\right)$ is given by,

$$-\frac{\delta a}{a} = \frac{a}{b}\left(\frac{\pi}{2}\right)^2 \frac{1}{2}\frac{\delta V}{V} = -1.23\frac{a}{b}\frac{\delta V}{V}. \tag{1}$$

In the above equations the symbol b is the effective range parameter. For the case of the $^1S_0$ n-p potential, $a/b = -23.8/2.7$, giving the result of $\delta a/a = -10.8(\delta V/V)$. Therefore, a 1% shift in the n-p potential strength results in about an 11% change in a_{np}. The modification of a_{np} in the n-d system is $\delta a_{np} = a_{np}^{2N} - a_{np}^{3N} = -0.3 \pm 0.8$ fm. The corresponding effective shift in the n-p

potential strength is $\delta V/V = 1.15 \times 10^{-3} \pm 3.07 \times 10^{-3}$. Setting the limit at the three-sigma level, the upper limit on the effective shift in the $^1S_0$ n-p potential strength due to 3N forces is $|\delta V/V| < 0.01$. This result is the first empirically determined limit on the relative strengths of the 2N and 3N forces. It is a factor of about five smaller than the value estimated by van Kolck [14] for the NN $^3S_1$ potential. Though these findings suggest an extremely small influence of 3N forces, caution is urged in the interpretation of these results. Since the relative contributions of 3N-force terms are likely to be strongly energy and geometry-dependent, the findings reported here are most valid under similar conditions of this study, that is, in the n-d continuum at energies around 9 MeV in the c.m. system.

5 Results for a_{nn}

Our resuls obtained for the n-n FSI cross sections at the four n-n pair production angles are shown in Fig. 4 in comparison to finite-geometry rigorous 3N calculations using the Bonn B NN potential with different values for a_{nn}. Results for a_{nn} were determined by means of χ^2 minimization of the Monte-Carlo calculated finite-geometry cross sections using three modern NN potential mo-

Table 2. The table on the left shows a_{nn} values extracted from each n-n FSI configuration using Monte-Carlo cross sections generated for our finite-geometry experimental arrangement by the Bonn B, CD Bonn and Nijmegen I NN potentials. The systematic uncertainties originate from a $\pm 5\%$ systematic error in the experimental absolute cross sections. The table on the right shows a_{nn} values and associated errors extracted from the shapes of three n-n FSI cross-section curves.

$\theta_{nn}(^\circ)$	a_{nn} (fm)	Δa_{nn}^{stat} (fm)	Δa_{nn}^{syst} (fm)
Bonn B			
20.5°	−18.9	±0.2	±0.6
28.0°	−18.8	±0.2	±0.6
35.5°	−17.7	±0.4	±0.6
43.0°	−18.8	±0.4	±0.7
CD Bonn			
20.5°	−18.9	±0.2	±0.7
28.0°	−18.6	±0.3	±0.7
35.5°	−17.8	±0.3	±0.6
43.0°	−18.6	±0.4	±0.6
Nijmegen I			
20.5°	−19.2	±0.2	±0.8
28.0°	−18.8	±0.3	±0.7
35.5°	−18.0	±0.3	±0.6
43.0°	−18.7	±0.4	±0.6

$\theta_{nn}(^\circ)$	a_{nn}^{shape} (fm)	Δa_{nn}^{shape} (fm)
Bonn B		
20.5°	−19.0	±0.7
28.0°	−18.3	±0.9
35.5°	−18.8	±1.2
CD Bonn		
20.5°	−19.1	±0.6
28.0°	−18.3	±0.9
35.5°	−18.8	±1.3
Nijmegen I		
20.5°	−19.1	±0.6
28.0°	−18.4	±0.9
35.5°	−18.9	±1.3

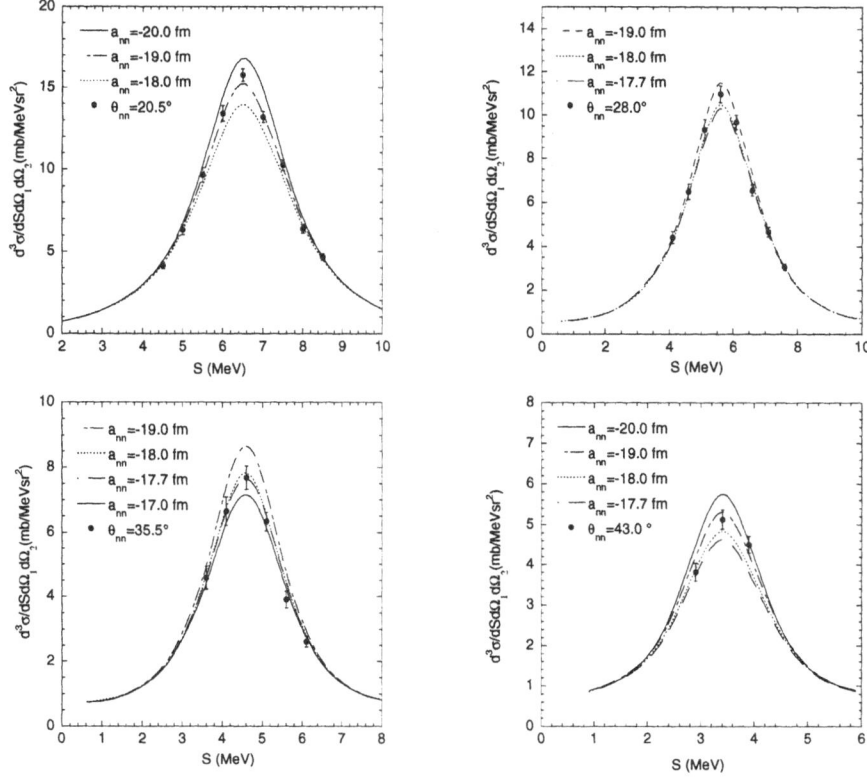

Figure 4. Measured and predicted (for the TUNL experimental setup) n-n FSI cross section at $\theta_{nn} = 20.5°$, $28.0°$, $35.5°$, and $43.0°$ (lab). The curves are based on different values for a_{nn}.

dels and the experimental absolute cross-section data. Like for a_{np}, a systematic angular dependence of a_{nn} was not observed, as can be seen from Table 2 (left side).

In addition, χ^2 minimizations of the Monte-Carlo cross-section calculations and the experimental FSI cross-section data with floating normalization were performed (see right side of Table 2). This shape analysis of the FSI peak gave an average value of $a_{nn} = -18.8 \pm 0.5$ fm, consistent with the result obtained from the analysis of the absolute FSI cross-section data. Our results for a_{nn} are in agreement with the average value of $a_{nn} = -18.6 \pm 0.5$ fm obtained from the π^--d capture reaction, thus clearly demonstrating that the influence of 3N forces is very small in the 3N configuration studied in the present work.

6 Summary

In summary, the present a_{np} and a_{nn} $^1S_0$ scattering lengths determinations from the 3N n-d breakup reaction and their comparison to the existing NN and quasi NN results, respectively, clearly show that 3N-force effects on the n-p and n-n FSI enhancements are considerably smaller than previously concluded from less accurate experimental data and theoretical analyses.

Acknowledgement. This work was supported in part by the U.S. Department of Energy, Office of High Energy and Nuclear Physics, under grant No. DE-FG02-97ER41033. The 3N calculations reported in this work were performed on the Cray T916 of the North Carolina Supercomputing Center at the Research Triangle Park, North Carolina.

References

1. W. Glöckle et al.: Phys. Reports **274**, 107 (1966)

2. A. Kievsky, M. Viviani, S. Rosati: Nucl. Phys. **A577**, 511 (1994); A. Kievsky, M. Viviani, S. Rosati: Phys. Rev. **C52**, R15 (1995)

3. V.G.J. Stoks et al.: Phys. Rev. **C49**, 2950 (1994)

4. R.B. Wiringa, V.G.J. Stoks, R. Schiavilla: Phys. Rev. **C51**, 38 (1995)

5. R. Machleidt, F. Sammarruca, Y. Song: Phys. Rev. **C53**, 1483 (1996)

6. G.A. Miller, B.M.K. Nefken, I. Šlaus: Phys. Reports **194**, 1 (1990)

7. C.R. Howell et al.: Phys. Lett., in press.

8. I. Šlaus, Y. Akaishi, H. Tanaka: Phys. Reports **173**, 257 (1989)

9. W. Tornow, H. Witała, R.T. Braun: Few-Body Syst. **21**, 97 (1996)

10. W. Tornow, H. Witała, R.T. Braun: Phys. Lett. **B318**, 281 (1993)

11. O. Dumbrajs et al.: Nucl. Phys. **B216**, 277 (1983)

12. S.A. Coon et al.: Nucl. Phys. **A317**, 242 (1979)

13. R. Machleidt: Adv. Nucl. Phys. **19**, 189 (1989)

14. U. van Kolck: Phys. Rev. **C49**, 2932 (1994)

Few-Body Systems Suppl. 10, 295–304 (1999)

Few-
Body
Systems
© by Springer-Verlag 1999

Searches for the H Dibaryon and the Pentaquark

Daniel Ashery*

School of Physics and Astronomy, Tel Aviv University, Tel-Aviv, Israel

Abstract. The pentaquark is described as a doublet of states: $P^0 = |\bar{c}suud\rangle$ and $P^- = |\bar{c}sddu\rangle$. The H dibaryon is described as: $H = |uuddss\rangle$. Both were predicted to be stable against strong decay or slightly unbound resonances. The considerations leading to these expectations are reviewed. The main properties of these particles: masses, decay modes, lifetimes and hadroproduction cross sections are discussed. Experimental efforts to search for these particles are reviewed.

1 Introduction

All the known Hadrons fall in two categories: the mesons, made out of quark-antiquark pairs and the baryons, made out of three quarks. Extensive efforts were made in order to find particles made out of more quarks. At this point we should define the nature of a multi-quark system. For example, a six-quark component in the deuteron will be heavier by about 350 MeV than the p-n component [1] . Such a component has therefore a de-stabilizing effect and has to be very small ($\sim 1\%$) so that we do not refer to the deuteron as a six-quark system. Only with a "proper mix" of flavors it is possible to find conditions where the multi-quark system is more bound than the color singlet sub-systems that have the same quark content.

Following these lines, Jaffe [2] predicted that a particle made out of [u,u,d,d,s,s] quarks should exist. It was named the "H" (Hexaquark) and a large effort is being made to find it experimentally. Following Jaffe's work, Lipkin [3] and Gignoux et al. [4] showed that the $[u, u, d, \bar{c}, s]$ and $[u, d, d, \bar{c}, s]$ combinations of quarks should also exist. These were named the "P" (Pentaquarks: P^0 and P^-). All this (and the following discussion) holds also for the corresponding antiparticles. Existence of the P, as for the H, means that it would be stable against strong decay and will have a lifetime typical to weakly decaying objects.

*Supported in part by the Israel Academy of Science and Humanities and by the US-Israel Binational Science Foundation

2 Binding Energy of the P and the H

The Color Hyperfine (CH) interaction between quarks of mass m, spin σ and color λ is given by:
$$V_{CH} = -\alpha \sum_{i,j} \lambda_i \lambda_j \sigma_i \sigma_j / m_i m_j.$$
The coefficient α is calculated from the Δ - N mass splitting of 300 MeV which is attributed to the CH interaction. The binding CH potential of a system is given by the difference between the CH interaction in the system and in the lightest color-singlet combination of quarks into which it can be decomposed. The wave function of the H, built in similar way to that of the deuteron, may be written schematically as:
$$\Psi_H = \alpha_1 \Psi_{6q} + \beta_1 \Psi_{(\Lambda - \Lambda)} + \gamma_1 \Psi_{(\Sigma - \Sigma)} + \delta_1 \Psi_{(\Xi - N)}.$$
The lightest color singlet combination is the $(\Lambda - \Lambda)$ system (2231 MeV). Assuming $SU(3)_F$ symmetry, the mass splitting is M($\Lambda - \Lambda$) - M(H) = $4(\alpha/m^2)$, half of the $\Delta - N$ mass splitting, making the H stable against strong decay. Similarly, the P^0 and P^- wave functions can be written as:
$$\Psi_{P^0} = \alpha_2 \Psi_{5q} + \beta_2 \Psi_{(D_S^- - p)} + \gamma_2 \Psi_{(\Sigma^+ - D^-)} + \delta_2 \Psi_{(\Lambda - \bar{D}^0)},$$
$$\Psi_{P^-} = \alpha_3 \Psi_{5q} + \beta_3 \Psi_{(D_S^- - n)} + \gamma_3 \Psi_{(\Sigma^- - \bar{D}^0)} + \delta_3 \Psi_{(\Lambda - D^-)}.$$
Here the lightest color singlet is the $(D_S - N)$ (N = p or n) system (2907 MeV) and similarly for the $\bar{P}$. Assuming $SU(3)_F$ symmetry and ignoring the CH due to the massive c quark the resulting mass splitting is M($D_S - N$) - M(P) = $4(\alpha/m^2)$, the same as for the H particle. These are the largest CH binding potentials obtained for 6q and 5q combinations. For both Pentaquark and H dibaryon we find a component of the wave function described as a multi-quark system where, unlike the case of the deuteron, it provides a gain in stability, no short-range repulsion so that this component may be significant. This is what makes us refer to these particles as multiquark systems.

These calculations have to be corrected for the $SU(3)_F$ symmetry breaking and for the finite mass of the c quark. The results of these corrections [4] are shown in Fig. 1, where $\delta = 1 - m_u/m_s$ and $\eta = m_u/m_c$. It can be seen that the P retains always a larger binding potential than the H and for reasonable values of the two parameters ($\delta < 0.5, \eta : 0.1 - 0.3$) the binding CH potential can be several tens of MeV.

The final binding energy is given by the total Hamiltonian:
$$H = \sum_i (p_i^2/2m_i) - \sum_{i,j} \lambda_i \lambda_j V_c(r_{ij}) - \sum_{i,j} (\lambda_i \lambda_j \sigma_i \sigma_j / m_i m_j) V_{ss}(r_{ij})$$
where $V_c(r_{ij})$ is the Color Electric central potential and $V_{ss}(r_{ij})$ is the spin-spin interaction. For the kinetic energy part, the massive c quark makes it much smaller in the P than in the H if the same form factor (internal momentum) is assumed. For 250 MeV/c the total kinetic energy term for the P is smaller by about 15 MeV than for the H.

The complete mass evaluation is, necessarily, more model dependent. Over 20 calculations of the H mass were performed using a variety of models such as Bag models, Chiral models, Instanton interaction, Skyrme model, lattice calculations and more [5] . While the results vary between very deep binding of over 1 GeV to a state unbound by 0.5 GeV, most of the results cluster around the $\Lambda\Lambda$ threshold. This indicates that most likely the H is either bound or unbound by a few MeV. Much fewer model calculations were done for the Pentaquark. These include Skyrme model calculations [6], Instanton interactions [7] and bag model calculations [8]. The results vary only by several tens of MeV but still allow for bound or unbound states. Calculations for both H and Pentaquarks were recently carried out using the Goldstone Boson Exchange model [9] concluding that both H and the charmed-strange Pentaquark are unbound but a non-strange Pentaquark is found to be bound. It can be therefore concluded that the available theoretical tools are insufficient for determination of the binding energies of the H and the P but it is likely that these systems may be bound by up to several tens of MeV or unbound by a similar amount and may be observed experimentally.

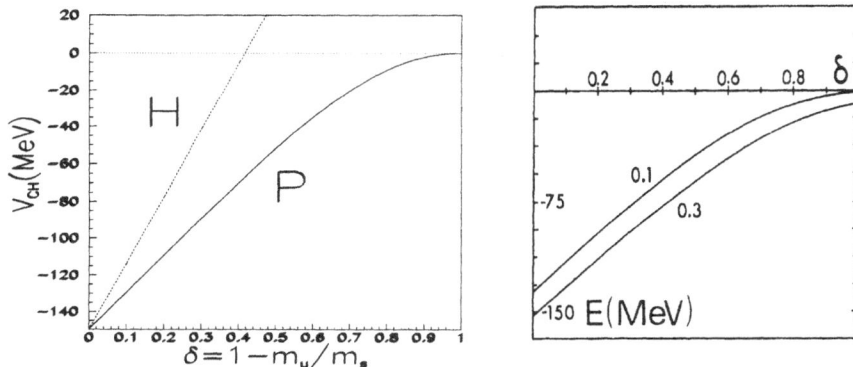

Figure 1. Color Hyperfine binding potential of the H and P corrected for the mass of the strange (left part) and charm (right part) quarks (ref. [4]).

3 Hadroproduction Cross Sections

There can be several ways in which to produce the P and the H. We estimate the hadroproduction cross sections as this process is used in experimental searches for the H and P. The first mechanism is the "coalescence production" [10] described by:

$$N_c = \int f(x_i, p_i) C(x_i, p_i) \prod_{i=1}^{N} \frac{2\theta(p_i)\delta(p_i^2 - m_i^2)d^4 p_i \cdot p_i^\mu d\sigma_\mu^i}{(2\pi\hbar)^3}. \tag{1}$$

Here the two color-singlet systems that compose the wave function (e.g. a D_S-p pair for the P or $\Lambda\Lambda$ pair for the H) are produced and coalesce into the final particle P or H. For this to happen the two systems must move within a small volume of phase space so that they can interact. The probability for that is given by Eq. 1 [10] where f is the distribution function of the coalescing particles in phase space, C is the "Coalescence Integral" which measures the probability for the particles to occupy the part of phase space that will allow the interaction and the rest is the phase space and momentum conservation.

Values of C were calculated [10] for production of the deuteron, H and P and are of the order of 10^{-2} to 10^{-3}. In fact it was found experimentally [11] that the cross section for high energy hadroproduction of deuterons is about 10^{-3} times the cross section for production of protons. We can therefore expect this order of magnitude for the cross section ratios for production of H relative to Λ and P relative to D_s. Similar results were obtained from estimates assuming a reaction mechanism of "full production" of the particle, in the same way as regular hadrons are produced [10, 12]. The estimated ratios are:

$$\sigma(P)/\sigma(D_s) \sim 10^{-2}, \quad \sigma(H)/\sigma(\Lambda) \sim 4 \cdot 10^{-3}.$$

It is better to leave the estimates in such relative terms rather than translate them to absolute cross sections so that their meaning is more general and they can be applied to a variety of hadronic interactions.

The production of the H was also considered in the $\Delta S=-2$ $^3He(K^-, K^+)Hn$ reaction. Aerts and Dover [13] calculated the matrix element:

$$M \sim \int \psi_{^3He} \, T(K^-p \to K\Xi) \, \Gamma(\Xi N \to H) \, d^3k_p \tag{2}$$

using a realistic ^{3}He wave function with two models for short range correlations, measured $K^-p \to K\Xi$ cross sections and a coalescence estimate for Γ. These calculations made it possible to plan experimental searches for the H and interpret the results by comparing with the expectations. Such comparisons were made also for searches where the only possible comparison was with coalescence calculations. However, the estimates of Eq. 2 are by nature more precise than those based on the coalescence model.

4   Lifetimes and Decay Modes

Some of the searches for the H and P are sensitive to the lifetime and decay mode of the particle. This by nature makes the search less sensitive as it depends on additional assumptions. Estimates of the H lifetime, depending on its binding energy, were made [14] and yielded a range of 10^{-9} to 10^{-6} seconds. The decay diagrams were based on a spectator model and, as the H becomes more bound, the reduced phase space and reduction of possible decay channels increases the lifetime. Assuming a binding energy of a few MeV, the shorter lifetimes are the more probable with spectator decay modes to $\Lambda N\pi$ or ΣN. For the pentaquark, as for most charmed baryons, the large number of decay channels is usually described as decay of the charmed quark while the rest of the system acts as spectator. One then can expect decay modes similar to those of the charmed mesons that contain the same quarks (Sect. 2). The lifetime is then expected to be similar to that of these charmed mesons, or somewhat longer due to the reduced phase space.

The decay through the color-octet component can open additional channels. This component would allow direct annihilation of the $\bar{c}$ quark on the s and d quarks via the Cabibbo allowed and suppressed transitions and other ways of quark final state interactions [15] . The effect on the P lifetime is to reduce it but it is hard to estimate the branching ratios to these decays and their effect on the lifetime. Consequently, the expected lifetime of the P can be either somewhat longer or shorter than that of the D_S in the range of 10^{-13} to 10^{-12} seconds.

5 Experimental Searches for the H

Most of this effort was done at Brookhaven National Laboratory. In addition to the $\Delta S = -2$ (K^-, K^+) reaction many experiments utilized hadroproduction with either light probes or relativistic heavy ions in which the presumed production mechanism is through coalescence. As for the detection method, most of the experiments tried to observe the decay of the H into $\Lambda N \pi$ or ΣN. These searches depend on knowledge of the decay mode and lifetime of the H. This is a disadvantage from which other experiments, that used associated production as signature, do not suffer. Some recent experiments will be discussed below.

We begin with three experiments that were completed and their negative results were published recently. BNL experiment E836 [16] studied the $^3He(K^-, K^+)Hn$ reaction using a 1.8 GeV/c K^- beam. The outgoing K^+ were detected by a magnetic spectrometer equipped with Čerenkov and time-of-flight detectors for particle identification and background suppression. The results show the strong peak from quasi-free Ξ^- production and a signature for the H should come from events with higher momenta. As there is no evidence for a peak in this region the authors determined upper limits for a mass range from about 2.05 to 2.2 GeV/c^2.

In order to understand their results and to compare them with those of other experiments, they are presented in Fig. 2 after being divided by the theoretical expectations for the production cross section in this reaction. These were taken from ref. [13] and corrected for the measured cross section for the $p(K^-, K^+)\Xi^-$ reaction [16]. A similar experiment was carried out at KEK (E224 [17]). The beam momentum was 1.66 GeV/c and the target was Carbon in scintillating fibers. This helped to reduce the background

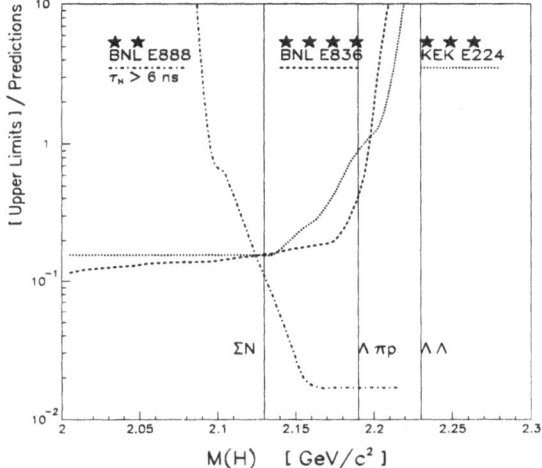

Figure 2. Upper limits for H production normalized to predictions. One or two stars are given to experiments dependent or independent on the H lifetime, respectively. One or two stars are given to experiments for which the predictions are from coalescence or customized calculations, respectively.

but no signal was observed above the quasi-free peak and the results are shown in Fig. 2 with the same normalization as for BNL E836. Common to these two experiments is that the results do not depend on the H lifetime and decay modes.

A quite different experiment, which does depend on the H lifetime and decay mode is BNL experiment E888 [18]. A 24 GeV/c proton beam hit a Cu target. Sweeping magnets downstream of the target removed charged particles and the remaining neutral particles were allowed to decay in a 10 m long vacuum tank. Using a downstream two-arm spectrometer the $H \to \Lambda n$ and $H \to \Sigma n$ decays were searched for by detecting the $\Lambda \to p\pi^-$ with $P_T(\Lambda) > 174\mathrm{MeV}/c$. Events pointing back to collimators were removed. The results are presented in terms of mass-dependent upper limits on $(d\sigma_H/d\Omega) \, / \, (d\sigma_\Lambda/d\Omega) \quad \theta = 48\mathrm{mr}$ and are based on an assumed lifetime of more than 6 ns for the H.

In order to interpret these results and put them on a common ground with the two previous experiments, they were divided by the cross section expected from coalescence calculations [19]; see Fig. 2. Because these results depend on the H lifetime and decay modes and because the coalescence calculations are less precise than the more dedicated calculations made for the two previous experiments, the uncertainties in these last results should be considered larger. The combination of the results from the three experiments puts an upper limit of about an order of magnitude below the theoretical expectations for H masses of less than about 2.15 GeV/c^2. The first two experiments cannot rule out the existence of the H with a mass between this value and the 2.23 GeV/c^2 threshold for strong decay.

In the last 20 years there have been several reports of observation of events that were attributed to the H particle. These events were observed in emulsions or bubble-chambers. Shahbazian et al. reported two $H \to \Sigma^- p$ events observed in bubble-chamber. One in 1988 with a mass of 2174 MeV [20] and one in 1993 with a mass of 2400 MeV [21]. Two more events with a mass near 2200 MeV were reported by Alekseev et al. [22]. While all these reports drew criticism also on technical grounds, it is certainly impossible to establish the existence of the H based on four events with different masses.

More recently, two BNL experiments published results with events and perhaps some structure in the bound H region. In E813 the $p(K^-, K^+)\Xi^-$ reaction was used. The Ξ^- particles were detected by a Si detector, slowed down and stopped in liquid deuterium to form a (Ξ^-, d) atom. This was expected to decay to an H and a monoenergetic neutron which is identified by a time-of-flight system. The spectrum from about half the data taken in this experiment shows some fluctuations [23] but no peak was observed and no claim was made. In E810 a 14.6×A GeV/c Si beam hit a Pb target and tracks were detected by the Multi-Particle Spectrometer [24]. Looking for $H \to \Sigma^- p$ and $H \to \Lambda \pi^- p$ events, a structure was observed corresponding to a H mass of about 2.21 GeV/c^2 [25]. These events had a decay time consistent with about half the Λ lifetime. The results from all these experiments are combined in Fig. 3. It should be noted that most of the observed events are in a mass region for which the published upper limits are relatively high compared with the expectations.

Recently, KEK experiment E224 searched for events with two Λ particles in the $C(K^-, K^+)$ $\Lambda\Lambda X$ reaction [26]. They used the scintillating fibers and CCD camera to identify two neutral vertices and the $\Lambda \rightarrow p\pi^-$ to identify $\Lambda\Lambda$ events. They observe an enhancement over phase space expectations just above the $\Lambda\Lambda$ threshold. These results are shown in Fig. 3 (not normalized to the other results).

Observation of double-Lambda hypernuclei has an impact on the search for the H. If it is strongly bound with a mass much

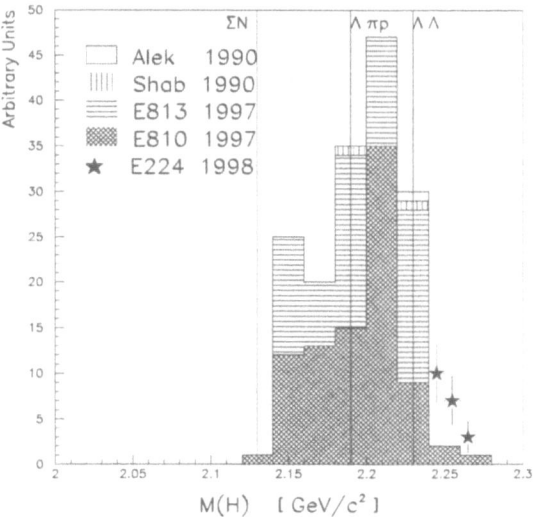

Figure 3. Observed events which may contain H candidates. The results of the various experiments are not normalized to each other.

smaller than twice m_Λ, a double-Lambda hypernucleus can decay strongly emitting an H. A positive identification of a double-Lambda hypernucleus weak decay puts limits on its mass:

$$2m_\Lambda - B_{\Lambda\Lambda}(A, Z) \leq m(H) \leq 2m_\Lambda$$

with $B_{\Lambda\Lambda}(A, Z)$ the binding energy of the two Lambdas in the nucleus. The value of this binding energy should come from the measurements. Based on binding energies of single-Lambda hypernuclei it can be estimated to be about 20 MeV. Therefore, like with the upper limits discussed earlier, also the observation of double-Lambda hypernuclei will not rule out a weakly bound H particle.

Two double-Lambda hypernucleus events observed in emulsion experiments were reported in the early sixties. They were interpreted as $^{10}_{\Lambda\Lambda}Be$ or $^{11}_{\Lambda\Lambda}B$ [27] and as $^{6}_{\Lambda\Lambda}He$ [28]. The identification was based on observation of two sequential Λ decays. It should be noted that it is not easy to establish whether such a decay came from a $_{\Lambda\Lambda}A$ or from $_H A$. Twenty-five years later a search for such events in emulsion was carried out at KEK with modern techniques and only one event identified as $^{10}_{\Lambda\Lambda}Be$ or $^{13}_{\Lambda\Lambda}Be$ was reported [29]. Obviously, if the first observations were correct one would have expected many more events in the KEK experiment. The existing data therefore contains no convincing evidence for observation of double-Lambda hypernuclei.

6 Search for the Pentaquark

As discussed in Sect. 3 the expected Pentaquark production cross section is of a few percent of that for the D_S. Previous charm hadroproduction experiments were able to observe only about 100 D_S events (over a similar background) in two decay modes ($\phi\pi$ and K^*K) combined together [30]. It was therefore necessary to use a very large statistics charm hadroproduction experiment in order to achieve the sensitivity needed for the Pentaquark search.

The search for the $P^0_{\bar{c}s}$ was carried out in experiment E791 [31] at Fermilab. The experiment recorded with a relatively loose trigger 2×10^{10} events from the interaction of a 500 GeV/c π^- beam in nuclear targets. Precision vertex and tracking information was provided by 23 silicon microstrip detectors upstream and downstream of the targets, ten proportional wire-chamber and 35 drift-chamber planes. Momentum was measured using two dipole magnets. Two multicell, threshold Čerenkov counters were used for π, K and p identification [32].

The pentaquark was searched for in its expected decay mode $P^0_{\bar{c}s} \to \phi\pi p$, where the ϕ subsequently decays to K^+K^- [33]. The sensitivity of the search was normalized to $D^\pm_s \to \phi\pi^\pm$ decays. The acceptance of the detector was calculated via Monte-Carlo simulation where the pentaquark was simulated with a lifetime of 0.4 ps and with two different masses, 2.75 and 2.86 GeV/c^2. The analysis process selected $P^0_{\bar{c}s}$ and $\bar{P}^0_{c\bar{s}}$ candidates equally.

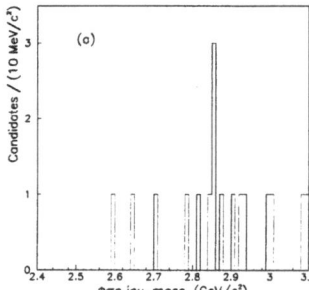 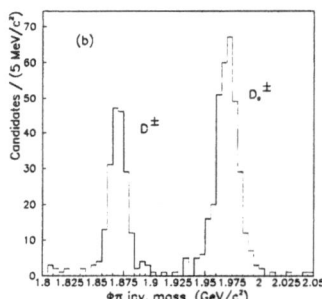

Figure 4. (a) The $\phi\pi p$ invariant mass spectrum for the final selection criteria. (b) $\phi\pi$ invariant mass spectrum showing the $D^\pm_s$ normalization sample. The left-most peak arises from Cabibbo-suppressed $D^\pm$ decays.

The $\phi\pi p$ invariant mass spectrum for the final analysis cuts is shown in Fig. 4(a). The $\phi\pi$ invariant mass spectrum for the $D^\pm_s \to \phi\pi^\pm$ normalization sample is shown in Fig. 4(b). This sample was selected using the same selection criteria (where relevant) as were used to select pentaquark candidates. In this manner the systematic error on the ratio of efficiencies for the two decay modes was minimized. The $\phi\pi p$ invariant mass spectrum (Fig. 4(a)) is roughly flat with no statistically significant structure. It is concluded that there is no evidence for $P^0_{\bar{c}s} \to \phi\pi p$ decays in these data.

The spectrum of Fig. 4(a) is used to obtain 90% C.L. upper limits on the product of the pentaquark production cross section and the pentaquark branching fraction to $\phi\pi p$, relative to that for $D_s^\pm \to \phi\pi^\pm$. The limits for two pentaquark masses: 2.75 GeV/c^2 and 2.86 GeV/c^2 are presented in Table 1. Assuming that the branching fractions of the $D_s^\pm \to \phi\pi^\pm$ and $P_{\bar{c}s}^0 \to \phi\pi p$ decays are similar, the resulting upper-limits approach the range of the estimated ratio between the pentaquark and $D_s^\pm$ production cross sections.

Table 1. Upper-limit (90% C.L.) for the ratio of cross section times branching fraction for $\sigma \cdot$ BR of the $P_{\bar{c}s}^0 \to \phi\pi p$ and $D_s^\pm \to \phi\pi^\pm$ decays and **preliminary** upper limits for the $P_{\bar{c}s}^0 \to K^* K p$ and $D_s^\pm \to K^* K^\pm$ decays. Results are given for two pentaquark masses and assuming a pentaquark lifetime of 0.4 ps.

	$P_{\bar{c}s}^0 \to \phi\pi p$		$P_{\bar{c}s}^0 \to K^* K p$	
M(P^0) GeV/c^2	2.75	2.86	2.75	2.86
90% CL Upper-Limit	0.023	0.040	0.04	0.02

A similar search is going on for another expected decay mode: $P_{\bar{c}s}^0 \to K^* K p$ with $K^* \to K\pi$ [34]. It is more difficult to study this mode because of the large width of the K* (50 MeV). This required application of tighter selection criteria which were optimized using the same procedure as described above. Preliminary results of this search are included in Table 1.

7 Summary

20 years after proposal of the H and 10 years after proposal of the Pentaquark we begin to have conclusive experimental results. The H dibaryon, if it exists, has most likely a mass of more than 2.15 GeV (bound by not more than 80 MeV). Suggestive signals seen for H with a mass of ~2.21 GeV, if confirmed, will be consistent with upper limits and with most observations of double-Lambda hypernuclei, if these are also confirmed (E885, E906). We need experimental studies sensitive to small binding energy (BNL E813, FNAL E791).

The Pentaquark was searched for via $\phi\pi p$ and $K^* K p$ decays achieving the desired sensitivity: $\sim 1\%$ of D$_s$ production. Within this sensitivity there is no convincing evidence for existence of the Pentaquark; it is not ruled out either. The Upper-Limit results approach the theoretical estimates and provide a good starting point for future experimental searches in high statistics charm experiments (E781, E831 and Charm 2000 at Fermilab, COMPASS at CERN).

References

1. H.J. Lipkin: Phys. Lett. **B198**, 131 (1987)
2. R.L. Jaffe: Phys. Rev. Lett. **38**, 195 (1977)
3. H.J. Lipkin: Phys. Lett. **B195**, 484 (1987)
4. C. Gignoux et al.: Phys. Lett. **B193**, 323 (1987)

304

5. B. Quinn et al.: In: *Proc. Baryon 92 International Conference*, ed. M. Gai, p. 278. World Scientific 1992

6. D.O. Riska and N.N. Scoccola: Phys. Lett. **B299**, 338 (1993)

7. S. Takeuchi, S. Nussinov, K. Kubodera: Phys. Lett. **B318**, 1 (1993)

8. S. Fleck et al.: Phys. Lett. **B220**, 616 (1989); S. Zouzou and J.-M. Richard: Few-Body Syst. **16**, 1 (1994)

9. F. Stancu: paper presented in this conference

10. J.I.Kapusta: Phys. Rev. **C21**, 1301 (1980); C.B.Dover et al.: BNL preprint TPR-90-60-BNL-45865 (1991); Phys. Rev. **C44**, 1636 (1991)

11. W. Bozzoli et al.: Nucl. Phys. **B144**, 317 (1978)

12. Casher, H. Neuberger and S. Nussinov: Phys. Rev. **D20**, 179 (1979); M.A. Moinester et al.: Z. Phys. **A356**, 207 (1996)

13. A.T.M. Aerts and C.B. Dover: Phys. Rev. **D28**, 450 (1983)

14. J.F. Donoghue et al.: Phys. Rev. **D34**, 3434 (1986)

15. H.J. Lipkin: private communication

16. R.W. Stotzer et al.: Phys. Rev. Lett. **78**, 3646 (1997)

17. J.K. Ahn et al.: Phys. Lett. **B378**, 53 (1996)

18. J. Belz et al.: Phys. Rev. Lett. **76**; 3277 (1996); Phys. Rev. **D53**, R3487 (1996)

19. B.A. Cole et al.: Phys. Lett. **B350**, 147 (1995)

20. B.A. Shahbazian et al.: Z. Phys. **C39**, 151 (1988)

21. B.A. Shahbazian et al.: Phys. Lett. **B316**, 593 (1993)

22. A.N. Alekseev et al.: Sov. J. Nucl. Phys. **52**, 1016 (1990)

23. W.D. Ramsay: In: *Proc. Int. Conf. on Particles and Nuclei*, Williamsburg (1996) p. 602. and G.B. Franklin, private communication

24. S.E. Eiseman et al.: Phys. Lett. **B287**, 44 (1992)

25. R.S. Longacre: In: *proc. 7th Int. Conf. on Hadron Spectroscopy*, eds. S.U. Chung and H.J. Willutzki, p. 219 (1997)

26. J.K. Ahn et al.: KEK preprint 98-24 (1998), submitted to Phys. Rev. Lett.

27. M. Danysz et al.: Nucl. Phys. **49**, 121 (1963)

28. P. Prowse: Phys. Rev. Lett. **17**, 782 (1966)

29. S. Aoki et al.: Prog. Theor. Phys. **85** 1287 (1991)

30. G.A. Alves et al.: Phys. Rev. Lett. **77**, 2388 (1996)

31. E. M. Aitala et al.: Phys. Rev. Lett. **76**, 364 (1996) and references therein

32. D. Bartlett et al.: Nucl. Instr. and Meth.

33. S. MayTal-Beck: PhD thesis, Tel Aviv University (unpublished) and E. M. Aitala et al.: Phys. Rev. Lett. **81**, 44 (1998)

34. G. Hurvits: PhD thesis, Tel Aviv University (unpublished)

Few-Body Systems Suppl. 10, 305–314 (1999)

Few-
Body
Systems
© by Springer-Verlag 1999

Few Body Systems and Electromagnetic Probes

Ingo Sick *

Dept. für Physik und Astronomie, Universität Basel, CH-4056 Basel, Switzerland

Abstract. We discuss recent experimental results on the electromagnetic structure of the neutron, the long- and short-range structure of the deuteron, the D-state and the role of meson exchange currents in the A=3 system.

1 Introduction

In this review, we address a number of topics on the structure of hadronic systems as investigated using the electromagnetic probe. For the three-nucleon system, the study of radiative capture has seen considerable activity recently, mainly as a consequence of the ability to now perform experiments using tensor-polarized particles, and the ability to treat the 3-nucleon (3N) continuum in an exact way. The structure of the two-body bound state, both in radial and momentum space, continued to receive attention. Much of the interest here is related to the mesonic degrees of freedom and the role of relativistic effects. The experimental study of the neutron structure has considerably progressed, leading to much more precise form factors suitable both for an understanding of neutron structure and the use of these form factors in calculating *nuclear* form factors.

2 Radiative capture

Much of the interest in p+d radiative capture originates from the ability to study higher order polarization observables when the process is induced by tensor-polarized deuterons. Tensor polarization observables are closely related to S–D transitions, quantities that provide direct information on the D-state components of nuclear wave functions and mesonic degrees of freedom. These more subtle effects now can be quantitatively studied as the 3N-continuum can be treated with a level of reliability that comes close to the one for the 2N-system.

E-mail address: Sick@ubaclu.unibas.ch

306

Recent data on p+d radiative capture have been taken by Schmid et al. at very low energy (E_d=0–80keV) and Anklin et al. (at 45MeV) [1, 2]. The low-energy region is of interest due to the fact that at E_d=0 the usually dominating E1 ($l = 0 \rightarrow l = 0$) transition is suppressed, such that the M1 term becomes the dominating one. In the M1 amplitude, meson exchange currents (MEC) play an important role, particularly in the vector analyzing power. At higher energy, the capture process becomes much more sensitive to the shorter-range structure and provides a direct measure of the A=3 D-state. Both experiments have been performed by inducing the reaction using vector- and tensor-polarized deuterons. In the case of the higher-energy experiment, where processes other than capture largely dominate, the photon and the ^{3}He nucleus have been detected in coincidence in order to cleanly identify the capture process. Calculations of the capture cross sections and analyzing powers

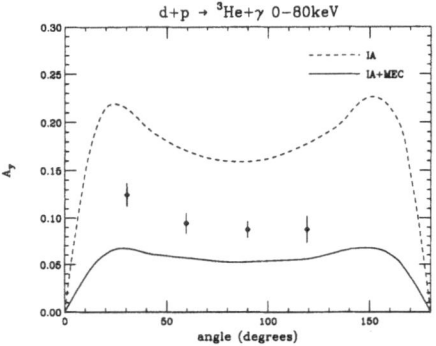

Figure 1. Vector analyzing power of p+d capture reaction at low energy.

have been performed using solutions of the Schrödinger equations for both the bound- and the continuum-states. The calculations of the Pisa group [1] employ the correlated hyperspherical approach, the Bochum group [2] uses the Faddeev approach. Both groups use modern N-N interactions, and include all the multipolarities (E1, M1, E2) relevant for the respective energies. MEC in the E-multipoles are included via Siegert's theorem, explicit MEC in the M1 amplitude are included for the low-energy data where M1 dominates.

Figures 1, 2 show the results for the vector analyzing power A_y at low energy, and the tensor analyzing power A_{yy} at 45MeV. A_y is quite sensitive to MEC, as seen from the difference between the dashed and solid curves; the difference between the full calculation and the data at present is not understood. At 45MeV the full calculation and experiment agree quite well, except for very large angles where M1 (which is known to receive an about equal contribution from MEC) is dominating. The dashed curve shows that the tensor analyzing power is indeed very sensitive to the A=3 (but not the deuteron) D-state. The dash-dot curve emphasizes that the exact treatment of the continuum state is indispensable; analyzing powers calculated in the plane-wave approximation have nothing to do with experiment. The sensitivity studies carried out also show that the data are not particularly sensitive to the N-N interaction employed. It is very important, however, to include multipolarities up to E2.

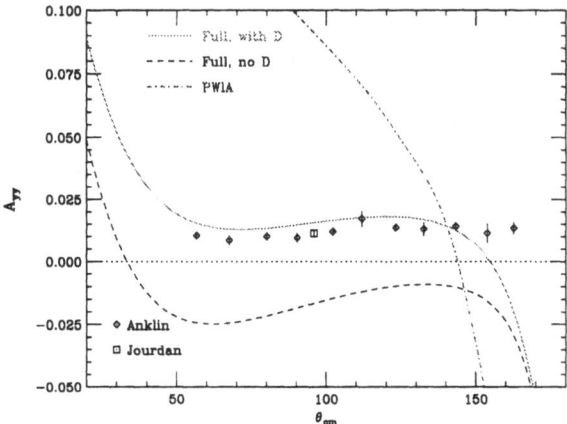

Figure 2. Tensor analyzing power of p+d radiative capture at 45MeV.

3 Deuteron radius

The deuteron is a fundamental system that has received extensive attention in the past, by both theory and experiment. The wave function of the deuteron can be calculated accurately for any given nucleon-nucleon (N-N) potential. The electromagnetic form factors, particularly at low q, can be predicted with high precision. Past work [4] has revealed one major deficit in our understanding: the deuteron root-mean-square (rms) radius calculated using many N-N potentials shows a tight linear correlation with the N-N triplet scattering length predicted by the potential employed. For the well known experimental scattering length this linear relation misses the experimental deuteron rms-radius derived from electron scattering by $0.019 \pm .003$fm. For a system as well understood as the deuteron such a discrepancy is most disturbing. Various corrections such as mesonic [3, 4, 5] and quark [6, 5] degrees of freedom have been investigated, without success.

We therefore have reanalyzed the *world data* using a model-independent approach. Particularly relevant for the determination of the radius are the data of [7, 8, 9]. We start from the original, unseparated cross sections, which allows to make, as part of the fit, the separation of the longitudinal and transverse structure functions using the full information available today. For cases where the deuteron data have been measured relative to the proton, we recalculate the cross sections using modern data for the proton [10]. In our analysis we also take into account the dominant systematic errors (normalization, energy,..) as quoted in the publications and thesis works. The data basis used includes a total of 329 points. We parameterize the deuteron form factors $F_{C0}(q)$, $F_{M1}(q)$, $F_{C2}(q)$ [or, alternatively, the structure functions $A(q)$ and $B(q)$] using the sum-of-gaussians form factors SOG [11]. The deuteron wave function exhibits a long tail towards large r due to the weak binding of the deuteron, a tail that affects the shape of the form factors at very low q. As the shape of the tail at large r is well known — it depends only on the binding energy of the deuteron — this shape is used as a constraint during the analysis of the data.

The fit of the *world* data (for references see [12, 13]) leads to a charge *rms*-radius of the deuteron of 2.112 ± 0.01 *fm*, where the error bar includes both statistical and systematic uncertainties of the data. This radius perfectly agrees with the one determined by Klarsfeld et al., despite the very different approach used. We have found, however, that the PWBA used in this analysis as well as all previous studies of electron-deuteron scattering, is not good enough. We therefore have calculated the Coulomb distortion using second order Born approximation as described in [14, 15]. For $Z\alpha \simeq 0.01$ this expansion in terms of powers of $(Z\alpha)$ is expected to be very accurate, and we estimate it to give more reliable results than the standard phase-shift codes developed for treating the Coulomb distortion for heavy nuclei.

After correcting the data for Coulomb distortion, the $F_{C0}(q)$, $F_{M1}(q)$, $F_{C2}(q)$ from the fit yield the radius listed in Table 1. We conclude from the

Table 1. Charge rms-radii from different sources. Corrections for dispersive effects (disp), meson exchange currents (MEC) and nuclear polarization (pol) have been applied to make the three last entries comparable.

Source	rms-radius (fm)	reference
(e, e) world data	2.130 ± 0.010	[12, 13]
(e, e) –disp. –MEC	2.123 ± 0.010	[13, 16, 5]
isotope shift –pol. –MEC	2.1316 ± 0.001	[5, 17, 18]
N–N scattering data	2.130 ± 0.002	[17]

comparison of the second, third and fourth entries of table 1 that there is good agreement between the radii determined from (e,e), N-N scattering and optical isotope shifts. A longstanding problem has been elucidated.

4 Deuteron form factors

In the course of the analysis of the world data, we have also obtained the various form factors up to $q=8fm^{-1}$, including their experimental error bars. When including the available polarization data (including the recent T_{20} data measured at AMPS), one also can separate the $F_{C0}(q)$, $F_{C2}(q)$ form factors. We show below the charge monopole and magnetic dipole form factors, the quantities that exhibit diffraction minima and are particularly sensitive to the ingredients of the theoretical calculations.

The form factors have been calculated using both relativistic [19, 20] and nonrelativistic [21, 22] approaches, and different N-N potentials (Paris, Bonn, AV18). The contribution of MEC is generally included in a consistent way, i.e. with the mesonic currents directly derived from the N-N potential employed. In order to exhibit the contribution of MEC, we include the IA prediction for the case of the Paris potential. Figures 3, 4 show that in general there is sur-

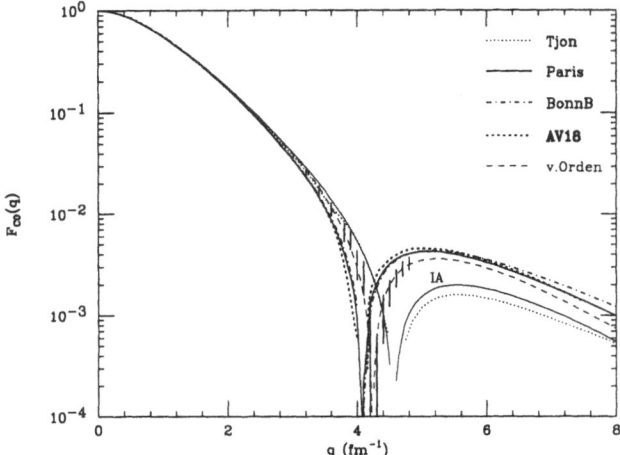

Figure 3. C0 form factor from world data (error bars) together with selected theoretical predictions identified by the N-N potential used or by author (for the relativistic calculations). The IA prediction is given for the Paris potential.

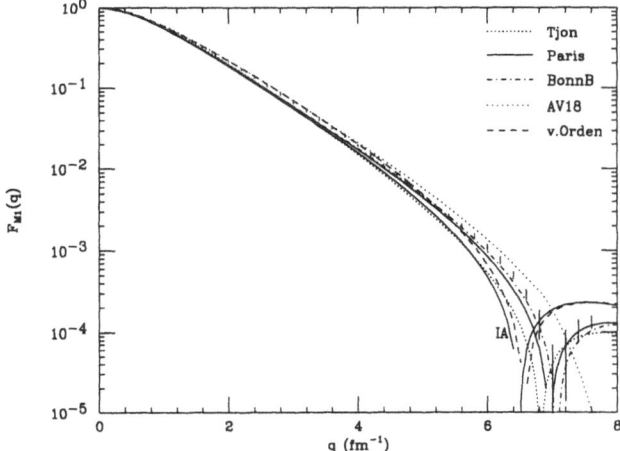

Figure 4. M1 form factor from world data (error bars) together with selected theoretical predictions.

prisingly good agreement between the predictions and the data. The biggest differences show up in the region of the diffraction minima. MEC do have a major effect, but they are smaller than for the (isovector) case of the A=3 form factors. There is no clear distinction between nonrelativistic and relativistic calculations; depending on the observable and the specific calculation, one or the other does better. Surprisingly, the nonrelativistic calculations do very well up to the largest momentum transfers, corresponding to wave function components of remarkably large momenta.

5   d(e,e'p) at large missing momenta

Recent advances in the experimental techniques have allowed us to significantly extend the measurements of the (e,e'p) cross sections. At NIKHEF, the use of large solid-angle hadron detectors in combination with the stretched electron beam made possible the measurement of much smaller cross sections. At MAMI, the combination of the high intensity CW beam and the large solid angle magnetic spectrometers led to an extension of the missing-momentum distribution all the way up to 900 MeV/c.

Figure 5 shows the (e,e'p) cross section as measured at NIKHEF [23]. The surprising feature of the distribution is the flattening out towards large missing momenta; according to the deuteron momentum distribution a continuing drop would be expected. Calculations by both Leidemann and Tjon can reproduce the data when allowing for 1) the charge-exchange process, and 2) excitation of an intermediary Δ via MEC processes. The importance of this latter process mainly results from the kinematics employed in this experiment which implies a large momentum-mismatch; the momentum transfer was *not* large as compared to the momenta studied.

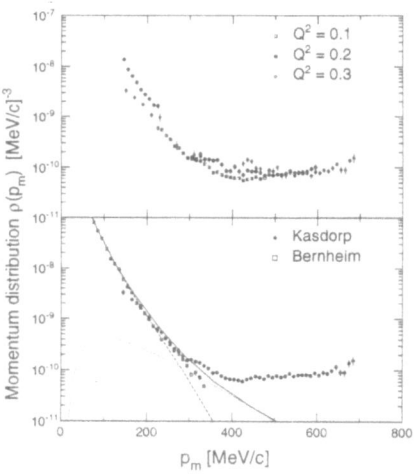

Figure 5. Cross section for d(e,e'p) as measured by Kasdorp et al.

The results of Boeglin et al. [24], measured at MAMI, are shown in Fig. 6. Under the kinematical conditions used here, the cross section keeps falling. While still present, the contribution of MEC, and the Δ in particular, is much smaller. The calculation of Arenhövel accounts quite well for the data. This experiment corresponds to kinematics that are closer to "parallel kinematics"; the systematic study of (e,e'p) cross sections measured on heavier nuclei [25] has shown that these conditions are much more favourable in suppressing non-IA contributions to the reaction mechanism. Recent (e,e'p) experiments on helium, also performed at NIKHEF, show similar features [26]. The spectra corresponding to large removal energies do show the behaviour expected from calculations of the correlated spectral function: a ridge at missing energy of $\sim k^2/2m_N$. The

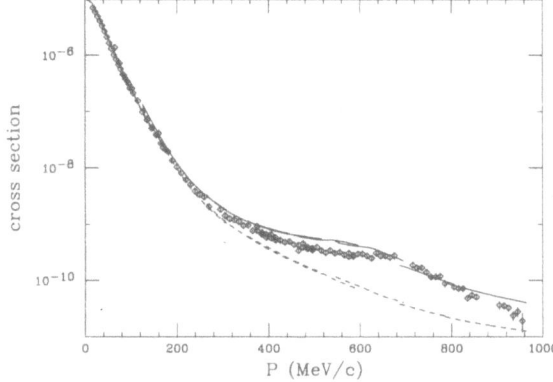

Figure 6. Cross section for d(e,e'p) as measured by Boeglin et al. Full calculation (solid), IA only (dashed).

cross section does however again exhibit rather large contributions due to non-IA processes, as a consequence of the low q employed.

In order to measure the spectral function at large missing energy and momenta, it will be necessary to suppress these non-IA contributions as much as possible, by working with kinematics involving large q, parallel kinematics and as little momentum mismatch as possible.

6   Neutron form factors

Precise knowledge of the nucleon form factors is important for both the understanding of nucleon structure and the prediction of nuclear form factors. Experiments aiming at the *neutron* form factors always have been hampered by the need to use nuclear targets. As a consequence, the neutron magnetic form factor G_{mn} is known to $\pm 10\%$ at best, while the neutron electric form factor G_{en} is known, over a rather small range of q, to $\pm 30\%$ at best. These uncertainties in general are *not* related to experimental uncertainties; they originate from the fact that the systematic theoretical uncertainties in the interpretation of the observables experimentally studied are too large.

Today's experimental tools allow us to make significant progress. In particular, for G_{mn}, CW electron beams make the study of the d(e,e'n) reaction under quasifree kinematics feasible; to extract G_{mn} from the experimentally measurable ratio d(e,e'n)/d(e,e'p) one needs very small theoretical corrections (of a few %) which can be calculated to $<1\%$. The high-intensity neutron beams available at today's hadron facilities allow us to produce *tagged, monoenergetic* neutron beams that can be used to accurately calibrate the neutron detector efficiency.

Such an experiment has recently been completed by the Basel/Mainz collaboration [27] who used the MAMI electron beam and the PSI neutron beam. Figure 7 shows the data. The upper panel gives the previously available form factors and displays the large scatter. The lower panel gives the new data in the form of a ratio to the dipole parameterization. G_{mn} now is determined to

±1.5% accuracy; data to higher q still are under analysis.

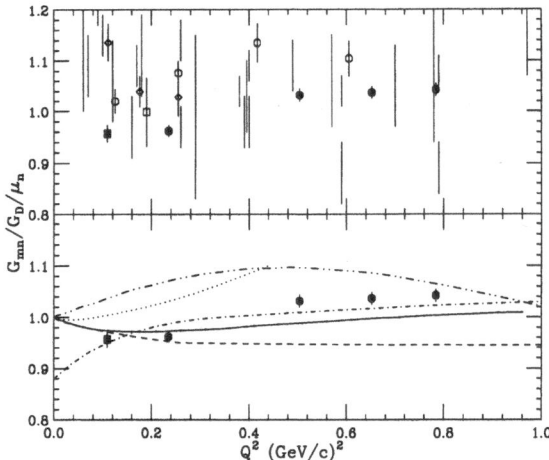

Figure 7. Neutron magnetic form factor: previous data (top), data from MAMI/PSI experiment with various model predictions (bottom).

The neutron *charge* form factor is even more poorly known; on top of the uncertainties introduced by the nuclear corrections comes the uncertainty arising from the subtraction of the dominant magnetic contribution to the cross section. In Fig. 8 we show the most accurate previous data, obtained from a very accurate (±1%) measurement of the electron-deuteron elastic cross section [7]. The large uncertainties result from the variance in the predictions of deuteron structure and MEC . When using e.g. the Paris potential, one obtains the experimental values of G_{en} and the fit labeled "Paris" in Fig. 8. When using a different N-N potential to predict deuteron structure, one obtains different values of G_{en} (not shown) and different fits (shown).

Today's technology allows us to employ scattering of *polarized* electrons from *polarized* neutrons, either $n(e, e')n$ or $n(e, e')n$. If the spin of the neutron is perpendicular to q, the polarization asymmetry is proportional to the interference term $G_{mn} \cdot G_{en}$. In order to avoid subtractions when using *nuclear* neutron targets such as polarized deuterons or ^{3}He, a coincident detection of the recoil neutron and electron is needed.

Such experiments have recently been carried out by several collaborations working at the Mainz Microtron [28, 29, 30, 31]. These collaborations either have measured the recoil neutron polarization in a double scattering experiment, or they have used a high pressure (several bar) polarized ^{3}He target. The resulting — preliminary — values for G_{en} are shown in Fig. 8. The experiments employing polarization observables are rapidly improving in quality; as their error bars are statistical in nature rather than systematic, important further improvements are to be expected. Experiments presently being set up at both MAMI and TJNAF will provide much more accurate data in the near future.

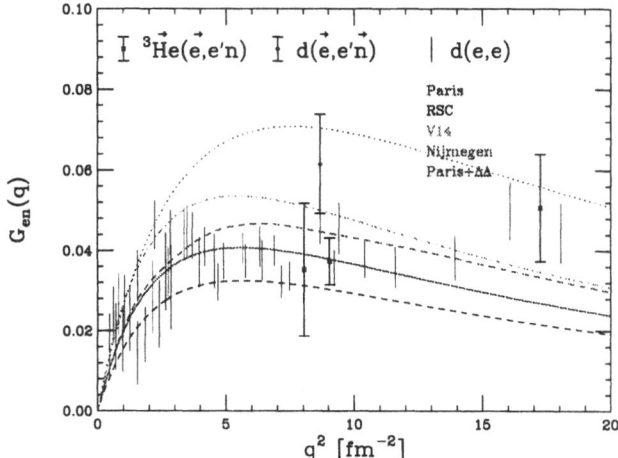

Figure 8. Neutron electric form factor: previous data (|) obtained using the deuteron body form factor and MEC for the Paris potential, together with fit (solid line). The fits obtained using different N-N models are also shown. The new data (preliminary) obtained by different collaborations working at MAMI are shown as points with error bars.

7 Conclusions

During the last years considerable progress has been made. This is due to 1) the availability of CW beams, which make feasible the measurement of better data, and the investigation of new observables, 2) higher momentum transfers, which lead to less momentum mismatch and a cleaner interpretation of the data , and 3) the measurement of polarization observables, which give access to the smaller wave function components and a better sensitivity to MEC.

At the same time considerable progress has been made in theory. This is mainly due to the achievement of an accurate description (for non-relativistic energies) of the 3N-continuum system; this now allows us to use the 3N system in the same way as previously the 2N-system, with the difference that the 3N-system is much richer in terms of observables accessible.

References

1. G.J. Schmit et al.: Phys. Rev. Lett. **76**, 3088 (1996)

2. H. Anklin et al.: Nucl. Phys. **A636**, 189 (1998)

3. M. Kohno: J. of Physics **G9**, L85 (1983)

4. S. Klarsfeld et al.: Nucl.Phys. **A456**, 373 (1986)

5. A. Buchmann, H. Henning and P.U. Sauer: Few Body Syst. **21**, 149 (1996)

6. C.W. Wong: Int. J. Mod. Phys. **E3**, 821 (1994)

7. S. Platchkov et al.: Nucl. Phys. **A510**, 740 (1990)

8. G.G. Simon, Ch. Schmitt and V.H. Walther: Nucl. Phys. **A364**, 285 (1981)

9. R.W. Berard et al.: Phys. Lett. **47B**, 355 (1973)

10. G. Hoehler et al.: Nucl. Phys. **B114**, 505 (1976)

11. I. Sick: Nucl. Phys. **A218**, 509 (1974)

12. I. Sick and D. Trautmann: Phys. Lett. **B375**, 16 (1996)

13. I. Sick and D. Trautmann: Nucl. Phys. **A** (in print)

14. R.R. Lewis: Phys. Rev. **102**, 537 (1956)

15. S.D. Drell and R.H. Pratt: Phys. Rev. **125**, 1394 (1962)

16. T. Herrmann and R. Rosenfelder: Eur. Phys. J. **A2**, 29 (1998)

17. J.L. Friar, J. Martorell and D.W.L. Sprung: Phys. Rev. **A56**, 5173 (1997)

18. K. Pachucki et al.: J. Phys. **B29**, 177 (1996)

19. E. Hummel and J.A. Tjon: Phys. Rev. **C49**, 21 (1994)

20. J.W. van Orden, N. Devine and F. Gross: Phys. Rev. Lett. **75**, 4369 (1995)

21. H. Henning: Private Communication (1997)

22. R. Schiavilla and D.O. Riska: Phys. Lett. **B244**, 373 (1990)

23. W.J. Kasdorp: Thesis. Univ. Utrecht 1997 (unpublished)

24. K.I. Blomqvist et al.: Phys. Lett. **B424**, 33 (1997)

25. I. Sick: In: *Proc. Elba Conf. Electron-Nucleus Scattering*, eds. O. Benhar, A. Fabrocini, p. 445. World Scientific 1997

26. J.J. van Leeuwe: Thesis. Univ. Utrecht 1996 (unpublished)

27. H. Anklin et al.: Phys. Lett. **B428**, 248 (1998)

28. M. Meyerhoff et al.: Phys. Lett. **B327**, 201 (1994)

29. J. Becker: Thesis. Univ. Mainz 1997 (unpublished)

30. M. Ostrick: Thesis. Univ. Mainz 1998 (unpublished)

31. D. Rohe: Thesis. Univ. Mainz 1998 (unpublished)

Few-Body Systems Suppl. 10, 315–322 (1999)

Few-
Body
Systems
© by Springer-Verlag 1999

Deuteron Electromagnetic Form Factors

E. J. Beise

representing the JLAB t_{20} collaboration *

University of Maryland, College Park, MD 20742 USA

Abstract. Detailed studies of the electromagnetic structure of the deuteron are fundamental to a global understanding of the strong interaction, either between two nucleons or between quarks embedded in hadronic matter. Recently there has been experimental progress on the determination of deuteron electromagnetic structure from elastic electron scattering measurements at Jefferson Lab and NIKHEF. In this paper a brief review of recent experiments is presented.

1 Introduction

The deuteron is the simplest nuclear system, and as a result is one of the best media in which to study the NN interaction. Elastic electron scattering allows one to investigate the ground state electromagnetic structure of the deuteron with high precision. At momentum transfers below $Q^2 \sim 0.4$ $(\text{GeV}/\text{c})^2$, the deuteron form factors are well determined by the non-relativistic impulse approximation (IA) with small theoretical uncertainties. In the region between 0.4-1 $(\text{GeV}/\text{c})^2$, various corrections to the IA become important, particularly isoscalar meson-exchange currents (MEC) and relativistic effects. Above 1 $(\text{GeV}/\text{c})^2$ where the single nucleon wave functions must have considerable overlap, it is of particular interest to try to identify whether conventional nucleon-based models begin to break down or whether models using quark-based degrees of freedom are required.

Experimentally, it continues to be a challenging problem to fully dissect the electromagnetic structure of the deuteron. Since the deuteron is a spin 1 object, it has three elementary form factors describing its charge monopole (G_C), charge quadrupole (G_Q) and magnetic dipole (G_M) distributions. With no polarization observables, the cross section for elastic e-d scattering can be written in terms of the elementary cross section for scattering from a spinless

---

*see *http://t20.jlab.org/t20_collab.html* or [1] for the full collaboration list

pointlike particle, $\left(\frac{d\sigma}{d\Omega}\right)_M$, multiplied by a function $\mathcal{S}$ which contains the details of the deuteron's internal structure

$$\frac{d\sigma}{d\Omega} = \left(\frac{d\sigma}{d\Omega}\right)_M \cdot \mathcal{S} = \left(\frac{d\sigma}{d\Omega}\right)_M \cdot \left[A(Q) + B(Q)\tan^2\left(\frac{\theta_e}{2}\right)\right]$$

The structure functions A and B depend only on the 4-momentum Q transfered from the electron to the deuteron. Using the traditional Rosenbluth method it is possible to independently extract A and B, and the magnetic form factor G_M is uniquely determined by $B(Q)$, but A contains a linear combination of all three elementary EM form factors:

$$
\begin{aligned}
A(Q) &= G_C^2(Q) + \frac{8}{9}\eta^2 G_Q^2(Q) + \frac{2}{3}\eta G_M^2(Q) \\
B(Q) &= \frac{4}{3}\eta(1+\eta)G_M^2(Q)
\end{aligned}
\tag{1}
$$

where $\eta = Q^2/4M_d^2$.

A third quantity is thus necessary to separate G_C and G_Q. Tensor polarization observables can accomplish this goal. Either the analyzing powers (T_{2i}) are extracted from asymmetries in elastic scattering of unpolarized electrons from a polarized deuteron target [2], [3], [4], [5], or the tensor moments (t_{2j}) are extracted through recoil polarization [6], [7]. The two methods give equivalent information, and the resulting deuteron tensor moments are:

$$
\begin{aligned}
t_{20} &= -\frac{1}{\sqrt{2}\,\mathcal{S}}\left[\frac{8}{3}\eta G_C G_Q + \frac{8}{9}\eta^2 G_Q^2 + \frac{1}{3}\eta\left[1 + 2(1+\eta)\tan^2\left(\frac{\theta_e}{2}\right)\right]G_M^2\right] \\
t_{21} &= \frac{2}{\sqrt{3}\,\mathcal{S}}\,\eta\left[\eta + \eta^2\sin^2\left(\frac{\theta_e}{2}\right)\right]^{1/2}G_M G_Q \sec\left(\frac{\theta_e}{2}\right) \\
t_{22} &= -\frac{1}{2\sqrt{3}\,\mathcal{S}}\,\eta G_M^2.
\end{aligned}
\tag{2}
$$

Figure 3 summarizes the previous and new data for t_{20}, A and the resulting G_C, as well as a selection of calculations, all of which are discussed below.

2 Polarized Target Experiments at NIKHEF

Four of the measurements of T_{20} shown in Fig. 3(a) used a tensor polarized deuteron target [2] (open circles), [3] (open squares), [4] (star), [5] (solid triangles). One of the primary sources of systematic error in the data prior to 1995 has been absolute determination of the target polarization while the target is in the beam. As a result, typically one data point was used as a normalization for the others in any given measurement. In the two recent NIKHEF experiments [4], [5], the deuteron polarization was directly measured, resulting in absolute normalization of all points and considerably smaller systematic errors.

Unpolarized electrons at 700 MeV from the AmPS storage facility were incident on a tensor polarized internal target. The polarized deuterium was prepared from an atomic beam source and fed into a T-shaped "dwell cell", resulting in an effective target thickness of 2×10^{13} atoms/cm^2. Scattered electrons were detected with an array of 60 thallium-doped cesium-iodide crystals, and the recoiling deuterons (and protons from deuteron breakup) were detected with a 16-layer scintillator range telescope. The experimental quantity of interest in the polarized target experiments is the asymmetry in scattering from deuterons with net polarization parallel or antiparallel to the direction of momentum transfer. The target polarization was determined using two polarimeters in conjunction: a Breit-Rabi polarimeter sampled the deuterium coming from the atomic beam source, and an ion-extraction system which monitored the relative deuterium polarization along the length of the target cell in the beam. Corrections to the data were made to account for the fact that the direction of the deuteron polarization was not constant over the length of the target, using calculated values of T_{21} and T_{22} from ref. [8]. The resulting values for T_{20} are the solid stars ([4]) and solid triangles ([5]) in Fig. 3(a).

3 Recent Results from Jefferson Lab

In 1997 two experiments on elastic e-d scattering were performed at the CEBAF accelerator of Jefferson Laboratory in Newport News, VA, USA. In Hall A, precise measurements of A and B were carried out. Unpolarized electrons of energies up to 4 GeV were scattered from a 15 cm liquid deuterium target, and the scattered electron and recoiling deuteron were detected in coincidence using the high resolution spectrometer pair. $A(Q^2)$ was measured over the range $0.5 < Q^2 < 6.5$ (GeV/c)^2, and $B(Q^2)$ over $0.5 < Q^2 < 1.4$ (GeV/c)^2. A more complete description of this experiment along with preliminary results can be found in [9].

In Hall C at CEBAF, a measurement of t$_{20}$ was carried out at $0.68 < Q^2 < 2.0$ (GeV/c)^2. A 110 μA beam of unpolarized electrons ranging from 1.4 to 4.0 GeV were scattered from a 12 cm liquid deuterium target. The scattered electrons were detected in the Hall C high momentum spectrometer (HMS), between the angles of 20.3° and 35.8°. The recoiling deuterons were focussed onto a polarimeter using a specially designed QQQD magnetic channel which also protected the polarimeter from direct view of the target. The fraction of recoiling deuterons collected by the QQQD channel varied from 1.0 to 0.5 over the kinematic range of the experiment.

The polarimeter (POLDER) [10] utilized the analyzing reaction $p(\boldsymbol{d},2p)n$, which has a large sensitivity to the deuteron's tensor polarization but no vector analyzing power. In kinematics where the recoiling neutron has low momentum, the $\boldsymbol{d}$-2p cross section and analyzing power are well described theoretically [11] by a simple charge-exchange process. The efficiency of the polarimeter can be written as (assuming no vector analyzing power)

318

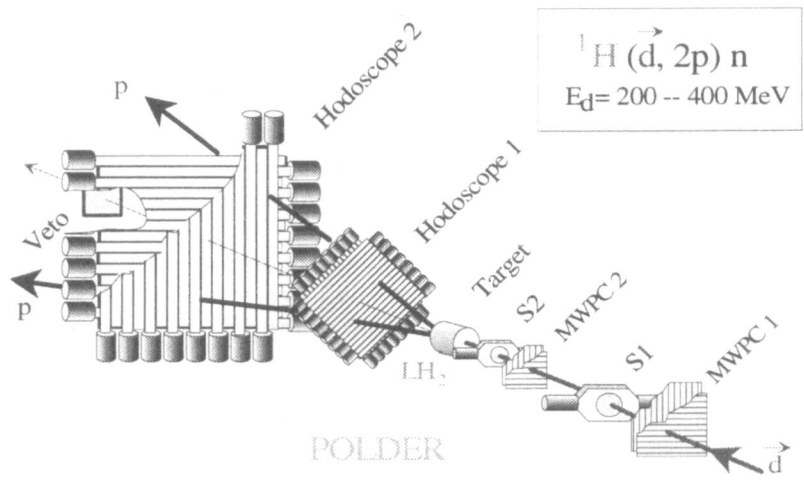

Figure 1. Schematic view of the POLDER polarimeter (see ref. [10] for a detailed description).

$$\epsilon_{pol}\,(\theta,\varphi) = \epsilon_0(\theta)\,[1 + t_{20}\,T_{20}(\theta) + 2\,\cos\phi\,t_{21}\,T_{21}(\theta) + 2\,\cos2\phi\,t_{22}\,T_{22}(\theta)]\quad(3)$$

where ϵ_0 is the unpolarized detector response and T_{2j} are the tensor analyzing powers of the *d*-2*p* reaction.

A schematic view of POLDER is shown in Fig. 1. The number and direction of incident deuterons were determined with two trigger scintillators and six planes of wires in two multiwire proportional chambers upstream of a 20 cm long liquid hydrogen target. The two protons emitted from the charge exchange reaction were detected with two pairs of scintillator hodoscopes. Using only timing information from the hodoscopes, the direction (θ,ϕ) of the center of mass of the two protons relative to the incident deuteron and the opening angle between them were reconstructed. The deuteron polarization was then deduced from modulations of the proton yield in θ and ϕ relative to the previously determined unpolarized yield.

In Fig. 2 is shown a two-dimensional plot of target mass as calculated by the electron arm only, and coincidence time between the deuteron channel and the HMS, for the data at 2.4 GeV incident electron beam energy. Coincident deuterons are clearly identified at M=1.87 (GeV/c^2). Coincident protons which were within the HMS momentum acceptance were removed from the raw data by adjustment of the coincidence timing electronics, and the remaining background of random protons was removed with software cuts.

Prior to its use at CEBAF, POLDER was calibrated using polarized deuteron beams from the SATURNE synchrotron, using deuteron kinetic ener-

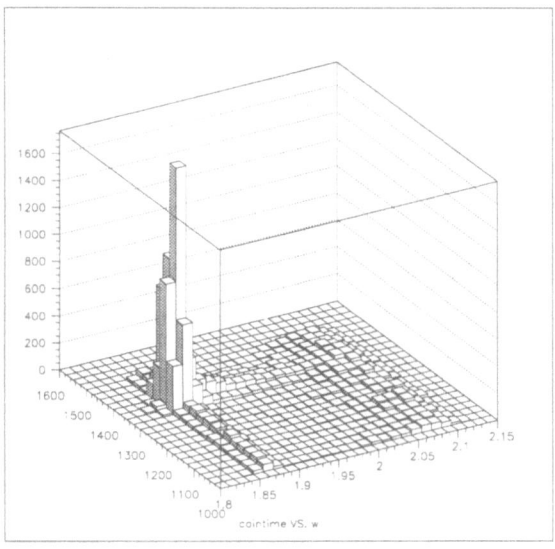

Figure 2. Reconstructed target mass vs. coincidence time (in raw TDC channels) for 2.4 GeV incident beam. The deuterons are located at M=1.87 (GeV/c^2).

gies from 140 to 520 MeV in steps of approximately 20 MeV. Both ϵ_0 and T_{ij} were determined with good precision, and the vector analyzing power T_{11} was measured to be 0.

Preliminary results for t_{20} are shown in Fig. 3(a), along with previous data and various models. The JLAB error bars include both statistical errors resulting from the number of elastic e-d events, and systematic errors, the largest of which comes from 3-4% variations in the relative normalization k between the SATURNE calibration data and the JLAB data. With further analysis, it is anticipated that the uncertainty in k can be reduced to 1%, which should result in a reduction in the errors by about a factor of two.

In addition to the t_{20} data, unpolarized cross sections were measured, which allowed the determination of $A(Q^2)$, shown in Fig. 3(b). These data are in good agreement with both the older SLAC data (solid stars)[12] and that from the recent JLAB Hall A experiment [9]. Earlier data from CEA [13] (open triangles) are also shown. Combined with the world's data set for $B(Q^2)$, the form factors G_C and G_Q can then be extracted. Shown in Fig. 3(c) is $G_C(Q^2)$ using t_{20} and A determined from only the Hall C experiment (solid circles): the errors are at present dominated by those of t_{20}. The node in G_C is at about 0.68 (GeV/c)^2, compared to 0.75 (GeV/c)^2 found in the 1991 Bates data (ref. [7], solid squares).

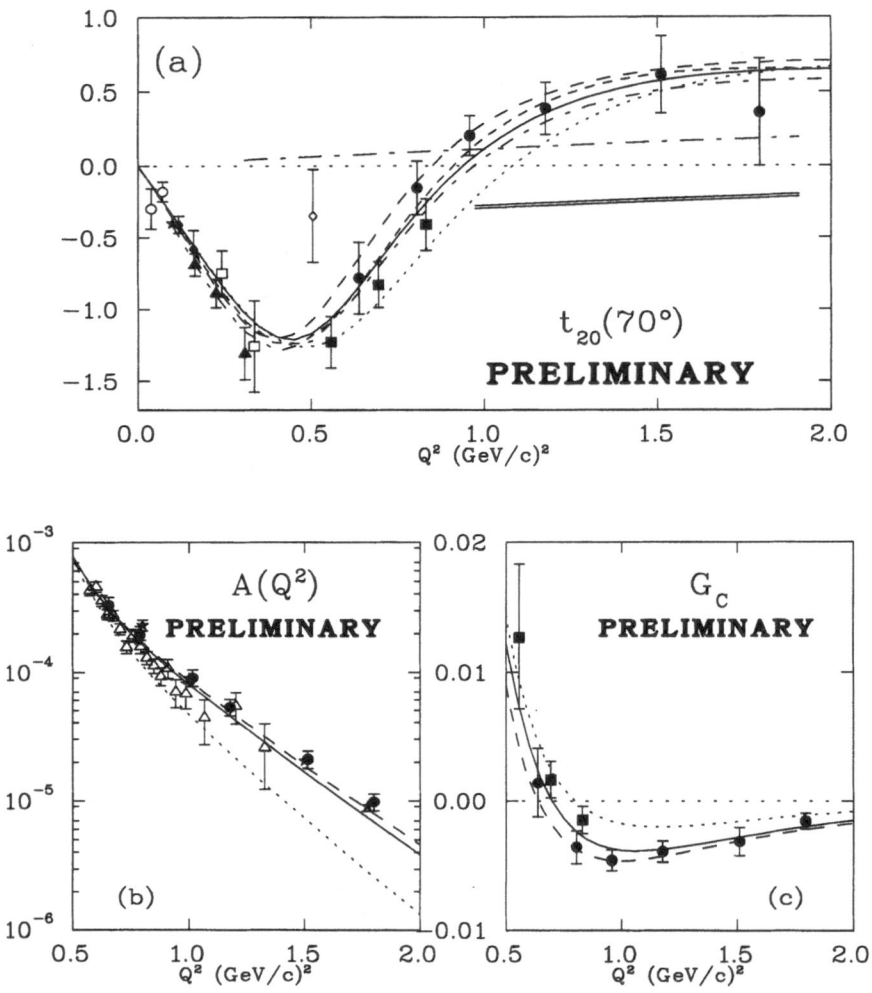

Figure 3. Preliminary results for t_{20}, A and G_C from the JLAB Hall C experiment (solid circles), shown with selected calculations and other data (see text for references).

4 Models of Elastic Form Factors

Numerous theoretical models of the deuteron elastic form factors exist in the literature and it is not possible to cover all models in this paper. Instead a few representative cases are shown in Fig. 3 along with the experimental data discussed below.

In ref. [8], Wiringa et al. calculated deuteron electomagnetic form factors using the non-relativistic Argonne v18 potential. A small charge-dependent term is included in the potential as a result of fitting both pp and np scattering data, as well as low energy nn scattering parameters. The long-dashed and dotted lines are the results of the calculation with and without meson exchange corrections included, respectively. With the meson exchange terms, the model gives good agreement with both the A and the JLAB t_{20} data, although B (not shown) is somewhat overpredicted (see [8]).

Several new relativistic calculations have recently become available, using different prescriptions but producing very similar results for deuteron electromagnetic observables. Van Orden et al. [14], (solid line) used a relativistically covariant impulse approximation within a quasipotential formalism, similar to the earlier calculation of Hummel and Tjon [15], but with the inclusion of additional diagrams where one nucleon is completely off-shell. The NN interaction used is fitted to Nijmegen energy dependent np phase shifts, with small P-wave components which originate from purely relativistic effects. The calculated node of B is very sensitive to P-wave interference terms. Phillips et al. [16] (short dashes) produced a similar calculation using equal-time formalism which results in a more symmetric inclusion of off-shell effects. Parameters from the Bonn-B potential are used for the interaction. Both calculations include a $\rho\pi\gamma$ meson exchange term. Carbonell et al. (dot-short dash) [17] used the light-front formalism (also with the Bonn potential) where meson exchange currents partially appear automatically in the impulse approximation part of the calculation. All three models give similarly good results for t_{20} and A but ref. [14] reproduces the minimum in B somewhat better than the other two; results for A and G_C from [14] are also shown in Figs. 3(b) and 3(c). Also included in Fig. 3(a) are representative perturbative QCD predictions for t_{20} ([18], dot-long dash), ([19], solid double line).

5 Summary and Outlook

Future directions in the study of deuteron electromagnetic form factors are two-fold. Improved precision in the region of the node of G_C, particularly with a different technique, could provide an important experimental check on the JLAB data. The next phase of the NIKHEF internal target measurements with the AmPS storage ring are planned for the very end of 1998, but may not take place due to the pending shutdown of the facility. Internal target experiments are also planned at the MIT-Bates laboratory using the BLAST detector [20], where measurements with comparable or better precision would be possible.

It would also be of great interest to extend the t_{20} data to higher momentum transfer. Although at present the highest Q^2 point from JLAB is consistent with at least one perturbative QCD calculation, the data taken as a whole are in much better agreement with "nucleons-only" models. The $A(Q^2)$ data from Hall A [9] appear to indicate a scaling trend consistent with quark-like degrees of freedom above about 4 $(GeV/c)^2$. Measurements of t_{20} at higher Q^2 could help elucidate this point, since the nucleon models and pQCD calculations display very different behaviour in this regime. A measurement at higher momentum transfer would however likely require either a new design of polarimeter based on some other analyzing reaction, or a polarized target combined with a high

322

luminosity, high energy electron beam [21].

References

1. C. Furget et al.: contributed to MESON98 (Cracow, Poland, June 1998)

2. V.F. Dmitriev et al.: Phys. Lett. **157B**, 143 (1985)

3. R. Gilman et al.: Phys. Rev. Lett. **65**, 1733 (1990)

4. M. Ferro-Luzzi et al.: Phys. Rev. Lett. **77** , 2630 (1996)

5. M. Ferro-Luzzi: contribution to the *7th International Conference on Polarized Gas Targets and Polarized Beams* (U. Ill., Aug 1997). Also M. Bouwhuis, Ph.D. thesis, University of Utrecht (1998), ISBN 90-393-1754-2

6. M.E. Schulze et al.: Phys. Rev. Lett. **C52**, 597 (1984)

7. I. The et al.: Phys. Rev. Lett. **67**, 173 (1991), and M. Garçon et al.: Phys. Rev. **C49**, 2516 (1994)

8. R.B. Wiringa, V.G.J. Stoks and R. Schiavilla: Phys. Rev. **C51**, 38 (1995)

9. J. Gomez: these proceedings

10. J.S. Réal: Thesis of the University of Grenoble I, ISN 94-05, unpublished; S. Kox et al.: Nucl. Inst. Meth. **A346**, 527 (1994)

11. D.V. Bugg and C. Wilkin: Phys. Lett. **B152** , 37 (1985) ; J. Carbonell, M. Barbaro and C. Wilkin: Nucl. Phys. **A529**, 653 (1991)

12. R. Arnold et al.: Phys. Rev. Lett. **35**, 776 (1975)

13. J. E. Elias et al.: Phys. Rev. **177**, 2075 (1969)

14. J.W. Orden, N. Devine and F. Gross: Phys. Rev. Lett. **75**, 4369 (1995)

15. E. Hummel and J.A. Tjon: Phys. Rev. **C49**, 21 (1994)

16. D.R. Phillips, S.J. Wallace, and N.K. Devine: Los Alamos preprint nucl-th/9802067 (also U Md preprint PP#98-093)

17. J. Carbonell, B. Desplanques, V.A. Karmanov and J.F. Mathiot: Phys. Reports **300**, 215 (1998)

18. S.J. Brodsky and J.R. Hiller: Phys. Rev. **D46**, 2141 (1992)

19. A. Kobushkin and A. Syamtomov: Phys. Rev. **D49**, 1637 (1994)

20. W. Turchinetz: MIT, private communication

21. see R. Gilman: in the proceedings of the *Workshop on Physics with 8 GeV Beams and Beyond*, Jefferson Lab, June 1998, and references therein

Few-Body Systems Suppl. 10, 323–326 (1999)

Few-
Body
Systems
© by Springer-Verlag 1999

The A_y puzzle and the nuclear force

D. Hüber

Theoretical Division, Los Alamos National Laboratory, M.S. B283, Los
Alamos, NM 87545, USA

Abstract. The nucleon-deuteron analyzing power A_y in elastic nucleon-
deuteron scattering poses a longstanding puzzle. At energies E_{lab} below approximately 30 MeV, A_y cannot be described by any realistic NN force. The
inclusion of existing three-nucleon forces does not improve the situation. Because of recent questions about the ^3P_J NN phases, we examine whether reasonable changes in the NN force can resolve the puzzle. In order to do this, we
investigate the effect on the ^3P_J waves produced by changes in different parts of
the potential (viz., the central force, tensor force, etc.), as well as on the 2-body
observables and on A_y. We find that it is not possible with reasonable changes
in the NN potential to increase the 3-body A_y and at the same time to keep
the 2-body observables unchanged. We therefore conclude that the A_y puzzle
is likely to be solved by new three-nucleon forces, such as those of spin-orbit
type, which have not yet been taken into account.

1 Introduction

The A_y puzzle is one of the main open problems in nuclear physics. The vector
analyzing powers A_y and iT_{11} are the only precisely measured observables in
low energy elastic nucleon-deuteron (Nd) scattering which cannot be described
by any of the modern NN potentials. The discrepancy is about 30% at energies of $E_{lab} = 3$ MeV or lower and becomes smaller with increasing energy,
until it disappears at about 30 MeV. The inclusion of today's three-nucleon
forces (3NFs) has either only very little effect, or, as in the case of the Tucson-
Melbourne (TM) 3NF, goes into the wrong direction. Attempts to solve this
puzzle by changes in the ^3P_j nucleon-nucleon (NN) forces (to which the vector
analyzing powers are most sensitive) did not lead to satisfactory results. For
an overview on the A_y puzzle, see [1].

In [2] we studied systematically which improvements in the description of
A_y can be made by changes in the NN potential. Contrary to earlier attempts
we did not change the NN potential in the ^3P_j channels as a whole, but rather
term by term (i.e., central force, tensor force, *etc.*). Also, since that question

Table 1. Possible changes in the NN potential which lead to an improved description of A_y. See text for explanation.

δ_C	δ_{l^2}	$\delta_{(ls)^2}$	δ_T	δ_{ls}	δ_{A_2}	δ_{A_y}
10	−10	0	52.0	32.0	3.6	54
10	−10	−10	49.4	30.4	5.1	55
22	−22	0	46.4	22.4	6.2	55
22	−22	−22	40.7	18.9	8.5	57
30	−30	0	42.7	16.0	7.4	56
30	−30	−30	34.9	11.2	9.6	57

was raised recently, we had a look into the question how much room is given for changes of the low energy ^3P_j NN phase-shifts. Our findings in ref. [2] are reported in the next section.

This study is carried through within the non-relativistic framework of the Lippmann-Schwinger and Faddeev equations.

We assume that any NN potential has to describe the NN data base. All modern NN potentials do that, and they all give the same answer for A_y, though their off-shell behaviour is quite different. We also assume that the conventional one-pion-exchange potential is correct. (For more details on our assumptions, we refer to [2].)

2 Results

As a working vehicle for this study, we chose the AV18 NN potential [3] due to its simplicity. This potential has a one-pion-exchange part (OPEP) for the long range, as well as phenomenological short range operators. In the ^3P_j channels these operators are 1 (central force), S_{12} (tensor force), $\boldsymbol{l} \cdot \boldsymbol{s}$ (spin-orbit force), $\boldsymbol{l}^2$ and $(\boldsymbol{l} \cdot \boldsymbol{s})^2$. The OPEP has a central and a tensor part. Whereas the tensor force is almost exclusively OPEP, the central force is dominated by the short range part.

We changed all the terms in the ^3P_j AV18 potential, one after the other, by $\pm 10\%$ and looked for the effects of these changes in the two-body and three-body nucleon analyzing powers. It turns out that if one wants to keep the two-body observable unchanged and wants to improve the three-body observable at the same time, one cannot avoid huge changes of 30-50% in the tensor force.

Table 1 shows some possible changes of the potential which lead to an improved description of A_y. Thereby δ_C, δ_{l^2}, $\delta_{(ls)^2}$, δ_T, and δ_{ls} are the changes of the corresponding terms in the AV18 potential in percent. δ_{A_2} and δ_{A_y} are the resulting changes in the two-body (A_2) and three-body (A_y) nucleon analyzing powers (the changes in A_2 are for $E_{lab} = 1$ MeV, the changes for A_y for $E_{lab} = 3$ MeV). Table 1 shows clearly that always a huge increase of the tensor force is needed in order to increase the prediction of A_y. Since the changes in the potential as given in Table 1 are only rough estimates, the prediction of A_2 is also increased for all examples. But the increase achieved for A_y is always much larger than for A_2, and much larger as the required 30% to describe the data. So one can imagine that it is possible with some fine tuning to keep the change in A_2 at an acceptable level and still have a large enough increase for A_y.

Unfortunately, as already noted above, the tensor force is almost exclusively OPEP. Thus an improvement in the description of A_y, and keeping the prediction for the two-body analyzing power A_2 unchanged at the same time, would require changes of 30-50% in the OPEP, which is impossible.

If one requires at the same time the predictions for the ^3P_j NN phase shifts to be unchanged, one is led to even more drastic changes in the potential, which are totally out of the question.

The next obvious question is then, are the NN phase-shifts determined correctly by the current phase-shift analyses (PSAs). Recently some doubt was put on whether the low energy ^3P_j NN phases are correct. We find it plausible that in a multi-energy PSA the solution for the phase-shifts is very likely to be unique. The constraints in a multi-energy PSA put onto the phases are continuity and analyticity. We show that just the requirement of the correct threshold behaviour allows no room for any changes in the low energy ^3P_j phase-shifts. The only additional assumption is that OPEP is correct. Indeed, OPEP is a very important input for PSAs. Our study repeats the earlier findings about the important role of OPEP.

Ambiguities for the ^3P_j phase-shifts, which are found in single-energy PSAs, are misleading and are known to be resolved by performing a multi-energy PSA and by putting in additional physics, which means OPEP in our case.

This finding is confirmed by a try to fit the AV18 NN potential [4] to NN phase-shifts which were modified at low energies. It turned out that a good fit to the phase-shifts is only possible by changing the pion-nucleon coupling constant significantly. Also, though the refitted potential cannot describe the modified phases at the lowest energies, its prediction for the phases at higher energies changes, too, and sometimes even fall outside the error bars of the Nijmegen PSA.

We can also exclude charge independence breaking (CIB) and charge symmetry breaking (CSB) as a solution for the A_y puzzle. An investigation of CIB and CSB effects as they are predicted by meson-exchange models [5] reveals that these effects are too small and partly have the wrong sign.

Long range electromagnetic corrections to the NN potential as a solution of the A_y puzzle are ruled out by ref. [6].

Thus we have ruled out any possibilities to solve the A_y puzzle on the two-body level. Therefore the only possible solution must be a 3NF. It must be a 3NF term which has not yet been tried, and due to the nature of the vector analyzing powers they must be of spin-dependent type, too. Such terms exist. Current 3NFs contain only such terms which were believed to be most important and least complicated. At the time the current 3NFs were developed it was not possible to test them in a realistic scattering calculation. So it is not unlikely that current 3NF models miss terms which are important in the vector analyzing powers.

We also comment on the fact that a 3NF is always equivalent to an off-shell ambiguity in the NN potential. Though this is a problem in principle, we show that it is not a practical one. Also, the fact that the Bonn potentials give almost the same prediction for A_y as all the other modern NN potentials, though the Bonn potentials have a totally different off-shell behaviour, shows that off-shell differences in the current NN potentials play an unimportant role for A_y.

3 Outlook

On a first view, it is striking that only the TM 3NF has a significant effect on A_y. The difference between the TM 3NF and the other 3NFs tried so far on A_y is that the latter ones do not have the so-called c-term of the TM 3NF. So a term like this c-term (with opposite sign) would be a likely candidate to solve the A_y puzzle. But as is shown in [7] the c-term in the TM 3NF should be dropped. On the other hand, terms similar to the c-term arise from chiral perturbation theory and are candidates for solving the A_y puzzle [7]. Work on these terms is underway.

Acknowledgement. This work was performed in part under the auspices of the U.S. Department of Energy. The work of D.H. was supported in part by the Deutsche Forschungsgemeinschaft under Project No. Hu 746/1-3. The numerical calculations have been performed on the Cray T90 of the Höchstleistungsrechenzentrum in Jülich, Germany.

References

1. W. Glöckle et al.: Phys. Rep. **274**, 107 (1996)

2. D. Hüber and J.L. Friar: Phys. Rev. **C58**, 674 (1998)

3. R.B. Wiringa, V.G.J. Stoks and R. Schiavilla: Phys. Rev. **C51**, 38 (1995)

4. R.B. Wiringa and V.G.J. Stoks: Private communication

5. G.Q. Li and R. Machleidt: submitted for publication; R. Machleidt: Private communication

6. V.G.J. Stoks: Phys. Rev. **C57**, 445 (1998)

7. J.L. Friar, D. Hüber and U. van Kolck: submitted for publication

Few-Body Systems Suppl. 10, 327–330 (1999)

Few-
Body
Systems
© by Springer-Verlag 1999

Measurements of the analyzing power of elastic pd-scattering at 190 MeV

M. Seip, R. Bieber, A.G. Drentje, M.J. van Goethem, M.N. Harakeh,
M. Hoefman, H. Huisman, N. Kalantar-Nayestanaki, H.R. Kremers,
J.G. Messchendorp, M. Volkerts, S.Y. van der Werf, H.W. Wilschut

Kernfysisch Versneller Instituut, Zernikelaan 25, 9747 AA Groningen, The
Netherlands

Abstract. A beam-polarimeter for medium energy protons and deuterons
(E=100-200 MeV) has been constructed. Its operation is based on the $p + d$,
$p + p$ and $d + p$ elastic scattering with coincident detection of both particles
in 4 independent planes. The 16 detectors of the polarimeter have a phoswich
construction to ensure particle identification. With this device the analyzing
power, A_y, of $p + d$ elastic scattering has been measured at 190 MeV for CM-
angles of 25° to 115° in two planes, monitored by $p + p$ elastic scattering
in the two other planes. The target was mixed CH_2/CD_2. This allowed for
simultaneous extraction of $A_y(p, d)$ and the beam polarizations. The analyzing
power is compared with calculations based on the Bonn-potential including
partial waves up to $J = 5$.

1   Experimental Setup

At KVI, the new superconducting cyclotron AGOR provides polarized proton beams in the energy range 120-200 MeV and vector and tensor polarized
deuteron beams in the energy range 70-200 MeV. To measure the degrees of polarization of these beams a beam-polarimeter has been constructed and tested.
A transection of its design is given in Fig. 1. The instrument consists of 8 arms,
45° apart in azimuthal angle. In each arm the polar angle at which detectors
can be placed ranges from 15° to 80°. In each of the 8 arms 2 phoswich detectors can be placed. The phoswich detectors are based on the $\Delta E - E$ principle,
i.e. each of them consists of 2 scintillator-layers, 2 or 4 mm NE102A and 10
cm NE115 so that particle identification can be obtained. Figure 2 illustrates
the particle identification obtained when 200 MeV α-particles are scattered off
a thin carbon target.

For the polarization-measurements only kinematically coincident events of
the $p + d$, $p + p$ and $d + p$ elastic reactions are taken into account.

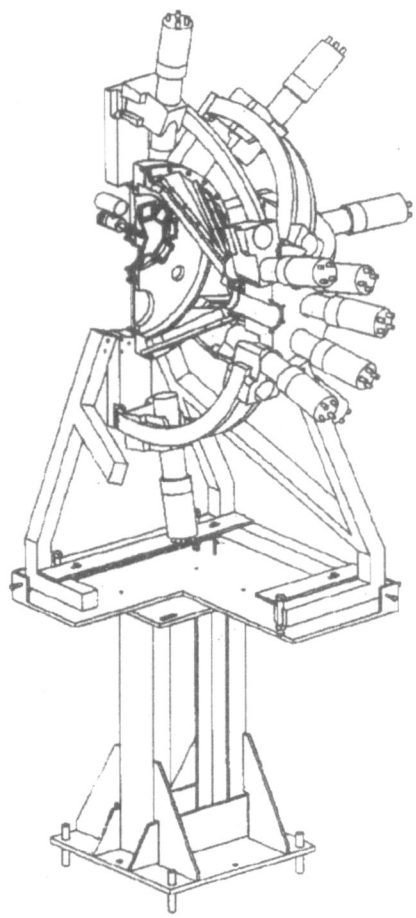

Figure 1. Transection of the beam polarimeter at KVI. The beam traverses from left to right.

2 Measuring polarizations and analyzing powers of the elastic $p+d$ reaction

The general equation for the cross section of a spin-$\frac{1}{2}$ particle on an unpolarized target is

$$\sigma(\theta, \phi) = I_\circ(\theta)[1 + p_z A_y \cos\phi] \tag{1}$$

where θ denotes the polar angle, ϕ the azimuthal angle and the z-axis is the spin-quantization axis (perpendicular to the beam). To extract analyzing powers a reference reaction with known analyzing powers at the energy measured needs to be monitored so that the degree of polarization p_z can be extracted.

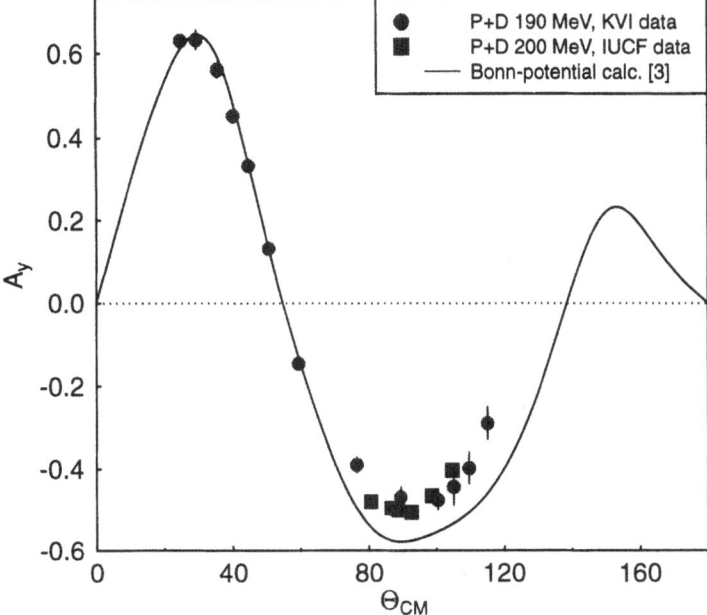

Figure 2. $\Delta E - E$ for one phoswich detector. Both axes are given in arbitrary units.

Using this polarization the analyzing power can be extracted from the measured asymmetry $p_z A_y$ of the monitored reaction with unknown analyzing power.

With the beam-polarimeter, data for the vector analyzing power A_y of $\boldsymbol{p}+d$ elastic scattering at 190 MeV were measured for CM-angles of 25° to 115° in the horizontal plane and one of the diagonal planes. Elastic scattering taken in the other diagonal plane at $\theta_{CM} = 39.7°$ and 48.0° served as a polarization monitor, using $\boldsymbol{p} + p$ analyzing powers obtained from the Nijmegen potential. Since the polarization axis is vertical, the yields collected in the vertical plane show no asymmetry. They serve to normalize different runs to each other. This allows for the simultaneous extraction of $A_y(\boldsymbol{p}, d)$ and the beam polarizations, $P^\uparrow$, $P^\downarrow$ and P_0. The latter is the rest polarization in the situation where both the weak- and strong-field transitions of the polarized-ion source were off. It is non-zero due to the relatively low magnetic field in the ECR ionizer. Its value is found to be $P_0 \simeq 0.07$. With the weak field- and the strong field units off, the source polarizations of $\simeq$ -60% and $\simeq$ +60% were obtained, respectively.

The preliminary data are shown in Fig. 3. Also shown are data taken at 200 MeV at IUCF [1]. Further data, taken with the SALAD setup at KVI [2], in the angular range $\theta_{CM} = 130°\text{-}160°$ are under analysis. The analyzing power is compared with calculations based on the Bonn-potential including partial waves up to $J = 5$ [3,4].

Up to CM-angles of 70° the data agree very well with the calculations. At higher angles the calculations disagree with both the IUCF and the KVI-data.

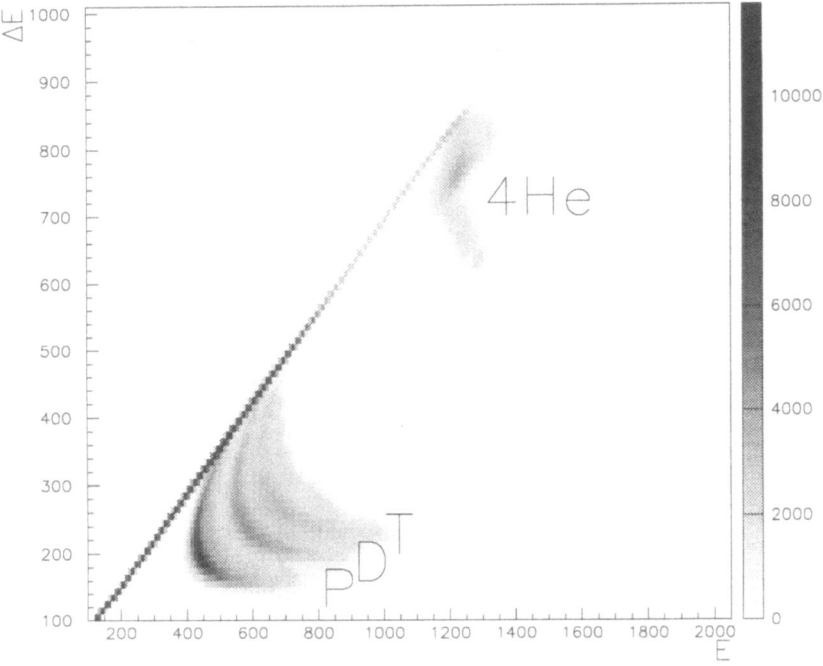

Figure 3. A_y for p+d elastic scattering at 190 MeV. The curve is based on a Bonn-potential calculation [3].

References

1. S.P. Wells et al.: Nucl. Instr. and Methods **A325**, 205 (1993)

2. N. Kalantar-Nayestanaki: Nucl. Phys. **A631** 242 (1998)

3. H. Witała: Private communication (1997)

4. W. Glöckle et al.: Phys. Reports **274**, 107 (1996)

Few-Body Systems Suppl. 10, 331–334 (1999)

Few-
Body
Systems
© by Springer-Verlag 1999

Proton-Deuteron Elastic Scattering for $E > 0$

E.O. Alt[1*], A.M. Mukhamedzhanov[2†**], A.I. Sattarov[3††***]

[1] Institut für Physik, Universität Mainz, D-55099 Mainz, Germany
[2] Cyclotron Institute, Texas A&M University, College Station, TX 77843, USA
[3] Physics Department, Texas A&M University, College Station, TX 77843, USA

Abstract. We report on the first reliable numerical results for proton-deuteron elastic scattering observables for energies above the deuteron breakup threshold, for the Paris potential. The calculations have been performed within the screening and renormalisation approach. The theoretical results are compared with recent experimental data.

1 Introduction

Theoretical investigations of the nucleon-deuteron (Nd) reaction are stimulated by the prospect that new information on various aspects of $2N$ and $3N$ forces can be extracted which is difficult or even impossible to obtain otherwise. Nowadays, for reactions with neutrons as projectiles (nd), calculations with 'realistic' nuclear potentials including $3N$ forces provide a good to very good reproduction of most experimental observables, with a few, as yet unexplained failures: A_y, iT_{11}, etc. Since experimental data are rather sparse and frequently lack the desired accuracy, nd calculations are usually compared with pd data which, from principle point of view, is clearly unsatisfactory.

For the proton-induced reaction (pd), data are abundant and of high precision. But similarly advanced and reliable calculations are as yet restricted to energies below the deuteron breakup threshold ($E < 0$) [1, 2]. Calculations have been performed also for $E > 0$, but only with such simple nuclear potentials [3] that their results can not be considered being more than semiquantitative.

*E-mail address: alt@dipmza.physik.uni-mainz.de
†Supported by DOE Grant DE-FG05-93ER40773
**E-mail address: akram@comp.tamu.edu
††Supported by Deutsche Forschungsgemeinschaft, Project 436 USB-113-1-0
***E-mail address: dior@phyacc.tamu.edu

What reasons are responsible for this lack of sophisticated and reliable *pd* calculations for $E > 0$? In the framework of integral equations for amplitudes, the presently only substantiated method is the screening and renormalisation approach; though being theoretically well founded [4, 5], it is very CPU-time consuming and, because of that, requires still one approximation (see below). For the differential equation approaches for wave functions, several fundamental questions have not yet found a satisfactory answer. For instance, what are the consequences of the lack of convergence of partial wave summations of Coulombian quantities, what of the assertion that the asymptotic behaviour of the infinitely angular-momentum coupled system of radial Faddeev equations differs from that of the partial-wave truncated finite system [6], or what of the incompleteness of the Coulomb boundary condition as imposed at present?

2 Complete Boundary Condition for the *ppn* Wave Function

To elaborate on the last question we formulate here the boundary condition to be imposed on the *ppn* wave function in all configuration space [7, 8]. The usual definition of Jacobi co-ordinates is used: $k_\alpha\,(r_\alpha)$ is the relative momentum (co-ordinate) between particles β and γ, and $q_\alpha\,(\rho_\alpha)$ the momentum (co-ordinate) of particle α relative to the center-of-mass of the pair (β, γ). Moreover, $\mu_\alpha = m_\beta m_\gamma / (m_\beta + m_\gamma)$ is the reduced mass, and $\eta_\alpha = e_\beta e_\gamma \mu_\alpha / k_\alpha$ the Coulomb parameter, for particle pair (β, γ). Define the following asymptotic regions:

$$\Omega_0: \quad r_\alpha, r_\beta, r_\gamma \to \infty, \text{ but not } r_\nu/\rho_\nu \to 0 \text{ for } \nu = \alpha, \beta, \gamma; \quad (1)$$

$$\Omega_\nu: \quad \rho_\nu \to \infty, \ r_\nu/\rho_\nu \to 0, \text{ for } \nu = \alpha, \beta, \text{ or } \gamma. \quad (2)$$

For definiteness let the neutron be particle α. Then one has in leading order

$$\Psi^{(+)}_{k_\alpha q_\alpha}(r_\alpha, \rho_\alpha) \stackrel{\Omega_0}{\approx} e^{i(k_\alpha \cdot r_\alpha + q_\alpha \cdot \rho_\alpha)} \, e^{i\eta_\alpha \ln(k_\alpha r_\alpha - k_\alpha \cdot r_\alpha)}: \quad p + p + n, \quad (3)$$

$$\stackrel{\Omega_\alpha}{\approx} e^{i(k_\alpha \cdot r_\alpha + q_\alpha \cdot \rho_\alpha)} \, \tilde\psi^{(+)}_{k_\alpha}(r_\alpha): \quad n + (p + p), \quad (4)$$

$$\stackrel{\Omega_{\beta \neq \alpha}}{\approx} e^{i(k_\beta \cdot r_\beta + q_\beta \cdot \rho_\beta)} \, \tilde\psi^{(+)}_{k_\beta(\rho_\beta)}(r_\beta): \quad p + (p + n), \quad (5)$$

where the (pp) and (pn) wave functions are defined as continuum solutions of

$$\left\{ \frac{k_\alpha^2}{2\mu_\alpha} + \frac{\Delta_{r_\alpha}}{2\mu_\alpha} - V_\alpha^N(r_\alpha) - V_\alpha^C(r_\alpha) \right\} e^{ik_\alpha \cdot r_\alpha} \tilde\psi^{(+)}_{k_\alpha}(r_\alpha) = 0, \quad (6)$$

$$\left\{ \frac{k_\beta^2(\rho_\beta)}{2\mu_\beta} + \frac{\Delta_{r_\beta}}{2\mu_\beta} - V_\beta^N(r_\beta) \right\} e^{ik_\beta(\rho_\beta) \cdot r_\beta} \tilde\psi^{(+)}_{k_\beta(\rho_\beta)}(r_\beta) = 0. \quad (7)$$

The (pp) wave function satisfies the usual Schrödinger equation (SE) (6) for a nuclear plus Coulomb interaction, while the one for the (pn) subsystem is solution of the two-body-like SE (7) where, instead of the (asymptotic) momentum k_β, a 'local momentum' $k_\beta(\rho_\beta) = k_\beta + a_\beta(\hat\rho_\beta)/\rho_\beta$ occurs which depends on the distance ρ_β of the (pn) pair from the other proton (for explicit definitions of a_β see [7]). This fact can be interpreted as source of a long-range, non-central, velocity-dependent three-body correlation between *ppn* [7, 8].

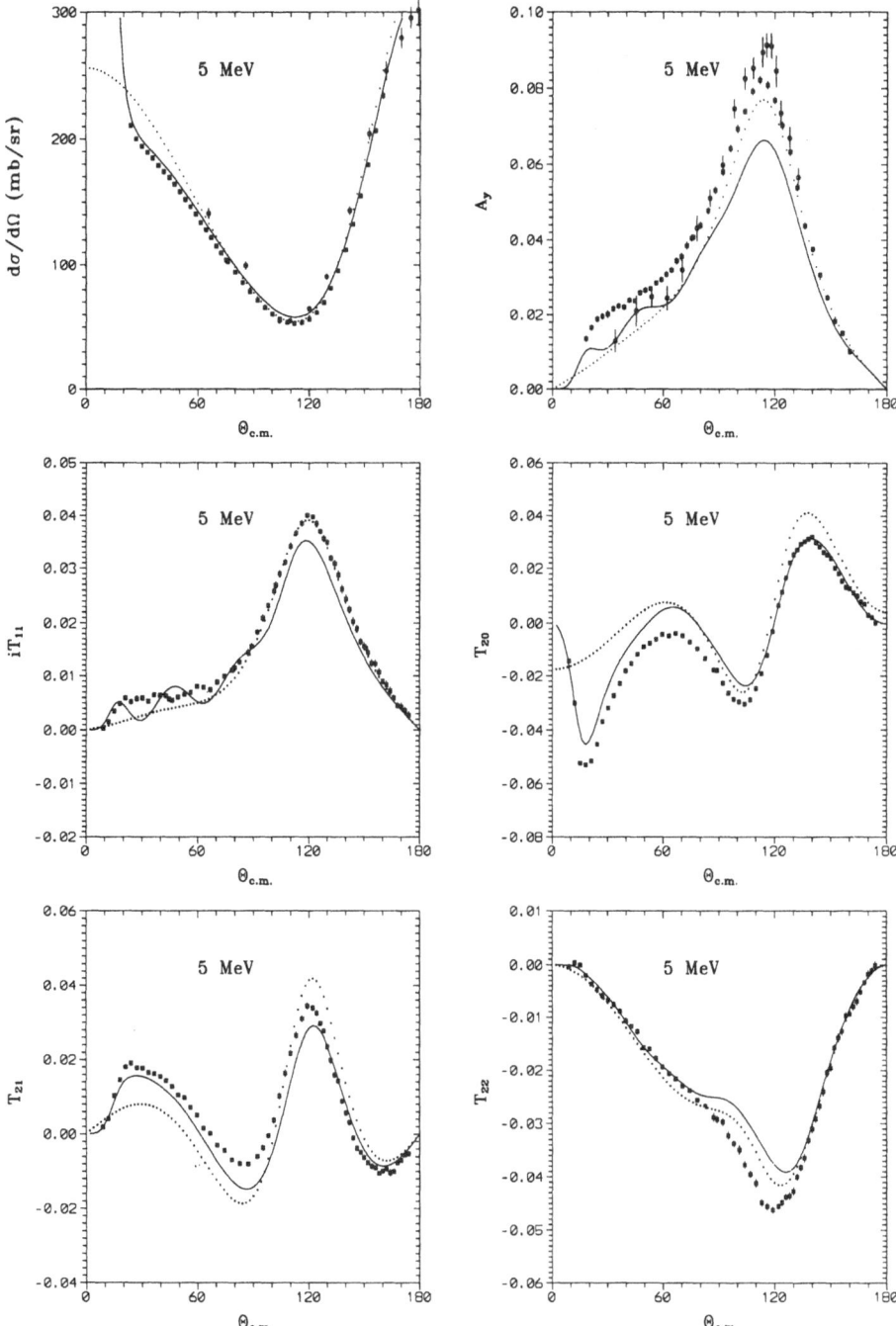

Figure 1. Differential cross section and polarisation observables for proton laboratory energy of 10 MeV versus c.m. scattering angle. Solid line: *pd*, dotted line: *nd*. Experiment: *pd* (■) from [10]; *nd* (○): from [11].

334

3 Results for pd Elastic Scattering Observables for $E > 0$

We use the screening and renormalisation approach (for a recent pedagogical review see [5]). As nuclear interaction the Paris potential (in fact, a separable representation [9] thereof) is taken, in the states $^3S_1 - {}^3D_1$, $^1S_0$, and in all P waves. The screening radius required for obtaining converged results for cross sections was $R = 100$ fm while for polarisation observables even $R = 625$ fm was not fully sufficient for c.m. angles $< 70°$. The Coulomb interaction was taken into account in the so-called Coulomb Born approximation which consists in approximating the pp-Coulomb T-matrix by the potential. The resulting error in the final pd scattering amplitudes was estimated to be less than 1%.

In the figures we compare our pd results with selected experimental data of Sagara et al. [10]. For comparison we also include the corresponding nd results. Inspection shows that on the average we find good but not perfect agreement of the calculated with measured quantities. However, the general trends, in particular also the relation between neutron (Tornow et al. [11]) and proton analysing powers, are well reproduced. The longstanding nd A_y- and iT_{11}-puzzles which also show up in our calculations and are clear indications that some aspects of the nuclear force are still not satisfactorily described in all 'realistic' potential models (cf. [12]), not surprisingly carry over to the pd case. For energies below the deuteron breakup threshold this has already been pointed out in [2]. A more extensive comparison with experiment is in preparation.

References

1. G.H. Berthold, A. Stadler, H. Zankel: Phys. Rev. **C41**, 1365 (1990)

2. A. Kievski: Contribution to this Conference and references therein

3. E.O. Alt, W. Sandhas, H. Ziegelmann: Nucl. Phys. **A445**, 429 (1985)

4. E.O. Alt, W. Sandhas, H. Ziegelmann: Phys. Rev. **C17**, 1981 (1978)

5. E.O. Alt and W. Sandhas: In: *Coulomb Interactions in Nuclear and Atomic Few-Body Collisions*, eds. F. Levin, D. Micha, p.1. New York: Plenum 1996

6. A.A. Kvitsinskii and D.M. Latypov: Sov. J. Nucl. Phys. **53**, 953 (1991)

7. E.O. Alt and A.M. Mukhamedzhanov: Phys. Rev. **A47**, 2004 (1993)

8. E.O. Alt: Invited Talk at this Conference

9. J. Haidenbauer and W. Plessas: Phys. Rev. **C30**, 1822 (1984)

10. K. Sagara et al.: Phys. Rev. **C50**, 576 (1994) and private communication

11. W. Tornow et al.: Phys. Lett. **B257**, 273 (1991)

12. W. Tornow, W. Witala, A. Kievski: Phys. Rev. **C57**, 555 (1998)

Few-Body Systems Suppl. 10, 335–338 (1999)

Few-
Body
Systems
© by Springer-Verlag 1999

Nucleon-Deuteron Scattering with Δ-Isobar Excitation

K. Chmielewski[1], S. Nemoto[1], A.C. Fonseca[2], and P.U. Sauer[1]

[1] Institut für Theoretische Physik, Universität Hannover, Appelstraße 2, D-30167 Hannover, Germany
[2] Centró Física Nuclear da U. L., Av. Prof. Gama Pinto, N°2, 1699 Lisboa, Portugal

Abstract. A two-baryon interaction with explicit Δ-isobar degrees of freedom is used for the description of nucleon-deuteron scattering. The Δ-isobar yields an effective three-nucleon force in three-nucleon systems. The three-particle scattering equations are solved by employing a separable expansion of the two-baryon transition matrix. The singularities in the three-particle scattering equations are treated by two independent numerical techniques; their equivalence is demonstrated. First results for nucleon-deuteron breakup scattering with Δ-isobar excitation are given.

1 Baryon-Baryon Interaction

The description is based on the coupled-channel two-baryon interaction depicted diagrammatically in Fig. 1. It allows for the virtual excitation of a nucleon into a Δ-isobar. In the three-nucleon medium, three-nucleon channels are coupled to those in which one nucleon is transformed into a single Δ-isobar. The virtual excitation of the Δ-isobar yields an effective three-nucleon force. Example processes which contribute to the three-nucleon force are shown in Fig. 2. The force model was developed in refs. [1, 2] and used there to calculate properties of the three-nucleon bound state.

We extent those calculations to nucleon-deuteron scattering and investigate whether the Δ-isobar is a remedy for some long-standing discrepancies in the theoretical description of nucleon-deuteron scattering like the A_y puzzle [3] and the so-called Sagara discrepancy [4].

2 Nucleon-Deuteron Elastic Scattering

The three-particle scattering equations are solved by means of a separable expansion of the two-baryon transition matrix developed in ref. [5]. First results of elastic nucleon-deuteron scattering with Δ-isobar excitation were reported in

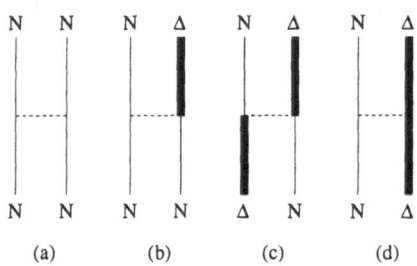

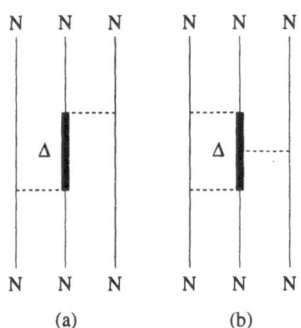

Figure 1. The baryon-baryon interaction employed. A thin solid lines indicates a nucleon (N), a thick solid line indicates a Δ-isobar (Δ). The instantaneous interaction is drawn as a dashed line. Processes (b)–(d) refer to isospin triplet partial waves only.

Figure 2. Example processes in the three-nucleon medium which give rise to effective three-nucleon forces. The legend of lines is the same as in Fig. 1.

ref. [6]. A more sophisticated study is given in ref. [7]. The beneficial influence of the Δ-isobar on the description of the cross section minimum is described in ref. [8] and in the contribution of Nemoto et al. [9] to this conference.

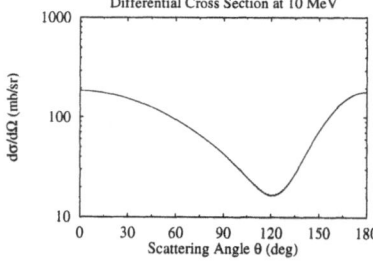

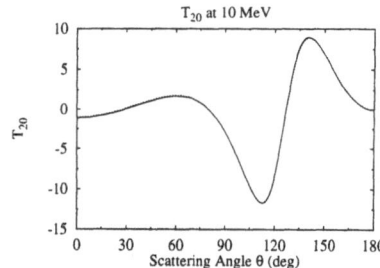

Figure 3. Differential cross section $\frac{d\sigma}{d\Omega}$ and deuteron analyzing power T_{20} of elastic nucleon-deuteron scattering as a function of the scattering angle θ at 10 MeV nucleon lab energy as calculated by the methods of real-axis integration (solid lines) and contour deformation (dashed lines). There are no visible differences.

Whereas refs. [5,7–9] used the technique of "contour deformation" to treat the singularities in the three-particle scattering equations, the present approach employs the technique of "real-axis integration." The details of the latter technique will be presented in ref. [10]. Both techniques are compared in Fig. 3 which shows two observables of elastic nucleon-deuteron scattering at 10 MeV nucleon lab energy. The agreement of the two independent calculations is perfect.

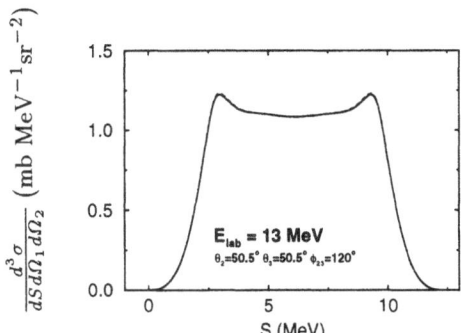

Figure 4. The nucleon-deuteron breakup cross section at 13 MeV nucleon lab energy as a function of the arc length S in the space star configuration. Results for the calculation with Δ-isobar (solid line) and the Paris potential (dashed line) are shown.

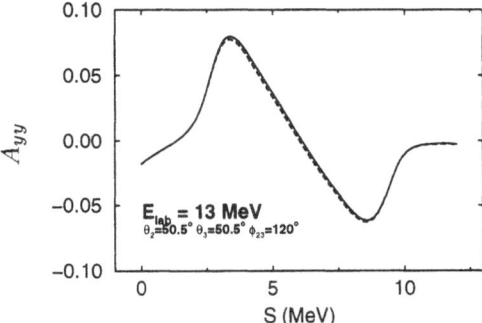

Figure 5. Deuteron analyzing power A_{yy} for nucleon-deuteron breakup scattering at 13 MeV nucleon lab energy as a function of the arc length S in the space star configuration. Results for the calculation with Δ-isobar (solid line) and the Paris potential (dashed line) are shown.

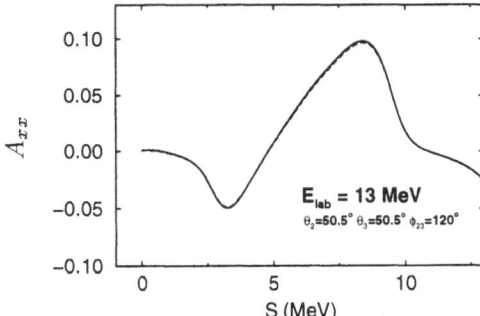

Figure 6. Same legend as in Fig. 5 but for the deuteron analyzing power A_{xx}.

3 Nucleon-Deuteron Breakup Scattering

Calculations for elastic nucleon-deuteron scattering based on the real-axis integration technique can be extended more or less straight-forward to nucleon-deuteron breakup scattering. At present calculations are under way which will explore the role of the Δ-isobar in breakup configurations. Preliminary results are shown in Figs. 4–6 where calculations for the purely nucleonic Paris potential are compared to calculations with virtual Δ-isobar excitation. Only two-baryon partial waves up to a total two-baryon angular momentum $I = 2$ are taken into account. The few observables considered do not exhibit a sizable Δ-isobar effect.

Acknowledgement. The work was supported by a graduate-student grant (K. C.) of the German state Lower Saxony and by a grant (S. N.) of the German Bundesministerium für Forschung und Technologie Contract No. 06 OH 741. Numerical calculations were performed at the Regionales Rechenzentrum für Niedersachsen.

References

1. C. Hajduk and P.U. Sauer: Nucl. Phys. **A322**, 329 (1979)

2. C. Hajduk, P.U. Sauer and W. Strueve: Nucl. Phys. **A405**, 581 (1983)

3. Y. Koike and J. Haidenbauer: Nucl. Phys. **A463**, 365c (1987)

4. K. Sagara et al.: Phys. Rev. **C50**, 576 (1994)

5. S. Nemoto et al.: Few-Body Systems (accepted for publication)

6. K. Chmielewski et al.: Few-Body Systems **Suppl. 8**, 394 (1995)

7. S. Nemoto et al.: Few-Body Systems. (accepted for publication)

8. S. Nemoto et al.: Phys. Rev. **C**. (accepted for publication)

9. S. Nemoto et al.: contribution to this conference

10. K. Chmielewski et al.: (in preparation)

Few-Body Systems Suppl. 10, 339–342 (1999)

Few-
Body
Systems
© by Springer-Verlag 1999

On the Sagara Discrepancy in the Cross Section Minimum of Elastic Nucleon-Deuteron Scattering

S. Nemoto[1], K. Chmielewski[1], S. Oryu[2] and P.U. Sauer[1]

[1] Institut für Theoretische Physik, Universität Hannover, Appelstraße 2, D-30167 Hannover, Germany

[2] Department of Physics, Faculty of Science and Technology, Frontier Research Center for Computational Science, Science University of Tokyo, Noda, Chiba 278, Japan

Abstract. A coupled-channel formulation of nucleon-deuteron scattering with Δ-isobar excitation is used. Δ-isobar excitation in the nuclear medium yields an effective three-nucleon force. The effect of Δ-isobar excitation on the spin-averaged differential cross section is studied. The discrepancy between theory and experiment in the diffraction minimum is reduced.

Virtual Δ-isobar excitation yields an effective three-nucleon force in the nuclear medium. Theoretically the Δ-isobar gets created through a two-baryon potential with channel coupling. Such a coupled-channel interaction is applied in ref. [1] to the three-nucleon bound state and in ref. [2] to elastic nucleon-deuteron scattering. Past calculations [3–6], based on traditional two-nucleon potentials, failed in the region of the diffraction minimum. This failure was pointed out in ref. [3] and therefore is often called Sagara discrepancy. We concentrate on the Δ-isobar effect for the diffraction minimum of the spin-averaged differential cross section here.

The calculational scheme is taken over from refs. [2,7]. The Hilbert space has a purely nucleonic sector and a sector in which a nucleon is turned into a Δ-isobar. The Δ-isobar is considered a stable baryon with mass 1232 MeV and with spin and isospin $\frac{3}{2}$; that treatment of the Δ-isobar is appropriate below the pion-production threshold; the considered scattering energies stay well below. The two-baryon potential is the parametrization A2 of ref. [7] and its purely nucleonic reference potential is the Paris potential [8]. The AGS scattering equations [9] with channel coupling are solved by a separable expansion of the two-baryon transition matrix. That separable expansion is judged in ref. [7] to remain valid for the considered scattering energies. The purely nucleonic partial wave $^3F_3$ in the isospin triplet $I = 3$ is used instead of the partial-wave coupled state since the effect of the two-baryon interaction in partial waves of total angular momentum $J = 3$ is small. The calculation takes three-baryon channels up to three-baryon total angular momentum $\frac{31}{2}$ into account. The calculation is done without Coulomb interaction. However, the experimental data to be described refer to proton-deuteron scattering. The Coulomb interaction is known to yield sizable corrections for the differential cross section at all scattering angles below 5 MeV proton lab energy and in forward scattering angles at all energies. The region of the diffraction minimum appears not to be influenced appreciably by the Coulomb interaction above 5 MeV proton lab energy.

Figure 1 shows our results for the differential cross section of nucleon-deuteron scattering at seven energies. Up to 10 MeV nucleon lab energy the effect of the Δ-isobar is invisible on the log-plot. At higher energies it appears beneficial for the theoretical prediction with Δ-isobar excitation. It decreases the Sagara discrepancy. The same effect was already seen in ref. [10]. The Δ-isobar improves the agreement between theoretical prediction and experimental data for backward angles, i.e., for $\theta > 90°$. Although the Δ-isobar worsens the agreement for forward angles, i.e., for $\theta < 90°$, this angular regime is still sensitive to Coulomb corrections at even higher energies.

In Fig. 2 we try to understand the role which the Δ-isobar plays for the results in more detail. We define the discrepancy

$$\Delta\sigma(\theta) \equiv \frac{\frac{d\sigma}{d\Omega}^{\text{calc}}(\theta) - \frac{d\sigma}{d\Omega}^{\text{exp}}(\theta)}{\frac{d\sigma}{d\Omega}^{\text{exp}}(\theta)} \times 100 \tag{1}$$

between theoretical prediction and experimental data as ref. [3]. Figure 2 shows that discrepancy $\Delta\sigma(\theta_{\min})$ at the minimum of the cross section for all considered energies. The experimental minimum position $\theta_{\min}$, the cross section value and its error are determined by fitting the available data points. The Δ-isobar effect in $\Delta\sigma(\theta_{\min})$ increases with energy in general. The crossing of zero for the result of the purely nucleonic potential is observed as in ref. [3]; with Δ-isobar excitation the crossing is shifted to lower energies. With Δ-isobar excitation the calculated differential cross section in the diffraction minimum remains always lower than the experimental one, except at 2 MeV. Up to 10 MeV the Δ-isobar gives constant contributions and at 10 MeV the Δ-isobar even worsens the agreement in the diffraction minimum.

The minimum of the cross section is improved by a complicated interference phenomenon, e.g., the $N\Delta$ partial waves coupled to the purely nucleonic ^3P_I and $^1D_2$ waves are most important yielding 75 % of the Δ contribution $d\sigma^{\text{calc}}/d\Omega(\theta_{\min})|_{N+\Delta} - d\sigma^{\text{calc}}/d\Omega(\theta_{\min})|_N$ at 135 MeV. We reported the details of the Δ-isobar effect in the diffraction minimum of the differential cross section for elastic nucleon-deuteron scattering in ref. [14].

We see a beneficial Δ-isobar effect in the diffraction minimum of the spin-averaged differential cross section for elastic nucleon-deuteron scattering at higher energies. There the Δ-isobar helps to remove the long-standing Sagara discrepancy. The found results, in their positive and negative implications, have still to be taken with some caution; it has to be checked that they are not acci-

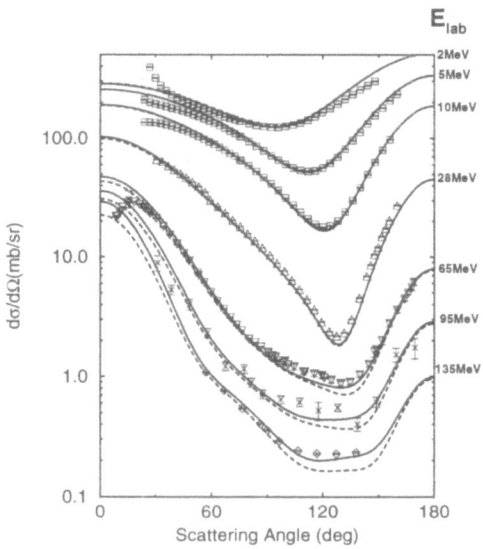

Figure 1. Differential cross sections of elastic nucleon-deuteron scattering as function of the c.m. scattering angle for 2 MeV, 5 MeV, 10 MeV, 28 MeV, 65 MeV, 95 MeV and 135 MeV nucleon lab energy. The differential cross sections derived from the coupled-channel potential with Δ-isobar excitation (solid curves) are compared with results based on the purely nucleonic reference potential (dashed curves). The experimental data refer to proton-deuteron scattering; they are taken at 2 MeV, 5 MeV and 10 MeV from ref. [3], at 28 MeV from ref. [15], at 65 MeV from ref. [16], at 95 MeV from ref. [17] and at 135 MeV from ref. [5].

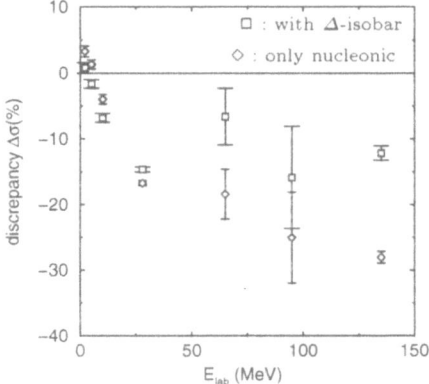

Figure 2. Discrepancy $\Delta\sigma$ between experimental data and theoretical prediction at the experimental minimum of the differential cross section as a function of energy according to Eq. (1). Results with ($\square$) and without ($\diamond$) Δ-isobar excitation are compared. The error bars reflect uncertainties in the experimental data only; no attempt is made to estimate the systematic uncertainties of the calculation.

dental to the particular form of our dynamic model: The Δ-isobar effect may possibly be changed by Coulomb corrections for proton-deuteron scattering as emphasized in refs. [11, 12] and by the influence of higher partial waves. Relativistic corrections are unlikely to be effective [4] at the energies considered. We also notice that a complementary calculation [13] on the same problem, using a traditional irreducible and not an effective three-nucleon force, arrives at similar conclusions.

Acknowledgement. This work was supported by a grant of the German Bundesministerium für Forschung und Technologie Contract No. 06 OH 741 for S. N. The numerical calculations were perfomed at Regionales Rechenzentrum für Niedersachsen and at Frontier Research Center for Computational Science in the Science University of Tokyo.

References

1. Ch. Hajduk, P.U. Sauer, W. Strueve: Nucl. Phys. **A405**, 581 (1983)

2. S. Nemoto et al.: Few-Body Systems, accepted for publication

3. K. Sagara et al.: Phys. Rev. **C50**, 576 (1994)

4. W. Glöckle et al.: Phys. Reports **274**, 107 (1996)

5. N. Sakamoto et al.: Phys. Lett. **B367**, 60 (1996)

6. Y. Koike, S. Ishikawa: In: *Contribution to the XVth International Conference on Few-Body Problems in Physics, Conference Handbook* (Groningen, 1997), p. 236

7. S. Nemoto et al.: Few-Body Systems, accepted for publication

8. M. Lacombe et al.: Phys. Rev. **C21**, 861 (1980)

9. E.O. Alt, P. Grassberger, W. Sandhas: Nucl. Phys. **B2**, 167 (1967)

10. K. Chmielewski et al.: In: *Contribution to the XVth International Conference on Few-Body Problems in Physics, Conference Handbook* (Groningen, 1997), p. 219

11. S. Oryu, H. Yamada: Phys. Rev. **C49**, 2337 (1994) and H. Yamada: Private communication

12. E.O. Alt, M. Rauh: Few-Body Systems **17**, 121 (1994), and references therein

13. H. Witała et al.: preprint, nucl-th/9801018, 1998

14. S. Nemoto et al.: Phys. Rev. (in print)

15. K. Hatanaka et al.: Nucl. Phys. **A426**, 77 (1984)

16. H. Shimizu et al.: Nucl. Phys. **A382**, 242 (1982)

17. O. Chamberlain et al.: Phys. Rev. **94**, 666 (1954)

Few-Body Systems Suppl. 10, 343–346 (1999)

Few-Body Systems
© by Springer-Verlag 1999

Channel Selection in Proton-Deuteron Bremsstrahlung at 190 MeV

M. Volkerts[1], J.C.S. Bacelar[1], M.J. van Goethem[1], H. Huisman[1], N. Kalantar-Nayestanaki[1], A. Kugler[2], H. Löhner[1], J.G. Messchendorp[1], R.W. Ostendorf[1], S. Schadmand[1]*, R. Simon[3], V. Wagner[2] and H.W. Wilschut[1]

[1] University of Groningen, Kernfysisch Versneller Instituut, Zernikelaan 25, 9747 AA Groningen, The Netherlands
[2] Nuclear Physics Institute, 250 86 Řež u Prahy, Czech Republic
[3] Gesellschaft für Schwerionenforschung, Planckstraße 1, 64291 Darmstadt, Germany

Abstract. For the first time an exclusive proton-deuteron bremsstrahlung experiment has been performed in which all the exit channels can be identified separately. Preliminary results indicate we are able to identify the different exit channels with high accuracy. For the 3 and 4-body final state, no high quality calculations are available at the present time.

1 Introduction

Nucleon-nucleon bremsstrahlung is the most fundamental reaction used in studying the off-shell behavior of the nucleon. It only involves two strongly interacting particles in the final state and the electromagnetic interaction, which is well known from QED. In addition, the cross sections for nucleon-nucleon bremsstrahlung serves as a basic input for calculations that deal with photon production in heavy-ion collisions. These calculations rely heavily on theoretical predictions of the nucleon-nucleon bremsstrahlung process, in which approximations (such as the Soft Photon Approximation, SPA) are made that contain, to leading order, no off-shell information.

At KVI we performed a series of experiments to study nucleon-nucleon bremsstrahlung starting with the proton-proton system[1]. In order to complete the picture of understanding nucleon-nucleon bremsstrahlung it is necessary to measure pn bremsstrahlung along with pp bremsstrahlung. Due to experimental difficulties this is usually done by measuring proton-deuteron bremsstrahlung.

Present address Universität Giessen, Heinrich-BUFF-Ring 16, 35392 Giessen, Germany

No free neutron targets exist and generally speaking, the neutron beams lack the high quality and/or intensity required for precise measurements. Most of the pdγ experiments in the past are inclusive in the sense that only the energy and angular distributions of the outgoing photons have been measured, implying an integration over all particles and reaction channels [2, 3]. Comparison with theoretical calculations based on a meson exchange potential model shows that the theory underestimates the data by up to 40% depending on the photon emission angle [4], demonstrating the need for more experimental and theoretical effort.

We have performed, for the first time, an exclusive experiment in which the 2-body final state γ³He (radiative capture), 3-body final state pdγ ("coherent" bremsstrahlung) and the 4-body final state ppnγ have been measured simultaneously. Cross sections and analyzing powers of these reactions were measured for all exit channels.

2 Experimental Setup

The experiment was performed using the 190 MeV polarized-proton beam from the AGOR cyclotron impinging on a LD$_2$ target [5] with a typical beam current of 1 nA. The outgoing hadrons were detected with the Small-Angle Large-Acceptance Detector, SALAD, specifically designed and built for these measurements. SALAD consists of two multi-wire proportional chambers [6] for determining the positions of the outgoing hadrons and a stack of plastic scintillators for measuring the energy of these particles, backed by a second layer used for vetoing elastically scattered protons up to 135 MeV. The detector has almost cylindrical symmetry with full azimuthal coverage between 6° and 19° and limited coverage for polar angles up to 26°. The large solid angle (about 500 msr) is mandated by the very small cross sections of the nucleon-nucleon bremsstrahlung process. To allow safe beam passage, the detector was equipped with a central hole.

For measuring the position and energy of the photons the Two-Arm Photon Spectrometer, TAPS [7], was used, in two different geometries. In the first phase of the experiment all the crystals where mounted in a hexagonal frame upstream from the target, keeping the cylindrical symmetry. This allows for adding corresponding bins, thereby increasing the statistical accuracy. This "supercluster" geometry covers a polar angular range of 125°-170°. In the second phase of the experiment the crystals were mounted in 6 blocks of 64 crystals each that were positioned around the target covering an angular range of 57°-176°. With this "block" geometry angular distributions could be measured. For more experimental details see [8] where also the analysis procedure is outlined.

3 Preliminary Results

In this section some preliminary results for the 3 and 4-body final state for the "supercluster" geometry will be presented. For preliminary results of the radiative capture process see [9]. In the present data for the 4-body final state, ppnγ, the two protons and the photon are fully identified and their 4 momenta measured. The 4 momentum of the neutron is inferred by momentum conservation.

For four different combinations of proton and deuteron scattering angles, the five-fold differential cross-section as a function of the photon polar angle (measured with respect to the outgoing proton) is shown in the four left panels of Fig. 1. Two calculations [10] are shown: a SPA calculations for the free pnγ process (dotted line, downscaled by a factor of 3) and a classical calculation for the pdγ process (solid line, arbitrary normalization). Neither of these two calculations is capable of explaining the data as one should expect since intermediate energy states in the deuteron and initial- and final- state effects are neglected in these calculations. In the four right panels of Fig. 1 preliminary

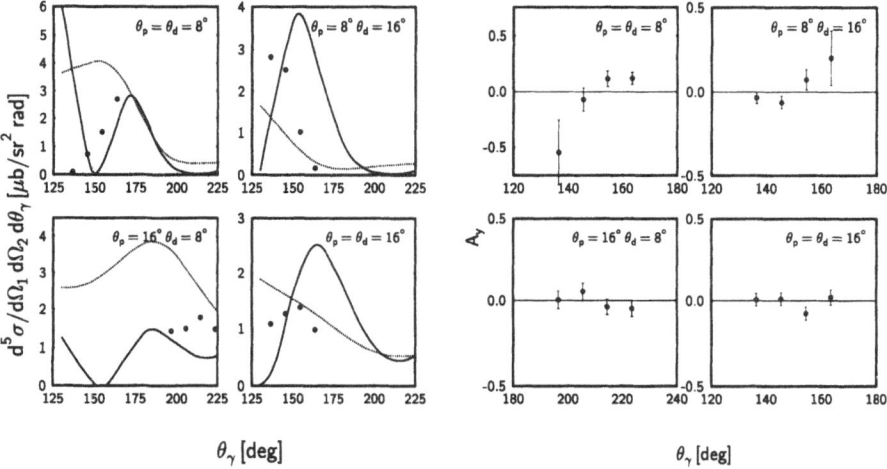

Figure 1. Preliminary cross sections for the 3-body final state (four left panels) and preliminary analyzing powers (four right panels), both as a function of θ_γ for 4 different angle combinations (given in the figures). See the text for explanation.

results for the analyzing power A_y are shown for the same combinations of angles. No calculation for this observable has ever been done.

For the 4-body final state one can select the kinematics for which the neutrons were spectators in the collision and carried near zero momentum and only the protons took place in the bremsstrahlung process. The Fermi motion of the protons will affect the kinematics considerably but the largest part of the cross sections will come from low-momentum part of the wave function. In Fig. 2 preliminary cross section for 4 different combinations of proton angles are shown. The calculation shown is a fully relativistic calculation for the free ppγ process [11]. One can clearly see the need for a complete calculation where the deuteron wave function is included and final-state effects are properly taken into account.

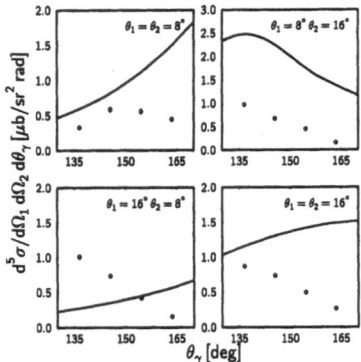

Figure 2. Preliminary cross sections for the 4-body final state where the neutron is assumed to be a spectator with zero momentum.

4　Summary

For the first time an exclusive proton-deuteron bremsstrahlung experiment has been performed in which the 2-body final state γ^3He, 3-body final state pdγ and the 4-body final state ppnγ have been measured simultaneously. Preliminary results indicate we are able to identify the different exit channels with high accuracy. For the 3 and 4-body final state, no high quality calculations are available at the present time.

References

1. H. Huisman et al.: these proceedings

2. J.A. Pinston et al.: Phys. Lett., **B249**, 402 (1990)

3. J. Clayton et al.: Phys. Rev., **C45**, 1810 (1992)

4. K. Nakayama et al.: Phys. Rev., **C45**, 2039 (1992)

5. N. Kalantar-Nayestanaki, J. Mulder and J. Zijlstra: submitted to Nucl. Inst. and Meth. A

6. M. Volkerts et al.: submitted to Nucl. Inst. and Meth. A

7. A.R. Gabler et al.: Nucl. Inst. and Meth., **A346**, 168 (1994)

8. N. Kalantar-Nayestanaki: Nucl. Phys. **A631**, 242c (1998)

9. *KVI Annual Report 1997*, KVI (1997)

10. O. Scholten: Private communication

11. G.H. Martinus, O. Scholten and J.A. Tjon: Phys. Rev. **C56**, 2945 (1997)

Few-Body Systems Suppl. 10, 347–350 (1999)

Few-
Body
Systems
© by Springer-Verlag 1999

Effects of Final State Interaction in the ³He Decay caused by Muon Capture.

R. Skibiński[1], W. Glöckle[2], J. Golak[1], H. Witała[1,2]

[1] Institute of Physics, Jagiellonian University, PL-30059 Cracow, Poland
[2] Institut für Theoretische Physik II, Ruhr-Universität Bochum, D-44780 Bochum, Germany

Abstract. The ³He decay caused by μ-capture leading to neutron, deuteron and μ-neutrino in outgoing channel is investigated. Three-nucleon Faddeev wave functions for the both initial and final states are calculated using the Bonn B and Paris nucleon-nucleon potentials. Large effects of interactions between nucleons in the final state on the decay rate are found and comparison to recent experimental data shows a nice agreement for full calculations.

1 Introduction

Recently a considerable progress in calculations of three nucleon (3N) continuum states have been achieved [1]. As was shown in a series of recent papers [2, 3, 4] the exact treatment of the interactions among the three nucleons in the continuum states is very important for correct analysis of electromagnetic processes, such as an inelastic electron scattering on ³He or radiative p-d capture. Here we would like to study the importance of final state interactions (FSI) among the nucleons in ³He decay caused by μ-capture. We restricted ourselves to the case

$$\mu^- + {}^3\text{He} \rightarrow \text{d} + \text{n} + \nu_\mu \ .$$

Recently the first measurement of energy spectra for deuteron and proton has been reported [5, 6].

2 Theoretical formalism

The lagrangian density has the form of the current-current interaction and both the initial and final states are taken as products of leptonic and hadronic wave functions. The leptonic part of the current is exactly known and the single nucleon current operator is parametrized in terms of single-nucleon weak-interaction form factors $g_i^{V,A}$ [7]. We use the nonrelativistic limit of the hadronic

current i^λ, consistent with our wave functions and evaluate
$I^\lambda = < \Psi^{out} \mid i^\lambda \mid \Psi_{^3He} >$. After introducing standard Jacobi momenta p, q one gets for matrix elements of i^λ in the laboratory system:

$$< pq \mid i^\lambda(Q) \mid p'q' > = \delta(p' - p)\delta(q' - (q - \frac{2}{3}Q))\mathcal{I}^\lambda(q, Q) ,$$

$$\mathcal{I}^0(q, Q) = \frac{3}{(2\pi)^3}\{g_1^V + g_1^A(\frac{\sigma\pi}{m} - \frac{\sigma\nu}{2m})\} , \tag{1}$$

$$\mathcal{I}(q, Q) = \frac{3}{(2\pi)^3}\{g_1^V(\frac{\pi}{m} - \frac{\nu}{2m}) - \frac{\nu}{2m}(g_1^V + g_2^V 2m)i(\sigma \times \hat{\nu}) +$$

$$+ g_1^A\sigma - g_2^A\frac{1}{2m}m_\mu\hat{\nu}(\sigma\nu)\} ,$$

with $\pi \equiv \frac{2}{3}\nu + q$; $\nu = -Q$.

The final scattering state $\mid \Psi^{out} > = \mid\Psi_{nd}^{(-)} >$ can be decomposed into Faddeev components and finally one gets

$$I^\mu = I_{PWIAS}^\mu + I_{rescatt}^\mu , \tag{2}$$

where I_{PWIAS}^μ (plane wave impulse approximation symmetrized) corresponds to the case when no interaction between outgoing nucleon and deuteron in the final state is present. The $I_{rescatt}^\mu$ term contains all rescatterings and can be written as [2] [3]

$$I_{rescatt}^\mu \equiv \frac{1}{\sqrt{3}} < \Phi_{nd} \mid P \mid U^\mu > , \tag{3}$$

with

$$\mid U^\mu > = t_1 G_0(1 + P)i^\mu(Q) \mid \Psi_{^3He} > + t_1 G_0 P \mid U^\mu > . \tag{4}$$

This is the same integral equation which one finds in the 3N continuum [1] and the same methods can be used to solve it for any NN interaction. We solve Eq. (4) in a partial wave decomposition and in momentum space. For details we refer to [8, 9]. The decay rate is calculated from the S-matrix in a standard way.

3   Results

The initial 3N bound state of ^{3}He used is always based on 34-channel Faddeev calculation. In treating the 3N continuum we included all NN force components for total NN angular momentum up to $j_{max} = 1$. The $j = 2$ contributions turned out to be practically negligible. As NN force we use the Bonn B and Paris potentials. In the case of a kinematically complete situation in the outgoing channel, we show in Fig. 1 the decay rate $\frac{d^2\Gamma}{dE_d dE_\nu}$ for the deuteron energy $E_d = 18$ MeV. There is a drastic increase in decay rate by a factor of ≈ 200, when interactions between neutron and deuteron in the final state are included. In order to compare our results to experimental data one has to integrate

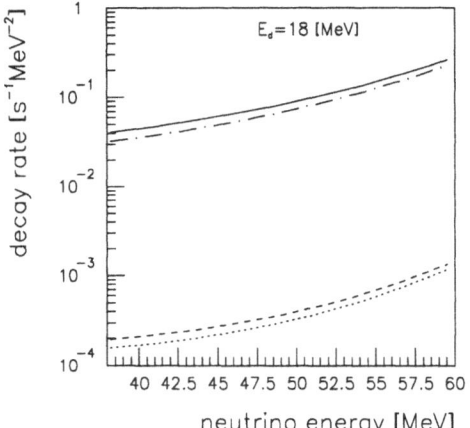

Figure 1. The decay rate $\frac{d^2\Gamma}{dE_d dE_\nu}$ at $E_d = 18$MeV. The dashed and dotted lines are the PWIAS predictions for the Bonn B and Paris potentials, respectively. The solid and dashed-dotted are the Bonn B and Paris potential predictions when in the final state all NN force components up to $j_{max} = 1$ are included.

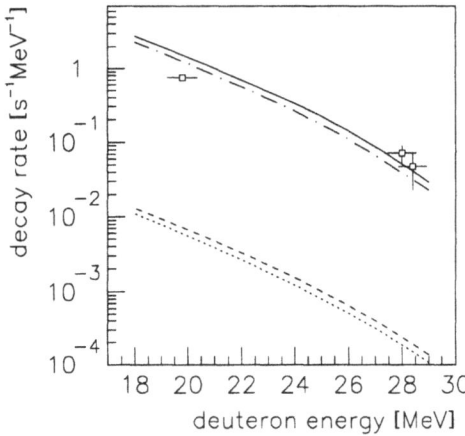

Figure 2. The decay rate $\frac{d\Gamma}{dE_d}$ for the case when the neutrino is not observed. For the description of lines see Fig.1. The squares are experimental points from [5, 6].

numerically the decay rate at every deuteron energy over the neutrino energy. This leads to the results presented in Fig. 2. It is clearly seen that while the PWIAS predictions drastically underestimate the data, the inclusion of FSI leads to a rather good agreement.

4 Summary

We calculated decay rates for muon capture on ^{3}He leading to a n+d+ν_μ channel. Realistic NN forces have been used (the Bonn B and Paris potentials). The bound and continuum 3N states were evaluated consistently solving the corresponding Faddeev equations. We used the most simple nonrelativistic single nucleon weak-current operator parametrized by the standard nucleon weak form factors. The predictions of the plane-wave impulse approximation differ drastically by a factor of about ~ 200 from the full result. The full calculations are in a nice agreement with the recent experimental data.

Acknowledgement. This work was supported by the Polish Committee for Scientific Research under Grant No. 2P03B03914, and by the Science and Technology Cooperation Germany-Poland under Grant No. X098.91. The numerical calculations were performed on the Cray T90 of the Höchstleistungsrechenzentrum in Jülich, Germany and on the Convex 3840 of the ACK-Cyfronet, Cracow, Poland under Grant No. KBN/C3840/UJ/020/1994.

References

1. W. Glöckle et al.: Phys. Reports **274**, 107 (1996)

2. S. Ishikawa et al.: Il Nuovo Cimento **107A**, 305 (1994); Phys. Lett. **B339**, 293 (1994); Phys. Rev. **C57**, 39 (1998)

3. J. Golak et al.: Phys. Rev. **C51**, 1638 (1995); Phys. Rev. **C52**, 1216 (1995)

4. W. Glöckle et al.: In: *Proceedings of the 2nd Workshop on Electronuclear Physics* (May 1998, MIT, USA)

5. W.J. Cummings et al.: Phys. Rev. Lett. **68**, 293 (1992)

6. S.E. Kuhn et al.: Phys. Rev. **C50**, 1771 (1994)

7. N. Tatara, Y. Kohyama, K. Kubodera: Phys. Rev. **C42**, 1694 (1990)

8. W. Glöckle: In: *The Quantum Mechanical Few-Body Problem.* Berlin: Springer Verlag 1983

9. H. Witała, W. Glöckle, Th. Cornelius: Few-Body Systems **3**, 123 (1988)

Few-Body Systems Suppl. 10, 351–354 (1999)

Few-
Body
Systems
© by Springer-Verlag 1999

Electron-induced two-proton knockout from ^{3}He

D.L. Groep[1], M.F. van Batenburg[1], Th.S. Bauer[1,2], H.P. Blok[1,3],
D.J. Boersma[1,2], E. Cisbani[4], R. De Leo[5], S. Frullani[4], F. Garibaldi[4],
W. Glöckle[6], J. Golak[7], P. Heimberg[1,3], W.H.A. Hesselink[1,3], M. Iodice[4],
D. Ireland[8], E. Jans[1], H. Kamada[6], L. Lapikás[1], G. Lolos[9], R. Perrino[10],
A. Scott[8], R. Starink[1,3], M.F.M. Steenbakkers[1,3], G.M. Urciuoli[4],
H. de Vries[1], L.B. Weinstein[11], H. Witała[7]

[1] NIKHEF, P.O. Box 41882, 1009 DB Amsterdam, The Netherlands

[2] Universiteit Utrecht, Utrecht, The Netherlands

[3] Vrije Universiteit, Amsterdam, The Netherlands

[4] Istituto Superiore di Sanità, Laboratorio di Fisica, INFN, Italy

[5] INFN Sezione di Bari, Dipartimento Interateneo di Fisica, Bari, Italy

[6] Institut für Theor. Physik II, Ruhr-Universität Bochum, Germany

[7] Institute of Physics, Jagellonian University, Cracow, Poland

[8] Department of Physics and Astronomy, University of Glasgow, UK

[9] Department of Physics, University of Regina, Canada

[10] INFN Sezione di Lecce, Lecce, Italy

[11] Physics Department, Old Dominion University, Norfolk, Virginia, USA

1 Introduction

Electron-induced two-nucleon knockout reactions at intermediate electron energies are driven both by one-body currents (coupling of the virtual photon to correlated nucleon pairs) and two-body currents (intermediate Δ-excitation and meson exchange currents (MEC)). In addition, final state interactions (FSI) contribute to the cross section. Recent experiments [1] have shown that the $^{16}O(e, e'pp)^{14}C$ reaction is dominated by the direct knockout of strongly correlated proton pairs via one-body currents.

Since continuum Faddeev calculations employing realistic NN potentials are currently available for the three-body system [2], an exclusive measurement of the three-body breakup of ^{3}He offers the opportunity to compare data to parameter-free model predictions. The relative importance of competing two-proton knockout mechanisms can be investigated by varying the energy and momentum of the virtual photon, as was done in the experiment described herein.

2 Experimental Setup

The experiment was performed in the EMIN endstation at NIKHEF, with the high duty-factor electron beam extracted from the pulse stretcher AmPS. The incident electron energy was 564 MeV and the duty factor and average current of the extracted beam were typically 80% and 3 μA, respectively. The target cell was a high-pressure (3 MPa) cryogenic (15 K) barrel cell with a diameter of 5 cm. The scattered electrons were detected in the QDQ magnetic spectrometer, and the emitted protons in the large acceptance HADRON3 and HADRON4 plastic scintillator arrays [3], covering solid angles of 240 and 550 msr, respectively. The coincidence time resolutions per detector pair were approximately 0.8 ns (FWHM). The experimental missing energy resolution amounted to 5 MeV (FWHM). Target thickness calibrations were at each kinematic setting obtained by measuring elastic electron scattering off ^{3}He.

3 Kinematics

Data were taken at momentum transfers (q) of 305, 375 and 445 MeV/c for an energy transfer (ω) of 220 MeV. At the central q value, data in the continuous ω-range from 170 to 290 MeV were collected. The experimental acceptances in (ω, q) are indicated in Fig. 1.

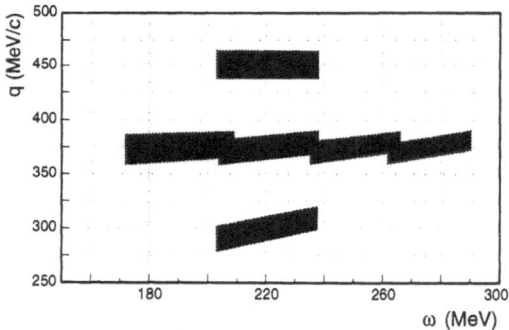

Figure 1. Experimental acceptances of the virtual photon in transferred energy ω and transferred three-momentum q.

Since the ^{3}He$(e, e'pp)$ reaction is kinematically complete, the outgoing neutron momentum can be identified with the missing momentum, defined as:

$$\boldsymbol{p}_{\rm m} = \boldsymbol{q} - \boldsymbol{p}'_1 - \boldsymbol{p}'_2 \equiv \boldsymbol{p}'_n, \tag{1}$$

in which $\boldsymbol{p}'_1$ and $\boldsymbol{p}'_2$ are the momenta of the forward and backward proton in the final state, respectively. In addition, the angle γ_1 between the three-momentum vector q and the momentum of the forward proton in the final state was used as an independent variable, while the cross section was averaged over the other kinematic quantities.

4 Results

At present, only the data taken at $\omega = 220$ MeV have been analyzed. These data were taken in three kinematic settings, covering largely overlapping regions in p_m and γ_1. The data shown below are still preliminary, but are not expected to change more than 10% in the complete analysis.

The data are compared to continuum Faddeev calculations [2] performed with the Bonn-B NN potential. In these calculations, two-body angular momenta up to $j = 2$ are accounted for, and both the ^{3}He ground state and the $3N$ breakup state are treated in the same framework, while all rescattering effects in the final state are accounted for up to infinite order.

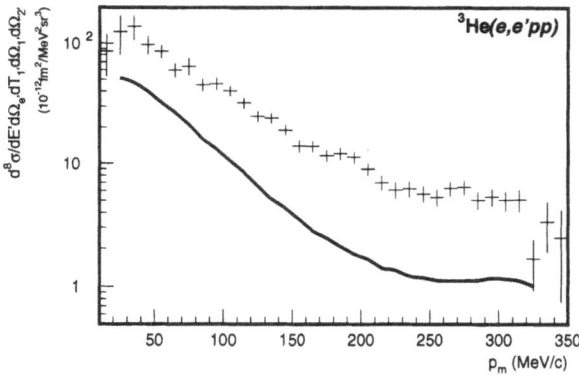

Figure 2. Preliminary eight-fold differential cross section of the ^{3}He$(e, e'pp)$ reaction as a function of the missing (neutron) momentum for the $(\omega, q) = (220 \text{ MeV}, 305 \text{ MeV}/c)$ kinematics. The solid curve corresponds to the result of continuum Faddeev calculations by Golak et al., averaged over the experimental detection volume.

Figure 2 shows the differential cross section for the $q = 305$ MeV/c and $\omega = 220$ MeV kinematics as a function of the missing momentum. In the low p_m-region, where the neutron momenta are small, the calculations fall short by about a factor of two. At higher missing momenta, the difference between the data and the theoretical prediction increases up to a factor five. In Fig. 3 the cross section is shown as a function of the transferred three-momentum q for a region in the (γ_1, p_m) phase space that is common to the detection volumes of the ω=220 MeV kinematic points. The measured cross section decreases by a factor of 5 between $q = 305$ and 445 MeV/c. This ratio is similar to the ratio predicted by the electron-proton cross section $K\sigma_{ep}$, which describes the coupling of a photon to an off-shell proton. This supports the picture of

354

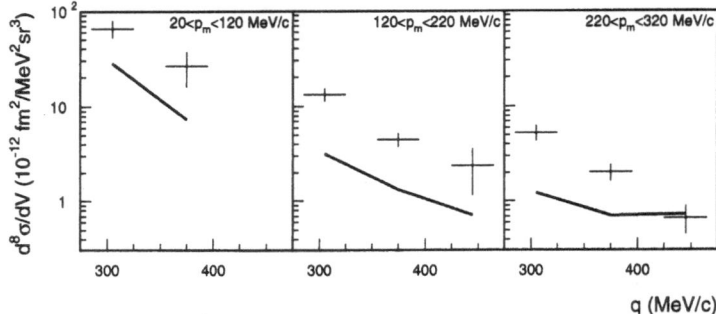

Figure 3. Preliminary eight-fold differential cross section for the ^{3}He$(e, e'pp)$ reaction for three values of three-momentum transfer q. The data have been averaged over an interval in γ_1 from 10 to 25 degrees in three bins for the neutron momentum p_m. For the setting at $q = 445$ MeV/c there is no acceptance for $p_m \leq 150$ MeV/c. The solid lines are the results of continuum Faddeev calculations evaluated within the experimental detection volume.

the photon coupling to one of the correlated protons, thus causing the second proton to be emitted as well, while the neutron acts as a spectator. At low p_m the results of the full Faddeev calculations show a similar decrease of the cross section as a function of q. Here, the underestimation of the data at $q = 375$ and 445 MeV/c is similar to the one observed at $q = 305$ MeV/c, as shown in Fig. 2. However, in the region $p_m \geq 250$ MeV/c the calculated q-dependence is significantly flatter than the experimental one.

Several ingredients of the calculations are currently being investigated as a possible source for the observed discrepancies: the NN potential used, meson-exchange and Δ contributions and the upper limit of the two-body angular momenta included. Three-body effects might play a role as well, because they influence the part of the ^{3}He wave function that contains three nucleons with high momentum.

Acknowledgement. This work is part of the research programme of the "Sticht-ing voor Fundamenteel Onderzoek der Materie (FOM)", which is financially supported by the "Nederlandse Organisatie voor Wetenschappelijk Onderzoek (NWO)". The support of the Science and Technology Cooperation Germany-Poland is gratefully acknowledged.

References

1. C.J.G. Onderwater et al.: Phys. Rev. Lett. **78**, 4893 (1997) and Phys. Rev. Lett. **81**, 2213 (1998)

2. J. Golak et al.: Phys. Rev. **C51**, 1638 (1995)

3. A.R. Pellegrino et al.: to be published in Nucl. Instr. and Methods

Few-Body Systems Suppl. 10, 355–358 (1999)

Few-
Body
Systems
© by Springer-Verlag 1999

Measurements of the virtual bremsstrahlung in the $p+p$ and $p+d$ systems

J.G. Messchendorp[1], J.C.S. Bacelar[1], M.J. van Goethem[1], M.N. Harakeh[1],
M. Hoefman[1], H. Huisman[1], N. Kalantar-Nayestanaki[1], A. Kugler[3],
H. Löhner[1], R.W. Ostendorf[1], S. Schadmand[1], R. Simon[2], M. Volkerts[1],
V. Wagner[3], H.W. Wilschut[1]

[1] KVI Groningen, The Netherlands

[2] GSI Darmstadt, Germany

[3] NPI Řež u Prahy, Czech Republic

Abstract.
   Virtual bremsstrahlung yields measured in $\vec{p}+p{\rightarrow}p+p+e^{+}+e^{-}$ and $\vec{p}+d{\rightarrow}{}^{3}\mathrm{He}+e^{+}+e^{-}$ are presented. The experiments were performed with a 190 MeV polarized proton beam obtained from the cyclotron AGOR at KVI in Groningen. Differential cross sections, response functions and analyzing powers were obtained for both reactions in exclusive measurements in which all outgoing particles were measured in a coincidence setup between SALAD and TAPS. The data are compared with gauge-invariant calculations using a NN T-matrix fitted to elastic phase-shifts. For $pp{\rightarrow}ppe^{+}e^{-}$ a reasonable agreement is found for all measured virtual-photon invariant masses, $M_{\gamma} >15$ MeV/c^{2} up to 80 MeV/c^{2}. For the $pd{\rightarrow}{}^{3}\mathrm{He}\,e^{+}e^{-}$ angular distributions, the calculations underestimate the data for $\theta_{CM} >100°$ similar to what is found in the real-photon capture reaction.

   The $p+p{\rightarrow}p+p+\gamma$ process is known to be an ideal probe to study the nucleon-nucleon (NN) interaction beyond the elastic region. The interest lies in studying the (half) off-shell behaviour of the NN potential. An advantage is that meson-exchange contributions are suppressed in this reaction. On the other hand, to improve the understanding of non-nucleonic degrees of freedom (mesons and resonances) in nuclei, reactions of the type $pd{\rightarrow}{}^{3}\mathrm{He}\gamma$ are favourable. The large momentum transfer in this process is expected to isolate and enhance meson-exchange contributions [1]. In the two-nucleon system ($np{\rightarrow}d\gamma$) the effects of meson-exchange can be adequately accounted for through the use of Siegert's theorem. The role of these contributions in heavier nuclei is so far not satisfactorily understood. To bridge our understanding of these effects from the two-nucleon process to that in the multinucleon system, a study of the three-nucleon system is imperative.

The study of the virtual bremsstrahlung process in few-body scattering is an important extension to the real-bremsstrahlung studies. A new aspect is the presence of a longitudinally polarized virtual photon. By measuring the momenta of both leptons arising from the virtual-photon decay, the $pp{\rightarrow}ppe^+e^-$ and $pd{\rightarrow}^3\text{He}\,e^+e^-$ reaction amplitudes can be decomposed into terms related to different virtual-photon polarizations [2]. In principle all response functions (W_i with i indicating the polarization: T for transverse, L for longitudinal and combinations for the interference terms) can be measured.

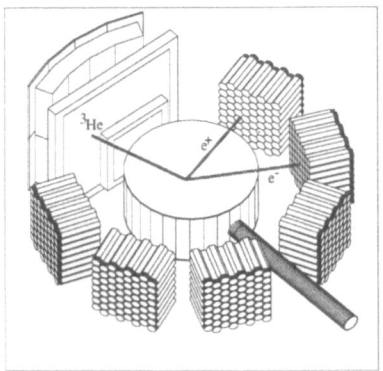

Figure 1. Schematic view of the detector system SALAD and of TAPS in block mode. For the $pp{\rightarrow}ppe^+e^-$ the two protons are detected in SALAD.

Two reactions were measured so far: $\vec{p}+p$, and $\vec{p}+d$. For both reactions, polarized protons of 190 MeV were used. The small cross section of the virtual bremsstrahlung reaction makes it necessary to use liquid targets (LH$_2$ and LD$_2$) and to perform exclusive measurements. We used 6 mm thick targets for both experiments.

The experiment was performed with a coincidence setup between two different detector systems [2], schematically depicted in Fig. 1. The Small-Angle Large-Acceptance Detector (SALAD) [3] measures the scattering angle ($6° < \theta_p < 21°$ with full azimuthal angle coverage, and up to $\theta_p = 26°$ with limited coverage) and the energy of the hadrons (protons, deuterons and ^{3}He) with an angular resolution of 0.5 degrees and an energy resolution of about 5%. The electrons and positrons are detected in the photon spectrometer TAPS (Two-Arm Photon Spectrometer) [4] consisting of 384 BaF$_2$ crystals placed around the target position. This results in a good coverage of the polar angle of the virtual photon and the possibility to detect virtual photons with large invariant masses up to the kinematical limit (93 MeV/c^2 for $pp{\rightarrow}ppe^+e^-$ and 130 MeV/c^2 for $pd{\rightarrow}^3\text{He}\,e^+e^-$ reaction).

For both reactions, the position of the leptons allowed us to decompose the reaction amplitude into the different response functions. Furthermore, the kinematical overdetermination gives the possibility to reduce the background to a negligible level [2].

The virtual-bremsstrahlung yields have been measured simultaneously with the real-bremsstrahlung yields allowing the study of its relative behaviour.

Theoretical calculations of the $pp{\rightarrow}ppe^+e^-$ channel are based on a low-energy theorem (LET) approximation which was derived [5] along the lines of the original theorem by Low for the real-photon production [6]. These calculations are performed by expanding the leading-order contributions (where the photon couples to external legs) around the on-shell kinematics. Furthermore, higher-order contributions are added in such a way that the current is conserved. These extra terms take rescattering and meson-exchange diagrams partially into account. Contributions resulting from Δ degrees of freedom and negative-energy states are not taken explicitly into account.

Calculations for the $pd{\rightarrow}{}^3$Heγ^* channel [7] are performed similarly to those of the $pp{\rightarrow}ppe^+e^-$ channel. Also here, leading-order contributions are calculated and a contact term is constructed in such a way to obey current conservation. The ^{3}He wave-function has been obtained from the Argonne V18 potential, in Plane Wave Approximation.

In the left part of Fig. 2 the results of the analysis of the $pp{\rightarrow}ppe^+e^-$ reaction is shown and compared to calculations. The differential cross section (for polar proton angles $6^\circ{<}\theta_p{<}26^\circ$ and invariant masses $M_\gamma{>}15$ MeV/c^2), transverse (W_T) and transverse-transverse (W_{TT}) response functions are plotted as a function of the virtual-photon invariant mass M_γ. The absolute normalisation of the cross sections has been obtained by analyzing the well known elastic channel $p{+}p{\rightarrow}p{+}p$ resulting in a systematical error of $\pm 5\%$. For all these observables a reasonable agreement with the calculations is observed. As predicted by the model, the longitudinal response is suppressed and only appear to be significant at large invariant masses. Unfortunately, here statistics becomes poor. Second generation experiments planned at KVI will obtain good statistical accuracy at these large invariant masses.

The results for the $pd{\rightarrow}{}^3$He$\,e^+e^-$ experiment are shown in the right part of Fig. 2. In the top panel, the capture cross-section is given as a function of the center-of-mass angle. The data are integrated over photon invariant masses larger than 10 MeV/c^2. The calculations underestimate the measured dilepton yield for scattering angles between $100^\circ - 130^\circ$. The same is observed in the real- capture yields [1], and therefore points to a deficiency of the calculations in the hadronic sector. Within the experimental errors the ratio of dilepton to real-photon yields is well predicted by the calculations as can be seen in the bottom panel of Fig. 2. The middle panel shows the analyzing power measured for the virtual-photon capture and a comparison with the measurements [1] performed for the real-photon capture.

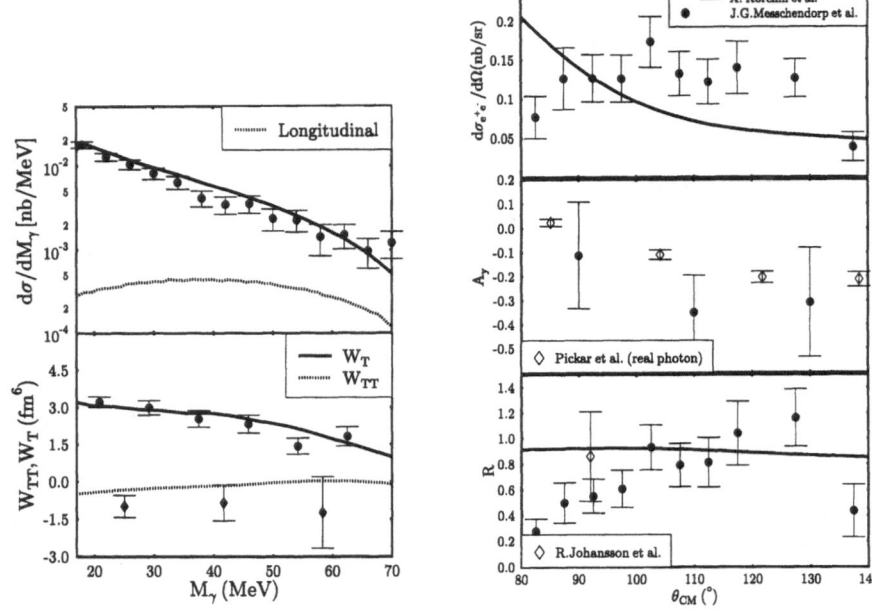

Figure 2. Left: Differential cross-section and response functions W_T and W_{TT} as a function of the invariant mass of the virtual photon for the reaction $pp \rightarrow ppe^+e^-$. Right: Differential cross-section, analyzing power and ratio $R = \frac{\sigma_{e^+e^-}}{\alpha\sigma_\gamma}$ as function of the center-of-mass angle θ_{CM}, for the reaction $pd \rightarrow {}^3He\,e^+e^-$.

References

1. M.A. Pickar et al.: Phys. Rev. **C35**, 37 (1987)

2. J.G. Messchendorp et al.: Nucl. Phys. **A631**, 618c (1998)

3. N. Kalantar-Nayestanaki: Nucl. Phys. **A631**, 242c (1998)

4. A.R. Gabler et al.: Nucl. Instr. Meth. Phys. Res. **A346**, 168 (1994)

5. A.Yu. Korchin and O. Scholten: Nucl. Phys. **A581**, 493 (1995)

6. F.E. Low: Phys. Rev. **110**, 974 (1958)

7. A.Yu. Korchin et al.: submitted to Phys. Lett. **B**

Few-Body Systems Suppl. 10, 359–362 (1999)

Few-
Body
Systems
© by Springer-Verlag 1999

Benchmark calculations for the n+t system

F. Ciesielski[1], J. Carbonell[1], C. Gignoux[1], A.C. Fonseca[2]

[1] Institut des Sciences Nucléaires, 53 Av. des Martyrs, F-38026 Grenoble
[2] Centro Física Nuclear da U.L., Av. Prof. Gama Pinto, 2, P-1699 Lisboa

Abstract. The results for the n+t reaction obtained by solving the Faddeev-Yakubovsky and AGS equations are presented. We show how the resonant behaviour, which is the main feature of the n+t elastic cross section, is missed by the realistic NN potentials when limited to the $j = 1$ waves.

The aim of this benchmark is to check the ability of the realistic NN potentials to reproduce the n+t elastic cross section, accurately measured in [1]. We have jointly used two very different formalisms in the present calculations: on the one hand, the Faddeev-Yakubovsky (FY) approach in configuration space and, on the other hand, the AGS equations based on a converged expansion of the 3N t-matrix and a rank one expansion of the NN potential. First results including references and methods have been presented before [2, 3].

It has been reported elsewhere [4] that the integrated cross section is well reproduced by the phenomenological MT I-III potential. The two considered approaches are in reasonable agreement when dealing with this potential as it is illustrated in Fig. 1. It is worth noticing the good description of the differential cross section at neutron laboratory energy $T_{lab} = 3.5$ MeV reached with this simple model interaction. In what follows we show that both methods lead to very similar results, even in the case of realistic NN interactions.

The results discussed below concern the scattering lengths and the cross sections obtained with a realistic NN potential, mainly the Argonne V14. We restrict ourselves to the energies below the first inelastic threshold n+n+d, i.e. $T_{\text{lab}} \leqslant 6.8$ MeV, and except when explicitly mentioned, only the $^1S_0$, $^3S_1$ and $^3D_1$ interaction components have been included.

Since in the AGS approach the 2+2 subamplitudes are treated exactly by convolution and the 3N t-matrix expansion is well under control, the difference between the two calculations may only be attributed to the use of rank one separable expansion for 2N t-matrix. In particular, this fact is responsible for the slight differences between the calculated singlet (a_0) and triplet (a_1) scat-

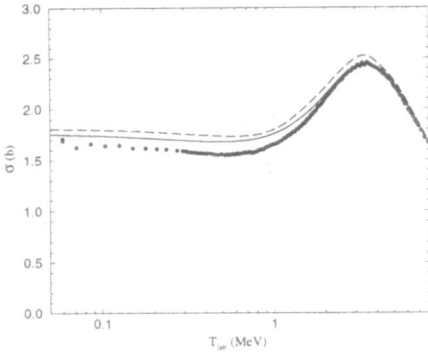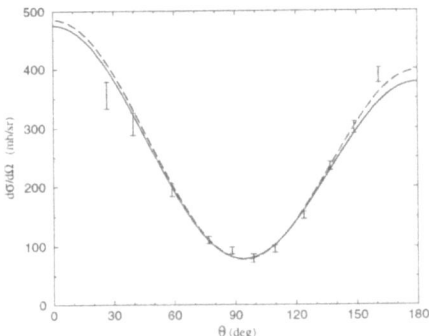

Figure 1. Experimental n+t total and differential ($T_{lab} = 3.5$ MeV) cross sections are well reproduced by the FY (solid line) and AGS (dashed line) calculations with MT I-III NN interaction.

tering lengths, as emphasized in Table 1. However, the more precise results of FY in this case lead to greatly overestimated zero-energy cross section and coherent scattering length. Concerning a_0 and a_1 individually, the discrepancies existing between the different experimental values prevent us from a reliable comparison. We notice furthermore that the last published results seem to be in disagreement with the older ones and with all the calculations. A definitive measurement of a_0 and a_1 is thus necessary.

Table 1. n+t zero-energy parameters with AV14. The singlet, triplet, coherent scattering lengths a_0, a_1, a_c (fm) and the zero energy cross section σ_0 (b) are overestimated, unless a three nucleon term is included.

		a_0	a_1	a_c	σ_0
exp.		?	?	3.65±0.08	1.70±0.03
AV14	(AGS)	4.39	3.89	4.01	2.03
AV14	(FY)	4.31	3.79	3.92	1.94
AV14+TNI	(FY)	3.99	3.53	3.65	1.68

The cross sections calculated with the FY (solid line) and AGS (dashed line) approaches are plotted in Fig. 2. They are very close each other in the whole energy domain, but far from the experimental points, not only in the very low-energy region (a direct consequence of the overestimated values of a_0 and a_1), but also in the resonant region. Since the physical origins of these two defects are very different they are to be discussed separately: i) the low-energy overestimate is due to the too small triton binding energy given by AV14. A strong correlation between this energy and the N+NNN scattering length has been observed and this problem can be simply overcomed by including an *ad hoc* 3N interaction [2, 5]. The resulting curve (dotted line) is plotted on the left part of Fig. 2, and the new scattering lengths shown in Table 1. Note that these values are compatible with those obtained by the more elaborate Urbana TNI [6]. ii) the bump in the elastic cross section at $T_{\text{lab}} \approx 3.5$ MeV is the consequence of the n+t P-waves resonant behaviour. Far from the zero-energy correction but still at small energies, the effect of a TNI is supposed to be negligible. This is indeed the case for the phenomenological TNI used, but could also be expected in view of the N+NN case [7].

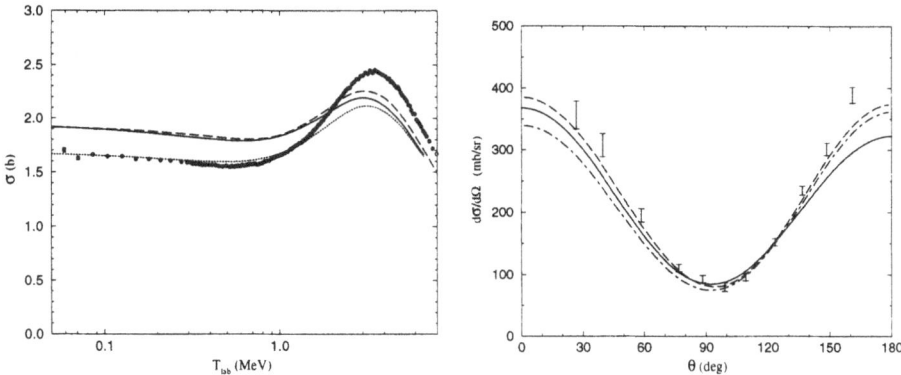

Figure 2. n+t integrated and differential cross sections at $T_{\text{lab}} = 3.5$ MeV. The experimental points are not well fitted by the FY (solid line) or AGS (dashed line) results with AV14 interaction. The dotted line shows the correction caused by an *ad hoc* three nucleon force. A differential cross section obtained with Bonn B potential and AGS approach (dot-dashed line) is included showing the same defect.

The differential cross section at $T_{\text{lab}} = 3.5$ MeV is plotted in Fig. 2 and illustrates a reasonable agreement between the FY (solid line) and AGS (dashed line) approaches at least concerning the total cross section. The results confirm the defect of AV14 in the peak region, where the n+t P-waves play an important role. Further calculations with other realistic interactions lead to similar conclusions [2, 5]. This behaviour does even not seem to be a characteristic of the local potentials, but has also been obtained in the AGS formalism for the Bonn-B potential (dot-dashed line in Fig. 2).

So far, neither the inclusion of the NN P-waves, closing the $V_{NN}^{j=1}$ shell, nor the inclusion of a model TNI are able to improve the performances of the realistic potentials in the peak region. The reason for such a defect may be due to a lack of convergence in the calculations or has to be found in the NN interaction itself. More complete calculations in both aproaches are in progress. We have already pointed out how a simple model (MT I-III) succeeds in the n+t challenge, both in the total and differential cross sections. The comparison between the n+t P-waves phaseshifts in Fig. 3 reveals an interesting feature. Whereas in the MT case, 3 partial waves (0^-, 1^-, 2^-) are degenerated and their contributions differ only by a statistical factors, in the AV14 case this degeneracy is broken. Unfortunately, the spreading of the original phaseshift is unsymmetrical and globally the 3 phaseshifts are all weakened. The origin of this could be the inadequacy of the present approach for the NN interaction, which seems to show already some limits in the so-called A_y puzzle [8] and in other 4N scattering observables [9].

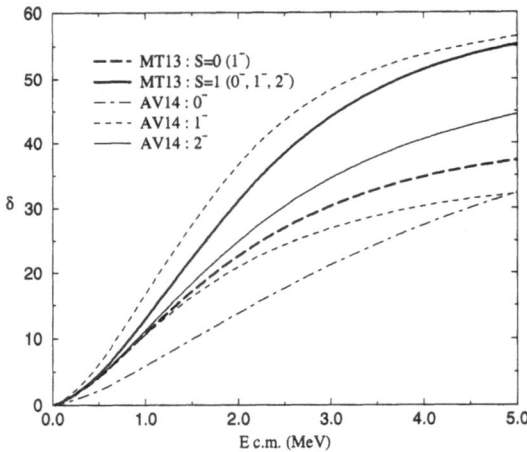

Figure 3. n+t P-waves phaseshifts. The $J^\pi = 0^-$ (dot-dashed thin line), $J^\pi = 1^-$ (dashed thin line), and $J^\pi = 2^-$ (solid thin line), obtained with AV14 are compared to those, i.e. $S = 0, J^\pi = 1^-$ (long-dashed thick line) and $S = 1, J^\pi = 0^-, 1^-, 2^-$ (solid thick line), obtained with MT I-III. The asymmetrical degeneracy breaking results in a weakening of the correlated contribution to the cross section.

Acknowledgement. The numerical calculations have been performed with the Cray-T3E of the CGCV (CEA) and IDRIS (CNRS).

References

1. T.W. Phillips, B.L. Berman, J.D. Seagrave: Phys. Rev. **C22**, 384 (1980); J.D. Seagrave, L. Cranberg, J.E. Simmons: Phys. Rev. **119**, 1981 (1960)

2. F. Ciesielski, J. Carbonell, C. Gignoux: Nucl. Phys. **A631**, 653c (1998) ; F. Ciesielski: Thèse Université J. Fourier (Grenoble) (1997)

3. A.C. Fonseca: Nucl. Phys. **A631**, 675 (1998); Phys. Rev. **C40**, 1390 (1989)

4. F. Ciesielski and J. Carbonell: Phys. Rev. **C58** 58 (1998)

5. F. Ciesielski, J. Carbonell, C. Gignoux: Phys. Lett. **B** (in press)

6. M. Viviani: Contribution to this conference

7. W. Glockle et al.: Physics Reports **274**, 107 (1996)

8. D. Hüber: Contribution to this conference

9. A.C. Fonseca: Contribution to this conference

Few-Body Systems Suppl. 10, 363–366 (1999)

Few-
Body
Systems
© by Springer-Verlag 1999

Neutron–^{3}H and Proton–^{3}He Low–Energy Scattering

M. Viviani

INFN, Piazza Torricelli, 2, 56100 Pisa, Italy

Abstract. The Kohn variational principle and the correlated Hyperspherical Harmonics technique are applied to study the $n - {}^3$H and $p - {}^3$He low energy scattering. The calculated $n - {}^3$H total cross section at zero energy agrees well with the measured value, while a small discrepancy is found for the coherent scattering length. For the $p - {}^3$He scattering process, a comparison between the calculated and measured observables is performed at 3 MeV, where accurate experimental data are available. A reasonable agreement is found for the differential cross section, whereas clear discrepancies are observed for the polarization observables.

The problem of four nucleon scattering has been considered for a long time (see ref. [1] and references cited therein). However, calculations with realistic nuclear hamiltonians have been performed only rather recently, by using either the resonating group method (RGM) [2], or the Faddeev–Yakubovsky (FY) method [3, 4], or the correlated Hyperspherical Harmonic (CHH) technique [5]. The present accuracy in these calculations is in some measure lesser than that achieved in the $N - d$ case. Nevertheless clear indications of rather large discrepancies between theoretical predictions and experimental data have been pointed out, in particular for the energy dependence of the $n - {}^3$H cross section [3] and in some $p - {}^3$H polarization observables [2].

In the present contribution, preliminary results for some $n - {}^3$H and $p - {}^3$He observables are reported. The scattering wave functions are expanded in terms of the CHH basis [6] and determined by using the Kohn variational principle. Such a technique has been successfully used in the study of the $N - d$ scattering below and above the deuteron breakup threshold [7, 8]. The present calculations follow exactly the same line as that described in ref. [7] for the $N - d$ case.

The calculated singlet and triplet $n - {}^3$H scattering lengths, corresponding to different potential models, are plotted versus the ^{3}H binding energy in Fig. 1. Their most recent experimental values [9, 10] are also reported. The models including only NN forces are the AV14 [11], AV8 [12] and AV18 [13]

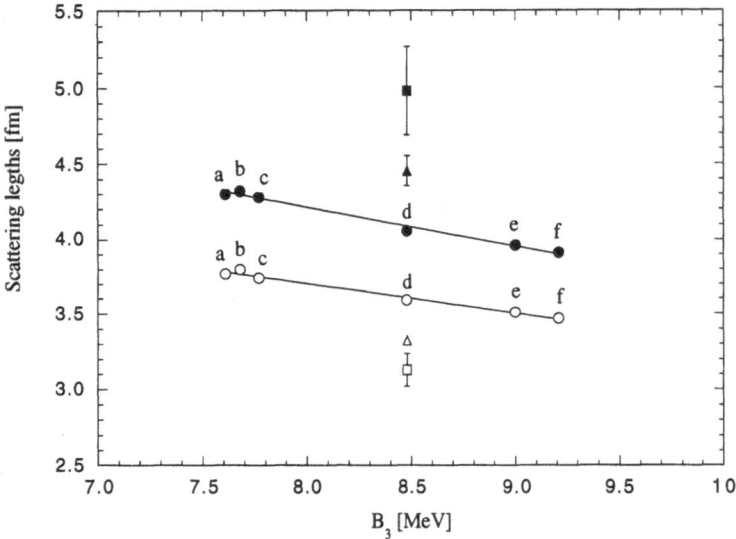

Figure 1. Singlet (full symbols) and triplet (open symbols) $n - {}^3\text{H}$ scattering lengths plotted against the ${}^3\text{H}$ binding energy. Circles labelled by a, b, c, d, e, f correspond to the AV18, AV14, AV8, AV18UR, AV14BR1 and AV14BR2 models, respectively. The AV14UR and AV18UR model predictions are almost coincident. The squares (triangles) are the experimental values of ref. [9] (ref. [10]). The straight lines are linear fits of the theoretical results.

potentials. The ones including also 3N forces are: the AV14+Urbana model VIII (AV14UR) [14], AV18+Urbana model IX (AV18UR) [15], AV14+Brazil with $\Lambda = 5.6 m_\pi$ (AV14BR1) and AV14+Brazil with $\Lambda = 5.8 m_\pi$ (AV14BR2) [16]. In the AV14UR and AV18UR models, one parameter of the 3N potentials was fixed in order to reproduce the experimental ${}^3\text{H}$ binding energy value $B_3 = 8.48$ MeV. The AV14BR1 and AV14BR2 models have been chosen to give slightly larger B_3 values. In the case of the AV14 interaction, the present estimate for the singlet (triplet) scattering length is 4.32 (3.80) fm, in good agreement with the value 4.31 (3.79) fm obtained in ref. [3] by solving the FY equations. By inspecting Fig. 1, it can be noticed that all the estimates for the singlet (triplet) scattering lengths fall essentially on a straight line. However, the experimental values extracted from the data do not lie on the theoretical curves. This disagreement is related to a rather small discrepancy between the calculated and measured coherent scattering length [5].

A preliminary study of the $p - {}^3\text{He}$ differential cross section and some polarization observables have been performed at $E_{c.m.} = 3$ MeV. At this energy, accurate experimental data are available [17, 18] and, therefore, detailed comparisons can be made. Only S- and P-wave phase shifts and mixing angle

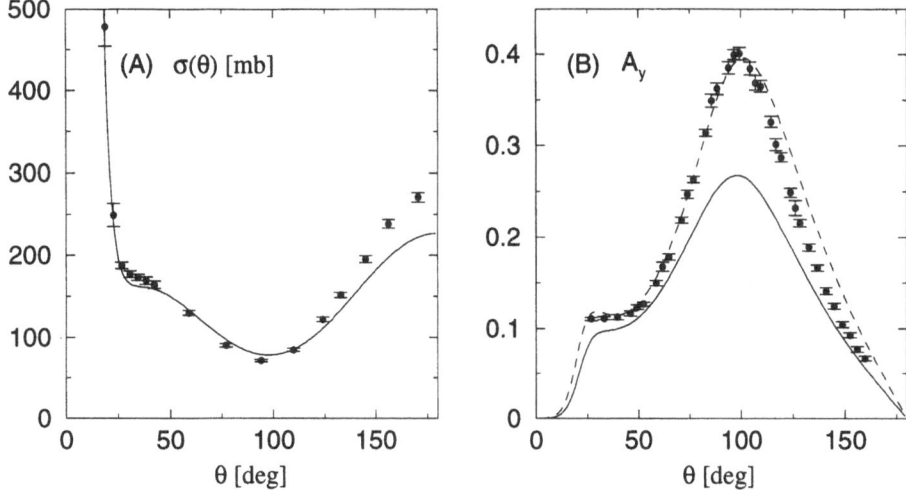

Figure 2. Preliminary $p - {}^3$He differential cross section (A) and proton analysing power (B) at $E_{c.m.} = 3$ MeV for the AV18 potential model (solid lines). The experimental differential cross section is taken from ref. [9] and the analysing power values from ref. [10]. The dashed line is the theoretical proton analyzing power calculated by using for the ${}^3P_2$ phase shift the experimental (PSA) value.

parameters have been taken into account in the theoretical calculations of the observables. That approximation is somewhat justified by the phase shift analysis (PSA) performed in ref. [17] (the D-wave phase shift are almost negligible for $E_{c.m.} = 3$ MeV). The calculated (solid lines) and measured (points with error bars) differential cross section σ and proton analysing power A_y are reported in Fig. 2. The interaction model used is the AV18, and the Coulomb interactions between the protons is fully taken into account. It is worthy to notice that the proper treatment of the Coulomb repulsion is necessary in order to have the correct behavior of the observables at small angles. As it can be seen from Fig. 2, a reasonable agreement is obtained for the differential cross section, in particular at small angles and in the minimum region. On the other hand, a rather large underprediction of A_y is observed. Such a discrepancy is mainly due to the small value of the calculated ${}^3P_2$ phase shift. In fact, the theoretical calculation predicts $\delta({}^3P_2) = 30°$, whereas the value $\delta({}^3P_2) \approx 37°$ was obtained in the PSA of ref. [17]. To stress the effect of this disagreement, the dashed line in Fig. 2 shows the value assumed by A_y if the PSA value of $\delta({}^3P_2)$, together with the theoretical estimates for all the other parameters, is used in calculating this observable. A much better agreement with the experimental data is now obtained. Investigations are in progress to test the connections of this disagreement with the similar underprediction of A_y in the low-energy $N - d$ scattering processes.

Acknowledgement. This work has been done in collaboration with S. Rosati and A. Kievsky. The author wishes to thank J. Carbonell and L.D. Knutson for valuable discussions and for providing their results prior to publication.

References

1. D.R. Tilley, H.R. Weller and G.M. Hale: Nucl. Phys. **A541**, 1 (1992)

2. H.M. Hofman and G.M. Hale: Nucl. Phys. **A613**, 69 (1997)

3. F. Ciesielski, J. Carbonell and C. Cignoux: In: *Proceedings of the XV International Conference of Few–Body Problems in Physics* (Groningen (Holland), 22–27 July 1997), Nucl. Phys. **A631**, 635c (1998)

4. A.C. Fonseca: In: *Proceedings of the XV International Conference of Few–Body Problems in Physics* (Groningen (Holland), 22–27 July 1997), Nucl. Phys. **A631**, 675c (1998)

5. M. Viviani, S. Rosati and A. Kievsky: Phys. Rev. Lett. **81**, 1580 (1998)

6. A. Kievsky, S. Rosati and M. Viviani: Nucl. Phys. **A551**, 241 (1993)

7. A. Kievsky, M. Viviani and S. Rosati: Nucl. Phys. **A577**, 511 (1994)

8. A. Kievsky, M. Viviani and S. Rosati, **C56**, 2987 (1997); A. Kievsky, M. Viviani and S. Rosati: in preparation

9. H. Rauch et al.: Phys. Lett. **165B**, 39 (1985)

10. G.M. Hale et al.: Phys. Rev. **C42**, 438 (1990)

11. R.B. Wiringa, R.A. Smith and T.L. Ainsworth: Phys. Rev. **C29**, 1207 (1984)

12. R.B. Wiringa (unpublished)

13. R.B. Wiringa, V.G.J. Stoks and R. Schiavilla: Phys. Rev. **C51**, 38 (1995)

14. R.B. Wiringa: Phys. Rev. **C43**, 1585 (1991)

15. B.S. Pudliner et al.: Phys. Rev. **C56**, 1720 (1997)

16. H.T. Coelho, T.K. Das and M.R. Robilotta: Phys. Rev. **C28**, 1812 (1983)

17. M.T. Alley and L.D. Knutson: Phys. Rev. **C48**, 1901 (1993)

18. D.G. McDonald, W. Haeberli and L.W. Morrow: Phys. Rev. **133**, B1178 (1964)

Few-Body Systems Suppl. 10, 367–370 (1999)

Few-
Body
Systems
© by Springer-Verlag 1999

Use of Realistic Interactions to Calculate Four-Nucleon Scattering Observables

A.C. Fonseca

Centro de Física Nuclear da Universidade de Lisboa, Av. Prof. Gama Pinto, N: 2, 1699 Lisboa Codex, Portugal

Abstract. The four-body equations of Alt, Grassberger and Sandhas are solved for a system of four nucleons, using realistic NN interactions in channels $^1S_0$, $^3S_1 - ^3D_1$, $^1P_1$, $^3P_0$, $^3P_1$ and $^3P_2$. The results of the calculation are compared with data for the reactions $dd \rightarrow dd$ and $dd \rightarrow p^3\text{H}$. The calculations indicate that the nucleon-nucleon p-waves have a strong effect on $4N$ observables, but one finds an overall disagreement with data that indicates a possible failure of known realistic interactions.

1 Introduction

In a recent review article on the three-nucleon continuum [1] one finds that $n - d$ elastic observables (cross sections, vector and tensor polarizations) are insensitive to the choice of realistic NN potential. Behond the persistent Ay discrepancy at low energy, the agreement between calculations and data is excellent in the energy range up to 65 MeV. In the present work we extend our understanding of realistic NN interactions by testing them in the four-nucleon continuum.

The starting point involves the solution of Alt, Grassberger and Sandhas equations [2] for the transition operators involving all $(2) + (2)$ and $(3) + 1$ channels. For local NN potentials such equations are three-vector variable integral equations which after partial wave decomposition reduce to a set of coupled equations in three continuous variables. Similar equations were recently solved for ^{4}He [3] and n^3H elastic scattering [4] below three-body breakup threshold, requiring more than one hundred hours of single processor super computing time to handle large dimension matrices ($n > 10^6$). Since scattering calculations require a great number of channels, one follows an approach based on the separable representation of subsystem amplitudes in order to reduce the equations to two (or one) continuous variable. Although the number of effective $1 + (3)$ channels increases by a factor of three or four, there is a net gain due to reduction in the dimensionality of the equations and the internal sum of two- and three-body subsystem channels in the kernel of the equations, leading to matrices that are three orders of magnitude smaller. The integral equations we use are the same as in ref. [5] and result from the AGS equations [6] after one has: (a) represented the original NN t-matrix by an operator of rank one; (b) represented the resulting $3N$ t-matrix by a finite rank operator and taken as many terms as needed for convergence. Since in the modified AGS equations [6] the $2 + 2$ subamplitudes are expressed in terms of a convolution integral involving two non-interacting pair-propagators, as first proposed by Fonseca [7], the sole approximation in this approach involves a rank one representation of the $2N$ t-matrix which may be obtained from the well-known method of Ernest, Shakin, and Taylor [8]. The multi-term representation of the $3N$ t-matrix is done using the EDPE method developed by Sofianos, McGurk and Fiedeldey [9]. This latest approximation for the $3N$ t-matrix is well under control since one may compare the finite rank approximation with the original t-matrix results for the $3N$ observables (cross sections and analyzing powers), and check the convergence rate of $4N$ observables for increasing rank of the $3N$ representation.

This method was first used in ref. [5] to calculate the binding energy of ^{4}He and later confirmed to be accurate by the exact work of Kamada and Glöckle [10]. More recently [11] the results of our calculations for n^3H elastic scattering were shown to agree with the results of the Grenoble group [4], for both Mafliet-Tjon and Argonne V14 potentials taken in $2N$ partial waves with $j \leq 1^+$ ($^1S_0$, $^3S_1 - ^3D_1$).

2 Results

The four-nucleon calculations we present here make use of the Paris, Bonn-A, Bonn-B and Argonne V14 potentials in channels $^1S_0, ^3S_1 - ^3D_1$, $^1P_1$, $^3P_0$, $^3P_1$ and $^3P_2$. The first two channels correspond to including all $2N$ partial waves $j \leq 1^+$, while the first five channels to $j < 1$. As mentioned before the sole approximation is the use of a rank one EST expansion of the respective $2N$ t-matrix in each partial wave. In all calculations we include all positive and negative parity three-nucleon ($3N$) subamplitudes with total angular momentum up to $J = 7/2^+$ as well as all underlying $3N$ channels corresponding to particle-pair orbital angular momentum $L \leq 3$. Likewise all $2+2$ subamplitudes that are consistent with the underlying $2N$ channels are included. Finally, all $4N$ observables are calculated using four-nucleon amplitudes with total angular momentum up to $\mathcal{J} = 6^\pm$ in all corresponding $1 + 3$ and $2 + 2$ channels with relative orbital angular momentum $\mathcal{L} \leq 5$.

Since preliminary work was first presented in ref. [12], we do not show here how the results converge with the rank in the EDPE expansion of $3N$ t-matrix, nor with the number of the $3N$ subamplitudes for increasing $3N$ total angular momentum J. All results are shown here for fixed rank "r" equal six for $3N$ subamplitudes with $J \leq 3/2^\pm$, four for $J = 5/2^\pm$ and two for $J = 7/2^+$.

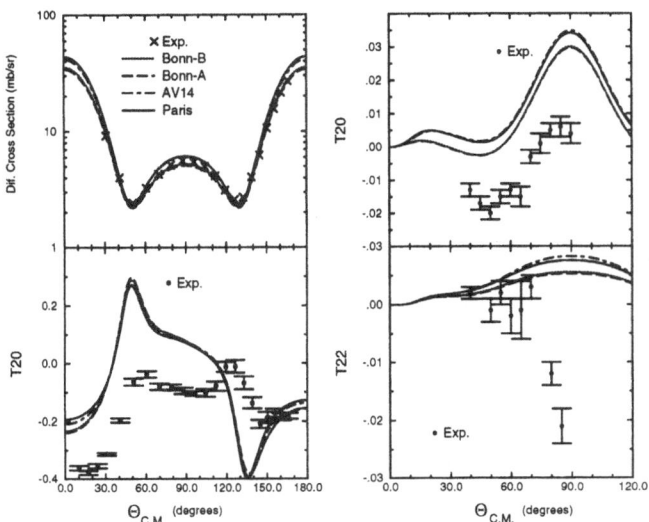

Figure 1. $dd \rightarrow p^3\mathrm{H}$ (left) and $dd \rightarrow dd$ (right) observables for different realistic interactions taken in $2N$ partial waves with $j \leq 1^+$. The data points are from ref. [13]. All $3N$ subamplitudes up to $J = 7/2^+$ are included with ranks $r = 6666442$ respectively.

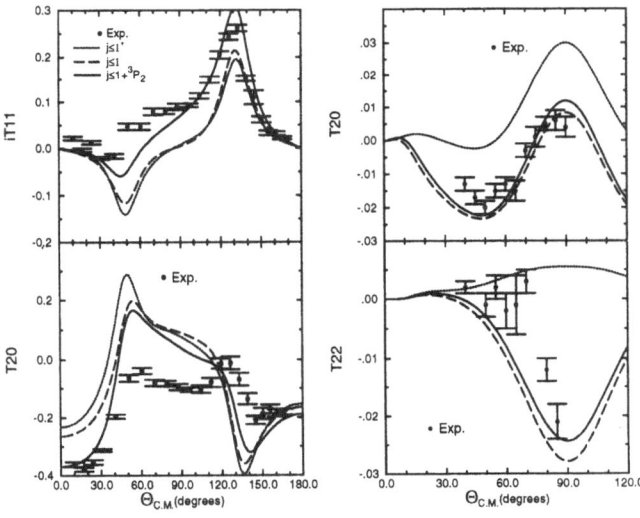

Figure 2. Same as in Fig. 1 for Bonn-B but including higher $2N$ partial waves.

In Fig. 1, we show both $dd \to p^3H$ and $dd \to dd$ observables and compare different realistic interactions. Only $2N$ partial waves with $j \leq 1^+$ are included. Although one finds $4N$ observables to be more sensitive to $2N$ input than $3N$ observables, there are no dramatic changes that may indicate a profound difference between potentials. Vis-a-vis the data the discrepancy is large and one is compeled to include higher $2N$ partial waves. This is done in Fig. 2 for Bonn-B. The outcome is impressive but concerning. The NN p-waves introduce a remarkable change in the observables, particularly iT_{11} for $dd \to p^3H$ and T_{20} and T_{22} for $dd \to dd$. Nevertheless the remaining discrepancies in the tensor observables for $dd \to p^3H$, of which T_{20} is an example, cannot be attributed to the use of rank one EST for the $2N$ t-matrix, given the benchmark agreement [11] with the Grenoble group for $n-^3H$ elastic scattering in the same energy region. The failure of realistic interactions in reproducing the $n-^3H$ elastic resonance shape is bound to have a counterpart in the $dd \to p^3H$ and $dd \to dd$ sector, particularly because the resonant behaviour is associated with negative parity $4N$ states that exist in both $\mathcal{T} = 0$ and $\mathcal{T} = 1$ reactions. Nevertheless one needs to study the effect of higher partial waves and realistic $3N$ forces before taking a final conclusion.

References

1. W. Glöckle et al.: Phys. Reports **274**, 107 (1996)

2. E.O. Alt, P. Grassberger, W. Sandhas: Phys. Rev. **C1**, 85 (1970)

3. H. Kamada and W. Glöckle: Phys. Lett. **B292**, 1 (1993)

4. F. Ciesielski, J. Carbonell, C. Gignoux: Nucl. Phys. **A631**, 653c (1998)

5. A.C. Fonseca: Phys. Rev. **C40**, 1390 (1989)

6. H. Haberzettl and W. Sandhas: Phys. Rev. **C24**, 359 (1981)

7. A.C. Fonseca and P.E. Shanley: Phys. Rev. **D13** 2255 (1976); ibidem: Phys. Rev. **C14**, 1343 (1976)

8. J. Ernest, C.M. Shakin and R.M. Thaler: Phys. Rev. **C8**, 46 (1973)

9. S. Sofianos, N.J. McGurk and H. Fiedeldey: Nucl. Phys. **A318**, 295 (1979)

10. H. Kamada and W. Glöckle: Nucl. Phys. **A548**, 205 (1992)

11. F. Ciesielski et al.: Contribution to this conference

12. A.C. Fonseca: Nucl. Phys. **A631**, 675c (1998)

13. W. Gruebler et al.: Nucl. Phys. **A193**, 129 (1972); ibidem **A193**, 149 (1972); ibidem **A369**, 381 (1981)

Few-Body Systems Suppl. 10, 371–374 (1999)

Few-
Body
Systems
© by Springer-Verlag 1999

A Study of the Deuteron D-state Probability in the $^3\overrightarrow{\text{He}}(\overrightarrow{d},p)^4$He Reaction

S. Oryu[1*], S. Gojuki[1], T. Hino[1†], E. Uzu[1], H. Kamada[2]

[1]Department of Physics, Faculty of Science and Technology,
  Science University of Tokyo, Noda, Chiba 278-8510, Japan
[2]Institut für Theoretische Physik II, Ruhr-Universität Bochum,
  D-44780 Bochum, Germany

Abstract. The deuteron D-state probability and the spin polarization correlation coefficient $C_{//}$ for the $^3\overrightarrow{\text{He}}(\overrightarrow{d},p)^4$He reaction were investigated at a deuteron lab energy $E_d = 270$ MeV in the $p+n+^3$He three-body Hilbert space. Here the Faddeev-Born Approximation was treated with the realistic NN and N-^{3}He interactions, because the observables involve the forward proton scattering cross section at high energy. We present theoretical predictions for thirteen different deuteron wave functions from realistic potentials. The calculated results of $C_{//}$ are about a factor of 2 larger than the experimental value. A physical interpretation of the discrepancies is presented based upon a five-body system picture.

The deuteron D-state probability is studied in the polarized $^3\overrightarrow{\text{He}}(\overrightarrow{d},p)^4$He reaction at a deuteron lab energy $E_d = 270$ MeV. The vector and tensor analyzing powers A_y, A_{xx}, A_{yy} and the spin vector polarization correlation coefficients C_{xx}, C_{yy} at a forward c.m. angle have been measured [1]. The spin polarization correlation coefficient $C_{//}$ is defined by those spin observables; *i.e.*,

$$C_{//}(\theta) = 1 + \frac{1}{4}(A_{yy} + A_{xx}) + \frac{3}{4}(C_{yy} + C_{xx}). \tag{1}$$

It is demonstrated that $C_{//}$ agrees with the cross section ratio at the forward angle $\theta_{\text{lab}} = \theta_{\text{c.m.}} = 0°$ and the backward angle $\theta_{\text{c.m.}} = 180°$; *i.e.*, $C_{//}(0°; 180°) = \sigma_{\text{pol}}(0°; 180°)/\sigma_0(0°; 180°)$, where $\sigma_0(\theta)$ is the differential cross section for the unpolarized reaction ^{3}He$(d,p)^4$He and $\sigma_{\text{pol}}(\theta)$ is the differential cross section for polarized $^3\overrightarrow{\text{He}}(\overrightarrow{d},p)^4$He reaction.

---

*E-mail address: oryu@ph.noda.sut.ac.jp
†Alternative address: Fujitsu Limited

On the other hand, in the Plane Wave Born Approximation (PWBA) one can also show:

$$C_{//}(0^{\circ}) \;=\; \frac{9}{4}\frac{w^2(k)}{u^2(k) + w^2(k)}, \tag{2}$$

$$\boldsymbol{k} \;=\; \boldsymbol{q}_p - \frac{1}{2}\boldsymbol{q}_d, \tag{3}$$

where $u(k)$ and $w(k)$ are the S- and D-state deuteron wave functions, respectively, and $\boldsymbol{k}$ is the internal momentum which is given by the initial momentum $\boldsymbol{q}_d$ of the deuteron and final momentum $\boldsymbol{q}_p$ of the proton in the reaction. Therefore, the polarization correlation coefficient $C_{//}$ obtained from the experimental data in Eq. (1) reflects directly the D-state probability of the deuteron wave function in the PWBA of Eq. (2).

We use the Faddeev equations to calculate the reaction $^3\mathrm{He}(d,p)^4\mathrm{He}$ as a three-body problem consisting of $^3\mathrm{He}$, p and n. We assume the Born term as the solution of the Faddeev equations (Faddeev-Born approximation), because contributions from higher terms should become small at high energy. We employ a separable RGM potential (SeRGM) for the N-$^3\mathrm{He}$ interaction [2] which is introduced by a technical method of subtracting the Pauli forbidden states from the RGM N-$^3\mathrm{He}$ potential and by the EST separable expansion [2], and for the NN interaction the PEST potential [3] as the first step. However, the Born term is essentially given by the initial deuteron wave function and the final $^4\mathrm{He}$ (or N-$^3\mathrm{He}$) wave function. Therefore we can utilize various deuteron wave functions. In Fig. 1 we show the polarization correlation coefficients $C_{//}$ as points generated using deuteron wave function obtained from various realistic NN potentials which have different D-state probabilities. The solid curve is the prediction obtained from the Faddeev-Born approximation using a PEST potential modified to reproduce various D-state probabilities. The experimental value is shown as a solid line with its error bar corresponding to the shaded region. The theoretical predictions of $C_{//}$ are about a factor of 2 larger than the experimental value; i.e., if the prediction derived from the Faddeev-Born approximation were correct, then the D-state probability should be 2 %, which is very different from conventional expectations.

For the final state, we can choose several $^4\mathrm{He}$ wave functions which are obtained not only from the SeRGM N-$^3\mathrm{He}$ potential, but also the phase-shift equivalent PEST-type separable N-$^3\mathrm{He}$ potential (SePSEQ), and by direct solution of the four-body Faddeev-Yakubovsky (F-Y) equations with the AV14-NN interaction. The calculated results for $C_{//}$ in the Faddeev-Born approximation and in the PWBA are compared as a function of the scattering angle in Fig. 2a. The solid curve corresponds to the result from the Faddeev-Born approximation and the dashed curve corresponds to the result from the PWBA. Both theoretical results completely agree in the forward scattering region $0^{\circ} \leq \theta_{\mathrm{c.m.}} \leq 90^{\circ}$. However, $C_{//}$ in the Faddeev-Born approximation oscillates for $\theta_{\mathrm{c.m.}} \geq 90^{\circ}$ resulting from the truncation of higher total angular momentum than $J = 39/2$ in the SeRGM potential case. The results using the F-Y wave function and those for the SePSEQ-potential for $J = 39/2$ are in very good agreement with

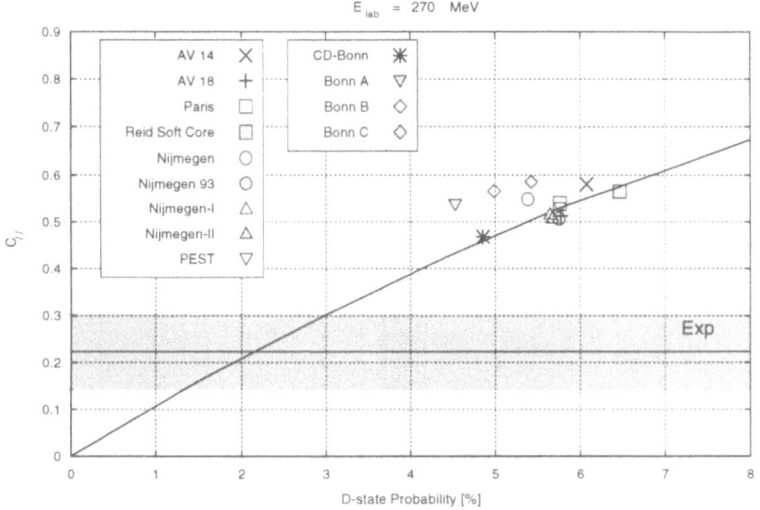

Figure 1. $C_{//}$ plotted versus D-state probability. The points and curve are produced using deuteron wave functions from various NN potentials. The experimental value is shown as a straight solid line with an error bar corresponding to the shaded region.

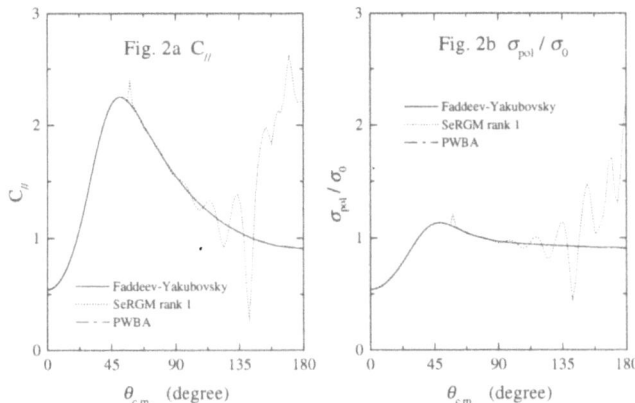

Figure 2. a: for $C_{//}$, and b: for $\sigma_{\mathrm{pol}}(\theta)/\sigma_0(\theta)$ are compared as a function of angle.

PWBA at all angles. We also present the cross section ratio $\sigma_{\mathrm{pol}}(\theta)/\sigma_0(\theta)$ in Fig. 2b. The two quantities at the forward and backward angles agree with those of $C_{//}$ except for the SePSEQ potential.

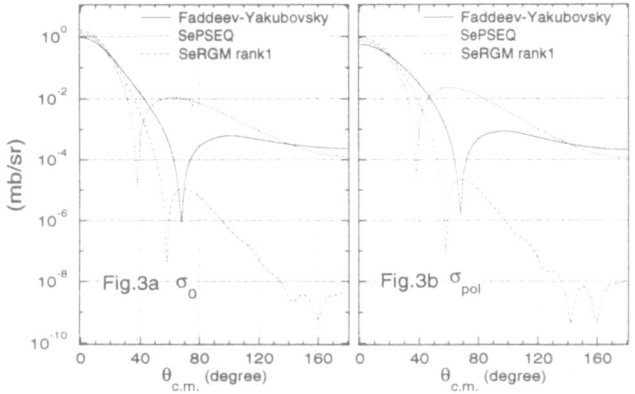

Figure 3. a: the usual differential cross section $\sigma_0(\theta)$, and b: the polarized differential cross section $\sigma_{\text{pol}}(\theta)$ are compared as a function of angle.

Figs. 3a and 3b illustrate the unpolarized and polarized differential cross sections, respectively. Cross sections for three different n-^{3}He wave functions are compared. It is found that the potential dependence of the cross section is large but the ratio and $C_{//}$ are almost independent of the final-state interaction. Therefore, we conclude that the quantity $C_{//}$ is a good physical observable with which to investigate the D-state probability of the incident deuteron. The small nodules around 58° and the oscillations at large angles in Figs. 2a and 2b are numerical errors which come from the ratio ($\sim 0/0$) due to the very small differential cross sections.

Finally, our results seem to imply that the Faddeev-Born approximation modeled as $p + n + ^3$He is not adequate to constrain the D-state probability quantitatively, but treating processes $p+d+d$ and $p+^3$H$+p$ as five-body systems would be necessary. However, if we restrict ourselves to the forward scattering of protons, then ^{3}He could be a good core object, and including a few higher order processes would improve our theoretical scheme.

Acknowledgement. This work was supported by a grant of the FRCCS projects in the Science University of Tokyo.

References

1. T. Uesaka: Doctoral Thesis of the University of Tokyo (1997)

2. S. Oryu et al.: Few-Body Systems **17**, 185 (1994); S. Oryu et al.: Nucl. Instr. & Meth. in Phys. Res. **A402**, 402 (1998)

3. J. Haidenbauer and W. Plessas: Phys. Rev. **C30**, 1822 (1984) and ibid **32**, 1424 (1985)

Few-Body Systems Suppl. 10, 375–378 (1999)

Few-
Body
Systems
© by Springer-Verlag 1999

Pole Structure of the $\frac{3}{2}^+$ Resonance in ^{5}He and $n + \alpha$ Scattering Observables

E. Simeckova, P. Bem, P. Vercimak

Nuclear Physic Institute, Academy of Sciences of the Czech Republic,
CZ-25068 Rez, Czech Republic

Abstract. Presently adopted $n + \alpha$ scattering phase shifts from long past parametrization of the $\frac{3}{2}^+$, 16.75 MeV resonant state in ^{5}He are found at variance with results of the present R-matrix analysis of observables in the ^{5}He system. It enables to dismiss the controversy considered in the recent dynamical microscopic analysis ot the two-pole structure of the $S(\frac{3}{2}^+)$-matrix associated with this state.

In past works, the well known $J^\pi = \frac{3}{2}^+$, ^{5}He* (16.75 MeV) resonance was parametrized in terms of various approximations to the Breit-Wigner formula. Consequently, different sets of resonance parameters were extracted from experimental data. It has been demonstrated by the Los Alamos group that an unambiguous parametrization of a resonance could be extracted from an asymptotic quantity such as the S-matrix [1, 2]. Authors have performed a single-level (SL) and two-level (TL) R-matrix analysis of $d + t$ reaction cross-section below 250 keV and also a multilevel (ML) analysis of wide database of observables in the ^{5}He system. Different sets of $R(\frac{3}{2}^+)$-matrix parameters have generated the pole of the $S(\frac{3}{2}^+)$-matrix at the nearly identical position on the unphysical Riemann sheet $(\mathrm{sgn}\,\mathrm{Im}k_n, \mathrm{sgn}\,\mathrm{Im}k_d) = (-,-)$, k_i is the wave number in $n + \alpha$ and $d + t$ channels, respectively. It is commonly known, that the Los Alamos group has also observed the other pole (so called shadow pole), which was found to lie on another unphysical sheet $(-,+)$, for the TL and ML sets [2].

An analysis of observed two-pole structure within the dynamical microscopic model has recently been reported by the Debrecen group [3]. Model calculations have successfully reproduced the constellation of poles on Riemann sheets as it has been generated from ML resonance parameters. However, the model prediction was found in controversy with resonant $D(\frac{3}{2}^+)$-wave phase shifts of $n + \alpha$ scattering (from analysis of a limited set of data) due to Hoop and Barschall [4] (H.B.). Namely, the predicted location [3] of a shadow pole on a sheet (-,+) implies the behaviour of phase shift that passes through the $\frac{\pi}{2}$ point (see full line in Fig. 1a), or - in term of Argand phase diagram - the resonance loop of scattering amplitude encompasses the $\frac{i}{2}$ point (full line in Fig. 1b). The curves calculated by H.B. exhibit the opposite behaviour (dashed lines in Fig. 1), which is just the matter of controversy in the theoretical treatment of the shadow pole. Any analysis concerning the relevance of the H.B. set of phase shifts with updated database was not yet referred to.

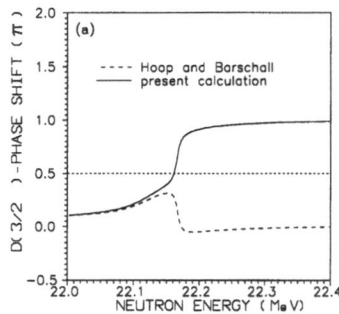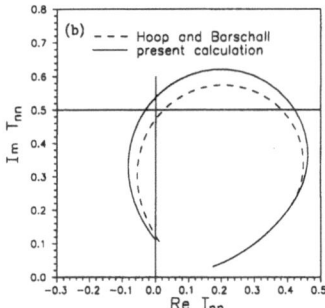

Figure 1. Calculated phase shift (a) and Argand diagram (b) of $D(\frac{3}{2}^+)$-wave in the $n + \alpha$ channel.

In our work, the two channel-multilevel R-matrix analysis of observables in $n + \alpha$ and $d + t$ channels was performed with the database up to the deuteron energy of 1 MeV and a neutron energy of 25 MeV. The database includes corresponding part of the data set cited in refs. [1, 2] and also present accurate data of the $d + t$ cross section reported recently [5]. The 380 data points were fitted with 35 free R-matrix parameters, describing the known ^{5}He states and the contribution from distant (background) levels. The fitting procedure has converged, having chi-square value of 1.17 per degree-of-freedom. Resulting $R(\frac{3}{2}^+)$-matrix parameters have generated the position and the width of the conventional $S(\frac{3}{2}^+)$-matrix pole, $E_M = 46.5$ keV and $\Gamma_M = 73.0$ keV, and the shadow pole (-,+), $E_S = 80.2$ keV and $\Gamma_S = 7.9$ keV, which agree well with results of the ML LASL analysis [2]. Further results and details of the present analysis will be referred to elsewhere. Here we compare present results and H.B. predictions for observables relevant to the resonating $\frac{3}{2}^+$ phase shift in the $n + \alpha$ channel.

In Fig. 1, the phase shifts (Fig. 1a) and the Argand diagram (Fig. 1b) of the scattering amplitude for the $D(\frac{3}{2}^+)$-wave in the $n + \alpha$ channel are shown, as they are calculated from the present analysis (full lines) and from the H.B. set of parameters (dashed line). Note that present curves, being identical with those calculated from ML set [2], are in accordance with dynamical microscopic calculations [3]. In Fig. 2, the $n + \alpha$ cross section observables from the Canberra experiment [6] are compared with the present fit and the calculations

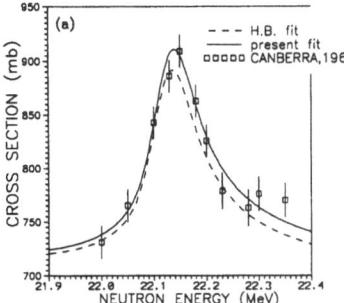 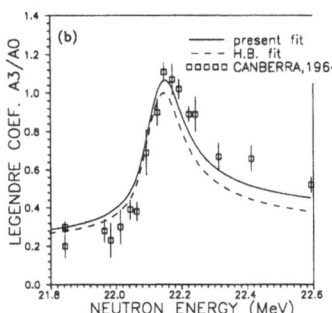

Figure 2. The total neutron cross section **(a)** and the Legendre-coefficient ratio A3/A0 of differential cross section **(b)** from ref. [6], compared with present and H.B. calculations. Calculated curves are averaged over the energy resolution of data.

based on H.B. phase shift parameters (only the total cross section was employed in H.B. analysis). Although the partial chi-square values (1.31 and 2.96 per point for present and H.B. calculations, respectively) prefer the present parametrization, the difference is of a little statistical significancy. Also, no statement could be made from an analysis of the other and lately measured $n + \alpha$ cross section (Karlsruhe, ref. [7]), the reason being in noticeable and not randomly distributed fluctuations of these data, see Fig. 3a. Actually, the whole

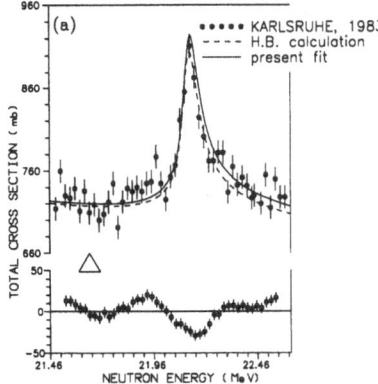 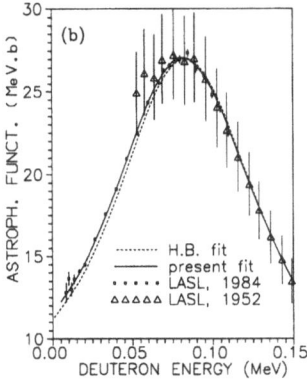

Figure 3. The total neutron cross section **(a)** from ref. [7] and the S-function of the integral $d + t$ reaction cross section **(b)** from ref. [1], (LASL, 1984) and ref. [9] (LASL, 1952), compared with present and H.B. calculations. For explanation of the symbol Δ in **(a)**, see text.

deviation Δ of experimental data from the present calculated curve (both averaged over some energy interval - five points in the showed case) illustrates that the resonance shape could be significantly distorted, making thus the determination of the resonance energy and the ordering $\Gamma_n < \Gamma_d$ (from single-level parametrization [7, 8] of only these data) less reliable. In Fig. 3b, the $d+t$ cross section from an earlier work of LASL [9] (used in H.B. analysis), is shown together with recent accurate LASL data [1]. The H.B. version gives acceptable fit to earlier (rather soft) data, but dramatically disagrees with recent ones, leading to the chi-square per point of 15 (the present fit gives the value of 1.1).

Results of the present work thus enable to conclude that the up to now adopted $D(\frac{3}{2}^+)$-wave phase shifts of $n + \alpha$ scattering from Hoop and Barshall [4] are disproved by the latest $d + t$ reaction cross section data. Further, the presently available $n + \alpha$ scattering observables do not contradict the dynamical treatment of the $\frac{3}{2}^+$ resonance in ^{5}He, formulated by the Debrecen group [3]. Finally, the parameters of the $\frac{3}{2}^+$, $E_n = 22133$ keV resonance, recommended [8] from the analysis of most recent $n + \alpha$ total cross section measurements [7] could partly be affected by systematical errors that we have ascertained in these data.

References

1. R.E. Brown, N. Jarmie, G.M. Hale: Phys. Rev. **C35**, 1999 (1987)

2. G.M. Hale, R.E. Brown, N. Jarmie: Phys. Rev. Lett. **59**, 763 (1987)

3. A. Csoto, R.G. Lovas, T. Kruppa: Phys. Rev. Lett, **70**, 1391 (1993)

4. B. Hoop, Jr., H.H. Barschall: Nucl. Phys. **83**, 65 (1966)

5. P. Bem et al.: Few-Body Systems **22**, 77 (1997)

6. R.E. Shamu and J.G. Jenkin: Phys. Rev. **135**, B99 (1964)

7. B. Haesner et al.: Phys. Rev. **C28**, 995 (1983)

8. F. Ajzenberg-Selove: Nucl. Phys. **A490**, 1 (1988)

9. V. Argo et al.: Phys. Rev. **87**, 612 (1952)

Few-Body Systems Suppl. 10, 379–382 (1999)

Few-
Body
Systems
© by Springer-Verlag 1999

Three-Body Coulomb Effects in the Direct Coulomb Breakup of ^{8}B into ^{7}Be + p in the Field of a ^{208}Pb Ion.

B.F. Irgaziev[1] *[†], E.O. Alt[2]**, A.M. Mukhamedzhanov[3]††***

[1] Physics Department, Tashkent State University, Tashkent 700095, Uzbekistan

[2] Institut für Physik, Universität Mainz, D-55099 Mainz, Germany

[3] Cyclotron Institute, Texas A&M University, College Station, TX 77843, USA

Abstract. The amplitude for the Coulomb breakup of a light nucleus in the field of a highly charged ion is considered in the framework of the distorted wave approach, with particular emphasis being laid on correctly taking into account the three-body Coulomb interactions in the final state. Numerical calculations have been performed for the double differential cross section for the reaction ^{208}Pb(^{8}B, ^{7}Be p)^{208}Pb. They clearly demonstrate the importance of long-range three-body Coulomb correlations in the astrophysically interesting regime when the ejectiles have the extremely small relative energies.

1 Introduction

Study of the dissociation ^{8}B → ^{7}Be + p in the Coulomb field of a fully stripped ^{208}Pb ion has attracted great interest from theorists as well as experimentalists. The reason lies in the possibility to extract the astrophysical S-factor for the inverse reaction, namely the ^{7}Be(p, γ)^{8}B radiative capture, for such low energies as are of interest in nuclear astrophysics but are very difficult to extract directly from experiment, due to the Coulomb barrier. A first experiment on the Coulomb breakup of ^{8}B has been carried out by Motobayashi et al. [1], with theoretical calculations performed by Typel and Baur [2]. In their analysis the possibility of Coulomb post-acceleration was not considered. Later

*Supported by Deutscher Akademischer Austauschdienst (DAAD).

†*E-mail address:* irgaz@iaph.silk.org

**E-mail address:* alt@dipmza.physik.uni-mainz.de

††Supported by DOE Grant DE-FG05-93ER40773

***E-mail address:* akram@comp.tamu.edu

on, post-acceleration effects were studied as higher-order dynamical processes in different approaches (see [3, 4] and references therein). However, a truly adequate treatment of the Coulomb final state interaction within the genuine three-body framework, which results in modifications of the three-body final state wave function, has not been attempted.

In fact, the well known form of the three-charged particle wave function in the appropriate asymptotic regions [5, 6] indicates that long-range three-body correlations occur, which manifest themselves by changing, for non-asymptotic distances between the three particles, the asymptotic relative momentum between the ejectiles into a "local" momentum. The latter depends, among other things, on the distance between the target nucleus and the center of mass of the ejectiles. Only if the separations between all three particles become comparably large asymptotically, the standard Redmond form of asymptotics [7] is valid. That these three-body correlations can give rise to considerable modifications of the standard two-body pictures of the Coulomb breakup has already been explored in [8].

The aim of the present investigation is to show that the aforementioned three-body Coulomb effects do, indeed, give important contributions to the double differential cross-section for the reaction ^{208}Pb(^{8}B, ^{7}Be p)^{208}Pb when the reaction fragments ^{7}Be and p have relative energies less than 500 keV. As a result, the S-factor extracted at very low relative energies of the ejectiles changes non-negligibly relative to its value extracted without three-body effects.

2 Amplitude for the Coulomb Breakup $A(a, bc)A$ in the Distorted-Wave Approach

Consider the reaction

$$a + A \rightarrow b + c + A, \tag{1}$$

where a is a cluster-like projectile which can separate into nuclei b and c, and A is a heavy, highly charged ion. The corresponding charge numbers are Z_b, Z_c, $Z_a = Z_b + Z_c$, and Z_A, respectively. We denote the radius vector between particles b and c by r, and the radius vector between ion A and the center of mass the system $(b + c)$ by ρ. As the Coulomb effects do not depend on spin, all particles are considered spinless.

The breakup amplitude pertaining to reaction (1) can be written as

$$M_{if} = \langle \Psi_f^{(-)} \mid \Delta U \mid \phi_a \Phi_i^{(+)} \rangle. \tag{2}$$

As shown in ref. [2] for the reaction ^{208}Pb(^{8}B, ^{7}Be p)^{208}Pb, the $M1$ transition contributes to the cross section in the resonance region only, being negligible for other relative energies of the fragments. Therefore, the assumption that the interaction ΔU is purely Coulombian is well justified in that case.

Let $U_i^C(\rho) = Z_a Z_A e^2 / \rho$ be the entrance channel Coulomb "optical" potential between particles A and a. Then, $\Phi_i^{(+)}(\rho)$ denotes the initial-state scattering wave function describing the relative motion of ion A and the projectile a

under the action of $U_i^C(\rho)$. Moreover, $\phi_a(\boldsymbol{r})$ is the overlap integral of the bound state wave functions of nuclei a and b with that of c, and $\Psi_f^{(-)}(\boldsymbol{r}, \boldsymbol{\rho})$ the full final state three-body wave function. The transition potential is given as $\Delta U = V_{bA}^C + V_{cA}^C - U_i^C$, with the Coulomb potentials $V_{bA}^C = Z_b Z_A e^2 / |\, \boldsymbol{\rho} + m_c/m_a \boldsymbol{r}\, |$ and $V_{cA}^C = Z_c Z_A e^2 / |\, \boldsymbol{\rho} - m_b/m_a \boldsymbol{r}\, |$, $m_a = m_b + m_c$.

Our calculations are based on the representation (2) of the breakup amplitude. Since we are going to study reactions of the type (1) in kinematical regions where they are highly peripheral, i.e. when the distance ρ between particles A and a is much larger than the sum of their nuclear radii, we use for the final state an approximate three-body wave function which is valid in the relevant region of configuration space. The latter is characterised by the fact that the variable r, because of its occurrence in the incoming bound state overlap function $\phi_a(\boldsymbol{r})$, is of limited variation while the variable ρ is large and eventually approaches infinity. Thus, we use for $\Psi_f^{(-)}(\boldsymbol{r}, \boldsymbol{\rho})$ the asymptotic three-charged particle wave function valid in this asymptotic region [6]. This wave function consists of four terms of which, however, only the first three ($\sim O(1/\rho)$) were used since the last one is $O(1/\rho^2)$ and was found to give negligible contributions to the reaction amplitude as compared to the other three terms. A multipole expansion can be carried out for the transition potential, resulting in

$$\Delta U = Z_A e^2 \sum_{l,m} \frac{4\pi}{(2l+1)} \frac{r^l}{\rho^{l+1}} \left[\left(\frac{m_c}{m_a}\right)^l Z_b + (-1)^l \left(\frac{m_b}{m_a}\right)^l Z_c \right] Y_{lm}^*(\hat{\rho}) Y_{lm}(\hat{r}). \quad (3)$$

In the special case of the reaction $^{208}\text{Pb}(^8\text{B}, {}^7\text{Be}\, p)^{208}\text{Pb}$, the Coulomb breakup amplitude (2) contains the overlap function ϕ_a of the bound state wave functions of ^8B and ^7Be. This overlap function is approximated by $\mathcal{S}^{1/2}\chi$, where $\mathcal{S}$ is the spectroscopic factor for the configuration $p + {}^7\text{Be}$ in ^8B, and χ is the bound state wave function of ^7Be and p in ^8B. There exist two possible configurations of $p + {}^7\text{Be}$ in ^8B, with the channel spin which is the sum of the spins of the p and of ^7Be being equal to 1 or 2. The dominant component is that with spin equal to 2 [9]. The function χ was calculated using the DWBA PTOLEMY code. The spectroscopic factor was taken so as to yield the correct vertex constant for the overlap function $\mathcal{S}^{1/2}\chi$ [9].

3   Results for the $^{208}\text{Pb}(^8\text{B}, {}^7\text{Be}\, p)^{208}\text{Pb}$ Coulomb Breakup

We have studied the dissociation of ^8B into $^7\text{Be} + p$ in the Coulomb field of a ^{208}Pb ion at a ^8B beam energy of $E_i = 46.5$ MeV/A, for scattering angles in the range from 1 to 10 degrees. Note that in the kinematical region of interest, the binding energy of the proton to the ^7Be nucleus is much smaller than the collision energy E_i. Using the above-mentioned final state wave function, the breakup amplitude is seen to consist of two types of contributions. That of the first type which arises with the help of the leading term of the asymptotic three-body wave function [5], factorises into an excitation factor of ^8B and the elastic scattering amplitude of ^8B in the field of the ^{208}Pb ion; those of the second type, however, which arise from the next-to-leading terms [6] do not. Obviously, if the contribution of the latter would be large then extraction of the astrophysical S-factor from the Coulomb breakup reaction would be impossible. Our calculations show that, indeed, the cross section is completely dominated by the first, leading term of the asymptotic wave function; the contribution of the further wave function terms to the breakup amplitude turns out to be less than 1%.

We note that the main contribution to the Coulomb breakup amplitude comes, as expected, from the dipole part in (3). The quadrupole term contributes less than 8% in the range $E_{^7\mathrm{Be}\,p} = 800$ - $1000\,\mathrm{keV}$ of relative kinetic energy of the ejectiles, its magnitude decreasing further with decreasing energy.

We find that the double differential cross section for the reaction $^{208}\mathrm{Pb}(^8\mathrm{B}, ^7\mathrm{Be}\,p)^{208}\mathrm{Pb}$ at a scattering angle of 2 degrees, when calculated with the proper (asymptotic) final state wave function, turns out to be larger by approximately 9% at the relative energy $E_{^7\mathrm{Be}\,p} = 300\,\mathrm{keV}$, by 6% at $E_{^7\mathrm{Be}\,p} = 600\,\mathrm{keV}$, by 1% at $E_{^7\mathrm{Be}\,p} \gtrsim 800\,\mathrm{keV}$, than that obtained by using only the Redmond asymptotic wave function. Note that the latter practically coincides with the factorised form of the wave function in the conventional DWBA for three particles in the final state. The observed difference in the small relative-energy region may be attributed to a significant increase of the local momentum over the value of the asymptotic (usual) momentum.

The conclusion to be drawn from these results is that the S-factor as extracted from a theory which takes into account three-body Coulomb effects will differ from that derived in the conventional way by the same percentages. Hence, future analyses of Coulomb breakup reactions with very low relative energy of the fragments should take care to include three-body Coulomb final state interactions.

References

1. T. Motobayashi et. al.: Phys. Rev. Lett. **73**, 2680 (1994)

2. G. Typel and G. Baur: Phys. Rev. **C50**, 2104 (1994)

3. C.A. Bertulani: Nucl. Phys. **A587**, 318 (1995)

4. C.A. Bertulani and M. Gai: Nucl. Phys. **A636**, 227 (1998)

5. E.O. Alt and A.M. Mukhamedzhanov: Phys. Rev. **A47**, 2004 (1993)

6. A.M. Mukhamedzhanov and M. Lieber: Phys. Rev. **A54**, 3078 (1997)

7. P.J. Redmond, cited in L. Rosenberg: Phys. Rev. **D8**, 1833 (1972)

8. E.O. Alt et al.: Phys. Atom. Nuclei **58**, 1860 (1995)

9. A.M. Mukhamedzhanov and N.K. Timofeyuk: Yad. Fiz. **51**, 679 (1990)

Few-Body Systems Suppl. 10, 383–386 (1999)

Few-
Body
Systems
© by Springer-Verlag 1999

Pionic decay of the hypertriton

H. Kamada[1*], J. Golak[2], K. Miyagawa[3], H. Witała[2], W. Glöckle[1]

[1] Institut für Theoretische Physik II, Ruhr Universität Bochum, D-44780 Bochum, Germany
[2] Institute of Physics, Jagellonian University, PL 30059 Cracow, Poland
[3] Department of Applied Physics, Okayama University of Science, Ridai-cho Okayama 700, Japan

Abstract. The pionic decay of the hypertriton is investigated. Three-body Faddeev equations are rigorously solved for the bound state and scattering states using realistic NN and YN interactions. The theoretically predicted decay rates agree well with experiment.

1 Introduction

The hypertriton is the lightest hyper-nucleus, consisting of a proton, a neutron and a hyperon. We calculated this bound state as a rigorous solution of the three-body Faddeev equation using the realistic potentials of the Nijmegen type for the YN [1] and the NN [2] systems. Our theoretical prediction of the binding energy agrees quite well [3] with data. Thereby in that force model the $\Lambda - \Sigma$ conversion plays an important role. Looking into the hypertriton wave function we find a large probability (98.7%) for a deuteron component [3]. Thus the hypertriton is to a large extent a system with a Λ weakly bound to a deuteron.

Because the Λ particle decays weakly and there is no kinematical restriction, the hypertriton also has to decay. The following processes occur:

$$
\begin{aligned}
{}^{3}_{\Lambda}He \rightarrow \quad & n + n + p \quad 1.5\% \\
\rightarrow \quad & d + n \quad 0.18\%
\end{aligned}
\tag{1}
$$

and

---

*Alternative address: Institut für Kernphysik, Fachbereich 5 der Technischen Hochschule Darmstadt, D-64289 Darmstadt, Germany

$$
\begin{aligned}
{}^{3}_{\Lambda}H \rightarrow \quad & {}^{3}H + \pi^0 & 12.8\% \\
\rightarrow \quad & {}^{3}He + \pi^- & 25.7\% \\
\rightarrow \quad & d + n + \pi^0 & 20.6\% \\
\rightarrow \quad & d + p + \pi^- & 41.3\% \\
\rightarrow \quad & n + n + p + \pi^0 & 0.32\% \\
\rightarrow \quad & n + p + p + \pi^- & 0.64\%
\end{aligned}
\tag{2}
$$

These are the so called *nonmesonic* decay and *mesonic* decay processses. In order to determine the decay rates we use rigorous Faddeev solutions for the hypertriton and the three nucleon continuum states. For those calculations we employ modern YN and NN potentials. In the case of the nonmesonic decay we need an interaction for the weak-strong NY $\rightarrow$ NN processes. It is based on the exchange of several mesons (π, η, K, ρ, ω and K^*) between Y and N [4]. Each meson contributes significantly and there is a strong interference, which leads to a result close to the one calculated by the π exchange only. We found a total nonmesonic decay rate of $6.39 \times 10^7 \mathrm{sec}^{-1}$[5]. For the processes (1) the percentages shown are the theoretical decay rates normalized to the free Λ decay rate (Γ/Γ_Λ).

In the mesonic decay mode, the rates are larger than in the nonmesonic decay mode. This situation is different to the case of heavier hypernuclei, where the mesonic decays are Pauli blocked. A phenomenological model was recently studied by Congleton [6] who used ${}^{3}_{\Lambda}H$ and ${}^{3}He$ wave functions found from variational calculations. We used three-body Faddeev equations and treated the elementary weak decay process ($\Lambda \rightarrow N + \pi$) consistently to the pionic part of the nonmesonic process. We obtained a lifetime of the hypertriton according to the total decay rate (mesonic and nonmesonic) as $\tau = 2.56 \times 10^{-10}$sec [7]. This compares well to the experimental value of $\tau^{exp.} = (2.64 + 0.92 - 0.54) \times 10^{-10}$sec [8]. In (2), we also show our theoretical individual mesonic decay rate in terms of the free Λ decay rates. A recent summary of our results can be found in[9].

In this contribution we would like to present additional calculations for the pionic decay.

2 Various decay rates

Many channels are open in the pionic decay. Here we would like to concentrate to the case: ${}^{3}_{\Lambda}H \rightarrow {}^{3}He + \pi^-$. The decay rate is given by

$$
\begin{aligned}
d\Gamma = \ & \frac{1}{2} \sum |\sqrt{3} \langle \Psi_{k_\pi}^{(-)} | \hat{O} | \Psi_{{}^{3}_{\Lambda}H} \rangle|^2 \frac{d\boldsymbol{k}_\pi}{8\pi^2 \omega_\pi} \\
\times \ & \delta(M_{{}^{3}_{\Lambda}H} - M_{{}^{3}He} - \omega_\pi - \frac{k_\pi^2}{2M_{{}^{3}He}})
\end{aligned}
\tag{3}
$$

with

$$
\hat{O} = i\sqrt{2} G_F m_\pi^2 \left(A_\pi + \frac{B_\pi}{2M} \boldsymbol{\sigma} \cdot \boldsymbol{k}_\pi \right)
\tag{4}
$$

Table 1. The decay rates of the process: $^3_\Lambda$H $\rightarrow$ ^{3}He $+ \pi^-$

NN-Potential	d-state probability	Γ/Γ_Λ
Nijmegen 93 [2]	5.39 %	25.66 %
Nijmegen -I [2]	5.68 %	25.79 %
CD-Bonn [10]	4.86 %	25.68 %
Argonne V18 [11]	5.78 %	25.77 %
Ruhr-Pot[12]	5.85 %	25.78 %
exp. [8, 6]	-	(25.15±5.13) %

Here $\Psi^{(-)}$ is the pion-^{3}He scattering state, $\Psi_{^3_\Lambda H}$ the hypertriton bound state, $\hat{O}$ the operator for $\Lambda \rightarrow \pi^- + p$, $M_{^3_\Lambda He}$ the mass of the hypertriton, $M_{^3He}$ the mass of ^{3}He, $\omega_\pi = \sqrt{m_\pi^2 + k_\pi^2}$ the pion energy, and $G_F m_\pi^2 = 2.21 \times 10^{-7}$ the weak coupling constant. The constants $A_\pi = 1.05$ and $B_\pi = -7.15$ measure the parity, violating and conserving parts [4]. In the nonrelativistic limit $\bar{M}$ is taken as the average mass of the nucleon and the Λ particle.

Using the Nijmegen-93 NN potential [2] and Nijmegen YN potential [1] we obtained the rate for that process as $\Gamma = 0.973 \times 10^{-9} \text{sec}^{-1}$. We show in Table 1 the insensitivity of Γ against choosing different modern NN potentials. They support different d-state probabilities, which are also given. No significant dependence is seen and the theoretical values lie well within the experimental error bars. We plan to test other YN potentials in forthcoming studies, which might be more sensitive to the observables. The ratio for the decay rate into this specific channel to the decay rate into all channels including a π^- is found to be 0.379, which is close to the experimental value of 0.35 $\pm$0.04 [8, 6].

For the channels $^3_\Lambda H \rightarrow d + N + \pi$ and $^3_\Lambda H \rightarrow 3N + \pi$ we determined differential decay rates. We plot [7] $d\Gamma/d\,k_\pi$ for the processes $^3_\Lambda H \rightarrow d+p+\pi^-$ and $^3_\Lambda H \rightarrow p+p+n+\pi^-$ in Fig. 1. Both individual rates peak near the p + p + n +π^- threshold around $k_\pi = 101.3$MeV. The rate for the p + d + π^- channel dominates. In [7], we also exhibited the Dalitz plot for $d\Gamma/dT_\pi dT_d$ in the p+d+ π^- decay where T_π and T_d are the kinetic energies for π and the deuteron, respectively.

Measurements of hypertriton decay properties would certainly be very useful to test these predictions based on modern dynamics.

Acknowledgement. This work was supported by the Research Contract No. 41324878(COSY-044) of the Forschungszentrum Jülich, the Deutsche Forschungsgemeinschaft and the Science and Technology Cooperation Germany-Poland under Grant No. XO81.91. The numerical calculations have been performed on the Cray T90 of the Höchstleistungsrechenzentrum in Jülich, Germany.

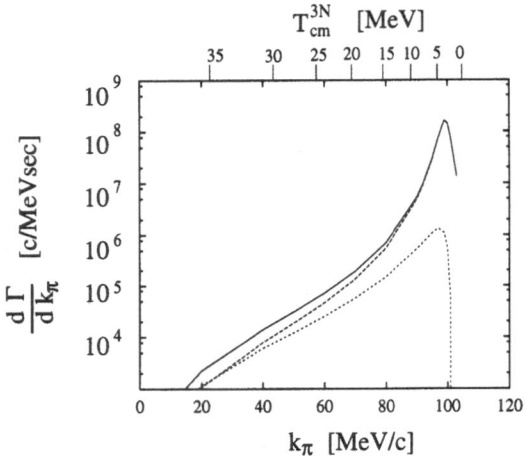

Figure 1. Differential decay rates $d\Gamma/dk_\pi$ for the processes : $^3_\Lambda H \to d + p + \pi^-$ (long dashed line), $^3_\Lambda H \to p + p + n + \pi^-$ (short dashed line) and sum of both (solid line).

References

1. P.M M. Maessen et al.: Phys. Rev. **C40**, 2905 (1989)

2. V.G.J. Stoks et al.: Phys. Rev. **C49**, 2950 (1994)

3. K. Miyagawa et al.: Phys. Rev. **C51**, 2905 (1995)

4. A. Parreño, A. Ramos, E. Oset: Phys. Rev. **C51**, 2477 (1995)

5. J. Golak et al.: Phys. Rev. **C55**, 2196 (1998); **56**, 2892 (1998)

6. J.G. Congleton: J. Phys. **G18**, 339 (1992)

7. H. Kamada et al.: Phys. Rev. **C57**, 1595 (1998)

8. Keyes et al.: Phys. Rev. **D1**, 66 (1970); Phys. Rev. Lett. **20**, 819 (1968); Nucl. Phys. **B67**, 269 (1973)

9. W. Glöckle et al.: In: *Proceedings of the International Conference on Hypernuclear and Strange Particle Physics*, eds. R.E. Chrien and D.J. Millener, Nucl. Phys. **A639** in print

10. R. Machleidt, F. Sammarruca and Y. Song: Phys. Rev. **C53**, R1483 (1996)

11. R. B. Wiringa, V.G.J. Stoks and R. Schiavilla: Phys. Rev. **C51**, 38 (1995)

12. D. Plümper, J. Flender and M.F. Gari: Phys. Rev. **C49**, 2370 (1994)

Few-Body Systems Suppl. 10, 387–390 (1999)

Few-
Body
Systems
© by Springer-Verlag 1999

Goldstone-Boson-Exchange Dynamics in the Constituent-Quark Model for Baryons

R.F. Wagenbrunn[1], L.Ya. Glozman[2], W. Plessas[1], K. Varga[3]

[1] Institute for Theoretical Physics, University of Graz, Universitätsplatz 5, A-8010 Graz, Austria

[2] Institute for Theoretical Physics, University of Tübingen, Auf der Morgenstelle 14, D-72076 Tübingen, Germany

[3] Theory Division, Argonne National Laboratory, Argonne, IL 60439, USA

Abstract. We describe light and strange baryons within the chiral constituent-quark model whose hyperfine interaction is provided by pseudoscalar meson exchange. We discuss the extension of the model to include vector and scalar meson exchanges.

In the development of effective models for low-energy QCD it is essential to observe the mechanism of spontaneous breaking of chiral symmetry. It makes the original QCD degrees of freedom, current quarks and gluons, to get changed into new effective ones, namely, constituent quarks and Goldstone bosons. Consequently, hadrons at low energies should be described by an effective Lagrangian involving just these degrees of freedom. In the valence-quark picture this means that, in particular, baryons should be described as systems of three constituent quarks Q whose dynamics is governed by confinement and a hyperfine interaction mediated by Goldstone-boson exchange (GBE) [1, 2].

As has been shown in explicit constructions of nonrelativistic [3, 4] and semirelativistic [5, 6] constituent-quark models along this line, a quite realistic description of light- and strange-baryon spectroscopy can be achieved. In particular, the notorious problem of correct level orderings of positive- and negative-parity states can be resolved in this manner. The success is connected with the characteristic symmetry in the spin-flavor operator in the most important spin-spin component of the pseudoscalar-exchange chiral potential. Specifically, a semirelativistic Hamiltonian of the following form [5, 6]

$$H = \sum_{i=1}^{3} \sqrt{\boldsymbol{p}_i{}^2 + m_i^2} + V_\chi + V_{\text{conf}} \qquad (1)$$

388

produces a spectrum as shown in Fig. 1. In Eq. (1) we have a relativistic kinetic-energy operator, a linear confinement potential, and a chiral potential involving the exchange of all pseudoscalar mesons (π, K, η, η') [6].

The GBE chiral CQM developed so far has still to be extended to include also the tensor forces of the pseudoscalar meson-exchange interaction. From

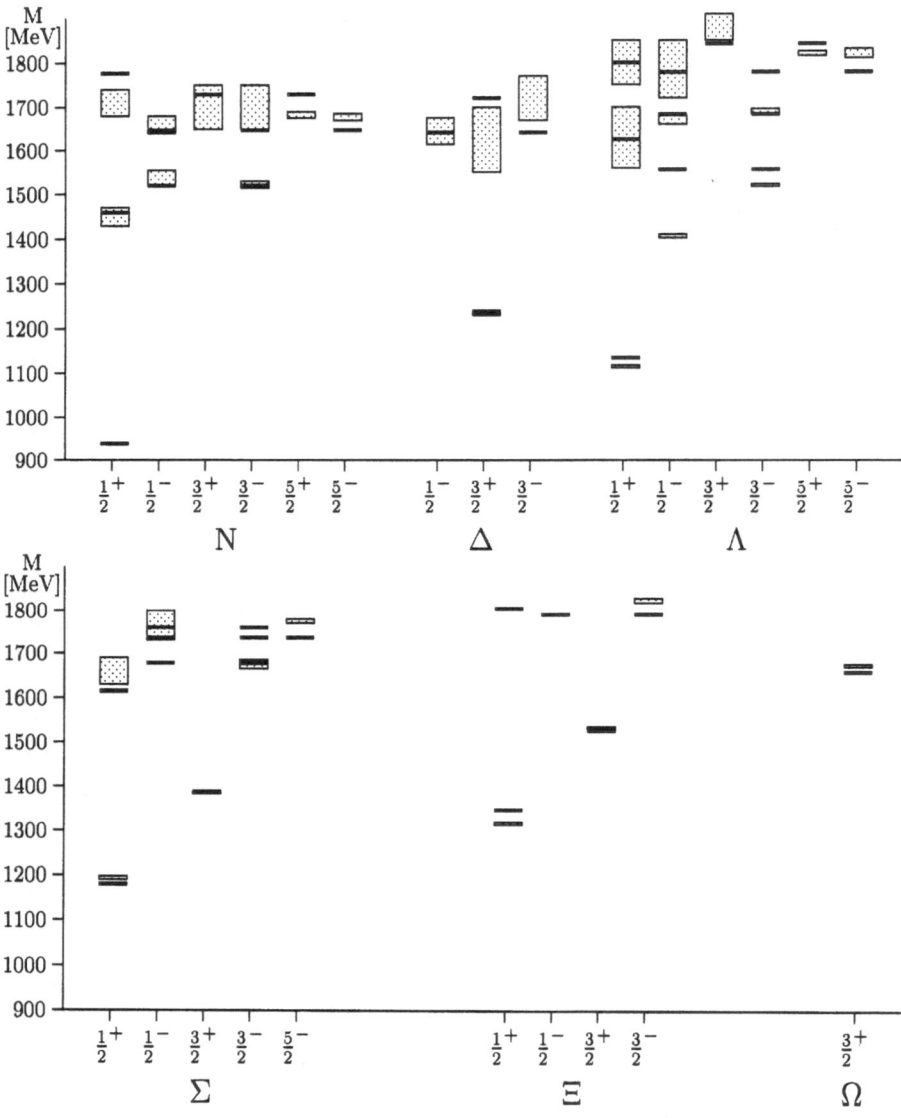

Figure 1. Energy levels of the lowest light and strange baryon states with total angular momentum and parity J^P for the model of ref. [6].

phenomenology we observe that the tensor-force effects must be minor, as the splittings within LS-multiplets like, e.g., $N(1535)$-$N(1520)$ are rather small. However, the tensor force of the pseudoscalar meson exchange is rather strong. It causes the level splittings as exemplified in Fig. 2b for the lowest-lying multiplets of the $\frac{1}{2}^-$-$\frac{3}{2}^-$ and $\frac{1}{2}^-$-$\frac{3}{2}^-$-$\frac{5}{2}^-$ nucleon resonances, i.e., for the $N(1535)$-$N(1520)$ and $N(1650)$-$N(1700)$-$N(1675)$ mulitplets, respectively.

In this situation we can think of multiple Goldstone-boson exchanges and introduce their effects by vector meson (ρ, ω, K^*, ϕ) and scalar meson (σ) exchanges. Indeed, we have found that in baryon spectra the tensor forces coming with vector meson exchange have an effect opposite to those stemming from the pseudoscalar exchange. So, the level splittings within the LS-multiplets are much reduced after the addition of vector meson exchange (see Fig. 2c).

The vector and scalar meson exchanges also give rise to spin-orbit forces. However, the corresponding contributions again roughly cancel each other. In particular, for a pair of quarks in a relative P-wave the spin-orbit forces from ρ and ω exchanges have different signs [8]. In this connection it is important to observe that, in deriving a nucleon-nucleon interaction, the two would add up

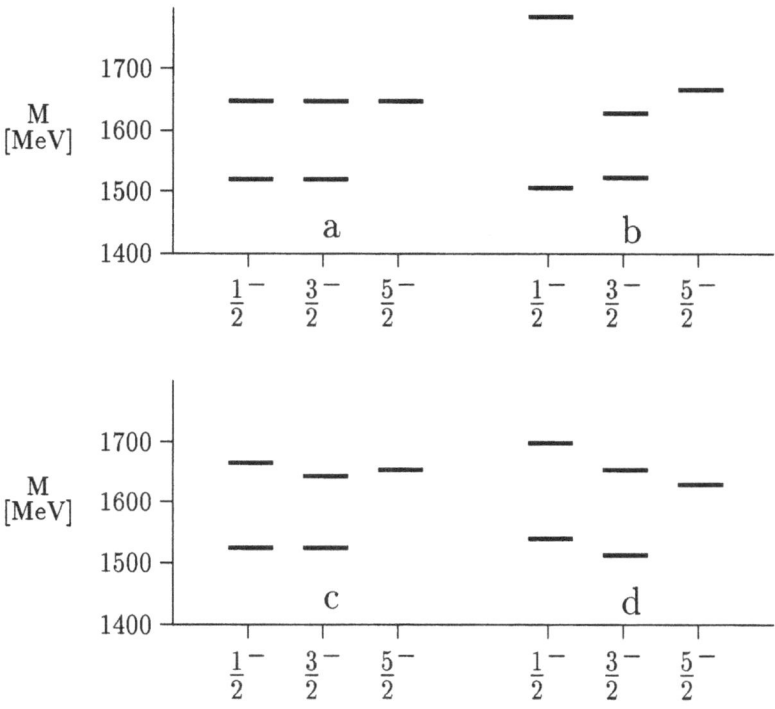

Figure 2. LS-multiplets of nucleon excitation levels for the GBE chiral CQM with a) no tensor forces (as in Fig. 1), b) tensor forces from pseudoscalar meson exchange included, c) addition of vector-meson-exchange tensor forces, and d) full model with all force components from pseudoscalar, vector, and scalar exchanges.

to give rise to a sizable spin-orbit force, just as it is needed there.

In the full model with all force components from pseudoscalar, vector, and scalar meson exchange included quite a reasonable description of the LS-multiplets is achieved (Fig. 2d). In total the level splittings remain small, certainly much reduced as compared to the case with pseudoscalar exchange only. (Fig. 2b). The most prominent role in the baryon spectra is still played by the spin-spin interaction, which is now provided by the pseudoscalar and vector meson exchanges.

At present, we have only a preliminary parametrization of the extended GBE chiral CQM [9]. We are still working to properly tune the various force ingredients in our model. We are confident that a fit with only a few open parameters will be possible providing a good description of the baryon spectra including the details of their fine structure.

Acknowledgement. R.F.W. is grateful to the Paul-Urban Foundation for providing support to attend this conference.

References

1. L.Ya. Glozman and D.O. Riska: Phys. Reports **268**, 263 (1996)

2. L.Ya. Glozman: In: *Perturbative and Nonperturbative Aspects of Quantum Field Theory* (Lecture Notes in Physics, vol. 479), eds.: H. Latal and W. Schweiger. Berlin: Springer 1997

3. L.Ya. Glozman, Z. Papp, and W. Plessas: Phys. Lett. **B381**, 311 (1996)

4. L.Ya. Glozman et al.: Nucl. Phys. **A623**, 90c (1997)

5. L.Ya. Glozman et al.: Phys. Rev. **C57**, 3406 (1998)

6. L.Ya. Glozman et al.: Preprint hep-ph/9706507; Phys. Rev. **D** (to appear)

7. Particle Data Group, C. Caso et al.: Eur. Phys. J. **C3**, 1 (1998)

8. L.Ya. Glozman: Preprint hep-ph/9805345

9. R.F. Wagenbrunn et al.: In: Proceedings of the Joint ECT*/JLAB Workshop on *N* Physics and Nonperturbative QCD*, Trento, 1998. To appear in Few-Body Systems Supplementum

Few-Body Systems Suppl. 10, 391–394 (1999)

Few-
Body
Systems
© by Springer-Verlag 1999

Hadronic Decays of Light and Strange Baryon Resonances in the GBE Constituent-Quark Model

A. Krassnigg[1], W. Plessas[1], L. Theußl[1,2], K. Varga[3], R.F. Wagenbrunn[1]

[1] Institute for Theoretical Physics, University of Graz, Universitätsplatz 5, A-8010 Graz, Austria

[2] Institut des Sciences Nucléaires, Université Joseph Fourier, Avenue des Martyrs 53, F-38026 Grenoble, France

[3] Theory Division, Argonne National Laboratory, Argonne, IL 60439, USA

Abstract. We report further results from a study of hadronic decays of light and strange baryon resonances within a recently proposed constituent-quark model whose hyperfine interaction is derived from Goldstone-boson exchange. Three-quark wave functions generated by solution of a semirelativistic Schrödinger-type equation with the stochastic variational method are employed for calculating partial decay widths along the elementary emission model. Complementing previous studies on π and η decays, we here concentrate on the strange K decays.

In the constitutent-quark model (CQM) of refs. [1, 2, 3], the hyperfine interaction is deduced from Goldstone-boson-exchange (GBE) dynamics [4, 5]. This is motivated by the spontaneous breaking of chiral symmetry in low-energy QCD. Already in the version where the GBE mechanism is constrained to be mediated by the spin-spin component of pseudoscalar meson exchange only, the chiral CQM reproduces the spectra of all light and strange baryons in close agreement with phenomenology; in particular, it yields the right level orderings of positive- and negative-parity excitations simultaneously in the light and strange baryon spectra [1, 2, 3].

We solve the semirelativistic GBE CQM for the three-quark system via the stochastic variational method [6]. The corresponding wave functions are then employed in the calculation of hadronic decay widths along the elementary emission model (EEM) (cf., e.g., ref. [7]). Its decay operator derives from the interaction Lagrangian

$$\mathcal{L} \sim i\, g_{QM}\, \bar{\psi}\, \gamma_\mu \gamma_5\, \boldsymbol{\lambda}^F \cdot (\partial_\mu \boldsymbol{\Phi})\, \psi \,, \tag{1}$$

where ψ and $\bar{\psi}$ are the constituent-quark fields, $\boldsymbol{\Phi}$ is the meson field vector, g_{QM} is the quark-meson coupling constant, and $\boldsymbol{\lambda}^F$ are the Gell-Mann flavor matrices. In the lowest-order nonrelativistic reduction one arrives at the following form of the EEM decay operator

$$\hat{O}_\alpha = 3\,i\,\frac{g_{QM}}{2m_k}\,X^\alpha\,e^{-i\mathbf{q}\cdot\mathbf{r}_k}\left[\left(1+\frac{\omega}{2m_k}\right)\boldsymbol{\sigma}\cdot\mathbf{q} - \frac{\omega}{m_k}\,\boldsymbol{\sigma}\cdot\mathbf{p}_k\right]\,, \quad (\alpha = 1,\dots,8). \tag{2}$$

It describes the resonance decay by the emission of a meson of type α (π: $\alpha = 1,2,3$; K: $\alpha = 4,5,6,7$; η: $\alpha = 8$) from a pointlike quark. In Eq. (2), ω and $\mathbf{q}$ are the meson energy and momentum, m_k, $\mathbf{p}_k$, and $\mathbf{r}_k$ are the mass, momentum, and position of quark k, $\boldsymbol{\sigma}$ are the Pauli spin matrices, and X^α is the flavor transition operator corresponding to the decay of type α (for more details see, e.g., ref. [8]). The first term in the square brackets is referred to as the "direct term" whereas the second one is usually called "recoil" term.

For the calculation of partial decay widths of resonance B going to the final-state baryon B' one works in the rest frame of B. We compare results obtained with the direct term only to the results including recoil corrections. Thereby one can learn about the structure dependence or independence of specific decay amplitudes. The direct term basically represents the overlap of the B and B' wave functions, while the recoil term includes a derivative and therefore introduces the spatial dependence of the wave functions more sensitively. The calculations have been performed in a similar spirit as in ref. [9] before, but without introducing any additional parameters in the decay operator. In the first step of our studies it is our primary aim to get insight into the behaviour of the three-Q wave functions rather than to provide a realistic description of the decay widths. For the case of the semirelativistic pseudoscalar-exchange chiral CQM [1, 2, 3], results of partial widths for the π- and η-decay modes of the light resonances N^* and Δ^* have already been reported in ref. [10]. The same decay modes of the strange resonances Λ^* and Σ^* have been discussed in ref. [11]. Here, we add the corresponding results for the strange K and $\bar{K}$ decays. In Table 1 partial decay widths for the transitions $\Lambda^* \to N\bar{K}$, $\Sigma^* \to N\bar{K}$, $\Sigma^* \to \Delta\bar{K}$, and $N^* \to \Lambda K$ are given.

The results with the direct term only appear unrealistic in all cases. Addition of the recoil term helps to improve the situation in some cases. So, at least the orders of magnitude of some decay widths come out right. Still, the overall quality of the results remains unsatisfactory. For the present status of our calculations this is no surprise, however. Certainly, a more realistic decay model has to be employed if one aims at explaining the experimental data. First of all, the quark-meson vertex evidently has a finite extension and this should be taken into account. Furthermore, the mechanism for meson emission from a baryon resonance should be considered more explicitly. This problem is intimately related with the setup of the CQM itself. Irrespective of the type of the effective $Q-Q$ interactions (e.g., one-gluon exchange vs. meson exchange) it is obviously not sufficient to describe the resonances just as $\{QQQ\}$ bound states with sharp energies. A more realistic description including their finite widths is surely needed. In the framework of the GBE CQM such an extension

Table 1. Kaon partial decay widths for the GBE CQM calculated with the operator $\hat{O}_\alpha$ of Eq. (2). In all cases the theoretical resonance energies of the CQM, given in the second column, have been employed (instead of the experimental ones). The theoretical N, Δ, and Λ ground-state masses entering the calculations are 939 MeV, 1240 MeV, and 1136 MeV, respectively. Experimental data are from the latest compilation of the PDG [12].

decaying resonance			Γ [MeV]		
$I(J^P_{LS})$	M_{th}	decay mode	direct	direct +recoil	Exp.
$0(\frac{3}{2}^-_{1\,\frac{1}{2}})$	1556	$\Lambda_{1520} \to N_{939}$	1.9	3	7
$0(\frac{1}{2}^+_{0\,\frac{1}{2}})$	1626	$\Lambda_{1600} \to N_{939}$	1.4	1.1	34 ± 11
$0(\frac{1}{2}^-_{1\,\frac{1}{2}})$	1683	$\Lambda_{1670} \to N_{939}$	0.6	17	7 ± 2
$0(\frac{3}{2}^-_{1\,\frac{1}{2}})$	1683	$\Lambda_{1690} \to N_{939}$	0.6	1.1	15 ± 3
$0(\frac{1}{2}^-_{1\,\frac{3}{2}})$	1779	$\Lambda_{1800} \to N_{939}$	0.01	0.2	98 ± 23
$0(\frac{1}{2}^+_{0\,\frac{1}{2}})$	1800	$\Lambda_{1810} \to N_{939}$	0.3	11	53 ± 23
$0(\frac{5}{2}^+_{2\,\frac{1}{2}})$	1842	$\Lambda_{1820} \to N_{939}$	2.2	3.8	48 ± 4
$0(\frac{5}{2}^-_{1\,\frac{3}{2}})$	1779	$\Lambda_{1830} \to N_{939}$	0	0.01	6 ± 3
$0(\frac{3}{2}^+_{2\,\frac{1}{2}})$	1842	$\Lambda_{1890} \to N_{939}$	2.2	81	28 ± 8
$1(\frac{1}{2}^+_{0\,\frac{1}{2}})$	1616	$\Sigma_{1660} \to N_{939}$	0	0.2	20 ± 10
$1(\frac{3}{2}^-_{1\,\frac{1}{2}})$	1678	$\Sigma_{1670} \to N_{939}$	0.2	0.3	6 ± 2
$1(\frac{1}{2}^-_{1\,\frac{1}{2}})$	1678	$\Sigma_{1750} \to N_{939}$	0.2	22	23 ± 14
$1(\frac{5}{2}^-_{1\,\frac{3}{2}})$	1736	$\Sigma_{1775} \to N_{939}$	2.6	4.7	48 ± 4
$1(\frac{1}{2}^-_{1\,\frac{1}{2}})$	1678	$\Sigma_{1750} \to \Delta_{1232}$	0	0	
$1(\frac{5}{2}^-_{1\,\frac{3}{2}})$	1736	$\Sigma_{1775} \to \Delta_{1232}$	0	0	
$\frac{1}{2}(\frac{1}{2}^-_{1\,\frac{3}{2}})$	1648	$N_{1650} \to \Lambda_{1116}$	0	0.1	11 ± 6
$\frac{1}{2}(\frac{5}{2}^-_{1\,\frac{3}{2}})$	1648	$N_{1675} \to \Lambda_{1116}$	0	0	< 1.5
$\frac{1}{2}(\frac{5}{2}^+_{2\,\frac{1}{2}})$	1729	$N_{1680} \to \Lambda_{1116}$	0.03	0.04	
$\frac{1}{2}(\frac{3}{2}^-_{1\,\frac{3}{2}})$	1648	$N_{1700} \to \Lambda_{1116}$	0	0	< 3
$\frac{1}{2}(\frac{1}{2}^+_{0\,\frac{1}{2}})$	1777	$N_{1710} \to \Lambda_{1116}$	1.2	0.1	15 ± 10
$\frac{1}{2}(\frac{3}{2}^+_{2\,\frac{1}{2}})$	1729	$N_{1720} \to \Lambda_{1116}$	0.03	22	12 ± 11

could naturally be achieved by introducing additional Fock components, e.g., of the type $\{QQQ\text{meson}\}$. We may then expect an even better reproduction of the light and strange baryon spectra, specifically for resonances lying close to decay thresholds. Furthermore, the wave-function behaviour will be changed considerably. Hopefully, this would then allow for an improved description of hadronic decays and other types of dynamic observables.

Acknowledgement. A.K., L.T., and R.F.W. are grateful to the Faculty of Natural Sciences of the University of Graz and the Paul-Urban Foundation, respectively, for providing support to attend this conference.

References

1. L.Ya. Glozman et al.: Phys. Rev. **C57**, 3406 (1998)

2. L.Ya. Glozman et al.: Preprint hep-ph/9706507, Phys. Rev. **D**, to appear

3. R.F. Wagenbrunn et al.: These Proceedings

4. L.Ya. Glozman, D.O. Riska: Phys. Reports **268**, 263 (1996

5. L.Yà. Glozman: Lecture Notes in Physics **479**, 363 (1997)

6. K. Varga, Y. Suzuki: Phys. Rev. **C52**, 2885 (1995); K. Varga, Y. Ohbayasi, Y. Suzuki: Phys. Lett. **B396**, 1 (1997)

7. A. Le Yaouanc et al.: In: *Hadron Transitions in the Quark Model.* New York: Gordon and Breach 1988

8. A. Krassnigg: Diploma Thesis, Karl-Franzens-Universität Graz, 1998

9. R. Koniuk, N. Isgur: Phys. Rev. **D21**, 1868 (1980)

10. L.Ya. Glozman et al.: Proceedings of the 4th N^* Workshop, Washington, DC, 1997. πN Newsletter **14**, 99 (1998)

11. W. Plessas et al.: Proceedings of the N^* Workshop, Trento, 1998, to appear in Few-Body Systems Supplements

12. Particle Data Group: Eur. Phys. J. **C3**, 1 (1998)

Few-Body Systems Suppl. 10, 395–398 (1999)

Few-
Body
Systems
© by Springer-Verlag 1999

Understanding the Low Energy Hadron Spectrum in a Chiral Quark Cluster Model

F. Fernández[1], P. González[2], A. Valcarce[1]

[1] Grupo de Física Nuclear, Universidad de Salamanca, E-37008 Salamanca, Spain

[2] Dpto. de Física Teórica e IFIC, Universidad de Valencia - CSIC, E-46100 Burjassot, Valencia, Spain

Abstract. The low energy N and Δ spectra are studied by means of a chiral quark cluster model. We solve the Schrödinger equation in the hyperspherical harmonic approach. The interacting potential includes Goldstone boson exchanges besides the usual one-gluon exchange. The predicted baryonic spectrum is quite reasonable. However, if consistency with the two-baryon sector is required, the observed inversion of the positive and negative parity excitations of the nucleon cannot be obtained. Alternative solutions are discussed.

1 Introduction

Since the advent of QCD, quark models, originally formulated from symmetry arguments plus a few simple assumptions, have evolved significantly. On the one hand, they have found a more sound theoretical justification, on the other hand they have incorporated other relevant features of the basic theory. In particular, since the 70's the one-gluon exchange (OGE) potential derived from QCD has been a key ingredient in most quark models. More recently chiral symmetry, spontaneously broken in QCD, has been taken into account in a non-relativistic quark model scheme through an effective quark-quark interaction mediated by the exchange of a pseudoscalar as well as a scalar field.

When these contributions are added to the kinetic energy and the usual confinement terms, the resulting Hamiltonian corresponds to the so-called chiral quark cluster model (CQCM).

2 The Chiral Quark Cluster Model

The form of the OGE interaction was derived long ago by Rújula et al. [1]. To explicitly derive the form of the chiral potential one proceeds to a non-

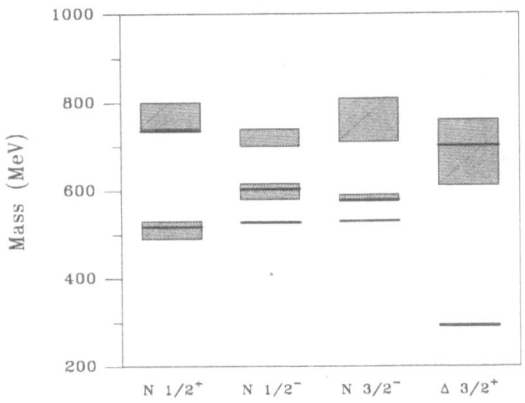

Figure 1. N and Δ low-energy spectrum without three-body forces

relativistic reduction of an effective Hamiltonian that mimics QCD in the regime between the scale of chiral symmetry breaking and the confinement scale. The emerging image is that of constituent quarks interacting through Goldstone modes associated to the spontaneous breaking of the symmetry [2]. The potentials obtained are given in ref. [3]. The CQCM has been applied to the two-baryon system explaining the deuteron observables, the N-N phase shifts and the main features of the N-Δ and Δ-Δ interactions [3, 4].

Its application to the baryon spectra should be done with care, because the contact interaction (Dirac delta) of the OGE potential has to be regularized in order not to get an unbound spectrum when solving the Schrödinger equation.

When regularization of the delta interaction is done by means of a spreading Yukawa function, a reasonable spectrum comes out [5]. It is important to note that the regularization depends on the model space in which the calculation is done and the parameter of this regularization (r_0) should not be understood as a true parameter of the model Hamiltonian. The regularization scheme chosen in ref. [5] consists of two steps: i) the model space is truncated leaving only up to K=2 excitations in the hyperspherical harmonic approach and ii) the δ is regularized by demanding that the matrix elements of the regularized form between harmonic oscillator wave functions are the same as the matrix elements of the exact δ. This prescription guarantees consistency between the spectrum and the NN calculation. The value obtained for the regularizing parameter should be considered within the model space one has used and not generalized to a different working scheme. To illustrate this point, let us note that in ref. [6], using the same potential and the same regularization as in ref. [5], but in a Faddeev approach, the results obtained are strikingly different. Obviously, this does not imply that the interaction used in ref. [5] is not appropriate, as

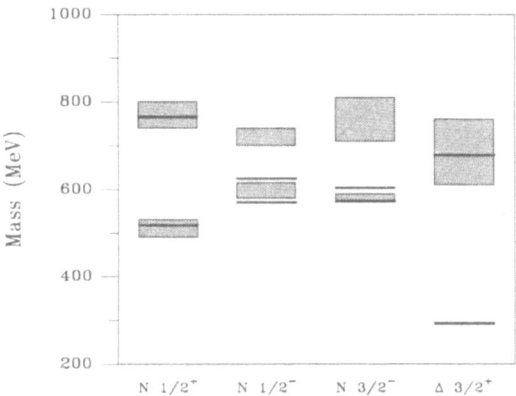

Figure 2. N and Δ low-energy spectrum including three-body forces

claimed in ref. [6], but that the regularization has not been properly done for the model space used.

3 Results and Discussion

At difference with the baryon-baryon interaction, the spectrum is very sensitive to the specific form of the confinement term. If a functional form $V_{conf} = -ar^n$ (n integer) is chosen, a linear confinement is preferred by data [5].

As in many other spectroscopy quark models, with the standard value of the parameters, the position of the Roper resonances (first radial excited states) and the corresponding first negative parity states are predicted inverted with respect to data. This is so in spite of the fact that the $(\sigma_i \cdot \sigma_j)(\tau_i \cdot \tau_j)$ structure of the pseudoscalar potential favors their correct location since it gives attraction (repulsion) for symmetric (antisymmetric) spin-isospin pairs.

The model can be forced to generate the inversion of the Roper and the negative parity excitations of the nucleon, by slightly increasing the strength of the pseudoscalar potential, $\Lambda = 5.4$ fm^{-1}. In this calculation, to avoid problems with the Dirac delta in the OGE, one can use a smoother regularization than in ref. [5], taken $r_0 = 0.8$ fm, just by increasing the strong coupling constant, $\alpha_s = 0.65$. For this value of r_0, the results obtained with the hyperspherical harmonic approach and a Faddeev calculation including up to four amplitudes are similar. Nonetheless, the spectra are of the same quality as those reported in ref. [5]. Using a confinement constant $a_c = 60.12$ MeV fm^{-1}, the calculated spectra are shown in Fig. 1.

However, by enhancing artificially the pseudoscalar potential the agreement for the two-baryon sector, especially the binding energy of the deuteron, is destroyed. Therefore the complete inversion of these two levels cannot be generated from the pseudoscalar potential if consistency with the two-baryon system is required.

A solution to this problem can be obtained through the consideration of a three-quark force as proposed some time ago by Desplanques et al. [7]. The specific form of the interaction, on a pure phenomenological basis, is

$$V = \frac{1}{2} \sum_{i \neq j \neq k \neq i}^{3} \frac{V_0}{m_q^3} \frac{e^{-m_0 r_{ij}}}{m_0 r_{ij}} \frac{e^{-m_0 r_{ik}}}{m_0 r_{ik}} \tag{1}$$

where V_0 is taken to be -50.51 GeV^{-2} fm^{-6} and $m_0 = 250$ MeV. Certainly it is not much to relocate two levels of nucleon and Δ adding two more free parameters (V_0 and m_0). However, the effect of this potential on the whole spectrum is positive, see Fig. 2, it does not spoil the two-baryon results and allows to choose a bigger value for the spreading range of the Dirac delta in the OGE potential ($r_0 = 2.$ fm) what eliminates in a great part its model space dependence, and also the cutoff for the pseudoscalar exchange that reproduces the deuteron binding energy, $\Lambda = 4.2$ fm^{-1}. The confinement constant is in this case $a_c = 49.80$ MeV fm^{-1}. Furthermore, three-body forces combined with Goldstone boson exchanges give baryon sizes bigger than those obtained in ref. [8], which may have beneficial effects on pionic strong baryon decay processes.

Then, we can conclude that it is possible to have a consistent precise description of the baryon spectrum and the baryon-baryon system in a chiral quark cluster model provided that a three-quark force has been incorporated.

Acknowledgement. This work has been supported by Dirección General de Investigación Científica y Técnica (DGICYT) under Contract No. PB94-0080.

References

1. A. de Rújula, H. Georgi, S.L. Glashow: Phys. Rev. **D12**, 147 (1975)

2. A. Manohar, H. Georgi: Nucl. Phys. **B234**, 189 (1984)

3. F. Fernández et al.: J. Phys. **G19**, 2013 (1993)

4. A. Valcarce et al.: Phys. Rev. **C50**, 2246 (1994); A. Valcarce et al.: Phys. Rev. **C52**, 38 (1995), Phys. Rev. **C56**, 3026 (1997)

5. A. Valcarce et al.: Phys. Lett. **B367**, 35 (1996)

6. L.Ya. Glozman et al.: Phys. Rev. **C57**, 3406 (1998)

7. B. Desplanques et al.: Z. Phys. **A343**, 331 (1992)

8. F. Cano et al.: Nucl. Phys. **A603**, 257 (1996)

Few-Body Systems Suppl. 10, 399–402 (1999)

Few-
Body
Systems
© by Springer-Verlag 1999

Exotic Hadrons

Fl. Stancu*

Institute of Physics, B.5, University of Liege, Sart Tilman, B-4000 Liege 1, Belgium

Abstract. Among the exotic hadrons here I discuss multiquark systems formed of more than three quarks and/or antiquarks. Special attention is paid to pentaquarks containing a heavy (charmed) antiquark. These are being searched experimentally at Fermilab. Their stability is studied within a Goldstone boson exchange model. The results are compared to those obtained from conventional models based on one gluon exchange.

1   Introduction

QCD inspired models predict the existence of exotic hadrons formed of more than three quarks and/or antiquarks ($q^m \bar{q}^n$ with $m + n > 3$). Most studies are devoted to systems described by the colour state $[222]_C$. These are the tetraquarks $q^2 \bar{q}^2$ [1], the pentaquarks $q^4 \bar{q}$ [2, 3] and the hexaquarks q^6 [1].

Here we are concerned with the study of pentaquarks containing a heavy antiquark. The reason is that pentaquarks with one strange quark and a charmed antiquark are presently being searched at Fermilab [4]. An account of this searching is given by D. Ashery at this conference [5].

The first theoretical studies of pentaquarks [2, 3] have been performed within a constituent quark model based on one gluon exchange (OGE) interaction. Starting from simplifying approximations it was found that the pentaquarks $uuds\bar{c}$ and $udds\bar{c}$ and their conjugates are stable against strong decays. Better approximations [6] lead to instability. But more sophisticated calculations [7] suggested several candidates for stability and especially those with strangeness $S = -1$ or -2. In particular the $uuds\bar{c}$ system was bound by - 52 MeV. On the other hand the nonstrange systems $uudd\bar{Q}$ (Q = c or b) were found unbound. So far, the ongoing experiments [4, 5] had as guidelines the predictions of OGE based models. Within the confidence level of the analyzed experiments, no convincing evidence for the production of pentaquarks with the flavour content $uuds\bar{c}$ and $udds\bar{c}$ has been observed [4, 5].

*E-mail address: fstancu@ulg.ac.be

2 Stability of pentaquarks

In the stability problem we are interested in the quantity

$$\Delta E = E(q^4\bar{q}) - E_T \tag{1}$$

where $E(q^4\bar{q})$ represents the pentaquark energy and E_T is the lowest threshold energy for dissociation into a baryon + a meson. A negative ΔE suggests the possibility of a stable compact system.

The theoretical predictions are model dependent. Here we are concerned with constituent quark models which simulate the low-energy limit of QCD. I discuss the stability of pentaquarks within the Goldstone boson exchange (GBE) model [8, 9]. It will be shown that this model gives results at variance with OGE models. In the GBE model the hyperfine splitting in hadrons is due to the short-range part of the Goldstone boson exchange interaction between quarks, instead of the OGE interaction. The GBE interaction is flavour-dependent and its main merit is that it reproduces the correct ordering of positive and negative parity states in all parts of the considered spectrum of nonstrange and strange baryons in contrast to any OGE model. Moreover, the GBE interaction induces a strong short-range repulsion in the Λ-Λ system, which suggests that a deeply bound H-baryon should not exist [10]. This is in agreement with the high-sensitivity experiments at Brookhaven [11] where no evidence for H production has been found. On the other hand, in its present form, the GBE model does not apply to hyperfine splitting in mesons. According to ref. [8] there is no meson exchange interaction between quarks and antiquarks. It is assumed that the $q\bar{q}$ pseudoscalar pairs are automatically included in the GBE interaction. Therefore a light quark and the heavy antiquark interact via the confinement potential only and the model Hamiltonian contains GBE interactions only between light quarks.

For a comparative discussion it is useful to consider a schematic OGE interaction between the light quarks

$$V_{cm} = -C_{cm} \sum_{i<j} \lambda_i^c . \lambda_j^c \sigma_i . \sigma_j \tag{2}$$

with $C_{cm} = 293/16$ MeV determined from the $\Delta - N$ splitting, and a schematic GBE interaction

$$V_\chi = -C_\chi \sum_{i<j} \lambda_i^F . \lambda_j^F \sigma_i . \sigma_j \tag{3}$$

with $C_\chi = 293/10$ MeV determined also from the $\Delta - N$ splitting.

At this stage it is useful to note that one can have either positive or negative parity pentaquarks, as it will be explained below. Reference [12] considered pentaquarks of negative parity, by analogy to those proposed in the OGE model [2, 3]. The parity is given by the intrinsic parity of the antiquark. The light quarks are assumed identical. The ground state orbital (O) wave function is symmetric under permutation of any two light quarks, i.e. it corresponds to the partition $[4]_O$. Thus the parity of the pentaquark is equal to the parity of the antiquark because the q^4 subsystem has no internal angular momentum and there is no angular momentum in the relative motion between this subsystem and the heavy antiquark either. Note that the q^4 subsystem must be in a colour state $[211]_C$. As we look for the lowest state we have to take the spin S of the q^4 subsystem equal to zero. Then the totally antisymmetric state of q^4 must have the composition

$$|[4]_O([211]_F[22]_S;[31]_{FS})[211]_C >$$ (4)

In this case the expectation value of (3) is $< V_\chi >= -16C_\chi$. Inner product rules [13] also allow a flavour-spin (FS) state $[31]_{FS} = [31]_F$ x $[22]_S$. But this has higher energy because $< V_\chi >= -28/3C_\chi$. Therefore the lowest state must contain strangeness, i.e. has flavour symmetry $[211]_F$. Reference [12] shows that negative parity pentaquarks in this state are unbound by the GBE interaction. This is at variance with OGE based models.

The comparison of operators (2) and (3) suggests that the lowest state in OGE models is also of type (4) but with C and F interchanged, i.e. strangeness is also required. In the CS coupling scheme one gets $< V_{cm} >= -16C_{cm}$.

Now, to introduce positive parity pentaquarks one can consider a state like (4) but with an orbital symmetry [31] and a flavour-spin symmetry [4]. In this case the q^4 state of spin S = 0 has the structure $[4]_{FS} = [22]_F$ x $[22]_S$, thus it does not necessarily contain strangeness. The orbital state $[31]_O$ must have nonzero angular momentum. A state with an internal angular momentum L = 1 has been constructed in [14]. The parity of such a state is P = $(-)^{L+1} = 1$. The lowest positive parity state is then

$$|[31]_O([22]_F[22]_S;[4]_{FS})[211]_C >$$ (5)

with an expectation value $< V_\chi >= -28C_\chi$. Thus the GBE interaction brings a larger attraction for this state than for (4) i.e. the positive parity state should be lower than the negative parity one.

3 Results

Within the realistic model [9] and based on a simple variational solution it has been shown in [14] that the nonstrange positive parity pentaquarks $uudd\overline{Q}$ are bound by ΔE = - 75.6 MeV and - 95.6 MeV for Q = c and b respectively. The strange pentaquarks of type $uuds\overline{Q}$ or $udss\overline{Q}$ described by (5) turned out to be unbound. The reason is that the GBE interaction is weaker for strange pentaquarks due to the factor $1/(m_i m_j)$.

As the $uudd\overline{Q}$ pentaquarks of negative parity are predicted to be unbound by a chromomagnetic interaction [3, 7], the same system but with positive parity is expected to be even more unstable due to the increase in the kinetic energy produced by the excitation of a quark to the p-shell [14]. While the GBE interaction overcomes this excitation, the OGE interaction (2) does not make a distinction between the flavour symmetries associated to $[4]_O$ and $[31]_O$ orbital states so that the $[31]_O$ state appears higher than $[4]_O$, due to a higher kinetic contribution, thus nonstrange positive parity pentaquarks are unstable in OGE models.

One can see that the OGE and the GBE interactions predict contradictory results for anticharmed pentaquarks: while the GBE interaction stabilizes a given system, the OGE interaction destabilizes it and vice versa. A similar situation occurs for other exotic hadrons [15].

In conclusion the GBE model predicts that *nonstrange positive parity* pentaquarks with anticharm or antibeauty are the best candidates for stability against strong decays. If bound, they should be lighter than the strange ones of the same parity and should also have a longer weak decay lifetime. Experimentalists are encouraged to search for such pentaquarks.

Acknowledgement. I am grateful to Danny Ashery for very useful discussions.

References

1. R.L. Jaffe: Phys. Rev. **D15**, 267 (1977)

2. C. Gignoux, B. Silvestre-Brac, J.-M. Richard: Phys. Lett. **B193**, 323 (1987)

3. H.J. Lipkin: Phys. Lett. **B195**, 484 (1987); Nucl. Phys. **A625**, 207 (1997)

4. E.M. Aitala et al. (Fermilab E791 collab.): Phys. Rev. Lett. **81**, 44 (1998)

5. D. Ashery: these proceedings

6. S. Fleck et al.: Phys. Lett. **220**, 616 (1989)

7. J. Leandri, B. Silvestre-Brac: Phys. Rev. **D40**, 2340 (1989)

8. L.Ya. Glozman, D.O. Riska: Phys. Rep. **268**, 263 (1996)

9. L.Ya. Glozman, Z. Papp, W. Plessas: Phys. Lett. **B381**, 311 (1996) ; L.Ya. Glozman et al.: Nucl. Phys. **A623**, 90c (1997)

10. Fl. Stancu, S. Pepin, L.Ya. Glozman: Phys. Rev. **D57**, 4393 (1998)

11. R.W. Stotzer et al.: Phys. Rev. Lett. **78**, 3646 (1997)

12. M. Genovese et al.: Phys. Lett. **B425**, 171 (1998)

13. Fl. Stancu: In: *Group Theory in Subnuclear Physics*, Sect. 4.8. Oxford: Clarendon Press 1996

14. Fl. Stancu: e-print hep-ph/9803442

15. Fl. Stancu: In: Proceedings of the workshop *N* Physics and Nonperturbative QCD*, Trento, Italy, May 18-29, to appear in Few-Body Systems Suppl.

Few-Body Systems Suppl. 10, 403–406 (1999)

Few-
Body
Systems
© by Springer-Verlag 1999

Describing the Nucleon Electromagnetic Form Factors at High Momentum Transfers

L. Theußl[1,2], B. Desplanques[2], B. Silvestre-Brac[2], K. Varga[3]

[1] Institute for Theoretical Physics, University of Graz, Universitätsplatz 5, A-8010 Graz, Austria

[2] Institut des Sciences Nucléaires, CNRS-UJF, 53 Avenue des Martyrs, F-38026 Grenoble, France

[3] Theory Division, Argonne National Laboratory, 9700 S. Cass Ave., Argonne, IL 60439, USA

Abstract. Electromagnetic form factors of the nucleon are calculated within the framework of a non-relativistic constituent-quark model. The emphasis is put on the reliability and accuracy of present day numerical methods used to solve the three-body problem. The high-q^2 behaviour of the form factors is determined by the form of the wave function at short distances and, due to the small absolute values that one deals with, an accurate solution is essential.

1 Introduction

Currently, there is some theoretical interest devoted to the nucleon form factor and, in particular, to its behaviour at high momentum transfers. From QCD in the perturbative regime, one expects the nucleon form factor to scale like q^{-4} up to log terms. Experimentally, this behaviour seems to be reached rather quickly, around $q^2 \sim 10\ \mathrm{GeV}^2$.

Simple descriptions in terms of constituent quarks have relied on the harmonic oscillator wave functions, where the calculated form factors drop exponentially to zero beyond $q^2 \sim 3\ \mathrm{GeV}^2$. Curiously, it has been deduced from this result that a constituent quark model does not lead to a power law behaviour of form factors at high q^2. As shown in details in a paper to appear [1], this is not true if one considers more realistic interquark potentials and their corresponding wave functions, however, due to the small values that one deals with, the problem becomes a considerable numerical task.

Structure calculations were performed with the same quark-quark force [2] in Grenoble (Faddeev equations) [3], Valencia (hyperspherical formalism) [4] and Graz (stochastic variational approach) [5]. The results show differences in the binding energy of a few MeV, so it is interesting to test the quality of the wave functions by looking at the predictions for the form factors as well. The results of the first two approaches are to be published elsewhere [1]. In the present contribution, we consider the form factors calculated with the wave functions obtained by the stochastic variational approach [5].

We will give a short reminder of the argument for a power law behaviour of the form factors and a brief introduction to the stochastic variational approach before presenting our results.

2 Power law expectation for the nucleon form factor

Electron scattering on the nucleon involves many diagrams that differ by their time ordering and include the exchange of at least two gluons. One of them is shown in Fig.1 where the virtual photon transfers some momentum q to a quark, which is shared with the other two quarks by the successive exchange of two gluons.

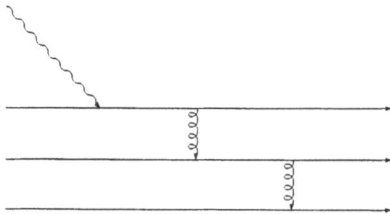

Figure 1. Electron scattering on the nucleon

Examination of this diagram provides a quick estimate of the behaviour of the form factor at high q^2. Each gluon propagator provides a factor $\frac{1}{q^2}$ and, in the non-relativistic limit, each intermediate quark also introduces a factor $\frac{1}{q^2}$. Therefore the form factor is expected to behave asymptotically as:

$$F_N^{n.r.}(q^2)_{q^2 \to \infty} \sim q^{-8}.$$

This behaviour is expected to be valid only in a non-relativistic framework and for quark-quark potentials that are at least as singular as a Coulomb potential at the origin (and therefore scale like $\frac{1}{q^2}$ at high q). It is however not clear a priori at which scale this asymptotic behaviour sets in, and it is therefore interesting to see whether present day calculational methods are able to confirm this theoretical expectation.

3 Description of the nucleon wave function within the stochastic variational approach

We employed in our calculations the Bhaduri et al. [2] quark-quark force,

$$V = \frac{1}{2} \sum_{i<j} \left(-\frac{\kappa}{r_{ij}} + \frac{r_{ij}}{a^2} - D + \frac{\kappa_\sigma}{m_i m_j} \frac{exp(-r_{ij}/r_0)}{r_0^2 r_{ij}} \boldsymbol{\sigma}_i \cdot \boldsymbol{\sigma}_j \right),$$

with $\kappa = 102.67$ MeV fm, $a = 0.0326$ MeV$^{-1/2}$ fm$^{1/2}$, $r_0 = 0.4545$ fm, $m = 337$ MeV and $D = 913.5$ MeV.

In the stochastic variational approach [5], the spatial part of the nucleon wave function is approximated as a sum of basis functions that are chosen to be of gaussian type:

$$\varphi_{\alpha j}(\boldsymbol{x}, \boldsymbol{y}) = x^{2\nu+\lambda} y^{2n+l} \exp(-\beta_j x^2 - \delta_j y^2 + \gamma_j \boldsymbol{x}.\boldsymbol{y}) \mathcal{F}_{\alpha_0}^{JM_J TM_T}(\hat{x}, \hat{y}),$$

where $\boldsymbol{x}$ and $\boldsymbol{y}$ are the internal Jacobi coordinates of the three quark system and $\mathcal{F}$ denotes the orbital-spin-isospin part of the wave function. The parameters $\nu, \lambda, n, l, \beta, \delta$ and γ are chosen randomly and are fixed by a standard routine that minimizes the energy (see ref. [5] for details).

There are two drawbacks of this procedure that are essential for our purpose. First the criterion of minimal energy does not necessarily imply a better quality of the wave function. Second, the specific form of the correlated gaussian basis functions implies that the derivative of the total wave function at the origin is always identically zero. Since the values of the form factors at high momentum transfers are determined by the short distance behaviour of the wave function, one might expect some troubles in this approach. It is however interesting to ask, just until where the obtained wave function can be regarded as a reliable approximation, especially in view of possible applications.

4 Results

In Fig. 2, we compare our results with those of three other approaches [1]: the Hyperspherical Harmonics formalism with the grand orbital momentum value K=0 and 2, and Faddeev calculations with 2 and 8 amplitudes retained in the expansion.

It was found that the hypercentral approximation in the hyperspherical formalism ($K = 0$) systematically leads to a q^{-7} behaviour. Solutions of the Faddeev-equations lead to a q^{-8} behaviour, but the absolute magnitude of the form factors depends on the number of partial waves retained in the Faddeev amplitude. The solutions obtained by the stochastic variational method agree with the Faddeev results when 8 amplitudes are retained. However, sizeable departures are observed at squared momentum transfers of ~ 50 GeV^2 and it turned out to be difficult to stabilize the results beyond this limit. This can be traced back to the difficulties of the stochastic variational approach as mentioned in the previous section: the derivative of the wave function at the origin is identically zero, whereas for the exact result it is not.

Summing up, we can say that constituent quark models do lead to a power law behaviour of form factors at high q^2, but one has to be careful about the numerical precision of the method that one employs.

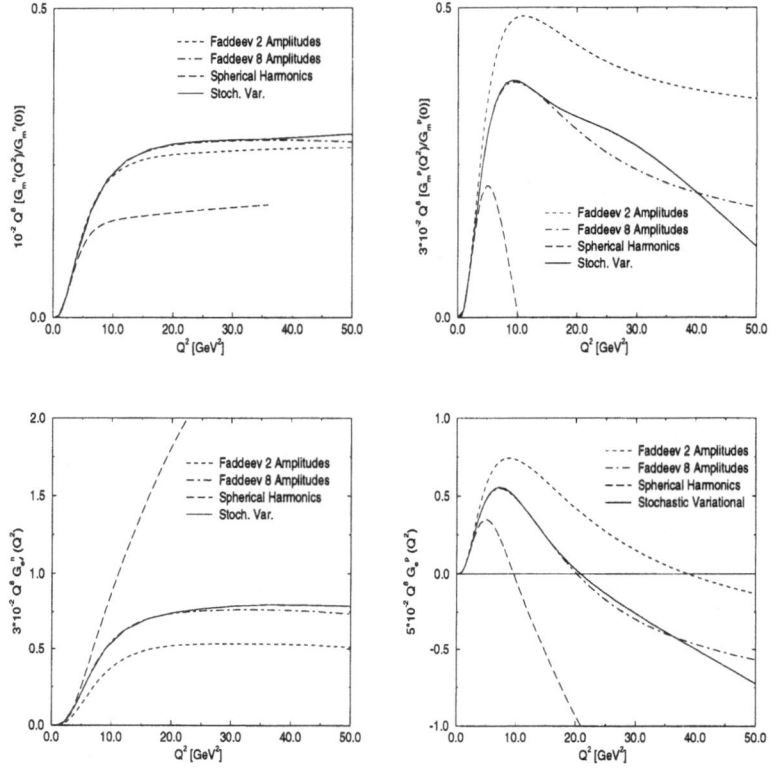

Figure 2. Electromagnetic form factors of the nucleon up to $q^2 = 50\,\mathrm{GeV}^2$

Acknowledgement. L.T. would like to thank the members of the Institut des Sciences Nucléaires in Grenoble for their hospitality during the time where this work was performed.

References

1. B. Desplanques et al.: to be published

2. R.K. Bhaduri, L.K. Cohler and Y. Nogami: Nuov. Cim. **65A**, 376 (1981)

3. B. Silvestre-Brac and C. Gignoux: Phys. Rev. **D32**, 74 (1985)

4. F. Cano: Doctoral Tesis, University of Valencia, 1997

5. K. Varga and Y. Suzuki: Phys. Rev. **C52**, 2885 (1995)

Few-Body Systems Suppl. 10, 407–410 (1999)

Few-
Body
Systems
© by Springer-Verlag 1999

Investigation of N-N* Electromagnetic Form Factors within a Front-Form CQM

G. Salmè [1], E. Pace [2] and S. Simula [3]

[1] Istituto Nazionale di Fisica Nucleare, Sezione di Roma I, P.le A. Moro 2, I-00185 Rome, Italy

[2] Dipartimento di Fisica, Università di Roma "Tor Vergata", and Istituto Nazionale di Fisica Nucleare, Sezione Tor Vergata, Via della Ricerca Scientifica 1, I-00133, Rome, Italy

[3] Istituto Nazionale di Fisica Nucleare, Sezione di Roma III, Via della Vasca Navale 84, I-00146 Rome, Italy

Abstract. The helicity amplitudes for the transitions $N - S_{11}$ and $N - S_{31}$ are presented. The amplitudes have been obtained within our front-form CQM model, based on hadron eigenstates of a relativistic mass operator and CQ current with Dirac and Pauli form factors.

Hadron electromagnetic (em) form factors have been recently investigated within the front-form constituent quark (CQ) model of [1] for space-like values of the four-momentum transfer. The main features of the model are: i) the use of hadron eigenfunctions of a relativistic mass operator, that includes an effective $q - q$ interaction and reproduces the hadron spectra for a large set of quantum numbers [2]; ii) the use of a one-body em current operator containing phenomenological Dirac and Pauli form factors for CQ's, which are determined by the request of reproducing the existing experimental data for the pion and nucleon elastic form factors (cf. [1]). Such a model has been already applied for obtaining a parameter-free prediction of the em form factors for the transitions to $N^*(1440)$ and $\Delta(1232)$, including the possible effects due to the D-wave components in the Δ wave function, [1].

In this contribution, we will present an analysis of transition form factors for $N \to S_{11}(1535)$, $N \to S_{11}(1650)$ and $N \to S_{31}(1620)$.

The current for negative-parity transition with $J_f = 1/2$ is given in terms of Dirac ($F_1^{f\tau}$) and Pauli-like ($F_2^{f\tau}$) form factors by (cf. [3])

$$\bar{\Psi}_f \, J^\mu \, \Psi_\tau = \bar{\Psi}_f \gamma^5 \left[\frac{p_f^\mu + p_i^\mu}{M_f - M_i} F_2^{f\tau} - \frac{M_f + M_i}{M_f - M_i} q^\mu F_1^{f\tau} + \gamma^\mu (F_1^{f\tau} + F_2^{f\tau}) \right] \Psi_\tau \quad (1)$$

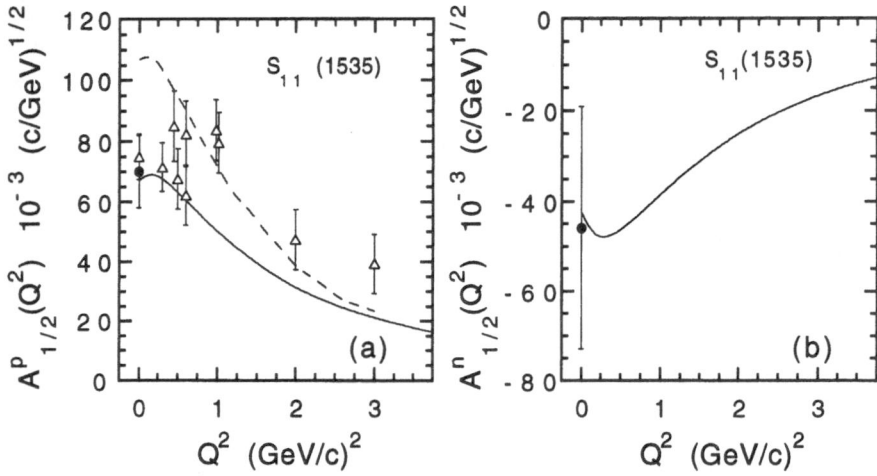

Figure 1. - (a) The transverse helicity $A_{1/2}$ for the transition $p \to S_{11}(1535)$ vs. Q^2. Solid line: $A_{1/2}$ from the hadron wave functions corresponding to the interaction of [2] and the nucleon em current with CQ form factors of [1]; dashed line: a non relativistic CQM calculation [4]. Solid dot: PDG '96 [5]; triangles: data analysis from [6]. - (b) The same as in Fig. 1(a), but for $n \to S_{11}(1535)$.

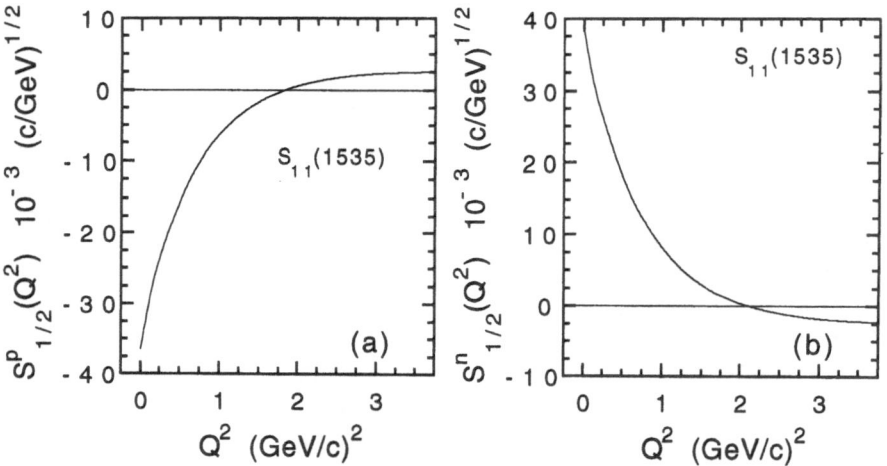

Figure 2. The same as in Fig. 1, but for the longitudinal helicity $S_{1/2}$.

where $\tau = p, n$. By using such a current, the helicities for negative-parity transition can be written as follows

$$S_{1/2}^{\tau}(Q^2) = \zeta \sqrt{\frac{2\pi\alpha}{k^*}} \sqrt{\frac{Q^+}{2M_iM_f}} \sqrt{\frac{Q^+Q^-}{4M_f}} \frac{M_f - M_i}{Q^2\sqrt{2}} \left[F_1^{f\tau} - \frac{Q^2}{(M_f - M_i)^2} F_2^{f\tau} \right]$$

$$A_{1/2}^{\tau}(Q^2) = -\zeta \sqrt{\frac{2\pi\alpha}{k^*}} \sqrt{\frac{Q^+}{2M_iM_f}} \left(F_1^{f\tau} + F_2^{f\tau} \right) \quad (2)$$

where ζ is the sign of the πN decay amplitude, $k^* = (M_f^2 - M_i^2)/2M_f$, $Q^{\pm} = (M_f \pm M_i)^2 + Q^2$. The invariant form factors in Eq. (2) can be obtained within the front-form CQ model following standard procedures (see, e.g., [1]), namely approximating the plus component of the transition current, $\mathcal{I}^+$, in terms of the sum of one-body CQ currents, containing CQ Dirac and Pauli form factors. In particular

$$F_1^{f\tau} = -\frac{1}{2}Tr\left(\sigma_z\mathcal{I}^+(\tau)\right) \qquad F_2^{f\tau} = -\frac{M_f - M_i}{2Q}Tr\left(\sigma_x\mathcal{I}^+(\tau)\right). \qquad (3)$$

where $\mathcal{I}^+_{\nu_f\nu_i}(\tau) = \bar{u}^f_{LF}(\nu_f)\sum_{j=1}^{3}\left(e_j\gamma^+ f_1^j(Q^2) + i\kappa_j\frac{\sigma^{+\rho}q_\rho}{2m_j}f_2^j(Q^2)\right)u^\tau_{LF}(\nu_i).$

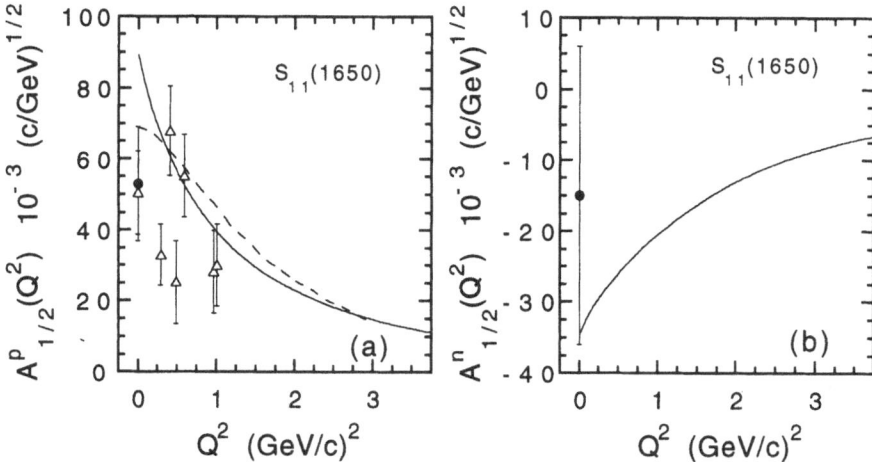

Figure 3. - (a) The transverse helicity $A_{1/2}$ for the transition $p \to S_{11}(1650)$ vs. Q^2. Solid line: $A_{1/2}$ from the hadron wave functions corresponding to the interaction of [2] and the nucleon em current with CQ form factors of [1]; dashed line: a non relativistic CQM calculation [4]. Solid dot: PDG '96 [5]; triangles: data analysis from [6]. - (b) The same as in Fig. 3(a), but for $n \to S_{11}(1650)$.

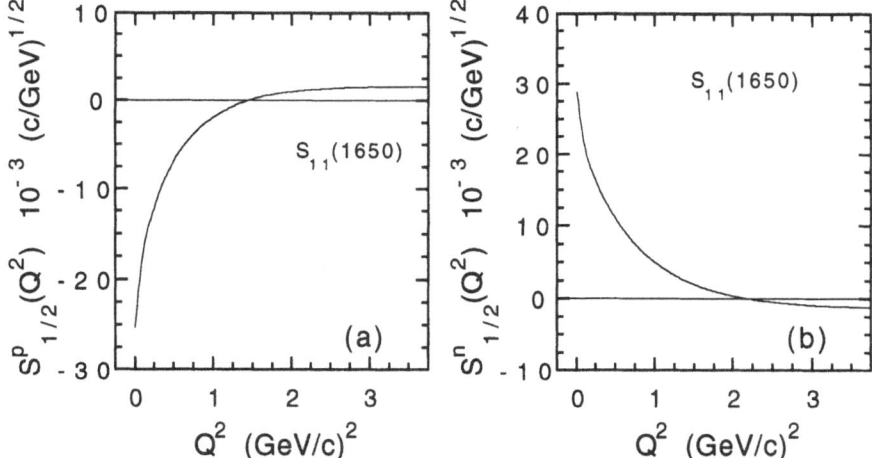

Figure 4. The same for Fig. 3, but for the longitudinal helicity $S_{1/2}$.

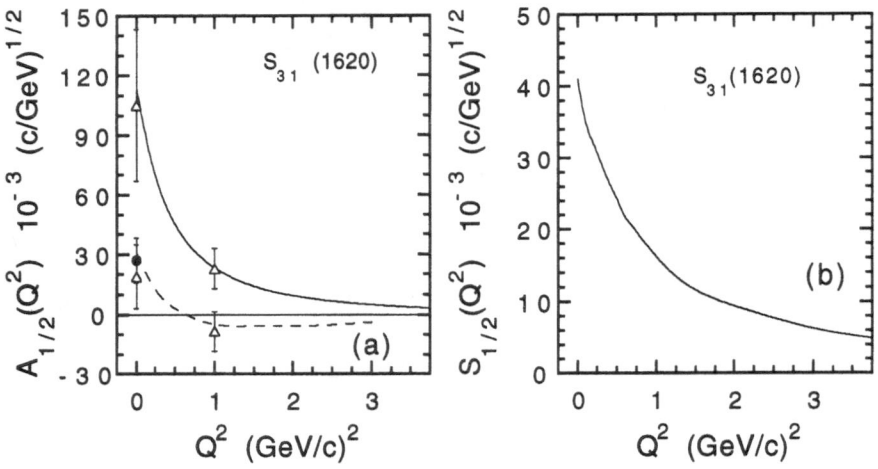

Figure 5. - (a) The transverse helicity $A_{1/2}$ for the transition $p \rightarrow S_{31}(1620)$ vs. Q^2. Solid line: $A_{1/2}$ from the hadron wave functions corresponding to the interaction of [2] and the nucleon em current with CQ form factors of [1]; dashed line: a non relativistic CQM calculation [4]. - (b) The same as in Fig. 5a, but for $S_{1/2}$.

In Figs. 1-5, our *parameter-free* evaluation of the helicity amplitudes, $A_{1/2}$ and $S_{1/2}$ are shown for $N \rightarrow S_{11}(1535)$, $S_{11}(1650)$ and $S_{31}(1620)$, respectively. In the case of $S_{31}(1620)$ the results for p and n coincides (as in the case of $P_{33}(1232)$), since only the isovector part of the CQ current is effective, given the isospin of the resonance.

The overall agreement between our predictions and the data is encouraging, though a most accurate set of data is necessary in order to reliably discriminate between different models. However, the sensitivity to relativistic effects for the P-wave resonances seems sizeable.

References

1. F. Cardarelli et al.: Phys. Lett. **B357**, 267 (1995); Few Body Systems Suppl. **8**, 345 (1995); Phys. Lett. **B371**, 7 (1996); Phys. Lett. **B397**, 13 (1997); Nucl. Phys. **A623**, 361c (1997)

2. S. Capstick and N. Isgur: Phys. Rev. **D34**, 2809 (1986)

3. R.H. Stanley and H.J. Weber: Phys. Rev. **C52**, 435 (1995)

4. M. Aiello, M.M. Giannini and E. Santopinto: J. of Phys. **G24**, 753 (1998)

5. Particle Data Group: Phys. Rev. **D54**, 1 (1996).

6. V. D. Burkert: Czech. J. Phys. **46**, 627 (1996) and private communications

Few-Body Systems Suppl. 10, 411–414 (1999)

Few-
Body
Systems
© by Springer-Verlag 1999

Nucleon structure functions in a constituent quark scenario *

Sergio Scopetta[1], Vicente Vento [1,2], Marco Traini [3]

[1] Departament de Fisica Teòrica, Universitat de València 46100 Burjassot (València), Spain
[2] Institut de Física Corpuscular, Consejo Superior de Investigaciones Científicas, Spain
[3] Dipartimento di Fisica, Università di Trento, I-38050 Povo (Trento), and INFN, Gruppo Collegato di Trento, Italy.

Abstract. Using a simple picture of the constituent quark as a composite system of point-like partons, we construct the polarized parton distributions by a convolution between constituent quark momentum distributions and constituent quark structure functions. We achieve good agreement with experiments in the unpolarized as well as in the polarized case, though a good description of the recent polarized neutron data requires the introduction of one more parameter. When our results are compared with similar calculations using non-composite constituent quarks, the accord of the present scheme with the experiments is impressive. We conclude that DIS data are consistent with a low energy scenario dominated by composite constituents of the nucleon.

At low energies, the so called naive quark model accounts for a large number of experimental observations. At large energies, QCD sets the framework for an understanding of the Deep Inelastic Scattering (DIS) phenomena beyond the Parton Model. However, the perturbative approach to QCD does not provide absolute values for the observables. The description based on the Operator Product Expansion (OPE) and the QCD evolution requires the input of non-perturbative matrix elements. We have developed an approach which uses model calculations for the latter ingredients [1]. Moreover, in order to relate the constituent quark with the current partons of the theory, a procedure, hereafter called ACMP, has been applied [2, 3]. Within this approach, constituent quarks are effective particles made up of point-like partons (current quarks (antiquarks) and gluons), interacting by a residual interaction described as in

*Supported in part by DGICYT-PB94-0080 and TMR programme of the European Commission ERB FMRX-CT96-008

a quark model. The hadron structure functions are obtained by a convolution of the constituent quark model wave function with the constituent quark structure function. This idea has been recently used to estimate the pion structure function [4]. We summarize here our application to the unpolarized [3] and polarized [5] DIS off the nucleon. It will be found that DIS data are consistent with a low energy scenario dominated by composite constituents.

In our picture the constituent quarks are themselves complex objects whose structure functions are described by a set of functions Φ_{ab} that specify the number of point-like partons of type b, which are present in the constituents of type a with fraction x of its total momentum [2, 3]. In general a and b specify all the relevant quantum numbers of the partons, i.e., flavor and spin. Let us discuss first the unpolarized case for the proton [3].

The functions describing the nucleon parton distributions omitting spin degrees of freedom are expressed in terms of the independent $\Phi_{ab}(x)$ and of the constituent probability distributions u_0 and d_0, at the hadronic scale μ_0^2 [1], as

$$f(x, \mu_0^2) = \int_x^1 \frac{dz}{z} [u_0(z, \mu_0^2) \Phi_{uf}(\frac{x}{z}, \mu_0^2) + d_0(z, \mu_0^2) \Phi_{df}(\frac{x}{z}, \mu_0^2)] \qquad (1)$$

where f labels the various partons, i.e., valence quarks (u_v, d_v), sea quarks (u_s, d_s, s), sea antiquarks ($\bar{u}, \bar{d}, \bar{s}$) and gluons g. The different types and functional forms of the structure functions for the constituent quarks are derived from three very natural assumptions [2]: i) The point-like partons are the quarks, antiquarks and gluons described by QCD; ii) Regge behavior for $x \to 0$ and duality ideas; iii) invariance under charge conjugation and isospin.

These considerations define the following structure functions [2]

$$\Phi_{qf}(x, \mu_0^2) = C_f x^{a_f}(1 - x)^{A-1} , \qquad (2)$$

where $f = q_v, q_s, g$ for the valence quarks, the sea and the gluons, respectively. Regge phenomenology suggests: $a_{q_v} = -0.5$ (ρ meson exchange) and $a_{q_s} = a_g = -1$ (*pomeron* exchange). The other ingredients of the formalism, i.e., the probability distributions for each constituent quark, are defined according to the procedure of ref. [1] and shown in [3]. Our last assumption relates to the hadronic scale μ_0^2, i.e., that at which the constituent quark structure is defined. We choose $\mu_0^2 = 0.34$ GeV^2, as defined in ref. [1], namely by fixing the momentum carried by the various partons. This choice of the hadronic scale determines all the parameters except one, which is fixed through the data [3]. To complete the process, the above input distributions are NLO-evolved in the DIS scheme to the experimental scale, where they are compared with the data.

We next generalize our previous discussion to the polarized parton distributions. As it is explained in ref. [5], using $SU(6)$ (spin-isospin) symmetry and other reasonable simplifying assumptions, it can be shown that

$$\Delta f(x, \mu_0^2) = \int_x^1 \frac{dz}{z} [u_0(z, \mu_0^2) \Delta\Phi_{uf}(\frac{x}{z}, \mu_0^2) + d_0(z, \mu_0^2) \Delta\Phi_{df}(\frac{x}{z}, \mu_0^2)] , \qquad (3)$$

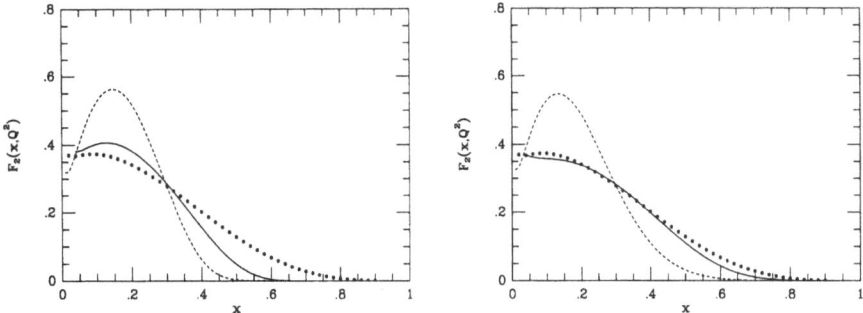

Figure 1. The proton $F_2(x, Q^2)$, obtained by NLO-evolution to $Q^2 = 10$ GeV^2 (full), compared to the data (dots) [10]. The result which would be obtained disregarding the constituent structure is also shown (dashed). Left (right) panel: constituent wave functions form ref. [8] (ref. [9]).

where f labels the various partons; it means that the $ACMP$ procedure can be extended to the polarized case just by introducing three additional structure functions for the constituent quarks: $\Delta\Phi_{qq_v}$, $\Delta\Phi_{qq_s}$ and $\Delta\Phi_{qg}$. In order to determine them we add two minimal assumptions: *iv)* factorization: $\Delta\Phi$ cannot depend upon the quark model used; *v)* positivity: the constraint $\Delta\Phi \leq \Phi$ is saturated for $x = 1$. In such a way we determine completely the $\Delta\Phi$'s. In fact, the QCD partonic picture, Regge behavior and duality imply that

$$\Delta\Phi_{qf} = \Delta C_f x^{-\Delta a_f}(1 - x)^{\Delta A_f - 1} \tag{4}$$

and $-\frac{1}{2} < \Delta a_f < 0$, for all $f = q_v, q_s, g$, as allowed by dominant exchange of the A_1 meson trajectory [6]. Moreover, the assumption that the positivity restriction is saturated for $x = 1$, in the spirit of ref. [7], implies that the $\Phi's$ and the $\Delta\Phi's$ have the same large x behavior, and that $\Delta C_f = C_f$, (the latter being introduced in (2)); it means that the partons which carry all of the momentum also carry all of the polarization. Let us stress that the change between the polarized functions and the unpolarized ones comes only from Regge behavior; as a matter of fact, it turns out that, *except for the exponent Δa_f shown above,* the $\Delta\Phi$'s, Eq. (4), are given by the unpolarized functions, Eq. (2). The other ingredients, i.e., the polarized distributions for each constituent quark, are defined according to the procedure of ref. [1] and they are shown in ref. [5]. Finally, the parton distributions at the hadronic scale are evolved to the experimental scale by performing a NLO evolution in the AB scheme [6]. Results are shown in Figs. 1 and 2. Figure 1 refers to the unpolarized case. The structure function $F_2(x, Q^2)$, obtained evolving the parton distributions Eq. (1), calculated using Eq. (2) for the Φ_{qf}'s and two different models for u_o and d_o, describes successfully the data. The agreement becomes impressive if compared with a similar calculation with non-composite constituents.

In the polarized case, it is found [5] that the constituent structure functions Eq. (4) give a good result for the proton, but they fail in reproducing the recent precise neutron data. This is to be ascribed to our naive input for the sea and to the symmetry for the u and d quarks [5]. In particular, it has been

414

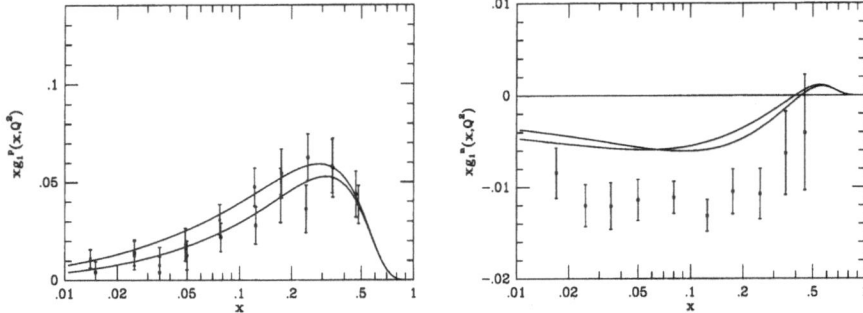

Figure 2. Left (Right): $xg_1(x, Q^2)$ for the proton (neutron) evolved at NLO to $Q^2 = 10\,(5)$ GeV^2, for the two extreme Regge behaviors mentioned in the text (full curves). The wave functions used are from ref. [9]. The data [10] are shown for comparison.

shown that, by redefining the sea $\Delta\Phi$, changing *only one* parameter so that the experimental sea polarization is recovered, also the neutron is rather well described. Figure 2 refers to this last scenario. The procedure is also able to predict successfully several observables, such as the nucleon axial charges [5]. It should be noticed that in this framework the *spin crisis*, as initially presented, does not arise.

Summarizing, low energy models seem to be consistent with DIS data when a structure for the constituent is introduced. The crucial role played by the sea in the polarized case, as well as the implementation of Chiral Symmetry Breaking in our procedure, have to be more deeply investigated. It will be the subject of future work.

References

1. M. Traini et al.: Nucl. Phys. **A614**, 472 (1997)

2. G. Altarelli et al.: Nucl. Phys. **B69**, 531 (1974)

3. S. Scopetta, V. Vento and M. Traini: Phys. Lett. **B421**, 64 (1998)

4. G. Altarelli, S. Petrarca and F. Rapuano: Phys. Lett. **B373**, 200 (1996)

5. S. Scopetta, V. Vento and M. Traini: Phys. Lett. **B** (1998) to appear

6. G. Altarelli et al.: Nucl. Phys. **B496**, 337 (1997)

7. R. Carlitz and J. Kaur: Phys. Rev. Lett. **38**, 673 (1977)

8. N. Isgur and G. Karl: Phys. Rev. **D18**, 4187 (1978)

9. R. Bijker, F. Iachello and A. Leviatan: Ann. Phys. **236**, 69 (1994)

10. EMC: Nucl. Phys. **B328**, 1 (1989); SMC, Phys. Rev. **D56**, 5330 (1997); E154: Phys. Rev. Lett. **79**, 26 (1997)

Few-Body Systems Suppl. 10, 415–418 (1999)

Few-
Body
Systems
© by Springer-Verlag 1999

The Strange Magnetic Moment of the Proton in the Chiral Quark Model

D.O. Riska*

Department of Physics, 00014 University of Helsinki, Finland

Abstract. The strange magnetic moment of the proton is small in the chiral quark model, because of a near cancellation between the quantum fluctuations that involve kaons and s-quarks and loops that involve radiative transitions between strange vector mesons and kaons.

1 Introduction

The observation that the strangeness magnetic moment of the proton may be positive [1] ($G_M^s(Q^2 = 0.1 \text{ GeV}^2) = 0.23 \pm 0.37$) was unexpected as most predictions had given negative values [2, 3, 4, 5]. The expectation of a negative value for $G_M^s(0)$ may be explained as follows: Consider a u quark with spin s_z projection $+\frac{1}{2}$. If this fluctuates into a kaon and an s quark, the probability that s_z of the s quark be $-\frac{1}{2}$ is twice that for the value $+\frac{1}{2}$. Hence, as the charge of the s quark is $-e/3$, the s quark should contribute a positive amount to the magnetic moment of the u quark. On the other hand following spin-flip the $\bar{s}$ quark in the kaon has orbital angular momentum $l_z = +1$, and thus as its charge is $+e/3$, it should also contribute a positive amount to the net magnetic moment of the u quark. As there are twice as many u quarks with $s_z = +\frac{1}{2}$ as with $s_z = -\frac{1}{2}$ in the proton it follows that this fluctuation will give a positive contribution to the magnetic moment of the proton. As G_M^s is defined as the matrix element of $\bar{s}\gamma^\mu s$ this matrix element is obtained by multiplication of the "conventional" magnetic moment by -3, and thus should be negative.

A positive value for $G_M^s(0)$ has to arise from transition couplings between strange mesons. It has recently been noted that a fluctuation into a loop with a K^* meson, which decays into kaon that is reabsorbed with the s quark, gives a positive value [6]. That work based on the nonrelativistic oscillator quark model gave the result $G_M^s(0) = 0.035$. It is shown here that the chiral quark model leads to a small negative value for $G_M^s(0)$.

E-mail address: riska@pcu.helsinki.fi

This calculation should be contrasted with those in refs. [7, 4], where the strangeness fluctuations were considered at the hadronic level. The consideration of the strangeness fluctuations of the constituent quarks is motivated by the fact that the magnetic moments of the nucleons are explained by the constituent quark model, but not by the hadronic constituent model [8]. The smaller meson-quark coupling constants imply loop corrections, which are sufficiently small so as not perturb the overall quark model description of the magnetic moments. The quark model approach automatically takes into account all baryonic intermediate states.

2 Strangeness Loops with Transition Couplings

Consider the (hidden) strangeness fluctuations $q \to K^*s \to K\gamma s \to q\gamma$ of a light u or d quark q. The key vertex in this loop diagram is the $K^* \to K\gamma$ vertex, which is described by the transverse current matrix element:

$$< K^a(k')|J_\mu|K_\sigma^{*b}(k) >= -i\frac{g_{K \cdot K\gamma}}{m_K^*}\epsilon_{\mu\lambda\nu\sigma}k_\lambda k_\nu' \delta^{ab}. \tag{1}$$

The coupling constant $g_{K \cdot K\gamma}$ is determined by the empirical decay widths for radiative decay of the K^* as $g_{K^{*+}K^+\gamma} = 0.75$ and $g_{K^{*-}K^-\gamma} = 1.14$.

Given the current matrix element (1) the contributions to the anomalous magnetic moment of the u and d quarks from the $K^*K\gamma$ loop diagrams take the following form (when expressed in terms of nuclear magnetons and after assigning the kaon line a "strangeness charge" of -1):

$$G_M^s(0)_{u,d} = -\frac{g_{Kqs}g_{K \cdot qs}g_{K \cdot K\gamma}}{2\pi^2}\frac{m_p}{m_K}\int_0^1 dx(1-x)$$

$$\cdot \int_0^1 dy\{m_u(m_s - m_u x)\{\frac{1}{G_1} - \frac{1}{G_2} - \frac{1}{G_3} + \frac{1}{G_4}\} - log(\frac{G_2G_3}{G_1G_4})\}. \tag{2}$$

Here m_p is the proton mass and m_u and m_s the constituent masses of the u, d and s quarks respectively for which we use $m_u = 340$ MeV, $m_s = 460$ MeV [9]. The quantities G_j above are defined as

$$G_1 = G(m_K, m_{K^*}), \ G_2 = G(\Lambda, m_{K^*}), \ G_3 = G(m_K, \Lambda^*), \ G_4 = G(\Lambda, \Lambda^*), \tag{3}$$

where the function G is defined as

$$G(m, m') = m_s^2(1 - x) - m_u^2 x(1 - x) + m^2 x(1 - y) + m'^2 xy. \tag{4}$$

In the case of the u quark the coupling $g_{K \cdot K\gamma}$ constant is that for the K^+, K^{*+} mesons and in the case of the d quark it is that for the K^0, K^{*0} mesons.

The π-quark coupling constant is determined by the πNN coupling $g_{\pi NN}$ as [9]: $g_{Kqs} = g_{\pi qq} = 3/5(m_q/m_N)g_{\pi NN} = 2.9$.

The corresponding vector meson coupling constant $g_{K \cdot qs}$ is related to the nonstrange vector meson (ρ) quark coupling strength as $g_{K \cdot qs} \simeq g_{\rho qq} =$

Table 1. Contributions to the strange magnetic moment of the proton (in nuclear magnetons) from the loops with $K^* \to K\gamma$ transition vertices and with diagonal $KK\gamma$ vertices as a function of the cut-off parameter Λ (in MeV).

Λ	$K^*K\gamma$	$KK\gamma$	Sum
800	0.0	-0.016	-0.016
1000	0.006	-0.031	-0.025
1200	0.026	-0.046	-0.020

$(3/5)(m_q/m_N)g_{\rho NN}$. With the usual value for the ρNN vector coupling constant $g^2_{\rho NN}/4\pi = 0.52$ and the including the factor $(1+\kappa)$, where $\kappa = 6.6$ ([10]) is the ρNN tensor coupling, the value $g_{K^*qs} = g_{\rho qq} = 4.2$ obtains.

As the K^*Ks loop diagrams are logarithmically divergent a cut-off is included in their evaluation by insertion of monopole form factors of the form $(\Lambda^2 - m_K^2)/(\Lambda^2 + k'^2)$ and $(\Lambda^{*2} - m_K^{*2})/(\Lambda^{2*} + k^2)$ at the K and K^* vertices respectively. The cut-off masses Λ and Λ^* are free parameters. We shall here take $\Lambda = \Lambda^*$ in the numerical calculations.

The contribution from these loops to the strange magnetic moment of the proton is obtained as

$$G^s_M(0)_p = \frac{4}{3}G^s_M(0)_u - \frac{1}{3}G^s_M(0)_d, \tag{5}$$

and thus will have a value close to the corresponding constituent quark values. For Λ values in the range 1–1.5 GeV the contributions are positive, but small (Table 1).

These values are smaller than the corresponding value obtained in ref. [6] with the harmonic oscillator quark model. The difference is due to the non-relativistic nature of the latter which implies an overestimate of the radiative widths of the vector mesons. With the usual static fermion currents of the constituent quark model the coupling constant $g_{K^*K\gamma}$ should be 2, which is twice too large.

3 Strange Fluctuations with Diagonal Couplings

The expressions for the contributions from the diagonal Ks loop contribution to the strange magnetic moments of the u and d quarks are similar to the corresponding expressions for the contribution of pionic fluctuation to the neutron magnetic moment [8]. These kaonic loops then give the same contribution to the anomalous strange magnetic moment of the u and d quarks. This may be written in the form

$$G_M^s(0) = -\frac{g_{Kqs}^2}{4\pi^2}m_p \int_0^1 dx(1-x)(m_s - m_u x)$$

$$\cdot\{(1-x)[\frac{1}{H_1} - \frac{1}{H_4} - x\frac{(\Lambda^2 - m_K^2)}{H_4^2}] + \int_0^1 dyx[\frac{1}{H_1} - \frac{1}{H_2} - \frac{1}{H_3} + \frac{1}{H_4}]\} \quad (6)$$

Here the quantities H_j are defined as

$$H_1 = H(m_K, m_K), \; H_2 = H(m_K, \Lambda), \; H_3 = H(\Lambda, m_K), \; H_4 = H(\Lambda, \Lambda),$$

$$H(m, m') = (m_s^2 - m_u^2 x)(1-x) + m^2 xy + m'^2 x(1-y). \quad (7)$$

By (5) the kaonic loop contribution (6) to the constituent quark equals that of the proton. For Λ below 1.2 GeV this negative contribution is larger in magnitude than the positive contribution of the K^*Ks loop contributions considered above.

4 Discussion

The result obtained here is that the strange magnetic moment of the proton is small because of a strong cancellation between strangeness fluctuations, which involve radiative transition couplings and such which do not. The chiral quark model calculation leads to a stronger cancellation than the non-relativistic quark model calculation in ref. [6].

Acknowledgement. This work was supported in part by the Academy of Finland under contract 34081 and is based on a collaboration with L.Ya. Glozman.

References

1. B. Mueller et al.: Phys. Rev. Lett. **78**, 3824 (1997)

2. M.J. Musolf et al.: Phys. Reports **239**, 1 (1994)

3. E.J. Beise et al.: Proc. SPIN96 symposium, eprint nucl-ex/9610011

4. L.L. Barz et al.: eprint hep-ph/9803221

5. S.J. Dong, K.F. Liu and A.G. Williams: eprint hep-ph/9712483

6. P. Geiger and N. Isgur: Phys. Rev. **55**, 299 (1997)

7. M.J. Musolf and M. Burkhardt: Z. Phys. **C61**, 433 (1994)

8. H.A. Bethe and F. de Hoffman: In: *Mesons and Fields, Vol. II.* Evanston: Row, Peterson (1955)

9. L.Ya. Glozman and D.O. Riska: Phys. Rep. **268**, 263 (1996)

10. G. Höhler and E. Pietarinen: Nucl. Phys. **B95** (1975) 216

Few-Body Systems Suppl. 10, 419–422 (1999)

Few-
Body
Systems
© by Springer-Verlag 1999

Electromagnetic Probe of Nucleon Strangeness

Stephen R. Cotanch[1*], Robert A. Williams[2†]

[1] Department of Physics, North Carolina State University, Raleigh, NC 27695-8202, USA
[2] Thomas Jefferson National Accelerator Facility, Newport News, VA 23606, USA

Abstract. We continue our strangeness studies with new calculations for the nucleon time-like form factors and probing lepton pair production process $N(\pi, e^+e^-)N$. Using a recent value for the ϕN coupling constant obtained from our analysis of limited $p(\gamma, \phi)p$ data, we again document a several order of magnitude e^+e^- production resonant cross section structure. This dramatic ϕ peak follows from vector meson dominance and provides an interesting experimental signature for OZI violation and the strangeness content of the nucleon.

1 Introduction

As detailed previously [1], we reported the utility of the $N(\pi, e^+e^-)N$ process for extracting the nucleon electric and magnetic form factors in the unmeasured time-like region $0 \leq q_\gamma{}^2 \leq 4M_N^2$. This energy interval spans the ρ, ω and ϕ vector meson masses which affords a decisive test of vector meson dominance (VMD) and, in turn, an opportunity to determine the poorly known coupling constant, $g_{\phi NN}$. The magnitude of $g_{\phi NN}$ is important for directly assessing the strangeness content of the nucleon, a topic of current interest [2].

In this work we revisit out $N(\pi, e^+e^-)N$ calculation utilizing a new value for $g_{\phi NN}$ obtained from our recent analysis of $p(\gamma, \phi)p$ data [3]. Using a consistent phenomenological framework based upon quantum hadrodynamics (QHD), we obtain $g_{\phi NN} = -.65$ by fitting the high t measurements. This is substantially smaller than our previous coupling and leads to a reduction in the $N(\pi, e^+e^-)N$ peak cross section by a factor of 4. Nevertheless, due to the dramatic resonant VMD mechanism, the ϕ peak remains a measurably distinct prediction clearly testable by experiment. Complete details are presented in the next section.

---

*_E-mail address:_ cotanch@ncsu.edu
†_E-mail address:_ bobw@cebaf.gov

2 Model Details and Results

We begin by briefly summarizing our low energy ϕ photoproduction analysis which provided a new value for $g_{\phi NN}$. Our model entails a consistent application of QHD and incorporates both diffractive (Pomeron exchange) and non-diffractive mechanisms. For a comprehensive description consult ref. [3]. In Fig. 1 four calculations, along with data, are compared for $|t|$ between 0 and 1 GeV^2. The curve legends are: dotted, only Pomeron exchange (t channel); dashed, Pomeron plus π^0 and η exchange (t channel); dot-dashed, Pomeron plus ϕNN coupling (s and u channel); solid, Pomeron plus π^0 and η exchange and ϕNN coupling. To describe the high $|t|$ data, where t channel exchanges are suppressed, requires $g_{\phi NN} = -.65$ which is smaller by a factor of 2 than the value we used in ref. [1]. More extensive analyses of approved future ϕ production measurements at Jefferson Lab should further improve this determination.

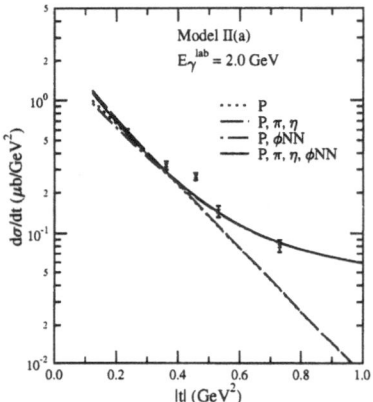
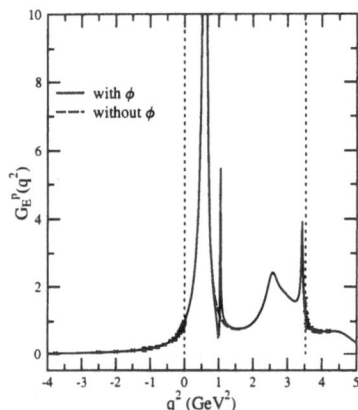

Figure 1. Model and data for $p(\gamma, \phi)p$.   **Figure 2.** p electric form factor.

With this value for the ϕN coupling we have re-computed the nucleon electric and magnetic form factors. As specified in ref. [4], we utilized a hybrid VMD model, implementing Sakurai's universality hypothesis, to obtain a good description of the baryon octet EM form factors and nucleon data. In Fig. 2 we display the electric proton form factor, $G_E^p(q^2)$, with and without the ϕ contribution. Equally interesting is the neutron magnetic form factor which is also predicted sensitive to $g_{\phi NN}$. This is shown in Fig. 3. Hence it is important to obtain both the proton and neutron form factors since they provide complimentary $g_{\phi NN}$ information.

We now present our new predictions for $p(\pi^-, e^+e^-)n$ corresponding to our revised form factors. As described previously [1], our QHD formalism includes covariance, gauge invariance, crossing symmetry and duality constraints. This approach provides a reasonable framework for comprehensively understanding pseudoscalar meson electromagnetic production and radiative capture (see ref. [5] for further details). We first applied this model to π photoproduction data using Born amplitudes (i.e. N, π and ρ graphs) including final state absorption effects and phenomenological hadronic vertex factors. A reasonable description of the data was obtained for several energies which validates the model parameterization. Then without further approximation or adjustment, the $p(\pi^-, e^+e^-)n$ cross section can be predicted from crossing symmetry as shown in Fig. 4. Notice that even with the reduced ϕ contribution to the nucleon form factors there is still over a two order of magnitude enhancement in the cross section. The dashed curve is for $g_{\phi NN} = 0$ and documents the effect from the $\rho\pi\phi$ coupling due to t channel ρ-exchange between π and N. The smaller, dashed curve peak near 4° represents the conventional, expected ϕN production which constitutes a realistic ϕN background for assessing s and u channel production governed by $g_{\phi NN}$.

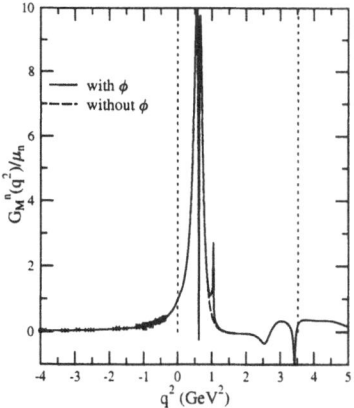

Figure 3. n magnetic form factor.

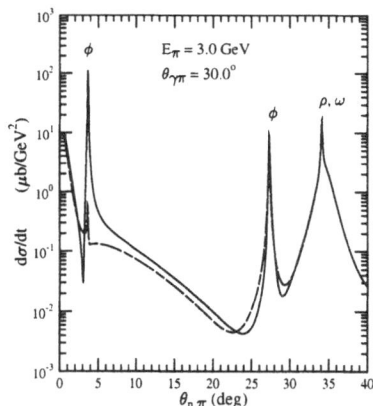

Figure 4. $p(\pi^-, e^+e^-)n$ predictions.

The novel dual peak structure follows from VMD and generates a sensitive energy and angular dependence. This should facilitate, through data analysis by diagrammatic VMD models, kinematically isolating ϕ production from the proton (u channel), ρ-exchange (t channel $\rho\pi\phi$ contribution) or the neutron (s channel). For example, in Fig. 4 u channel diagrams dominate since $s \gg u$, indicating such measurements will mainly provide proton form factor information. Similarly for these kinematics the $n(\pi^+, e^+e^-)p$ reaction should essentially provide neutron form factor information. This can be obtained from $d(\pi^+, e^+e^-)pp$ experiments. Related, we have also derived a high t theorem for the ratio of ϕ to ω peak cross sections, $R = d^3\sigma(q^2 = M_\phi^2)/d^3\sigma(q^2 = M_\omega^2) = g_{\phi NN}^2/g_{\omega NN}^2 f$, with the same s and t. Here f is a known kinematical factor and the triple cross section has only a single peak as a function of q^2. For our vector meson couplings we predict $R = .035f$ which agrees well with the actual numerical calculation of $R \simeq .04f$. Note this is roughly an order of magnitude larger than the OZI prediction $R = tan^2(\delta)f' = 4.2.10^{-3}f'$, where $\delta = 3.7°$ is the small deviation from ideal mixing of the u, d quark components in the ϕ and f' is another known factor of order unity.

Finally, it should be stressed that OZI evading kaon loops can also contribute to $g_{\phi NN}$. Interestingly, however, two independent analyses [6, 7] report such loop diagrams effectively cancel and ref. [7] calculates a remnant value $g^2_{\phi NN}/4\pi \simeq .0046$. This is substantially smaller than our value of .034 and if used in this analysis would result in a further reduction of the $4°$ resonance peak in Fig. 4 by almost an order of magnitude. Accordingly, should loop suppression prove valid, any experimental determination of $g^2_{\phi NN}/4\pi \gg .005$ would directly document either a significant OZI violation (with no nucleon strangeness) or, and more likely, an appreciable $s\bar{s}$ component in the nucleon.

3 Summary

We reaffirm that the $N(\pi, e^+e^-)N$ process is a viable method for obtaining the unknown nucleon time-like form factors in the low-lying vector meson region. Such information, especially when combined with improved high t ϕ photoproduction measurements, should enable a more definitive determination of $g_{\phi NN}$. Another related experiment would be Compton virtual scattering, $N(\gamma, e^+e^-)N$, in this same energy regime. Together, these complimentary measurements should provide significant insight into the strangeness content of the nucleon.

Acknowledgement. This work is supported by DOE grant DE-FG02-97ER41048 and NSF grant HRD-9633750.

References

1. R.A. Williams and S.R. Cotanch: Phys. Rev. Lett. **77**, 1008 (1996)

2. J. Ellis, E. Gabathuler and M. Karliner: Phys. Lett. **217B**, 173 (1989)

3. R.A. Williams: Phys. Rev. **C57**, 223 (1998)

4. R.A. Williams and C.P. Truman: Phys. Rev. **C53**, 1580 (1996)

5. R.A. Williams, C.-R. Ji and S.R. Cotanch: Phys. Rev. **D41**, 1449 (1990) Phys. Rev. **C43**, 452 (1991); *ibid.* **C46**, 1617 (1992); *ibid.* **C48**, 1318 (1993)

6. N. Isgur and P. Geiger: Phys. Rev. **D55**, 299 (1997)

7. Ulf-G. Meiβner et al.: Report No. JLAB-THY-97-02; hep-ph/9701296

Few-Body Systems Suppl. 10, 423–426 (1999)

Few-
Body
Systems
© by Springer-Verlag 1999

Bloom-Gilman duality of the nucleon structure function and the elastic peak contribution

S. Simula[1] *, G. Ricco, M. Anghinolfi, M. Ripani, M. Taiuti[2]

[1] Istituto Nazionale di Fisica Nucleare, Sezione Roma III, Italy
[2] Physics Dept., University of Genova and INFN, Sezione di Genova, Italy

Abstract. The occurrence of the Bloom-Gilman duality in the nucleon structure function is investigated by analyzing the Q^2-behavior of low-order moments, both including and excluding the contribution arising from the nucleon elastic peak. The Natchmann definition of the moments has been adopted in order to cancel out target-mass effects. It is shown that the onset of the Bloom-Gilman duality occurs around $Q^2 \sim 2 \ (GeV/c)^2$ if only the inelastic part of the nucleon structure function is considered, whereas the inclusion of the nucleon elastic peak contribution leads to remarkable violations of the Bloom-Gilman duality.

The investigation of inelastic lepton scattering off nucleon (and nuclei) can provide relevant information on the concept of parton-hadron duality, which deals with the relation among the physics in the nucleon-resonance and Deep Inelastic Scattering (DIS) regions. As is well known, well before the advent of QCD, local parton-hadron duality was observed empirically by Bloom and Gilman [1] in the proton structure function $F_2(x, Q^2)$ measured at $SLAC$ (where $x \equiv Q^2/2m\nu$ is the Bjorken scaling variable, m the nucleon mass and Q^2 the squared four-momentum transfer). More precisely, they found that the smooth scaling curve measured in the DIS region at high Q^2 represents a good average over the resonance bumps seen in the same x region at low Q^2.

In ref. [2] we addressed the specific question whether and to what extent the Bloom-Gilman duality already observed in the proton occurs also in the structure function of a nucleus. To that end all the available experimental data for the structure functions of proton, deuteron and light complex nuclei were analyzed in terms of low-order moments in the Q^2 range from 0.3 to 5 $(GeV/c)^2$. If only the inelastic parts of the structure functions are considered, we found

---

*E-mail address: simula@hpteo1.roma3.infn.it

that in case of the proton and the deuteron the Bloom-Gilman duality is fulfilled starting from $Q^2 \sim 2$ $(GeV/c)^2$, whereas in case of complex nuclei, despite the poor statistics of the available data, the onset of the local parton-hadron duality is clearly anticipated. Besides these interesting findings, we observed also that the inclusion of the contribution arising from the nucleon elastic peak leads to remarkable violations of the local parton-hadron duality for all the targets considered. In this contribution we present an improvement of the work of ref. [2] about the failure of local parton-hadron duality around the nucleon elastic peak.

Following the works of refs. [3, 4], the analysis of ref. [2] was carried out using the Cornwall-Norton definition of the moments, viz.

$$M_n^{(CN)}(Q^2) \equiv \int_0^1 d\xi \; \xi^{n-2} \; F_2(\xi, Q^2) \tag{1}$$

where $\xi \equiv 2x/[1 + \sqrt{1 + 4m^2x^2/Q^2}]$ is the Nachtmann variable. The Q^2 behavior of Eq. (1) was compared with the one of the moments $A_n^{(CN)}(Q^2)$ of the leading-twist (dual) structure function, viz.

$$A_n^{(CN)}(Q^2) \equiv \int_0^1 d\xi \; \xi^{n-2} \; F_2^{(dual)}(\xi, Q^2) \tag{2}$$

with

$$
\begin{aligned}
F_2^{(dual)}(\xi, Q^2) =\; & \frac{x^2}{(1 + \frac{4m^2x^2}{Q^2})^{3/2}} \frac{F_2^{(LT)}(\xi, Q^2)}{\xi^2} \\
& + 6\frac{m^2}{Q^2}\frac{x^3}{(1 + \frac{4m^2x^2}{Q^2})^2} \int_\xi^1 d\xi' \frac{F_2^{(LT)}(\xi', Q^2)}{\xi'^2} \\
& + 12\frac{m^4}{Q^4}\frac{x^4}{(1 + \frac{4m^2x^2}{Q^2})^{5/2}} \int_\xi^1 d\xi' \int_{\xi'}^1 d\xi'' \frac{F_2^{(LT)}(\xi'', Q^2)}{\xi''^2}
\end{aligned} \tag{3}
$$

where the ξ-dependence as well as the various integrals appearing in the r.h.s. account for target mass effects, which have to be included in order to cover the low Q^2 region [3]. In Eq. (3) $F_2^{(LT)}(x, Q^2)$ represents the leading-twist (LT) nucleon structure function, fitted to high Q^2 proton and deuteron data, and extrapolated down to low values of Q^2 by means of the Altarelli-Parisi evolution equations. In the DIS region one gets $F_2^{(LT)}(x, Q^2) = \sum_f e_f^2 x \left[\rho_f(x, Q^2) + \bar\rho_f(x, Q^2)\right]$, with $\rho_f(x, Q^2)$ being the quark distribution of flavor f.

However, Eqs. (1-3) suffer from a well known mismatch, because $\xi(x = 1) \equiv \xi_{max} = 2/(1 + \sqrt{1 + 4m^2/Q^2}) < 1$. Therefore, while the evaluation of the integral of Eq. (1) stops at the physical threshold $\xi_{max} < 1$, corresponding to the elastic end-point $x = 1$, the r.h.s. of Eq. (2) requires the values of the dual function (3) in the unphysical region $\xi_{max} \le \xi \le 1$. Consequently, the comparison of the experimental $M_n^{(CN)}(Q^2)$, obtained using all the available data sets and including the nucleon elastic peak contribution, with the dual moments $A_n^{(CN)}(Q^2)$ should be handled with care. Since the mismatch originates from target-mass corrections (i.e., from the fact that $m \neq 0$) we now adopt a different definition of the moments, which avoids completely the mismatch problem, namely we consider the Natchmann definition of moments [5], given by

$$M_n(Q^2) \equiv \int_0^1 dx \, \frac{\xi^{n+1}}{x^3} \, F_2(x, Q^2) \, \frac{3 + 3(n+1)r + n(n+2)r^2}{(n+2)(n+3)} \tag{4}$$

where $r \equiv \sqrt{1 + 4m^2 x^2/Q^2}$ ($\xi = 2x/(1+r)$). Since the r.h.s. of Eq. (4) projects out the contributions of spin-n operators [5], the Q^2 behavior of $M_n(Q^2)$ has to be compared with the one of the LT moments $A_n(Q^2)$, defined as

$$A_n(Q^2) \equiv \int_0^1 dx \, x^{n-2} \, F_2^{(LT)}(x, Q^2) \tag{5}$$

where the LT structure function $F_2^{(LT)}(x, Q^2)$ does not include any target-mass corrections by definition. Now, the evaluation of Eq. (5) requires the values of the structure function $F_2^{(LT)}(x, Q^2)$ only in the physical region $0 \leq x \leq 1$.

As pointed out in ref. [4] and described in ref. [2], the Bloom-Gilman duality should manifest itself in the dominance of the LT moments $A_n(Q^2)$ in the Q^2 behavior of low-order moments $M_n(Q^2)$ starting from a value $Q^2 \simeq Q_0^2$ almost independent of the order n of the moment (for high values of n the moments of the structure functions $F_2(x, Q^2)$ and $F_2^{(LT)}(x, Q^2)$ should differ because of the rapidly varying behavior of the nucleon-resonances peaks). In other words we expect that the low-order residual moments $\Delta M_n(Q^2) \equiv M_n(Q^2) - A_n(Q^2)$ become a small fraction of the LT moments $A_n(Q^2)$ for $Q^2 \gtrsim Q_0^2$. This is exactly what we see for $Q^2 \gtrsim Q_0^2 \sim 2 \, (GeV/c)^2$ in Fig. 1(a), where some low-order residual moments $\Delta M_n(Q^2)$ have been evaluated including only the inelastic data. However, when the contribution from the nucleon elastic peak is added, the dominance of $A_n(Q^2)$ starts from values of Q^2 which strongly depend upon n for $n > 2$. As for $n = 2$, since the second moment $M_2(Q^2)$ corresponds to the area of the structure function, the dominance of the LT second moment $A_2(Q^2)$ corresponds to the *global* parton-hadron duality, which holds with and without the inclusion of the nucleon elastic peak contribution (see Fig. 1). Our results imply that the parton-hadron duality holds for local averages of the nucleon structure function over the resonance bumps, but the elastic peak. Note that the applicability of the concept of local parton-hadron duality in the region around the nucleon elastic peak was found to be critical also in refs. [1, 4]. Finally, let us point out that results of the same quality as those shown in Fig. 1 hold as well in case of the deuteron structure function.

In conclusion, the occurrence of the Bloom-Gilman duality in the nucleon structure function has been investigated by analyzing the Q^2-behavior of low-order moments, both including and excluding the contribution arising from the nucleon elastic peak. The Natchmann definition of the moments has been adopted in order to cancel out target-mass effects. It has been shown that the

426

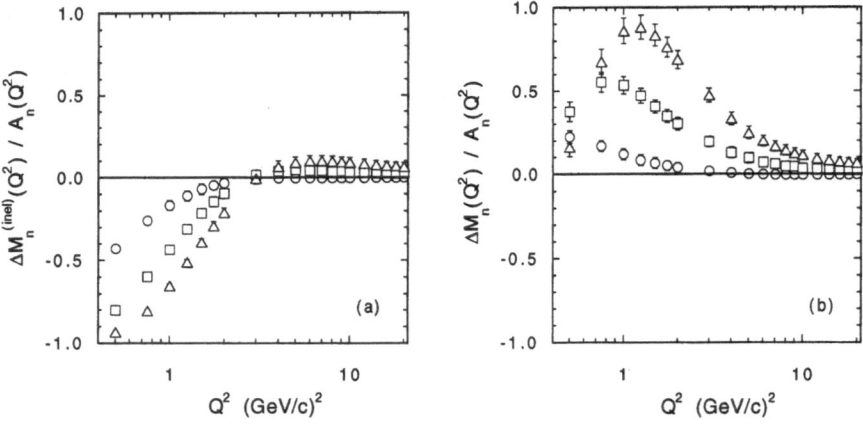

Figure 1. Ratio of the residual moments $\Delta M_n(Q^2) \equiv M_n(Q^2) - A_n(Q^2)$ to the LT moments $A_n(Q^2)$ (see Eqs. (4-5)) vs. Q^2. Dots, squares and triangles correspond to $n = 2, 4$ and 6, respectively. For the calculation of $A_n(Q^2)$ (Eq. (5)) the parton distributions of ref. [6] have been adopted. In (a) only the inelastic part of the proton structure function is considered, whereas in (b) also the proton elastic peak contribution is included in the determination of Eq. (4). For the experimental data set adopted for the evaluation of Eq. (4) see ref. [2].

onset of the Bloom-Gilman duality occurs around $Q^2 \sim 2 \ (GeV/c)^2$ if only the inelastic part of the nucleon structure function is considered, whereas the inclusion of the nucleon elastic peak contribution leads to remarkable violations of the Bloom-Gilman duality.

References

1. E. Bloom and F. Gilman: Phys. Rev. Lett. **25**, 1140 (1970); Phys. Rev. **D4**, 2901 (1971)

2. G. Ricco et al.: Phys. Rev. **C57**, 356 (1998)

3. H. Georgi and H.D. Politzer: Phys. Rev. **D14**, 1829 (1976)

4. A. De Rujula, H. Georgi and H.D. Politzer: Ann. of Phys. **103**, 315 (1977)

5. O. Natchmann: Nucl. Phys. **B63**, 237 (1973)

6. M. Gluck, E. Reya and A. Vogt: Z. Phys. **C53**, 127 (1992); **C67**, 433 (1995); H. Plothow-Besch: preprint /CERN-PPE/15-3-1995 (PDF Library) and Comp. Phys. Comm. **75**, 396 (1993)

Few-Body Systems Suppl. 10, 427–430 (1999)

Few-
Body
Systems
© by Springer-Verlag 1999

Front-form calculation of the deuteron EM form factors

V.A. Karmanov[*][1], J. Carbonell[†] [2]

[1] Lebedev Physical Institute, Leninsky Prospekt 53, 117924 Moscow, Russia
[2] Institut des Sciences Nucléaires, 53 av. des Martyrs, 38026 Grenoble, France

Abstract. The deuteron form factors are calculated in the framework of the relativistic nucleon-meson dynamics. The relativistic effects change considerably the S- and D-waves of the deuteron, result in the dominating extra component in the deuteron wave function and generate the contact interaction between the nucleon, mesons and photon. The prediction for the polarization observable t_{20} is in good agreement with the recent data of TJNAF/CEBAF.

1 Introduction

We calculate the deuteron form factors in the framework of the explicitly covariant version of light-front dynamics (see [1] for review). In this approach one can use in full measure the knowledge of the nonrelativistic wave functions and incorporate selfconsistently the relativistic effects. We assume that the nucleons interact with each other by exchanging the same mesons that are incorporated in the NN potential, however we don't use any nonrelativistic potential approximation.

2 Wave function

In the explicitly covariant formulation of light-front dynamics the wave function is defined in the space-time on the plane formally given by the light-front equation $\omega \cdot x = 0$, where ω is the four-vector with $\omega^2 = 0$. This restores the relativistic covariance in comparison to the standard light-front approach. The latter is obtained as a particular case for $\omega = (1, 0, 0, -1)$. The wave functions depend on the orientation of the light front, as they should. This dependence is covariantly parametrized in terms of the four-vector ω.

The explicit covariance allows to construct the general form of the light-front wave function for a system with given spin. The relativistic deuteron

---

[*]e-mail: karmanov@sci.lebedev.ru
[†]e-mail: carbonel@isn.in2p3.fr

wave function on the light front contains six spin components, in contrast to two components – S and D-waves – in the nonrelativistic case. Its general form was found in [2]. In our calculation we keep only the three dominating components f_1, f_2 and f_5.

The relativistic deuteron wave function was calculated in [3] by a perturbative way incorporating exact relativistic OBEP kernel found in light-front dynamics, and the nonrelativistic wave function as the starting point. For the kernel the set of six mesons $(\pi, \eta, \rho, \omega, \delta$ and $\sigma)$ and the parameters corresponding to the Bonn model [4] were used.

In the nonrelativistic region of k the components f_1 and f_2 found by this way turn into the usual S- and D-waves, whereas other components become negligible. However, starting from $k \approx 0.5$ GeV/c the component f_5 dominates over all other components, including f_1 and f_2.

The physical meaning of this dominating extra component has been clarified [5] by comparing, in $1/m$ approximation, the analytical expression for the amplitude of the deuteron electrodisintegration near threshold with the nonrelativistic one, including meson exchange currents. For the isovector transition, in the region, where the so called pair term with the pion exchange dominates, this component (together with a similar component in the scattering state) automatically incorporates 50% of the contribution of the pair term and therefore dominates too. Another 50% is given by the contact (instantaneous) interaction (see below).

3 Electromagnetic vertex

A peculiarity of the light-front dynamics is the presence of the so called instantaneous (or contact) interaction $NNB\gamma$, where B is the meson π, ρ, etc. This is a trace of disappeared diagrams corresponding to vacuum fluctuations. The fermion fields in this interaction are sandwiched with the factor $\hat{\omega} = \omega_\mu \gamma^\mu$ that in the standard approach is the well known γ^+ (in addition to the standard factors of the nucleon-meson Lagrangian, like γ_5, γ_μ, etc.). Therefore the full electromagnetic vertex which we use to calculate the deuteron form factors corresponds to the sum of the impulse approximation and of the contact interaction. It partially takes into account the many-body currents. For the contact interaction we take into account the sum over all six mesons contributing to the Bonn potential, with the parameters of the Bonn model [4].

4 Finding form factors

The next step is to extract the deuteron form factors from the electromagnetic vertex. The physical amplitudes should not depend on the orientation of the light-front plane. However in practice, because of the incompatibility between the transformation properties of the approximate e.m. current and the wave function, the nonphysical ω dependence survives in the on-energy shell deuteron electromagnetic vertex. Due to covariance, the general form of this dependence can be found explicitly and then separated from the physical form factors. The explicit formulas extracting the physical form factors $\mathcal{F}_1, \mathcal{F}_2, \mathcal{G}_1$ from the nonphysical contributions were found in refs. [6].

The polarization observable t_{20} measured in experiment is a combination of the form factors (see e.g. [1]). Besides momentum transfer it depends also explicitly on $\tan^2 \frac{1}{2}\theta$.

5 Numerical results and conclusions

Our prediction [1, 7] for the polarization observable t_{20} together with the most recent measurements are shown in figure 1. The calculation is in close agreement with the existing experimental data in all the momentum range, including the new results obtained at TJNAF/CEBAF [8] which corresponds to $\theta = 70°$.

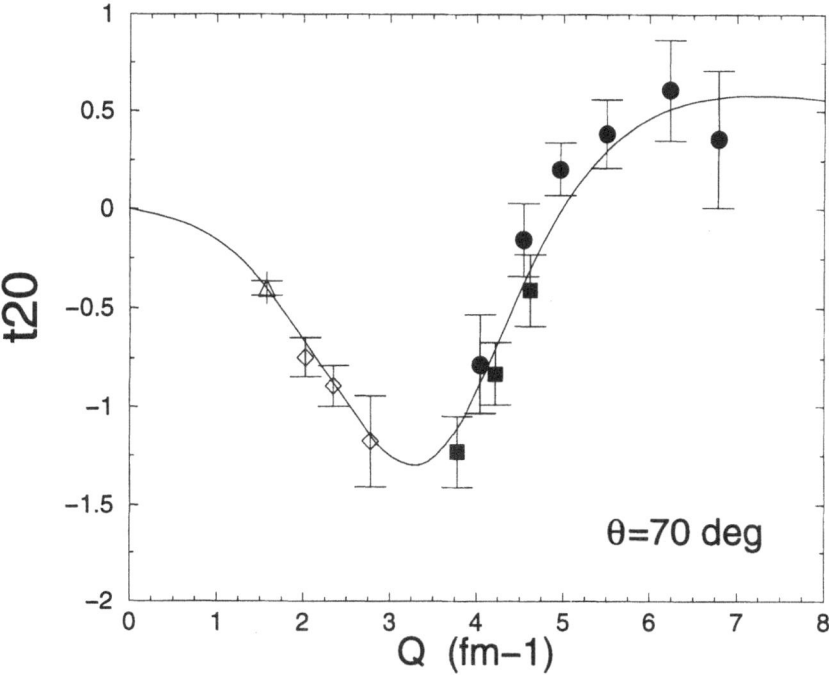

Figure 1. The deuteron tensor polarization t_{20} at $\theta = 70°$ compared to most recent experimental data: triangle from [9], diamonds from [10], squares from [11], circles from [8]

This coincidence means that the deuteron structure at small distances is understood rather well. It is given by the relativistic nucleon-meson dynamics. In its turn, the relativistic nucleon-meson dynamics is well described theoretically, in the framework of the phenomenological nucleon-meson Lagrangian. The parameters of this Lagrangian are found phenomenologically. Their calculation in the framework of QCD is a separate problem.

This allows us to make the general conclusion that: *in a large extent, the relativistic nuclear physics can be developed independently of its derivation from QCD.*

However, these results can be improved in many respects. An exact solution of the equation for the deuteron is indeed feasible. It implies the redefinition of the parameters of the NN kernel in the framework of the light-front equations. This necessitates also a careful treatment of higher order contributions to the kernel. Finally, the calculation of electromagnetic observables should include higher Fock states ($NN\pi$, etc.) and the meson exchange currents corresponding to the direct interaction of the photon with the intermediate mesons.

Acknowledgement. The authors are sincerely grateful to B. Desplanques and J.-F. Mathiot for many useful discussions, valuable remarks and helpful advices.

References

1. J. Carbonell et al.: Phys. Reports **300**, 215 (1998)

2. V.A. Karmanov: Nucl. Phys. **A362**, 331 (1981)

3. J. Carbonell and V.A. Karmanov: Nucl. Phys. **A581**, 625 (1995)

4. R. Machleidt, K. Holinde and Ch. Elster: Phys. Reports **149**, 1 (1987)

5. B. Desplanques, V.A. Karmanov and J.-F. Mathiot: Nucl. Phys. **A589**, 697 (1995)

6. V.A. Karmanov and A.V. Smirnov: Nucl. Phys. **A546**, 691 (1992); **A575**, 520 (1994)

7. J. Carbonell and V.A. Karmanov: Preprint ISN-98.54, to be published

8. S. Kox et al.: European Conf. on El.-Mag. Int. with Nucleons and Nuclei, Santorini (Greece), October 1997;
   E.J. Beise et al.: 16th European Conf. on Few-Body Problems in Physics, Autrans (France), June 1998

9. M. Ferro-Luzzi et al.: Phys. Rev. Lett. **77**, 2630 (1996)

10. M. Ferro-Luzzi et al.: Nucl. Phys. **A631**, 190 (1998)

11. I. The et al.: Phys. Rev. Lett. **67**, 163 (1991)

Few-Body Systems Suppl. 10, 431–434 (1999)

Few-
Body
Systems
© by Springer-Verlag 1999

Measurements of the Deuteron Elastic Structure Function $A(Q^2)$ at the Jefferson Laboratory

J. Gomez* †

Thomas Jefferson National Accelerator Facility, Newport News, Va, USA

Abstract. The deuteron elastic structure function $A(Q^2)$ has been extracted in the range $0.7 \leq Q^2 \leq 6.0$ $(\text{GeV/c})^2$ from cross section measurements of elastic electron-deuteron in coincidence.

Measurements of the elastic deuteron electromagnetic form factors offer unique opportunities to test models of short-range aspects of the nucleon-nucleon interaction, meson-exchange currents, isobaric configurations and, quark degrees of freedom. The elastic electron-deuteron cross section is given by $d\sigma/d\Omega = \sigma_M \left[A(Q^2) + B(Q^2) \tan^2(\theta/2) \right]$ where θ is the electron scattering angle, $\sigma_M = \alpha^2 E' \cos^2(\theta/2)/[4E^3 \sin^4(\theta/2)]$ is the Mott cross section, α is the fine-structure constant, E and E' are the incident and scattered electron energies and $Q^2 = 4EE' \sin^2(\theta/2)$ is the four-momentum transfer squared. The deuteron elastic structure functions $A(Q^2)$ and $B(Q^2)$ are given in terms of the charge, quadrupole and magnetic form factors $F_c(Q^2)$, $F_q(Q^2)$ and $F_m(Q^2)$ by $A(Q^2) = F_c^2(Q^2) + (8/9)\tau^2 F_q^2(Q^2) + (2/3)\tau F_m^2(Q^2)$ and $B(Q^2) = (4/3)\tau(1+\tau)F_m^2(Q^2)$ with $\tau = Q^2/4M_d^2$. M_d is the deuteron mass. The aim of the experiment reported here was to extend the previously measured kinematical range of $A(Q^2)$ and to resolve inconsistencies in previous data sets [1, 2, 3] by measuring elastic electron-deuteron (e-d) cross sections for $0.7 \leq Q^2 \leq 6.0$ $(\text{GeV/c})^2$.

The experiment was conducted in one of the experimental areas (Hall A) of the Thomas Jefferson National Accelerator Facility (JLab). Incident electron beams with energies from 3.2 to 4.4 GeV, currents from 5 to 120 μA and, 100% duty-factor were used in this experiment. Beam current and energy uncertainties were estimated to be $\pm 2\%$ and $\pm 0.2\%$, respectively. Uncertainties due to beam position and angle at the target are negligible. The target system consisted of two 15 cm long cylindrical cells: one filled with liquid hydrogen, the other with liquid deuterium. Measured beam-induced density changes were

---

*Representing the JLab Hall A collaboration.
† E-mail address: gomez@jlab.org

~2% at 120 μA. A 15 cm long "empty" target was used to measure possible contributions from the full cell end-caps to the measured cross sections. They were found to be negligible.

The Hall A experimental facility at JLab consists of two, magnetically identical, QQDQ High Resolution Spectrometers (HRS) of 4 GeV/c maximum momentum. One HRS deflected negatively charged particles into the focal plane (electron HRS) while the other was set for positive particles (recoil HRS). Both HRS have two planes of plastic scintillators for triggering and timing and a pair of drift chambers for track reconstruction. In addition, the electron HRS has a gas Čerenkov and a Pb-glass calorimeter for electron identification. The trigger logic was set to accept all electron-recoil spectrometer coincidences as well as samples of single-arm triggers (for detector efficiency studies). The coincidence trigger efficiency ranged from 98% to 100%.

Coincidence elastic electron-proton (e-p) cross sections were measured in this experiment to check our understanding of spectrometer optics and double-arm acceptance. The e-p kinematics were selected such that the electron-recoil solid angle jacobian for e-p was the same as for e-d. The e-p data were taken with and without solid-angle defining collimators in front of the spectrometers.

In the data analysis, electron events were required to have a minimum pulse height in the Čerenkov counter and energy deposition in the calorimeter consistent with the momentum determined from the drift chamber track. Coincident events were identified using the relative time-of-flight (TOF) between the electron and recoil triggers. Contributions from random coincidences were in general negligible.

The elastic e-p and e-d cross sections were calculated according to $d\sigma/d\Omega = N_{ep(ed)} C_{eff}/[N_i N_t F \, \Delta\Omega]$ where $N_{ep(ed)}$ is the number of e-p(e-d) elastic events, N_i is the number of incident electrons, N_t is the number of target nuclei/cm^2, $\Delta\Omega$ is the effective double-arm solid angle including the spectrometer acceptance dependent part of the radiative corrections, F is the portion of the radiative corrections that depends only on Q^2 and target thickness, and C_{eff} is the product of corrections such as detector and trigger inefficiency in both electron and recoil spectrometers (1-3%), computer dead time (typically 5%) and, proton (~2%) and deuteron (~4%) absorption losses in the target and detectors.

The effective double-arm solid angle $\Delta\Omega$ was evaluated with a Monte Carlo computer program that simulated elastic e-p and e-d scattering under identical conditions as our measurements. The program ray-traced scattered electrons and recoil nuclei from the target to the detectors through models representing the magnetic characteristics, physical apertures and alignment of each HRS. The effects from ionization energy losses and multiple scattering in the target and vacuum windows were taken into account for both electrons and recoil nuclei. Bremsstrahlung radiation losses for both incident and scattered electrons in the target and vacuum windows as well as internal radiative effects were also taken into account. Details on this simulation method can be found in ref. [4].

The measured elastic e-p cross sections, with and without collimators, agree within ±6% with the values calculated from a recent parametrization [5] of

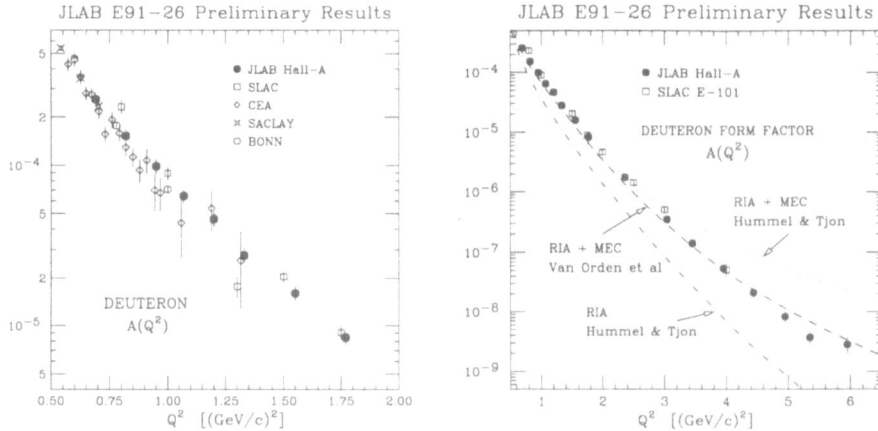

Figure 1. The left panel shows our data in the "low" Q^2 region. The previous measurements tend to show two long-standing diverging trends, one supported by the SLAC data [2] and the other by the CEA [1] and Bonn [3] data. Our data confirm the trend of the SLAC data. The right panel shows all of our data together with previous SLAC data. The two data sets agree well in the range of overlap. Our data continue to exhibit a smooth fall-off with Q^2. Theoretical calculations by Van Orden, Devine and Gross (VDG) [6] and Hummel and Tjon (HT) [7] are also shown. In the HT case, relativistic impulse approximation (RIA) calculations with and without meson-exchange currents (MEC) are shown. Clearly, at large Q^2, the RIA calculation alone lacks enough strength to account for the data, and the model becomes very sensitive to the inclusion of MEC. In the HT model, the $\rho\gamma\pi$ and $\omega\varepsilon\gamma$ MEC are included with form factors given by the Vector Dominance Model (VMD). Although not shown, the VDG model has a similar behavior: the RIA alone lacks enough strength, and inclusion of a $\rho\gamma\pi$ MEC with VMD form factors overshoots the data. The VDG model shown includes a $\rho\gamma\pi$ MEC with form factors given by quark models [8, 9].

proton world data. Values of $A(Q^2)$ were then extracted from the measured e-d cross sections under the assumption that $B(Q^2)$ does not contribute in any sizeable way to the cross sections (supported by the existing $B(Q^2)$ data). The extracted $A(Q^2)$ values are presented in Figs. 1 and 2. The error bars represent statistical and systematic uncertainties added in quadrature. The statistical error ranged from $\pm 1\%$ to $\pm 30\%$. The systematic error has been estimated to be $\sim \pm 8\%$ and is dominated by the uncertainty in the double-arm solid angle ($\pm 6\%$).

In summary, we have measured the elastic deuteron structure function $A(Q^2)$ in the range $0.7 \leq Q^2 \leq 6$ $(GeV/c)^2$. The results have clarified inconsistencies in previous low Q^2 data. The precision of our data will provide severe constraints on theoretical calculations of the electromagnetic structure of the two-body nuclear system. The results are consistent with meson-nucleon calculations based on the relativistic impulse approximation augmented by meson-exchange currents (although with softer form factors than those ob-

434

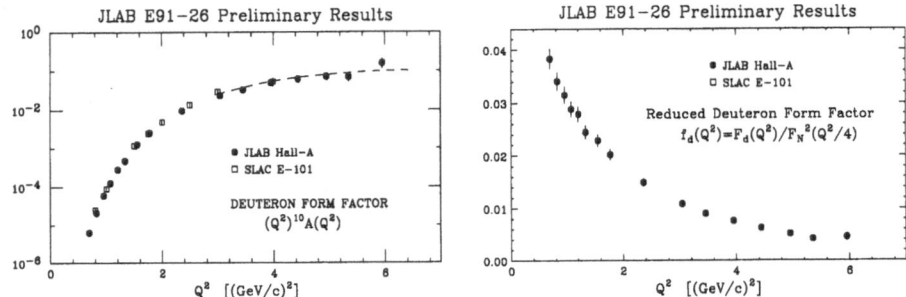

Figure 2. Deuteron models based on dimensional scaling [10, 11] and perturbative QCD [12] expect the "deuteron form factor" $F_d(Q^2)$ ($\equiv \sqrt{A(Q^2)}$) to fall as $(Q^2)^{-5}$. Consequently, the quantity $A(Q^2) \times (Q^2)^{10}$ should scale. Our data exhibits a scaling behavior compatible with those expectations (left panel). The right panel shows values for the "reduced" deuteron form factor $f_d(Q^2) \equiv F_d(Q^2)/F_N^2(Q^2/4)$ where the two powers of the nucleon form factor $F_N(Q^2) = (1 + Q^2/0.71)^{-2}$ remove in a minimal way the effects of nucleon structure [13].

tained from the VMD model). The results are also consistent with predictions of dimensional quark scaling and perturbative QCD. Future $A(Q^2)$ and $B(Q^2)$ measurements would be crucial for testing the apparent scaling behavior at large Q^2.

References

1. J.E. Elias et al.: Phys. Rev. **177**, 2075 (1969)

2. R.G. Arnold et al.: Phys. Rev. Lett. **35**, 776 (1975)

3. R. Cramer et al.: Z. Phys. **C29**, 513 (1985)

4. A.T. Katramatou et al.: Nucl. Instr. Meth. **A267**, 448 (1988)
   A.T. Katramatou: SLAC-NPAS-TN-86-08, 1986 (unpublished)

5. P.Y. Bosted: Phys. Rev. **C51**, 409 (1995)

6. J.W. Van Orden, N. Devine and F. Gross: Phys. Rev. Lett. **75**, 4369 (1995)

7. E. Hummel and J.A. Tjon: Phys. Rev. **C42**, 423 (1990)

8. K.L. Mitchell: PhD thesis, Kent State University, 1995 (unpublished)

9. F. Cardarelli et al.: Phys. Lett. **B359**, 1 (1995)

10. S.J. Brodsky and G.R. Farrar: Phys. Rev. Lett. **31**, 1153 (1973)

11. V. Matveev et al.: Nuovo Cimento Lett. **7**, 719 (1973)

12. S.J. Brodsky et al.: Phys. Rev. Lett. **51**, 83 (1983)

13. S.J. Brodsky and B.T. Chertok: Phys. Rev. Lett. **37**, 269 (1975)

Few-Body Systems Suppl. 10, 435–438 (1999)

Few-
Body
Systems
© by Springer-Verlag 1999

The deuteron isobaric component and the neutron charge form factor

A. Amghar[1], B. Desplanques[2]

[1] Dépt des Sciences Fondamentales, INHC, 35000 Boumerdes, Algeria
[2] ISN, 53 Avenue des Martyrs, F-38026 Grenoble Cedex, France

Abstract. The contributions to the deuteron charge form factors due to exchange currents involving the $\Delta\Delta$ component with π and ρ exchange are studied. They originate from the usual pair current, but at the quark level. They are found to provide a substantially large contribution to the deuteron charge form factor. In this case, it comes just second in order of importance after that of the pion pair current. Taking into account this new contribution implies a modification of the neutron charge form factor, G_E^n, that has been derived from the measurement of the structure function, $A(q^2)$. The previous fit of this form factor is improved, what could be further checked in a near future, using accurate determination of $A(q^2)$ soon available at higher values of q^2.

At high momentum transfers, contributions of exchange currents due to the excitations of the nucleon to higher energy states should be considered. They were not taken into account in previous studies [1, 2, 3, 4, 5] because they were thought to be negligible. In this study, we show that some of them are far from being small. For moderate momentum transfers, excitations that may have some relevance are essentially the resonance $\Delta(1232\text{MeV})$ and the Roper resonance $N^*(1440\text{MeV})$. Some of these new contributions have an origin similar to those coming from usual exchange currents (pair current of π, ρ and ω and the coupling $\rho\pi\gamma$), but at the quark level, making possible a modification of the spin and isospin structure of the nucleon.

In this study, we concentrate on a contribution involving the Δ excitation with π- and ρ-exchanges (Fig. 1), respectively denoted $\Delta\Delta(\pi)$ and $\Delta\Delta(\rho)$. The corresponding electromagnetic current due to the π- and ρ-exchanges are determined in the constituent quark model [6]. The expression of the isoscalar contribution of the first one to the charge density is given by:

$$\rho^{\Delta\Delta(\pi)} = \frac{G_M^S(\boldsymbol{q}^2)}{M_N}\left(\frac{f_{\pi N\Delta}}{m_\pi}\right)^2 \frac{\boldsymbol{S}_1 . \boldsymbol{q}\ \boldsymbol{S}_2 . \boldsymbol{k}_\pi}{\boldsymbol{k}_\pi^2 + m_\pi^2}\ K_{\pi N\Delta}^2(\boldsymbol{k}_\pi^2)\ \boldsymbol{T}_1 . \boldsymbol{T}_2 + 1 \leftrightarrow 2. \quad (1)$$

We notice that the above expression of $\rho^{\Delta\Delta(\pi)}$ is analogous to that of the usual

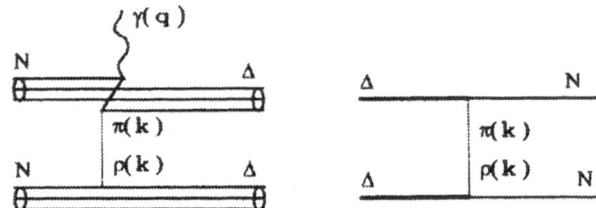

Figure 1. Representation of the total meson exchange contribution involving the $\Delta\Delta$ component in the intermediate state with the meson exchange current part on the left and the transition potential on the right

pion pair current. The spin and isospin Pauli matrices appearing in this one, σ and τ, are respectively replaced by transition matrices from states with spin or isospin $1/2$ to states with spin and isospin $3/2$, S and T [1]. The contribution due to the ρ-exchange, $\Delta\Delta(\rho)$, is obtained from Eq. (1) using standard changes.

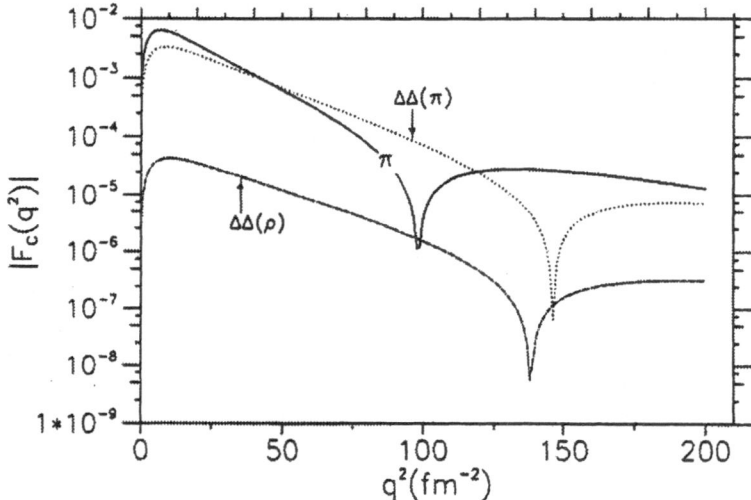

Figure 2. The contribution of $\Delta\Delta(\pi)$ and $\Delta\Delta(\rho)$ to the deuteron charge form factor. The Paris deuteron wave function is used

In Fig. 2, we show the contribution to the deuteron scalar charge form factor due to the $\Delta\Delta(\pi, \rho)$ component . The result so obtained can be compared to the contribution of the pion pair current. This last one is generally considered as the most important contribution to the exchange currents. We notice, that the $\Delta\Delta$ contribution with exchange of the pion is very important. It competes with that of the pion pair current at low momentum transfers and becomes more important beyond a squared momentum transfer of 50fm^{-2}. This is a direct consequence of the short range nature of the deuteron $\Delta\Delta$ component. As this one cannot be considered as well determined, the results for the largest momentum transfers should be taken with some caution. The $\Delta\Delta$ contribution to the deuteron charge form factor with ρ-exchange is much less important than that due to $\Delta\Delta(\pi)$, due to the large mass of this meson. This contribution cannot have any significant effect except at very high momentum transfer, what is due to the very short range character of its contribution. For the quadrupole form factor, the effect of these new contributions is rather small, and essentially concentrated at low momentum transfer [6].

The contributions of the Roper $N^*(1440\text{MeV})$ and of the $\rho\pi\gamma$ coupling with the excitation of the NN component to the $\Delta\Delta$ one have also been calculated [6, 7, 8], but their effect is negligible compared to the $\Delta\Delta(\pi)$ contribution.

The aim of the present study is to have a better determination of the neutron charge form factor, G_E^n, which remains poorly known until now. In absence of neutron target, the deuteron is used. This determination of G_E^n with a good precision is however limited by the experimental difficulties (relative errors on G_E^n can reach 20% at $q^2 = 15\text{fm}^{-2}$ [5]) and also by uncertainties in theoretical calculations. Since G_E^n is extracted from electron-deuteron elastic scattering, more precisely from the structure function $A(q^2)$ [5], a modification of this one by the new contributions of isobaric exchange currents induces a variation in the neutron charge form factor G_E^n [5]. This variation in G_E^n is given by the equation:

$$\frac{\delta G_E^n}{G_E^S} \simeq \frac{\delta A(q^2)}{2\,A(q^2)}. \tag{2}$$

We suppose that the proton charge and magnetic form factors, and the neutron magnetic form factor are well known. The newly determined neutron charge form factor is given by:

$$G_E^{n\,(\text{new})} = G_E^{n\,(\text{old})} - \delta G_E^n. \tag{3}$$

Beside the new contributions due to isobaric exchange currents, we may also correct for off-energy shell effects when determining G_E^n [6, 7, 9]. It follows that the analysis of Saclay's experimental results [5] using Paris wave function, see Fig. 3a, should be modified. Thus the value of G_E^n is increased for momentum transfers between 0 and 14 fm^{-2} and reduced outside the above interval. In Fig. 3b, we show results concerning G_E^n obtained in the present study. It is clearly shown that a fit better than that of reference [5] is achieved between our calculation and experimental data. The solid line in Fig. 3b is the new fit obtained for the neutron charge form factor with fit parameters $a = 1.25$ and $b = 14$ (Platchkov parametrization [5] is adopted). The value of $\chi^2/43$ is 39.3 for this specific fit. The dashed line represents the new fit using Galster parametrization with a parameter $p = 9$ [10]. The value of $\chi^2/43$ in the latter case is around 47.8.

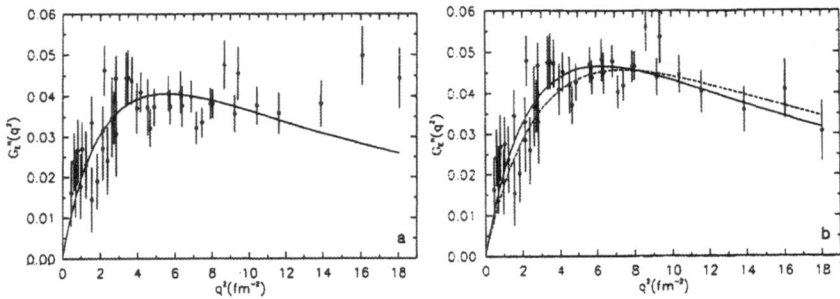

Figure 3. Neutron charge form factor, G_E^n: earlier fit (a), with Platchkov's parametrization [5] and new fit (b) with Platchkov's parametrization (solid line) and Galster's parametrization (dashed line) [10]. The data are taken from [5]

The disposal soon of new precise measurements of $A(q^2)$ at higher momentum transfer [11, 12] should allow one to extent the present analysis and see whether the better fit we got for G_E^n is founded. Indeed, this result essentially relies on the two last points in Fig. 3. On the other hand, the first measurements of this quantity using polarized electrons evidence a large spread [13], but support to some extent the increase of the fitted form factor found here.

References

1. M. Gari and H. Hyuga: Nucl. Phys. **A264**, 409 (1976)

2. H.J. Weber and H. Arenhövel: Phys. Reports **36**, 277 (1978)

3. G.H. Niephaus et al.: Phys. Rev. **C20**, 1096 (1979)

4. M. Hyuga and M. Gari: Nucl. Phys. **A274**, 333 (1976)

5. S. Platchkov et al.: Nucl. Phys. **A510**, 740 (1990)

6. A. Amghar et al.: Eur. Phys. J. **A1**, 85 (1998)

7. A. Amghar: Thèse de Doctorat de l'UJF, Grenoble I, (1993)

8. N. Aissat: Thèse de magistère de l'USTHB, Alger (1997)

9. A. Amghar and B. Desplanques: Nucl. Phys. **A585**, 657 (1995)

10. S. Galster et al.: Nucl. Phys. **B32**, 221 (1971)

11. J. Gomez: contributed paper presented at this conference

12. E.J. Beise: plenary talk given at this conference

13. J. Jourdan: review paper presented at INPC98, Paris, 24-28 August (1998)

Few-Body Systems Suppl. 10, 439–442 (1999)

New Tensor Force

V.I. Kukulin[1]*, V.N. Pomerantsev[1]*, S.G. Cooper[2], R. Mackintosh[2]

[1] Institute of Nuclear Physics, Moscow State University, Moscow 119899, Russia

[2] Physics Department, Open University, Milton Keynes, UK

Abstract. The tensor interactions between composite particles are studied by examples of $d + {}^4\text{He}$, $d + d$ and N + N considered as three-quark clusters. We discovered that the tensor mixing of nodal S-wave and nodeless D-wave has a few new interesting features which can explain some long-standing puzzles in the field.

1 Introduction

To date, two main types of tensor interactions have been studied. The first is the well known NN tensor force, mediated by one-pion or one-rho exchange, which results in deuteron D-state admixture and mixing in NN scattering. Another well studied type is a tensor interaction between deuterons or ${}^6\text{Li}$ and other nuclei. Our main interest here is tensor interactions between composite particles. Such interactions have many specific features caused by particle exchange and/or excitation effects. We will show that both these effects may result in very specific tensor mixing in such systems as $d + d$, $d + {}^4\text{He}$, ${}^6\text{Li} + {}^4\text{He}$ etc. and also in the NN system if the nucleons are treated as composite particles, i.e. three-quark clusters. It is most intriguing that one observes in all these systems one particular type of tensor mixing, i.e. $2S - 2D$ mixing (in h.o. notation) as will be explained below. This type of mixing is characterised by specific features unlike traditional tensor force effects. One such feature is the (relatively) large contribution of the short-range part of the interaction. This short range tensor force determines the sign of the ratio of asymptotic constants $\rho_2 = A_D/A_S$ for the S- and D-components of a bound state wave function and also the negative quadrupole moment in ${}^6\text{Li}$[1]. Using this new pattern for tensor mixing it is possible to resolve a recent dispute between two groups[2] about large space shell-model calculations.

*This work was supported in part by the Russian Foundation for Basic Research (grant no.97-02-17265

2 $2S - 2D$ Tensor Mixing and its Specific Features

We will discuss the new type of mixing by means of a model which can explain the long standing puzzle of the *negative* quadrupole moment Q of the ^{6}Li ground state. As many groups have found, it is difficult to explain the well established small *negative* value for the ^{6}Li quadrupole moment by large space calculations within the three-body model $\alpha + n + p$ which takes account of total antisymmetrization of six nucleons. All such calculations carried out with a large basis[1] give a large *positive* Q, even exceeding the deuteron quadrupole moment.

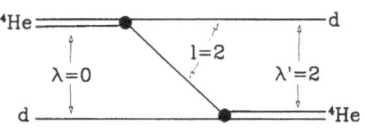

Figure 1. The d-exchange graph illustrating Majorana tensor force.

To resolve this puzzle it was conjectured [3] that this negative quadrupole moment arises from the D-component of the ^{4}He-cluster constituting ^{6}Li by means of d-exchange of "inner" deuteron (from α-cluster) and the "valence" deuteron (moving outside the ^{4}He core) – see Fig. 1.

It is well known that $l = 2$ component of ^{4}He with $S = 2$ and Young tableaux $f = [22]_x$ is composed from two deuterons with parallel spins and relative motion angular momentum 2. Further, due to the r^2 factor in the Q operator, the main contribution to Q comes from the peripheral region. Thus negative Q implies the opposite signs of asymptotic tails of S- and D-wave function components. This means, in turn, that the asymptotic ratio $\rho = A_D/A_S$ is also negative, which seems to be confirmed by some experiments although others appear to favour *positive* ρ and to date there is no final answer [4].

Nevertheless one can find [1] a tensor force which quantitatively describes the negative ^{6}Li quadrupole moment, the negative asymptotic mixing ratio ρ and even the energy behaviour of $^3S_1 - ^3D_1$ mixing parameter $\varepsilon_1(E)$ in $d + {}^4$He system. The potential model has a long range central component and a short range tensor force (see Table II in ref. [1]). The conventional $d - {}^4$He tensor force would have much the same radius as the central force. The short range strong tensor force most likely arises from the d-exchange effect depicted in Fig. 1 due to the small radial scale of a α-particle and its high binding energy in $d + d$ channel (~ 22.4 MeV). There are additional strong arguments in favour of just this exchange mechanism responsible for the tensor mixing in the ^{6}Li ground state: (i) The asymptotic mixing ratio ρ_2 in ^{4}He for $d + d$ channel is also negative [4] and this minus sign in the $d + d$ D-wave component of ^{4}He should pass to ^{4}He - d relative motion by d-exchange; (ii) the second from refs. [2] showed that Q for ^{6}Li is insensitive to the ^{6}Li binding energy in the $\alpha - d$ channel. This would be anomalous for a normal tensor force (cf. the NN system where the deuteron quadrupole moment is extremely sensitive to the deuteron binding energy). Noting that the S-wave component of the ^{6}Li ground state has a node (it is $2S$) while the D-wave component is nodeless ($2D$) and noting

---

[1]There have been a few RGM-like six-nucleon calculations on a very restricted basis whose authors claimed also the agreement for ^{6}Li quadrupole moment. However, due to the very restricted basis of such calculations, their results are by no means definitive.

that the range of the short range tensor force is less than the $2S$-node position (~ 2 fm), we conclude that there is a specific tensor mixing in this case. If the tensor force with radius $r_T < r_n$ is attractive, S- and D-components have the same sign (negative) at short $\alpha - d$ distances $r < r_n$ and thus, at short distances, the mixing character is quite similar to the NN case.

However, the $2S$-wave function changes its sign at $r > r_n$ and becomes positive in the asymptotic region while the nodeless $2D$-component remains negative. The interplay between the asymptotic signs of S- and D-components was checked by increasing the range of the tensor force. We were unable to fit Q, ρ_2 and ε_1 simultaneously with the larger r_T values. Thus, at least for the ^{6}Li ground state and ^{4}He - d low energy scattering, the short-range tensor force model with specific $2S - 2D$ tensor mixing *quantitatively* explains the experimental data. Unfortunately uncertainties remain because we do not yet know what local tensor force approximates the short range *nonlocal d-exchange* tensor force. Thus this requires further study. By contrast to the even parity tensor force, that for odd parity has been found to be much weaker and longer ranged. The tensor force in the ^{4}He - d system is therefore, like other exchange forces, *strongly parity dependent*.

Another place to study this $2S - 2D$ tensor mixing is the ^{4}He D-state component where the asymptotic mixing parameter ρ_2 is also negative. This negative ρ_2-value is anomalous because here the tensor mixing arises from the conventional NN tensor force where $\rho_2 > 0$ for deuterons. The negative ρ_2 in the D-component of ^{4}He has still not been interpreted (to the authors' knowledge). Our interpretation here is based on the observation that the dominating shell model configurations in ^{4}He are: $\alpha|(0s)^4[4]_x, LST = 000 > +\beta|(0s)^2(1p)^2[4]_x, LST = 000 > +\gamma|(0s)^2(1p)^2[22]_x, LST = 220 > +...$ where the first term includes zero h.o. quanta of excitation ($0\hbar\omega$) while two other terms correspond to the two-quanta excitations ($2\hbar\omega$). When rewritten as $d+d$ relative motion wave functions, the three components correspond to nodeless $0S$, nodal $2S$ and nodeless $2D$ wave functions respectively. If we switch on the NN tensor force, the strongest tensor mixing should occur between the second and third terms while the mixing between the first and third terms is likely to be much less. This is because the mixing of second and third terms occurs in the same $2\hbar\omega$ shell while the mixing of the first and third terms includes *inter-shell* mixing and so is smaller. Hence in this case we also have $2S - 2D$ tensor mixing. By analogy with the $d - ^4$He case we could conclude that the well known negative asymptotic ratio in ^{4}He is also a consequence of an *exchange interaction* between two deuterons in ^{4}He in the states with relative angular momenta $L = 2$ and parallel spins and $L = 0$ with anti-parallel spins.

3 Application to the NN System

The traditional approach to the NN system involves tensor mixing between nodeless S- and nodeless D-wave deuteron components. Also the range of the OPE tensor force (μ_π^{-1}), which is attractive, is the same as for the central part.

442

As a result, the asymptotic ratio ρ_2 in the deuteron is positive as is Q. However, the tensor force contribution in the deuteron and in low-energy NN scattering (in ε_1-behaviour) is still a little bit too low; e.g. the deuteron quadrupole moment Q as predicted by current potentials is too low. However, there is another model for NN interaction, known in the literature as the Moscow NN potential model [5]. In this model, certain orthogonality conditions to symmetric six-quark bag states give an additional short range node in the S-wave functions in place of the usual repulsive core, while the D-wave function has no node, another case of $2S - 2D$ tensor mixing.

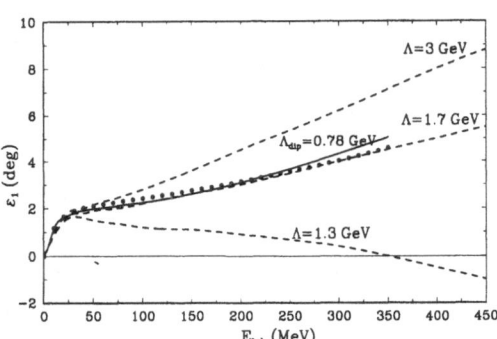

Figure 2. The ε_1 energy dependence for the conventional and Moscow NN potentials.

We now show that the $2S - 2D$ tensor mixing can consistently explain the behaviour of ε_1. In fact, the conventional NN potential models, because of the insufficient tensor force contribution mentioned above, are obliged to use rather unrealistically high values of the cut-off parameter $\Lambda \simeq 1.3 \div 1.7$ GeV/c to describe D-wave observables like Q, ρ_2 and ε_1. In the Moscow model, however, due to a different interference between the central and tensor parts of the interaction, a much smaller and quite realistic $\Lambda \simeq 0.7 \div 0.8$ GeV/c provides a good description for ε_1-behaviour in NN-scattering, see Fig. 2. In this figure we compare the ε_1-behaviour for different NN-potential models both traditional (dashed lines) and Moscow type (solid line). Note that the traditional NN-model predictions displayed in Fig. 2 employ a monopole cut-off factor for OPE-contribution while our model uses the dipole factor and $\Lambda_{\text{dipole}} = \sqrt{2}\Lambda_{\text{monopole}}$, so that if they would employ dipole type truncation the traditional models must use even larger Λ values.

References

1. S.G. Cooper et al.: Phys. Rev. **C57**, 2462 (1998)

2. A. Csoto, R.G. Lovas: Phys. Rev. **C53**, 1444 (1996); D.C. Zheng, J.P. Vary, B.R. Barrett: Phys. Rev. **C53**, 1447 (1996)

3. V.I. Kukulin: In: *Proc. 14-th Symp. on Nucl. Phys., Quernavaka, Mexico, 1991*, p.115. Singapore-New-Jersey-L.-Hong-Kong: World Scientific 1991

4. H.R. Weller, D.R. Lehman: Ann. Rev. Nucl. Part. Sci. **38**, 563 (1988)

5. V.I. Kukulin, V.N. Pomerantsev: Progr. Theor. Phys. **88**, 159 (1992)

Few-Body Systems Suppl. 10, 443–446 (1999)

Few-
Body
Systems
© by Springer-Verlag 1999

The negative-energy dependence of the n-p mixing parameter and the pion cut

W. van Dijk[1,2*] and M.W. Kermode[3†]

[1] Redeemer College, Ancaster, Ontario, Canada L9G 3N6

[2] Department of Physics and Astronomy, McMaster University, Hamilton, Ontario, Canada, L8S 4M1

[3] Department of Mathematical Sciences (Theoretical Physics Division), University of Liverpool, Liverpool, UK, L69 3BX

Abstract. In order to study the effect of the pion cut on the $^3S_1 - {}^3D_1$ n-p mixing parameter, we consider the negative-energy behaviour of the quantity $\tau = \tan(2\epsilon)/(2k^2)$, where ϵ is the eigen-mixing parameter. The results are consistent with our earlier finding that the experimental data can be fitted without including effects due to the pion cut. However, τ is a sensitive function of energy near the pion-exchange branch point of the **S** matrix.

1 Introduction

In order to propose a shape-independent expansion for the $^3S_1 - {}^3D_1$ mixing parameter, Adhikari et al. [1] considered the function

$$\tau(k^2) \equiv \frac{\tan 2\epsilon}{2k^2} = \frac{K_{02}/k^2}{K_{00} - K_{22}}, \tag{1}$$

where ϵ is the eigen-mixing parameter [2]. As seen in Eq. (1), $\tau(k^2)$ can also be expressed in terms of $K_{ll'}$ which is the ll'-channel element of the on-shell n-p **K** matrix. The authors make an effective range-type expansion of $\tau(k^2)$, which includes a pole term to approximate the nearest branch point of the one-pion-exchange (OPE) cut of the **K** (and also the **S**) matrix along the negative real axis in the complex energy plane.

Subsequently we showed that the OPE pole has negligible effect on the positive energy behaviour of $\tau(k^2)$ [3]. In fact, one can achieve a better fit to

---

*E-mail address: vandijk@quark.physics.mcmaster.ca

†E-mail address: kermode@liv.ac.uk

the data assuming $\tau(k^2)$ to be a rational function of k^2, without imposing the condition of a pole at $k^2 = -\frac{1}{4}m_\pi^2$, where m_π is the pion mass.

We wish to investigate further the dependence (or lack thereof) of τ on the pion cut. We note that the independence of τ on the pion cut is similar to that obtained in the analysis of Kermode et al. [4, 5] regarding the effective-range function $y = k \cot \delta$, where δ is the phase shift at energy k^2. Adopting an effective S-wave interaction, they studied the function y at negative energies, using the negative-energy phase equation with a OPE (Yukawa) potential which was cut off at a large distance R in coordinate space. The **S** matrix has no OPE branch cut unless $R = \infty$. As R was increased, it was found that y had an interesting negative-energy behaviour with the effect of the pion cut clearly seen from the limiting behaviour of the curves. However, the positive-energy dependence of y was essentially unchanged as R was varied (even to very large values).

The determination of τ at negative energies is not as simple as the determination of y for a single channel because a pair of coupled equations has to be solved, rather than a first-order non-linear differential equation for which the rapidly increasing exponential solution of the Schrödinger equation at all negative energies is well controlled. There is no simple phase equation for the determination of τ, so one has to ensure that the decreasing exponential solution is not completely dominated numerically by the increasing one.

2 One-pion-exchange potential model

Early work by Glendenning and Kramer [6] showed that the $J = 1$ phase parameters and deuteron properties could be reasonably well approximated by using an OPE potential with a hard core. In that spirit we use a class of potentials considered recently [7], but cut off at some R,

$$V = V_C + V_T S_{12},\tag{2}$$

where

$$\begin{aligned}V_C(r) &= V_0, \quad r < r_c \\ &= -f^2 \mu c^2 e^{-\mu r}/(\mu r), \quad r_c < r < R \\ &= 0, \quad r > R,\end{aligned}\tag{3}$$

and

$$\begin{aligned}V_T(r) &= 0, \quad r < r_c \\ &= -f^2 \mu c^2 e^{-\mu r}/(\mu r)[1 + 3/(\mu r) + 3/(\mu r)^2], \quad r_c < r < R \\ &= 0, \quad r > R.\end{aligned}\tag{4}$$

We use the same values of the parameters as in ref. [7], i.e., $f^2 \mu c^2 = 11.156184$ MeV, and $\mu^{-1} = 1.415$ fm and $2m/\hbar^2 = 0.0244113235$ MeV^{-1} fm^{-2}, where m is the reduced mass of the n-p system. Notice that inside the core radius r_c, the potential is purely central. Although we considered several of the cases investigated by Sprung et al. [7], we report the results for only one of them, namely the "hard-core" potential with $V_0 = \infty$ and $r_c = 0.48151186$ fm. This core radius was chosen to give a deuteron binding energy of 2.224575 MeV (when $R = 15$ fm).

3 Determination of τ at negative energies

We determine τ at negative energies by evaluating the **M**-matrix effective range function [8, 9]. We consider two real negative-energy $(E = -\hbar^2\gamma^2/2m < 0)$ solutions in the $^3\mathrm{S}_1 - {}^3\mathrm{D}_1$ state with the asymptotic forms

$$\left.\begin{array}{ll} u_\alpha(r) & \sim \quad -\mathcal{N}_0(i\gamma r) + \gamma^{-1}M_{00}\bar{\mathcal{J}}_0(i\gamma r) \\[2mm] w_\alpha(r) & \sim \quad -\gamma^{-3}M_{02}\bar{\mathcal{J}}_2(i\gamma r) \end{array}\right\}, \tag{5}$$

$$\left.\begin{array}{ll} u_\beta(r) & \sim \quad -\gamma^{-3}M_{20}\bar{\mathcal{J}}_0(i\gamma r) \\[2mm] w_\beta(r) & \sim \quad -\mathcal{N}_2(i\gamma r) + \gamma^{-5}M_{22}\bar{\mathcal{J}}_2(i\gamma r) \end{array}\right\}, \tag{6}$$

where $\bar{\mathcal{J}}_l(x) = -ixj_l(x)$ and $\mathcal{N}_l(x) = xn_l(x)$ for $l = 0, 2$, and j_l and n_l are the usual spherical Bessel functions. By integrating the Schrödinger equation for two linearly independent solutions outward and matching these solutions and their derivatives at $r = R$ to the asymptotic forms we obtain the **M**-matrix elements, $M_{ll'}$. For negative energies (just as for positive energies [3]) τ can be related to the **M**-matrix elements through the equation

$$\tau(-\gamma^2) = \frac{M_{02}}{M_{00}\gamma^4 - M_{22}}. \tag{7}$$

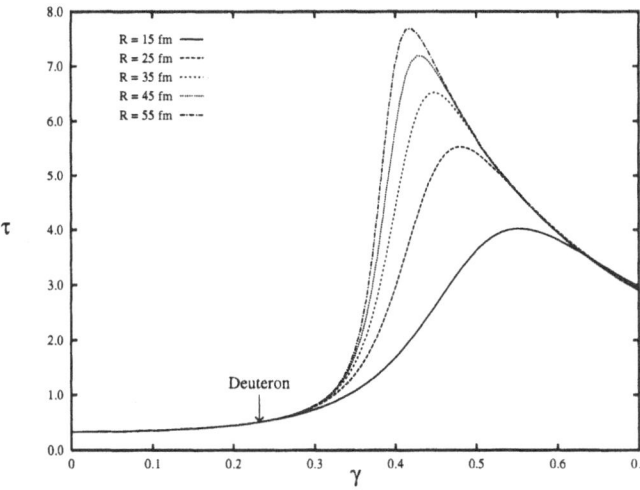

Figure 1. The quantity $\tau(-\gamma^2)$ in fm^2 as a function of γ in fm^{-1}.

The negative energy behaviour of τ for different cut-off distances R is displayed in Fig. 1.

4 Discussion and conclusion

Previously we showed that τ at positive energies showed very little sensitivity to the pion cut [3]. We conclude that this insensitivity persists down to and slightly below the deuteron energy. When γ ranges from 0.0 to 0.25 fm^{-1} the curves of τ are identical irrespective of the value of the cut-off radius.

For the deuteron bound state (with energy $E = -\hbar^2\alpha^2/(2m)$) the ratio of the asymptotic amplitudes ($\eta = A_D/A_S$) of the components of the wave function is $\eta = -\tan\epsilon|_{\gamma^2=\alpha^2} \approx \alpha^2\tau(-\alpha^2)$ for small η (or small ϵ). To be specific, for the deuteron $\alpha = 0.2316$ fm^{-1} and $\tau(-\alpha^2) = 0.5056$ fm^2, and we obtain $\eta = 0.0271$ for all values of R, which is also in agreement with the value obtained by solving for the bound state directly. Thus in the experimentally accessible region of the n-p system, τ is insensitive to whether the OPE branch point is there or not.

However, for negative energies not far below the deuteron bound-state energy, τ becomes quite sensitive to the cut-off distance. In fact the greatest sensitivity occurs near the OPE branch point, $\gamma = 0.35$ fm^{-1}. The curves of τ as a function of γ in Fig. 1 have a maximum which increases and shifts to the left (i.e., moves towards the value $\gamma = 0.35$ fm^{-1}) as R increases. For larger values of γ the curves for different R merge again. Other potentials of the class described in Sect. 2, e.g., soft-core potentials, give similar results.

Acknowledgement. We are grateful to EPSRC for support under research grant GR/L77409, and to NSERC Canada under grant OGP00-8672.

References

1. S.K. Adhikari et al.: Phys. Lett. **318B**, 14 (1993)

2. J.M. Blatt and L.C. Biedenharn: Phys. Rev. **86**, 399 (1952)

3. W. van Dijk, M.W. Kermode, D.J. Beachey: J. Phys. G: Nucl. Part. Phys. **21**, 651 (1995)

4. M.W. Kermode et al.: J. Phys. G: Nucl. Part. Phys. **10**, 773 (1984)

5. M.W. Kermode, S.G. Cooper, A. McKerrell: Phys. Lett. **142B**, 5 (1984)

6. N.K. Glendenning and G. Kramer: Phys. Rev. **126**, 2159 (1962)

7. D.W.L. Sprung et al.: Phys. Rev. **C49**, 2942 (1994)

8. M.H. Ross and G.L. Shaw: Ann. Phys. (NY) **13**, 147 (1961)

9. D.W.L. Sprung, M.W. Kermode, S. Klarsfeld: J. Phys. G: Nucl. Part. Phys. **8**, 923 (1982)

Few-Body Systems Suppl. 10, 447–450 (1999)

Few-
Body
Systems
© by Springer-Verlag 1999

Hadron Probing of the Deuteron Structure at Short Distances in Deuteron Breakup Reactions

A.P. Kobushkin[1*], A.P. Kostyuk[1] and E.A. Eliseev[2]

[1] Bogolyubov Institute for Theoretical Physics, 252143, Kiev, Ukraine
[2] Physics Department, Kiev National University, Kiev, Ukraine

Abstract. We discuss deuteron $A(d,p)X$ breakup and cumulative pion production $A(d,\pi)X$ in the framework of a constituent quark model of the deuteron. We demonstrate that consideration of the Pauli principle at the quark level, as well as multiple scattering, affect drastically cross section and polarization observables of these reactions and provide good description of the experimental data.

During last 15 years the deuteron structure was studied in a wide region, from that where description is given in terms of nucleons and mesons to that where quarks and gluons should be explicitly used for the deuteron description. Here we will analyze data on the deuteron breakup $A(d,p)X$ [1-11] and cumulative pion production $A(d,\pi)X$ [12, 13] and show that both of them can be understood in the framework of calculations which take into account the deuteron structure generated by the Pauli principle at the constituent quark level. In the present paper this was done using the simplest one-channel approach of the Resonating Group Method (RGM) [14]

$$\psi^d(1,2,\ldots,6) = \hat{A}\left\{\varphi_N(1,2,3)\varphi_N(4,5,6)\chi(\mathbf{r})\right\}, \tag{1}$$

where $\hat{A}$ is the quark antisymmetrizer and φ_N are wave functions of the nucleon three quark $(3q)$ clusters; $\chi(\mathbf{r})$ is the RGM distribution function and $\mathbf{r}$ stands for the relative coordinate between the centers of mass of the $3q$ bags.

Due to the presence of the antisymmetrizer in Eq. (1) the deuteron wave function (DWF), being decomposed into $3q \times 3q$ clusters, includes, apart from the standard pn component, nontrivial $\Delta\Delta$, NN^*, N^*N and $N^*N^{*\prime}$ components which correspond to all possible baryon resonance states (see ref. [14]). Most of the 1/2-isospin isobars have negative parity and thus generate effective P waves of the DWF [14, 15]. It should be mentioned that any known quark

---

*E-mail address: akob@ap3.bitp.kiev.ua

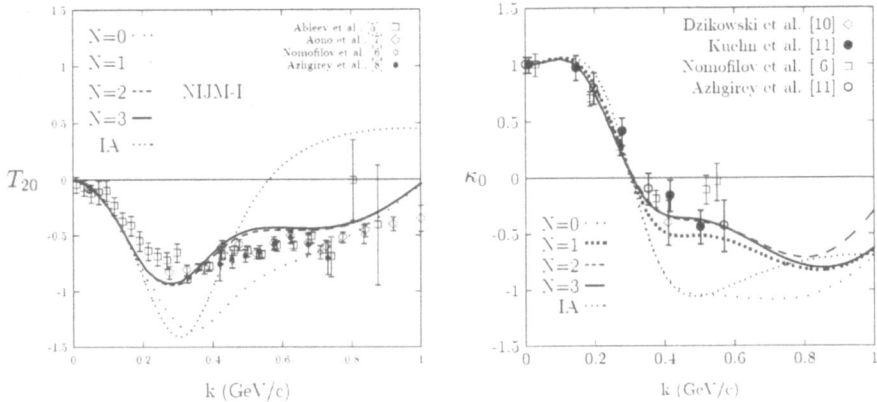

Figure 1. The tensor analyzing power T_{20} (left panel) and the polarization transfer κ_0 (right panel) in the $0°$ inclusive $^{12}C(d,p)$ breakup as a function of the relativistic internal momentum in the deuteron k. N is the maximum number of excitation quanta in the N^* resonance in the framework of oscillator quark model. The difference between IA and $N = 0$ demonstrates how important is multiple scattering.

approach to the DWF results in admixture of P components in the DWF [14, 15, 16] and study of effects connected with them should give important information about the deuteron structure at short distances.

Following Glozman-Kuchina paper (see ref. [14]) we choose $\chi(\mathbf{r})$ as a conventional DWF, $\chi_{NN}(\mathbf{r})$, modified by the RGM renormalization condition of ref. [17].

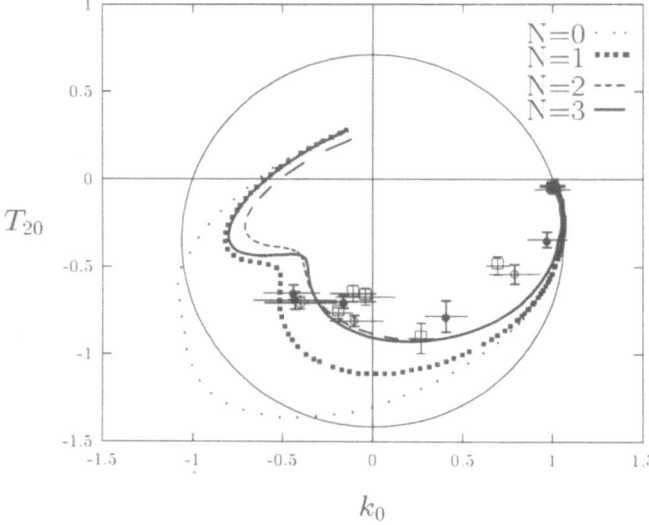

Figure 2. The polarization observables of Fig. 1 on the KPS plot [20].

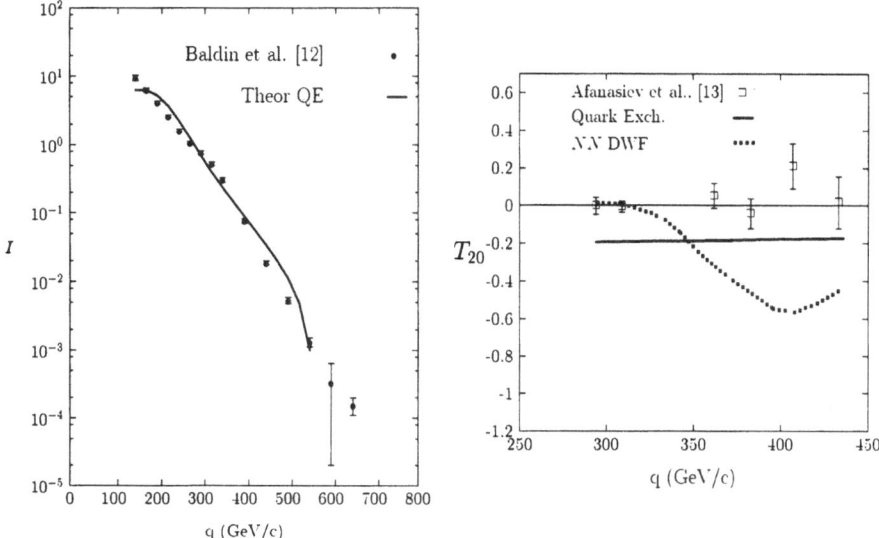

Figure 3. The differential cross section $I \equiv \frac{E}{2}\frac{d^3\sigma}{d^3p}$ (left panel) and the tensor analyzing power T_{20} (right panel) of the $0°$ cumulative pion production $(A(d,\pi)X)$; q is the pion momentum in the deuteron rest frame.

Multiple scattering effects were incorporated in the framework of the Bertocchi-Treleani model [18]. They appear to play an important role especially at high internal momenta in the deuteron.

In Figs. 1 and 2 we compare results of our calculations with the experimental data for the differential cross section, T_{20} and κ_0 in the $A(d,p)X$ breakup. The results were obtained with Nijm-I potential [19].

In spite of the fact that the $\Delta\Delta$ component has the largest probability among the non-nucleonic components of the deuteron, it does not contribute to (d,p) breakup. Nevertheless it may play significant role in cumulative pion production $A(d,\pi)X$. In our calculations we use the following model: the deuteron decays into a virtual $\Delta\Delta$ pair, then one of the Δ-s is absorbed by the target and the other one yields the cumulative pion and a nucleon. The results are shown in Fig. 3.

We conclude that effects related to the Pauli principle at the level of constituent quarks in the deuteron, as well as multiple scattering effects, play crucial role in $A(d,p)X$ breakup, especially at high deuteron internal momenta k. There is an evidence that the $\Delta\Delta$ component in the deuteron provides a dominant contribution into the cumulative pion production mechanism.

Acknowledgement. One of us (A.P. Kobushkin) is grateful to Organizing Committee of FB'98 for invitation and financial support during his stay in Autrans. We thank C. Perdrisat for important comments.

References

1. V.G. Ableev et al.: Nucl. Phys. **A393**, 491 (1983); **A411**, 591 (E) (1984); Pis'ma v ZhETF **37**, 196 (1983); JINR Rap. Comm. **1[54]**, 10 (1992)

2. C.F. Perdrisat et al.: Phys. Rev. Lett. **59**, 2840 (1987)

3. V.G. Ableev et al.: Pis'ma v ZhETF **47**, 558 (1988)

4. V. Punjabi et al.: Phys. Rev. **C39** 608 (1989)

5. V.G. Ableev et al.: JINR Rap. Comm. **4[43]-90**, 5 (1990)

6. A.A. Nomofilov et al.: Phys. Lett. **B325**, 327 (1994)

7. T. Aono et al.: Phys. Rev. Lett. **74**, 4997 (1995)

8. L.S. Azhgirey et al.: Phys. Lett. **B387**, 37 (1996)

9. N.E. Cheung et al.: Phys. Lett. **B284**, 210 (1992)

10. T. Dzikowski et al.: In: *Proc. of Int. Workshop 'Dubna Deuteron'97'*, p.181, Dubna, JINR, E2-92-25, 1992

11. B. Kuehn et al.: Phys. Lett. **B334**, 298 (1994); L.S. Azhgirey et al.: JINR Rep. Comm. **3[77]-96**, 23 (1996)

12. A.M. Baldin et al.: Preprint JINR E1-82-472 (1982)

13. S. Afanasiev et al.: Nucl. Phys. **A625**, 817 (1997)

14. L.Ya. Glozman et al.: Phys. Lett. **B200**, 983 (1988); L.Ya. Glozman and E.I. Kuchina: Phys. Rev. **C49**, 1149 (1994)

15. A.P. Kobushkin, A.I. Syamtomov and L.Yu. Glozman: Yad. Fiz. **59**, 833 (1996)

16. Fl. Stancu, S. Pepin and L.Ya. Glozman: Phys. Rev. **C56**, 2779 (1997)

17. K. Wildermuth and Y.C. Tang: In: *A Unified Theory of Nucleus.* Braunschweig: Vieweg 1977; M. Oka and K. Yazaki: Prog. Theor. Phys. **66**, 556 (1981)

18. L. Bertocchi and D. Treleani: Nuov. Cim. **36A**, 1 (1976)

19. W.G. Stoks et al.: Phys. Rev. **C49**, 2950 (1994); J.J. de Swart, C.P.F. Terheggen and W.G. Stoks: Preprint THEF-NYM-11; e-Preprint Archive: nucl-th/9509032

20. B. Kuehn, C.F. Perdrisat and E.A. Strokovsky: Yad. Fiz. **58**, 1898 (1995)

Few-Body Systems Suppl. 10, 451–454 (1999)

Few-
Body
Systems
© by Springer-Verlag 1999

First results on the tensor analyzing power A_{yy} in deuteron inclusive breakup at large transverse momenta of protons

V.P. Ladygin[1]*, S.V. Afanasiev[1], V.V. Arkhipov[1], L.S. Azhgirey[1],
V.K. Bondarev[1], E.V. Chernykh[1], M. Ehara[2], V.P. Ershov[1], G. Filipov[1,3],
V.V. Fimushkin[1], S. Fukui[2], S. Hasegawa[2], T. Hasegawa[4], N. Horikawa[2],
A.Yu. Isupov[1], T. Iwata[2], T. Kageya[2], V.A. Kashirin[1], M. Kawano[4],
A.N. Khrenov[1], A.D. Kirillov[1], V.I. Kolesnikov[1], A.G. Litvinenko[1],
A.I. Malakhov[1], T. Matsuda[4], I.I. Migulina[1], H. Nakayama[2], A.S. Nikiforov[1],
A.A. Nomofilov[1], E. Osada[4], V.N. Penev[1,3], Yu.K. Pilipenko[1],
S.G. Reznikov[1], P.A. Rukoyatkin[1], A.Yu. Semenov[1], I.A. Semenova[1],
V.I. Sharov[1], G.D. Stoletov[1], L.N. Strunov[1], N. Takabayashi[2], A. Wakai[2],
N.P. Yudin[5], S.A. Zaporozhets[1], V.N. Zhmyrov[1], L.S. Zolin[1]

[1] JINR, 141980 Dubna, Russia

[2] Nagoya University, 464-01 Nagoya, Japan

[3] INPNE, 1784 Sofia, Bulgaria

[4] Miyazaki University, 889-21 Miyazaki, Japan

[5] Moscow State University, 117234 Moscow, Russia

Abstract. The tensor analyzing power A_{yy} in inclusive breakup of 9 GeV/c deuterons on carbon has been measured at 85 mrad of the detected proton angle using polarized beam of the Synchrophasotron of the Laboratory for High Energies of JINR. The analyzing power A_{yy} remains positive at the highest measured momentum of the proton in definite contradiction with the predictions based on the standard nucleon-nucleon potentials.

1 Introduction

Recent experimental results on the tensor analyzing power T_{20} in deuteron inclusive breakup on hydrogen [1] and carbon [2] with the emission of the proton at a zero angle have shown that the traditional picture of the deuteron as a bound state of a neutron and a proton fails for short distances between them.

*E-mail address: ladygin@sunhe.jinr.ru

Neither account of mechanisms in addition to relativistic impulse approximation [3], nor including of additional components in the deuteron wave function (DWF) due to relativistic effects allow the T_{20} data be described at internal nucleon momenta ≥ 600 MeV/c. A failure of such attempts to describe the polarization observables for deuteron breakup at 0° within the framework of the traditional approaches [3, 4] can be considered as an indication of the manifestation of the non-nucleon degrees of freedom in the deuteron. An enlargement of the kinematical region to measure these observables is very desirable for a closer examination of this problem.

Here we present first results on the tensor analyzing power A_{yy} in deuteron inclusive breakup reaction on carbon target, $^{12}C(d,p)X$, at a 9 GeV/c initial deuteron momentum and a 85 mrad secondary proton emission angle in the laboratory.

2 Experiment and results

The experiment has been performed using a tensorially polarized deuteron beam of the Synchrophasotron of the Laboratory for High Energies of JINR. The tensor and vector polarizations, p_{zz} and p_z, were $p_{zz}^+ = 0.624 \pm 0.029(stat) \pm 0.025(sys)$, $p_{zz}^- = -0.722 \pm 0.022(stat) \pm 0.029(sys)$, $p_z^+ = 0.162 \pm 0.017(stat) \pm 0.003(sys)$ and $p_z^- = 0.209 \pm 0.013(stat) \pm 0.004(sys)$, respectively.

A slowly extracted 9 GeV/c deuteron beam with a typical intensity of $\sim 2 \cdot 10^9$ d/spill was directed onto a carbon target with the thickness either 6.5 g/cm^2 or 27.2 g/cm^2. The data were taken at 6 values of secondary particle momentum between 4.5 and 7.0 GeV/c. The polar angle acceptance of the setup was $\Delta\theta \approx \pm 8$ mr. The momentum acceptance depending on momentum varied between $\Delta p/p \approx \pm 0.02$ and ± 0.03.

The time-of-flight (TOF) information with a base line of ~ 28m was used in an off-line analysis for particle identification. The TOF resolution was better than 0.2 ns (1σ). The residual background was completely eliminated by the requirement that particles are detected at least in two prompt TOF windows.

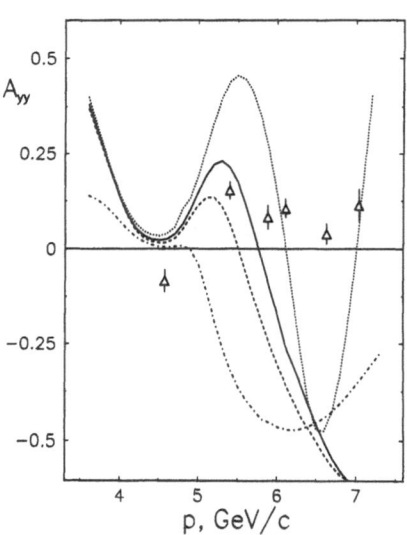

Figure 1. Tensor analyzing power A_{yy} at a 9 GeV/c and a 85 mrad of the proton emission angle. The solid, long-dashed, dotted and dash-dotted lines represent the results of calculations performed in in the framework of RHS model [5] with Paris [6], RSC [7], Bonn [8] and Moscow [9] DWFs, respectively.

In Fig. 1 the A_{yy} data are plotted versus proton momentum p in comparison with the predictions of the RHS model [5] for the $H(d, p)X$ reaction. The calculations are made using the conventional DWFs obtained from realistic nucleon-nucleon potentials [6-9] taking into account the finite acceptance of the setup. One can see that these calculations fail to reproduce the behavior of the experimental data, especially at high momenta of the protons.

Figure 2 shows the A_{yy} data from the present experiment along with the T_{20} data obtained at a zero transverse momentum of the proton on carbon target [1, 2, 10] plotted versus momentum of proton in the rest frame of the deuteron, q, (at a zero angle, $T_{20} = -\sqrt{2}A_{yy}$). The values of A_{yy} are positive for all the data at large momenta q. These values are incompatible with the predictions using DWFs from any reasonable nucleon-nucleon potential. On the other hand, attempts to take account of the non-nucleon degrees of freedom in the deuteron seem to be successful. An asymptotic positive value of $A_{yy} \sim 0.2$ for deuteron breakup at 0° was obtained in the framework of the QCD motivated approach [11]. Such a behavior of A_{yy} in the $^{12}C(d, p)X$ reaction at 0° was obtained in the calculations taking into account multiple scattering and the Pauli principle in the constituent quark level [12].

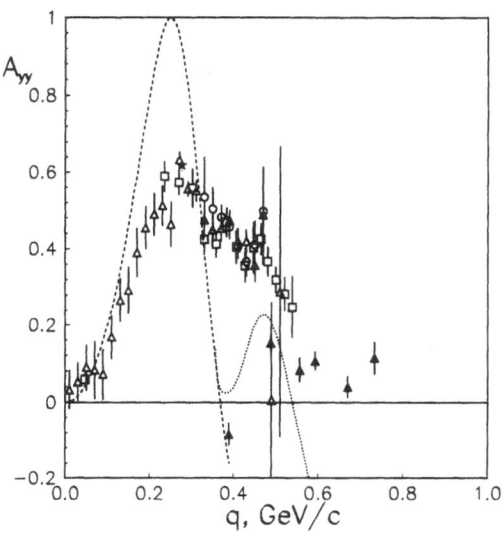

Figure 2. A_{yy} data from this experiment (filled triangles) as compared with the data obtained at 0° on carbon target from ref. [10] (open triangles), ref. [2] (open squares) and ref. [1] (open circles) versus proton momentum in the rest frame of the deuteron q. The dashed and dotted lines are the results of calculations using Paris DWF [6] for 0 and 85 mrad proton emission angles, respectively.

3 Conclusions

- The first results on the tensor and vector analyzing powers, A_{yy} and A_y, in inclusive deuteron breakup on carbon at large transverse momenta of protons have been obtained. The range of measurements corresponds to transverse proton momenta p_T between 390 and 600 MeV/c or to proton momenta in the rest frame of the deuteron up to ~ 730 MeV/c.

- We observe that A_{yy} remains positive up to the largest momenta of the proton. Such a behavior of A_{yy} definitely contradicts the predictions of the relativistic hard scattering model [5], while the cross section data obtained in the same experiment are satisfactorily described in framework of this model using conventional DWFs [6-9] This is apparently the consequence of the fact that the polarization data are much more sensitive to the details of the deuteron structure at short distances between the constituents.

- The new A_{yy} data obtained at large transverse momenta closely resemble A_{yy} data in deuteron inclusive breakup [1, 2] and dp backward elastic scattering [13] obtained at a zero transverse momentum. This similarity may be caused by the common reason, namely, by the manifestation of the non-nucleonic degrees of freedom in the deuteron.

- The non-zero value of the vector analyzing power A_y obtained simultaneously with A_{yy} can be interpreted as an important role of the spin-dependent part of the nucleon-nucleon amplitude.

Acknowledgement. One of the authors (V.P.L.) is grateful to the Organizing Committee of the 16-th European Conference on Few-Body Problems in Physics and Russian Foundation for Basic Research (grant N^o $98-02-26768$) for the financial support.

References

1. L.S. Azhgirey et al.: Phys. Lett. **B387**, 37 (1996)

2. T. Aono et al.: Phys. Rev. Lett. **74**, 4997 (1995)

3. G.I. Lykasov and M.G. Dolidze: Z. Phys. **A336**, 339 (1990)
   G.I. Lykasov: Part. and Nucl. **24**, 140 (1993)

4. L.S. Azhgirey et al.: Yad. Fiz. **48**, 87 (1988) : Z. Phys.

5. L.S. Azhgirey and N.P. Yudin: Yad. Fiz. **57**, 160 (1994)

6. M. Lacombe et al.: Phys. Lett. **B101**, 139 (1981)

7. R.V. Reid: Ann.Phys. (N.Y.) **50**, 411 (1968)

8. R. Machleidt et al.: Phys. Reports **149**, 1 (1987)

9. V.M. Krasnopol'sky et al.: Phys. Lett. **B165**, 7 (1985)

10. V.G. Ableev et al.: Pis'ma Zh. Eksp. Teor. Fiz. **47**, 558 (1988);
    JINR Rapid Comm. 4[43]-90, 5 (1990)

11. A.P. Kobushkin: J.Phys. G: Nucl.Part.Phys. **19**, 1993 (1993)

12. A.P. Kobushkin: Phys. Lett. **B421**, 53 (1998)

13. L.S. Azhgirey et al.: Phys. Lett. **B391**, 22 (1997); Yad. Fiz. **61**, 494 (1998)

Few-Body Systems Suppl. 10, 455–458 (1999)

Few-
Body
Systems
© by Springer-Verlag 1999

Measurement of the Tensor Analyzing Power T_{20} in the Fragmentation of 9 GeV Tensor Polarized Deuterons into Cumulative Pions

L. Zolin[1]*, S. Afanasiev[1], V. Arkhipov[1], V. Bondarev[1,2], G. Filipov[1,3],
M. Ehara[4], S. Fukui[4], S. Hasegawa[4], T. Hasegawa[5], N. Horikawa[4],
A. Isupov[1], T. Iwata[4], V. Kashirin[1], M. Kawano[5], T. Kageya[4], A. Khrenov[1],
V. Kolesnikov[1], V. Ladygin[1], A. Litvinenko[1], A. Malakhov[1], T. Matsuda[5],
H. Nakayama[4], A. Nikiforov[1], E. Osada[5], Yu. Pilipenko[1], S. Reznikov[1],
P. Rukoyatkin[1], A. Semenov[1], I. Semenova[1], N. Takabayashi[4], A. Wakai[4]

[1] JINR, 141980 Dubna, Russia

[2] St.-Petersburg State University, St.-Petersburg, 198350, Russia

[3] INPNE, 1784 Sofia, Bulgaria

[4] Nagoya University, 464-01 Nagoya, Japan

[5] Miyazaki University, 889-21 Miyazaki, Japan

Abstract. Tensor analyzing power T_{20} of the reaction $d^\uparrow A \to \pi^-(0°)X$ was measured in the fragmentation of 9 GeV deuterons into pions with momenta from 3.5 to 5.3 GeV/c on hydrogen, beryllium and carbon targets. This momentum range corresponds to the region of cumulative production with the cumulative variable x_c from 1.08 to 1.76. The values of T_{20} were found to be small and consistent with positive values in contradiction with the predictions based on a direct mechanism assuming NN collision between a high momentum nucleon in the deuteron and a target nucleon ($NN \to NN\pi$).

1 Introduction

An interest in study of reactions induced by polarized deuterons is caused by high sensitivity of spin observables both to the details of the deuteron structure and to the mechanism of the reactions. The data on the momentum distribution of nucleons in the deuteron, obtained in the deuteron breakup reaction [1] and in ed inelastic scattering [2], show good agreement confirming that both

---

*E-mail address: zolin@moonhe.jinr.ru

hadron and electron probes give reliable information on the deuteron structure. At the same time, both of them revealed marked deviation from Impulse Approximation (IA) calculations with standard two-component (S,D) deuteron wave functions [3] at the internal momenta of $k > 0.25$ GeV/c. Some of the models, accounting non-spectator graphs [4], could give a satisfactory explanation of this deviation. However, the measurements of T_{20} in $d^\uparrow A \to pX$ [5] and in $d^\uparrow p \to pd$ [6] showed that deviations of T_{20} from IA are considerable at the same $k > 0.25$ GeV/c (Fig. 1) and cannot be explained by any of the current models based on IA. A different character of the deviations in these two reactions indicates that non-nucleonic degrees of freedom can manifest themselves in a different way at the high momentum part of DWF. This example illustrates also that comparison of behaviour of observables (including the spin ones) in a series of different reactions is highly desirable to identify features of short range correlations (SRC) of nucleons in nuclei.

Another reaction to probe SRC is cumulative hadron production in fragmentation of nuclei [7]. The mechanism of cumulative hadron production is not clarified yet. Two alternative approaches are used. One is based on the hypothesis of the high momentum nucleon component in nuclei, the use of IA leads to assumption of the dominating so-called direct machanism. The other approach is based on consideration of SRC as multi-quark configurations, in this case the cumulative pion is produced by hadronization of quark-spectator at fragmentation of such a configuration [8]

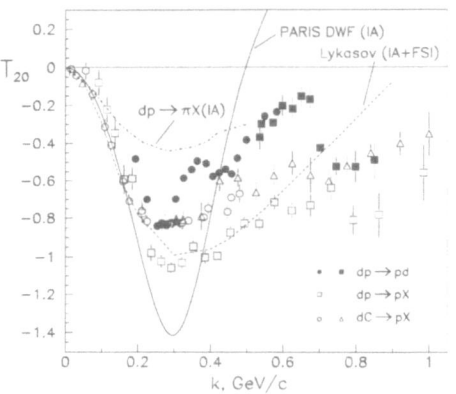

Figure 1. The T_{20} data for reactions induced by tensor polarized deuterons in GeV region: deuteron breakup $H, C(d,p)X$ [5] and $dp \to pd$ [6] (comments to the curves and the star point are in the text).

One of the motivations to study T_{20} in fragmentation of polarized deuterons into cumulative pions was to verify applicability of IA to this process.

Calculations based on the direct mechanism [9] predict a similar $T_{20}(k)$-dependance in $dA \to pX$ and $dA \to \pi X$ (Fig. 1). We present here the data of T_{20} vs two variables x_c and k_{min}. Cumulative number x_c is to be equal to a minimal effective number of nucleons which should take part in collision to produce a cumulative particle [7]. In the framework of the direct mechanism, x_c can be related to an internal momentum k, which means that the cumulative pion of x_c is to be produced on the nucleon of internal momentum $k \geq k_{min}(x_c)$.

The first measurement of T_{20} in the cumulative pion production at zero angle $d + A \to \pi^\pm(0°) + X$ was performed with the carbon target [9]. The value of T_{20} was determined by measuring the yield of pions $(N^{1,2})$ at two deuteron tensor polarizations $(p_{zz}^{1,2})$: $T_{20} = 2\sqrt{2}(N^1 - N^2)/(N^1 p_{zz}^2 - N^2 p_{zz}^1)$.

It was found that the value of $|T_{20}|$ is much smaller than that predicted by IA. However, this difference could be related to deviation caused by the C-target. Moreover, due to large errors it was difficult to draw a definite conclusion concerning the sign of T_{20}.

In this paper, we report the new data for $d^\uparrow A \to \pi^-(0°)X$ obtained with $H, Be,$ and C targets.

2 Experiment and results

The measurements at the Dubna Synchrophasotron were carried out using the setup (Fig. 2) described in [9]. π^- momentum changed from 3.5 to 5.3 GeV/c at the fixed beam momentum of p_d=9.0 GeV/c to range x_c from 1.08 to 1.76. The points at 3.5, 4.0, 4.5 and 5.0 GeV/c were taken with H-target (7 g/cm^2). To check the A-dependance effect 5.0 GeV/c-point was repeated with Be-target (36 g/cm^2), and, finally, point at 5.3 GeV/c was taken with C-target (55 g/cm^2). π^-s were selected by time of flight between the $F_{56_{1-4}}$ and $F_{6_{1-3}}$ counters (Fig. 2).

To control beam polarization stability during the run, we measured T_{20}^p of $dC \to p(0°)X$ at the fixed ratio $p_p/p_d = 2/3$. In all, there were 24 measurements during the run. They showed that $|p_{zz}^+| + |p_{zz}^-|$ remained stable within ±2%. T_{20}^p averaged over the run time is marked by the star in Fig. 1. The new results together with earlier data [9] are presented in Fig. 3.

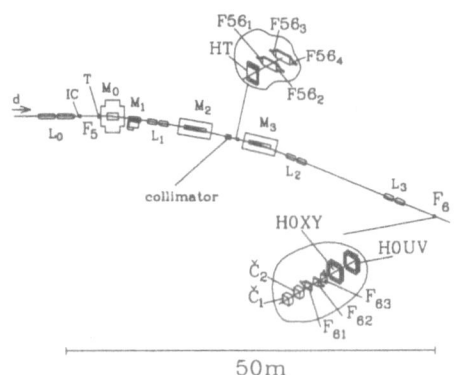

Figure 2. A layout of the setup on the 4V beam line in the Synchrophasotron experimental hall. T is target, and IC is monitoring ionization chamber. M_i denote bending magnets and L_i quadru- poles. F_i, C_i are scintillation and Cherenkov counters; $HT, HOXY,$ and $HOUV$ are scintillation hodoscopes.

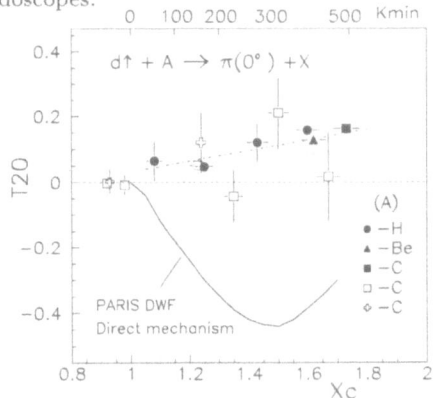

Figure 3. T_{20} vs x_c and k_{min} for $d^\uparrow + A \to \pi^-(0°) + X$ at fragmentation of 9 GeV deuterons on H,Be, and C targets. The previous data [9] with C-target marked by open squares (π^-) and crosses (π^+). The dashed line is a linear fit of the data. The solid curve shows calculations with Paris DWF [9].

As shown in Fig. 3, the T_{20} of π^- is of positive sign. It is nearly zero at x_c=1.0 and increases up to $\simeq 0.15$ at $x_c = 1.7$. Comparing the new T_{20} data obtained on $H, Be,$ and C targets, it was found that $T_{20}(x_c)$ does not depend on the target mass number.

The curve in Fig. 3 shows the prediction of the direct production mechanism made with the Paris DWF [9]. These culculations fail to reproduce $T_{20}(x_c)$ in any part of the cumulative region $(x_c > 1, k_{min} > 0)$ in contradiction with T_{20}-behaviour in the deuteron breakup $d^\uparrow A \to p(0°)X$ where quantitative agreement with IA calculations takes place up to $k = 0.25$ GeV/c.

3 Conclusion

The tensor analyzing power T_{20} of the reaction $d^\uparrow H(Be, C) \to \pi(0°)X$ at the deuteron momentum of 9 GeV/c and pion momenta from 3.0 to 5.3 GeV/c has been measured. As for the direct mechanism of pion production, this momentum region corresponds to internal momenta up to $k \leq 500$ MeV/c. The sign of T_{20} is opposite to that determined with the standard DWF assuming the direct mechanism. The above data show that the hypothesis of the direct $NN \to NN\pi$ process dominating is not appropriate in the fragmentation of deuterons into cumulative pions. Attention should be payed to a quark scenario of cumulative particle formation considering the deuteron core as a multiquark configuration.

Acknowledgement. One of authors (L.Z.) is grateful to Organizing Committee of FB'98 for invitation and financial support.

References

1. V.G. Ableev et al.: Nucl. Phys. **A393**, 491 (1983)

2. P. Bosted et al.: Phys. Rev. Lett. **49**, 1380 (1995)

3. M. Lacombe et al.: Phys. Lett. **B101**, 139 (1981); R.V. Reid: Ann. Phys. (N.Y.) **50**, 411 (1968)

4. G.I. Lykasov: Phys. Part. Nucl. **24**, 59 (1993)

5. C.F. Perdrisat et al.: Phys. Rev. Lett. **59**, 2840 (1987); V. Punjabi et al.: Phys. Rev. **C39**, 608 (1989); T. Aono et al.: Phys. Rev. Lett. **74**, 4997 (1995); L.S. Azhgirey et al.: Phys. Lett. **B387,** 37 (1996)

6. V. Punjabi et al.: Phys. Lett. **B350**, 178 (1995); L.S. Azhgirey et al.: Phys. Lett. **B391**, 22 (1997)

7. A.M. Baldin: Nucl. Phys. **A434**, 695c (1985); V.B. Kopeliovich: Phys. Reports **139**, 52 (1986)

8. V.V. Burov et al.: Phys. Lett. **B67**, 46 (1977); V.K. Lukyanov, A.I. Titov: Phys. Part. Nucl. **10**, 815 (1979)

9. S. Afanasiev et al.: Nucl. Phys. **A625**, 491 (1997); M.V. Tokarev: In: *Proc. Int. Workshop "Dubna Deuteron-91"*, JINR, E2-92-25, Dubna, 1992, p.84

Few-Body Systems Suppl. 10, 459–462 (1999)

Few-
Body
Systems
© by Springer-Verlag 1999

Polarization Transfer Observables in Pion-Deuteron Elastic Scattering

G. Suft[1], P. Amaudruz[6], W. Beulertz[1], A. Bock[1], E. Boschitz[2], B. van den Brandt[4], B. Brinkmöller[2*], M. Frank[1], A. Glombik[1†], W. Grüebler[7], P. Hautle[4], J. Hey[1], J.A. Konter[4], B. Kowalzik[1*], W. Kretschmer[1], S. Mango[4], R. Meier[2**], G. Mertens[3], S. Merz[1], H. Meyer[1], L. Sözüer[1], R. Tacik[5], R. Weidmann[1†]

[1] Physikalisches Inst. der Univ. Erlangen-Nürnberg, 91058 Erlangen, Germany

[2] Inst. für Exp. Kernphysik, Univ. Karlsruhe, 76131 Karlsruhe, Germany

[3] Physikalisches Institut, Universität Tübingen, 72076 Tübingen, Germany

[4] Paul Scherrer Institut, 5232 Villigen, Switzerland

[5] University of Regina, Saskatchewan, Canada

[6] TRIUMF, 4004 Wesbrook Mall, Vancouver, British Columbia, Canada

[7] ETH Zürich, ETH-Hönggerberg, 8093 Zürich, Switzerland

Abstract. Measurements of polarization transfer observables in $\pi^+ d$ elastic scattering have been performed at two pion energies, one below and one at the $\Delta(3,3)$ resonance. The results are compared to different predictions from the SAID phase shift analysis and Faddeev calculations.

1   Introduction

Pion deuteron elastic scattering has been studied experimentally and theoretically for almost two decades. During this time there have been exciting motivations for such studies like searching for charge symmetry breaking effects, dibaryon resonances and possible short range $N\Delta$ interactions, but the most important underlying question has been the thorough testing of sophisticated few body theories [1]. For that reason measurements have been performed at several energies across the $\Delta(3,3)$ resonance for the observables: $d\sigma/d\Omega$ and

---

*Present address: SAP AG, 69185 Walldorf, Germany

†Present address: debis Systemhaus GmbH, 80995 München, Germany

**Present address: Physikalisches Institut, Universität Tübingen, 72076 Tübingen, Germany

iT_{11} over almost the entire angular range, T_{20}, T_{21}, T_{22} and t_{20} in the backward hemisphere [2]. When comparing this data set with the existing few body theories it appeared that none of them was able to predict the cross section data at large angles and the vector analyzing power data at intermediate angles in the energy region above 180 MeV correctly. Efforts to improve the theoretical calculations failed. Only recently a complete new and very promising approach has been proposed [3] and is currently being developed, however it may take a long time, until numerical predictions will become available. For this reason measurements of polarization transfer observables are important to provide - at least in some angular regions - seven or more observables which are required for a model independent amplitude analysis.

2 Experimental setup

Measurements of polarization transfer observables in $\pi^+ d$ elastic scattering are extremely difficult because they require a high pion flux, a suitable radiation resistant polarized deuteron target plus a well calibrated deuteron polarimeter which allows to measure the different vector and tensor deuteron polarizations [4]. These conditions have become available only recently at PSI. In the polarization transfer experiment, pions were scattered at large angles from a vector polarized deuteron target. The recoil deuterons were focused onto the deuteron polarimeter by a quadrupole triplet to increase the accepted solid angle. The pion flux was monitored by an ionization chamber and by the $\pi d \longrightarrow pp$ reaction at $\theta_{c.m.} = 90°$, where the cross section is insensitive to the target polarization. For the material of the polarized deuteron target, $^{14}ND_3$ had to be used instead of ^6LiD because of the high deuteron background coming from the breakup of ^6Li.

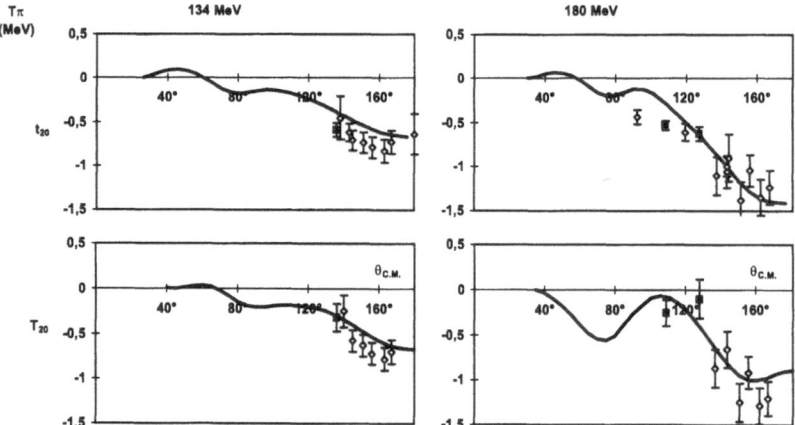

Figure 1. Angular distribution of t_{20} and T_{20} at $T_\pi = 134$ and 180 MeV in comparison with SAID SM94 calculations. The open diamonds are from measurements at TRIUMF and LAMPF, the closed squares are the data points of the present experiment.

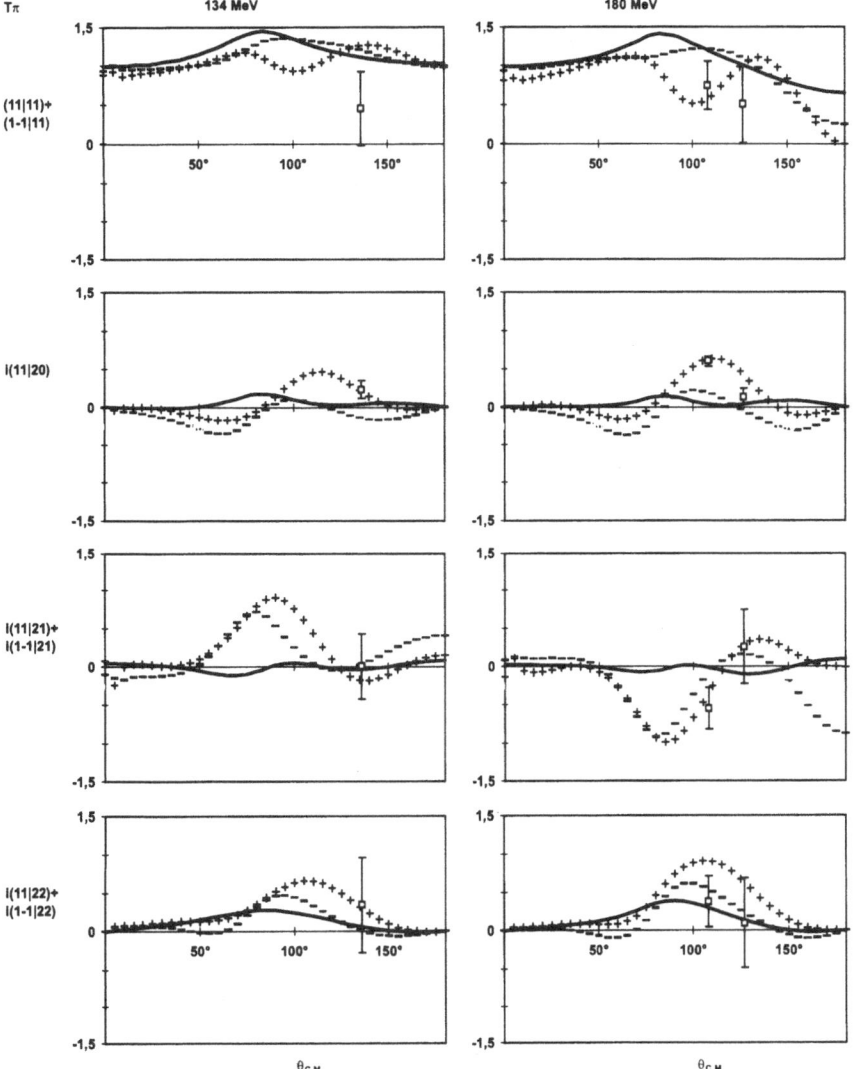

Figure 2. Comparison of the measured polarization transfer observables (squares) with Faddeev calculations (solid) and different SAID predictions (dashed C500, crosses SM94).

The target polarization was continuously monitored by the dynamical NMR signal, and by measuring the elastically scattered pions at an angle where reliable iT_{11} data are available. So the deuteron target polarization could be determined and compared with the NMR measurement. The vector and tensor polarization of the recoil deuterons was measured by the $^3He(d,p)^4He$ reaction in the polarimeter at $\theta = 0°, 25°$ and $45°$ at $\phi = 0°, 90°, 180°, 270°$. Due to the very low count rate of the protons of the $^3He(d,p)^4He$ reaction many on- and off-line tests were necessary to insure the correctness of the data reduction. The most crucial test has been the reproduction of earlier t_{20} measurements at LAMPF, and T_{20} measurements at TRIUMF [5], where $T_{20} = f(t_{20}, t_{21}, t_{22})$ can be calculated from our measured quantities in the polarimeter. As can be seen there is excellent agreement between our data and the previous measurements, Fig. 1.

3 Results and discussion

In Fig. 2 we present our results for the different polarization transfer observables together with a comparison between Faddeev calculations by Garcilazo [6] and different results from the SAID phase shift analysis SM94 (elastic πd channel) and C500 (couple channels $d(\pi, \pi)d$, $p(p, p)p$ and $d(\pi, p)p$) [7]. So far our results are not included in these databases. All measured polarization transfer observables are particularly in agreement with the SAID SM94 solutions within the statistical uncertainties. The data for 180 MeV and $\theta_{\pi, C.M.} = 108°$ are the combined results coming from two different data-taking periods. The single results from these periods agree very well with each other, especially for the observable $i(11|20)$ (first measurement [8]: 0.595 ± 0.088, second measurement: 0.613 ± 0.098). This is important because there is strong disagreement with the predictions of the Faddeev calculations for $i(11|20)$.

Acknowledgement. This work would have been impossible without the generous help and considerable skills of the staff of the PSI. It has been partially funded by the German Federal Ministry for Research and Technology (BMFT) under the Contracts No. 06 KA 6541 and 06 ER 3585.

References

1. H. Garcilazo, T. Mizutani: In: πNN *Systems.* Singapore: World Scientific 1990

2. M. Weßler et al.: Phys. Rev. **C51**, 2575 (1995) and references there in

3. A.N. Kvinikhidze, B. Blankleider: Nucl. Phys. **A574**, 788 (1994) and earlier references

4. W. Grüebler et al.: NIM **A262**, 307 (1987)

5. E. Ungricht et al.: Phys. Rev. Lett. **52**, 333 (1984); G.R. Smith et al.: Phys. Rev. **C38**, 251 (1988)

6. H. Garcilazo: Phys. Rev. **C35**, 1804 (1987) and private communication

7. R.A. Arndt, I.I. Strakovsky, R.L. Workman: Phys. Rev. **C50**, 1796 (1994); C.-H. Oh et al.: Phys. Rev. **C56**, 635 (1997)

8. G. Suft et al.: Phys. Lett. **B425**, 19 (1998); G. Suft: Dissertation, University Erlangen-Nürnberg, 1997, unpublished

Few-Body Systems Suppl. 10, 463–466 (1999)

Few-
Body
Systems
© by Springer-Verlag 1999

Determination of quark-antiquark potentials and meson spectra

C. Semay[1]*, B. Silvestre-Brac[2]

[1] Université de Mons-Hainaut, 20 Place du Parc, B-7000 Mons, Belgique
[2] Institut des Sciences Nucléaires, 53 Avenue des Martyrs, F-38026 Grenoble-Cedex, France

Abstract. We determine two different sets of quark-antiquark potential to be used in a Schrödinger or a Salpeter type of equation. The central part contains QCD inspired components: the Coulomb like interaction, the confinement potential and an instanton interaction. It is supplemented with phenomenological relativistic corrections. The constituent quarks are characterized by a colour charge density. The parameters are determined by a minimization procedure on representative samples of mesons and the full spectra are calculated.

1 Introduction

The purpose of this work is to obtain a good and simple quark-antiquark potential by improving a previous model [1]. "Good" means that the meson spectra and other properties of all mesons could be satisfactorily reproduced. "Simple" means that this interaction could be applicable without serious difficulties to baryons and multiquark systems. Contrary to usual models, the annihilation processes between quarks is taken into account by considering an instanton induced interaction which relies on very sound QCD properties [2]. We have developed a non-relativistic and a semi-relativistic models in order to compare the merits of the two kinematics. Our results are also compared with the very impressive work of Godfrey and Isgur (GI) on meson properties [3].

2 The model

The Hamiltonian is the sum of a kinetic energy term and a potential energy term. The non-relativistic expression is the usual kinetic part of Schrödinger

---

*Chercheur qualifié F.N.R.S.

equation while the semi-relativistic expression is the kinetic part of the spinless Salpeter equation.

The constituent quarks in our model are not considered as point-like particles but as effective degrees of freedom dressed by the virtual cloud of gluons and quark-antiquark pairs. For simplicity we assume, as in the GI work, that the quark density is given by a Gaussian shaped distribution, characterized by two parameters σ_0 and δ, with a size depending on the mass. Accordingly, a potential, which is a two-body operator, is transformed by convoluting the bare interaction with an effective two-body Gaussian quark distribution whose size is given in ref. [1]. Here we differ from the GI work which chooses an unnatural very complicated form for the size parameter.

The bare confinement interaction is the usual linear potential $ar + C$ where a is the string tension and C a constant which appears to be necessary for a good description of the spectra. The dressed and non-dressed confinement potentials differ only for very small values of r. We find convenient to use the simplest, that is to say the non-dressed expression. The results are practically identical and the numerical job is much easier [1].

A large contribution of the short range part of the potential stems from the one-gluon exchange process. The running coupling constant $\alpha_s(Q^2)$ is parametrized by a sum of three Gaussian shaped forms like in the GI work. But contrary to that work, the parameters of this function are determined by a fit to experimental data. The three Gaussian functions are related to two adjustable parameters: α_s^c (value of $\alpha_s(0)$) and β, a number introduced to take into account experimental errors. In configuration space the running coupling constant is then written as the sum of error functions divided by r. The effect of dressing is simply to modify the range of this effective one-gluon exchange potential.

The instanton interaction is a short range potential derived from QCD Lagrangian, which acts only for pseudoscalar mesons [2]. Its form depends on the flavour content of the meson. It can be parametrized with a simple Gaussian shaped function with a range r_0. It also depends on two dimensioned coupling constants: g and g'. In the isoscalar sector, the potential mixes $n\bar{n}$ and $s\bar{s}$ states. The effect of dressing is simply to modify the r_0 value. The pseudoscalar mesons can be well described in the framework of the instanton induced interaction without introducing completely phenomenological interactions as it is done in the GI work. This is the first main difference between their work and ours.

The spin-spin (or hyperfine) term is an essential ingredient of the total interaction. The hyperfine term is a relativistic correction to the central interaction. Its form factor is given by the Laplacian of the central part times a mass dependent term which depends on the Lorentz structure of the corresponding potential. The Lorentz structure of the confining term is not known precisely and this fact does not allow to give a definite expression for the real hyperfine term. Consequently, we choose to describe the hyperfine interaction in a completely phenomenological way [1]. This is the second main difference between the GI work and our model. We verified that a good fit can be obtained by using a sum of two Gaussian forms, one with a short range η_{01} and the other with a longer range η_{02}. Two strength parameters κ_1 and κ_2 are introduced to multiply the Gaussian functions. Again, the effect of dressing is simply to modify the range of the potentials. In the GI work the mass dependent factor is replaced by a complicated energy dependence function whose purpose is to take into account relativistic effects. We prefer to keep a simple mass dependence to allow an easier extension of our model to baryons and multiquarks systems. All the formulae and the corresponding formalim can be found in ref. [1]. The novelties of this study concern other relativistic corrections which are taken into account here.

Different types of spin-orbit terms exist. Keeping the same philosophy as the one developed for the hyperfine interaction, we find natural to express them in terms of Gaussian functions. To avoid a too large number of parameters we keep the same range parameters as those appearing in the hyperfine term. The most general form of the spin-orbit interaction is the sum of a symmetric term and an antisymmetric one. Taking into account the fact that the symmetric term is characterized by two possible different mass dependent factors, four new strength parameters κ_3 to κ_6 must be introduced in front of the corresponding Gaussian functions. The symmetric potential term is responsible for the splitting of ^3L_J states with $J = L - 1, L, L + 1$. The antisymmetric term couples states $^1L_{J=L}$ and $^3L_{J=L}$, and depends only on κ_3 and κ_4.

The radial part of the tensor potential is also given by two Gaussian shaped functions with the same range parameters as before. Again, two new strength parameters κ_7 and κ_8 are introduced.

3 Numerical methods

The meson wave functions are expanded in a harmonic oscillator basis up to 30 quanta [1]. The calculation of potential matrix elements are performed with Talmi's integrals and the semi-relativistic kinetic energy matrix elements are calculated with the method developed in ref. [4]. All results are verified by calculation of meson masses with the three-dimensional Fourier grid Hamiltonian method developed in ref. [5].

The parameters are determined numerically by minimizing a χ_S^2 function. A careful analysis has been performed in order to identify the internal quantum numbers of the mesons chosen to determine parameters. Based on a sample of 50 mesons, we have determined two different sets of parameters: one with non-relativistic kinematics (NR) and one with semi-relativistic kinematics (SR). 30 supplementary mesons have been considered to calculate a χ_E^2 value with an enlarged set of mesons.

4 Results

The parameters found for our two best models are given in Table 1. As usual the non-relativistic quark masses are about 200 MeV above the semi-relativistic masses. Other parameters have reasonable values and do not differ strongly in the two models. This is not true for the phenomenological parameters κ_k which define the strength of the hyperfine, spin-orbit and tensor potentials. In Table 2, we can see that, for the enlarged set of mesons, the semi-relativistic model gives a spectrum better than the non-relativistic one, and it is as good as the GI model. The complete spectra are reasonably well reproduced and will be presented in a forthcoming paper [6].

Table 1. Values of the parameters with units in powers of GeV.

Parameter	NR	SR	Parameter	NR	SR
m_n	0.400	0.230	η_{01}	0.481	0.112
m_s	0.630	0.433	η_{02}	1.488	1.501
m_c	1.860	1.668	κ_1	1.214	0.848
m_b	5.243	5.054	κ_2	0.927	-1.001
α_s^c	0.793	0.471	κ_3	-6.905	0.213
β	1.100	1.100	κ_4	1.427	-0.091
a	0.161	0.197	κ_5	5.429	-2.130
C	−0.661	−0.531	κ_6	8.484	2.651
g	11.54	13.99	κ_7	9.802	1.872
g'	4.963	5.787	κ_8	3.382	-0.468
r_0	2.668	3.348	σ_0	3.820	3.027
			δ	1.318	0.716

Table 2. Values of χ^2.

	NR	SR	GI
χ_S^2	29.8	44.7	51.4
χ_E^2	79.4	54.8	52.2

References

1. C. Semay and B. Silvestre-Brac: Nucl. Phys. **A618**, 455 (1997)

2. C.R. Münz et al.: Nucl. Phys. **A578**, 418 (1994)

3. S. Godfrey and N. Isgur: Phys. Rev. **D32**, 189 (1985).

4. L.P. Fulcher: Phys. Rev. **D50**, 447 (1994)

5. F. Brau and C. Semay: J. Comput. Phys. **139**, 127 (1998)

6. C. Semay and B. Silvestre-Brac: submitted for publication

Few-Body Systems Suppl. 10, 467–470 (1999)

Few-
Body
Systems
© by Springer-Verlag 1999

Interactions of Pion-like Particles from Lattice QCD *

H. Markum[1], H.R. Fiebig[2], A. Mihály[3], R. Pullirsch[1], K. Rabitsch[1]

[1] Institut für Kernphysik, Technische Universität Wien, A-1040 Vienna, Austria
[2] Physics Department, FIU-University Park, Miami, Florida 33199, USA
[3] Department of Theoretical Physics, Lajos Kossuth University, H-4010 Debrecen, Hungary

Abstract. An approximate local potential for the residual $\pi^+ - \pi^+$ interaction is computed. We use an $O(a^2)$ improved action on a coarse $9^3 \times 13$ lattice with spacing $a \approx 0.4$fm. We attempt extrapolation of the $\pi^+ - \pi^+$ potential to the chiral limit.

1 Introduction and Theoretical Method

The construction of improved lattice actions allows to work with lattice volumes large enough to accommodate systems of two hadrons, with manageable computational effort [1]. We take advantage of this development to study the residual effective interaction of two pseudo-scalar mesons on the lattice.

The current computation allows extrapolation of the extracted $\pi-\pi$ potential to the chiral limit. The progress in lattice simulations enables a comparison of lattice-based scattering phase shifts, computed with the potential, to experimental results [2] in the isospin $I = 2$ channel.

An $L^3 \times T = 9^3 \times 13$ lattice was used with an $O(a^2)$ tree-level and tadpole improved action with next-nearest neighbor couplings [3]. At $\beta = 6.2$ the corresponding lattice constant is $a \approx 0.4$fm or $a^{-1} \approx 500$MeV [4]. We considered six different values for the hopping parameter κ of Wilson fermions which determine the quark mass. The critical value for κ^{-1} is ≈ 5.5.

An outline of the theoretical framework may be found in [1, 5], here we mention only the essential points. Suitable operators for the $\pi^+ - \pi^+$ system, having isospin $I = 2$, are

$$\Phi_{\boldsymbol{p}}(t) = \phi_{-\boldsymbol{p}}(t) \, \phi_{+\boldsymbol{p}}(t) \, , \tag{1}$$

*Supported in part by NSF PHY-9700502, by OTKA T023844, and by FWF P10468-PHY

where

$$\phi_{\boldsymbol{p}}(t) = L^{-3} \sum_{\boldsymbol{x}} e^{i\boldsymbol{p}\cdot\boldsymbol{x}} \bar{\psi}^{\mathrm{d}}(\boldsymbol{x},t)\,\gamma_5\psi^{\mathrm{u}}(\boldsymbol{x},t) \tag{2}$$

describes single mesons with lattice momenta $\boldsymbol{p}$ consisting of an up and down quark. The correlation matrix for the π^+–π^+ system

$$C_{\boldsymbol{p}\boldsymbol{q}} = \langle \Phi_{\boldsymbol{p}}^{\dagger}(t)\,\Phi_{\boldsymbol{q}}(t_0)\rangle = \bar{C}_{\boldsymbol{p}\boldsymbol{q}} + C_{I,\boldsymbol{p}\boldsymbol{q}} \tag{3}$$

is a sum of a free, $\bar{C}$, and a residual-interaction contribution, C_{I}. The correlator $\bar{C}$ of two free pions is diagonal in p, q. Since here we are interested in the $\ell = 0$ (s-wave) interaction we make use of the irreducible representation A_1 of the lattice symmetry group $O(3, \mathcal{Z})$. The eigenvalues of the corresponding reduced correlators decrease exponentially with increasing time t

$$\begin{aligned}
\bar{C}^{(A_1)}(t,t_0) &\sim e^{-\bar{E}_p(t-t_0)} \quad \text{free} \\
C^{(A_1)}(t,t_0) &\sim e^{-E_n(t-t_0)} \quad \text{interacting} .
\end{aligned} \tag{4}$$

We also implement link variable fuzzing and operator smearing [6] at the sink. We define an effective interaction through

$$H_{\mathrm{I}} = -\frac{\partial}{\partial t} \ln(\bar{C}^{-1/2}\, C\, \bar{C}^{-1/2}) . \tag{5}$$

Matrix elements of H_{I} are obtained from linear fits to the logarithm of the eigenvalues of the correlators $\bar{C}$ and C. At this point, only the diagonal elements of C were utilized. Momentum-space matrix elements $(\boldsymbol{p}\,|H_{\mathrm{I}}|\boldsymbol{q})$ are computed in a truncated basis of small lattice momenta. The Fourier transform to coordinate space contains a local potential which in the s-wave projection reads

$$V(r) = \sum_{q} j_0(2qr)(q|H_{\mathrm{I}}^{(A_1)}|q) , \tag{6}$$

with j_0 a spherical Bessel function.

2 Results and Conclusion

Figure 1 shows the energy spectra for the free and interacting two-pion system according to (4) for $\kappa^{-1} = 5.804$ (left plot) and $\kappa^{-1} = 5.972$ (right plot), respectively. The spectra reveal that the residual interaction lowers the large-momentum levels of the free system, thus indicating attraction at short relative distances. Small momentum levels are shifted slightly upward. We take this as an indication of a repulsive component of the interaction at larger separations.

Chiral extrapolation of the potential was done by linear fits of $V(r)$ versus the squared pion mass m^2 for a fine (plot-grade) mesh of values r, fixed one at a time. Using sets of 3 through 6 data points, corresponding to the smallest available values of m^2, gives very similar results. The subsequent analysis was performed with the 3 smallest mass points.

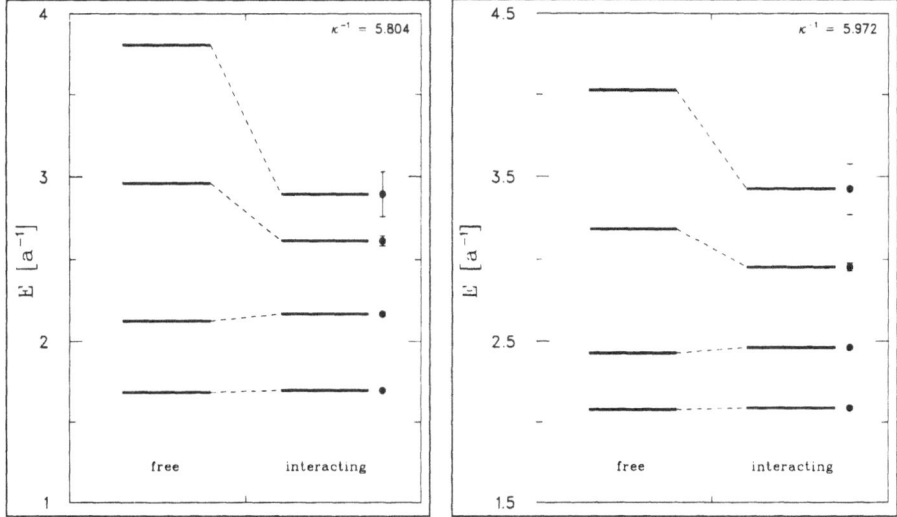

Figure 1. Energy levels for the free and interacting pion-pion system for $\kappa^{-1} = 5.804$ and $\kappa^{-1} = 5.972$, respectively.

The extrapolated potential $V(r)$ is shown in the left plot of Fig. 2 as a dashed line. The oscillations of $V(r)$ in the region $r > 2$ are due to the Fourier transform of the truncated momentum sums. The wave length is indicative of the lattice resolution at the current truncation to 4 momentum points.

A parametric fit to $V(r)$ with

$$V^{(\alpha)}(r) = \alpha_1 \frac{1 - \alpha_2 r^{\alpha_5}}{1 + \alpha_3 r^{\alpha_5 + 1} e^{\alpha_4 r}} + \alpha_0 \,, \tag{7}$$

at $\alpha_5 = 2$ fixed, was applied to the extrapolated potential. The result is shown in the left plot of Fig. 2 as a solid line. It suggests attraction at short distances followed by a repulsive barrier.

We have used $V^{(\alpha)}(r)$ in a Schrödinger equation to calculate s-wave scattering phase shifts $\delta_{\ell=0}^{I=2}(p)$, see the right plot of Fig. 2. The pion mass was set to multiples of the experimental value, corresponding to $m_\pi = 0.28$ in units of a^{-1}. The repulsive nature of the phase shifts is due to the hump of $V^{(\alpha)}(r)$ around, and extending beyond, $r \approx 1$, see left plot of Fig. 2. The data points in the right plot of Fig. 2 are experimental results compiled from [2].

To summarize, the steady conceptual and technical progress in lattice QCD made the calculation of few-quark correlation functions feasible. Relying on an improved action, a non-relativistic potential was extracted and scattering phase shifts for the $I = 2$ channel π–π system were computed. The range of the extracted potential is short compared to the current spatial resolution. The latter is determined by the somewhat large value of the lattice constant, and by the limitations imposed by the momentum truncation for the correlator matrices. This situation makes it difficult to reliably extract details of $V(r)$.

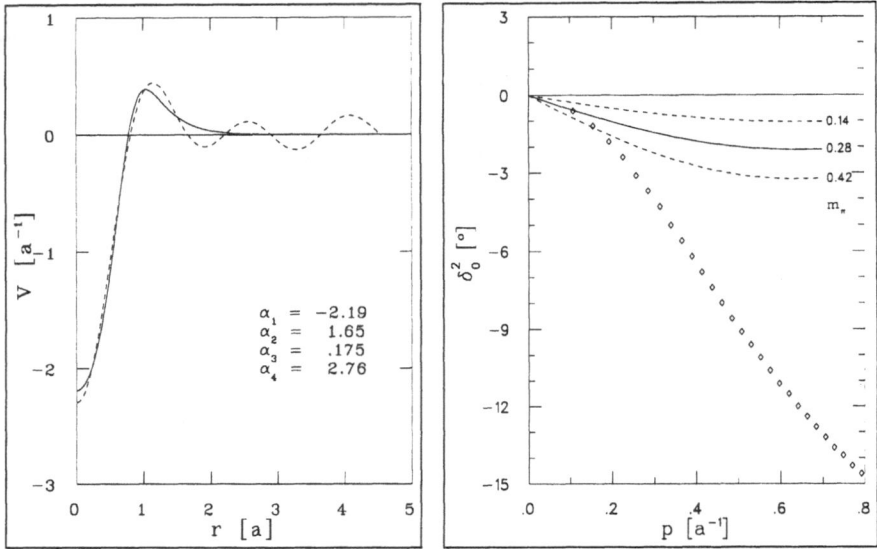

Figure 2. Result $V(r)$ of the chiral extrapolation shown by the dashed line. A parametric fit with $V^{(\alpha)}(r)$, see (7), gives the solid line. Scattering phase shifts $\delta_0^2(p)$ computed from $V^{(\alpha)}$ with three (ad hoc) values of m_π. The experimental pion mass corresponds to $m_\pi = 0.28$. Experimental data ($\diamond$) are from [2].

In addition to $V(r)$ there is a nonlocal potential [1] present which has not yet been computed.

The scattering phase shifts obtained from the underlying lattice study show repulsive behavior in the low-momentum region and, in this respect, compare favorably to experimental findings. Quantitatively, the computed phase shifts are too small by a sizeable factor. Relativistic corrections are at the 40% level at $p \approx 0.6a^{-1}$. The physics community comes now into a position to investigate hadron-hadron interactions from first principles of QCD.

References

1. H.R. Fiebig et al.: Nucl. Phys. **B** (Proc. Suppl.) **63A-C**, 188 (1998)

2. C.D. Froggatt, J.L. Peterson: Nucl. Phys. **B129**, 89 (1977)

3. H.W. Hamber et al.: Nucl. Phys. **B225**, 475 (1983); T. Eguchi, N. Kawamoto: Nucl. Phys. **B237**, 609 (1984)

4. H.R. Fiebig, R.M. Woloshyn: Phys. Lett. **B385**, 273 (1996)

5. J.D. Canosa, H.R. Fiebig: Phys. Rev. **D55**, 1487 (1997)

6. C. Albanese et al.: Phys. Lett. **B192**, 163 (1987); C. Alexandrou et al.: Nucl. Phys. **B414**, 815 (1994)

Few-Body Systems Suppl. 10, 471–474 (1999)

Few-
Body
Systems
© by Springer-Verlag 1999

Semi-relativistic RGM calculations of pion-pion scattering

R. Ceuleneer, C. Semay*

Université de Mons-Hainaut, Place du Parc 20, B-7000 Mons, Belgique

Abstract. Pion-pion scattering is investigated in the framework of the Resonating Group Method. The wave function of an isolated pion, described as a quark-antiquark system interacting through a potential composed of a central and a spin-spin part, is determined using the spinless Salpeter equation. Given these ingredients the kernel of the integro-differential equation governing the relative motion of the colliding pions is calculated using relativistic kinematics. The corresponding S- and D-wave pion-pion phase shifts and their Galilean counterparts are compared with experiment.

Recently we have carried out one-channel RGM calculations of pion-pion scattering in terms of the underlying quark dynamics expanding the wave function of an isolated pion, described as a quark-antiquark system, in harmonic oscillator bases extended up to $N\hbar\omega$ excitation energy with $0 \leq N \leq 8$ [1]. The exchange kernel of the RGM equation

$$T_{\boldsymbol{r}}\ \psi(\boldsymbol{r}) + \int K(\boldsymbol{r}, \boldsymbol{r}')\psi(\boldsymbol{r}')d^3r' = E\ \psi(\boldsymbol{r}) \qquad (1)$$

was evaluated without any simplifying assumption using usual shell-model techniques. The corresponding pion-pion scattering phase shifts were extracted from Eq. (1) numerically. The theoretical results obtained in this way were compared to those obtained by representing the pion wave function by a superposition of Gaussian-shaped functions [2]. In these works, devoted essentially to some numerical aspects of the RGM approach to hadron physics, we have emphasized the inconsistencies resulting from the use of Galilean kinematics. Our purpose is to make our previous results consistent, to some extent, with the requirements of special relativity by substituting the nonrelativistic kinetic energy operator of the ith particle by its relativistic counterpart. Thus the present calculations rely on the hamiltonian

$$H = T_1 + T_{\bar{1}} + T_2 + T_{\bar{2}} + V_{1\bar{1}} + V_{2\bar{2}} + V_{1\bar{2}} + V_{2\bar{1}} + V_{12} + V_{\bar{1}\bar{2}} \qquad (2)$$

---

*Chercheur qualifié F.N.R.S.

with

$$T_i = \sqrt{m^2 + \boldsymbol{p}_i^2} - m, \qquad m = m_u = m_d, \tag{3}$$

and

$$V_{ij} = -\frac{3}{16}\tilde{\lambda}_i.\tilde{\lambda}_j (v_{ij} + w_{ij}\, \boldsymbol{s}_i.\boldsymbol{s}_j). \tag{4}$$

The matrices $\tilde{\lambda}_i$ are the SU(3) color generators of the ith particle ($-\tilde{\lambda}_i^*$ for antiparticles). The dependence of V_{ij} upon the interparticle distance r_{ij} is taken of the form

$$v_{ij} = -\frac{A}{r_{ij}} + Br_{ij} - C, \quad w_{ij} = V_g \exp\left(-\frac{r_{ij}^2}{r_0^2}\right). \tag{5}$$

The wave function of an isolated S-wave pion is given by

$$\psi(r_{q\bar{q}}) = \phi(r_{q\bar{q}}) \left[s^{\frac{1}{2}}(q)s^{\frac{1}{2}}(\bar{q})\right]^0 \left[\tau^{\frac{1}{2}}(q)\tau^{\frac{1}{2}}(\bar{q})\right]^1 \tag{6}$$

where $s^{\frac{1}{2}}$ and $\tau^{\frac{1}{2}}$ denote the spin and isospin wavefunctions of a single quark or antiquark. The square brackets in (6) and in subsequent expressions stand for angular momentum and isospin coupling. From our nonrelativistic RGM calculations it appears that the extension of the harmonic oscillator expansion of the pion wave function beyond $N = 0$, which is essential to describe correctly the pion binding energy, has negligible effects upon the pion-pion scattering phase shifts. Therefore the calculations reported in the present work are carried out using a one-Gaussian approximation to describe the individual pions; thus the spatial wave function is approximated by

$$\phi(r_{q\bar{q}}) = b^{-\frac{3}{2}} \left(\frac{2}{\pi}\right)^{\frac{1}{4}} \exp\left(-\frac{r_{q\bar{q}}^2}{4b^2}\right) \tag{7}$$

where $b = (m\omega)^{-\frac{1}{2}}$ is the oscillator length parameter. The value of b and the corresponding theoretical pion mass are obtained by minimizing the expectation value of the relevant spinless Salpeter hamiltonian [3]

$$m_\pi = \min_b \int_0^\infty \phi(r_{q\bar{q}}) \left[2\sqrt{m^2 + \boldsymbol{p}_{q\bar{q}}^2} + v_{q\bar{q}} - \frac{3}{4}w_{q\bar{q}}\right] \phi(r_{q\bar{q})})r_{q\bar{q}}^2 dr_{q\bar{q}}. \tag{8}$$

The RGM wave function describing the two-pion system reads

$$\psi = \Phi(1\bar{1}2\bar{2}) - \Phi(1\bar{2}2\bar{1}) \tag{9}$$

with

$$\Phi(1\bar{1}2\bar{2}) = \sum_{L'M'I'} \phi_{L'M'}^{I'}(1\bar{1}2\bar{2}) \frac{f_{L'M'}^{I'}(r)}{r}, \quad L' + I' \quad \text{even} \tag{10}$$

and

$$\phi_{LM}^{I}(1\bar{1}2\bar{2}) = \mathcal{C}(1\bar{1})\mathcal{C}(2\bar{2})\phi(r_{1\bar{1}})\phi(r_{2\bar{2}})S(1\bar{1}2\bar{2})I(1\bar{1}2\bar{2})Y_{LM}(\hat{r}) \tag{11}$$

where $\mathcal{C}(1\bar{1})$ and $\mathcal{C}(2\bar{2})$ represent color singlets. The vector $\boldsymbol{r}$ is the relative separation of the pions $(1\bar{1})$ and $(2\bar{2})$. $S(1\bar{1}2\bar{2})$ and $I(1\bar{1}2\bar{2})$ are the total spin and isospin wavefunctions.

The partial wave $f_{LM}^I(r)$ satisfies the equation

$$\langle \phi_{LM}^I(1\bar{1}2\bar{2})|H - \mathcal{E}|\psi \rangle = 0, \tag{12}$$

in which the integration is carried out over the color, spin, isospin and spatial variables keeping r constant. Using expression (11) of ψ this equation splits into a direct and an exchange term thus

$$D_{LM}^I - E_{LM}^I = 0. \tag{13}$$

Owing to the color dependence of the interparticle interaction, only $V_{1\bar{1}}$ and $V_{2\bar{2}}$ contribute to the direct matrix element. Consequently, in the center-of-mass frame of the two-pion system, D_{LM}^I describes the relative motion of two free pions. Therefore, Eq. (1) reduces to

$$\left[2\sqrt{m_\pi^2 + \boldsymbol{p}^2} - 2m_\pi - E \right] \frac{f_L^I(r)}{r} + \frac{1}{r} \int_0^\infty K_L^I(r, r') f_L^I(r') dr' = 0 \tag{14}$$

where $\boldsymbol{p}$ and $E = \mathcal{E} + 4m - 2m_\pi$ are the relative momentum and energy of the colliding pions, respectively. The kernel of this equation is given by

$$K_L^I(r, r') = -\frac{48\,m\,r\,r'}{\pi^2 b^3} \left\{ \begin{matrix} \frac{1}{2} & \frac{1}{2} & 1 \\ \frac{1}{2} & \frac{1}{2} & 1 \\ \frac{1}{2} & \frac{1}{2} & 1 \\ 1 & 1 & I \end{matrix} \right\}$$

$$\times \left[T_L + \frac{\pi\sqrt{\pi}}{8m}(V_L - \mathcal{E}\delta_{L0}) \exp\left(-\frac{r^2 + r'^2}{2b^2} \right) \right], \tag{15}$$

with

$$T_L = \int_0^\infty dk \int_0^\infty dk' (kk')^2 I_1^L(k, k') j_L\left(\frac{kr}{b} \right) j_L\left(\frac{k'r'}{b} \right) \exp\left(-\frac{k^2 + k'^2}{2} \right) \tag{16}$$

in which,

$$I_1^L(k, k') = \int_{-1}^{+1} d\zeta P_L(\zeta) \frac{\exp(-\kappa^2)}{\kappa}$$

$$\times \int_0^\infty dK_1 \left\{ \sqrt{1 + (2mb)^{-2}K_1^2} - 1 \right\} K_1 \exp(-K_1^2) \sinh(2\kappa K_1). \tag{17}$$

The calculation of V_L is outlined in ref. [2].

From the numerical results relying on this model, it appears that the relativistic and Galilean exchange kernels of the RGM integro-differential equation are quite similar, which indicates that the relativistic effects are to a large extent taken into account, in the Galilean kernel, through the constituent quark mass and the parameters defining the interquark potential. Furthermore, our calculations show that these kernels bear closely upon the relativistic nature of the pion. Consequently, the large relativistic effects exhibited by the corresponding phase shifts originate not only from the relative motion of the colliding pions but also from the relativistic dynamics governing their internal structure.

We found that the $I = 2$ phase shifts obtained using relativistic kinematics are in excellent agreement with experiment, contrary to those relying on Galilean kinematics whose absolute values are too large. On the other hand the isoscalar phase shifts are too small whatever the procedures are used. However, the relativistic shifts increase smoothly with increasing $m_{\pi\pi} = 2m_\pi + E$, as the observed ones, whereas the Galilean phase shifts present an undesirable bump around 0.5 GeV. We found also that the exchange kernels are considerably smaller for $L = 2$ than for $L = 0$. Accordingly, the corresponding phase shifts are extremely small: up to $E = 1$ GeV they do not exceed 1 degree, which compares satisfactorily with experiment.

References

1. R. Ceuleneer and C. Semay: In: *Proceedings of the International Conference MESONS and LIGHT NUCLEI* (Stráž pod Ralskem, Czech Republic, 3-7 July 1995), Few-Body Systems Suppl. **9**, 303 (1995)

2. R. Ceuleneer, C. Semay and B. Silvestre-Brac: J. Phys. **G22**, 1395 (1996)

3. W. Lucha and F.F. Schöberl: In: *Proceedings of the International Conference on Quark Confinement and the Hadron Spectrum* (Como, Italy, 20-24 June 1994) p. 100, eds. N. Brambilla and G.M. Prosperi. World Scientific 1995

Few-Body Systems Suppl. 10, 475–478 (1999)

Momentum Dependent Nucleon-Nucleon Potentials and the Off-Shell Dependence of the Triton Binding Energy

H. Leeb[1]*, F. Korinek[1], M. Braun[2], S.A. Sofianos[2]

[1] Institute of Nuclear Physics, Technische Universität Wien, Wiedner Hauptstraße 8-10/142, A-1040 Vienna, Austria
[2] Department of Physics, University of South Africa, P.O. Box 392, Pretoria 0003, South Africa

Abstract. An inverse scattering scheme for nucleon-nucleon potentials in coupled channels is presented which includes a term proportional to the square of the momentum. The method is used to construct a set of fully phase equivalent potentials having a different momentum dependence. A characteristic dependence of the triton binding energy on the off-shell behaviour of these potentials is found.

1 Introduction

The nucleon-nucleon (NN) interaction is of central importance in nuclear physics and much effort has been devoted to its determination. Nevertheless, we are still not able to give a quantitative description in terms of QCD, the generally accepted theory of the strong interaction. Nowadays, the so-called realistic NN-potentials [1, 2, 3], are guided by the boson exchange picture originally introduced by Yukawa [4]. They reproduce quite well the scattering phase shifts and mixing parameters below the pion threshold and are extensively used for the evaluation of few- and many-nucleon observables.

An important issue of recent investigations is the off-shell dependence of the NN-interaction. This question is still not settled because there is no clear theoretical constraint and two-nucleon scattering data alone cannot clarify this point. Nevertheless, the corresponding differences in evaluated few- and many-nucleon observables cannot be attributed to off-shell effects alone because at higher energies these potentials exhibit different on-shell properties as well whose effect on few- and many-body observables is not well known.

---

*E-mail address: leeb@kph.tuwien.ac.at

In this contribution we propose a systematic way to study off-shell effects using fully on-shell equivalent NN-potentials with different momentum dependence which naturally arises in relativistic calculations. In this work we consider the simplest linearized form,

$$V(r) = V_a(r) + \frac{\mathbf{p}^2}{m} V_b(r) + V_b(r) \frac{\mathbf{p}^2}{m} \,, \tag{1}$$

which has been used for the parametrization of realistic NN-potentials [1, 2]. A first calculation performed by varying the momentum dependent part of the Nijmegen-I potential [1] revealed a significant off-shell dependence.

2 Formalism

Inverse scattering techniques [5] provide an elegant way for the construction of NN-potentials from scattering data. Although the methods were given already in the sixties first quantitative constructions of NN-potentials from phase shift data were presented only in the last decade [6]. These methods, however, were limited to local potentials having a fixed off-shell dependence. Recently, we developed inverse scattering methods which include also momentum dependent parts of the form of Eq. (1). Our methods are based on Darboux transformations and details of the formalism for uncoupled channels can be found in [7]. Here, we outline a non-trivial generalization of this scheme to coupled-channel systems.

Similarly to uncoupled channels [7] there exists a one-to-one relationship between a momentum dependent potential (1) and a linearly energy dependent potential. Thus we restrict ourselves to Darboux transformations of the coupled Sturm-Liouville equation

$$\left(\frac{\mathrm{d}^2}{\mathrm{d}x^2} \mathbf{1} - D(x) - U_0(x) \right) \Psi_0(k, x) = k^2 h(x) \Psi_0(k, x) \tag{2}$$

where k is the wave number, $h(x)$ is a scalar function which contains the energy dependent part of the potential (see [7]), $U_0(x)$ is a quadratic potential matrix and $D(x)$ is a diagonal matrix containing the centrifugal term. Assuming that all solutions ζ_0 associated with the potential $U_0(x)$ are known, then it can be shown [8] that the functions

$$\zeta_2(\alpha, x) = \zeta_0(\alpha, x) \left(1 - A \int_x^c \mathrm{d}s\, h(s) \bar\zeta_0(\alpha, s) \zeta_0(\alpha, s) \right)^{-1}, \tag{3}$$

$$\Psi_2(k, x) = \Psi_0(k, x) + \frac{\zeta_2(\alpha, x)}{\alpha^2 - k^2} A\, W\left[\zeta_0(\alpha, x); \Psi_0(k, x) \right] \tag{4}$$

with

$$W\left[\zeta_0(\alpha, x); \Psi_0(k, x) \right] = \bar\zeta_0(\alpha, x) \left(\frac{\mathrm{d}}{\mathrm{d}x} \Psi_0(k, x) \right) - \left(\frac{\mathrm{d}}{\mathrm{d}x} \bar\zeta_0(\alpha, x) \right) \Psi_0(k, x) \tag{5}$$

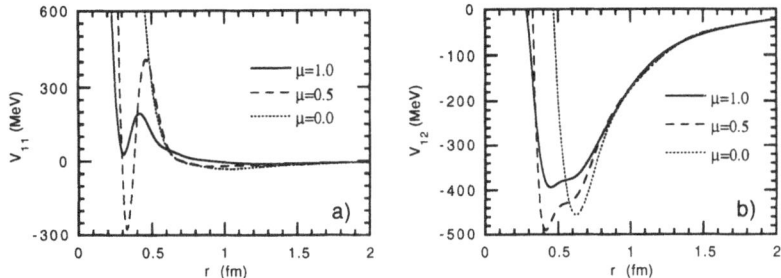

Figure 1. $V_a(r)$ of the 11-component (a) and the 12-component (b) of the NN-potential matrix in the $^3S_1-^3D_1$-channel for different μ-values.

are solutions of the same Sturm-Liouville equation with the potential

$$U_2(x) = U_0(x) - 2\sqrt{h(x)}\frac{\mathrm{d}}{\mathrm{d}x}\left(\sqrt{h(x)}\zeta_2(\alpha, x)A\bar{\zeta_0}(\alpha, x)\right). \qquad (6)$$

which depends on the chosen solution ζ_0, the complex value α and the symmetric matrix A. The bar means transposition of the vector solution. Korinek [8] developed a generalisation of this transformation (4,6) associated with a set of functions $\zeta_0(\alpha_i, x)$ and constant matrices A_i. Thus a hierarchy of potentials is obtained where the wave functions are known in closed form. It can be shown [8] that the α_i correspond to poles of the S-matrix and the corresponding matrices A_i are related to the residues at $k = \alpha_i$.

To construct NN-potentials with this novel scheme we followed the procedure of the Hamburg group [6] and approximated the phase shift and mixing parameters by rational expressions via Padé approximants. Then the poles and residues are used to determine the parameters α_i and A_i. The constructed set of NN-potentials for the $^1S_0$ and $^3S_1-^3D_1$ partial waves are on-shell equivalent to the Nijmegen-I potential up to at least 2 GeV (see Fig. 1). These potentials have different strength $V_b(r)$, i.e. we use $V_b(r) = \lambda V_b^{Nij-I}(r)$ for the $^1S_0$ and $V_b(r) = \mu V_b^{Nij-I}(r)$ for the $^3S_1-^3D_1$ partial waves with different λ and μ.

3   Triton binding energy

We used this set of on-shell equivalent NN-potentials in a five-channel calculation of the triton binding energy, E_t. The results obtained, summarized in Table 1, clearly demonstrate the sensitivity of E_t on off-shell variations of the NN-interaction stemming from uncertainties in its momentum dependence. This is in agreement with the finding of Jäde et al. [6] that there is no significant difference in the evaluated value of E_t obtained with the Paris-potential ($\lambda = \mu = 1$) and with its phase equivalent local potential ($\lambda = \mu = 0$). Our results, however, indicate that this apparent insensitivity of E_t is only due to a

Table 1. Triton binding energy E_t as a function of the strengths (λ, μ) of the momentum dependent part of the NN-interaction

	$\lambda = 0.0$	$\lambda = 0.5$	$\lambda = 1.0$
$\mu = 0.0$	-7.475 MeV	-7.585 MeV	-7.644 MeV
$\mu = 0.5$	-7.443 MeV	-7.560 MeV	-7.625 MeV
$\mu = 1.0$	-7.242 MeV	-7.358 MeV	-7.423 MeV

compensation of the effects of the momentum dependent terms in the $^1S_0$ and the $^3S_1$-$^3D_1$ channel.

4 Conclusions

The newly developed inverse scattering scheme for coupled partial waves provides a way to construct phase equivalent potentials with different momentum dependence. This enable us to study the corresponding off-shell effects on the triton binding energy. The significant variations in E_t found suggest that systematic studies with such sets of fully phase equivalent potentials must be undertaken. Such investigations that include also scattering observables are in progress.

Acknowledgement. Work supported by Fonds zur Förderung der Wissenschaftlichen Forschung, Österreich, project number P10467-PHY, the Foundation for Research and Development and the University of South Africa.

References

1. V.G.J. Stoks et al.: Phys. Rev. **C49**, 2950 (1994)

2. M. Lacombe et al.: Phys. Rev. **C21**, 861 (1980)

3. R. Machleidt, K. Holinde and Ch. Elster: Phys. Rep. **149**, 1 (1987)

4. H. Yukawa: Proc. Phys.-Math. Soc. Japan **17**, 48 (1935)

5. K. Chadan and P.C. Sabatier: In: *Inverse Problems in Quantum Scattering Theory.* New York: Springer 1989

6. L. Jäde, M. Sander and H.V. von Geramb: In: *Inverse and Algebraic Quantum Scattering Theory* (Lecture Notes in Physics, vol. 488), p.124. Berlin: Springer 1997

7. F. Korinek et al.: Nucl. Phys. **A607**, 123 (1996)

8. F. Korinek: Thesis. Technische Univ. Wien 1998

Few–Body Systems Suppl. 10, 479–482 (1999)

Few–
Body
Systems
© by Springer-Verlag 1999

Low–momentum effective theory for nucleons using the method of unitary transformation

E. Epelbaoum[1,2]*, W. Glöckle[1]†, Ulf-G. Meißner[2] **

[1] Ruhr-Universität Bochum, Institut für Theoretische Physik II, D-44870 Bochum, Germany
[2] Forschungszentrum Jülich, Institut für Kernphysik (Theorie), D-52425 Jülich, Germany

Abstract. We have shown explicitly for the case of a Malfliet–Tjon type model potential that it is possible to construct a low–momentum effective theory for two nucleons. To that aim we decouple the low and high momentum components of this two–nucleon potential using the method of unitary transformation. The corresponding unitary operator is parametrized in terms of an operator A. The requirement of decoupling these two spaces leads to a nonlinear integral equation, which we have solved numerically. We also discuss another possibility of finding a solution to this nonlinear equation using the two–nucleon t-matrix.

1 Introduction

Chiral perturbation theory for two and more nucleons became a subject of a great research interest in the past few years, see e.g. the pioneering work in [1, 2]. One hopes to be able to clarify the structure of nuclear forces in this way. However, only the low–momentum matrix elements of nuclear forces may be systematically treated in this approach since it is based on a consistent power counting of small momenta and pion masses compared to the typical hadronic scale of $\Lambda_{\text{had}} \simeq 1$ GeV. A natural problem arises due to the appearance of shallow nuclear bound states indicating the breakdown of perturbation theory. Furthermore, solving the Lippmann–Schwinger equation for constructing the deuteron necessarily involves momenta $|\boldsymbol{p}| > \Lambda_{\text{had}}$. For such momenta, the chiral effective potential constructed according to the conventional power counting rules is no longer applicable. One might therefore question the validity or usefulness of such an approach alltogether. For recent discussions of

---

*E-mail address: evgeni.epelbaum@hadron.tp2.ruhr-uni-bochum.de
†E-mail address: walter.gloeckle@hadron.tp2.ruhr-uni-bochum.de
**E-mail address: Ulf-G.Meissner@fz-juelich.de

this subject see [3] and a different power counting scheme has been presented in [4]. A similar problem arises in standard few– and many–body calculations based on realistic nucleon–nucleon (NN) forces. The potentials, if derived from meson–exchange diagrams, are generally based on a non–relativistic expansion in powers of momenta over the nucleon mass and are then used in various types of bound state equations. These usually involve integrations over a much larger range of momenta as used in the construction of the NN potentials. In this general context the question of the existence and the properties of a low–momentum effective theory for nucleons are thus of great importance. We show here that it is indeed possible to construct an effective two–nucleon potential from a given realistic potential which involves only low momenta, i.e. momenta below a chosen momentum cut–off, but which gives exactly the same results for bound and scattering states.

2 Formalism

In this study we consider the momentum space Hamiltonian for the two–nucleon system of the form $H(p, p') = H_0(p)\delta(p - p') + V(p, p')$, where H_0 stands for the kinetic energy and V is a momentum–space Malfliet–Tjon [5] potential

$$V_{\mathrm{MT}}(q_1, q_2) = \frac{1}{2\pi^2}\left(\frac{V_R}{t + \mu_R^2} - \frac{V_A}{t + \mu_A^2}\right) , \tag{1}$$

with $t = (q_1 - q_2)^2$. We choose the parameters as given in [6], $V_R = 7.29$, $V_A = 3.18$, $\mu_R = 614$ MeV and $\mu_A = 306$ MeV. From here on, we only consider the S-wave part of this potential which can be obtained analytically. Although this potential is quite simple, it captures essential features of the NN interaction, in particular, it supports exactly one bound state at $E = -2.23$ MeV. We introduce the projection operators $\eta(\lambda) = \int d^3p \, |p\rangle\langle p|$, $|p| \leq (>) \Lambda$, where Λ is a momentum cut–off which separates the low from the high momentum regions. Apparently $\eta^2 = \eta$, $\lambda^2 = \lambda$, $\eta\lambda = \lambda\eta = 0$ and $\eta + \lambda = 1$. In order to decouple the Schrödinger equation for the low and high momentum components

$$\begin{pmatrix} \eta H \eta & \eta H \lambda \\ \lambda H \eta & \lambda H \lambda \end{pmatrix} \begin{pmatrix} \eta|\Psi\rangle \\ \lambda|\Psi\rangle \end{pmatrix} = E \begin{pmatrix} \eta|\Psi\rangle \\ \lambda|\Psi\rangle \end{pmatrix} , \tag{2}$$

we introduce the unitary transformation U, $H \longrightarrow H' = U^\dagger H U$, such that

$$\eta H' \lambda = \lambda H' \eta = 0 . \tag{3}$$

We choose a parametrisation of U given by Okubo [7],

$$U = \begin{pmatrix} (1 + A^\dagger A)^{-1/2} & -A^\dagger(1 + AA^\dagger)^{-1/2} \\ A(1 + A^\dagger A)^{-1/2} & (1 + AA^\dagger)^{-1/2} \end{pmatrix} , \tag{4}$$

where A has to satisfy the condition $A = \lambda A \eta$. It is then straightforward to recast the conditions Eq. (3) in a form of the following nonlinear equation for the operator A

$$V(\boldsymbol{p},\boldsymbol{q}) \quad - \quad \int d^3q'\, A(\boldsymbol{p},\boldsymbol{q}')V(\boldsymbol{q}',\boldsymbol{q}) + \int d^3p'\, V(\boldsymbol{p},\boldsymbol{p}')A(\boldsymbol{p}',\boldsymbol{q})$$

$$- \quad \int d^3q'\, d^3p'\, A(\boldsymbol{p},\boldsymbol{q}')V(\boldsymbol{q}',\boldsymbol{p}')A(\boldsymbol{p}',\boldsymbol{q})$$

$$= \quad (E_{\boldsymbol{q}} - E_{\boldsymbol{p}})\, A(\boldsymbol{p},\boldsymbol{q}) \quad . \tag{5}$$

Here we denoted the momenta from the η and λ–spaces by $\boldsymbol{q}$ and $\boldsymbol{p}$, respectively. Once A and thus U have been determined, the effective Hamiltonian in the η–space can easily be obtained. Another way of determining the operator A using the scattering states to the original Hamiltonian H is exhibited in refs. [8, 10].

3 Results

We have solved the nonlinear Eq. (5), by discretizing using Gauss–Legendre quadrature points. To avoid the singularities of the operator $A(\boldsymbol{p},\boldsymbol{q})$ when both $|\boldsymbol{p}|$ and $|\boldsymbol{q}|$ go to the cut–off Λ we provide a regularization scheme [9]. To be specific, we redefine the original potential $V(\boldsymbol{p},\boldsymbol{p}')$ by multiplying it with some smooth functions $f(\boldsymbol{p})$ and $f(\boldsymbol{p}')$ which are zero in some neighborhood of the point $|\boldsymbol{p}| = \Lambda$ and one elsewhere. The regularization is chosen mild enough so that it has no effect on the observables. In Fig. 1, we compare the original

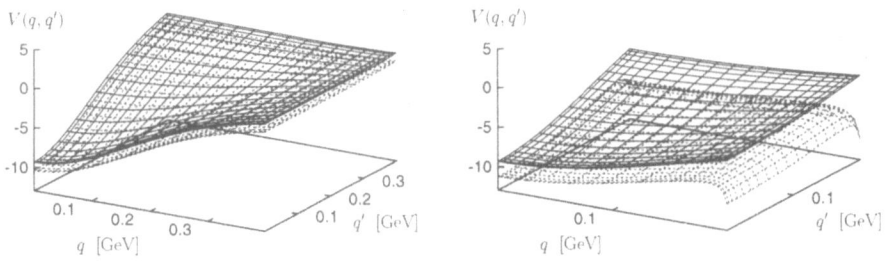

Figure 1. Effective two–nucleon potential (dashed lines) in comparison with the original potential (solid lines). Left (right) panel: Λ=400 (200) MeV.

potential with the effective one. The latter is defined via $V_{\mathrm{eff}} = \mathcal{H}' - \mathcal{H}_0$. In the range of the small momenta, the potentials are very similar for Λ =400 MeV. The main effect of integrating out of high momentum components at the level of the potential seems to be given in this case just by an overall shift. However, one finds significant differences between the effective and the original potential when the cut–off is chosen very small, $\Lambda \leq 200$ MeV. This is shown in Fig. 1. The solution of the effective LS equation is now very simple since the integration is confined to $q \leq \Lambda$. Using 40 quadrature points the resulting binding energy agrees within 9 digits with the result gained from the

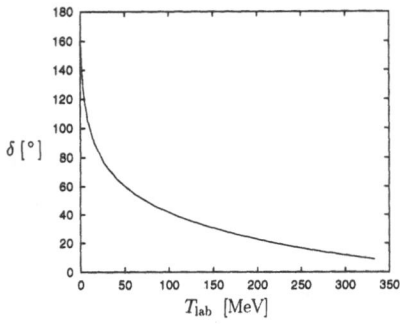 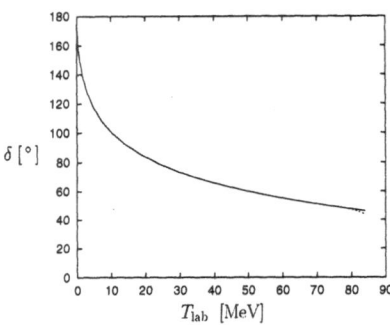

Figure 2. Phase shifts from the effective potential (dashed line) and the original potential (solid line) as a function of the kinetic energy in the lab frame. Left (right) panel: $\Lambda = 400$ (200) MeV.

corresponding homogeneous equation driven by the original potential V and defined in the whole momentum range. Furthermore, the S—wave phase shifts agree perfectly well solving either the Lippmann–Schwinger equation in the full momentum space or in the space of low momenta only. This is shown in Fig. 2. Note that in both figures we have not shown the corresponding functions in the region of regularization.

In summary, we have shown how to construct an effective low energy theory for nucleons based on the method of unitary transformations. For a simple S—wave potential, we have shown that the theory projected onto the subspace of momenta below a given cut–off reproduces exactly the features of the original one.

References

1. S. Weinberg: Phys. Lett. **B251**, 288 (1990); Nucl. Phys. **B363**, 3 (1991)

2. C. Ordóñez, L. Ray and U. van Kolck: Phys. Rev. Lett. **72**, 1982 (1994); Phys. Rev. C53, 2086 (1996); U. van Kolck: Phys. Rev. **C49**, 2932 (1994)

3. S.R. Beane, T.D. Cohen and D.R. Phillips: Nucl. Phys. **A632**, 445 (1998)

4. D.B. Kaplan, M.J. Savage and M.B. Wise: Phys. Lett. **B424** , 390 (1998); nucl-th/9802075

5. R.A. Malfliet and J.A. Tjon: Nucl. Phys. **A127**, 161 (1969)

6. Ch. Elster, J.H. Thomas and W. Glöckle: Few Body Syst. **24**, 55 (1998)

7. S. Okubo: Prog. Theor. Phys. **12**, 603 (1954)

8. K. Suzuki and R. Okamoto: Prog. Theor. Phys. **70**, 439 (1983)

9. E. Epelbaoum, W. Glöckle and Ulf-G. Meißner: nucl-th/9804005; Phys. Lett. **B** (1998) in print

10. E. Epelbaoum, W. Glöckle and Ulf-G. Meißner: in preparation.

Few-Body Systems Suppl. 10, 483–486 (1999)

Few-
Body
Systems
© by Springer-Verlag 1999

N-Particle Effective Generators of the Poincaré Group Derived from a Field Theory

A. Krüger, W. Glöckle

Institut für Theoretische Physik II, Ruhr-Universität Bochum, D-44780 Bochum, Germany

Abstract. In quantum mechanics the principle of relativity is guaranteed by unitary operators being associated with inhomogeneous Lorentz transformations ensuring that quantum mechanical expectation values remain unchanged. In field theory the ten generators of inhomogeneous Lorentz transformations can be derived from a scalar Lagrangian density describing the physical system of interest. They obey the well known Poincaré Lie algebra. For interacting systems some of the generators become operators allowing for particle production or annihilation so that the generators act on the full Fock space.

However, given a field theory on the whole Fock space we prove that it is possible to construct generators acting on a subspace with a finite number of particles by one and the same unitary transformation of all generators leaving the Poincaré algebra valid. In this manner it is in principle possible to derive a relativistically invariant theory of interacting particles on a Hilbert space with a finite number of particles from a field theoretical Lagrangian.

1 Introduction

The principle of relativity requires that quantum mechanical matrix elements remain invariant under inhomogeneous Lorentz transformations (Poincaré transformations):

$$|\langle\Phi|\Psi\rangle| \xrightarrow{T} |\langle\Phi'|\Psi'\rangle| \overset{!}{=} |\langle\Phi|\Psi\rangle| \tag{1}$$

where T is a Poincaré transformation. The Ts form a ten parameter Lie group with

$$T = T(\boldsymbol{\theta}, \boldsymbol{\zeta}, a^\mu) \tag{2}$$

where $\boldsymbol{\theta}$ is the angle of a rotation, $\boldsymbol{\zeta}$ is the rapidity of a boost and a^μ is a four dimensional translation. In the case of proper orthochronous Lorentz transformations we need the Ts to be represented by unitary operators U to guarantee (1). We employ a symbolical notation and write

$$U(\boldsymbol{\theta}, \boldsymbol{\zeta}, a^{\mu}) = \mathrm{e}^{-\mathrm{i}(a_{\mu} P^{\mu} + \boldsymbol{\theta} \cdot \boldsymbol{J} + \boldsymbol{\zeta} \cdot \boldsymbol{K})}. \tag{3}$$

Here P^{μ}, J_n, K_n are hermitian operators called *generators* which can be identified as being

$$H \equiv P^0 \quad \text{Hamilton operator} \tag{4}$$

$$P_n \qquad \text{momentum opepator} \tag{5}$$

$$J_n \qquad \text{ang. mom. operator} \tag{6}$$

$$K_n \qquad \text{boost operator.} \tag{7}$$

The group properties of Poincaré transformations require a set of commutation relations of the generators the so called *Lie algebra* of the Poincaré group being characteristic for Poincaré transformations:

$$[O_i, O_j] = f_{ijk} O_k \tag{8}$$

where

$$O_i = P^{\mu}, J_n, K_n, \quad \text{and } i = 1, \ldots 10. \tag{9}$$

In a standard manner one can find a representation of the generators from field theory starting from a scalar Lagrangian density:

$$\mathcal{L} = \mathcal{L}_0 + \mathcal{L}_I. \tag{10}$$

We concentrate on the case of a theory of interacting particles where $\mathcal{L}_I \neq 0$. In Dirac's instant form of dynamics we get:

$$\begin{aligned} H &= H_0 + H_I \\ \boldsymbol{K} &= \boldsymbol{K}_0 + \boldsymbol{K}_I \\ \boldsymbol{P} &= \boldsymbol{P}_0 \\ \boldsymbol{J} &= \boldsymbol{J}_0. \end{aligned} \tag{11}$$

In general the terms H_I and $\boldsymbol{K}_I$ will allow for particle creation and annihilation so that the set (11) acts on the full Fock space with an infinite number of particles. However we may be interested in a space with a fixed number of particles. Due to Okubo [1] it is possible to transform the generator of time translations H by a unitary operator $\mathcal{U}$ in a way that the transformed operator $\tilde{H}$ is an *effective* operator, which means that $\tilde{H}$ doesn't mix the space of interest with a finite number of particles with the rest of the Fock space.

The question arises whether this operator $\mathcal{U}$ will also transform the other generators O_i into *effective* operators $\tilde{O}_i$.

2 Existence proof of effective generators

We prove [2, 3] that Okubo's operator $\mathcal{U}$ is indeed transforming *all the generators* into effective generators at the *same time*. We want to write down $\mathcal{U}$ and then give the basic assumptions and the spirit of the proof.

To start with, we define a projector η on the subspace of interest with a finite number of particles and a projector $\Lambda \equiv 1 - \eta$ on the rest of the Fock space. Then any state in Fock space can be written as:

$$|\chi\rangle = \begin{pmatrix} \eta\chi \\ \Lambda\chi \end{pmatrix} \tag{12}$$

and the generators look in general like:

$$O_i = \begin{pmatrix} \eta O_i \eta & \eta O_i \Lambda \\ \Lambda O_i \eta & \Lambda O_i \Lambda \end{pmatrix} \quad i = 1, \ldots 10. \tag{13}$$

Using Dirac's instant form of dynamics we have:

$$\eta O_i \Lambda \neq 0 \neq \Lambda O_i \eta \qquad O_i = H, K_n \tag{14}$$

and

$$\eta O_i \Lambda = 0 = \Lambda O_i \eta \qquad O_i = P_n, J_n. \tag{15}$$

Our aim is to find a transformation $\mathcal{U}$ giving

$$\tilde{O}_i = \begin{pmatrix} \eta \tilde{O}_i \eta & 0 \\ 0 & \Lambda \tilde{O}_i \Lambda \end{pmatrix} \quad i = 1, \ldots 10. \tag{16}$$

We use Okubo's form of a unitary operator:

$$\mathcal{U} = \begin{pmatrix} (1 + A^\dagger A)^{-\frac{1}{2}} \eta & (1 + A^\dagger A)^{-\frac{1}{2}} A^\dagger \\ -(1 + AA^\dagger)^{-\frac{1}{2}} A & (1 + AA^\dagger)^{-\frac{1}{2}} \Lambda \end{pmatrix} \tag{17}$$

where

$$A = \Lambda A \eta. \tag{18}$$

Then the diagonalization of a generator O_i requires a nonlinear condition on A_i:

$$\Lambda([O_i, A_i] + O_i - A_i O_i A_i)\eta = 0 \quad i = 1, \ldots 10 \tag{19}$$

To solve these equations we make the following assumptions:

$$A_i = \sum_{\nu=1}^{\infty} g^\nu A_{i\nu} \tag{20}$$

$$\text{and} \quad H_I \sim g, \quad K_{In} \sim g. \tag{21}$$

Using the form (20) of A_i we can give formal solutions of A_i, $i = 1, \ldots 10$ by solving Eqs. (19) order by order in the parameter g.

To start our proof we solve for H or $i = 1$ respectively, i.e. we find $A_H \equiv A_1$. The result [3] looks like:

$$A_{H1} = -\int d\Lambda d\eta \frac{1}{E_\Lambda - E_\eta} |\Lambda\rangle\langle\Lambda|H_I|\eta\rangle\langle\eta|$$

$$A_{H2} = \ldots \tag{22}$$

$$\vdots$$

Here the integration is over the η and Λ spaces spanned by eigenstates of the momentum operator and E_η and E_Λ are the corresponding eigenvalues of the free Hamilton operator.

Given $A_{H\nu}$ we can show easily ([3], Eqs. (53)-(57)) that A_H is also a solution of Eqs. (19) for $O_i = P_n$ and $O_i = J_n$. However a lengthy but straight forward calculation is needed ([3], Eqs. (66)-(82)), to show that A_H is also a solution of Eqs. (19) for $O_i = K_n$. Consequently one and the same operator $A \equiv A_H$ is solution of Eqs. (19) for all the generators O_i, $i = 1, \ldots 10$.

Hence we proved the existence of one unitary operator $\mathcal{U}$ transforming the ten generators into effective generators at the same time.

3 Conclusions

We conclude that because of the unitarity of $\mathcal{U}$ the Lie algebra of the Poincaré group (8) is still valid for the effective generators on the η space:

$$
\begin{aligned}
[O_i, O_j] &= f_{ijk} O_k \\
\Rightarrow \quad [\eta \tilde{O}_i \eta, \eta \tilde{O}_j \eta] &= f_{ijk} \eta \tilde{O}_k \eta.
\end{aligned}
\tag{23}
$$

It follows that a nontrivial representation of the generators of Poincaré transformations is found which acts on a space with a finite number of particles. Hence expectation values on this subspace are still Lorentz invariant.

Of interest for practical calculations is that the original Schrödinger equation decomposes like

$$
H|\chi\rangle = E|\chi\rangle \longrightarrow \left\{ \begin{array}{l} \eta \tilde{H} \eta |\tilde{\chi}\rangle = E\eta|\tilde{\chi}\rangle \\ \Lambda \tilde{H} \Lambda |\tilde{\chi}\rangle = E\Lambda|\tilde{\chi}\rangle \end{array} \right.
\tag{24}
$$

Here the eigenvalue E remains unchanged. Being interested in a system of two particles for instance one chooses η to project on the subspace of these two particles. Okubo showed that if there are incident particles only in the η space then the component $\Lambda|\tilde{\chi}\rangle$ is actually zero so that we just need to solve

$$
\eta \tilde{H} \eta |\tilde{\chi}\rangle = E\eta|\tilde{\chi}\rangle.
\tag{25}
$$

References

1. S. Okubo: Prog. Theor. Phys. **12**, 603 (1954)

2. W. Glöckle and L. Müller: Phys. Rev. **C23**, 1183 (1981)

3. A. Krüger and W. Glöckle: nucl-th/9712043

Few-Body Systems Suppl. 10, 487–490 (1999)

Few-
Body
Systems
© by Springer-Verlag 1999

Self-similar Properties of the Light Nuclei Interactions

V.V. Glagolev[1], G. Martinská[2], M.S. Nioradze[3], T. Siemiarczuk[4], J. Urbán[2]

[1] Joint Institute for Nuclear Research, Dubna, Russia
[2] University of P.J. Šafarik, Košice, Slovak Republic
[3] High Energy Physics Institute, Tbilisi State University, Tbilisi, Georgia
[4] Institute for Nuclear Research, Warsaw, Poland

Abstract. A new way for representating nuclei fragmentation data is suggested. The self-similar behaviour of these processes called out by the kinematics is demonstrated. The convenience of working in accelerated nuclei is underlined, particularly for the determination of the binding energy of a wide class of nuclei fragments.

1 Introduction

Nowadays fairly large samples of experimental data on the light nuclei $(d, {}^3He, {}^4He)$ proton interactions have been accumulated in the full solid angle conditions of the JINR LHE 1m hydrogen bubble chamber. The different fragmentation reactions, including both mesonless and pion containing channels have been reliably identified [1].

It is now possible to compare the characteristics of the reactions at different energies and with different fragment masses. One of the classical data representation – the Chew–Low plot [2] has usually been used for the $2 \to 3$ processes and for multiparticle processes.

However in the case of nuclear collisions the masses of the produced fragments A' may considerably differ from that of the initial nucleus A. The comparison of the reactions in terms of the Chew–Low plots become inappropriate at least for the great differences in the four-momentum tranfers which lead to a " dispersion " of events containing different fragments.

2 A new fragmentation data representation

Changing nothing on the generality of the proposed fragmentation data representation, let us examine the simple case of the proton-nucleus interaction: $p + A \to A' + X$.

The use of the relative four-velocity squared $b_{AA'}$, introduced by Baldin [3]

$$b_{AA'} = -(\frac{P_A}{M_A} - \frac{P_{A'}}{M_{A'}})^2 \mid_{\boldsymbol{p_A}=0} = 2(\frac{E_{A'}}{M_{A'}} - 1) \qquad (1)$$

instead of $|t|$ is the basic difference between the Chew–Low plot and the proposed way. Here and from now on P stands for the four-momentum. This variable seems to be the most suitable for the processes in the relativistic nuclear physics and it is widely used.

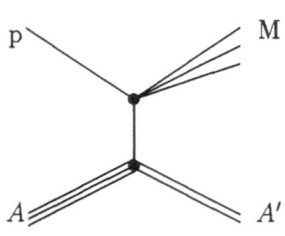

The second difference lies in the normalization of the missing mass squared MM^2 to the studied fragment A'.

In the notation of the diagram (Fig. 1), M can be interpreted as a product of the interaction of the incoming proton with an off-mass-shell part of the nucleus A, denoted R (recoil). We use the quantity S', expressed in the recoiled on-mass-shell "R" rest frame as:

Figure 1. Schematical representation of the fragmentation process.

$$S' = (P_p + P_{"R"})^2 = M_p{}^2 + M_{"R"}{}^2 + 2E_p M_{"R"},$$

for normalization purposes. One can easily get an expression for the contour line of the plot:

$$\frac{MM^2}{S'} = \frac{(M_A - M_{A'})^2 + 2E_p(M_A - M_{A'}) + M_p^2}{S'} - \frac{M_{A'}(M_A + E_p)}{S'}b_{AA'}$$
$$\pm \frac{2M_{A'}|\boldsymbol{p_p}|\sqrt{b_{AA'} + \frac{b_{AA'}{}^2}{4}}}{S'}, \qquad (2)$$

for which the lower bound is given by $M_X{}^2/S'$ where M_X is the sum of masses in the upper vertex of the diagram in Fig. 1.

As an example in Fig. 2 the full contour line for the ^{4}Hep$\rightarrow^3$Hpp at 2.15 A GeV/c is shown.

We demonstrate the use of the introduced representation on the fragmentation process, characterized by small values of $b_{AA'} \ll 1$. Examples of the contour lines in the fragmentation region for the discussed reaction are displayed in Fig. 3 at several incoming proton momenta from 2 to 100 GeV/c. In addition to the contours the figure also shows the central lines, corresponding to the two first terms of Eq. (2).

In this region one can see the characteristic behaviour of the limiting curve width. The width Γ is defined as the distance between the lower and upper branches of the graph at a given value of $b_{AA'}$. From Eq. (2), the expression for the width comes out:

$$\Gamma = \frac{4M_{A'}|\boldsymbol{p_p}|\sqrt{b_{AA'} + \frac{b_{AA'}{}^2}{4}}}{S'}.$$

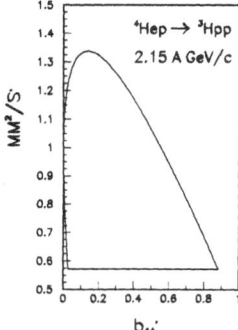

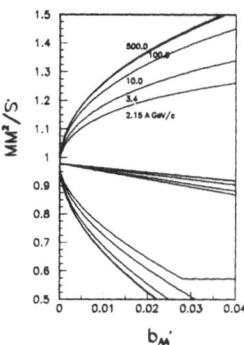

Figure 2. The contours of the $(\frac{MM^2}{S'}$ vs $b_{AA'})$ plots for the reaction $^4\text{Hep}\rightarrow^3\text{Hpp}$ at 2.15 A GeV/c.

Figure 3. Limiting contours for the $^4\text{Hep}\rightarrow^3\text{Hpp}$ reaction in the fragmentation region at different incoming proton momenta.

The ratio of the widths at two energies for a given type of reaction (fixed A and A') does not depend on $b_{AA'}$ and, with increasing energy, tends to 1. In this way the self-similar behaviour of the fragmentation process has been demonstrated.

3 Experimental results

Figure 4 displays the $^4\text{Hep}\rightarrow^3\text{Hpp}$ reaction data and the contour lines of the proton momentum 3.4 A GeV/c. The given experimental points, as it can be seen, are mainly inside the corresponding contour lines. A similar situation takes place also for the $^3\text{Hep}\rightarrow\text{dpp}$ reaction at 4.5 A GeV/c.

We would like to emphasize an additional feature of the plot, suitable for a supplementary use. At $b_{AA'}=0$ the branches of the contour lines meet at

$$\frac{MM^2}{S'} = 1 - 2\epsilon\frac{(M_{''R''} + E_p)}{S'} + \frac{\epsilon^2}{S'}, \qquad (3)$$

where ϵ is the binding energy of the fragment A' in the parent nucleus A.

For the special case of quasi NN scattering, this expression, with an accuracy of 10^{-4}, turns into

$$\frac{MM^2}{S'} \simeq 1 - \frac{\epsilon}{M_p}, \qquad (4)$$

which allows to determine the binding energy of different fragments, including the unstable ones. For this reason the plot of the, e.g. $^4\text{Hep}\rightarrow^3\text{Hepn}$ reaction at small $b_{AA'}$ is examined.

Figure 5 shows these data in the region of $b_{AA'} < 5 \cdot 10^{-4}$. Projecting the experimental data onto the y-axis and fitting a Gaussian

490

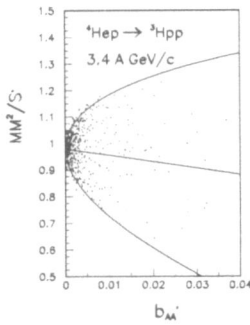

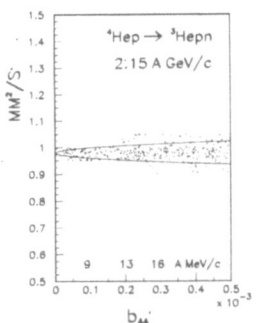

Figure 4. The MM^2/S' vs $b_{AA'}$ plot for the ^{4}Hep$\rightarrow^3$Hpp reaction at 3.4 A GeV/c .

Figure 5. The plot for the ^{4}Hep$\rightarrow^3$Hepn reaction at 2.15 A GeV/c in the $b_{AA'} < 5 \cdot 10^{-4}$ region.

to the obtained MM^2/S' distribution, one estimates the mean value $\left\langle \frac{MM^2}{S'} \right\rangle = 0.9780 \pm 0.0013$. Hence, using Eq. (4), we obtain for the binding energy of ^{3}He in ^{4}He $\epsilon = (20.6 \pm 1.3)$MeV/$c^2$, beeing in good agreement with the value $\epsilon_t = 20.57$MeV/c^2, evaluated from the known masses of ^{4}He, ^{3}He and n. Using the exact expression (3), one obtains $\epsilon = (20.7 \pm 1.3)$MeV/$c^2$.

In addition to $b_{AA'}$ scale (Fig. 5) also the momentum scale with respect to A nucleus rest frame is present, from which can be seen, that due to small values of the fragment momenta, the proposed method of binding energy determination cannot be applied to the fragmentation of the nuclei at rest (target). To work in the beams of accelerated nuclei with precise measurements of the fast fragments momenta in the region of $b_{AA'} \rightarrow 0$ seems to be unvoidable.

4   Conclusions

A new fragmentation data representation is introduced. By its aid the self-similar behaviour of the light nuclei fragmentation processes at small $b_{AA'}$ is shown. A method is proposed to determine the binding energy of the fragments in the parent nucleus.

References

1. V.V. Glagolev: Nucl. Phys. (Proc. Suppl.) **36**, 509 (1994)

2. G.F. Chew and F.E. Low: Phys. Rev. **113**, 1640 (1959)

3. A.M. Baldin: Nucl. Phys. **A447**, 203c (1985)

Few-Body Systems Suppl. 10, 491–494 (1999)

Few-
Body
Systems
© by Springer-Verlag 1999

The Forward Spectrometer for Exclusive Study of the Roper Resonance at Saturne II Accelerator (LNS, CE Saclay)*

A.N. Prokofiev[†], G.D. Alkhazov, V.V. Astashin, A.G. Atamanchuk,
V.V. Baublis, V.Ya. Gerzenstein, V.V. Golubev, V.L. Golovtsov,
A.V. Khanzadeev, B.G. Komkov, A.V. Kravtsov, L.G. Kudin, E.M. Oristchin,
B.V. Razmyslovich, V.M. Samsonov, V.V. Vikhrov, S.S. Volkov,
A.A. Vorobyov, An.A. Vorobyov, A.A. Zhdanov
for $SPES4\pi$ Collaboration

Petersburg Nuclear Physics Institute, 188350, Gatchina, RUSSIA

Abstract. This is a brief description of the forward spectrometer which was used as a part of $SPES4\pi$ installation for an exclusive investigation of the $p(\alpha, \alpha')X$ and $p(\vec{d}, d')X$ reactions at the SATURNE-II accelerator (CE Saclay). The spectrometer was designed to register the products of the recoil particle decay and to select the specific channels of the reaction. It consists of a set of multiwire drift chambers and scintillation counter hodoscope and was especially adapted for the operation close to a high intensity direct beam of the accelerated particles. The main parameters of the spectrometer and the results of its test are presented.

The inclusive reaction $p(\alpha, \alpha')X$ was recently studied at the SATURNE II accelerator using the $SPES4$ spectrometer for measurements of missing mass spectra of the scattered α-particles [1]. The results indicated the existence of a strong isoscalar excitation of the target nucleons at the energy close to the energy of the Roper resonance $P_{11}(1440 \text{ MeV})$. This one may be interpreted as a radial nucleon excitation and can be used for the nucleon compressibility determination. A serious problem of this experiment is that both the target and the projectile can be excited and may give contributions to the inclusive spectrum. Therefore, it is crucial to separate the different processes and to reduce the physical background connected mainly with the Δ-isobar production. In order to separate the different decay channels it was necessary to perform

*This work was sponsored by CEA, CNRS/IN2P3 and INTAS/RFBR (Grant 95-1345).
[†] *E-mail address:* prokan@lnpi.spb.su

the coincidence (exclusive) measurements. Such measurements were performed during 1995-97 at LNS (CE Saclay) where inelastic $p(\alpha, \alpha')X$ and $p(\vec{d}, d')X$ reactions were studied in an exclusive experiment using new $SPES4\pi$ installation [2] (Fig. 1).

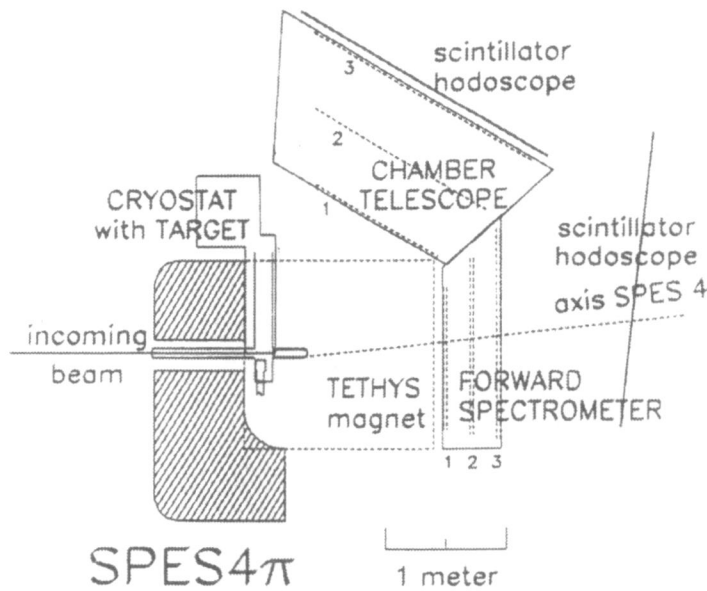

Figure 1. Schematic view of the $SPES4\pi$ installation involving a non-focusing magnetic spectrometer for the Roper resonance exclusive investigation.

The scattered particles α' and d' were registered, as in the single arm inclusive experiment [1], by the $SPES4$ spectrometer [3,4]. The resonance decay products were detected by means of a large acceptance non-focusing magnetic spectrometer consisting of two independent parts: Forward (FS) and Side (SS) spectrometers. The FS had the aim to detect the protons arising from the decay of the Roper resonance, to select them from pions and other background particles, and to measure the proton kinematical parameters. A distribution of the proton momentum versus angle of the recoil for inelastic α-particle scattering and Roper resonance excitation is presented in Fig. 2. The simulation was made for a fixed α' scattering angle $\theta_{\alpha'} = 0.8°$. According to the simulation results, the recoil proton momentum is restricted by $p \leq 1.0$ GeV/c and its incoming angle by $|\theta_p| \leq 50°$. The protons are expected to be separated from the pions using the time-of-flight and energy loss measurements by the scintillation counters hodoscope. The momenta of particles are measured by the drift chamber telescope, which determines the trajectory of the particles. The requirement for the accuracy in measurement of the proton kinematical parameters stems from the necessity to obtain good enough missing mass resolution, so that one could distinguish between 1π and 2π modes of the Roper resonance decay.

The FS consists of 6×2 layers of multiwire hexagonal drift chambers and of a scintillation counter hodoscope. The specific feature of FS drift chambers is the use of a hexagonal cell as a basic element of the whole chamber structure. The drift chamber structure as well as the result of the electric field

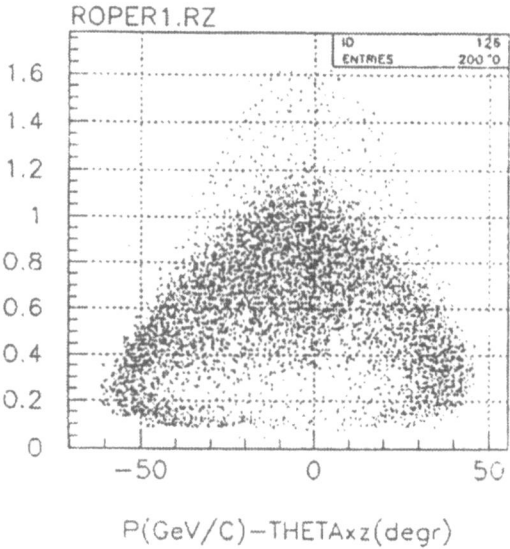

Figure 2. The $p_p - \theta_x$ plot for the protons from the Roper resonance decay (Monte-Carlo simulation).

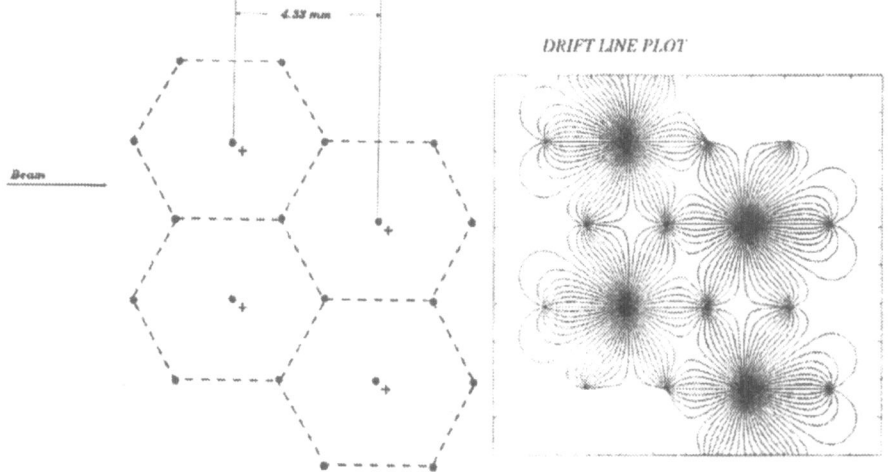

Figure 3. Left: the structure of the drift chambers. Right: the electric field map calculation for honeycomb drift chambers structure.

topography calculation are presented in Fig. 3. Every chamber has a hole in the middle of the plane for passing of unscattered or scattered at small angle particles.The use of the honeycomb structure without walls allowed us to reduce the density of the materials inside the chambers. A reliable operation of the FS was achieved at the beam intensity of about 2.10^8 particles per bunch which corresponds to an intensity of about 10^6sec^{-1} of the secondary particles crossing the chamber plane. The spatial resolution achieved was better than 1mm for multiwire drift chambers. The time-of flight and the energy loss resolution of the scintillation hodoscope were 1nsec and 1.5 MeV respectively. The obtained resolution allowed us to distinguish between different type of particles and specifically to identify the deuterons from "genuine elastic" pd-backward scattering. The main parameters of the FS are presented in the table below.

The Forward Spectrometer main parameters

The drift chambers

Number of the chambers: 6 $(X_1, Y_1, U, V, Y_2, X_2)$.
U and V chamber wires inclination: 8°.
Size of the chambers: from 0.5×1.2 m^2 to 0.7×1.7 m^2.
Sensitive (anode) wires diameter: 0.025 mm gold plated tungsten.
Cathode wires diameter: 0.090 mm.
Gas mixture: $50\% Ar + 50\% C_2 H_6$.
High voltage ≈ 2.2 kV.
Typical spatial resolution ≤ 1 mm.

The scintillation counter hodoscope

Size of the hodoscope: 1.2×2.5 m^2 (including the hole for the direct beam passing 0.2×0.2 m^2).
Scintillator bar thickness: 15 mm.
Number of scintillator bars: 26.
Scintillator NE-110 with the attenuation length of 4 m and pulse rise time ≈ 1.0 nsec.
Typical resolution: $\Delta T_{TOF} \approx 1$ nsec, $\frac{dE}{dx} \approx 1.5$ MeV.

References

1. H.P. Morsh et al.: Phys. Rev. Lett. **69**, 1336 (1992)

2. R. Dahl et al.: In: *Nouvelles de Saturne* **No21**, 69 (1997)

3. E. Grorud et al.: NIM **188**, 549 (1981)

4. M. Bedjidian et al.: NIM **A257**, 132 (1987)

Few-Body Systems Suppl. 10, 495–498 (1999)

Few-
Body
Systems
© by Springer-Verlag 1999

Study of Delta and Roper resonances excitation in light nuclei induced reactions

E.A. Strokovsky[1]*, L.S. Azhgirey[1], L.V. Malinina[1], N.M. Piskunov[1],
I.M. Sitnik[1], M. Boivin[2], T. Hennino[2], R. Kunne[2], M. Kagarlis[2],
P. Radvanyi[2], E. Tomasi-Gustaffson[2], J.-L. Boyard[3], L. Farhi[3],
J.C. Jourdain[3], B. Ramstein[3], M. Roy-Stephan[3], G.D. Alkhazov[4],
A.V. Kravtsov[4], V.A. Mylnikov[4], E.M. Orichtchin[4], A.N. Prokofiev[4],
B.V. Razmyslovich[4], I.I. Tkach[4], S.S. Volkov[4], A.A. Zhdanov[4], R. Dahl[5],
M. Drews[5], C. Ellegaard[5], C. Gaarde[5†], J.S. Larsen[5], M. Skousen[5],
P. Morsch[6], W. Augustiniak[7], P. Zupranski[7], C.F. Perdrisat[8], V. Punjabi[9]

[1] JINR, 141980, Dubna, Moscow region, Russia

[2] LNS CEA/DSM and CNRS/IN2P3 CE, Saclay, 91191 Gif-sur-Yvette
Cedex, France

[3] CNRS/IN2P3 IPN and University Paris-sud, 91406 Orsay Cedex, France

[4] PNPI RAS, 188350 Gatchina, Russia

[5] The Niels Bohr Institutet, DK-2100 Copenhagen, Denmark

[6] KFA-Juelich, D-52425 Juelich, Germany

[7] Soltan Institute for Nuclear Studies, Warsaw, Poland

[8] The College of William and Mary, Williamsburg, Virginia 23185, USA

[9] Norfolk State University, Norfolk, Virginia 23504, USA

Abstract. The excitation mechanisms and properties of broad hadronic resonances in nuclei, which are intimately related with nuclear medium response on high energy excitations, have attracted much attention during last decade. Recently, a new step in these investigations was taken: inelastic $p(d, d')$ scattering of polarized deuterons at momentum of 3.73 GeV/c was studied in an exclusive-type experiment at Saclay (exp. 278C), using the SPES-4π spectrometer. First (preliminary) results of this experiment are reported in this talk as well as data of inclusive-type experiments performed in Dubna. Data on the tensor analyzing power T_{20} of genuine elastic $p(d, p)d$ backward (c.m.) scattering are also reported.

*E-mail address strok@sunhe.jinr.ru
†deceased

In many cases, in particular in studies of structure of nuclei at short distances, a coupling between "external" and "internal" degrees of freedom (d.o.f.) of constituent nucleons cannot be ignored[1]: the closer nucleons are to each other the more important this coupling, which results in excitation of a nucleon, ought to be. Several examples of such coupling were presented at this conference, for example in talks [1, 2], where attention was payed to the existence and the role of Δ and higher nucleon resonances for the problem of structure of light nuclei at short distances. In particular, some non-standard components in the deuteron wave function (DWF) (see ref. [2]) arise from these couplings.

Unavoidably, in any search for direct signatures for such DWF components, the question arises, how one may distinguish between nucleon resonances that "pre-existed" in the deuteron and those "created" during its interaction with a target. A necessary pre-requisite to answer this question is a good understanding of the mechanisms of nucleon excitation in NN collisions, as well as a good knowledge of the properties (widths, coupling constants etc) of relevant resonances, like the Roper $N(1440)$, and negative parity N^* resonances, like $N(1535)$. At present this understanding is not satisfactory and needs new experimental data.

Combining different energy dependence of mechanisms of resonance excitations and specific combination of quantum numbers in initial states, it is possible to separate contributions of the different partial mechanisms (see, for example, refs. [3]). Especially useful are reactions with isoscalar projectiles, such as α or d, where "isospin-filtering" takes place. Dedicated $p(\alpha, \alpha')X$ experiments were performed at SATURNE-II [4] at kinetic energy of 1 GeV/nucleon; $p(d, d')X$ and $C(d, d')X$ experiments with polarized deuterons with momenta from 4.2 to 9 GeV/c were done in Dubna [5][2]. Tensor analysing power T_{20} was measured for inelastic (d, d') scattering at 0° (Fig. 1). No significant difference in T_{20} for $p(d, d')X$ and $C(d, d')X$ was observed. The data display an approximate scaling when plotted versus t: 4-momentum transfer squared.

To continue the study of N^* excitations in $p(\alpha, \alpha')$ and $p(d, d')$ reactions in exclusive-type experiments, the SPES-4π setup was installed at Saclay. Data were taken in experiment 278 with SATURNE-II beams during 1996 and 1997. The first preliminary data from this experiment, as obtained with polarized deuteron beam with momentum of 3.73 GeV/c, are reported here.

The SPES-4π setup [7] includes two main components: i) the high resolution magnetic spectrometer SPES-4, ii) the new, non-focusing magnetic spectrometer consisting of wide aperture magnet TETHYS and two sets of coordinate detectors (proportional and drift chambers) and scintillation hodoscopes: the "Lateral Spectrometer" (LS) and the "Forward Spectrometer" (FS). The target station was located inside magnetic field of TETHYS. The scintillation hodoscopes of both arms were used for measurements of time of flight and dE/dx of detected particles and provided the logical signals for the main trigger.

---

[1] Here "external" refers to the movement of a nucleon as a whole in a nucleus, while "internal" d.o.f. are related with states of constituents of the nucleon.

[2] Previously obtained (d, d') data were, as a rule, by-products of other experiments (see, for example, refs. [6]).

Figure 1. T_{20} data for $p(d, d')X$ inelastic scattering at 0° from ref. [5]. Stars: 9 GeV/c; open squares: 5.532 GeV/c; open circles: 4.495 GeV/c; full circles: data from ref. [8] for momenta 4.24 – 6.55 GeV/c.

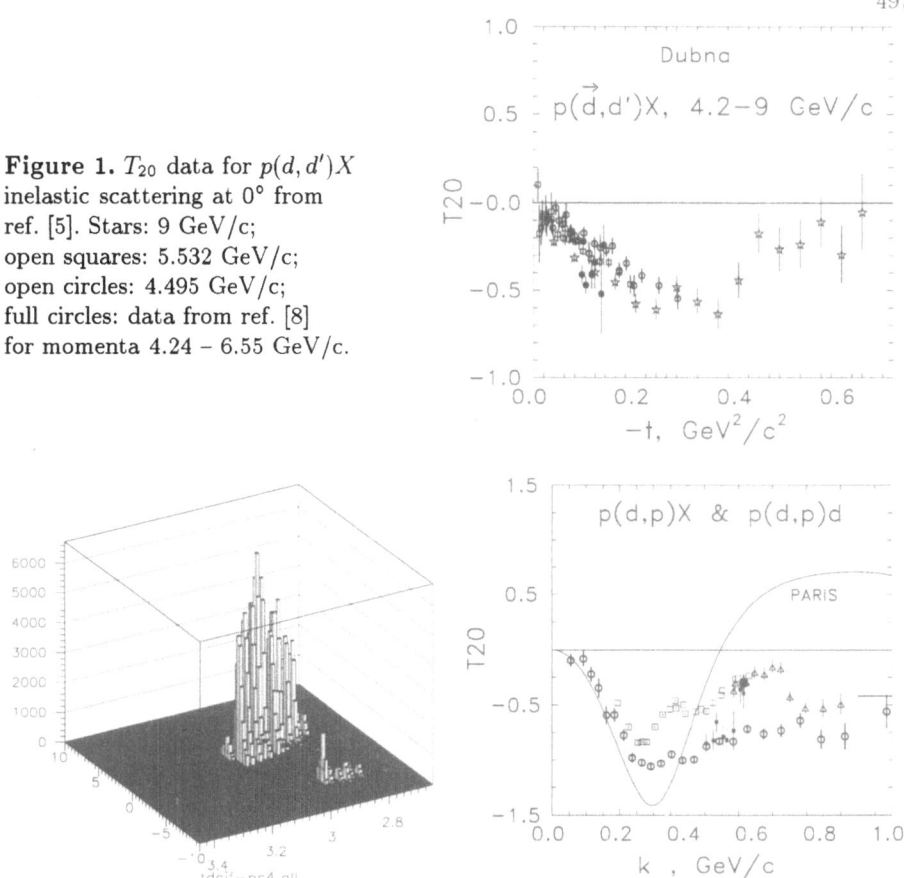

Figure 2. *Left*: event distribution on the plot of "momentum in SPES-4" (GeV/c) – "time of flight in SPES-4" (ns). Deuterons in SPES-4 from inelasic $p(d, d')$ reaction (the big "mountain"), protons from $p(d, p)d$ (the small peak), and $p(d, p)pn$ (the tail of the small peak) reactions are separated clearly. *Right*: data on T_{20} for $p(d, p)d$ backward (c.m.) elastic scattering (open squares and open triangles) from refs. [8, 10] and for $p(d, p)X$ breakup at 0° at $T_d = 7.4$ GeV/c from ref. [11] (open circles). Full black circle: data from this experiment for $p(d, p)d$ at $\vartheta_{c.m.} = 180°$ when both final particles are detected in coincidences. Small full squares: data from this experiment for $p(d, p)pn$ at $\vartheta_{c.m.} = 180°$. As in refs. [8, 10, 11] the data are plotted versus the Light Front variable k (the nucleon momentum in deuteron).

Deuterons inelastically scattered in 6 cm of liquid hydrogen at $\vartheta_{lab} \sim 0.4°$–0.7° were detected by SPES-4; charged particles (mostly protons and pions) produced in the reaction were detected by the LS or the FS. Particle identification was based on information from momentum, time of flight and ionization losses of each detected particle. Beam polarization was changed in the usual burst-to-burst mode; states 5,6,7,8 (in the SATURNE-II notations, [9]) were used. The beam polarization was measured using SATURNE's Low-Energy Polarimeter. The tensorial beam polarization $|\rho_{20}|$ in this experiment was ~0.64.

A coincidence of signals from SPES-4 and at least one of the arms triggered the setup. Events taken with this trigger and the polarized deuteron beam at 3.73 GeV/c come mostly from two reactions (Fig. 2, left plot): (i) the inelastic $p(d, d')X_{ch}$ (the deuteron was detected in SPES-4) and (ii) the elastic backward $p(d, p)d$ scattering. The latter reaction was unambiguously identified by *detecting the recoil proton in SPES-4* (with momentum ~ 2.93 GeV/c) *in coincidence with the scattered deuteron in FS* (with momentum ~ 0.8 GeV/c).

The $p(d, d')X_{ch}$ events are actually quasi-inclusive because the trigger requires at least one charged particle in the (FS, LS) arms, restricting the available phase space for the X_{ch} system. Nevertheless, T_{20} for this reaction is in reasonable agreement with data presented in Fig. 1.

New data for $p(d, p)d$ backward elastic scattering are presented in Fig. 2 (right plot) and agree perfectly with existing data obtained in experiments with detection of recoil protons only. In this experiment, for the first time for energies above ~ 1 GeV, *both* scattered particles were registered.

Acknowledgement. This work was supported in part by INTAS-RFBR grant $N°$ 95-1345. Two participants acknowledge support from the US Department of Energy (V.P., grant $N°$ DE-FG05-89ER40525) and from the US National Science Foundation (C.F.P., grant $N°$ 97-04502).

References

1. S. Nemoto et al.: talk at this Conference

2. A.P. Kobushkin: talk at this Conference

3. P. Fernandez de Cordoba, E. Oset: Nucl. Phys. **A544**, 793 (1992); P. Fernandez de Cordoba et al.: Phys. Lett. **B319**, 416 (1993); Nucl. Phys. **A586**, 586 (1995); E.A. Strokovsky, F.A. Gareev: YaF **58**, 1404 (1995); S. Hirenzaki et al.: Phys. Lett. **B378**, 29 (1996); Phys. Rev. **C53**, 277 (1996)

4. H.P. Morsch et al.: Phys. Rev. Lett. **69**, 1336 (1992); Nucl. Phys. **A553**, 645c (1993)

5. L.S. Azhgirey et al.: JINR Rapid Comm. **2[88]**, 17 (1998) (final data tables are presented); see also Phys. Lett. **B361**, 21 (1995)

6. J. Banaigs et al.: Phys. Lett. **45B**, 535 (1973); V.G. Ableev et al.: YaF **37**, 348 (1983); L.S. Azhgirey et al.: YaF **48**, 1758 (1998)

7. Exp. 278 Proposal (unpubl.); A.N. Prokofiev et al.: talk at this Conference.

8. L.S. Azhgirey et al.: Phys. Lett. **B391**, 22 (1997); YaF **61**, 494 (1998)

9. J. Arvieux et al.: Nucl. Phys. **A431**, 1613 (1984)

10. V. Punjabi et al.: Phys. Lett. **B350**, 178 (1995)

11. L.S. Azhgirey et al.: Phys. Lett. **B387**, 37 (1996)

Few-Body Systems Suppl. 10, 499–502 (1999)

Few-
Body
Systems
© by Springer-Verlag 1999

Pionic decay of possible d' dibaryon and models of quark-pion and NN interaction

I. Obukhovsky[1]* and A. Obukhovsky[2]

[1] Institute of Nuclear Physics, Moscow State University, Moscow 119899, Russia

[2] Institute for Nuclear Research, Russian Academy of Science

Abstract. The decay width $\Gamma_{\pi N d'}$ of a possible d' dibaryon is calculated on basis of a $^3P_0$ quark pair creation model for two alternative models of NN interaction (OBE and Moscow potentials).

1 Introduction

A possible d' dibaryon ($M_{d'}$=2065 MeV, $\Gamma_{d'}$=0.5 MeV, $J^P = 0^-$, T=0) proposed earlier (see e.g. [1] and references therein) for interpretation of a resonance-like behaviour of the pionic double-charge exchange (DCX) scattering on nuclei has been considered in refs. [2, 3] as a six-quark resonance. Such consideration runs into an obstacle: the evaluation of the d' mass in the non-relativistic quark model (NRQM) shows that the experimental value of $M_{d'}$ is too small to be explained with the standard qq interaction fitted to the baryon spectrum. It does not necessarily mean that the d' resonance in nuclei is not of quark origin, but it becomes obvious that more sophisticated approaches taking into account other interactions (e.g. the interaction of the dibaryon with outer hadronic fields) are desirable. The necessity of such approaches is at least seen in attempts to calculate the d' decay width [4].

Another aspect of the problem is that the d' mass, calculated in the NRQM, could be modified in a nuclear medium. The well known Δ-isobar mass shift (30-40 MeV) in nuclei [5, 6] implies a virtual decay $\Delta \rightarrow \pi+N$ and propagation of the virtual pion in a nuclear medium. It is applicable to the d' too, but the role of the virtual decay $d' \rightarrow \pi NN$ could be more important. The d' resonance peak in a nuclear medium is only 50 MeV higher than the πNN threshold and the decay width should be strictly dependent on the phase space volume of

*This work was supported in part by the Russian Foundation for Basic Research (grants no.96-02-18071 and 96-02-18072)

the final πNN state (recall that the quantum numbers of the d' prevent the decay into the NN channel). Then even a small mass shift of the proposed d' dibaryon could lead to a large variation of the decay width. This would make the d' almost impossible to be observable out of a nuclear medium if its mass becomes close to the calculated value $M_{d'} \approx$2.3-2.5 GeV [2, 3]. Against the d' existence outside of a nuclear medium are the recent experimental observations of the CHAOS collaboration [7] concerning the DCX (π^+, π^-) reaction on ^4He, which are unable to confirm the d' signal with a sufficient statistical accuracy in processes on light systems p+p and ^4He. Therefore, an evaluation of the dependence of the d' decay width on its mass would be important.

The first evaluations of the d' decay width $\Gamma_{d'}$ have shown that $\Gamma_{d'}$ is strictly dependent on the short range nucleon-nucleon dynamics [4]. Here we show that the value of $\Gamma_{d'}$ is very sensitive to the choice of the quark-pion coupling. It calls for some comments. As the d' mass is very close to the πNN threshold, the standard pseudoscalar (PS) or pseudovector (PV) quark-pion coupling models lead to very different results for the d' decay width in the lowest order of the $\frac{v}{c}$ expansion. At this order, the PS coupling is proportional to the $\boldsymbol{\sigma}_q \cdot \mathbf{k}$ operator (here $\mathbf{k} = \mathbf{p_q} - \mathbf{p'_q}$ is the momentum of the pion) that leads to the k^2 dependence of the transition matrix element [4]. In the limit $k \to 0$ the matrix element goes to zero [4, 8]. On the other hand, the PV coupling already in the zero order of $\frac{v}{c}$ leads to the Galilean-invariant vertex $\sim \boldsymbol{\sigma}_q \cdot \mathbf{k} - \frac{\omega_\pi}{2m_q}\boldsymbol{\sigma}_q \cdot (\mathbf{p_q} + \mathbf{p'_q})$ and the second (gradient) term of this vertex $\sim \frac{1}{2i}\boldsymbol{\sigma}_q \cdot (\boldsymbol{\nabla}_q - \overleftarrow{\boldsymbol{\nabla}}_{q'})$ gives rise to a constant non-zero matrix element of the $d' \to \pi$NN transition at $k \to 0$. The matter is that the contribution of the gradient term to the transition amplitude depends on off-shell properties both of the initial and final states.

Another state-dependent contribution to the πNN decay width could be due to the $q\bar{q}$ structure of the pion. As a rule the pion is considered as a structureless (point-like) Goldstone boson. However, it seems reasonable to say that the pion is not exactly a Goldstone boson. This is not only because of its non-zero mass, but also because there are processes where the inner $q\bar{q}$ structure of the pion plays a principle role, e.g. in the πN decay of the Roper resonance [9]).

For these reasons, an extension of the previous evaluations of the d' decay width [4] to more realistic quark-pion coupling models in this energy region would be useful for a solution of the d' problem.

We consider here the $\pi d'$ coupling and calculate the d' decay width in the framework of the phenomenologically successful $^3P_0$ quark-pair-creation model (QPCM, see e.g. ref. [9] and references therein). In the limit of the zero $q - \bar{q}$ pion radius this model goes to the standard PV coupling. Therefore, varying the π-radius we have obtained the results for the PV coupling too. Moreover, in this limit we can normalize the strength of the (phenomenological) constant for $q\bar{q}$ pair creation to the PV pion-nucleon coupling constant.

2   Models of NN interactions and two modes of d' decay

Two decades ago it was pointed out [10, 11] that excited six-quark configurations $s^4p^2[42]_X$ and $s^5p[51]_X$ are the most important for NN interaction
at short distance (in the lowest partial waves L=0 and 1 correspondingly).
They are compatible with all possible color-spin (CS) states in inner products
$[2^3]_C \circ [42]_S$ and $[2^3]_C \circ [3^2]_S$ and a full NN wave function at short distance could
be represented as a superposition of these configurations. But in this case, the
NN wave function has a node at $r \approx b_3$, where $b_3 \approx$0.5-0.6 fm is a radius of
the quark core of the nucleon. On the contrary, non-excited "bag-like" configurations $s^6[6]_X$ and $s^5p[51]_X$ (maximally symmetric in X-space) are compatible
with only one (in each channel) CS state and are not properly NN states (they
could be represented as superpositions of a few baryon-baryon states). They
are dibaryon candidates. In line with this analysis, the Moscow NN potential
model (MP) was proposed a decade ago [12]. It was postulated that the NN
wave function is orthogonal to the "bag-like" configurations and, as a consequence, the MP wave function has a node at $r_c \approx$0.5-0.6 fm instead of the
repulsive core. The MP and standard repulsive-core potentials (OBEP, Paris,
etc.) are phase equivalent but off-shell different. It reveals especially in short
range effects. The d' decay is one of such effects.

The d' dibaryon as a "naturally prepared" six-quark configuration
$$d' = |s^5p(b_6)[51]_X([321]_{CS})LST=110, J^P=0^->$$
has only two modes of decay with emission of the S-wave pion: (i) the de-
excitation mode $d' \to d_0 + \pi$ and (ii) the excitation mode $d' \to d_f + \pi$ ($d_2 + \pi$),
where d_0, d_f and d_2 are the six-quark configurations
$$d_0 = |s^6(b_6)[6]_X([2^3]_{CS})LST=001, J^P=0^+>,$$
$$d_f = |s^4p^2(b_6)[42]_X([2^3]_{CS})LST=001, J^P=0^+>,$$
$$d_2 = |s^5p(b_6)[51]_X([f_{CS}])LST=001, J^P=0^+>,$$
most important at short distance in the final $^1S_0$ NN state. The probabilities
of both modes have been calculated algebraically with f.p.c. techniques [4] and
with projecting onto final $^1S_0$ NN states in two alternative NN interaction
models: OBEP [13] and MP [14].

3   Results and conclusions

We have obtained that the transition to the "bag-like" configuration d_0 dominates (about 80%) the transition to the full superposition "NN-like" configurations $d_2 + \sum d_f$. Therefore, the d' decay width $\Gamma_{d'}$ strictly depends on the quark
structure of the NN wave function at short distance: it is large for standard
repulsive-core models (OBEP) and it is anomalously small for the MP model,
which uses a condition of orthogonality to the "bag-like" state d_0.

The results on the d' decay width $\Gamma_{d'}^{tot}$ are shown in Fig. 1 for two values
of the π-radius: $b_\pi=0$ (the curves labeled with PV) and $b_\pi=0.5b_3=0.3$ fm (the
curves labeled with QPCM). The PS coupling results of ref. [4] are also shown
(the curve labeled with PS). One can see that (i) the decay width strictly
depends both on the NN interaction at short distance and the choice of quark-

502

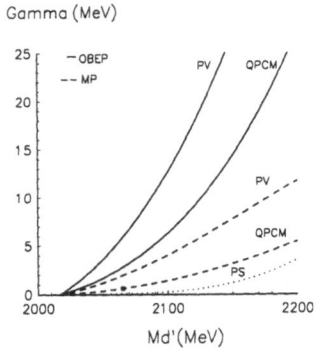

Gamma (MeV)

Figure 1. Dependence of the d' decay width on the d' mass.

pion coupling, (ii) the dependence on the d' mass is very sharp, what confirms our assumption that at the d' dibaryon mass of about 2.3-2.5 GeV the d' could not be seen as a narrow resonance.

References

1. R. Bilger, H. Clement et al.: Nucl. Phys. **A596**, 586 (1996)

2. G. Wagner, L.Ya. Glozman et al.: Nucl. Phys. **A594**, 263 (1995)

3. A.J. Buchmann, G. Wagner and A. Faessler: Phys. Rev. **C57**, 3340 (1998)

4. I.T. Obukhovsky, K. Itonaga et al.: Phys. Rev. **C56**, 3295 (1997)

5. M. Betz and T.-S.H. Lee: Phys. Rev. **C23**, 375 (1981)

6. V.G. Ableev et al.: Pis'ma JETP **40**, 35 (1984)

7. CHAOS Collaboration (Gräter et al.): Phys. Lett. **B240**, 37 (1998)

8. A. Samsonov and M. Schepkin: Nucl-th/9712079

9. F. Cano, P. Gonzàlez et al.: Nucl. Phys. **A603**, 257 (1996)

10. I.T. Obukhovsky, V.G. Neudatchin et al.: Phys. Rev. Lett. **88B**, 231 (1979)

11. M. Harvey: Nucl. Phys. **A352**, 301 (1981); **A352**, 326 (1981)

12. V.I. Kukulin et al.: Phys. Lett. **153B**, 7 (1985)

13. T. Ueda et al.: Prog. Theor. Phys. **95**, 115 (1996)

14. V.I. Kukulin, V.N. Pomerantsev et al.: Phys. Rev. C **57**, 535 (1998)

Few-Body Systems Suppl. 10, 503–506 (1999)

Few-
Body
Systems
© by Springer-Verlag 1999

Energy-dependent partial wave analysis of the reaction $\pi^- pp(^1S_0) \to pn$

D. Bosnar[1], B. Blankleider[2], P. Salvisberg[3*], H.-J. Weyer[4]

[1] Physics Department, University of Zagreb, HR-10000 Zagreb,Croatia
[2] Department of Physics, The Flinders University of South Australia, Bedford Park, SA 5042, Australia
[3] Institute of Physics, University of Basel, CH-4056 Basel, Switzerland
[4] Paul Scherrer Institute, CH-5232 Villigen PSI, Switzerland

Abstract.

We have performed an energy-dependent partial wave analysis of the reaction $\pi^- pp \to pn$ where the two initial-state protons are in a $^1S_0$ state . In contrast to a previous single-energy analysis, we find our partial wave amplitudes to be consistent with theoretical calculations involving just meson and baryon degrees of freedom.

1 Introduction

The study of pion absorption on two nucleons is of fundamental importance. The relatively large amount of momentum transfer involved, at least 367 MeV/c, implies that one should be sensitive to the short-range part of the nucleon-nucleon force, and hence to possible quark degrees of freedom. Particularly interesting is absorption on the pp-system, where an intermediate s-wave $N\Delta$ state is forbidden. Such observations have motivated intensive experimental studies of the $\pi^- pp(^1S_0) \leftrightarrow pn$ reaction [1, 2]. Most theoretical descriptions have used purely meson and baryon degrees of freedom [3] although in one notable exception [4] quark degrees of freedom were used. A distinguishing feature of these two types of calculation is the relative size of the amplitudes corresponding to $^3S_1$ and $^3D_1$ pn final states. In the quark description the $^3S_1$ final state amplitude dominates while the meson-baryon calculations give the $^3D_1$ amplitude as the dominant one. To ascertain which type of calculation is favoured by experimental data, Piasetzky et al. [5] performed a partial wave

*Present address: Swisscom Ltd., Innere Margarethenstrasse 4, CH-4002 Basel, Switzerland

analysis at the single energy of 62.5 MeV. Their results favour the calculation with quark degrees of freedom.

By contrast we have performed an energy-dependent partial wave analysis based on the total sum of currently available data. The amplitudes are taken to be smooth functions of energy. This requirement imposes an important constraint on the amplitude determination.

2 Formalism and Analysis

We expand both cross sections and polarisation observables in terms of Legendre polynomials. For the differential cross section we write

$$d\sigma/d\Omega = \sum_{l=0}^{\infty} A_l(E) P_l(\cos\theta). \tag{1}$$

It has been found that all cross section data up to 210 MeV pion lab energy can be fitted well with a Legendre expansion of just second order [2]. So far the only polarisation experiments available are for the analysing power A_{0y} in reaction $\vec{p}n \rightarrow \pi^- pp(^1S_0)$ [6, 7]. This we expand as

$$A_{y0} = \frac{\sigma_{y0}}{\sigma_{00}} \quad \text{where:} \quad \sigma_{y0} = \sum_l B_l P_{l1}(\cos\theta) \tag{2}$$

where $\sigma_{00} \equiv d\sigma/d\Omega$ is the spin averaged differential cross section as in Eq. (1).

The coefficients A_l and B_l can be expressed in terms of partial wave amplitudes [8]. An important assumption in this work is that absorption takes place only through s and p-waves. This restricts the number of partial wave amplitudes to three, T_a, T_b and T_c, so that

$$A_0 = K\left(\frac{1}{4}|T_a|^2 + \frac{1}{2}|T_b|^2 + \frac{1}{2}|T_c|^2\right), \quad A_1 = K\left[\frac{1}{\sqrt{2}}\mathrm{Re}(T_aT_b^*) - \mathrm{Re}(T_aT_c^*)\right],$$

$$A_2 = K\left[-\sqrt{2}\mathrm{Re}(T_bT_c^*) + \frac{1}{2}|T_c|^2\right] ; \tag{3}$$

$$B_1 = K\left[\frac{1}{\sqrt{2}}\mathrm{Im}(T_aT_b^*) + \frac{1}{2}\mathrm{Im}(T_aT_c^*)\right] \quad \text{and} \quad B_2 = K\frac{1}{\sqrt{2}}\mathrm{Im}(T_bT_c^*), \tag{4}$$

where K is a kinematical factor.

To impose a smooth energy dependence on the partial wave amplitudes, we express each as a polynomial in the pion c.m. momentum k. Angular momentum considerations restrict the low-energy behaviour of a partial wave amplitude to the form k^l, where l is the pion's orbital angular momentum. The high energy behaviour of amplitudes is not known a priori. To estimate the order of polynomial in k needed to describe $\pi^- pp(^1S_0) \rightarrow pn$ amplitudes below 250 MeV, we have performed a few-body calculation using the $NN-\pi NN$ model [9]. This gave us the following "minimal" momentum dependence of the amplitudes:

$$T_a(k) = t_0^a + t_1^a k^2 + t_2^a k^3, \quad T_b(k) = t_1^b k + t_2^b k^2, \quad T_c(k) = t_1^c k + t_2^c k^2 \tag{5}$$

where the coefficients t_i^α are complex parameters.

The fitting procedure involves minimising the χ^2 function which we define as:

$$\chi^2 = \sum_{i=1}^{n_E} \frac{1}{n_\theta} \sum_{j=1}^{n_\theta} \frac{[\sigma_{00}(ij) - \sigma_{00}^{exp}(ij)]^2}{\Delta^2 \sigma_{00}^{exp}(ij)} + \sum_{i=1}^{m_E} \frac{1}{m_\theta} \sum_{j=1}^{m_\theta} \frac{[A_{y0}(ij) - A_{y0}^{exp}(ij)]^2}{\Delta^2 A_{y0}^{exp}(ij)} \quad (6)$$

where n_E is the number of energies for which experimental differential cross sections σ_{00}^{exp} are available, and n_θ the number of angular points at a given energy (m_E and m_θ are the corresponding quantities for the asymmetry A_{y0}).

In the fitting procedure we have used data from all reactions that yield information about $\pi^- pp(^1S_0) \to pn$; namely, π^- absorption on ^{3}He [2], polarisation measurements of A_{y0} [6, 7] and $np \to d\pi^0$ data [10] (scaled by an empirical factor) from which we extracted A_0 for $\pi^-(pp) \to pn$ at very low energies.

3 Results and discussion

With the above parametrisation of amplitudes we obtain a unique solution that minimises the χ^2 function of Eq. (6). The obtained momentum dependence of A_0, A_1, A_2, using Eq. (3), is shown in Fig. 1(a). In Fig. 1(b) we show the reconstruction of A_{y0} using Eq. (2) and Eq. (4). The relative size of the final

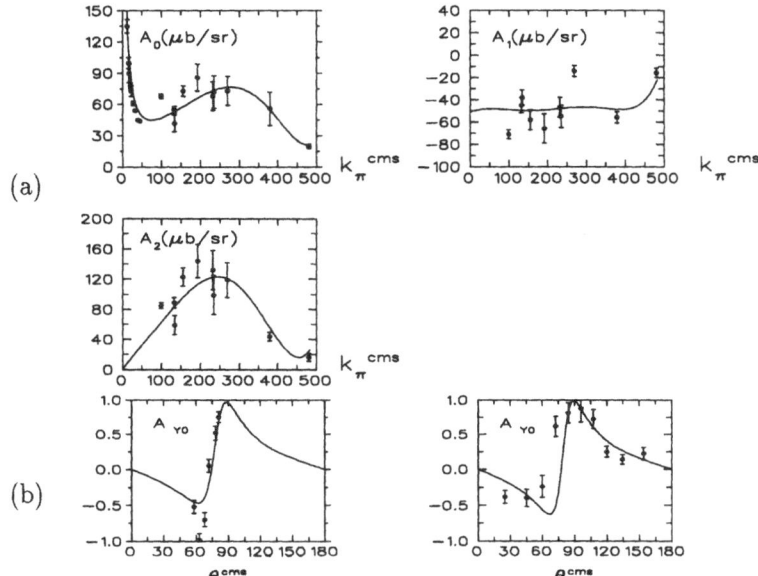

Figure 1. (a) Results of the fit (solid lines) compared to data: (a) Legendre coefficients A_1, A_1, A_2, (b) Analysing power A_{y0} at the two energies T_p=400 MeV (left) and T_p=443 MeV (right).

506

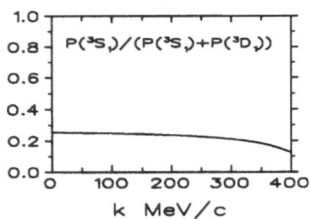

Figure 2. Relative size of the $^3S_1$ and $^3D_1$ final state contributions.

state $^3S_1$ and $^3D_1$ contributions is expressed in terms of amplitudes as:

$$\frac{P(^3S_1)}{P(^3S_1) + P(^3D_1)} = \frac{|T_b|^2}{|T_b|^2 + |T_c|^2}. \tag{7}$$

Using the fitted amplitudes in Eq. (7), we display the momentum dependence of this ratio in Fig. 2. We find that the $^3D_1$ contribution is clearly the dominant one. This result, in contrast to the previous single-energy analysis of ref. [5], favours theoretical calculations involving meson and baryon degrees of freedom.

Acknowledgement. D.B. acknowledges support from research contract 119250 of the Croatian Ministry of Science.

References

1. H.-J. Weyer: Phys. Rep. **195**, 295 (1990)

2. The most recent list of exp.: H. Hahn et al.: Phys. Rev. **C53**, 1074 (1996)

3. L.L. Kiang, T.-S.H. Lee and D.O. Riska: Phys. Rev. **C50**, 2703 (1994); T.-S.H. Lee and K. Ohta: Phys. Rev. Lett. **49**, 1079 (1982); R.R. Silbar and E. Piasetzky: Phys. Rev. **C29**, 1116 (1984); **C30**, 1365(E) (1984); H. Toki and H. Safarian: Phys. Lett. **B119**, 285 (1982); O.V. Maxwell and C.Y. Cheung: Nucl. Phys. **A454**, 606 (1986); J.A. Niskanen: Phys. Rev. **C43**, 36 (1991)

4. G.A. Miller and A. Gal: Phys. Rev. **C36**, 2450 (1987)

5. E. Piasetzky et al.: Phys. Rev. Lett. **57**, 2135 (1986)

6. C. Ponting et al.: Phys. Rev. Lett. **63**, 1792 (1989)

7. M.G. Bachman et al.: Phys. Rev. **C52**, 495 (1995) and private communication

8. B. Blankleider: In: *Particle Production Near Threshold* (AIP Conference Proceedings No. 221, p. 150), Nashville, Indiana October 1990, eds. H. Nann and E. J. Stephenson. New York: AIP 1991

9. I.R. Afnan and B. Blankleider: Phys. Rev. **C22**, 1638 (1980)

10. D. Hutcheon et al.: Phys. Rev. Lett. **64**, 176 (1990)

Few-Body Systems Suppl. 10, 507–510 (1999)

Few-
Body
Systems
© by Springer-Verlag 1999

The Reaction $D(e, pp)e'\pi^-$ on Polarized Deuteron at High Proton Momenta

V.N. Stibunov[1], A.Yu. Loginov[1], D.M. Nikolenko[2], A.V. Osipov[1], I.A. Rachek[2], A.A. Sidorov[1], D.K. Toporkov[2]

[1] Nuclear Physics Institute at Tomsk Polytechnical University, Tomsk, Russia
[2] Budker Institute of Nuclear Physics, Novosibirsk, Russia

Abstract. The differential cross section and target asymmetry components of the reaction $D(e, pp)e'\pi^-$ on polarized deuteron were measured. The kinetic energies of the protons were measured within 55-180 MeV and 46-265 MeV and the acceptance angles in lab. frame are $\Theta_{1,2} = 64^0 - 82^0$, $\Delta\phi_{1,2} = 32^0$. The sharp peak of the tensor a_{20}-component of the target asymmetry is found near the invariant mass of the $pp\pi$-system $M_{pp\pi} = 2300\ MeV/c^2$. The performed calculations of the differential yield and the tensor target asymmetry do not describe the obtained experimental results.

The interest to study π^--meson production on a deuteron for high polar angles and large momenta of both protons proceeds from an opportunity to acquire a new information on the dynamics of NN-interaction at short internucleon distances. In the region of proton momenta larger than the Fermi-momentum, the quasifree mechanism of the π^--meson production appears to be suppressed. The relative contribution of more complex reaction mechanisms grows in this kinematic area and these reactions require new models to describe the nucleon systems and hadron interactions. It is for these reasons the previous experiments in Hamburg [1] , Saclay [2] and Bonn [3, 4] chose to search for dibaryon states and observe ($\Delta\Delta$)-states as their main subject.

Our experiment was focused on the region of an even higher opening angle and larger values of the invariant mass than before. Also, the use of a polarized deuteron target enabled us to consider a number of polarization observables.

The measurements reported here were conducted simultaneously with the experiments performed [5, 6], which used an internal tensor-polarized deuterium target in the VEPP-3 storage ring at 2 GeV electron energy. The particle-detection system consisted of two identical two-arm apparatus to detect the protons in coincidence [7, 8]. Each arm of the detector was placed symmetrically around the electron-beam axis at a polar angle of 75^0 with respect to the beam line. The proton telescope included a drift chamber and the thin and thick scintillator counters. Each proton arm detected particles within the range of angles $\theta = 68° - 82°$ and $\Delta\varphi = 32°$. The kinetic energy of the protons which deposit all their energy was reconstructed combining the values of the energy deposition in the detector layers. In these measurements the direction and sign of the target polarization were changed periodically during the data acquisition [6]. The integrated luminosity and the average value of the tensor target polarization were determined from electron-deuteron elastic scattering [5].

The collected data were processed in a few consecutive stages [8, 9], which resulted in momentum vectors for both protons and reconstructed the vertex coordinates for the events. Computation of the pion momentum and photon energy was done on an assumption of a zero angle of electron scattering. The selected events were used to determine the yield of the reaction to each detector for two signs and two directions of the guide magnetic field.

The components of the experimental target asymmetry are defined as the counting rate combinations [6] :

$$a_{11} = \frac{1}{N} \sum_{i=1,2} (-1)^{\delta_{i1}} \left[N^i_{1+} + N^i_{2-} \right], \quad a_{20} = \frac{1}{N} \sum_{i,j=1,2} (N^i_{j+} - N^i_{j-}),$$

$$a_{22} = \frac{1}{N} \sum_{i,j=1,2} (-1)^{\delta_{ij}} [N^i_{j-} - N^i_{j+}], \quad (1)$$

where N^i_{jk} is the counting rate in the detector system i with the magnetic guide field index j, and the sign of deuteron tenzor polarization degree P_{zz} given by k and N is the total counting rate.

We used (see ref. [8]) the connection between the yield of the reaction summed over i, j and k into a 6-D phase space volume of the momenta, V_6 and differential cross section of the reaction

$$Y(V_6) = \int_{V_6} \epsilon L \frac{d^6\sigma}{d^3p_1 d^3p_2} d^3p_1 d^3p_2, \quad (2)$$

where p_1 and p_2 are the momenta of the protons, ϵ is the total detection and selection efficiency of the pp- events, L is the integral luminosity obtained from the measured elastic ed-scattering.

The dependences on the pp-system invariant mass that we obtained at this experiment for cross sections and analyzing power components of the reaction were presented in ref. [8, 9]. Here we present the first results as a function of the $pp\pi^-$-system mass, $M_{pp\pi}$. The differential yield of the reaction is shown in Fig. 1. and the tensor a_{20}-component of the target asymmetry in Fig. 2.

The calculations of the cross section of the investigated process were made in a few theoretical models. The cross section of the process initiated by electrons was expressed in the terms of the cross section of a reaction induced by the virtual photons. We used Dalitz-Yennie's virtual photon spectrum. For NEWGAM-code [8] one nucleon pion photoproduction operator has been taken

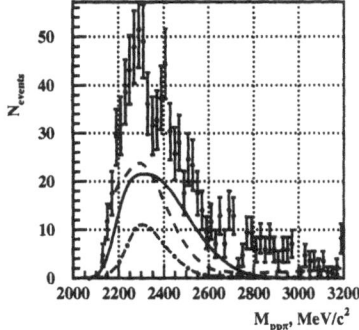

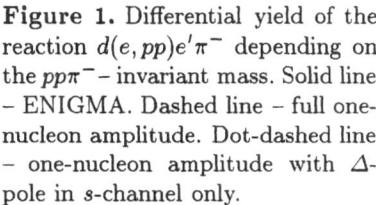

Figure 1. Differential yield of the reaction $d(e, pp)e'\pi^-$ depending on the $pp\pi^-$ – invariant mass. Solid line – ENIGMA. Dashed line – full one-nucleon amplitude. Dot-dashed line – one-nucleon amplitude with Δ-pole in s-channel only.

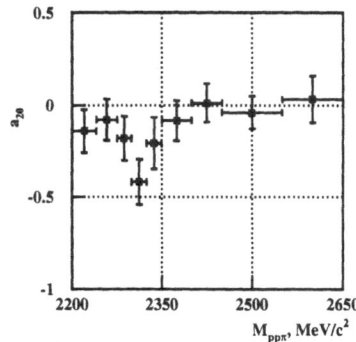

Figure 2. Tensor a_{20} – component of the target asymmetry as a function of the $pp\pi^-$ invariant mass.

from the phenomenological analysis [10] and the deuteron wave function was obtained using the Paris N-N potential. Also we used the ENIGMA-code which was developed for the exclusive pion electroproduction on nuclei [11]. The calculations of the polarization observables and cross section of the reaction were made within the spectator model using the elementary pion photoproduction amplitude discussed in ref. [12]. The Born terms of this amplitude are determined in pseudovector πN-coupling, the Δ-resonance is considered both in the s- and the u-channels and the ρ and ω-mesons exchange are considered in the t-channel. This amplitude is useful for the studies of the Δ-resonance. In addition, we studied the role of various dynamic effects in the $\Delta(1232)$-isobar photoproduction on a polarized deuteron using the relativistic impulse approximation in the nucleon-spectator model [13].

The experimental and some calculated dependences of the differential yield of the reaction on the mass of the $pp\pi$-system can be see in Fig. 1. The solid curve shows the result of the ENIGMA-code, the result of the NEWGAM-code is slightly different from this result. The dashed line corresponds to the calculation based on the total one-nucleon amplitude of the π-meson photoproduction [12], whereas the dot-dashed line shows the result of the calculation based on the one-nucleon amplitude including the Δ-isobar in the s-channel only. One can see that the experimental spectrum is peaked at $M_{pp\pi} = 2300 \ MeV/c^2$. It is clear from this figure that the experimental yields of the reaction are much higher than their calculated counterparts.

Figure 2 plots the behavior of the tensor target a_{20} –asymmetry versus the invariant mass of the $pp\pi^-$-system. Here one can see a peculiar feature - an sharp rise in the range of masses $M_{pp\pi}=2300 \ MeV/c^2$. Note that events from

510

this range correspond to $\Delta^0(1232)$-isobar production. This could be seen from the distribution of the invariant mass of the pion and the fastest of the two protons, $M_{p\pi}$ - which exhibits a clean peak at 1232 MeV/c^2. The calculated values of the a_{20}-component of the target asymmetry is below 0.6 in the mass region near 2300 MeV/c^2.

We keep on working on further analysis of obtained results. These results allow us to make two conclusions. The behavior of the differential yield and a_{20} -component of the target asymmetry near $M_{pp\pi}{=}2300\ MeV/c^2$ is associated with the excitation of $\Delta^0(1232)$-isobar on the deuteron. The noticeable difference near $M_{pp\pi}{=}2300\ MeV/c^2$ between the experimental values and the calculated tensor target a_{20} asymmetry and the reaction yield may be related with an excitation of a dibaryon resonance state and its following decay into proton and $\Delta^0(1232)$-isobar.

Acknowledgement. This work was supported by the Russian Foundation for Fundamental Research grants No.98-02-17993, No.98-02-17949 and by the IN-TAS grant No.96-0424.

References

1. P. Benz and P. Soding: Phys. Lett. **52B**, 367 (1974)

2. P.E. Argan, G. Audit, A. Bloch et al.: Phys. Rev. Lett. **41**, 86 (1978)

3. B. Bock, W. Ruhm, K. Althoff et al.: Nucl. Phys. **A459**, 573 (1986)

4. B. Ruhm, B. Bock, K. Althoff et al.: Nucl. Phys. **A459**, 557 (1986)

5. R. Gilman et al.: Phys. Rev. Lett. **65**, 1773 (1990)

6. S.I. Mishnev et al.: Phys. Lett. **B302**, 277 (1993)

7. L.G. Isaeva et al.: Nucl. Instr. and Meth. **A325**, 16 (1993)

8. A.Yu. Loginov, A.V. Osipov, A.A. Sidorov et al.: Pis'ma ZETF **67**, N10, 736 (1998); JETF Letters **67**, N10, 770 (1998)

9. A. Osipov, A. Sidorov, V. Stibunov et al.: In: *Proc. of the 14th Int. Conference on Particle and Nuclei*, p. 276, ed. C. Carlson. WSP 1996

10. W.J. Metcalf, R.L. Walker: Nucl. Phys. **B76**, 253 (1974)

11. J.L. Visschers: In : *Proc. of MC93 Int. Conf. on Monte Carlo Simulation in High Energy and Nucl. Phys*, p. 350, ed. P. Dragovitsch. WSP 1994

12. R.M. Davidson et al.: Phys. Rev. **D43**, 71 (1991)

13. A.Yu. Loginov et al.: Izvestiya Vischikh Uchebnikh Zavedenii. Fizika. **6**, 31 (1998), ibid. **12**, 240 (1998)

Few-Body Systems Suppl. 10, 511–514 (1999)

Few-
Body
Systems
© by Springer-Verlag 1999

Results from proton-proton bremsstrahlung measurements at 190 MeV

H. Huisman[1], J.C.S. Bacelar[1], M.J. van Goethem[1], M.N. Harakeh[1],
M. Hoefman[1], N. Kalantar-Nayestanaki[1], H. Löhner[1], J.G. Messchendorp[1],
M. Seip[1], R.W. Ostendorf[1], S. Schadmand[1,4], R. Turrisi[1,5], M. Volkerts[1],
H.W. Wilschut[1], A. van der Woude[1], R. Holzmann[2], R. Simon[2], A. Kugler[3],
K. Tcherkashenko[3], V. Wagner[3]

[1] KVI Gröningen

[2] GSI Darmstadt

[3] NPI Řež u Prahy

[4] Physikalische Institut Giessen

[5] GANIL, Caen

Abstract.
    A series of bremsstrahlung measurements have been performed at KVI
with the 190 MeV polarized proton beam of the new superconducting cyclotron, AGOR. The aim of these measurements is to investigate the dynamics of interactions involving intermediate off–shell nucleons. Nucleon-nucleon
bremsstrahlung is the most fundamental process used for such studies as it
involves the strong interaction between two nucleons and the well-known electromagnetic interaction. The present experiment provides the most accurate
experiment to date on the proton-proton system. The accuracy is high enough
to investigate higher order effects such as the virtual Δ-isobar and meson exchange currents. Surprisingly, none of the microscopic calculations is able to
describe the data both in magnitude and in shape.

1   Introduction

Nucleon-nucleon bremsstrahlung (NNγ) is the most fundamental reaction with
which one can study off-shell effects in the NN interaction. The most simple
type of models used to describe bremsstrahlung are the so called Soft Photon
Approximations (SPA), which are based on a low energy theorem due to Low
[1]. These calculations use elastic scattering data only as input. Another type of
calculations are microscopic calculations, which include the off-shell dynamics
of the intermediate nucleons. These microscopic calculations can be extended

with higher-order corrections, examples of which are the virtual Δ-isobar and the meson-exchange currents (MEC) [2]. In this contribution, we report on the ppγ cross sections and analyzing powers measured with high accuracies at a beam energy of 190 MeV. This allows to investigate, in detail, calculations with or without higher order effects, such as the Δ-isobar and meson-exchange currents.

2 Experiment and Data analysis

To detect the outgoing protons at forward angles, the Small-Angle Large-Acceptance Detector, SALAD [3], was designed and built. The detection system has a full 2π azimuthal coverage between 6° and 19° polar angle. It can measure the energy of protons lying between 20 and 135 MeV with a resolution of about 2.5 MeV. The upper limit of the range of the polar angles increases to about 26° for a smaller region of the azimuthal angles. The detector was designed with a large solid angle (about 500 msr) in order to obtain high statistics for the very small cross sections of the ppγ process. To allow the beam free passage, our detector was designed with a central hole. To determine the coordinates of the charged particles, we used two wire chambers (MWPC) [4] the first of which with three planes ($x, y,$ and 45°). The second chamber has two planes (x and y). The angular resolution obtained with these chambers is about 0.5°. Plastic scintillators are used to measure the energy of the particles. A thick layer consists of 24 elements, detecting the particles originating from bremsstrahlung events. This layer is backed by a second thin layer, consisting of 26 elements, only detecting protons with an energy high enough to punch through the first layer. Protons of this high an energy originate from elastic events and are thus rejected by a hardware trigger. We used a 6 mm thick liquid hydrogen target for these measurements [5].

For the measurement of the position and the energy of the photons, we used the Two-Arm Photon Spectrometer, TAPS [6] consisting of 390 BaF$_2$ crystals. These crystals were mounted at backward angles in a large hexagon surrounding the beam-pipe. The angular range was 125° to 170°. This cylindrical symmetry allows to add up counts for the whole azimuthal range and is thus essential for high statistics. Another ppγ experiment was performed to look at the angular distribution of the reaction. The crystals were redistributed in 6 rectangular frames. These frames were positioned on both sides of the beam line surrounding the target. The results of this experiment will be discussed elsewhere. In both geometries, TAPS covered more than 20% of the full 4π solid angle.

The ppγ events of interest are extracted from the data in an off-line analysis. The low-energy particles detected in TAPS have a longer time of flight than photons and can thus easily be discriminated against. For a further reduction of the background the overdetermined kinematics of ppγ is used. The 3-body final state has 9 kinematic variables of which only 5 are free because of energy and momentum conservation. All 9 variables are measured in the experiment, providing 4 redundant variables, used for background reduction.

Starting with the polar and azimuthal angles of the two outgoing protons and the polar angle of the photon, the rest of the kinematics is reconstructed : the two proton energies, the photon energy and the azimuthal angle of the photon. Since background events will in general produce a kinematically forbidden combination of scattering angles, reconstruction provides a major reduction of the background. The combination of time cuts and kinematical reconstruction reduces the background to the level of 5%. The remaining background is removed by requiring that the measured energy of one of the protons does not differ by more than 15 MeV from its reconstructed energy.

3   Results

In Fig. 1 the data are plotted along with three different theoretical predictions. The dashed curve is the result of a SPA calculation [7], which uses the phase-shifts of ref. [8] as input. This SPA calculation is relativistic and gauge invariant. The dotted curve is the result of a fully relativistic microscopic calculation by Martinus et al. [2] based on the Fleischer-Tjon NN potential [9]. It incorporates the off-shell dynamics of the intermediate nucleons and re-scattering contributions. The solid curve is the result of a calculation by the same group and contains in addition the magnetic meson-exchange currents and the virtual Δ-isobar. One can see from Fig. 1 that the cross sections are in almost perfect agreement with the SPA but not with the more sophisticated microscopic calculations. Part of the reason could reside in the fact that the NN-potential of

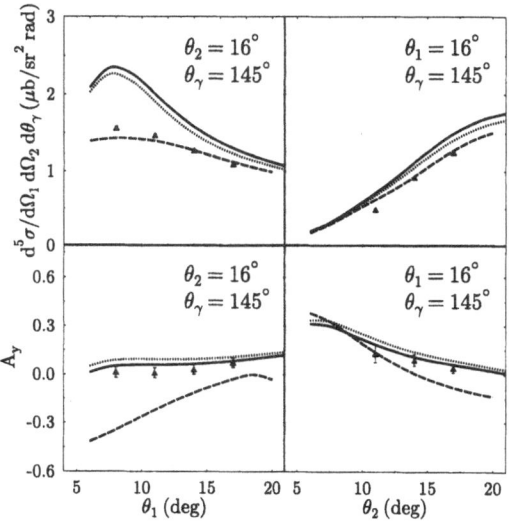

Figure 1. Preliminary ppγ cross sections and analyzing powers measured at 190 MeV incident energy (explanations in the text).

ref. [9] does not provide a high-quality fit to the present-day NN database. In addition, not all higher order corrections such as the invariant mass dependence of the NNγ vertex are explicitly included. In contrast with the cross sections, the analyzing powers are much better described by the calculations of Martinus et al. Other microscopic calculations yield similar results [10, 11].

4 Conclusions

In summary a high-precision proton-proton bremsstrahlung measurement of cross section and analyzing power has been performed at an incident energy of 190 MeV. Both protons were measured at small forward angles in coincidence with the photon. The statistical error on the measurements is superior to any prior measurement of this process. Theoretical calculations for the presented kinematics have been performed with a relativistic and gauge-invariant SPA and with a microscopic model. Surprisingly, the SPA of ref. [7] describes the measured cross section better than the microscopic model. The analyzing powers, however, are best described by the microscopic calculation of ref. [2]. Since the microscopic model disagrees both in magnitude and shape with the cross sections presented here, we conclude that there is to this date no high-quality NN model calculation consistent with the data and that more theoretical work needs to be done.

References

1. F.E. Low: Phys. Rev. **110**, 974 (1958)

2. G.H. Martinus, O. Scholten and J.A. Tjon: Phys. Rev. C **56**, 2945 (1997); G.H. Martinus, O. Scholten and J.A. Tjon: accepted for publication in Phys. Rev. C

3. N. Kalantar-Nayestanaki: Nucl. Phys. A **631**, 242c (1998)

4. M. Volkerts et al.: submitted to Nucl. Instr. and Meth. (1998)

5. N. Kalantar-Nayestanaki, J. Mulder and J. Zijlstra: accepted for publication in Nucl. Instr. and Meth. (1998)

6. A.R. Gabler et al.: Nucl. Instr. and Meth. A **346**, 168 (1994)

7. M.K. Liou, R.G.E. Timmermans and B.F. Gibson: Phys. Rev. C **54**, 1574 (1996); Phys. Lett. B **345**, 372 (1995)

8. J.R. Bergervoet et al.: Phys. Rev. C **41**, 1435 (1990)

9. J. Fleischer and J. Tjon: Nucl. Phys. A **84**, 375 (1974); Phys. Rev. D **15**, 2537 (1977); Phys. Rev. D **21**, 87 (1980)

10. V. Herrmann and K. Nakayama: Phys. Rev. C **45**, 1450 (1992)

11. K. Nakayama: private communication

Few-Body Systems Suppl. 10, 515–518 (1999)

Few-
Body
Systems
© by Springer-Verlag 1999

Kaon Photoproduction with Form Factors in a Gauge-invariant Approach

H. Haberzettl[1], C. Bennhold[1], T. Mart[1,2], T. Feuster[3]

[1] Center for Nuclear Studies, Physics Department, The George Washington University, Washington, DC 20052, U. S. A.
[2] Jurusan Fisika, FMIPA, Universitas Indonesia, Depok 16424, Indonesia
[3] Institut für Theoretische Physik, Universität Gießen, D-35392 Gießen, Germany

Abstract. The general gauge-invariant photoproduction formalism given by Haberzettl is applied to kaon photoproduction off the nucleon at the tree level, with form factors describing composite nucleons. We demonstrate that, in contrast to Ohta's gauge-invariance prescription, this formalism allows electric current contributions to be multiplied by a form factor, i.e., they do not need to be treated like bare currents. Numerical results show that Haberzettl's gauge procedure, when compared to Ohta's, leads to much improved χ^2 values. Moreover, predictions for the new Bonn SAPHIR data for $p(\gamma, K^+)\Lambda$ are given.

Gauge invariance is one of the central issues in dynamical descriptions of how photons interact with hadronic systems (see ref. [1], and references therein). For the simple example of $\gamma p \to n\pi^+$ with pseudoscalar coupling for the πNN vertex, one finds already at the tree level (see Fig. 1) that the corresponding amplitude violates gauge invariance if the baryon structure is described by form factors. This amplitude may be written as [2]

$$
\epsilon \cdot \widetilde{M}_{fi} = \sum_{j=1}^{4} \widehat{A}_j \bar{u}_n \left(\epsilon_\mu M_j^\mu \right) u_p - 2ge\bar{u}_n\gamma_5\epsilon_\mu \left[p'^\mu \frac{\widehat{F} - F_s}{s - m^2} + q^\mu \frac{\widehat{F} - F_t}{t - \mu^2} \right] u_p \ , \quad (1)
$$

with individually gauge-invariant currents,

$$
\begin{aligned}
M_1^\mu &= -\gamma_5\gamma^\mu \ k \cdot \gamma \ , & (2) \\
M_2^\mu &= 2\gamma_5 \left(p^\mu \ k \cdot p' - p'^\mu k \cdot p \right) \ , & (3) \\
M_3^\mu &= \gamma_5 \left(\gamma^\mu \ k \cdot p - p^\mu \ k \cdot \gamma \right) \ , & (4) \\
M_4^\mu &= \gamma_5 \left(\gamma^\mu \ k \cdot p' - p'^\mu \ k \cdot \gamma \right) \ , & (5)
\end{aligned}
$$

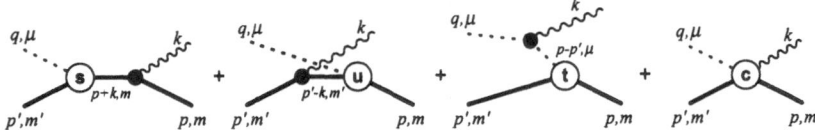

Figure 1. Tree-level photoproduction diagrams. Time proceeds from right to left. The form factors F_s, F_u, and F_t in the text describe the vertices labeled s, u, and t, respectively, with appropriate momenta and masses shown for their legs. The right-most diagram corresponds to the contact current M_c^μ required to restore gauge invariance.

with coefficient functions

$$\widehat{A}_1 = \frac{ge}{s-m^2}(1+\kappa_{\rm p})F_s + \frac{ge}{u-m^2}\kappa_n F_u \ , \tag{6}$$

$$\widehat{A}_2 = \frac{2ge}{(s-m^2)(t-\mu^2)}\widehat{F} \ , \tag{7}$$

$$\widehat{A}_3 = \frac{ge}{s-m^2}\frac{\kappa_p}{m}F_s \ , \tag{8}$$

$$\widehat{A}_4 = \frac{ge}{u-m^2}\frac{\kappa_n}{m}F_u \ , \tag{9}$$

and a gauge-violating term given by the last term in Eq. (1).

With all external legs on-shell in Fig. 1, the respective form factors are only functions of one of the Mandelstam variables, s, u, or t, i.e.,

$$F_s = F_s(s) = f\big((p+k)^2, m'^2, \mu^2\big) \ , \tag{10}$$

$$F_u = F_u(u) = f\big(m^2, (p'-k)^2, \mu^2\big) \ , \tag{11}$$

$$F_t = F_t(t) = f\big(m^2, m'^2, (p-p')^2\big) \ , \tag{12}$$

where $f(p^2, p'^2, q^2)$ is a general πNN form factor depending on the squared four-momenta of its three hadronic legs.

The function $\widehat{F}$ appearing here cancels out in Eq. (1), and hence it is undetermined. Introducing this free function here allows us to write Eq. (1) so that the gauge-invariant limit of having no form factors, *viz.*

$$\text{Point-like nucleons:} \quad F_s = F_u = F_t = \widehat{F} = 1 \ , \tag{13}$$

immediately provides for vanishing of the gauge-violating contribution to the amplitude (1).

For extended nucleons, and without a detailed dynamical treatment of the compositeness of nucleons [1], any prescription for restoring gauge invariance amounts to introducing an additional contact current M_c^μ (generically depicted by the fourth diagram in Fig. 1), with on-shell matrix elements cancelling exactly the gauge-violating term in Eq. (1). Apart from purely transverse components, for the present example this contact current is essentially given by the term in the square brackets of Eq. (1). Adding this contact contribution to

Eq. (1), one then obtains a gauge-invariant amplitude,

$$\epsilon \cdot \widehat{M}_{fi} = \sum_{j=1}^{4} \widehat{A}_j \, \bar{u}_n \left(\epsilon_\mu M_j^\mu \right) u_p \ ,$$ (14)

which *does* depend on $\widehat{F}$ now via $\widehat{A}_2$ of Eq. (7). In other words, we may use $\widehat{F}$ to distinguish between different choices for repairing gauge invariance.

One of the most popular prescriptions for restoring gauge invariance is due to Ohta [3]. Using analytic continuation and minimal substitution, Ohta finds that the required $\widehat{F}$ is constant,

$$\text{Ohta:} \quad \widehat{F} = f(m^2, m'^2, \mu^2) = 1 \ ,$$ (15)

determined by the normalization condition for the form factor in the unphysical region where all three legs are on-shell. This corresponds precisely to what one obtains for $\widehat{F}$ in the structureless case (13) and therefore the electric term $\widehat{A}_2$ of Eq. (7) is treated as in the bare case, thus effectively freezing all degrees of freedom arising from the compositeness of the πNN vertex.

The general meson photoproduction theory of ref. [1] provides another, more flexible, way of choosing $\widehat{F}$. Haberzettl's formalism allows one to take $\widehat{F}$ as a linear combination of all form factors appearing in the problem, i.e.,

$$\text{Haberzettl:} \quad \widehat{F} = a_s F_s(s) + a_u F_u(u) + a_t F_t(t) \ ,$$ (16)

where the coefficients are restricted by $a_s + a_u + a_t = 1$ in order to provide the proper limit for vanishing photon momentum (see ref. [2] for details).

We have tested the relative merits of both prescriptions for repairing gauge invariance for the kaon photoproduction reactions $\gamma p \to \Lambda K^+$ and $\gamma p \to \Sigma^0 K^+$. In both cases, one can take over Eqs. (6)-(9) and (14) by replacing the pion by K^+ and the neutron by the respective hyperon. The underlying resonance model we use is the one of ref. [4]. For simplicity, we employ here the same hadronic form factor for all resonances, parameterized as

$$f(p'^2, p^2, q^2) = \Lambda^4 / \left(\Lambda^4 + (p^2 - m^2)^2 + (p'^2 - m'^2)^2 + (q^2 - \mu^2)^2 \right) \ ,$$ (17)

where Λ is some cutoff parameter.

One of the main numerical results is summarized in Fig. 2. The upper panel shows χ^2 per data point as a function of one of the leading Born coupling constants, $g_{K\Lambda N}/\sqrt{4\pi}$, for the two different gauge prescriptions by Ohta and Haberzettl ($g_{K\Lambda N}$ was chosen here because χ^2 shows very little sensitivity on the other leading coupling constant, $g_{K\Sigma N}$ [2]). Clearly, Haberzettl's method provides χ^2 values better than Ohta's by at least a factor of two, which, moreover, are almost independent of $g_{K\Lambda N}$, in stark contrast to Ohta's. In the fits the form factor cutoff Λ was allowed to vary freely. As is seen in the lower panel of Fig. 2, in the case of Haberzettl's method, the cutoff decreases with increasing $K\Lambda N$ coupling constant, leaving the magnitude of the *effective* coupling, i.e., coupling constant times form factor, roughly constant. Since Ohta's

518

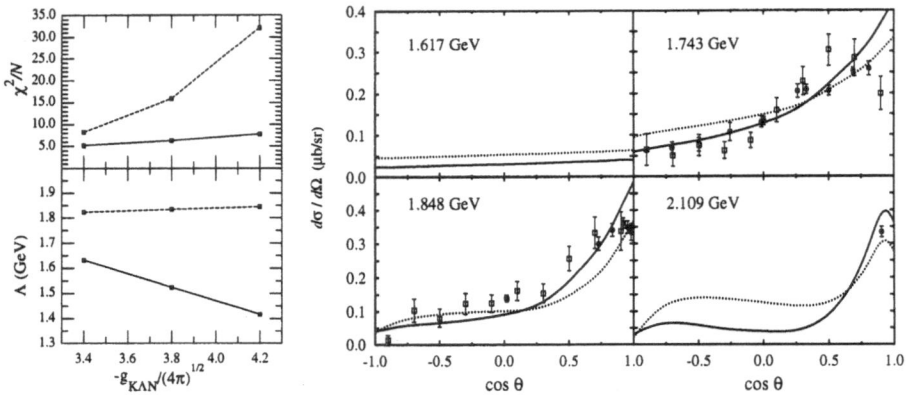

Figure 2. χ^2/N and cutoff parameter Λ as functions of coupling constant $-g_{K\Lambda N}$ (solid lines: Eq. (16); dashed lines: Eq. (15)).

Figure 3. Differential cross sections for $p(\gamma, K^+)\Lambda$ (solid lines: Eq. (16); dashed lines: Eq. (15); experimental points: old Bonn SAPHIR data [5]).

method does not involve form factors for electric contributions [cf. Eqs. (7) and (15)] no such compensation is possible there, and as a consequence the cutoff remains insensitive to the coupling constant (see ref. [2] for more details).

Figure 3 shows differential cross sections for $p(\gamma, K^+)\Lambda$ for four energies for which new, as yet unpublished, Bonn SAPHIR data exist. In the figure, we show the old SAPHIR data [5]. The new data have *not* been included in our fit and the curves shown in Fig. 3 are, therefore, *predictions*. As will be seen when they become available publicly, the new data clearly favor the gauge-invariance prescription by Haberzettl.

Our overall conclusion from the present findings is that Ohta's approach seems too restrictive to account for the full hadronic structure while properly maintaining gauge invariance, whereas the method put forward in ref. [1] seems well capable of providing this facility.

This work was supported in part by Grant No. DE-FG02-95ER40907 of the U.S. Department of Energy.

References

1. H. Haberzettl: Phys. Rev. **C56**, 2041 (1997)

2. H. Haberzettl et al.: Phys. Rev. **C58**, R40 (1998)

3. K. Ohta: Phys. Rev. **C40**, 1335 (1989)

4. T. Mart, C. Bennhold and C.E. Hyde-Wright: Phys. Rev. **C51**, R1074 (1995)

5. M. Bockhorst et al.: Z. Phys. **C63**, 37 (1994)

Few-Body Systems Suppl. 10, 519–522 (1999)

Few-Body Systems
© by Springer-Verlag 1999

Beam asymmetry Σ in meson photoproduction at GRAAL

F. Renard[1], J. Ajaka[2], V. Bellini[3], J.P. Bocquet[1], M. Breuer[4], M. Capogni[2],
M. Castoldi[5], L. Ciciani[2], A. D'Angelo[4], J.-P. Didelez[2], R. Di Salvo[2],
M.A. Duval[2], C. Gaulard[6], F. Ghio[7], B. Girolami[7], M. Guidal[2], E. Hourany[2],
I. Kilvington[8], V. Kouznetsov[9], A. Lapik[9], P. Levi Sandri[6], A. Lleres[1],
D. Moricciani[4], V. Nedorezov[9], L. Nicoletti[10], C. Perrin[1], D. Rebreyend[1],
N.V. Rudnev[9], C. Schaerf[10,4], A. Turinge[11], A. Zucchiatti[5]

[1] IN2P3, Institut des Sciences Nucléaires, 38026 Grenoble, France

[2] IN2P3, Institut de Physique Nucléaire, 91406 Orsay, France

[3] INFN, Laboratori Nazionali del Sud and Università di Catania, 95123, Italy

[4] INFN, Sezione di RomaII, I-00133 Roma, Italy

[5] INFN, Sezione di Genova, I-16146 Genova, Italy

[6] INFN, Laboratori Nazionali di Frascati, I-00044 Frascati Italy

[7] INFN, Sezione di Roma I and Istituto Superiore di Sanità, I-00161 Roma, Italy

[8] European Synchrotron Radiation Facility, F-38026 Grenoble, France

[9] Institute for Nuclear Research, RU-117312 Moscow, Russia

[10] Universitá di Roma "Tor Vergata", I-00133 Roma, Italy

[11] I. Kurchatov Institute of Atomic Energy, RU-123182 Moscow, Russia

Meson photoproduction on the nucleon is a major tool for the investigation of nucleon resonances. In particular eta photoproduction selectively probes certain resonances which are difficult to explore with pions. The low energy behaviour of the eta production is dominated by the $S_{11}(1535)$ resonance and the influence of other resonances (like $D_{13}(1520)$) is difficult to detect. Polarization observables enhance the sensitivity to the contributions of the less pronounced resonances since they depend mostly on the interference between the dominant E_{0+} multipole and the smaller ones (like $E_{2-} + M_{2-}$ for the $D_{13}(1520)$)[1].

The Graal facility provides a polarized and tagged photon beam by the backward Compton scattering of laser light on the high energy electrons circulating in the ESRF storage ring [2]. Using the UV line (351 nm) of an Ar-Ion

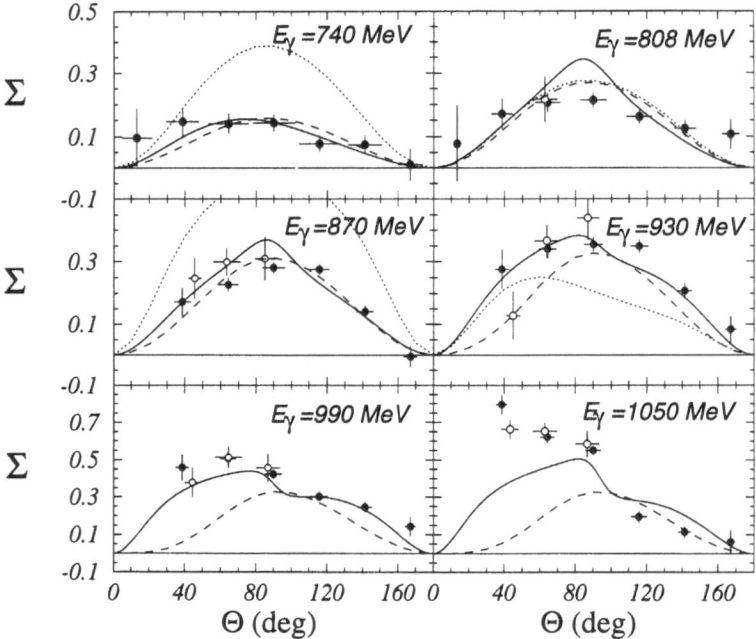

Figure 1. Beam Asymmetry Σ for η photoproduction at different incident gamma-ray energies as a function of the η angle Θ in the c.m.. The full circles are the results when the 2 or 6 decay photons are all detected in the BGO ball. The open circles are the results for one photon in the BGO and the other one in the shower detector. The error bars include statistical and systematic errors while the horizontal bars indicate the angular resolution. Predictions from [5] are represented by dotted curves; from [6] by dashed curves and from a nodal approach [8] by full curves.

laser we have produced a gamma-ray beam with an energy from 550 to 1470 MeV. Its polarization is 0.96 at the maximum photon energy and the energy resolution has been measured to be 16 MeV (FWHM). Using the green line of the same laser we have measured the beam polarization asymmetries Σ in the photoproduction of η [3], π^o and π^+ in the energy region 550-1100 MeV.

The photon energy is measured by an internal tagging system, located after the dipole at the end of the intersection line in the ring lattice. This detector consists of 10 plastic scintillators and a solid state Silicon microstrip detector with 128 channels and a pitch of 0.3 mm. The position of the microstrip traversed by the scattered electron (momentum analysed by the dipole) gives the energy of the gamma-ray. The photon energy resolution of the tagging system is 16 MeV (fwhm).

The Graal detector consists of a BGO [4] calorimeter which can analyse the high energy gamma-rays produced between 25° and 155° with an energy resolution of 3% at 1GeV. The charged particles are detected by two cylindrical wire chambers (in the same angular range) and a barrel made of 32 strips of

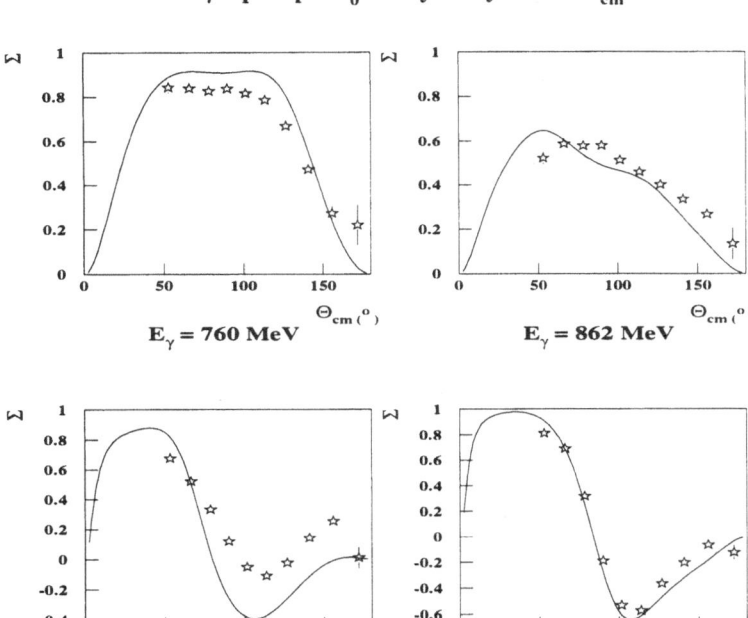

Figure 2. Beam Asymmetry Σ for π^o photoproduction, full curve is from reference [2].

plastic scintillator parallel to the beam axis. Particles moving at angles smaller than 25° encounter two plane wire chambers, two walls of plastic scintillator bars 3 cm thick and a shower wall made by a sandwich of four layers of Lead and plastic scintillators 4 cm thick. Two disks of plastic scintillator separated by a disk of Lead close the backward cone.

π^o events are detected when the two decay photons enter the BGO ball and the proton is in the BGO or the forward walls. The two body kinematics condition is sufficient to reduce the background to less than 1%. For the η we have added the events for which one photon is in the BGO and the other in the shower wall. The symmetry of the apparatus, around the beam axis, provides directly the azimuthal distribution of the events and therefore the beam polarization asymmetry. However the data are collected with two orthogonal directions of the beam polarization: horizontal and vertical. The addition of the two states of polarization reconstitutes an unpolarized gamma-ray beam which is used to correct for instrumental asymmetries. For each interval of gamma-ray energy and meson centre of mass angle the product ΣP is derived from the experimental data with a best fit procedure:

$$\frac{2N_{ver}}{N_{hor} + N_{ver}} = 1 + \Sigma P \cos(2\varphi) \tag{1}$$

N_{hor} and N_{ver} are respectively the number of events collected with the beam polarization in the Horizontal and Vertical planes normalized to the respective numbers of gamma-rays impinging on the target; P is the degree of polarization of the photon beam and Σ is the beam asymmetry. The energy dependence of P has been calculated by QED, from the laser light polarization at the collision point, and a Monte Carlo simulation showed that the effect of collimation was negligible in the energy range covered by the tagging.

The asymmetries for η photoproduction are indicated in Fig. 1. The most striking results are the large asymmetries at small angles which appear at high energies. This feature clearly demonstrates the effect of interference of higher energy resonances with the S_{11} channel and will allow to determine η branching ratios down to very low values. These data can be combined with the target asymmetry and the differential cross sections to obtain model insensitive constraints on the electro-strong parameters for the excitation and decay of the $D_{13}(1520)$ resonance [9, 10].

A preliminary set of selected measured asymmetries for π^o are indicated in Fig. 2 and compared with the SAID model [7]. The agreement seems poor at the energy of 961 MeV corresponding to a centre of mass energy of 1638 MeV. The Graal data improve very significantly the accuracy of the existing asymmetries and though the situation is more complex for π^o than for η it demonstrates again the sensitivity of the polarization observables.

References

1. B. Saghai and F. Tabakin: Phys. Rev. **C55**, 917 (1997); Phys. Rev. **C53**, 66 (1996); F. Fasano, F. Tabakin and B. Saghai: Phys. Rev. **C46**, 2430 (1992)

2. Graal collaboration: Nucl. Phys. **A622**, 110c (1997)

3. Graal collaboration: Phys. Rev. Lett., **81**, 1797 (1998)

4. F. Ghio et al.: Nucl. Inst. and Meth. **A404**, 71 (1998); P. Levi Sandri et al.: Nucl. Inst. and Meth. **A370**, 396 (1996); A. Zucchiatti et al.: Nucl. Inst. and Meth. **A317**, 492 (1992)

5. C. Bennhold et al.: Nucl. Phys. **A530**, 625 (1991); L. Tiator et al.: Nucl. Phys. **A580**, 455 (1994)

6. G. Knöchlein, D. Drechsel and L. Tiator: Z. Phys. **A352**, 327 (1995)

7. R. Arndt et al.: Phys. Rev. **C47**, 2759 (1993); Phys. Rev. **C52**, 2759 (1995)

8. B. Saghai: private communication

9. N.C. Mukhopadhyay and N. Mathur: to be published

10. L. Tiator, G. Knöchlein and C. Bennhold: to be published

Few-Body Systems Suppl. 10, 523–526 (1999)

Few-
Body
Systems
© by Springer-Verlag 1999

Virtual Compton Scattering at MAMI $\gamma^*p \to \gamma'p'$

S. Kerhoas[1], P. Bartsch[4], J. Berthot[2], P.Y. Bertin[2], V. Breton[2],
W.U. Boeglin[7], R. Böhm[4], N. d'Hose[1], T. Caprano[4], S. Derber[4],
N. Degrande[3], M. Distler[4], J.E. Ducret[1], R. Edelhoff[4], I. Ewald[4],
H. Fonvieille[2], J. Friedrich[4], J.M. Friedrich[4], R. Geiges[4], Th. Gousset[1],
P.A.M. Guichon[1], H. Holvoet[3], Ch. Hyde-Wright[5], P. Jennewein[4],
M. Kahrau[4], M. Korn[4], H. Kramer[4], K.W. Krygier[4], V. Kunde[4], B. Lannoy[3],
D. Lhuillier[1], A. Liesenfeld[4], C. Marchand[1], D. Marchand[1], J. Martino[1],
H. Merkel[4], K. Merle[4], P. Merle[4], G. De Meyer[3], J. Mougey[1], R. Neuhausen[4],
E. Offermann[6], Th. Pospischil[4], G. Quemener[2], O. Ravel[2], Y. Roblin[2],
J. Roche[1], G. Rosner[4], D. Ryckbosch[3], P. Sauer[4], H. Schmieden[4], S. Schardt[4],
G. Tamas[4], M. Tytgat[3], M. Vanderhaeghen[4], L. Van Hoorebeke[3], R. Van de
Vyver[3], J. Van de Wiele[8], P. Vernin[1], A. Wagner[4], Th. Walcher[4], S. Wolf[4]

[1] DAPNIA/SPhN, CEA/Saclay, F-91191 Gif-sur-Yvette Cedex
[2] LPC, Univ. Blaise Pascal, IN2P3 Aubière, France
[3] University of Gent, Belgium
[4] Institut für Kernphysik, Universität Mainz, Germany
[5] Old Dominium University, Virginia, U.S.A
[6] C.E.B.A.F, Virginia, U.S.A
[7] Florida International University, Miami, Florida, U.S.A
[8] IPN, IN2P3 Orsay, France

1 Introduction

The virtual Compton scattering (VCS) is the electron scattering on a proton
which radiates a real photon before being detected. The new observables, called
Generalized Polarizabilities (GP), extracted from this VCS at threshold can be
understood as the deformation of the charge and current distributions of the
proton [1]. These GP are functions of the mass of the virtual photon Q^2. In real
Compton scattering ($Q^2 = 0$), some polarizabilities of the nucleon are already
measured [2]. With the VCS, we will generalize these observables by measuring
them at different values of Q^2.

2 Formalism to extract Polarizabilities from cross sections

If we subtract from our experimental cross section ($d^5\sigma = \Phi\mathcal{M}$) the known cross sections[1] calculated from the Bethe Heitler (where the photon is emitted by the electrons BH) and Born (where the photon is emitted by a proton in the intermediate state) processes we obtain an interesting development in the final photon energy q' [1]

$$\frac{d^5\sigma^{exp} - d^5\sigma^{BH+Born}}{\Phi} = \mathcal{M}_0 - \mathcal{M}_0^{BH+Born} + (\mathcal{M}_1 - \mathcal{M}_1^{BH+Born})q' + ... \quad (1)$$

The extrapolation at q'=0 of our measurement gives us the constant term $\mathcal{M}_0 - \mathcal{M}_0^{BH+Born}$ (function of $Q^2, \epsilon, \theta, \phi$). This structure dependent term beyond the BH and Born processes was analyzed in terms of a multipole expansion which gives 5 independent Generalized Polarizabilities [1, 3]. This parametrization can be written as follows:

$$\frac{\mathcal{M}_0 - \mathcal{M}_0^{BH+Born}}{v_{LT}} = \frac{v_{LL}}{v_{LT}}(P_{LL}(Q^2) - \frac{1}{\epsilon}P_{TT}(Q^2)) + P_{LT}(Q^2) \quad (2)$$

where v_{LL}, v_{LT} are known kinematical coefficients (function of $Q^2, \epsilon, \theta, \phi$) and P_{LL}, P_{TT} and P_{LT} are linear combinations of 5 GP. With one value of ϵ, we can extract only two combinations of 5 GP, $(P_{LL} - \frac{1}{\epsilon}P_{TT})$ and P_{LT}.

3 Experimental results

Experimentally we performed the reaction $p(e, e'p')\gamma$ in two high resolution spectrometers in the A1 hall in Mainz. With a missing mass reconstruction we select our VCS events. For the first experiment [4], we decided to measure the GP at fixed $Q^2 = 0.33\,\text{GeV}^2$. The other fixed kinematical quantities are ϵ=0.62 and φ=0,180° (in-plane measurement). Figure 1 presents five differential cross sections as a function of the angle θ between the two photons, for five photon energies q'. At low q', our data are in agreement with the QED calculation. When q' increases, we can see a deviation from this calculation which proves the effect of the polarizabilities. By using Eq. (1) at 15 different bins in θ, we extract the constant term in the development in q'. The 15 values of $\mathcal{M}_0 - \mathcal{M}_0^{BH+Born}$ should verify Eq. (2), where the slope and the ordinate at origin are the unknown values that we want to extract from our data. Figure 2 presents this expected line and Table 1 the extracted values. For each observable, the first error is the statistical one coming out from the linear fit, and the second one is due to the method of extraction of the constant term in $q' = 0$. This systematic error will decrease very soon when the final analysis is finished. These experimental values are compared with different theoretical predictions [5, 6, 7, 8]. For this linear combination of GP, the best agreement is obtained for the Heavy Baryon Chiral Perturbation Theory, but the way to constraint really these models is to measure the GP independently. This can be the next generation of VCS experiments at threshold.

---

[1] QED+ Elastic Form Factors

$$d^5\sigma = d^5\sigma^{BH+Born} + \Phi\, q'\, (M_0 - M_0^{BH+Born}) + \Phi\, q'^2\, ...$$

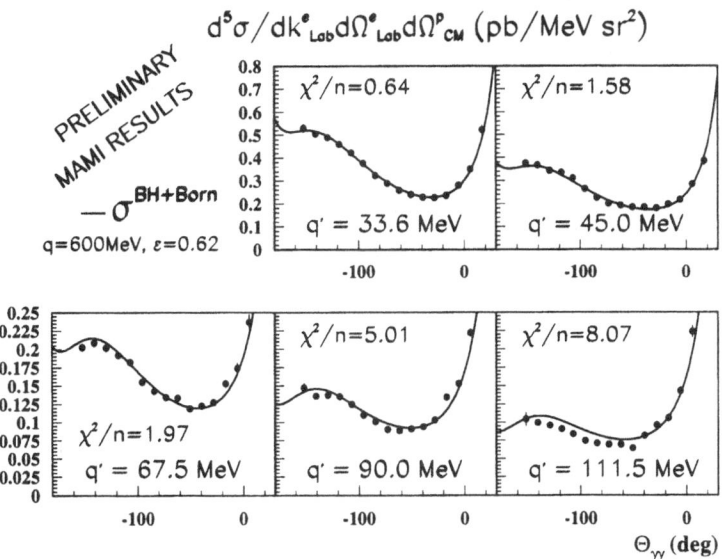

Figure 1. Five differential cross sections as a function of θ(angle between the 2 photons) for fixed $q\,(\text{or}Q^2), \epsilon, \varphi$ and five different values of q'. The agreement between the data and the curve is evaluated with a χ^2. When q' increases χ^2 increases also. This is the proof of the effect of the polarizabilities

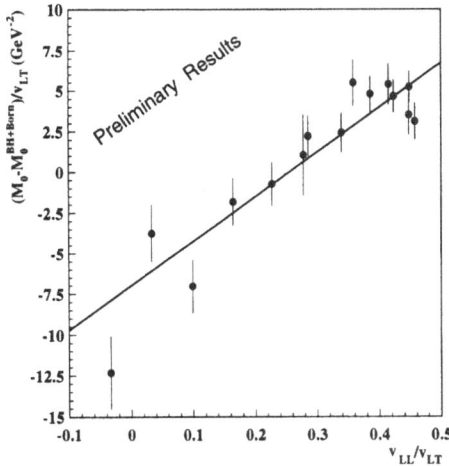

Figure 2. Parametrization of the structure term $\mathcal{M}_0 - \mathcal{M}_0^{BH+Born}$ beyond the LET, according to Eq. 2

Table 1. Experimental results are compared to theoretical ones (HBChPT for Heavy Baryon Chiral Perturbation Theory, LSM for Linear Sigma Model, NRQCM for Non Relativistic Quark Constituent Model and ELM for Effective Lagrangian Model)

$Q^2 = 0.33$ $(\mathrm{GeV/c})^2$	$(P_{LL}(Q^2) - \frac{1}{\epsilon}P_{TT}(Q^2))$ GeV^2	$P_{LT}(Q^2)$ GeV^2
This experiment	$27 \pm 3\ (\pm\ 12)$	$-7 \pm 1\ (\pm\ 4)$
HBChPT	26.3	-5.7
LSM	10.9	0.
ELM	5.9	-1.9
NRQCM	17.0	-1.7

4 Conclusion

Virtual Compton Scattering at Mainz demonstrates that it is possible to measure 2 linear combinations : $P_{LL} - \frac{1}{\epsilon}P_{TT}$ and P_{LT} of 5 Generalized Polarizabilities. Two other experiments will soon give other values of the same linear combinations : at CEBAF with $Q^2 = 1$ GeV^2 and $Q^2 = 2$ GeV^2, and at MIT-BATES with $Q^2 = 0.05$ GeV^2. The dependence of the GP on the momentum transfer is predicted quite differently in various models so this measurement at different Q^2 will give strong constraints for describing the non-perturbative structure of the proton. To measure independently the 6 Generalized Polarizabilities it is necessary to perform a double polarized experiment [9].

References

1. P. Guichon et al.: Nucl. Phys. **A591**, 606 (1995)

2. B.E. MacGibbon et al.: Phys. Rev. **C52**, 2097 (1995)

3. D. Drechsel et al.: Phys. Rev. **C55**, 424 (1997)

4. N. D'Hose and T. Walcher: In: *Nucleon Structure Study by Virtual Compton Scattering*, MAMI proposal, 1994

5. G.Q. Liu, A.W. Thomas and P. Guichon: Aust. J. Phys. **49**, 905 (1996)

6. M. Vanderhaeghen: Phys. Lett. **B368**, 13 (1996)

7. A. Metz and D. Drechsel: Mainz Report No. MKPH-T-96-17

8. T.R. Hemmert, B.R. Holstein, G. Knöchlein and S. Scherer: Phys. Rev. **D55**, 2630 (1997); Phys. Rev. Lett. **79**, 22 (1997)

9. P.A.M. Guichon and M. Vanderhaeghen: Progress in Particle and Nuclear Physics **41**, (1998)

Few-Body Systems Suppl. 10, 527–530 (1999)

Few-
Body
Systems
© by Springer-Verlag 1999

One, two, three, ..., infinity

R. Guardiola*

Departamento de Física Atómica, Molecular i Nuclear, Universidad de Valencia, Avda. Dr. Moliner 50, E-46100 Burjassot, Spain

Abstract. As concluding remarks to the European Few-Body Conference, the author presents a parallelism between the Few-Body and the Many-Body theories along the last years

1 Introduction

I was much pleased when the organizers of the conference asked me to give a brief talk, at the end of the meeting, as a farewell address. They suggested to me that the talk *should not be* a resumé of the conference, as far as each of the session chairpersons had done this task for each of the specific topics.

Facing the question of selecting an adequate subject for this talk, I decided to present some comments on the interplay between Few-Body and Many-Body physics, and at the same time to borrow the title from the excellent book of George Gamow [1], in such a form that the small numbers (1, 2, 3, ...) will represent the few-body systems and, obviously, infinity will refer to extended systems. In this interplay between two extremes, there are many facts in common. The physical systems have the same constituents, the same interactions and, up to some extent, almost the same theories (more precisely, theories with the same underlying concepts).

In some cases the extended system is simply a limiting case of the systems with a finite number of constituents, like in the case of nuclear matter, defined as the limit of the known properties of natural nuclei, or the strongly interacting quantum liquids, which may be considered as limits of drops of few atoms. In other cases the relation is not so simple. For example, the properties of the atomic-like quantum dots do not bear any resemblance with those of the electron gas system nor with the usual atoms. Finally, the interest in the two extremes has not always been simultaneous, sometimes because they were not known experimentally or the theories could not be applied equally well to both extremes.

---

*E-mail: Rafael.Guardiola@uv.es

2 Cross–fertilization

There has been a continuous exchange of ideas and methodologies between these two extremes, as well as with other branches of Physics, specially with Statistical Physics. The history is really shocking, because in some cases theories are born in, say, few body physics and after developing in many body physics they go back to their roots.

A special case corresponds to the Quantum Monte Carlo methods. The two main theories of relevance in nuclear systems, exact within statistical errors, were designed to deal with systems of few particles. From one side, there is the Domain Green Function Monte Carlo (GFMC), invented by Kalos in 1962 [2] for the study of the ^{4}He nucleus. On the other side, Anderson [3] proposed in 1975 the short–time approximation to the Green Function Monte Carlo method applied to the imaginary–time Schrödinger equation, currently known as Diffusion Monte Carlo (DMC) for its analogy with the classical diffusion problem, for the description of few–electron atomic and molecular systems. Both methods acquired their full state of development in the application to extended systems, and finally they went back to the motivating few–body applications.

This is not the only case of mutual interaction between the two extremes. The Coupled Cluster method (CCM) of Coester and Kümmel was designed for infinite systems, and applied afterwards to atoms, molecules and nuclei. Analogously, Hypernetted Chain methods (FHNC), which were extended from statistical physics to describe infinite fermionic and bosonic quantum systems, are being nowadays applied to the study of nuclei. Even specific few–body theories have found their place in the description of extended systems, like the case of Faddeev equations modified (Bethe-Faddeev) for their application in the nuclear media.

There has been also feedback from the experimental point of view. Just a simple example: the success in determining properties of liquid ^{4}He led to the interest in studying both theoretically and experimentally the properties of small aggregates (drops) of such atoms.

3 Present Status of the Few–Body Problem

As seen from the perspective of three years, from the previous FB95 conference at Peníscola [4] up to now, there has been an impressive development of both theory and experiment in the field of three– and four–particle systems. Some of these advances could be foreseen already in 1995, but it is nevertheless very gratifying that they have already taken place.

In this context, and focusing in nuclear systems, it is very important to realize that theoretical calculations based in very different grounds and technologies, have converged to essentially the same results and with a very high degree of accuracy. And not only in the description of ground-state properties, but also in the study of elastic scattering and break–up reactions. The degree of accuracy of theoretical calculations is only obscured by the very high precision measurements of collisions at very low energies with polarized projectiles and/or targets.

The theory/experiment comparison is revealing a series of new facts. From one side, there is an overall agreement, or, in other words, theoreticians are able to explain experimental measurements, when using modern realistic NN interactions. Nevertheless, very carefully determined polarization properties (like the tensor analyzing power) or very low energy elastic scattering measurements (like the determination of the n–^{3}H scattering length) resist themselves to the theoretical explanation.

Here is a good moment to recall the familiar *Coester band* [5] in nuclear matter studies of the seventies: two–body realistic NN interactions of very different nature gave for the equilibrium properties of nuclear matter (energy per nucleon and saturation density) predictions located along a narrow band far away from the saturation point. The conclusion at that time was that the usual nuclear matter calculations could not discriminate among the usual, basically phase-shift equivalent, potentials. Therefore, the reason of the discrepancy should be elsewhere: unadequacy of the theoretical calculations, the need of a three–body force, the importance of mesonic degrees of freedom, and so on. In this conference I was recalled of this fact by the calculations presented by the Pisa group of the nucleon–triton scattering length *versus* the binding energy of triton for several phase-shift equivalent potentials, giving rise to a series of almost aligned points which could be called the *Pisa band*. This statement is ilustrated in Fig. 1.

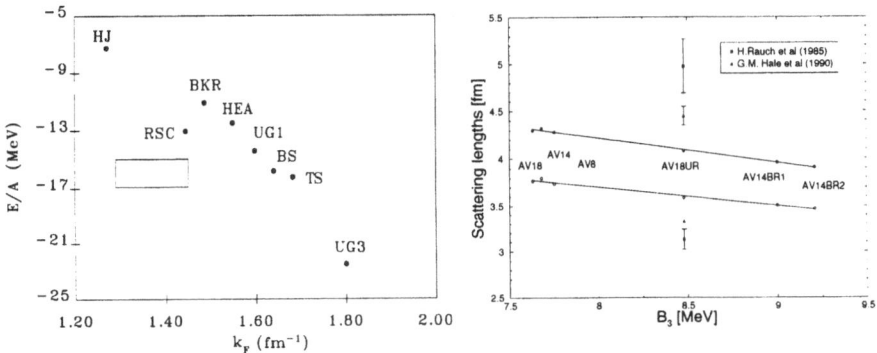

Figure 1. The left figure shows the *Coester band* [5] spanned by several nuclear matter calculations around the seventies for the usual NN interactions at that time (Hamada-Johnston, Reid soft core, Bressel-Kerman-Rouben, Holinde-Erkelenz-Alzetta, Ueda-Green, Bryan-Scott and de Tourreil-Sprung. The box indicates the experimental saturation point. The figure on the right, presented by A. Kievsky to this Conference, shows calculated and measured nucleon-triton scattering lengths versus the binding energy of the triton, for several modern NN potentials with and without three–body interactions (*Pisa band*)

One could easily guess which is the conclusion I want to draw from the present theory/experiment relationship: given the quality of the experimental measurements, and the accuracy reached by the present few–body theories, we are in the privileged situation of trying to answer questions of fundamental nature, such as the validity of a non–relativistic description of nuclear systems, or the representability of nuclear interactions by means of potentials. In this context it is important to insist in the fact that present theory/experiment confrontations go beyond static nuclear parameters, and may selectively enhance the role of specific partial waves of the interaction.

4 Mots Finaux

Nous sommes arrivés à la fin de la Seizième Confêrence Européenne de Physique sur le Petit Nombre de Corps. Et je voudrais finir en français, la très belle langue de Molière, Paul Valéry et Georges Brassens.

J'aimerais exprimer, au nom de tous les présents, notre profonde satisfaction pour l'excellent travail de l'équipe organisatrice de la Conférence, dirigée par *Bertrand Desplanques* avec l'inestimable collaboration de *J. Carbonell, C. Gignoux, S. Kox, L. Lévy, K. Protasov, D. Rebreyend, J.M. Richard, B. Silvestre-Brac, P. Valiron* et *L. Wiesenfeld* dans le travail scientifique, et *Jocelyne Riffault* et *Anne Marie Guglielmini* dans les aspects administratifs.

Ils ont été capables de nous fournir les moyens et l'ambiance qui nous a permis d'échanger nos idées et nos connaisances, non seulement sur des matières qui nous sont familières, mais aussi sur d'autres matières que nous considèrons aujourd'hui comme *exotiques* mais qui dans peu de temps seront aussi familières que les autres.

Je crois qu'une seule parole espagnole suffira pour exprimer notre appréciation de leur travail: *Olé!*

Acknowledgement. This work is supported by DGICyT under contract No. PB92–0820 and by the EEC network No. ERBCHRXCT940456. The author is particularly grateful to Michele Viviani for having provided the nucleon-triton part of Fig. 1, and to Mariela Portesi for a critical reading of the manuscript.

References

1. George Gamow, *One, two, three ... infinity*. New York: The Viking Press 1947

2. M.H. Kalos: Phys. Rev. **128**, 1791 (1962)

3. J.B. Anderson: J. Chem. Phys. **63**, 1499 (1975)

4. *Few–Body Problems in Physics '95*, ed. R. Guardiola, Few-Body Syst. Suppl. **8** (1996)

5. F. Coester et al.: Phys. Rev. **C1**, 769 (1970)

List of Participants

ADAMOWSKI Janusz
Faculty of Physics
and Nuclear Techniques
Technical University (AGH)
al. Mickiewicza 30
PL-30059 Kraków
Poland
Adamowski@novell.ftj.agh.edu.pl

ALT Erwin
Institut für Physik
Universität Mainz
D-55099 Mainz
Germany
alt@dipmza.physik.uni-mainz.de

AMGHAR Abdelhamid
Dept. de Physique Fondamentale
INHC
35000 Boumerdes
Algeria
amghar@ist.cerist.dz

ASHERY Daniel
School of Physics and Astronomy,
Tel Aviv University,
Tel-Aviv
Israel
ashery@silly.tau.ac.il

BACELAR J.C.
Kernfysisch Versneller Instituut
Rijks Universiteit Gröningen
Zernikelaan 25
NL-9747 AA Gröningen
The Netherlands
bacelar@kvi.nl

BAKKER Bernard
Dept. of Physics and Astronomy
Vrije Universiteit
De Boelelaan 1081
NL-1081 HV Amsterdam
The Netherlands
blgbkkr@nat.vu.nl

BALL Jacques
DAPNIA, SPhNHE
F-91191 Gif-sur-Yvette Cedex
France
BALL@phnx7.saclay.cea.fr

BARNEA Nir
ECT*
Strada delle Tarabelle 286
I-38050 Villazano (Trento)
Italy
barnea@ect.unitn.it

BAWIN Michel
Institut de Physique B5
Université de Liège au Sart-Tilman
B-4000 Liège
Belgium
michel.bawin@ulg.ac.be

BEISE Elizabeth J.
Department of Physics
University of Maryland
College Park, MD 20742-4111
USA
beise@physics.umd.edu

BELYAEV Vladimir
RCNP, Osaka University
Mihogaoko 10-1, Ibaraki
Osaka 567, Japan
belyaev@miho.rcnp.osaka-u.ac.jp
belyaev@pib1.physik.uni-bonn.de

BEYER Michael
Fachbereich Physik
Universität Rostock
Universitätsplatz 1
D-18051 Rostock, Germany
beyer@darss.mpg.uni-rostock.de

BIJTEBIER Jacques
Theoretische Natuurkunde
Vrije Universiteit Brussel
Pleinlaan 2,
1050 Brussel, Belgium
jbijtebi@vub.ac.be

BLOKHINTSEV Leonid
Institute of Nuclear Physics
Moscow State University
119899 Moscow
Russia
blokh@srdlan.npi.msu.su

BOCQUET Jean Paul
Institut des Sciences Nucléaires
53, Avenue des Martyrs
F-38026 Grenoble Cedex
France
bocquet@isn.in2p3.fr

BOGDANOVA Ludmila
ITEP
B.Cheremushkinskaya ul., 25
117218 Moscow
Russia
bogdanova@vxitep.itep.ru
ludmila@mucatex.msk.ru

BOSNAR Damir
Physics Department
University of Zagreb
HR-10000 Zagreb
Croatia
bosnar@plan.phy.hr

BOUDARD Alain
DAPNIA, SPhNHE
F-91191 Gif-sur-Yvette Cedex
France
boudard@phnx7.saclay.cea.fr

BRAU Fabian
Université de Mons-Hainaut
Place du Parc 20
B-7000 Mons
Belgium
fabian.brau@umh.ac.be

CARBONELL Jaume
Institut des Sciences Nucléaires
53, Avenue des Martyrs
F-38026 Grenoble Cedex
France
carbonel@isnhp4.in2p3.fr

CARLSON Joseph A.
T-5, MS B283,
Los Alamos, NM 87545
USA
carlson@qmc.lanl.gov

CEULENEER Rene
Université de Mons-Hainaut,
Place du Parc 20,
B-7000 Mons
Belgium
RENE.CEULENEER@UMH.AC.BE

CHMIELEWSKI Karsten
Institut für Theoretische Physik
Universität Hannover
D-30167 Hannover
Germany
chmiele@itp.uni-hannover.de

CIESIELSKI Frederic
Institut des Sciences Nucléaires
53, Avenue des Martyrs
F-38026 Grenoble Cedex
France
F.Ciesielski@isn.in2p3.fr

COESTER Fritz
Physics Division
Argonne National Laboratory,
Argonne IL 60439-4843
USA
coester@theory.phy.anl.gov

COTANCH Stephen
Department of Physics
North Carolina State University
Raleigh, NC 27695-8202
USA
cotanch@ncsu.edu

DANILIN Boris
Kurchatov Institute
Kurchatov sq. 1
123182 Moscow
Russia
danilin@cerber.polyn.kiae.su

DESPLANQUES Bertrand
Institut des Sciences Nucléaires
53, Avenue des Martyrs
F-38026 Grenoble Cedex
France
desplanq@isn.in2p3.fr

EPELBAOUM Eugeni
Ruhr-Universität Bochum
Institut für Theoretische Physik II
D-44870 Bochum, Germany
evgenie@hadron.tp2.ruhr-uni-
bochum.de

FABRE de la RIPELLE Michel
Institut de Physique Nucléaire
F-91406 Orsay Cedex
France

FERNANDEZ Francisco
Nuclear Physics Group
University of Salamanca
Plza. Merced s/n
E-37008 Salamanca, Spain
fdz@rs6000.usal.es

FONSECA Antonio
Centró Física Nuclear da U.L.
Av. Prof. Gama Pinto, 2
P-1699 Lisboa, Portugal
fonseca@alf4.cii.fc.ul.pt

FREDERICO Tobias
Dep. de Física
Instituto Tecnológico de Aeronáutica
Centro Técnico Aeroespacial
12.228-900 São José dos Campos
São Paulo, Brazil
FREDERICO@alpher.npl.washington.edu
tobias@fis.ita.cta.br

FRIAR James L.
T-5, MS B283
Los Alamos National Laboratory
Los Alamos, New Mexico 87545
USA
FRIAR@SUE.LANL.GOV

GIGNOUX Claude
Institut des Sciences Nucléaires
53, Avenue des Martyrs
F-38026 Grenoble Cedex
France
C.Gignoux@isn.in2p3.fr

GIRAUD Bertrand
Service Physique Théorique
DSM, C.E. Saclay
F-91191 Gif-sur-Yvette
France
giraud@spht.saclay.cea.fr

GLAGOLEV Viktor
Joint Institute for Nuclear Research
Laboratory of High Energy
141980, Dubna, Moscow Region
Russia
glagolev@sunhe.jinr.ru

GOMEZ Javier
Physics Division
TJNAF
Newport News, Virginia 23606
USA
gomez@jlab.org

GUARDIOLA Rafael
Dept. Física Atómica
Molecular i Nuclear
Universidad de Valencia
Avda. Dr. Moliner, 50
46100 Burjasot (Valencia)
Spain
Rafael.Guardiola@uv.es

HABERZETTL Helmut
Physics Department
George Washington University
Washington, DC 20052, USA
helmut@gwis2.circ.gwu.edu

HEGERFELDT Gerhard C.
Institut für Theoretische Physik
Universität Goettingen
Germany
hegerf@Theorie.Physik.UNI-
Goettingen.DE

HUEBER Dirk
Theoretical Division
Los Alamos National Laboratory
M.S. B283
Los Alamos, NM 87545
USA
hueber@t5.lanl.gov

HUISMAN Harry
Kernfysisch Versneller Instituut
Rijks Universiteit Gröningen
Zernikelaan 25
NL-9747 AA Gröningen
The Netherlands
huisman@kvi.nl

HUMBERSTON John
Dept. of Physics and Astronomy
University College London
Gower Street
London WC1E 6BT
UK
j.humberston@ucl.ac.uk

IRGAZIEV Bakhodir
Theoretical Physics Department
Tashkent State University
700095 Tashkent
Uzbekistan
irgaz@iaph.silk.glas.apc.org
irgaz@univer.tashkent.su

JANS Eddy
National Institute for Nuclear
and High-Energy Physics, NIKHEF
P.O. Box 41882
NL-1009 DB Amsterdam
The Netherlands
eddy@nikhef.nl

JENSEN A. S.
Institute of Physics and Astronomy
Aarhus University
DK-8000 Aarhus C, Denmark
asj@dfi.aau.dk

JI Chueng-Ryong
Department of Physics
North Carolina State University
Raleigh, N.C. 27695-8202
USA
crji@unity.ncsu.edu

KAMADA Hiroyuki
Institut für Kernphysik
Fachbereich 5 der Technischen
Hochschule Darmstadt
D-64289 Darmstadt
Germany
kamada@tp2lin48.tp2.ruhr-uni-
bochum.de

KARMANOV Vladimir
Lebedev Physical Institute
Leninsky Prospekt 53
117924 Moscow, Russia
karmanov@sci.lebedev.ru

KARTAVTSEV Oleg
Bogoliubov Laboratory
of Theoretical Physics
Joint Institute for Nuclear Research
141980 Dubna, Moscow region
Russia
oik@thsun1.jinr.ru

KERHOAS Sophie
CEA Saclay - DSM/DAPNIA/SPHN
Bat. 703 Orme des Merisiers
F-91191 Gif-sur-Yvette Cedex
France
skerhoas@cea.fr
sofy@phnx7.saclay.cea.fr

KIEVSKY Alejandro
Istituto Nazionale di Fisica Nucleare
Dipartimento di Fisica
Piazza Torricelli 2
I-56100 Pisa, Italy
kievsky@pisa.infn.it

KOBUSHKIN Alexander
Bogolyubov Institute
for Theoretical Physics
252143 Kiev, Metrologicheskaya str.
14-B Ukraine
kobushkin@gluk.apc.org
akob@ap3.gluk.apc.org

KONYA Balazs
Institute of Nuclear Research
P.O. Box 51
H-4001 Debrecen, Hungary
konyab@indigo.atomki.hu

KOVACS Zoltan
Institute for Theoretical Physics
Eötvös University
Budapest, Hungary
kz@poe.elte.hu

KRASSNIGG Andreas
Institute for Theoretical Physics
University of Graz
Universitätsplatz 5
A-8010 Graz, Austria
andreas.krassnigg@kfunigraz.ac.at

KRIKEB Ali
Institut de Physique Nucléaire
Université Claude Bernard
43, bld du 11 novembre 1918
F-69622 Villeurbanne Cedex
France
krikeb@ipnl.in2p3.fr

KRIVEC Rajmund
Stefan Institute
Department of Theoretical Physics
Jamova 39, P.O. Box 3000 1001
Ljubljana
Slovenia rajmund.krivec@ijs.si

KRUEGER Andreas
Institut für Theoretische Physik II
Ruhr-Universität Bochum
D-44780 Bochum, Germany
andreask@hadron.tp2.ruhr-uni-
bochum.de

KUKULIN Vladimir
Institute of Nuclear Physics
Moscow State University
119899 Moscow, Russia
KUKULIN@anna19.npi.msu.su

LADYGIN Vladimir
Joint Institute for Nuclear Researches
141980, Dubna, Moscow Region
Russia
ladygin@sunhe.jinr.ru
ladygin@moonhe.jinr.ru

LAHIFF Andrew
Department of Physics
Flinders University
G.P.O. Box 2100
Adelaide, S.A. 5001
Australia
phadl@apollo.ph.flinders.edu.au

LARSEN Sigurd
Physics Deptartment (009-00)
Temple University
Philadelphia, PA 19122
USA
syl@astro.ocis.temple.edu

LEEB Helmut
Institut für Kernphysik
Technische Universität Wien
Wiedner Hauptstrasse 8-10/142
A-1040 Vienna
Austria
leeb@kph.tuwien.ac.at

LEIDEMANN Winfried
Dipatimento di Fisica
Universita di Trento
I-38050 Povo
Italy

LU Lanchun
Institut des Sciences Nucléaires
53, Avenue des Martyrs
F-38026 Grenoble Cedex
France
lulc@isnhp6.in2p3.fr

MANDELZWEIG Victor
Physics Department
Hebrew University
Jerusalem
Israel
VICTOR@vms.huji.ac.il

MARKUM Harald
Institut für Kernphysik Technische
Universität Wien Wiedner Haupt-
strasse 8-10 A-1040 VIENNA
Austria
markum@as1.kph.tuwien.ac.at

MARKUSHIN Valeri
Paul Scherrer Institute
CH-5232 Villigen PSI
Switzerland
markushin@psi.ch

MARTIN Andre
Theory Division, CERN
CH-1211 Geneva
Switzerland
martina@mail.cern.ch

MATVEENKO Alexander V.
Bogoliubov Laboratory
of Theoretical Physics
Joint Institute for Nuclear Research
141980 Dubna, Moscow
Russia
amatv@thsun1.jinr.ru

MELDE Thomas
University of Manitoba
Winnipeg, MB R3T 2N2
Canada
tmelde@theory.uwinnipeg.ca

MESSCHENDORP J.G.
Kernfysisch Versneller Instituut
Zernikelaan 25
9747 AA Gröningen
The Netherlands
messchendorp@kvi.nl

MOSZKOWSKI Steven
Dept. of Physics and Astronomy
UCLA
Los Angeles, CA 90095
USA
stevemos@ucla.edu

MOTOVILOV Alexander K.
Laboratory of Theoretical Physics
Joint Institute for Nuclear Research
141980 Dubna, Moscow region
Russia
motovilv@thsun1.jinr.ru

MOUGEY Jean
Institut des Sciences Nucléaires
53, Avenue des Martyrs
F-38026 Grenoble Cedex
France
mougey@isnhp6.in2p3.fr

MUKHAMEDZHANOV Akram
Cyclotron Institute
Texas A@M University
College Station
Texas, 77843
USA
AKRAM@CYCOMP.TAMU.EDU

NEMOTO Shino
Institut für Theoretische Physik
Appelstr. 2
D-30167 Hannover
Germany
nemoto@itp.uni-hannover.de

NIELSEN Esben
Institute of Physics and Astronomy
University of Aarhus
Ny Munkegade
DK-8000 Aarhus C
Denmark
simlo@dfi.aau.dk

NOGGA Andreas
Institut für Theoretische Physik II
Ruhr-Universität Bochum
D-44780 Bochum
Germany
Andreas.Nogga@ruhr-uni-bochum.de

NOGUERA Santiago
Departamento de Fisica Teòrica
Universidad de València
E-46100 Burjasot, València
Spain
noguera@evalvx.ific.uv.es

OBUKHOVSKY I.T.
Institute of Nuclear Physics
Moscow State University
119899 Moscow, Russia
obukh@nucl-th.npi.msu.su

ORR Nigel
Lab. de Physique Corpusculaire
IN2P3 − CNRS
ISMRA et Université de Caen
F-14050 Caen Cedex, France
orr@caeas3.in2p3.fr

ORYU Shinsho
Professor of Department of Physics
Science University of Tokyo
2641 Yamazaki, Noda-city
Chiba 278-8510, Japan
oryu@ph.noda.sut.ac.jp

OVCHINNIKOV Serge
Theoretical Physics
The University of Tennessee
200 South College
Knoxville, TN 37996, USA
serge@orph28.phy.ornl.gov

PACE Emanuele
Dip. di Fisica
Universita di Roma
"Tor Vergata" and INFN
Sezione Tor Vergata
Viadella Ricerca Scientifica 1
I-00133 Roma, Italy
PACE@ROMA2.INFN.IT

PACHECO Joao
Groupe de Physique Théorique
Institut de Physique Nucléaire
F-91406 Orsay Cedex
France
pacheco@ipno.in2p3.fr

PAPP Zoltan
Institute of Nuclear Research
P.O. Box 51
H-4001 Debrecen
Hungary
pz@indigo.atomki.hu

PASCALUTSA Vladimir
Institute for Theoretical Physics
University of Utrecht
Princetonplein 5,
NL-3584 CC Utrecht
The Netherlands
pascalu@fys.ruu.nl

PASQUINI Barbara
Institut für Kernphysik
Johannes Gutenberg-Universität
Johann-Joachim-Becher-Weg 45
D-55099 Mainz
Germany
pasquini@kph.uni-mainz.de

PLESSAS Willibald
Institute for Theoretical Physics
University of Graz
Universitätsplatz 5
A-8010 Graz, Austria
plessas@edvz.kfunigraz.ac.at

PONOMAREV Leonid
RRC "Kurchatov Institute"
pl. Kurchatova 1
123128 Moscow, Russia
leonid@mucatex.msk.ru

POPOV Yuri
Nuclear Physics Institute
Moscow State University
Moscow 119899, Russia
POPOV@srdlan.npi.msu.su

PORTESI Mariela
Universidad de Valencia
E-46100 Burjasot, Valencia
Spain
Mariela.Portesi@uv.es

PROKOFIEV Alexander
Petersburg Nucl. Phys. Inst.
18835, Gatchina, Russia
prokan@lnpi.spb.su

PROTASOV Konstantin
Institut des Sciences Nucléaires
53, Avenue des Martyrs
F-38026 Grenoble Cedex
France
protasov@isn.in2p3.fr

RAKITIANSKI S. A.
Department of Physics
University of South Africa
P. O. Box 392, Pretoria
South Africa
rakitsa@kiaat.unisa.ac.za

REBREYEND Dominique
Institut des Sciences Nucléaires
53, Avenue des Martyrs
F-38026 Grenoble Cedex
France
D.Rebreyend@isn.in2p3.fr

RICHARD Jean-Marc
Institut des Sciences Nucléaires
53, Avenue des Martyrs
F-38026 Grenoble Cedex
France
jmrichar@isnnx2.in2p3.fr

RISKA Dan Olof
Department of Physics
University of Helsinki
FIN-00014 Helsinki
Finland
riska@science.helsinki.fi

ROBERTS Winston
Department of Physics
Old Dominion University
Norfolk, VA 23529
TJNAF
MS 12H2, 12000 Jefferson Avenue
Newport News, VA 23606
USA
winston@physics.odu.edu

ROSATI Sergio
University of Pisa
Physics Department
Piazza Torricelli 2
I-56100, Pisa
Italy
rosati@ibmth.difi.unipi.it

ROSINA Mitja
Department of Theoretical Physics
J. Stefan Institute, PO Box 3000
SI-1001 Ljubljana
Slovenia
Mitja.Rosina@ijs.si

ROST Jan Michael
Theoretische Quantendynamik
Universität Freiburg
D-79104 Freiburg
Germany
briggs@sun2.ruf.uni-freiburg.de

SALME Giovanni
INFN,Sezione Sanita
Viale Regina Elena 299
I-00161 Roma
Italy
GSLM@ISS.INFN.IT

SANDHAS Werner
Physikalisches Institut
Universität Bonn
Nussallee 12
D-53115 Bonn
Germany
unp064@ibm.rhrz.uni-bonn.de

SAUGE Sebastien
Laboratoire d'Astrophysique
Observatoire de Grenoble, B.P. 53
F-38041 Grenoble Cedex 9
France
Sebastien.Sauge@obs.ujf-grenoble.fr

SCOPETTA Sergio
Departamento de Fisica Teòrica
Universidad de València
Av.da Dr. Moliner
E-46100 Burjassot (València)
Spain
scopetta@titan.ific.uv.es

SEMAY Claude
Université de Mons-Hainaut
Place du Parc 20
B-7000 Mons
Belgium
Claude.Semay@umh.ac.be

SICK Ingo
Department of Physics
University of Basel
CH-4056 Basel
Switzerland
sick@ubaclu.unibas.ch

SILVESTRE-BRAC Bernard
Institut des Sciences Nucléaires
53, Avenue des Martyrs
F-38026 Grenoble Cedex
France
silvestre@isnhp1.in2p3.fr

SIMECKOVA Eva
Nuclear Physics Institute
Academy of Sciences of the Czech
Republic
CZ-250 68 Řež, Czech Republic
simeckova@vax.ujf.cas.cz

SIMULA Silvano
Istituto Nazionale di Fisica Nucleare
Sezione Sanita
I-00161 Roma
Italy
simula@hpteo2.iss.infn.it

SITNIK Igor
Joint Institute for Nuclear Research
141980 Dubna, Moscow region
Russia
sitnik@sunhe.jinr.ru

SKIBINSKI Roman
Jagiellonian University
Institute of Physics
4 Reymonta St.
PL-30059 Krakow
Poland
skibinsk@mezon.if.uj.edu.pl

STANCU Ica
Institut de Physique, B.5
Sart Tilman
B-4000 Liège 1
Belgique
stancu@baryon.theo.phys.ulg.ac.be

STIBUNOV Victor
Institute for Nuclear Physics
Tomsk Politekhnical University
prospekt Lenina 2A
634050 Tomsk
Russia
stib@npi.tpu.ru

STROKOVSKY Eugene
Lab. for Particle Physics
Joint Institute for Nuclear Research
141980 Dubna, Moscow region
Russia
strok@sunhe.jinr.ru

SUFT Gaby
Physikalisches Institut
Erwin-Rommel Str. 1
D-91058 Erlangen
Germany
Gaby.Suft@Physik.Uni-Erlangen.de

SULTANOV Renat A.
Instituto de Física Teórica
UNESP
01405-900 São Paulo, SP
Brazil
sultanov@ift.unesp.br

THEUSSL Lukas
Institute for Theoretical Physics
University of Graz
Universitätsplatz 5
A-8010 Graz
Austria
lut@physik.kfunigraz.ac.at

TORNOW Werner
Department of Physics
Duke University
and Triangle Universities
Nuclear Laboratory
Durham, North Carolina 27708-0308
USA
Tornow@tunl.duke.edu

VALIRON Pierre
Laboratoire d'Astrophysique
Observatoire de Grenoble, B.P. 53
F-38041 Grenoble Cedex 9
France
Pierre.Valiron@obs.ujf-grenoble.fr

Van DIJK Wytse
Dept. of Physics and Astronomy
McMaster University
Hamilton, Ontario
Canada L8S 4M1
vandijk@quark.physics.mcmaster.ca

Van REETH Peter
Dept. of Physics and Astronomy
University College London
Gower Street
London WC1 6BT
UK
pvr@phys.ucl.ac.uk

VARGA Kalman
Theory Division
Argonne National Laboratory
9700 S. Cass Ave.
Argonne, IL 60439
USA
VARGA@THEORY.PHY.ANL.GOV

VIVIANI Michele
INFN, Sezione di Pisa
Piazza Torricelli 2
I-56100 Pisa, Italy
viviani@galileo.pi.infn.it

VOLKERTS Marcel
Kernfysisch Versneller Instituut
Rijks Universiteit Gröningen
Zernikelaan 25
NL-9747 AA Gröningen
The Netherlands
VOLKERTS@KVI.nl

Von WITSCH Wolfram
Institut für Strahlen- und Kernphysik
University of Bonn
Nussallee 14-16
D-53115 Bonn, Germany
vwitsch@iskp.uni-bonn.de

VORONIN Alexei
P.N. Lebedev Physical Institute
53, Leninsky prosp.
117924 Moscow
Russia
alexei@voronin.msk.ru

WAETZOLD Lothar
Institut für Strahlen-
und Kernphysik
der Universität Bonn
Nussallee 14-16
D-53115 Bonn, Germany
waetzold@iskp.uni-bonn.de

WAGENBRUNN Robert
Institute for Theoretical Physics
University of Graz
Universitätsplatz 5
A-8010 Graz, Austria
rfw@physik.kfunigraz.ac.at

WEBER Christian
Institut für Strahlen-
und Kernphysik
der Universität Bonn
Nussallee 14-16
D-53115 Bonn, Germany
cweber@iskp.uni-bonn.de

YAKOVLEV Serguei L.
Dept. of Math. and Comp. Physics
St.Petersburg University
Petrodvorets, Ulyanovskaya Str. 1
198904, St.Petersburg, Russia
yakovlev@snoopy.phys.spbu.ru
yakovlev@mph.niif.spb.su

YAMAMURA Hisahiko
Department of Applied Physics
Okayama University of Science
Ridai-cho Okayama 700
Japan
yamamu-s@dap.ous.ac.jp

YAMAZAKI Toshimitsu
University of Tokyo
5-3-1 Koji-machi, Chiyoda-ku
Tokyo, 102 Japan
yamazaki@nucl.phys.s.u-tokyo.ac.jp

ZOLIN Leonid
Joint Institute for Nuclear Research
Dubna, Joliot Curie 6
141980 Dubna, Moscow Region
Russia
zolin@moonhe.jinr.ru

List of Authors

SpringerPhysics

V. D. Burkert, N. C. Mukhopadhyay, B. Saghai, S. Simula (eds.)

N* Physics and Nonpertubative Quantum Chromodynamics

Proceedings of the Joint ECT*/Jefferson Lab Workshop, Trento, Italy, May 18–29, 1998

1999. XV, 372 pages. 130 figures.

Hardcover DM 160,–, öS 1120,–, sFr 144,–

Reduced price for subscribers to "Few-Body Systems":

Hardcover DM 144,–, öS 1008,–, sFr 131,50

(all prices are recommended retail prices)

ISBN 3-211-83299-8

Few-Body Systems, Supplement 11

The workshop was devoted to a summary of recent experimental and theoretical research on N* physics. Special emphasis was given to the information that photo- and electro-production of nucleon resonances can provide on the nonperturbative regime of quantum chromodynamics. Discussions among experimentalists and theoreticians were stimulated in order to pursue the interpretation of the huge amount of forthcoming data from several laboratories in the world. This volume contains both the invited lectures and the contributions. On the main topics, like single and double pion production, pi- and K-meson production, the GDH sum rule, and the spin of the proton.

 SpringerWienNewYork

Sachsenplatz 4–6, P.O.Box 89, A-1201 Wien, Fax +43-1-330 24 26

e-mail: books@springer.at, Internet: http://www.springer.at

New York, NY 10010, 175 Fifth Avenue • D-14197 Berlin, Heidelberger Platz 3

Tokyo 113, 3–13, Hongo 3-chome, Bunkyo-ku